Science, Theory and Clinical Application in Orthopaedic Manual Physical Therapy

Scientific Therapeutic Exercise Progressions (STEP): The Back and Lower Extremity

5547 South 4015 West #7
Taylorsville, UT 84118

Publisher: The Academy of Graduate Physical Therapy, Inc.
First Published 2009

ISBN: 978-0-578-01558-3

Project Managers: Ola Grimsby and Jim Rivard
Illustrators: Visual Health Information and Ola Grimsby
Print Editor: Asha Hossain
Text Editor: Dana Grant
Photographer: Jim Rivard
Cover Design: Asha Hossain

Contributors

Ola Grimsby PT, DMT, FFAAOMPT
Founder, Chairperson of the Ola Grimsby Institute
www.olagrimsby.com

Bill Hayner PT, FAAOMPT
MTI Physical Therapy
Ola Grimsby Institute Instructor

Alec Kay MSPT, DMT, OCS, ATC, FAAOMPT
Co-Owner of United Physical Therapy
Ola Grimsby Institute Instructor

Micheal Maninang GR, DPT, MHS, MOMT, OCS, FAAOMPT
Physical Therapy Associates
Ola Grimsby Institute Instructor

Sarah Olson MSPT, OMT, FAAOMPT
MTI Physical Therapy
Ola Grimsby Institute Instructor

Brian Power PT, DPT, MOMT, FAAOMPT
Co-owner MTI Physical Therapy
Ola Grimsby Institute Instructor

Jim Rivard PT, MOMT, OCS, FAAOMPT
Co-owner MTI Physical Therapy
Instructor and Vice President of Curriculum Development for
The Ola Grimsby Institute

Arthur Schwarcs PT, PhD, ATC, MNMST, CEAS
Associate Professor at Husson Univeristy
Ola Grimsby Institute Instructor

Rob Tillman PT, MOMT, FAAOMPT
President of Orthopedic Rehabilitation and Specialty Centers
Ola Grimsby Institute Instructor

Dan Washeck PT, OCS, FAAOMPT
MTI Physical Therapy
Ola Grimsby Institute Instructor

Science, Theory and Clinical Application in Orthopaedic Manual Physical Therapy

Scientific Therapeutic Exercise Progressions (STEP): The Back and Lower Extremity

Editors: Jim Rivard PT, MOMT, OCS, FAAOMPT
Ola Grimsby PT, DMT, FFAAOMPT

Resources

The Ola Grimsby Institute
Orthopaedic Manual Therapy (OMT) courses for evaluation, treatment, exercise and residency training.
www.olagrimsby.com

Nutrition and Health
Didrik Sopler, PhD:
www.tissuerecovery.com

Exercise Equipment
www.lojer.com
www.rehabpropulleys.com
www.cardonrehab.com

Home Exercise Pulley Equipment
www.qtekproducts.com

Exercise Software
www.vhikits.com

Foreword

It has been my good fortune to have my physical therapy education complemented by an introduction to a broader paradigm of exercise rehabilitation. In the late 1980's I attended a manual therapy course taught by Ola Grimsby. Presented with clinical application of scientific and clinical research, my rationale for all aspects of patient care changed significantly. Most traumatized was my belief system of how exercise was used in orthopedic rehabilitation and how it differed from research related to training healthy subjects. Ola Grimsby presented the fundamentals of Medical Exercise Therapy (MET) as taught in his home country of Norway. The MET curriculum originated in the 1960's under the creative wings of Oddvar Holten. As a manual therapist, he combined his experiences as clinician, coach and athlete to bring training principles to patient care. MET laid out a foundation of exercise dosage and progression that began with tissue repair, joint mobilization, pain inhibition and resolution of edema. Specific training dosage was described for influencing motor function as it relates to coordination, timing, endurance, speed, volume strength and power. Being exposed to such a wide range of specific training parameters to influence such a larger list of potential training goals inspired me to better understand training principles as they relate to patient care.

My manual therapy education, under the guidance of Brad Jordan and Ola Grimsby, included training in exercise from several of the MET instructors from Norway. One of these MET instructors, Anders Myklebust, provided my first clinic training in MET through The Ola Grimsby Institute's (OGI) residency program. Though he overwhelmed my capacity to absorb and digest information, I was nonetheless inspired to seek out more training in this exercise specialty. Another MET instructor, Ronnie Stensnes, not only progressed my exercise instruction through his involvement in residency teaching for the OGI, but was also kind enough to allow me to assist him with teaching weekend seminars in MET. His gentle approach with such a young mind help nurture me through the frustrating years of becoming clinically proficient. I continued to be a student of MET classes for many years. Finally having the opportunity to take a class directly from Oddvar Holten was not only an inspiration but a privilege. Rolf Leirvik must also be acknowledged as a instructor for this course, demonstrating applications of the MET principles for both the spine and extremities.

The OGI eventually began teaching exercise courses outside the MET banner. The courses were renamed Scientific, Therapeutic, Exercise Progressions, or STEP. Becoming an instructor for these courses, I apprenticed not only under Ola Grimsby, but also under Rick Hobusch, one of only a few US born MET certified instructors, and Torhild Kvarekval, a Norwegian born and trained manual therapist instructing for the OGI. This mentorship as a clinician and instructor was invaluable. As most educators understand, learning is just beginning when you start to teach. Continuing to instruct STEP courses with Ola Grimsby has taken me all over the United States and to 10 other countries. His grasp on the principles of STEP continue to broaden my understanding. Despite my best efforts as an instructor, each class seemed to end with a sense of learning more than I taught. Each student brings a clinical and life experience that, when applied to the principles of STEP, creates and ever growing set of exercises, modifications and applications. I am grateful to the countless residency and course participants that continue to teach and serve as my inspiration to remain a student for life. Leading the battle for the removal of modalities for pain control as the primary form of orthopaedic physical therapy in the early 1970's, Ola Grimsby has been a driving force in shaping the face of orthopedic manual therapy. He fought for the incorporation of histology and tissue repair with exercise in the late 1970's, the inclusion neurophysiology for motor facilitation and pain inhibition in the early 1980's, the use of diet and supplementation for tissue maintenance/repair in the late 1980's and the push for residency training in manual therapy in the early 1990's as one of the Founding Fellows of the American Academy of Orthopaedic Manual Physical Therapists. Attempting to capture Ola's words in print, and to capture the clinical examples of STEP that have evolved with each course, has been the driving force behind the creation of Volumes 2 and 3 in this text series. The impact that the experiences of decades of clinical instructors, students and patients have been poured onto these pages is immeasurable.

Exercise rehabilitation remains one of the most difficult aspects of patient care for the orthopaedic manual physical therapist. The main thrust of these texts is to provide a scientific foundation for exercise design, dosage and progression. With this foundation exercises can be customized to each patient, to train a specific issue with an exact dosage. The exercises themselves are not as important as the concepts behind their design, dosage and progression. It is my hope that we have laid out a thought process, with specific dosage parameters, to arm the clinician with the ability to create, rather than to simply attempt to fit their patient into the design of a certain exercise, protocol or research article. These books should not only provide new and exciting exercise examples for the spine and extremities but also enhance the design and dosage of all exercises. New research continues to sharpen the focus of these concepts and we look forward to the continued evolution of this material.

— Jim Rivard

Table of Contents

Acknowledgments

Every book has two stories; one on the page and one unseen. As a work of non-fiction, the story in this book is easy to understand. However, the untold story involves students, teachers, clients, therapists and that relentless quest for knowledge that so invigorates the human mind. For many years, in classrooms and clinics and training sessions, those connected to the STEP curriculum have asked for such a book. Students from around the globe have sought such a publication. Their energy, desire, and emerging commitment to their work provided inspiration for all. Gratefully, we acknowledge the countless thousands of individuals that have in one way or another contributed to this publication. Those whose experience, knowledge, study, research and wisdom have enabled this book to evolve into a publication that has long-term value, inherent meaning and practical applications. Thanks to all who were involved with the books, course books and lectures from which this text is drawn. A token reference at the end of a sentence seems to fall short of acknowledging the time, dedication and inspiration of each author(s) of the hundreds of references cited in this text. Each articles represents the efforts of so many people, including subjects in the studies. Without the contributions of the Norwegian Medical Exercise Therapy (MET) group, in creating the foundation and sharing their knowledge and experience, this text would not have been possible. We also wish to thank the following for their insights and assistance with writing and editing this text: Torhild Kvarekval, Ben Grotenhuis, Laura Markey, Brian Power, Robin Schoenfeld, Asha Hossain and Dana Grant. Also of note is the exercise models and patients that provided their time to be photographed for this text. Thanks to all those who gave of their time simply because they cared. Thanks to those who shared their ideas because they believed. Thanks to those who saw the promise of so many possibilities. And most of all, thanks to those who will make the ink on these pages come to life.

Preface

Volume 3 of this text series is just that, Volume 3. This textbook can stand-alone as a reference for evidence related to motor dysfunction and exercises to restore function. Hundreds of examples for each joint system—from the cervical spine, thoracic spine and TMJ, to the shoulder, elbow and wrist—are demonstrated. But a deeper meaning of the text is lost if the reader simply thumbs through the pages searching for an interesting looking exercise to apply to a patient. This textbook is a companion text for Volume 1 (*Science, Theory and Clinical application in Orthopaedic Manual Physical Therapy: Applied Science and Theory*), which covers the integration of research in basic sciences and work physiology. Prior to initiating exercise a basic understanding of how to incorporate the fundamentals of—histology, neurophysiology, work physiology, exercise dosage and exercise progression—is required. It is not the exercises themselves but the design, dosage and progression chosen that determines the outcome.

This text, in a general way, will discuss the concepts of exercise dosage and progression with application to specific joint systems. More in depth discussions are provided in Volume 1 on all aspects of exercise design. References for cellular and tissue responses to mechanical stimulus are given, providing more clarity to early stage exercises shown. Descriptions of exercise equipment and types of exercise assist in making choices for exercise design. Detailed chapters on dosing exercise to achieve specific outcomes are provided, arming the clinician with the knowledge to modify almost any exercise to achieve almost any result. Knowing where to start is one thing, but understanding when and how to progress is another. Also outlined in Volume 1 is a logical approach to progressing exercise from the earliest stages of tissue repair all the way to returning to normal function.

The reader is encouraged to keep Volume 1 handy to review detailed discussion on the fundamentals of exercise design, dosage and progression. The following pages provide a limited summary of this information to assist with the dosage and progression parameters discussed in each chapter of this textbook, but this information cannot replace the contents of Volume 1.

STEP Principles: Summary of Resistance Training Recommendations

Quality	Muscle Action	Selection	Sequence	Resistance	Volume	Rest Intervals	Velocity	Frequency
Tissue Repair Pain & Edema Reduction	ECC & CON	SJ	Away from pain, or not into pain	From assisted, to 0% of 1RM, up to 50% of 1RM	1–5 set, 10–60 reps	≥1 minute	S	1–5x daily 7x/weekly
Tendinopathy	ECC only	Tendon specific	Mid to shortened range	As heavy as is pain free progress to mild soreness	1 set, 8–15 reps	None	S	1–2x daily 7x/weekly
Mobilization	ECC into restriction	SJ	Distraction then glides	Assisted to 50% of 1RM	1–5 set, 10–60 reps 3–5 reps if as a stretch	None	S	1–3x daily 7x/weekly
Coordination	CON emphasis ECC	SJ & MJ ex	Balance/Vestibular Planar Tri-planar/PNF	Assisted or 0%–50% of 1RM	2–5 sets, 30–50 reps	1 min.	Progress from S-M-F.	1–3x daily 7x/weekly
Vascularity	CON emphasis ECC	SJ & MJ ex	Into guarded pattern first then opposite	Pathology: 55% to 65% of 1RM Normals: 50–70% of 1RM for 10–15 reps (Kraemer)	1–3 sets, 25–30 reps	Acute: Until respiration returns to steady state Subacute: <30 sec.	Respiration rate	1x daily 6–7x/week
Endurance			For all groups:			For all groups:	For all:	
Novice	ECC & CON	SJ & MJ ex.	A variety in sequencing is recommended	50–70% of 1RM	1–3 sets, 24–30 reps	1–2 min. for high rep sets	S–MR	2–3x/week
Intermediate	ECC & CON	SJ & MJ ex.		50–70% of 1RM	Multiple sets, 24–30 reps	<1 min. for 10–15 reps		2–4x/week
Advanced	ECC & CON	SJ & MJ		30–80% of 1RM—PER	Mult. Sets, 25 reps or more—PER		M–HR	4–6x/week
Hypertrophy			For all groups:					
Novice	ECC & CON	SJ & MJ ex.	Large < Small	60–70% of 1RM	1–3 sets, 8–12 reps	1–2 min.	S, M	2–3x/week
Intermediate	ECC & CON	SJ & MJ ex.	MJ < SJ	70–80% of 1RM	Mult. Sets, 6–12 reps	1–2 min.	S, M	2–4x/week
Advanced	ECC & CON	SJ & MJ	HI < LI	70–100% 1RM with emphasis on 70–85%–PER	Mult. Sets, 1–12 reps Emphasis on 6–12PER	2–3 min. – VH; 1–2 min.—L–HM	S, M, F	4–6x/week
	ECC only			>1RM	1–3 reps only/1 set			1x per week per muscle
Strength			For all groups			For all Groups		
Novice	ECC & CON	SJ & MJ ex.	Large < Small	60–70% of 1RM	1–3 sets, 8–12 reps	2–3 min. for core,	S, M	2–3x/week
Intermediate	ECC & CON	SJ & MJ ex.	MJ < SJ	70–80% of 1RM	Multi. Sets, 6–12 reps	1–2 min. for others	M	2–4x/week
Advanced	ECC & CON	SJ & MJ ex. Emphasize MJ	HI < LI	1RMPER. > 1RM	Multi. Sets, 1–12 reps— PER		US–F	4–6x/week
Power		For all groups:	For all Groups:	For all Groups:		For all groups:		
v	ECC & CON ECC & CON ECC & CON	Mostly MJ	Large < Small Most complex < Least complex HI < LI	Heavy Loads >80%— strength Light Loads 30–60%— Velocity—PER > 1RM	Train for strength 1–3 sets, 3–6 reps 3–6 sets, 1–6 reps – PER 1–3 reps only/1 set	2–3 min. for core muscles 1–2 min. for other muscles	M F F	2–3x/week 2–4x/week 4–6x/week 1x per week per muscle

KEY: ECC, Eccentric; CON, concentric; ISO, isometric; SJ, single-joint; MJ, multiple-joint; ex., exercises; HI, high intensity; LI, low intensity; 1RM, 1-repetition maximum; PER., periodized; VH, very heavy; L-MH, light-to-moderately-heavy; S, slow; M, moderate;, US, unintentionally slow; F, fast; MR, moderate repetitions; HR, high repetitions. Modified from: Kraemer et al. (2002), Kraemer and Newton (2000) and Deschenes and Kraemer (2002).

STEP Concepts: Stage 1

- 3–5 Exercises
- Many repetitions/Minimal resistance, primarily for coordination, but also for:
  - < 50% of 1RM (mobilization).
  - < 60% of 1RM (local endurance).
  - < 50% of 1RM (tissue repair, edema reduction and range of motion).
  - < 40% of 1RM if atrophy (hypertrophy)
- Low speed (for improved coordination):
  - Deformity of collagen with slow stretch.
  - Concentric facilitation improved.
- Range of training:
  - Hypomobile joint: train in the outer range of available normal physiological motion.
  - Hypermobile: train in the mid to inner range of available normal physiological motion.
- Selective tissue training (STT): provide the optimal stimulus for repair of bone, muscle, collagen, cartilage, disc, etc.
- Neurophysiological influences of training: pain inhibition and coordination.
- Normalization of breathing patterns and posture.

Options

- Joint in resting position, start contraction from length tension position.
- Pure Concentric work for improved vascularity:
  - Use a pulley with specific wheel to remove the eccentric phase.
  - Weight stack rests between repetitions.
  - Avoid isometric work.
- Joint locking may be required to avoid motion around a non-physiological axis.

STEP Concepts: Stage 2

- Increased repetitions with an increased number of exercises (5–10 exercises).
- Increase repetitions with additional sets (endurance).
- Increase speed/not weight (strength/endurance).
- Combined concentric and eccentric work for further tissue tension accommodation.
- Isometrics to fix strength in shortened position of muscles that work eccentrically into the pathological range of motion.
- Body/limb position changes from recumbent to more dependent.
- Integration of balance training and normalization of ocular reflexes.
- Planar motions with exercise to full range/partial range tri-planar (diagonal patterns).
- Remove locking or change to less aggressive type.
- Histological influence of increased lubrication with increased speed.

STEP Concepts: Stage 3

- Increase weight (60–80% of 1RM), decrease repetitions (strength).
- Change work order: eccentric to concentric work to stabilize into the newly gained range.
- Isometrics to fix strength in gained range closure to the pathological range.
- Tri-planar motions in available range (diagonal patterns).
- Progression toward more functional motions and exercise application.

STEP Concepts: Stage 4

- Tri-planar motion through full range of motion around the physiological axis (coordination, endurance, strength and hypertrophy).
- Endurance, hypertrophy, speed, strength and power (80–90% of 1RM).
- Functional exercises for retraining of activities of daily living, sport and job activities.

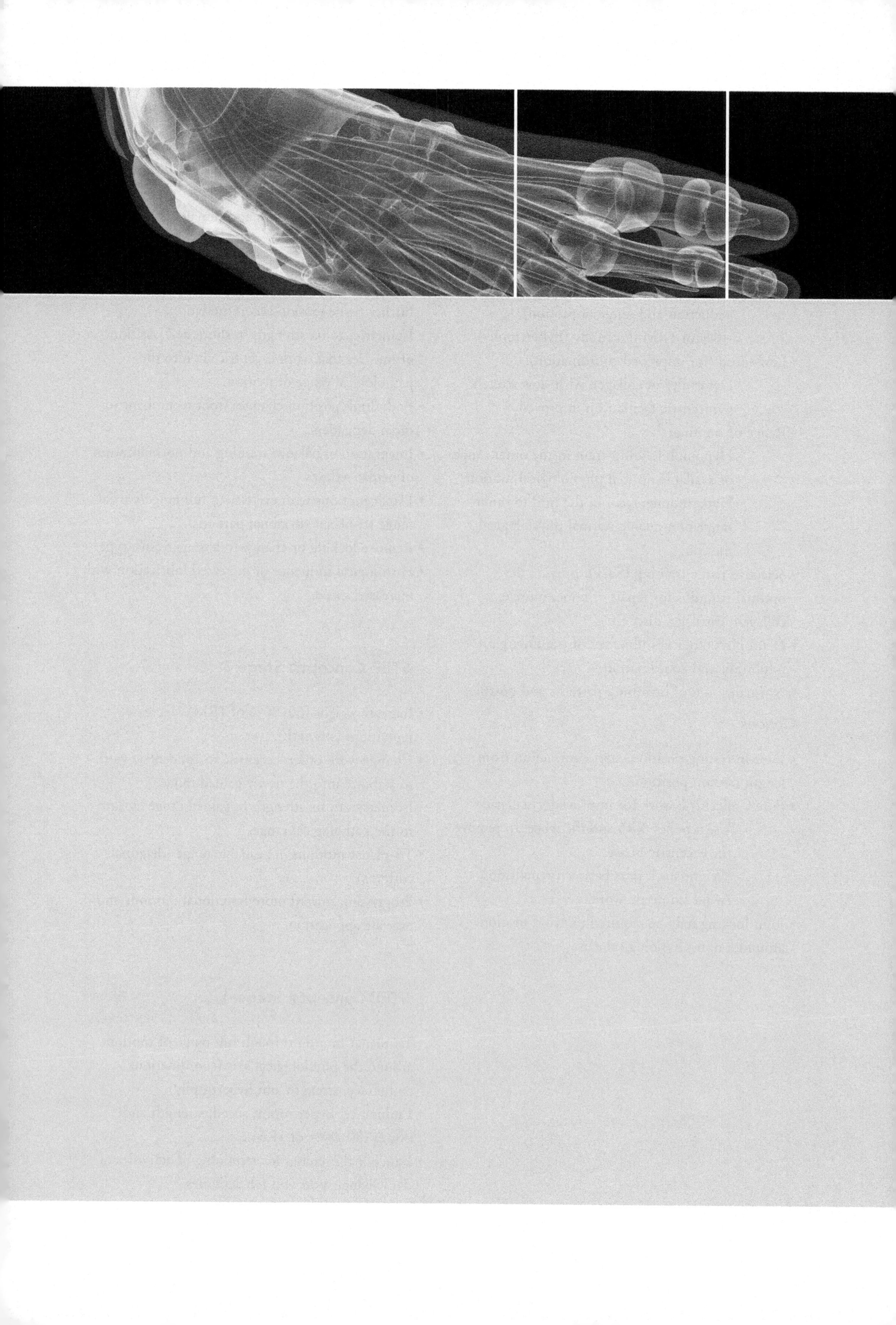

Authors: Jim Rivard, Dan Washeck, Micheal Maninang, Ola Grimsby

CHAPTER 1

Exercise Rehabilitation of the Ankle and Foot

Introduction

In this chapter:

The intent of this chapter is but to provide a logical treatment model for progression and dosage of exercise in the rehabilitation of common foot and ankle pathology. It is assumed the reader is familiar with anatomy, biomechanics and pathology associated with the lower limb. In a basic sense, the foot and ankle provide the stable contact points of the lower extremities to allow normal weight bearing and locomotion functions to occur. This stability is required not only to provide weight bearing support but also to provide leverage for propulsion; however, these are not the only functions of the foot and ankle. They also function to provide the lower extremity with shock absorption capability and an ability to adapt to different surfaces during gait. For these two latter functions, mobility is needed. For these purposes the foot and ankle need to be stable and mobile at the same time to allow for a normal gait pattern. The complexity of the foot and ankle is indicated by the number of bones and joints involved: 26 bones, two to five or even more sesamoid bones, seven principal joints and a number of lesser joints.

As with all joint systems, the source of pain can come from a multitude of tissues. Acute tissue injury in the ankle complex will alter the distal lower extremity's ability to perform, affecting entire body function. Joint restriction will limit mobility and function. Hypermobile joint motion will allow excessive motion, further stressing tissue. Even with non-injured tissue and normal joint mobility, muscles need to coordinate and control motion to attenuate ground reaction forces transmitted into joint structures. Without proper motor control from the lumbar spine to the foot, motions at the foot and ankle can occur through too far a range and/or too fast. Biomechanical restriction for any reason in the chain of joints from the lumbar spine to the foot needs to be addressed.

Segmental dysfunction in the lower lumbar spine can lead to inhibition, or altered motor patterns, of muscles innervated at these levels. For example, gluteal weakness from a pain free lumbar segmental dysfunction at L4 can alter force attenuation at the hip leading to excessive forces transmitted through the foot. An L5 dysfunction may inhibit peroneal function leading to a primary ankle sprain. The orthopedic manual physical therapist addresses not only the biomechanical chain of joint systems, but also the neurological chain of segmental innervation when addressing orthopedic impairment. The acute tissue injury is addressed but underlying impairments for the root cause must also be identified and addressed. The reader may need to integrate concepts outlined in the knee, hip, pelvis and lumbar chapters of this text to address the consequences of more complex ankle dysfunction.

Basic rehabilitation involves addressing primary tissue repair, normalizing joint mobility in the biomechanical chain, providing external support when needed, retraining the motor system and finally returning to normal function. Manual interventions of soft tissue massage and joint mobilization cannot be overlooked as a basis for treatment, but will not be covered in the context of this chapter. Exercise is the primary intervention to improve tissue repair and restore function.

Early exercise has been demonstrated to be safe and effective for acute ankle sprains (Brooks et al. 1981, Weinstein 1993, Eiff et al. 1994, Glasoe et al. 1999). Immobilization has been extensively documented to decrease protein synthesis for ligament repair, limit effective matrix organization and reduce collagen strength. Early mobilization, compared to casting, has been associated with an earlier return to work, return to activities of daily living, better mobility and less muscle atrophy (Eiff et al. 1994, Klein et al. 1993). This chapter will attempt to outline more common patterns of dysfunction for the ankle, as well as exercise-based interventions to correct them. Specific exercise examples are given, but the intension is not to provide the reader with a few exercises to use for ankle pathology. The primary goal is to provide evidence for a exercise dosage and design, along with a logical progression from acute injury to function. Integration of basic sciences with work physiology provide a framework for exercise rehabilitation.

Section 1: Stage 1 Exercise Progression Concepts for the Ankle and Foot

Training Goals

The basic components of an initial exercise program are to 1) normalize joint motion in the biomechanical chain, 2) provide tissue repair stimulus, 3) normalize motor patterns including segmental dysfunction in the neurological chain, 4) restore function motion including balance and gait 5) and finally to elevate the overall training level, preferably to a greater level then prior to injury. Emphasis is on pain inhibition, providing tissue stimulus for repair, facilitating muscle for restoring coordination and improving range of motion. Several training goals may be trained in the initial sessions. Typically restoring joint mobility is the first goal, but with a more acute trauma, tissue repair training may precede addressing mobility issues. The order in which these more acute impairments are addressed is patient dependent.

Passive intervention prior to exercise is typically necessary. Soft tissue work to muscles and fascia will reduce pain, decrease joint compression forces, improve coordination by reducing abnormal spasm and assist in the lubrication of tissues. Joint mobilization can serve to improve motion through plastic deformity of tight capsules, increasing lubrication of collagen and cartilage, inhibit pain, inhibit muscle guarding and facilitate joint receptors for normal muscle recruitment and proprioception.

Initial exercises used for the ankle and foot depend on the type of injury, whether or not it is macrotraumatic (sprain) or microtraumatic (tendonitis/tendinopathy). For macrotraumatic injuries early emphasis may be on edema control, pain inhibition, normalizing muscle guarding and select tissue training for repair. For microtraumatic injuries or chronic pain and dysfunction, biomechanical and gait analyses are prioritized to remove factors leading to overuse and cumulative injury. Proper evaluation of the entire lower quarter and lumbar spine should not be overlooked when considering microtraumatic ankle injury. Orthotics and shoe modifications are important issues that may need to be addressed with issues of foot deformity, gross hypermobility (grade 5 and 6) or with severe motor dysfunction. When possible, joint mobility and motor impairment should be addressed first, prior to utilization of orthotics. Normalizing these impairments may eliminate the need for external devices. The use of orthotics will not be addressed in this chapter.

Joint / Myofascial Mobilization

Joint mobility must be restored prior to attempting exercise to normalize motor function. Passive joint mobilization is the most specific intervention to evaluate and restore joint motion, and may be necessary prior to attempting exercise. Passive mobilization has been shown to be an effective adjunct to active care (Green et al. 2001, Pellow and Brantingham 2001, Collins et al. 2004, Esenhart et al. 2003, Young et al. 2004). Restoring joint mobility not only allows for normal range of motion, but also normalizes afferent feedback from soft tissue structures that aides in motor pattern recruitment and proprioception. Restoring joint motion, through passive techniques or exercises designed to produce arthrokinematic distraction or glides, can produce a neurological improvement in motor performance prior to direct training.

Exercise for tissue repair may also be required with reduced joint play and limited range. Evaluation procedures should determine whether limited active range is due to collagen restriction (hypomobile model) or from muscle guarding (hypo- and hypermobile models). Repetitive motion will improve collagen elasticity and cartilage lubrication, as well as creating afferent input to reduce pain and normalize muscle guarding, while static holds of arthrokinematic motion creates plastic deformity of joint capsules. Mobilization exercises are dosed between 30–50% 1RM at 30–50 repetitions for elasticity and neurological influences or sustained holds are performed for 15–30 seconds for plastic deformity of collagen.

The ankle and foot is a complex system of articulations that work together. Passive mobilization of these articulations may be necessary to allow for active exercise. The manual therapist can perform mobilizations to all of these articular surfaces but the patient may also be taught to perform simple techniques in the ankle, forefoot and toes. Below are a few examples of self mobilizations exercises.

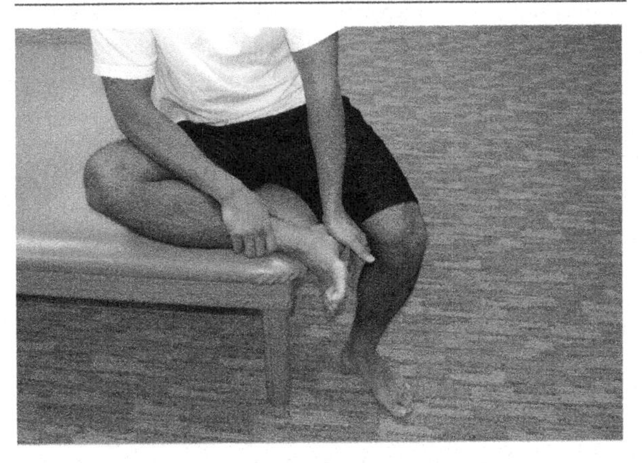

Figure 1.1: Self-mobilization for medial and lateral glide of the calcaneus on the talus. In sitting the lower limb is externally rotated to place the heel on the edge of a chair. The fixation hand holds the talus anterior and provides a pad between the ankle and the chair. The mobilizing hand holds the calcaneus performing medial and lateral glides for either oscillatory or stretch techniques.

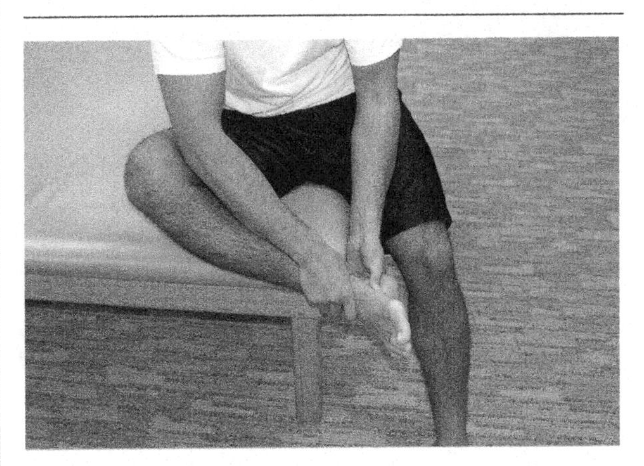

Figure 1.2: Self-mobilization of the subtalar joint or talocrural joint for anterior and posterior translation. The foot is crossed over resting on the opposite thigh. The ankle is gripped with a hand posterior on the calcaneus and anterior on the talus. Squeezing the hands together creates a relative posterior glide of the talus on the calcaneous. The hand placement can be shifted for an anterior glide, or moved to the talocrural joint for similar translatoric glides.

Figure 1.3: Self-mobilization of the great toe into distraction and dorsal glides to promote extension are performed in sitting with involved leg crossed on top of the uninvolved thigh. The patient can be instructed in self mobilization of the tarsometatarsal (TMT) and interphalangeal (IP) articulations for distraction, translatoric glides (anterior, posterior, medial and lateral) and spin (internal and external rotation).

Active mobilization with external forces improving distraction or translatoric glides of arthrokinematic motions can be performed as stretch articulations in non-weight bearing as well as more functional positions. Using pulley systems with straps, an external force moment is created attempting to replace the hands of the manual therapist in for self mobilization exercises. Though not as specific as manual mobilization, self mobilizations provide more independent options for joint mobilization.

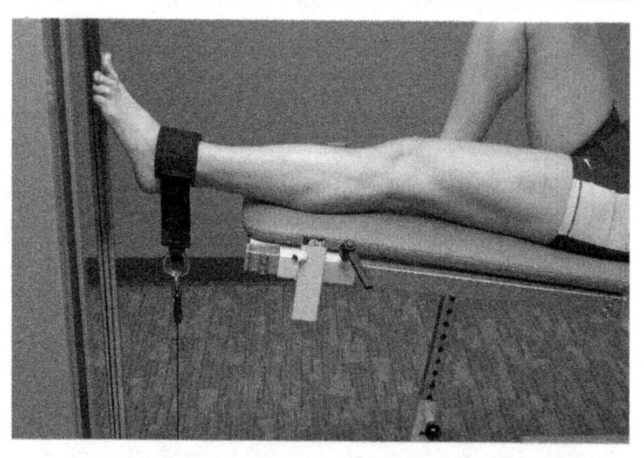

Figure 1.4: Non-weight bearing posterior glide of the tibia and fibula on the talus to mobilize plantarflexion. In supine, the foot rests on a wall, bolster or with the foot braced on a wall pulley (pictured). A vertical force moment from a wall pulley is placed as close to the joint line of the talocrural as possible. Mobilization with movement can be performed with active plantarflexion by pushing the foot into the pulley support. Passive sustained stretch can also be performed by relaxing the ankle allowing the pulley strap to perform the posterior glide of the tibia on the talus.

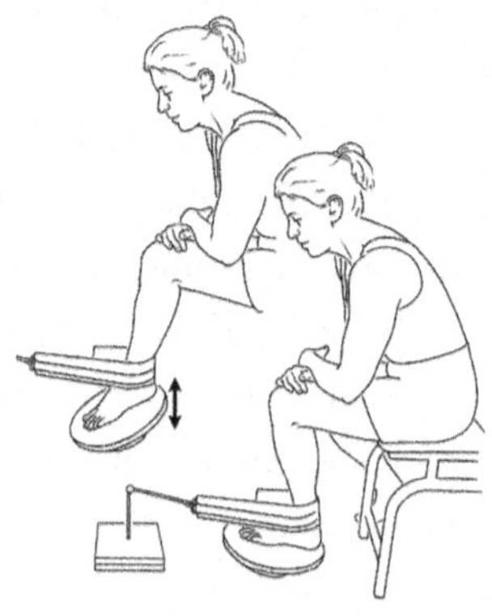

Figure 1.5: Partial weight bearing anterior glide of the tibia and fibula to mobilize dorsiflexion in sitting. An anterior force moment is placed on posterior at the joint line of the talocrural joint with the heel on the edge of a block while performing dorsiflexion.

Figure 1.6: Full weight bearing anterior glide of the tibia and fibula to mobilize dorsiflexion in standing. An anterior force moment is placed on posterior at the joint line of the talocrural joint with the heel on the edge of a block create slight plantarflexion placing the joint in loose packed position and avoid mobilization in a closed packed position. With limited dorsiflexion the closed packed position will occur much earlier in the range than in the normal ankle. A squat is performed to move the tibia on the talus for dorsiflexion in weight bearing.

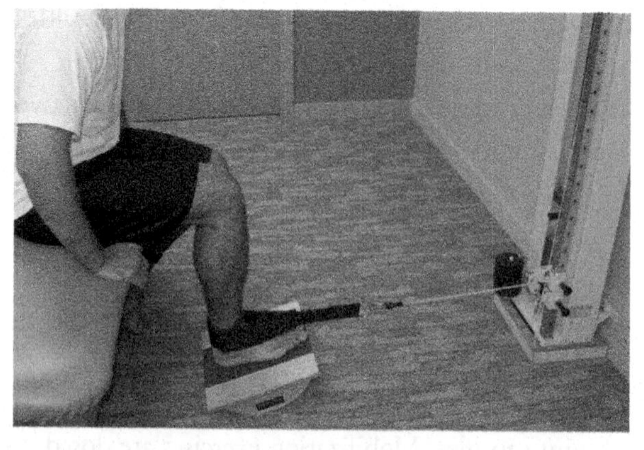

Figure 1.7: Antterior glide of the tibia and fibula on talus to actively mobilize dorsiflexion. A posterior force moment is created as close to the talocrural joint line as possible while

the patient actively dorsiflexes the ankle. The heel is placed on the edge of a step to allow the forefoot to freely move down. The same technique can be performed in sitting to reduce body weight through the ankle.

Sustained Stretching versus Dynamic Stretching

Passive testing of the joints may reveal normal joint motion but range of motion being limited by myofascial restriction. Passive stretching is commonly avoided in Stage 1 as pain and muscle guarding define the Stage 1 tissue state. Attempting to stretch muscle in guarding mediated by neurological influences of pain can increase tone and irritability. With the initial training goals of rehabilitation being coordination, endurance and eventually strengthening, passive muscle stretching might even hinder proper healing as it weakens newly synthesizing collagen bonds. Early training is focused on attaining normal patterns of coordinated muscle recruitment. Stretching may have a negative impact on this goal, but even if the effect is minimal, stretching provides no significant positive goal in early training. Stretching has also been shown to inhibit muscle performance, temporarily reducing endurance, strength and power.

Marek et al. 2005 found that both static and contract-relax proprioceptive neuromuscular facilitation stretching caused similar deficits in strength, power output and muscle activation at both slow (60 degrees per second) and fast (300 degrees per second) velocities. Nelson et al. (2005) recommended that heavy static stretching exercises be avoided prior to any performances requiring maximal muscle strength and endurance, finding a reduction of up to 28 percent. These findings were consistent with previous studies on static stretching (Nelson and Kokkonen 2001) and ballistic stretching (Kokkonen et al. 1998). A dynamic stretching approach that involves active motion dosed to increase local circulation to tonic muscles in guarding is more successful in restoring normal resting tone and increasing range of motion. Dosing training at 60% of 1RM with three sets of 24 repetitions will maximize local circulation for this effect. Involved muscles work eccentrically toward the lengthened range to improve mobility.

Figure 1.8: Dorsiflexion mobilization: anterior glide of the tibia on talus during active squatting. An anterior glide of the tibia is performed with a belt around the distal tibia and fibula. The ankle is placed in slight plantarflexion by standing on foam wedges. The plantarflexed position places the talocrural joint in its new relative resting position. Attempting to mobilize an anterior translation of the tibia too close to the end range of available dorsiflexion will be unsuccessful, as this represents the new closed packed position in which no gliding can occur. The patient is instructed to squat down allowing the knees to move forward, thus creating dorsiflexion at the ankle. Anterior ankle pain is avoided throughout the procedure. Often this mobilizing exercise is used to resolve an anterior impingement of the talocrural joint during squatting on a level surface.

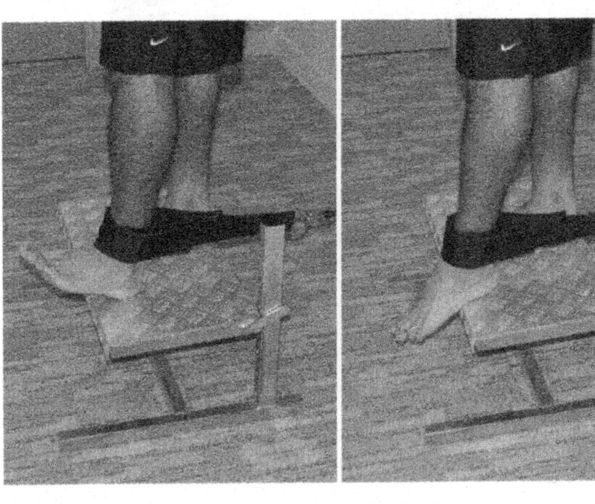

Figure 1.9a,b: Plantarflexion mobilization during active plantarflexion in a partial weight bearing position. The patient is standing on a step with a mobilizing force moment from a wall pulley providing a posterior moment of the distal tibia and fibula on the talus. The patient performs an active plantarflexion with the forefoot off the edge of the step. The pulley increases the relative posterior glide of the tibia and fibula necessary for full plantarflexion. This mobilization is somewhat more difficult to coordinate then the previous anterior glide with squatting, as the foot is free with only partial weight bearing. The patient may require upper quarter stabilization to reduce the balance challenge during the exercise. The motion is performed slowly with emphasis on the end range of plantarflexion.

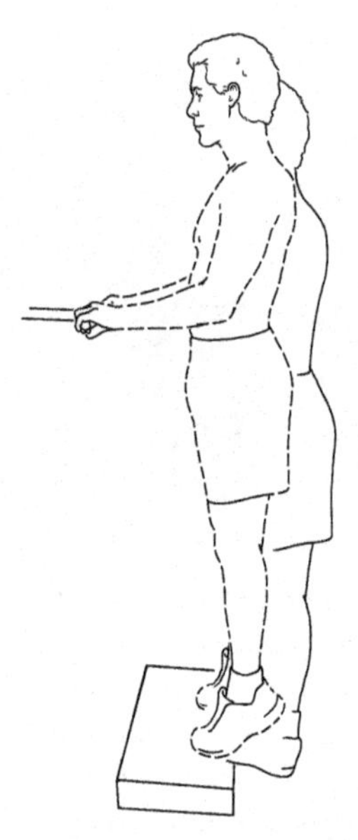

Figure 1.10: Standing calf raise. Dorsiflexion range can be improved with active training of the plantarflexors (60% of 1RM for three sets of 24). Eccentric motion is performed to the end range to dynamically lengthen the muscle. Performing contractions slowly over five seconds may improve the lengthening effect (Bandy et al. 1998).

Figure 1.11a,b: Anterior lunge with anterior glide of tibia. A combination of joint mobilization and dynamic muscle lengthening can be performed with the anterior lunge. The anterior force moment at the tibia mobilizes the talocrural joint for dorsiflexion while the lunge motion eccentrically lengthens the gastrocnemius and soleus. The heel must maintain contact with the floor and the knee remains in extension to achieve full muscle lengthening. Deceleration of the tibia on the talus is a more functional approach to training the plantarflexors, as opposed to performing standing plantarflexion, as this is their function during gait.

Tissue Repair / Edema Resolution / Pain Inhibition

Early tissue training of the ankle is an effective way to reduce pain, provide lubrication and nutrition to tissue, normalize fluid pressures and reduce muscle guarding. The selective tissue training model involves applying the optimal stimulus for the primary tissue(s) in lesion to stimulate cellular metabolism for the products of fiber (type I collagen) and glycosaminoglycan (GAG). New collagen production will assist in repairing the damaged tissue and increase the tensile force properties. GAG production will assist with nutrition and increase elasticity, which also improves tensile force properties. For repair, collagen requires modified tension in the line of stress. This concept is true for all collagen including the tendons, ligaments and joint capsules of the ankle and foot. Low resistance repetitive motion, active or passive, is performed in the line of stress of the injured collage with the joint system moving around a physiological axis. Pain is avoided at all times to avoid excessive tensile forces signaling collagen damage. Motion is also performed without symptoms of impingement, suggestive of the joint moving around a non physiological axis creating tissue irritation.

With acute collagen injury, overuse tendinopathy or surgical repair, all early intervention should focus on improving the ability to tolerate tension. Repetitive motion improves elasticity, repair and improved tensile strength, while immobilization, heating and stretching are associated with reduced tensile strength (see chapter 3). A specific tendon or ligament injury dictates a very specific direction of motion to train. The ankle capsule, however, has fiber orientation in all planes. Injury may be associated with one specific direction, such an in inversion sprain, but motion in all planes will provide tension to the collagen to promote cellular activity for tissue repair.

The cartilage in the joint articulations of the foot and ankle requires compression and decompression with gliding to provide the optimal stimulus for repair. This is achieved with the normal roll gliding in the joint with active or passive movement. The key is creating motion around a normal axis with normal compressive forces, not impingement forces.

The compressive force mechanically stimulates the chondrocytes to produce fiber (type II collagen) and GAG. The gliding component assists in making the synovial fluid less viscous to increase joint lubrication and improve nutrition transport through the chondrosynovial membrane.

Bone will benefit from early biomechanical energy in the line of force. This can occur through modified weight bearing training and/or muscle contraction forces.

Following lateral malleolar fracture, Hedstrom et al. (1994) found no benefit from early active range of motion exercise, compared to a control group using a walking cast. But only non-weight bearing, unloaded training, was performed. Early motion can reduce pain and inflammation, but combined with progressive controlled weight bearing, improved fracture healing can occur. Sondernaa et al. (1986) found improvements with active movements compared to immobilization up to six weeks, equalizing after one year. Finsen et al. (1989) found little difference at two years comparing immobilization to early movement. Long-term outcome is likely to be more similar, but early healing can be enhanced with motion. Early motion has been shown to be beneficial in ankle fracture healing and functional outcomes (Vasili 1957, Burwell HN, Charnley 1965, Segal et al. 1985).

Tissue repair emphasis requires a low level of resistance, from zero to no more than 50% of 1RM, to allow for thousands of repetitions daily (Holten and Grimsby 1986). Kern-Steiner et al. (1999) utilized this concept by applying 90–100 repetitions of unloaded weight bearing squats and heel rises to promote ligament healing after an inversion sprain. The type IV mechanoreceptor is the key measure of "modified" tension or compression. Caution should be used with cryotherapy and medication used to block the pain signals, as this may allow for excessive stress or strain to injured tissues during exercise. Tissue training is best when pain signals can be fully perceived to serve as a gauge for the range of safe "modified" tension in the line of stress.

Tissue Repair, Pain inhibition, Edema Reduction and Joint Mobilization

- *Sets*: 1–5 sets or training sessions daily
- *Repetitions*: 2–3 sets of 40 or more
- *Resistance*: <25% of 1RM
- *Frequency*: Multiple times daily (2–4 hours)

Coordination

- *Sets*: 1–5 sets or training sessions daily
- *Repetitions*: 2–3 sets of 30 or more
- *Resistance*: Zero to <50% of 1RM
- *Frequency*: One to three times daily

Vascularity / Local Endurance

- *Sets*: 2–3 sets or training sessions daily
- *Repetitions*: 2–3 sets of 24
- *Resistance*: 50–60% of 1RM
- *Frequency*: One time daily

Muscle in Atrophy

- *Sets*: 2–3 sets
- *Repetitions*: 2–3 sets of 40 or more
- *Resistance*: 30–40% of 1RM
- *Frequency*: One time daily

Direction and Motions

Differential diagnosis is required to identify the tissue in lesion, or at least the directions of tissue injury. The more specific the diagnosis the more specific exercise dosage can be to provide the optimal stimulus for tissue repair. The name of the pathology is not always as important as the tissue involved. As an example, a lateral collateral ligament strain involves collagen as the tissue in lesion. Rather than performing random "ankle alphabet" motions often prescribed with acute ankle sprains (Glasoe et al. 1999), specific directions and ranges are chosen to protect the injured tissue while providing the optimal motions for tissue repair. For collagen the optimal stimulus is modified tension in the line of stress. In severe injury or post surgical situations, motion may begin in a plane other than the one of collagen injury. Motion in the plane of injury begin by moving away from the direction of pain keeping motion in the mid to beginning range of collagen tension but never taking it to the end range tension. As the collagen begins to repair, the range of motion can move further toward the end range but is never taken to a range of pain.

Sequencing of Direction of Collagen Tissue Training

- *Movement is initiated in the pain free planes of motion*
- *Movement in painful plane of motion, but away from pain*
- *Movement in painful plane of motion, toward pain but not into painful range*

Cartilage pathologies are trained in a similar manner by emphasizing pain free planes of motion and moving away from pain in the pathological plane. Cartilage also frequently requires unloading to perform modified compression and decompression with gliding. This may require non-weight bearing training or an unloading device.

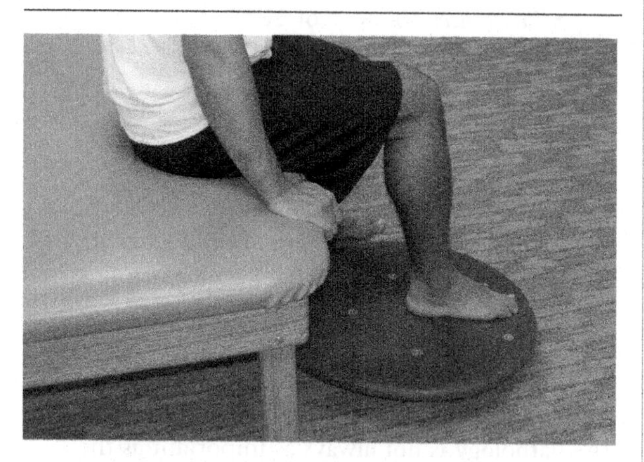

Figure 1.12: Sitting Biomechanical Ankle Platform Systems (BAPS®) board for tissue training. Sitting, in partial weight bearing, modifies the tissue strain while the direction of training can be specifically controlled. A smaller ball on the bottom of the board will limit the excursion of range. The direction of training can also be controlled, to prevent end range tissue strain, as in full inversion with a lateral collateral ligament strain, place a block or towel roll under the lateral side of the BAPS®.

Example of Initial Exercises for a Lateral Collateral Ligament Strain

All exercises are performed from zero to 50% of 1RM in the pain free range with three to five sets of 50 repetitions. When possible, training in elevation will help to control edema by reducing the hydrostatic pressure of the blood while the active motion will assist in venous return. Training five to ten times daily in the first week will have a greater impact on tissue repair, pain control and resolving edema than once a day for a longer length of time.

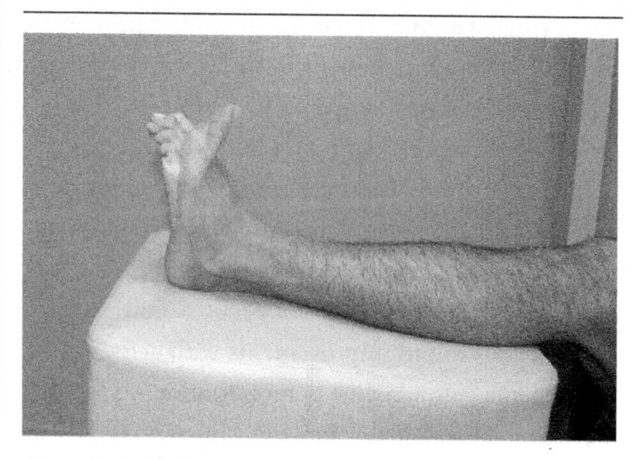

Figure 1.13: Supine active toe curl. The patient is in non-weight bearing with the foot elevated. The digits are alternately flexed and extended in the pain free range. The motion is performed slowly to avoid pain but to also emphasize elongation of restricted tissues.

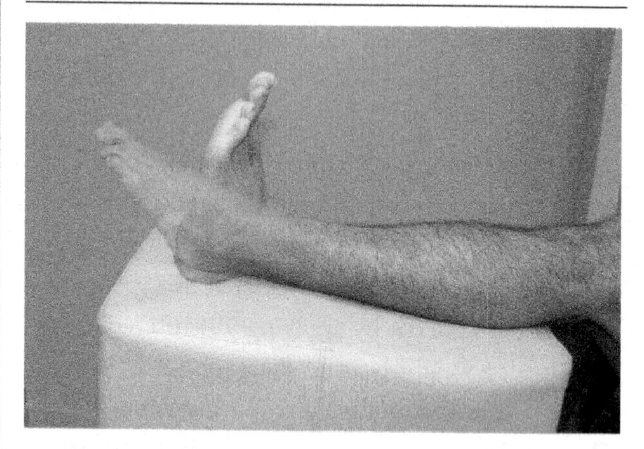

Figure 1.14: Supine ankle dorsiflexion and plantarflexion in the pain free range. The patient is in non-weight bearing with the foot elevated. The ankle is alternately dorsiflexed and plantarflexed in the pain free range of motion. The patient is instructed to perform the motion slowly to avoid pain and to emphasize elongation of restricted tissues. Combining motion of the digits will further emphasize movement of fluid in the region. If the patient is comfortable with the heel not supported, a greater range of motion can be achieved with the heel free to move off the edge of a bolster.

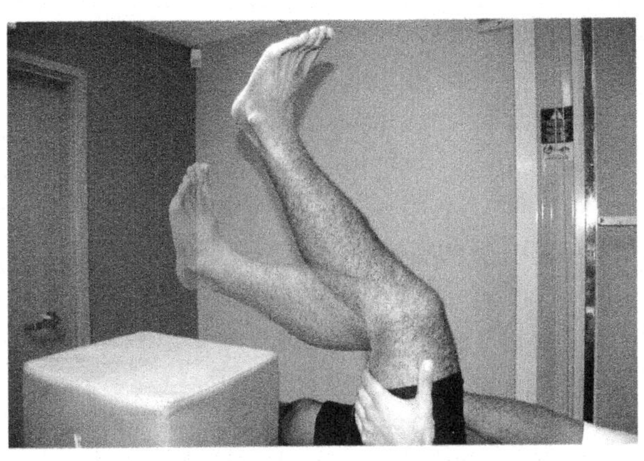

Figure 1.15: Supine knee extension to promote circulation in the lower limb. The patient is lying in supine with the leg supported vertically by the arms. The knee is alternately flexed and extended, emphasizing quadriceps function, to move fluid through the lower limb with gravity assisting the venous pump created by muscle contraction.

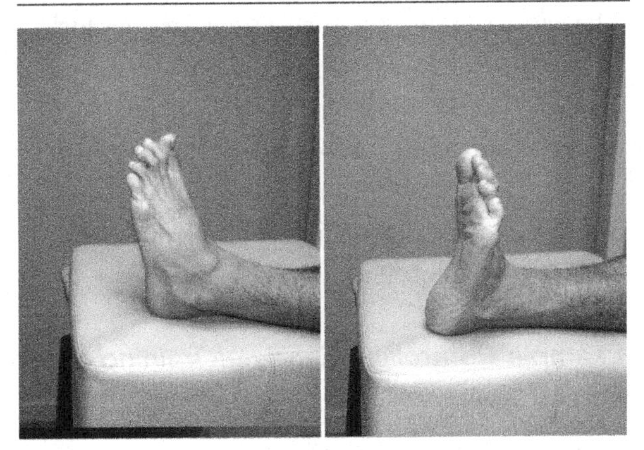

Figure 1.16a,b: Supine inversion and eversion of the ankle, perpendicular to gravity. The ankle is supported in elevation with active inversion and eversion while gravity is eliminated. For ligamentous injury such as an inversion sprain to the lateral collateral ligament, emphasis is on moving away from the direction of tissue injury into eversion, avoiding the end range position of inversion.

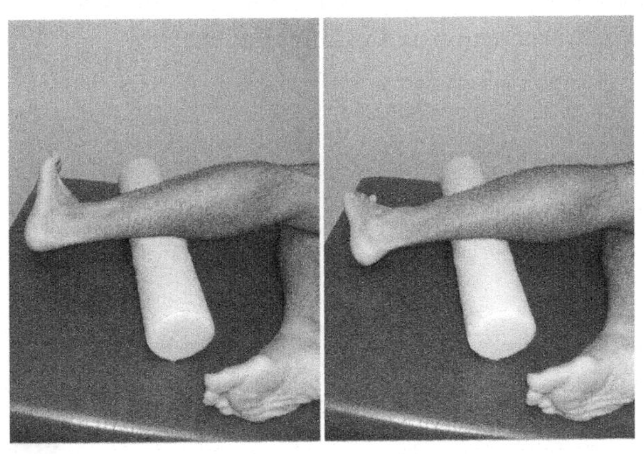

Figure 1.17a,b: Sidelying eversion against gravity. The patient is sidelying with the lower leg supported on a bolster. The table blocks the inversion range, artificially blocking the painful range of inversion.

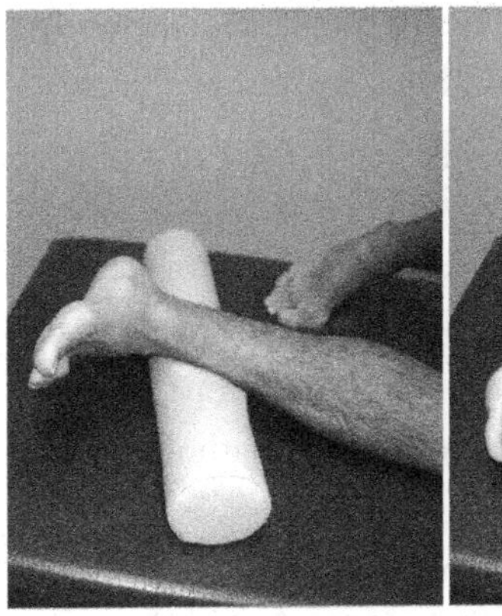

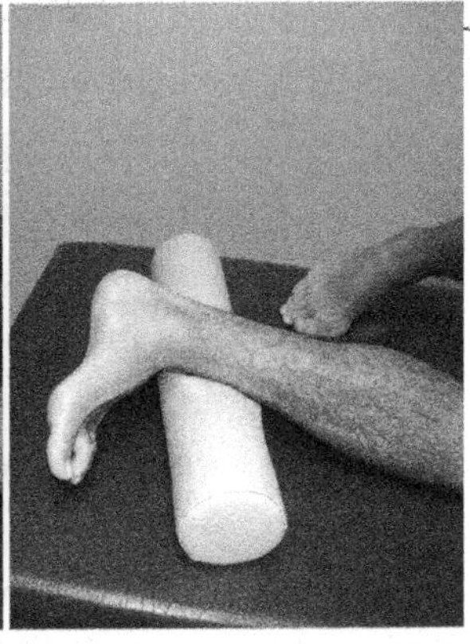

Figure 1.18a,b: Side lying inversion against gravity from a full eversion range to neutral, avoiding the inversion range.

The initial movement directions of the Stage 1 exercises are chosen based on contraindications, tissue presentation, presence of pain, direction of joint restriction or hypermobility, muscles in guarding and/or muscle performance issues. Selecting specific exercises and directions should follow a logical thought process. A thorough evaluation should make these choices self-evident. Initial exercises focus on joint mobilization and tissue repair, both of which lead to reduced pain levels. Next muscles in guarding are normalized with an emphasis on training for local vascularity into the pattern of guarding. Once joint motion is reestablished—through joint mobilization, resolution of edema and/or resolution of tonic muscle guarding—training for motor performance can begin. Specific theory outlined in Volume 1, Chapter 8 in this book series on Exercise Prescription, can be summarized as:

1) Train away from pain. Chose movements that involve non-injured joints, other planes of movement or in the same plane but away from the direction of pain.
2) Tissue Injury. Provide the optimal stimulus for the tissue injured, up to but not into pain. Collagen injuries are trained from mid to shortened range of collagen tension while bone and cartilage injuries are trained with reduced loads in the pain free range.
3) Joint Hypomobility. Provide external resistance to duplicate arthrokinematic glides or joint distraction with sustained stretch for greater than

10 seconds to deform collagen. Osteokinematic motions are used with high repetitions toward the restriction for collagen elasticity.

4) Joint Hypermobility. Train concentrically away from the hypermobile directions to provide tissue stimulus in the shortened range and begin coordination training of the muscles that function eccentrically toward the pathological range.
5) Muscle Spasm. Train tonic muscle guarding into the guarded pattern with 60% of 1RM for one to three sets of 30 to 24 repetitions, respectively.
6) Muscle Performance (coordination – facilitation). Identify motions, or specific muscles, lacking in recruitment, timing, coordination and/or endurance and train in simple patterns.
7) Weight bearing. Train at the highest level of weight bearing within the range of tissue tolerance and coordination.

Concentric Emphasis

Traditionally, isometric work has been used first in many rehabilitation protocols to avoid movement that may potentially injure tissue or for strengthening. Strength is one of the last functional qualities needed for motor performance, at least after coordination and endurance have been established. Isometric work should be avoided in the initial stages of rehabilitation as it fails to provide motion for tissue repair, reduction in effusion, pain inhibition and coordination training. All of these issues are positively influenced with specifically designed, dosed and controlled dynamic exercises. Tonic muscle guarding will reduce mobility and contribute to what may be clinically tested as a joint restriction, capsular pattern or a "short" muscle. In the foot and ankle this is most commonly seen in the deep predominantly tonic plantarflexors such as the soleus, reducing dorsiflexion range. Exercise into the guarded pattern, or away from pain, increases the eccentric range of the muscles improving flexibility. Tonic muscle guarding of the soleus muscle leading to lost dorsiflexion can be addressed with concentric plantarflexion, using the solues, providing oxygen to the muscle to assist in resolve guarding. The eccentric return allows for controlled repetitive motion toward dorsiflexion to improve muscle extensibility and collagen elasticity in tendons and joint capsules. As listed previously, the dosage to most efficiently improve local circulation for vascularity is 55–65% of 1RM or one to three sets of 24–30 repetitions.

Training pure concentric work, removing the eccentric phase, is used to emphasize local circulation and development of capillaries in the more significantly low trained state or with post surgical states. Pure concentric training increases capillary density, which is not associated with eccentric work (Hather et al. 1991). Pure concentric work is most easily performed clinically with an eccentric-stop wheel attached to a pulley system. The wheel spins for the concentric phase but locks during the eccentric phase removing the resistance and lowering the weight for the patient. The eccentric stop wheel can be replaced by allowing the weight stack or foot to come to rest between each repetition to decrease the intramuscular pressure and allow the muscle to fill with blood.

Example Controlling Range of Training for Lateral Collateral Sprain

Collagen injury with a lateral collateral ligament sprain requires modified tension in the line of stress, in the pain free range. End range inversion is avoided with all exercises, as this position places maximal tension on the lateral collateral ligament. All exercises are dosed at < 50% of 1RM to allow for 30–50 repetitions, emphasizing tissue repair, edema reduction and pain inhibition.

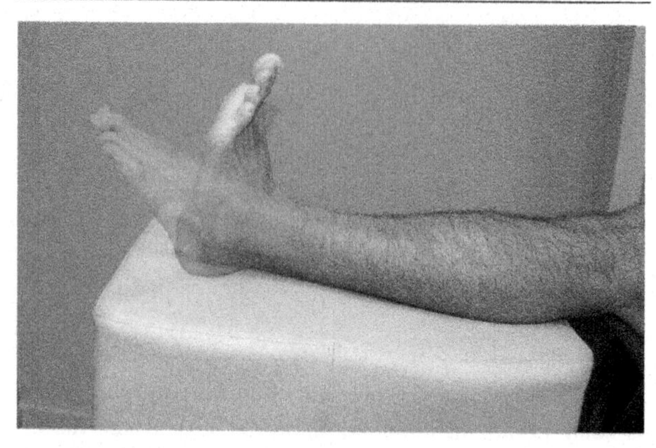

Figure 1.19: Toe flexion and extension. Movement of the

toes, without ankle motion, provides muscle contraction to move fluid and create afferent input for pain inhibition.

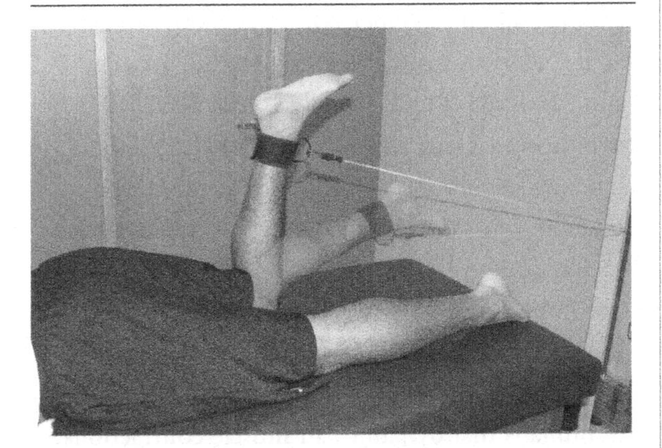

Figure 1.20: Knee flexion and extension. Similar to the previous exercise, knee flexion and extension without ankle motion provides large muscle contraction to move fluid in the lower limb. Extension can be performed in supine to maintain foot elevation, with flexion performed in prone or standing.

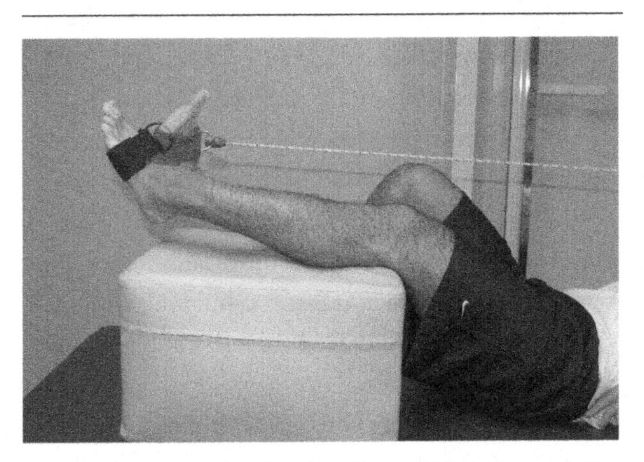

Figure 1.21: Non weight bearing plantarflexion. Non-weight bearing reduces the load to tissues and reduces the level of coordination required, allowing for a greater pain free range of motion. Pulley resistance allows for specific dosage of 50% of 1RM. Acute muscle/tendon strains or surgical repairs can also be specifically dosed for resistance, working up to 60% of 1RM to emphasize local vascularity.

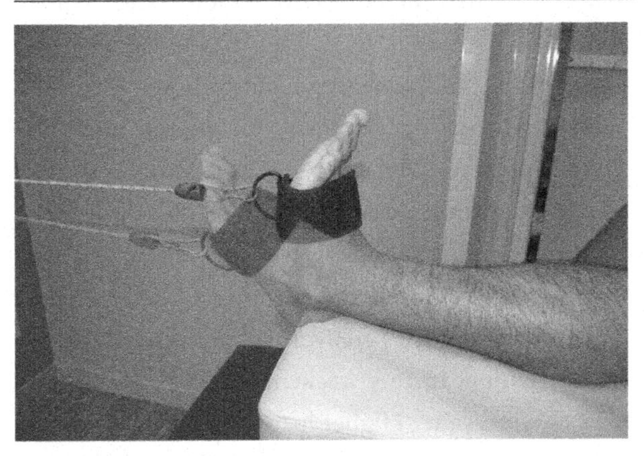

Figure 1.22: Ankle dorsiflexion in sitting or supine. If resistance is necessary, a free weight or a pulley strap around the forefoot can be applied. Motion is performed into dorsiflexion and slight eversion. Pain may be caused in end range plantarflexion, as this movement is typically combined with slight inversion. Maintaining slight eversion during the movement may be enough to control symptoms, or limiting full plantarflexion range may be necessary by blocking the forefoot from this range.

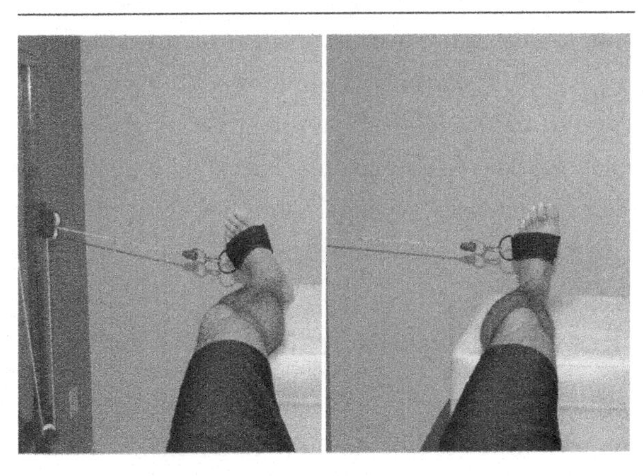

Figure 1.23: Ankle inversion with pulley resistance. Modified tension in the line of stress is achieved by avoiding the inversion range. Range of motion is performed from neutral to full eversion, avoiding the neutral to inversion.

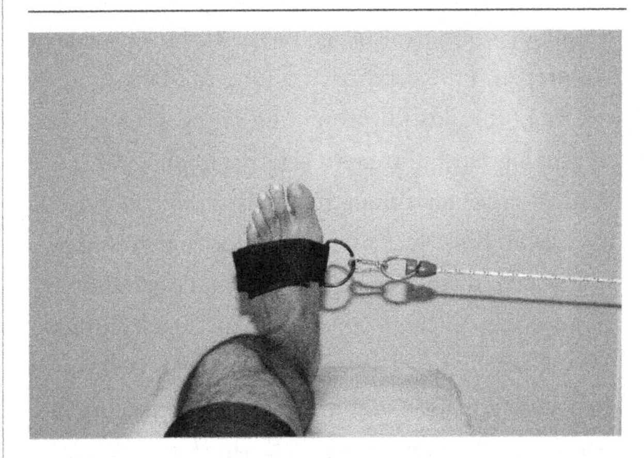

Figure 1.24: Ankle eversion with pulley resistance. Modified tension in the line of stress is achieved by avoiding the inversion range. Range of motion is performed from neutral to full eversion, avoiding the neutral to inversion.

Elastic resistance is not emphasized in early rehabilitation, as the length-tension relationship of muscle performance is not as closely replicated as when using free weights and pulleys. Kaminski et al. (2003) failed to show improvements in strength with a strengthening protocol (sets of 10 repetitions) and proprioception ankle program in subjects with ankle instability, training three times per week for six weeks.

Not all exercise approaches for early tissue repair need to be as complicated in theory or design. A simple stationary bike also influences concentric work, with little eccentric resistance, to provide a safe environment for early high repetition training for vascularity and tissue repair. Rather than repetitions and resistance, time is used for improving endurance and tissue elasticity. Hoiness et al. (2003) was able to improve peak eversion torque, reduce figure-8 running time, single leg stance balance and maximum speed with a program of high-intensity cycling 45 minutes, three times per week for six weeks.

Eccentric Tissue Training

The clinical benefits of eccentric training are numerous for rehabilitation from a functional standpoint, however, pure eccentric training can also be used for addressing collagen repair in tendinopathy and surgical repair. While concentric training emphasizes capillary density, eccentric training emphasizes collagen hypertrophy associated with repair. Pure eccentric training can be an effective stimulus for promoting collagen repair and improving tensile strength (Hather et al. 1991). With overuse tendinopathy the basic initial concept is to avoid directly training the muscle involved, as not to take further energy out of a muscle/tendon trying to repair in an "overuse" situation. Training for vascularity in all associated muscles is performed to bringing oxygen and nutrition to the healing structures. All motions are performed other than the primary motion of the tendon involved.

The primary motion of the tendon involved is then trained with eccentric work. Eccentric work uses very little energy, 70–75 percent less oxygen than concentric work (Knuttgen et al. 1971), which makes eccentric training ideal for early rehabilitation of overuse tendinopathies. Avoiding the concentric phase of training will prevent continued overuse in the tendinopathy during the acute phase of training. Pure eccentric training, using little oxygen, can then provide the appropriate cellular stimulus for collagen repair without adding to the over use condition.

Exercise specific to collagen training has been utilized in rehabilitation of Achilles tendinopathy (Mafi et al. 2001, Silbernagel et al. 2001, Roos et al. 2004) and patellar tendinopathy (Jonsson and Alfredson 2005). Langberg et al. (2006) demonstrated that chronically injured Achilles tendons respond to 12 weeks of eccentric training by increasing collagen synthesis rate, while collagen metabolism in healthy control tendons seems not to be affected. Active training can improve intrinsic properties of tendon healing and speed recovery. Langberg et al. (1998) demonstrated that in humans that dynamic calf muscle contractions results in increased peritendinous blood flow at the Achilles tendon. Hannukainen et al. (2005) measured an increase in glucose uptake in the Achilles tendon with exercise, associated with increased metabolism. Compared to muscle, the uptake was significantly less and did not increase with increased intensity of training. The concept of straining collagen with eccentric training can be applied to most tendon injuries, surgical repairs and muscle tears. With higher peak torques than concentric work, eccentric training allows for higher tensile forces, which stimulates cellular activity of fibroblasts, improves collagen density and elasticity.

Concepts for training collagen with pure eccentric work are different than typical weight training for muscular strength or coordination. The emphasis with eccentrics is on creating as high a tensile force as possible without pain or loss of coordination. Pain not only signals a gross level of deformity of collagen but may also alter normal recruitment patterns. Warren et al. (1993) demonstrated that with repetitive strain to rat tails, tissue damage significantly increased after eight to nine repetitions. The number of repetitions has an impact on the level of delayed onset muscle soreness (DOMS) created with pure eccentric training and with plyometric training that emphasizes high speed eccentrics. The number of repetitions used clinically is set between 10–15 to avoid a repetitive strain to the tendon, further weakening an already pathological state. Higher load is emphasized over higher repetitions. Langberg et al. (2006) utilized three sets of 15, two times daily with a backpack containing 20 percent of the body weight. As the exercise is addressing collagen and not muscle, only one set of high tensile force should be performed during the exercise session, but can be performed multiple times daily. The patient will dictate, by positive or negative tissue responses, how many sets and training sessions are performed daily. Progressing to one to two sets, four to five times daily may be necessary in acute post surgical cases, with caution in loading to avoiding any negative tissue responses. As the tendon heals over the course of several days, the number of training sessions daily is reduced, with emphasis on increasing the range of motion and speed in which the tendon is strained.

Basic Tendinosis Rehab Concepts

- *Initially concentric training emphasis for vascularity for all motions in the region of the tendon, except the primary motion of the tendon (3 sets of 24 at 60% of 1RM).*
- *Eccentric work one to two sets of 10–15 repetitions with resistance as heavy as possible to allow for coordinated motion up to, but not into pain*
- *Avoid excessive rest/immobilization, stretching, NSAIDs, corticosteroids, heating with ultrasound.*
- *Progress to full eccentric range and increase speeds, prior to increasing resistance.*

Example Controlling Range and Dosage of Training for Acute Achilles Tendon Repair

Non weight bearing training can begin with the heel on the ground for active dorsiflexion and plantarflexion training in the pain free range only, no added resistance, two to three sets of 50 performed hourly during the first week. Caution should be taken when performing exercises after icing, or with pain medication, as normal pain signals for excessive tension are blocked or reduced. The concept of "modified tension in the line of stress" requires a gauge for when tissue tolerance is exceeded and damage is occurring. The type IV mechanoreceptor provides this gauge, as long as the pathway is intact and not blocked with medication or icing.

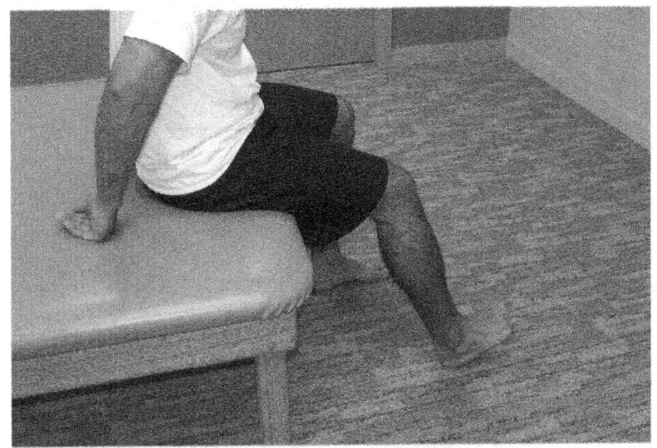

Figure 1.25: Active dorsiflexion in sitting. Dorsiflexion is performed in partial weight bearing with the heel on the ground. The heel is placed further forward so the ankle is in full plantarflexion. Dorsiflexion is performed slowly until tension is felt through the repair without pain reproduction. The position is not sustained to avoid any plastic deformity that would weaken the collagen matrix of the tendon.

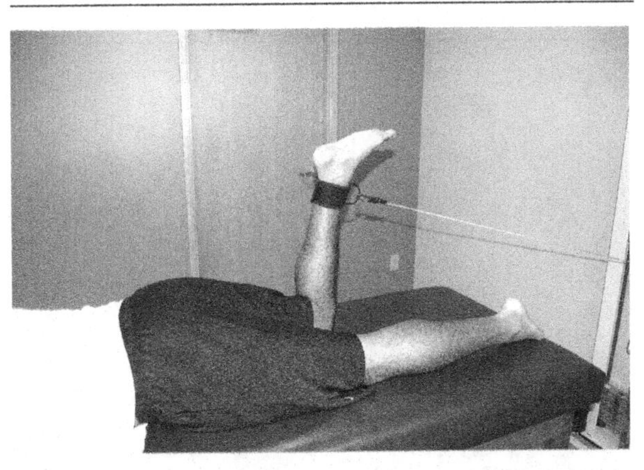

Figure 1.26: Prone knee flexion. Performing knee flexion without resistance emphasizes the hamstrings, but the gastrocnemius also contracts placing proximal tension through the Achilles tendon.

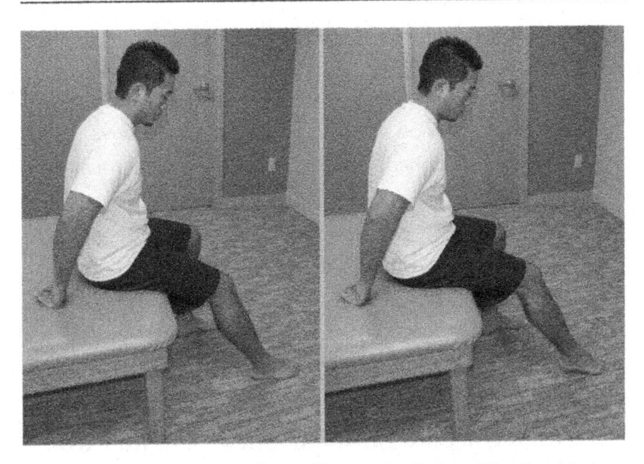

Figure 1.27a,b: Active plantarflexion in sitting. The heel is placed only as far back as possible toward dorsiflexion until tension is felt through the repair, but no pain is reproduced (a). Active plantarflexion is performed from mid range, slowly moving the tendon to the shortened range at full

plantarflexion (b). An assist with the hands lifting the thigh can reduce the resistance to allow for high repetition training without pain or fatigue.

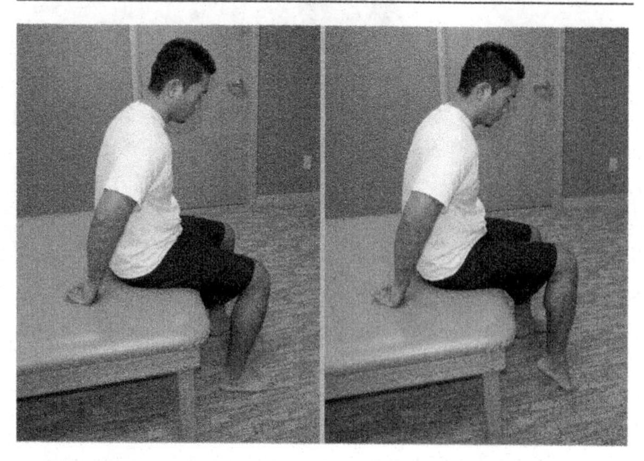

Figure 1.28a,b: Progression of the range of motion into further dorsiflexion. As the collagen elasticity improves the heel can be moved back further for both active dorsiflexion and plantarflexion. Range is still controlled to avoid pain.

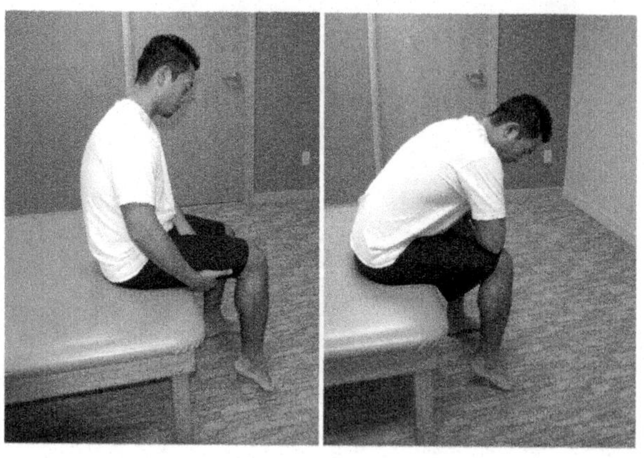

Figure 1.29a,b: Pure eccentric training of Achilles repair. Pure eccentric training allows for a greater load than performed with concentric work. The same exercise of plantarflexion is performed as previous, but the concentric phase is assisted by the upper extremities lifting the thigh (a), with full weight of the leg for eccentric training. Resistance is added by leaning the body weight over the knee during the eccentric phase of plantarflexion (b), with no resistance or active assistance for the concentric return. Pure eccentric training is performed one time daily, initially with one to two sets of 10–15 only to prevent repetitive train. Daily frequency is increased as pain and range of motion improve.

Any post surgical rehabilitation of a surgically repaired tendon should follow guidelines set by the surgeon. Generally, however, a trend toward early immobilization is excessive and can lead to further collagen atrophy, prolonged inflammation and reduced function. Providing the optimal collagen healing environment requires slow controlled movements, with modified tension in the line of stress, without pain blocking medication or modalities. Comparing early mobilization of injured Achilles tendons in rats (immediate versus starting after seven days), Godbout et al. (2006) found better mechanical properties at 28 days in the group that avoided early mobilization, beginning to apply stress after seven days. Immediate training was associated with the presence of neutrophils and increased the concentration of macrophages. The authors speculated that the persistence of the inflammatory response and undue mechanical loading led to fibrosis and reduced tendon function. The authors concluded that early mobilization of the tendon would be detrimental.

Many surgeons that have the best intensions in mind for their patients share this opinion. Studies such as this only reinforce their caution in early mobilization. Their clinical experience may even include cases in which early, uncontrolled movement led to detrimental effects. The issue may not be a concern for damage to the surgical procedure, but a belief that nothing positive will result. Murrell et al. (1998) did not find detrimental effects but found no improvement in healing of rat Achilles tendons surgically transected, after 15 minutes of swimming for 15 days.

The clinical message here is not to avoid early mobilization, but to avoid excessive mobilization that maintains or increases the inflammatory response. This previous study was performed on rats exercising on a wheel. This is hardly a controlled environment for range, resistance, repetitions, rest breaks and pain avoidance. Correctly dosed training should involve a reduction in pain with motion and palpation, reduction in temperature, resolution of muscle stiffness/guarding, increased range of motion, reduction in size of the inflamed tendon, reduced stiffness in the tendon after prolonged rest and improvements in functional status. If a negative response is noted in any of these variables, the exercise is modified to reduce the level of tendon strain, or overall amount, until a more positive result is obtained. Designing a more specific post surgical rehabilitation program, in conjunction with the physician and dosed specifically for each individual, should provide the most efficient and effective outcome. One protocol cannot address the needs of every patient.

Vascularity / Coordination / Motor Learning

As tissue tolerance, joint mechanics and pain improves, emphasis shifts towards restoring coordination of basic movement patterns. Coordination and motor learning exercises are the first functional quality related to muscle performance, which are required as building blocks for progressing to training endurance, strength and power.

The number of repetitions to retrain a coordinated movement is higher than 5,000–6,000 (Tropp 1985). Initial attempts to facilitate muscles might require isolation non-weight bearing techniques, but this should be progressed in later stages to closed chain functional movements utilizing groups of muscles in patterns. For example, Kulig et al. (2004) demonstrated selective activation of the posterior tibialis in closed chain resisted foot adduction. If intervention simply stopped here in rehabilitating muscle function, coordination and endurance of the muscle would occur in this one plane, but functionally the muscle will still be inadequate during complex weight bearing activities in all planes of movement.

Dosage for functional qualities related to motor performance requires low resistance, high repetition of dynamic motion. Isometric training is not preferred, as little coordination is developed and circulation is reduced with sustained contraction. Pain is avoided with motor performance training to prevent guarding patterns from creating abnormal movement patterns. In the presence of significant tonic fiber atrophy resistance is set at <40% of 1RM to prevent quick fatigue and compensatory motor recruitment, as well as minimize the intramuscular pressure that reduces local circulation at higher RM percentages. Coordination is a function of repetitions for neurological adaptation. Resistance at 50% of 1RM or less is recommended to allow for a high number of repetitions without fatigue. Emphasis on local muscle circulation and endurance is most efficiently achieved at around 60% of 1RM. Specific repetitions, resistance and rest breaks ensures enough energy is left in the system for protein synthesis for repairing tissues. The closer the resistance is to zero the more repetitions that can be performed. A high number of repetitions with each exercise also provides the neurological influence of firing mechanoreceptors for pain inhibition and resolution of tonic muscle guarding.

Tissue Repair, Pain Inhibition, Edema Reduction and Joint Mobilization

- *Sets*: 1–5 sets or training sessions daily
- *Repetitions*: 2–10 hours during the day
- *Resistance*: Assisted exercise up to 25% 1RM
- *Frequency*: Multiple times daily

Coordination

- *Sets*: 1–5 sets or training sessions daily
- *Repetitions*: 20–50 during the day
- *Resistance*: Assisted exercise up to 50% 1RM
- *Frequency*: 1–3 times daily

Vascularity / Local Endurance

- *Sets*: 2–3 sets or training sessions daily
- *Repetitions*: 25
- *Resistance*: Assisted to 50–60% 1RM
- *Frequency*: One time daily

Muscle in Atrophy

- *Sets*: 2–3 sets or training sessions daily
- *Repetitions*: 40 or more
- *Resistance*: Assisted to 30–40% 1RM
- *Frequency*: One time daily

Abnormal Myokinematics

The ankle will present with classic patterns of motor dysfunction that should be identified with a proper evaluation and retrained. Non-weight bearing testing of the foot and ankle may elicit gross weakness that can be addressed specifically. If these tests are pain free and strong, weight bearing testing will provide a higher level of load, as well as a higher functional demand, which may identify more subtle impairments. Proximal joint and motor performance, as previously discussed, must also be ruled out at the knee, hip, pelvis and lumbar spine. Often more proximal impairments are overlooked, as emphasis placed on the region of pain and primary complaint. The Manual Therapy Lesion concept of injury and impairment outlines are more global effect of pathology and the need for treatment to look beyond the region of chief complaints. The spinal influence on lower limb motor performance is commonly overlooked. Reduced proximal stability from lower lumbar and pelvic dysfunction affects the motor output in the trunk and lower limb patterns.

Basic Stages of the Manual Therapy Lesion (Grimsby 1988)

1. Collagen/tissue Trauma:

Acute injury, surgery, degenerative joint disease, repetitive strain, postural strain, hypo/hypermobile joint. Onset may be acute and painful or a slow progression.

2. Receptor Damage:

Structural damage to receptor, cut in neural pathway, imbedded in non-mobile capsules. The type I receptor is more easily damaged than type II, as it is located more superficial in the joint capsule. The pain free restricted joint will have a reduction in afferent input from lack of collagen elasticity changing motor performance within its motor field.

3. Reduced Muscle Fiber Recruitment:

Abnormal central processing of afferent signals from cutaneous, articular and muscle/tendon receptors. Altered central feedforward facilitation to alpha motor neurons (extrafusal muscle fibers) and gamma motor neurons (Intrafusal muscle fibers). End results include altered reflex responses for proprioception and kinesthesia, motor weakness, motor delay and/or poor timing. Central spinal segmental dysfunction, both with pain and without, alters central programming for distal performance.

4. Tonic Fiber Atrophy:

Reduced motor recruitment results in initial lack of recruitment, leading to atrophy over time. Tonic muscle fiber are more affected by type I mechanoreceptor loss and influences of the type IV mechanoreceptor (pain) system. Phasic muscle fiber atrophy occurring with more significant tissue damage and reduced activity.

5. Reduced Anti-Gravity Stability:

Reduced recruitment of the tonic system results in loss of dynamic control of joint motion, static postural stability/alignment and central balance mechanisms. Central stability at the lumbar spine and pelvis is reduced, which alter lower quarter motor patterns.

6. Motion Around Non-physiologic Axis:

Loss of dynamic control increases the neutral zone of function for the joint, increasing the range of the instantaneous axis of motion and altering mechanics for the biomechanical chain of the lower limb. Reduced tonic function creates an abnormal relationship between tonic (arthrokinematic) and phasic (osteokinematic) muscles. Compensatory motor recruitment occurs leading to abnormal mechanics.

7. Trauma/Acute Locking/Degeneration

Altered axis of motion leads to abnormal tissue stress/strain, resulting in further tissue damage and receptor damage. Hypermobile joints are prone to acute locking.

8. Pain/Guarding/Fear-anxiety of movement

The type IV mechanoreceptor responds to tissue damage with pain signals and tonic reflexogenic muscle guarding. Psychological influences of pain may lead to altered motor patterns and reduced effort. Overall effect is reduced tissues tolerance, abnormal joint/tissue loading, reduced afferent feedback, altered feedforward efferent drive and reduced function.

Treatment addressing the manual therapy lesion must first reduce pain and guarding. Numerous passive treatments and modalities can be used to this end but removing pain and guarding is not really treatment, only a necessary step before treatment. Treatment must focus on reduced tissue tolerance to stress and strain, abnormal joint mechanics, reduced mechanoreceptor function altering motor patterns and the resulting functional loss. Exercise provides the repetitive motion for tissue repair. Progressive training of the arthrokinematic tonic muscle system attempts to sensitize local muscle spindles to provide adequate afferent feedback to normalize motor patterns. The use of cutaneous receptors, through taping techniques, will also increase afferent input to improve motor patterns for both central programming and distal motor performance.

Treatment that seeks only to resolve pain does not address the primary biomechanical and neurophysiological issues related to reduced performance. As an example, residual pain and function deficits after lateral ankle sprain can affect as many as 55–72 percent of patients six weeks to 18 months after the initial injury (Gerber et al. 1998, Braun 1999). Fallet et al. (1998) concluded that the frequency of complications and breadth of longstanding symptoms after ankle sprain has led to the suggestion of a diagnosis of the "sprained ankle syndrome." Verhagen et al. (1995) concluded that there is no such thing as a simple ankle sprain, but within the contexts of the manual therapy lesion, ankle sprains are not different than any other joint injury. If the emphasis of therapy is only to reduce pain, restore range of motion and gross motor strength, then the potential for chronic symptoms or reoccurrence exists, resulting in the "syndrome" categorization. Rehabilitation must consider not only local issues, but also influences of the entire biomechanical chain, as well as neurophysiological issues of local mechanoreceptor damage and alterations in central motor programming and balance. It has been estimated that as many as 55 percent of individuals suffering ankle sprains do not seek treatment from a health care professional (Smith and Reischl 1986). Even as pain resolves through self-care, the underlying joint and motor pattern issues predispose the person to reoccurring ankle injuries, or injuries up the chain to the lumbar spine.

Athletes experiencing a prior ankle injury are up to five times more likely to reinjure their ankles. This has been shown in recreational basketball players (McKay et al. 2001), for soccer players (Arnason et al. 1996, Ekstrand et al. 1990), volleyball players (Bahr and Bahr 1997 and Bahr et al. 19994) and with military training (Milgrom et al. 1991). It might be concluded that rehabilitation is often incomplete from the first sprain, resulting in increased risk for re-injury. Treatment must go beyond resolution of pain and restoration of range of motion and strength.

Normalizing joint hypomobility is often overlooked in foot and ankle sprain injuries, as the assumption is that a hypermobility exists from the acute trauma. With the inversion sprain of the ankle, the talocrural joint is often most affected resulting in a hypermobility. But the subtalar joint often becomes hypomobile from the acute inflammation, particularly if excessive immobilization occurred as part of the initial treatment. The fibula can become anteriorly and inferiorly subluxated at the proximal or distal tibiofibular joint, preventing the normal excursion of the fibula and limiting posterior translation of the talus in relation to the mortise during dorsiflexion (Vitale and Fallat 1990). The cuboid may rotate medial and inferiorly subluxate, restricting forefoot motion and inhibiting the motor response of the peroneal muscles. The instantaneous axis of rotation (IAR) for joint motion moves toward any abnormal restrictive barrier, and if significantly restricted will become the new IAR, changing the normal movement pattern of the joint (Sammarco et al. 1973, Grimsby 1991, Konradsen and Magnusson 2000). Dynamic stabilization of the hypermobile joint cannot occur until mobility is restored at the hypomobile joints in the biomechanical chain. Passive joint mobilization will also temporarily increase the afferent response in joint mechanoreceptors, improving the neurological adaptation from training. Normalizing the neurological chain will not only restore gross strength of local muscles, but will also improve motor coordination of the entire limb, including postural righting and balance reflexes.

General Treatment Outline

- *Reduce pain and guarding: manual techniques and modalities.*
- *Provide optimal stimulus for tissue repair: repetitive motion.*
- *Normalize hypomobile structures locally and proximally to the lumbar region: joint articular, myofascial techniques and mobilizing exercise.*
- *Sensitization of muscle spindles to replace afferent loss due to mechanoreceptor damage: progressive exercise of tonic (arthrokinematic systems), joint mobilization/ manipulation before exercise, taping/bracing and visual feedback.*
- *Normalize motor function: resolve muscle guarding, motor timing and coordination.*
- *Improve performance: exercise for endurance, strength and function.*

Rehabilitation of the manual therapy lesion model emphasizes retraining of muscle spindle sensitivity to replace the afferent input lost from mechanoreceptor damage due to pathological changes. Capsular loss of mechanoreceptors reduces the afferent feedback for end range tension on collagen (arthrokinematic motion) associated with gross end ranges of motion (osteokinematic). With hypermobile joint motion the relative amount of joint play is increased, increasing the neutral zone, with reduced afferent input for proprioception and kinesthesia. Muscle spindles are most sensitive to the rate of length change and can substitute for the loss of joint mechanoreceptor input. Porter et al. (2002) found that the stretch-shortening reflex, via the muscles spindle, remains unaltered in chronic ankle instability. This suggests, that despite joint mechanoreceptor damage with trauma, muscle spindle function remains unaltered to provide afferent feedback. It is surmised that neuromuscular adaptation results in a lower threshold setting for the muscle spindles in response to the perturbation from motion. The sensitivity of the muscle spindle is under central nervous system control and can be modified. Neural input provided by the dynamic fusimotor neurons located in the spinal cord is believed to be control threshold setting (Leonard 1998). The dynamic fusimotor neurons control the primary group Ia afferents, making them more sensitive to dynamic stretch during motion (Leonard 1998). Given a lower threshold setting muscles responded with greater amplitude. These neuromuscular adaptations are thought to be primarily the result of the influence of the central nervous system at the spinal cord and not supraspinal centres (Cordova and Ingersoll 2003).

Mechanoreceptor damage in the foot and ankle alter the normal amount and timing of muscle recruitment to lower limb motor patterns. These types of alterations have been identified in pathology having a negative effect on static balance, dynamic balance and functional tasks. Normalizing peripheral reflex responses, as well as central programming, are crucial for a long-term outcome and full return to activity.

The peroneal muscles and posterior tibialis muscles, though working on opposite sides of the axis for pronation and supination, should be considered synergists in controlling the range and speed of joint mobility during functional weight bearing movements. Weakness of either groups results in a breakdown of normal foot mechanics. Reduction of peroneal muscle response/facilitation has been shown to increase overall ankle plantarflexion/ inversion range of motion (Lynch et al. 1996, Wilkerson et al. 1997). Acutely, weakness in the muscles of eversion can lead to lateral ligament sprains due to the uncontrolled posterior tibialis pull and lack of the restraining passive peroneal stiffness and eccentric control. Muscle weakness is likely more associated with inhibition from pain. Edema within the joint capsule will also alter normal afferent responses. Palmieri et al. (2004) simulated ankle joint effusion with injection, resulting in facilitation of the soleus, peroneus longus and tibialis anterior motoneuron pools. This is consistent with a protective guarding mechanism to stabilize the foot/ankle complex in order to maintain posture and/or function. In chronic ankle instability, after resolution of acute pain and edema, gross weakness may not be the issues as much as timing of muscle recruitment for stability. Direct muscle injury, in association with a sprain of the foot and ankle may also be a primary impairment requiring direct intervention. For example, inversion ankle sprains have been associated with direct longitudinal tearing of the peroneus brevis tendon, which may even require surgical intervention (Karlsson and Wiger 2002).

In examination of subjects with chronic ankle instability, differences in eversion strength from affected and non-affected sides has been identified (Bonnin 1950, Bosein et al. 1955, Staples 1972, Tropp 1986), while other authors have not found a difference (Lentell et al. 1990/1995, Wilkerson et al. 1997, Bernier et al. 1997, Kaminski et al. 1999). Ryan (1994) found weakness in invertors, not evertors. Porter et al. (2002) not only failed to find a difference in concentric strength of evertors, but also found that the stretch-shortening reflex remain unaltered in chronic ankle instability. Willems et al. (2002) concluded that the possible cause of chronic ankle instability is a combination of diminished evertor muscle weakness and combined with diminished proprioception.

Timing, or a delay in peroneal reaction time, may be more significant in chronic ankle instability, and has been identified (Konradsen et al. 1991, Lofvenberg et al. 1995, Karlsson and Andreasson 1991). Other studies, though, have failed to find this delay (Vaes et al. 2002, Isakov and Mizrahi 1991, Ebig et al. 1997, Nawoczenski et al. 1985). Peripheral mechanoreceptor or direct nerve injury may play a role in motor dysfunction in some subjects, while others may have an alteration in centrally derived motor program, or combinations of both. Subjects with ankle instability demonstrate a decreased ability to decelerate supination, with and early, sudden and presumably passive slowdown of ankle supination in the standing position (Vaes et al. 2001). With inadequate posterior tibialis muscle performance resulting dysfunctions include hindfoot valgus, forefoot abduction and pes planus; due to the uncontrolled and unchecked stress from the peroneus brevis (Den Hartog 2001). Muscle fatigue can significantly reduce motor performance, balance and have the potential for increase risk of injury (Johnston et al. 1998). In the presence of pathology, fatigue may occur much earlier due to muscle inhibition, overuse with muscle guarding or muscle compensation. Fatigue may then relate to a continued degradation of normal motor patterns.

Alterations in afferent feedback due to trauma, and secondary changes in feedforward mechanisms from the central nervous system, cannot be over emphasized. Duncan and McDonagh (1997) demonstrated an alteration in feedforward programming with reduced and delayed pre-activation of ankle musculature prior to ground contact during landing from a jump. Bullock-Saxton et al. (1994) demonstrated feedforward muscle activation deficits with mechanoreceptor damage at the ankle affecting proximal muscle performance; reduced hip extension strength in subjects with ankle sprains. Receptor damage at the hip not only affects the local motor performance, but the entire motor pattern for the pelvis and lower limb. Influences are noted bilaterally as well, not just on the involved limb.

Movement at the hip is important for controlling lateral displacement about a fixed foot, but control about the ankle also contributes significantly to maintaining postural stability (Rietdyk et al. 1999). A lateral shift of the pelvis over the foot, moving the center of mass laterally, drives the unstable ankle into inversion. Reduced eversion strength, predisposing the ankle to sprain, has not been shown in chronic ankle instability (Baumhauer et al. 1995). Controlling ankle inversion is not only a function of the ankle evertors, but the invertors eccentrically stabilize the lateral displacement of the tibia (Wilkerson et al. 1994/1997). Eccentric weakness of the invertors has been identified in chronic ankle instability (Munn et al. 2003), though other studies have not found eccentric weakness (Bernier et al. 1997). Reduced concentric inverter performance may also identify muscle imbalance. An elevated torque ratio of the concentric evertor to concentric invertor has been identified as one of the few predictors of ankle sprain (Baumhauer et al. 1995). Closed chain testing and treatment are more effective in the identification of instability from lack of more proximal control of the tibia, hip or pelvis.

Abnormal myokinematics may also be related to neurological issues. Absent of lumbar nerve injury, segmental facilitation in the spine (sensitization) may alter normal firing patterns and levels of recruitment. Peripheral nerve injury may directly contribute to de-efferentation of local muscles. For example, vulnerability of peripheral nerve injury with inversion trauma may lead to neurological deficit contributing to weakness after inversion ankle sprain. A high percentage of subjects with acute grade III ankle sprains have been shown to have peroneal (86 percent) and tibial (83 percent) nerve injuries (Nitz et al. 1985). Reduced conduction velocities of the deep peroneal nerve three weeks after an inversion sprain have been identified, though conduction velocities returned to baseline measures by five weeks (Kleinrensik et al. 1994). Neurological assessment of peripheral and spinal nerves, including screening for segmental facilitation, is necessary prior to treatment planning.

Speed

Initially the speed of training is kept at a slow to moderate rate. Slower speeds assist with coordination and prevents the risk of excessive tissue stress and strain. Generally speaking for motor performance, an increase in speed will produce an increased challenge to coordination, but coordination training is speed specific. The speed of training should eventually match the functional demand. Slower speeds are associated with greater concentric work peak torque with an emphasis on recruiting normal movement patterns, and increasing in the number of motor units recruited during the exercise. As coordination is improved, the patient will naturally increase the speed of performance, serving as a signal that the patient is ready to be progressed on that particular exercise (Stage 2 training).

Low speed may also mean slow enough to allow for collagen deformity, as in exercise for joint capsule mobilization or muscle/tendon lengthening. To improve plastic deformity of collagen, slow speed with a hold time of 10–15 seconds (joint capsule) or up to 30 seconds (muscle) is required for each repetition. A normal exercise may require two to three seconds to complete one cycle of concentric and eccentric work. Increasing this time to five to 15 seconds allows more time for tissue deformity in muscle to gain range of motion. For lengthening muscle this slow approach to each repetition is effective for both concentric work of the antagonist (Bandy et al. 1998) as with eccentric work of the agonist (Nelson and Bandy 2004).

The need for extensibility of the plantarflexors and Achilles tendon is most common in the ankle and foot. Passive stretching weakens the collagen of the muscle/tendon unit, which is not recommended in most pathological conditions. Repetitive motion with eccentric lengthening into dorsiflexion creates a thicker and more elastic tendon, rather than a longer and weaker tendon through passive stretching. Training plantarflexion with eccentric emphasis to the end of dorsiflexion initially requires a reduced load to allow for slow repetitions of up to 10 seconds for completion.

Slowing speed down, below that of a functional level, can also be a challenge to coordination. Slow speed changes are typically associated with weight bearing balance exercises and early agility training in Stage 2.

Bracing / Taping

Some clinicians presume that long-term application

of an ankle brace may cause the supporting structures of the ankle to weaken resulting in dependence on a support device. Healthy ankles are thought to develop weakness in the surrounding muscles if preventative bracing is used long term, causing a delay the leg musculature's ability to respond to external stimulus or perturbation, diminishing neuromuscular function and potentially placing the ankle-foot complex at risk for injury. These clinical assumptions are nothing more than assumptions. Cordova et al. (2000) demonstrated that long term preventative ankle bracing found that the duration of the peroneus longus stretch reflex (latency) is neither facilitated nor inhibited with extended use of an external ankle support. Proprioceptive input provided by the muscle spindles within the peroneus longus was not compromised.

Cordova and Ingersoll (2003) provide evidence that peroneus longus amplitude, in response to sudden inversion perturbation, is facilitated immediately after the application of a lace up style ankle brace. The peroneus longus amplitude was shown to increase after an eight-week application of a semi-rigid style ankle brace. The increased reflex response with an immediate application and extended use of external ankle support is effective in enhancing the normal neuromotor responses from the primary musculature that dynamically stabilizing the ankle. Cordova and Ingersoll (2003) also found that, after acute application, the lace up brace resulted in greater muscle spindle stretch reflex amplitude of the peroneus longus than the semi-rigid brace. The authors hypothesized that this may be due to increased afferent information provided to the central nervous system primarily by cutaneous mechanoreceptors, and perhaps other joint mechanoreceptors. The lace-up brace covers more surface area than a semi-rigid brace, potentially stimulating more receptors. Nishikawa and Grabiner (1999) also reported increased normalized peroneus longus Hoffman reflex (H-reflex) amplitude after application of a semi-rigid ankle brace. Their studied involved electrically stimulated peroneus longus group Ia afferent nerve fibres percutaneously, and not through deformation of the muscle spindles by movement of the ankle, as in the study by Cordova and Ingersoll (2003).

Braces can have an excitatory effect on the muscle responses. The peroneus longus was tested in these previously mentioned studies, but it must be remembered that muscles fire in patterns, not in isolated reflex arcs. Bracing and taping procedures, likely through sensitization of both cutaneous and joint mechanoreceptors, assist in recruitment of muscle, allowing for neurological adaption to occur with rehabilitative exercises.

The influence of taping and bracing on kinesthesia in ankles with chronic instability is inconclusive. Positive results have been shown in some studies (Gross 1987, Refshauge et al. 2000, Hubbard and Kaminski 2002), while other have shown little to negative responses (Glencross and Thornton 1981, Lentell et al. 1995). Much of this testing was performed in non-weight bearing, which may not reflect the overall influence of afferent input, combined with taping and bracing, in the weight bearing position. Whether bracing helps with kinesthesia in the individual patient is somewhat irrelevant in the clinical setting. Bracing and taping is typically applied as a protective measure to allow for training activities that will restore proprioception and kinesthesia.

Prophylactic ankle support is used to restrict frontal plane motion occurring at the subtalar joint; however, movement in the sagittal plane is constrained as well, which may interfere in the execution of functional tasks. Although ankle braces and taping have shown to be beneficial in the prevention of ankle injury (Rovere et al. 1988, Sitler and Horodyski 1995, Surve et al. 1994, Tropp et al. 1985), athletes often avoid wearing them due to perceived hindrance in performance (Homar et al. 2004). It is important to determine whether an external ankle supports substantially hinders the individual's ability to carry out sport-specific tasks. Cordova et al. (2000) did confirm that taping and lace-up style ankle braces restricted plantarflexion range of motion 10.5 and 9.3 degrees before any bout of continuous exercise, respectively. Taping may provide more customized and specific stabilization by directing restraints in vectors more associated with instability. The semirigid brace also does not allow for the maximum amount of inversion available, potentially slowing the change of direction from occurring. In a subsequent meta-analysis, Cordova et al. (2005) analyzed the utilization of external supports on performance. As external supports reduce range of motion, studies were assessed that looked at potential changes in sprint speed, agility speed and vertical jump. Overall the analysis showed that in healthy subjects, the average effects of external ankle support on sprint, agility running and vertical jump performances ranged from trivial to small in subjects who are not elite athletes. Although these performance effects are quite minimal in the general healthy athletic population, it is reasonable to assume these effects could have greater effects on the elite athlete (Hopkins et al. 1999). Cordova et al. (2002) performed a more specific literature review on the effects of several different types of braces and taping techniques on joint kinetics, sensorimotor function and functional performance, with emphasis on healthy subjects.

A compromise for health care practitioners between the prophylactic value of external ankle support versus the potential compromise in lower-extremity functional performance relates to the stage of healing and exercise progression. Early in rehabilitation the semi-rigid brace is preferred to prevent further injury and a safe return to higher level weight bearing and balance exercises. The use of external bracing and taping can be effective interventions to allow for a higher level of function and exercise. A rigid brace may be effective in preventing further abnormal damage to injured or post surgical tissues. A semi-rigid ankle brace is superior then a lace-up style brace in limiting rearfoot angular displacement and angular velocity, though even a lace-up brace is better than no support (Cordova et al. 2007). Progression involves reducing the level of external support to lace-up or taping. Under constraints of a semi-rigid brace the athlete is unable to achieve the maximum amount of dorsiflexion needed to provide the explosive propulsion required for push off for a change in direction. In these cases, the benefits of using lace-up ankle brace or ankle taping is a better option. The lace-up style ankle braces or ankle taping affords appropriate protection against frontal plane motion (Cordova et al. 2000), while allowing the athlete to achieve the necessary dorsiflexion to complete the motor skill. Softer braces and taping can provide cutaneous afferent input, improving proprioception, kinesthesia, reducing pain and improving performance.

Weight Bearing Training

Stage 1 might require gravity assisted or gravity eliminated weight bearing to obtain appropriate motion around a physiological axis. Side lying positioning, suspension slings or a weight bearing harness might be used initially to achieve the desired functional quality. Reducing the body weight for weight bearing training has been presented as a "new approach" in the literature (Kelsey and Tyson 1994), but the concept has been utilized in rehabilitation for some time. Medical Exercise Therapy in the 1970's, developed in Norway, utilized unloading principles for a tissue based exercise program designed to promote repair and restore function (Holent/Faugli 1996). Gardiner (1954) described the biomechanical aspects of slings and unloading for many different extremity exercises. The modern day Total Gym® is an extension of the original design, referred to as Goldie's Exerciser (McKenzie 1910). Unloading principles have been around for some time, with more recent application in rehabilitation of acute inversion ankle sprains (Kern-Steiner et al. 1999).

It must also be remembered that not all patients will have to start in Stage 1 in non-weight bearing, and may be ready to being training with partial or full weight bearing for some of the exercises. The previous examples of sitting plantarflexion, with the knee flexed, is an example of partial weight bearing training. This may be an adequate starting position to train weight bearing tissues such as cartilage or bone pathologies, when the emphasis is on tissue tolerance of weight bearing. In collagen and muscle pathologies, coordination in weight bearing, with the knee normally extended, may be the emphasis. To allow for training in this position the body weight must be reduced in a more functional position with the entire lower limb involved in motor pattern retraining. Primary emphasis in Stage 1 is non-weight bearing training, but partial weight bearing may be tolerated at significantly reduced levels using unloading options.

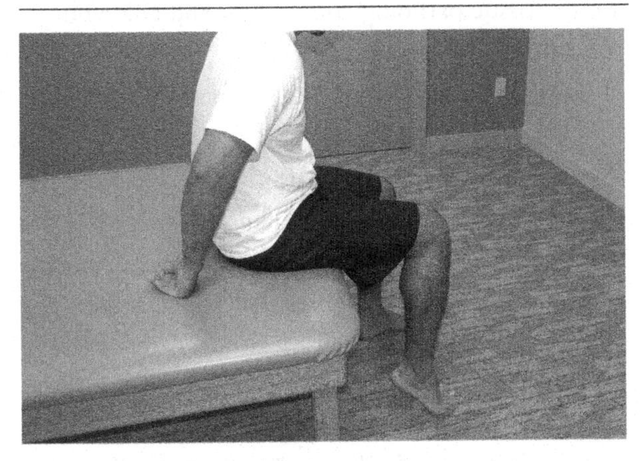

Figure 1.30: Sitting plantarflexion—partial weight bearing. Sitting reduces the body weight through the lower leg allowing for an increase in range of safe and pain free training. Plantarflexion can be performed in a limited pain free range with the heel hitting the ground at zero degrees dorsiflexion. Range of motion can be progressed by sliding the heel back to allow for an increase in dorsiflexion range.

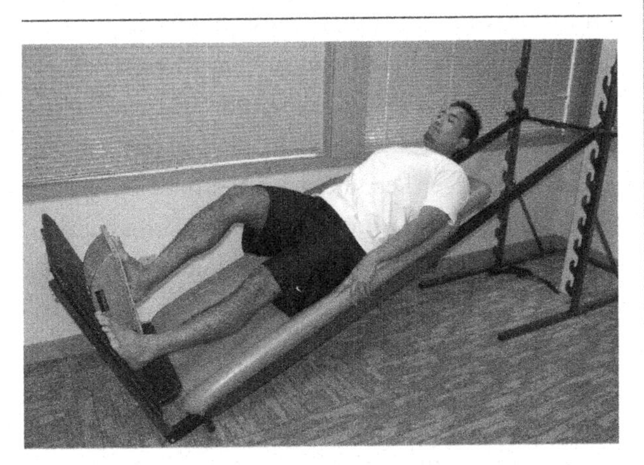

Figure 1.31: Partial weight bearing plantarflexion and dorsiflexion on a wobble board with an inclined position. Incline machines, or horizontal sleds, can allow for a significant reduction in body weight through the foot, and little coordination demand, to be tolerated early in the rehabilitation process. The wobble board emphasize mobilization for range of motion, while the incline bench allows for a gradual increase in weight bearing forces.

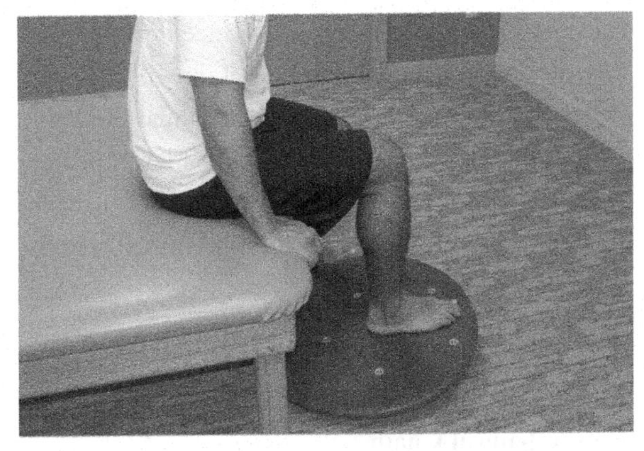

Figure 1.32: Wobble board or BAPS for proprioception training with reduced body weight in sitting. Coordination of cardinal plane motion, as well as combined circumduction, can be performed with partial weight bearing in sitting. Emphasis is placed on tissue training through different planes of motion and simple coordination.

Myofascial Stretching

As previously discussed, static stretching is not emphasized in Stage 1 to improve muscle elongation. Passive stretching produces longer and weaker collagen through plastic deformity, reducing the tensile tolerance of collagen in tendons. Reduced muscle length is more commonly associated with muscle guarding and pain in the acute stage. Muscle guarding is addressed through repetitive exercise to increase local muscle circulation. Collagen elasticity is improved through repetitive motion to become thicker and more resistant to tensile forces. Passive stretching is not recommended, and often contraindicated, early in pathologies directly related to collagen degradation. This includes tendinopathy, muscle strain and surgical repair of tendons.

Passive PNF approaches, such as hold-relax stretching, provide afferent input using a neurological approach to reducing tone. In dealing with collagen injury, these types of techniques are preferred, rather than static stretching. These techniques can avoid collagen plasticity with emphasis on brief lengthening of a few seconds following an active contraction. Muscles with reduced flexibility can also be treated indirectly with manual techniques or exercise to the level of spinal innervation. Facilitated spinal segments, or sensitized spinal segments, can reduce flexibility in any muscle innervated by the level due to increased neural drive altering resting tone. Manual therapy treatment to the L5 and S1 segmental levels of the spine may significantly reduce tonic muscle tone in the plantarflexors changing not only muscle length but also performance.

Adjunct Treatments

Biomechanical Chain

No joint system can be treated in isolation but must be viewed in the greater whole of the biomechanical chain. Nowhere is this point more clear than considering the foot and ankle at the bottom of a weight bearing chain involving the entire lower limb. Limitations in mobility of the lumbar spine, pelvis, hip or knee can alter the normal loads on the ankle, forefoot and toes leading to injury. For example, van Mechelen et al. (1992) found that restricted hip range of motion was present in runners who had sustained a running injury. Treatment, both passive and active, must address all proximal issues in adjunct to the foot and ankle treatment. Compensatory motion in the foot for proximal dysfunction can cause motion to occur too far, too fast or with poor timing, resulting in tissue breakdown. Excessive foot motion in and of itself is not deterministic of lower extremity problems (Donatelli et al. 1999, Viel et al. 1989, Rome et al. 2001). Joint specific hypermobilities of the foot can, however, lead to excessive stress along the joints' capsule and ligaments and place excessive demand on contractile tissues. An example of this is the overuse of the posterior tibialis muscle in a more lengthened position in an attempt to control excessive hindfoot pronation resulting in pain, tendinopathy and/or possible rupture.

Neurological Chain

The neurological chain refers to the efferent and afferent pathways from the lumbar spine through the lower limb. Segmental dysfunction in the lumbosacral region cannot only refer pain into the foot and ankle but can also alter the normal motor patterns. A complete evaluation involves not only nerve conduction testing and neural tension testing, but also screening for facilitated segments (sensitized segments) that can contribute to alterations in motor patterns and pain sensitization of distal structures. For example, segmental facilitation at the L5/S1 segment results in increased neurologically mediated tone in the gastroc-soleus. The muscle complex will test as "short" or "tight" due to this tone, altering mechanics of the foot, ankle and knee. The maintained elevation in tone can also contribute to primary pathologies including muscle tears, Achilles tendinopathy and plantar fasciitis. Primary treatment still focuses on the tissue pathology of the foot and ankle but spinal treatment is added to resolve mechanical dysfunction and/or segmental facilitation.

Neurological changes may also be associated with collagen injury and mechanoreceptor damage impacting foot/ankle motor facilitation and proprioceptive responses (see The Manual Therapy Lesion in chapter 8). An overstretched or strained ankle capsule and/or ligaments will easily damage type I mechanoreceptors found in the superficial layers of collagen. In a lateral collateral ligament sprain, damaged mechanoreceptors in the lateral collateral ligaments will reduce feedback for proprioception and kinesthesia resulting in decreased muscular responses. Impairments in balance, proprioception, coordination, endurance and strength result from the loss of afferent feedback. In the Biomechanical model exercises are directed towards providing modified tension in the line of stress of the lateral collateral ligament to promote collagen repair. In the Neurological model emphasis is placed on retraining the neurological recruitment of inhibited muscles. Joint mobilization to the joints in the region will influence motor recruitment. Cutaneous receptors, through taping and bracing, will also assist in increasing the motor facilitation. Progressive training for balance and coordination will incorporate more central motor pattern reflexes. Sensitizing the muscle spindles in the peroneals, with progressive training, will assist in providing feedback for facilitation of eversion when the ankle moves toward inversion and the damaged collagen.

Myokinematic Chain

As with the biomechanical model, where adjoining joint dysfunction can place abnormal stresses on the ankle, muscle imbalances from the knee, hip or lumbar spine can have similar effects. All muscles in the lower quarter must be assessed, preferable in weight bearing functional testing, to identify patterns of weakness that place excessive stress to ankle and foot structures. When specific muscle weakness is identified, this should be assessed in conjunction with lumbar spine segmental dysfunction or facilitation inhibiting segmentally innervated patterns of muscles. Structures of the foot compensate for all mechanical and motor dysfunction. For example, weakness or poor endurance of the gluteus medius, gluteus minimus and tensor fascia latae reduces the eccentric deceleration of internal rotation. The resultant internal rotation moment of the lower extremity in weight bearing may occur through too far a range or occur too quickly placing excessive demand on the tissues. The collapse into internal rotation leads to poor shock absorption and ineffective proximal control of pronation (Sahrmann 2002). The eccentric collapse of the lower quarter into pronation will be even more dramatic when performing high endurance or fatiguing activities. The excessive load on the foot structures — such as the posterior tibialis, plantar fascia and the Achilles tendon — can lead to secondary tissue pathologies when the primary impairment occurred more proximally at the hip. The mechanical chain concept of the lower limb, whether looked at biomechanically, neurologically or muscularly is not new. Freeman et al. (1965) theorized that functional instability of the foot and ankle resulted from reduced muscular coordination as consequence of a rupture of afferent nerve fibers in damaged ankle joint ligaments. Even earlier studies simulated the loss of afferent mechanoreceptor input on proprioception with anesthetization of joint capsules and pericapsular structures in the metatarsophalangeal joints of the great toe and index finger (Browne et al. 1954, Provins 1958). More recently, reduced hip extension strength has been measured in subjects with ankle sprains (Bullock-Saxton et al. 1994). The Manual Therapy Lesion, outlined in chapter 8, describes pathological mechanisms of tissue injury, mechanoreceptor damage and the resulting motor deficits leading to instability. The theoretical approach to rehabilitation is also discussed, with an emphasis on sensitizing muscle spindle afferent input to substitute for the loss of afferent input from damaged mechanoreceptors. Konradsen et al. (1993) demonstrated this concept by showing that reduced afferent input from anaesthetized ankle collateral ligaments could be replaced with afferent input from an active calf muscle.

Clinical Application: Plantar Fasciitis

The literature has little evidence comparing different forms of conservative treatment for plantar fasciitis. Cornwall and McPoil (1999) recommend incorporating 3 treatment strategies: reducing inflammation, reducing tissue stress and improving strength and flexibility. Not all of these approaches are supported in the literature, nor do they apply to every patient. Treatment planning may need to involve an integration of the biomechanical chain model, neurological chain model and the myokinematic chain model. As an example, plantar fasciitis may be the primary symptom but the underlying cause could be related to one or a combination of these three models. Biomechanically, plantar fasciitis could be looked at as having a low medial longitudinal arch height. It would seem logical to relate a low-arched individual as being more prone towards plantar fasciitis, as this would put excessive strains on the local tissues. Biomechanically, either hypermobile low arches or hypomobile high arches can cause excessive loads on the plantar fascia and lead to pathology (Cornwall 2000, Aquino et al. 1999). The biomechanics of the individual joints of the foot, ankle, knee hip and lumbar spine should be analyzed for optimal efficiency of motion throughout the lower limb. A high arched individual may need emphasis on improving joint mobility with passive mobilization, active exercise and the use of inserts and shoes to provide cushioning to assist in absorbing ground reaction forces. A low arched individual may require stabilization training of local muscles, orthotics or specific shoes to provide a component of motion control and stability.

According to the neurological model, plantar fasciitis may result from a primary facilitation, or sensitization, of the S1 spinal level. Somatic tissue sensitization of the collagen of the plantar facia may lead to pain, even in the absence of lumbar symptoms. Treatment would then involve not only direct intervention to the foot but also address issues in the lumbar region leading to facilitation.

Finally, the myokinematic chain would assess the motor patterns of the entire lower extremity beginning in the lumbar spine. Weakness, or poor motor timing, in motor patterns from the lumbar spine to the foot may result in a collapse of the foot into pronation. Excessive pronation is considered a common risk factor for the development of heel pain (Kwong 1988, Barrett and O'Malley 1999, Crosby and Humble 2001), although excessive pronation has not been associated with increased risk of plantar fasciitis or in other lower extremity abnormality (Donatelli et al. 1999, Reischl et al. 1999, Powers et al. 2002). Plantar fasciitis results from the duration of motion and not merely from the motion itself (Chandler and Kibler 1993, Kwong et al. 1988).

In the treatment of plantar fasciitis it can be seen how evaluation must involve a thorough assessment of joint systems, neurological systems and the motor system. Neither functions independently and all may be involved in the pathological presentation. Treatment cannot only focus on the primary tissue in lesion, but also in controlling forces placed on these tissues. The plantar fascia is primarily made of collagen. Chapter 3 discussed issues related to treating collagen pathology as it relates to increasing forces related to repair and reducing forces related to degeneration.

Reducing abnormal forces to the plantar fascia may be addressed biomechanically by normalizing joint mobility through the entire lower quarter. A hypomobile joint is compensated somewhere along the biomechanical chain, and the foot, being the last in line, may absorb the majority of the abnormal forces. In the hypermobile foot, orthotics may be necessary to prevent excessive range, or range that occurs too fast. Harty et al. (2005) described an increase in hamstring tightness, which may induce prolonged forefoot loading and, through the windlass mechanism, be a factor that increases repetitive injury to the plantar fascia. Improving muscle coordination, endurance and strength will reduce the forces transmitted to the foot, as they will be partially absorbed more proximally. Addressing the neurological chain may reduce the sensitization of tissues, but also improve normal motor facilitation. Reduced ankle dorsiflexion, obesity and work-related weight bearing appear to be independent risk factors for plantar fasciitis (Riddle et al. 2003). Reduced ankle dorsiflexion was described as one the most important risk factors. Given the breadth of issues that may relate to plantar fasciitis, it is easy to see how Crosby and Humble (2001) could conclude that no one modality of treatment for plantar fasciitis has been shown to be effective in all instances. An individualized treatment must address risk factors, clear proximal issues and address the primary tissue pain, all at the same time.

Stretching has been prescribed as a treatment for plantar fasciitis (Digiovanni et al. 2006, Quaschnick 1996) and the Achilles tendon (Poter et al. 2002, Davis et al. 1994). With collagen pathology stretching should be viewed with caution, or even as a contraindication, as stretching breaks bonds in collagen and weakens it tensile properties, compared to a dynamic eccentric training approach to lengthen and strengthen. Static stretching is also not a primary stimulus for repair, as repetitive motion can be when applied correctly. If the cause is found to be from a hypermobile medial arch due to subtalar/midtarsal instability or faulty mechanics leading to excessive pronation, then the last stimulus the collagen needs is more stretching and lengthening. A more appropriate treatment might include eccentric loading exercises to promote collagen repair. Even if stretching of the plantar fascia is not destructive, it may not be the most efficient form of intervention. Digiovanni et al. (2006) did find a non-weight bearing stretch of the plantar fascia over an eight weeks period to be superior to stretching the Achilles tendon. At two years, no difference was noted between the two groups. But a further understanding of the histological changes with plantar fasciitis, and the histological effects of intervention may assist in a more patient specific treatment design.

Plantar fasciitis has been described as a repetitive microtrauma overload injury of the attachment of the plantar fascia at the inferior aspect of the calcaneus (Harty et al. 2005). Wearing et al. (2004) identified that neither an abnormal foot shape, nor movement of the arch, are associated with chronic plantar fasciitis. However, once the condition is present, arch mechanics may influence the severity of plantar fasciitis. A common description is related to a pes planus foot type and lower-limb biomechanics that results in a lowered medial longitudinal arch creating excessive tensile strain within the fascia, microscopic tears and chronic inflammation. However, contrary to these clinical doctrines, histological evidence provides little support for this concept, with inflammation rarely observed in chronic plantar fasciitis (Wearing et al. 2006). This is consistent with descriptions of Khan et al. (2002), stating in the case of overuse tendinitis, it is unlikely that the patient is suffering from an inflammatory 'tendinitis' (Khan et al. 2002). Collagen breakdown, the most commonly believed theory, has been shown not to be consistent with the histological changes (Khan and Cook 2000). Even acute pathologies of several weeks are more correctly described as a 'tendinosis' (Khan et al. 1999). Initial changes occurring in the cell is a fundamental shift in the understanding of the histological process leading to tendinopathy. The order of breakdown appears to be 1) tenocyte abnormality, 2) matrix (ground substance) abnormality, 3) neovascularization followed by 4) collagen breakdown. It is unclear if similar mechanisms are in process leading to plantar fasciitis, but the intervention of modified tension in the line of stress is still consistent for the histological presentation. Pure eccentric work with higher tensile load for Achilles tendinopathy has been shown to be a superior training method, compared with training with combined concentric-eccentric work at lower forces (Mafi et al. 2001, Silbernagel et al. 2001, Roos et al. 2004).

The simple clinical approach is to perform eccentric plantarflexion on the involved foot, with the concentric phase performed by the opposite foot only. The load is as high as allows for coordinated motion in a relatively pain free environment. As described previously for Achilles tendinopathy, this collagen based treatment can be integrated into the overall treatment approach, performed with only one to two sets of 10–15 repetitions, multiple times daily. Specific taping techniques for plantar fasciitis may be helpful in reducing pain levels, allowing for increased exercise (Hyland et al. 2006). One daily training session with multiple sets and rest breaks is associated with training muscle. Training collagen does not require oxygen recovery, and can be performed multiple times per day, as collagen responds to the positive strain with production of fiber and glycosaminoglycan. A combination of dynamic stretching to the Achilles tendon can also be incorporated, with repetitive plantarflexion, emphasizing eccentric motion toward end range (Nelson and Bandy 2004).

Summary for Stage 1

Stage 1 training is generally not focused on training muscle, meaning it is not focused on improving muscle volume or strength. The initial emphasis is on providing the optimal stimulus for tissue repair, reducing pain, resolving edema, reducing muscle guarding and normalizing mobility. Facilitating inhibited muscle and coordinating basic movement pattern is the primary emphasis related to muscle performance. The training influence is directed more toward neurological adaptation, as coordination relates more to firing of motor patterns as opposed to direct influence on muscle tissue itself. As motor patterns improve greater force output is achieved prior to structural changes in the muscle tissue. Muscle hypertrophy will occur over time but will also require an increase in the loads more associated with Stage 2 and Stage 3 progression concepts.

Indications for Progression to Stage 2

All exercises progress at their own rate, not in a staged protocol progression. The stages reviewed in this chapter serve to organize the theoretical material in a logical order. As each patient presents with a specific tissue injury, previous training state, stage of healing and impairment list, individualized exercises and training goals should be matched to the patient. A specific functional quality, or goal, is set for each exercise. As this goal is achieved the exercise is discontinued or progressed to achieve a higher level functional quality. Some basic examples of responses indicating the patient is ready to progress from stage one concepts includes:

- *Pain levels decrease at a predetermined percentage by the therapist based on experience and the source of pain. Pain is not constant but related to specific motions or higher level activity.*
- *Range of motion in cardinal planes is improved to 10–25 percent of normal, depending upon joint and type of pathology.*
- *Joint mobility has normalized.*
- *Muscle guarding resolves.*
- *Tissue tolerance to stress and strain is improved demonstrated through a decrease in pain and edema with an increase in range of motion, weight bearing status and speed of exercise.*
- *Weight-bearing tolerance improves.*
- *Edema is reduced/resolved with resolution of joint temperature and/or girth.*
- *Speed of exercise naturally increases as coordination qualities improve. Change in speed is a sign that the patient's training state is improving and other aspects of dosing the exercises may be progressed as well.*
- *Perception of fatigue is reduced with training. The patient is completing the exercise program in a shorter period of time, requires shorter rest breaks between sets and has reduced heart rate during training. Post exercise symptoms of pain or muscle soreness are no longer present.*
- *The primary functional quality for an exercise is achieved.*
- *Re-testing demonstrates an improved performance level.*

Section 2: Stage 2 Exercise Progression Concepts for the Ankle and Foot

Training Goals

The four stages that are outlined in the STEP progression are not meant to be followed in a step-by-step fashion. The stages are merely an attempt to logically order the training sequence of functional qualities. A clinical training program may have exercises that conceptually fit into several of the stages at the same time. This is because each exercise is designed to address a specific tissue and/or functional quality. Exercises must be dosed specific to the patient's training level and tissue tolerance. As the goal of an exercise is achieved, the exercise is discontinued or progressed. Each exercise is progressed individually, not the entire program. Each visit should have some progression or change in the program, no matter how small. The patient should always be improving and pushed to their maximum safe level of performance.

During Stage 2, the overall training level is increased with five to 10 exercises of two to three sets each. The previous exercises of Stage 1 might continue, but the functional quality will change as the speed, sets and range of motion increases. As tissue tolerance improves, weight bearing is also increased to challenge balance and to restore functional gait patterns.

Increase Speed not Weight

The common emphasis on exercise progression is to increase the resistance used, but increasing resistance is the focus for strength training, which is not the primary functional quality being trained at this stage. Emphasis remains focused on coordination and endurance training through the entire range of motion. Rather than increasing weight, the speed of performance is increased. Speed requires overcoming inertia with acceleration and deceleration. Inertia is a type of force; therefore resistance is added to the exercise by adding speed. An increase in speed may naturally occur as the coordination level improves. If not, the patient can be cued to increase speed with an exercise.

Speed will also serve as a coordination challenge. Exercises performed below a functional level are increased toward a more normal speed. Fast coordination is typically more functionally related, as opposed to continuing to perform slowly with increased weight.

Increase Repetitions

The general progression concepts for Stage 2 are to increase the number of repetitions to further increase tissue training qualities and to further enhance coordination. Repetitions are not increased with each set, as this would require lowering the resistance and change the functional quality being trained. More repetitions are achieved by adding more sets or more exercises. The overall time of training will increase as the program is changed from three to five exercises to seven to ten. Some exercises may progress from one to two sets, to three to five sets. Additional exercises may address deficits in motor patterns involving the knee, hip, or lumbopelvic regions. Coordination of the entire lower quarter pattern in weight bearing, or partial weight bearing, is a primary baseline prior to progressing to endurance, strength, power and agility training.

Isometric Work

Isometric work (IW) was avoided in Stage 1 due to the presence of muscle guarding, as circulation is reduced with significant increases in intramuscular pressure. It also fails to provide the repetitive motion necessary for early tissue repair, pain modulation and coordination training. With resolution of muscle guarding and increasing joint mobility, isometrics are added to assist in fixing strength within the newly gained range of motion.

The primary use of isometric work in Stage 2 is to sensitize the muscle spindle to stretch. As the manual therapy lesion described earlier, mechanoreceptor damage in the joint reduces the normal afferent reflex responses for motion. Sensitizing the spindle increases the afferent input to assist in normalizing movement patterns, replacing some of the afferent input lost with the mechanoreceptor damage. As the muscle lengthens toward the hypermobile range, the spindle responds to the tension from the changing length of the muscle. Afferent input from the spindle will assist in the recruitment of alpha motor neuron pools for normal motor recruitment. The primary muscles that would eccentrically stabilize the hypermobile range are trained isometrically in their shortened range. The shortened range is a joint position opposite of the hypermobility. In this range isometric work can be safely performed to begin the process of sensitizing the spindles. In later stages, as a progression, the isometric work can be performed in a range closer to the hypermobile range. A basic clinical example of this concept is an ankle with an inversion sprain. The evertors would function eccentrically to decelerate the ankle toward inversion. Damage to the ligaments and joint capsule would damage mechanoreceptors and reduce muscle recruitment for lower limb motor patterns. Stage 1 began concentric work of the evertors from a mid to shortened range of the muscle, or from a neutral ankle to full eversion. End range inversion is avoided. Stage 2 begins isometric training of the evertors with higher resistance, but with the ankle in full eversion. As the spindle becomes more sensitive to stretch into the

1. Exercise for the Ankle and Foot

lengthened position, they will fire more easily when the moving toward inversion. The spindle will fire recruiting alpha motor neuron pools for improved recruitment of the evertors to decelerate the ankle from full eversion. Improved spindle sensitivity will also assist in positioning the ankle in neutral when at rest, rather than in slight eversion, associated with feedforward deficits in lateral ankle sprains.

Isometric Training Goals

- *Sensitize the muscle spindles in the shortened range of the muscle.*
- *To fix strength.*

Endurance

Stage 1 training emphasized coordination and motor learning to improve performance through neurological adaptation. Stage 2 training progresses to a more direct influence on muscle tissue by increasing the level of resistance by adding speed. Endurance is a functional quality related to local muscle circulation, which in this section refers to local muscular endurance, not cardiovascular endurance. Exercises can be dosed to increase endurance to tonic musculature of the ankle. Similar exercises performed in Stage 1, as with tissue repair and vascularization training may be utilized, with progressed dosage for endurance.

Dosage for Endurance

Vascularity / Local Endurance

- *Sets*: 2–3 sets or training sessions daily
- *Repetitions*: 25
- *Resistance*: Assisted to 50–60% 1RM
- *Frequency*: One time daily

Muscle in Atrophy

- *Sets*: 2–3 sets or training sessions daily
- *Repetitions*: 40 or more
- *Resistance*: Assisted to 30–40% 1RM
- *Frequency*: One time daily

Training for local muscle endurance may require non-weight bearing training of basic ankle movement with cuff weight or pulley resistance. All cardinal planes can be trained with easy control of the range of motion, as necessary to protect hypermobile ranges or mobilize hypomobile ranges. Free weights and pulley systems do allow for continued coordination training, with emphasis on the proper length tension curve from the resistance. Elastic resistance does not allow for proper length tension relationship with muscle training. Resistance continues to increase with elastic resistance, as the muscle torque decreases in the shortened range.

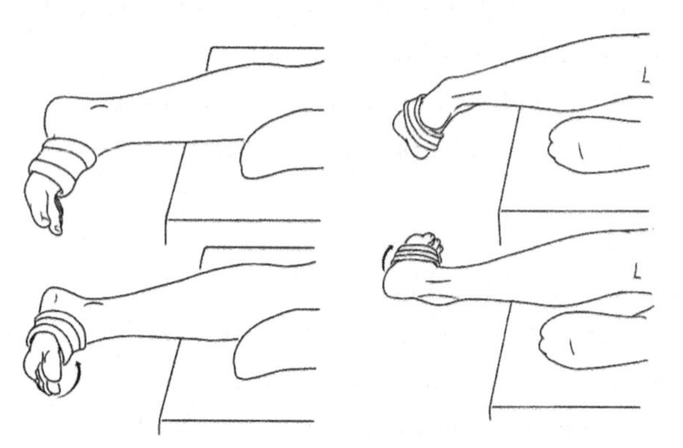

Figure 1.33a,b: Non-weight bearing training with free weights or pulleys can address local muscle endurance. Weight-bearing exercises may be added to address tissue tolerance and coordination, but specific muscle training in non-weight bearing may still be required. All cardinal planes can be trained, set specific to the individual patient presentation.

Full range of motion in non-weight bearing and full weight bearing should be attained prior to progressing to Stage 3 concepts. All exercises are progressed with each training session to the full available pain free range that can be coordinated around a physiological axis. A loss of weight bearing tolerance is a sign of excessive weight bearing stress and tissue inflammation, either from the training activities or daily activities that exceed tissue tolerance. With all exercise external resistance, or body weight, must be controlled to allow for coordinated repetitions. Weight bearing status is gradually progressed in this phase of improving tissue tolerance, further developing coordination for functional and athletic activities.

Partial-Full Weight Bearing Training

In Stage 2, weight bearing should be progressed from gravity eliminated positions to assisted weight bearing using upper extremity support or an unloading device. Gradual increases in weight bearing will allow increased tissue tolerance to compression and also provide increased proprioceptive feedback from joint receptors.

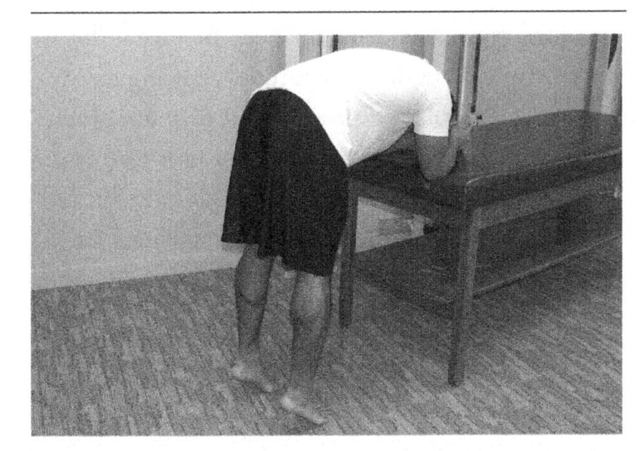

Figure 1.34: Partial weight bearing bilateral plantarflexion with the trunk supported. Reducing body weight can also be achieved by supporting the weight of the trunk on a counter or table. The uninvolved limb may also assist the motion training bilaterally, with progression to unilateral as coordination improves. Body weight is adjusted in this manor enough to prevent pain, improve coordination and to allow for 25 or more repetitions.

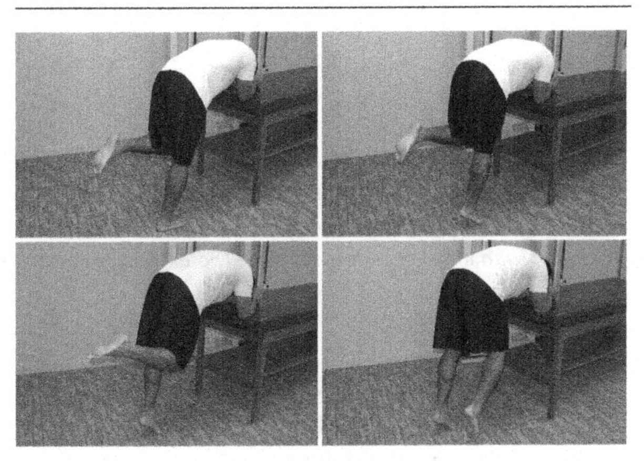

Figure 1.35a,b,c,d: "Achilles Shuffle"—Partial weight bearing pure eccentric training of the left Achilles for emphasis on collagen repair in tendon pathologies or repair. The patient stands on the right (uninvolved) limb with the body weight supported on a table (a). The concentric lift phase is performed by the right ankle (b). Weight is shifted onto the involved ankle while in full plantarflexion (c). Finally the partial body weight is lowered by the left ankle (involved tendon). The sequence is repeated for 10–15 repetitions. Range of motion and speed are increased prior to increasing the amount of body weight.

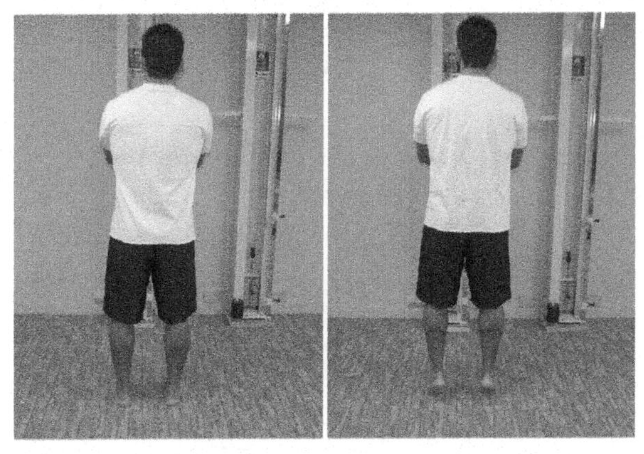

Figure 1.36: A progression from the previous exercises of partial weight bearing is a progression to full weight bearing plantarflexion with the trunk supported to reducing body weight. Speed is slowed down with this change to full weight bearing. As improvements in performance are made, range of motion is the next variable to progress, performing off the edge of a step. Next speed is increased to challenge both coordination and increase the relative load.

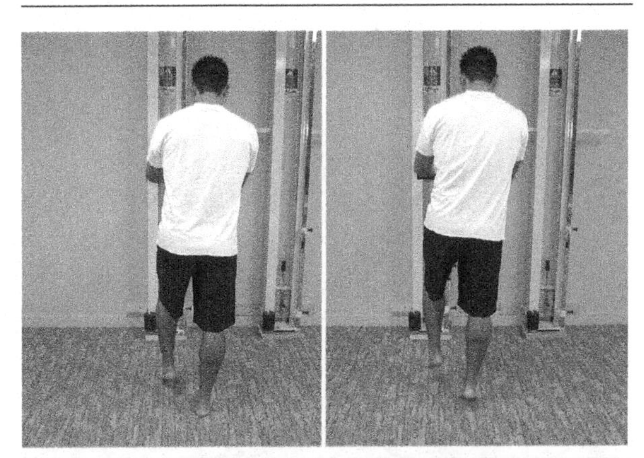

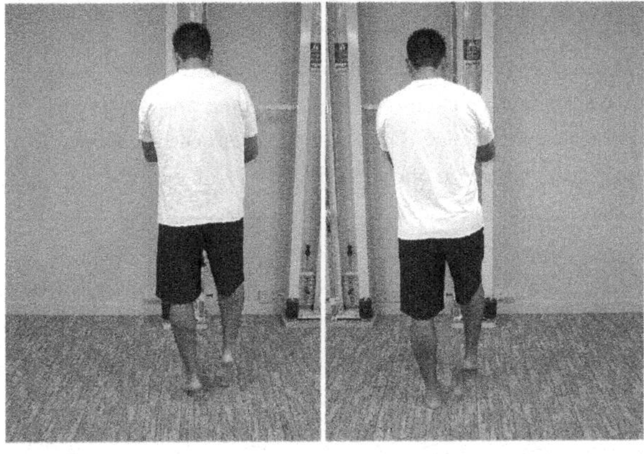

Figure 1.37a,b,c,d: "Achilles Shuffle"—Full weight bearing pure eccentric training of the left Achilles for emphasis on collagen repair in tendon pathologies or repair. The patient stands on the right (uninvolved) limb with hand support for balance control (a). The concentric lift phase is performed by the right ankle (b). Weight is shifted onto the involved ankle while in full plantarflexion (c). Finally the body weight is lowered by the left ankle (involved tendon). The sequence is repeated for 10–15 repetitions. Range of motion and speed are increased prior to increasing the amount of body weight.

1. Exercise for the Ankle and Foot

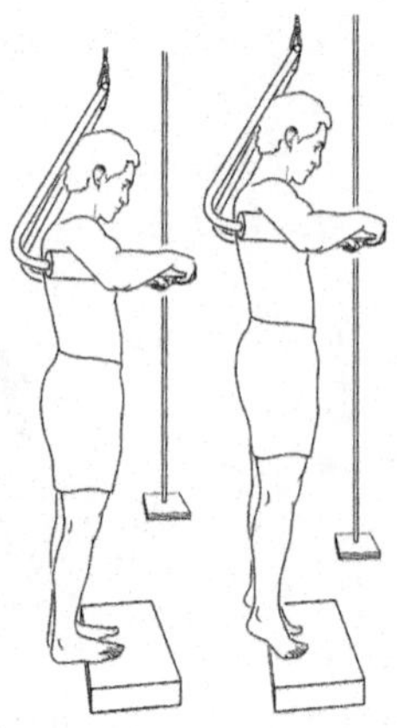

Figure 1.38: Partial weight bearing plantarflexion in standing. Reducing body weight with a pull down machine allows for partial weight bearing, stimulating more functional patterns for coordination and proprioception training. The weight is set to reduce body weight enough to prevent pain, improve coordination and to allow for 25 or more repetitions.

Figure 1.39: Partial weight bearing squats. Reducing body weight for squats assists with mobilizing ankle dorsiflexion, allowing for a challenge for tissue training and early coordination training. Body weight is reduced to allow for pain free, high repetition training.

Figure 1.40: Unloaded step training. Performing step training can emphasize concentric step-ups or eccentric step-downs, depending upon the functional needs of the patient.

Speed of training has a direct effect on muscle performance. When performing concentric activities, increasing the speed decreases the overall amount of time to recruit motor units, thus slow concentric work will improve performance. Eccentric work is the opposite situation, as faster speeds allow for a greater peak torque, and may improve coordination training. For the lower limb, this concept is most commonly applied with an eccentric step down performance. Increasing the speed will elevate the peak torque, allowing for lowing of the full body weight that may have been painful at slower speeds. To progress this type of eccentric training, speed is reduced to increase the challenge. An eccentric speed assist may increase the level of weight bearing tolerance, particularly in home training without unloading devices.

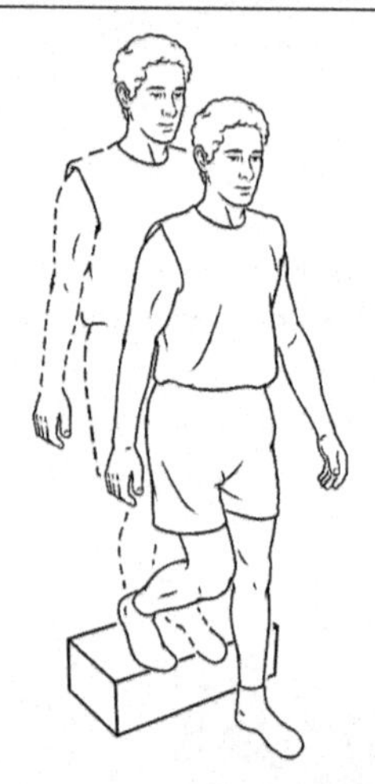

Figure 1.41: Step Downs. Once the patient is able to tolerate full weight bearing, fast eccentric step-downs (almost falling down on the leg) will allow a greater amount of load to be tolerated with initial eccentric training. The exercise can be progressed by slowing the speed.

More general aerobic activities can also be modified to allow pain free weight bearing. The ability to continue a regular aerobic conditioning program is important for most patients, both physically and psychologically. If a patient has a heel bruise or pain with calcaneal weight bearing, then treadmill walking might be initiated with an increased gradient, decreasing peak pressure and peak force in the heel (Grampp et al. 2000). This gradient could then be lowered as the tissue tolerance is increased and the patient remains pain free. Similar modifications of position can also be applied to other activities to allow earlier return to exercise.

Locking Techniques with Training

Locking concepts are more related to spinal treatment, but the concept of locking is simply to protect tissues from a particular range of motion. Performing heel raise exercises on the floor is a locking technique preventing full dorsiflexion, which may be associated with pain from excessive lengthening of the posterior ankle tissues or compression of the anterior joint structures. A similar concept can be applied to frontal plane motions of inversion and eversion. Placing a heel wedge under the foot for even one to two degrees change in position may be enough to allow for an increase in range of motion without pain during unloaded squats, step or lunge exercises. The challenge is only in designing the exercise for a safe application of an inversion or eversion moment at the heel. Assist with locking is gradually removed as tissue tolerance and coordination allow.

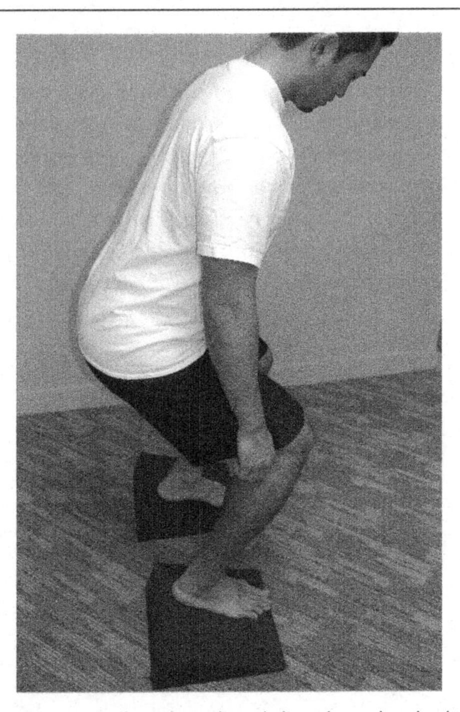

Figure 1.42: A small wedge placed directly under the heel will place the ankle in slight plantarflexion. The anterior talocrural joint may be painful in squatting past zero degrees dorsiflexion due to cartilage compression, or anterior joint capsule impingement. Placing the ankle in slight plantarflexion will increase the pain free range, allowing for a more functional squat performance.

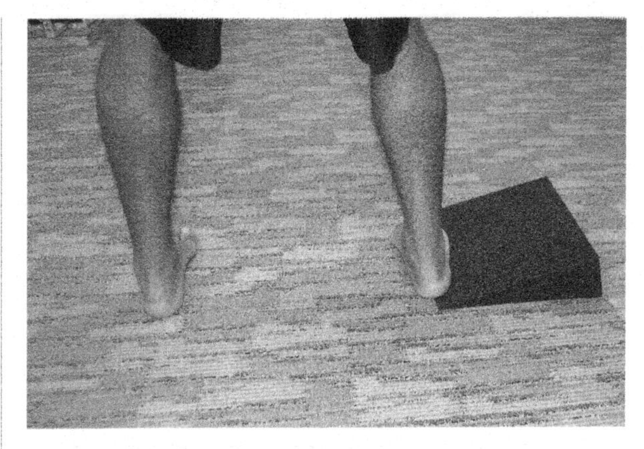

Figure 1.43: A small wedge placed under the lateral calcaneus places the heel in slight eversion, passively shorten the lateral ankle ligaments. Placing the wedge on the medial side will unload the medial joint structures

Proprioception

Proprioception is defined as the combined afferent neural input to the central nervous system from mechanoreceptors in joint capsule, skin, muscles and ligaments. Proprioceptive deficits of ankle ligaments are believed to weaken reflexive muscle contractions, reducing the ability to stabilize body sway (Freeman et al. 1965, Ryan 1994, Konradsen and Ravn 1991, Matsusaka et al. 2001, Rozzi et al. 1999). Proprioceptive activities should be initiated early using unloading weight bearing activities as pain allows.

Gradual increases in pain-free weight bearing while performing task specific activities have been shown to accelerate the return to function in lateral ankle sprains (Kern-Steiner R et al. 1999). Along with providing optimal stimulus to joint cartilage, as weight bearing is progressed mechanoreceptor recruitment will also increase, further enhancing motor performance and neurological adaptation. Exercise for proprioception can be performed separately with emphasis on many repetitions or time performed, but they also can be built into all weight bearing exercise. This type of training is of greater importance after trauma or surgery that results in tissue damage and subsequent damage to afferent feedback for proprioception. Despite the mechanism of injury or diagnosis, proprioception should be tested with impairments addressed.

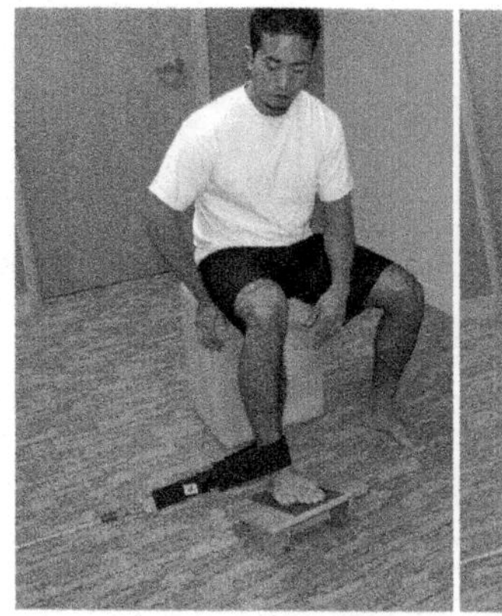 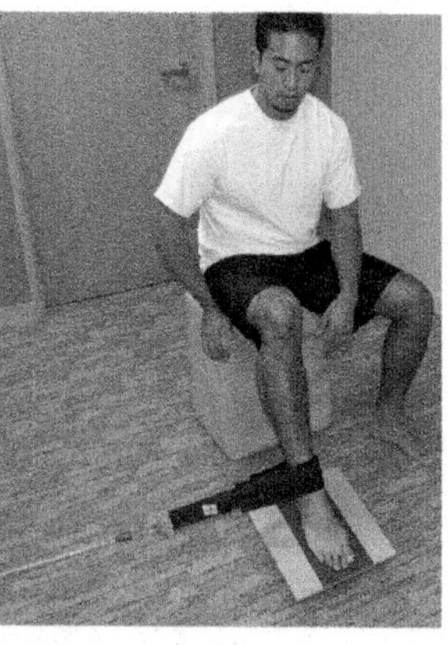

Figure 1.44a,b: Partial weight bearing plantarflexion and dorsiflexion with a secondary varus force from a pulley or band lateral to the ankle. The patient is instructed to maintain a neutral ankle from inversion/eversion during the motion. Initially visual feedback will assist with the coordination but should eventually be removed from the exercise. From a tissue training approach, the varus force may initially be used to prevent eversion from occurring, straining tissues injured in an eversion injury. Alternatively, for an inversion sprain, a varus force moment provides a greater challenge to the muscles of eversion to maintain a neutral ankle and protect the medial structures. The fixed axis of the wobble board makes this initial approach very safe from creating excessive inversion to reinjury sprained lateral structures. Later progressions will involve allowing motion into the range of inversion.

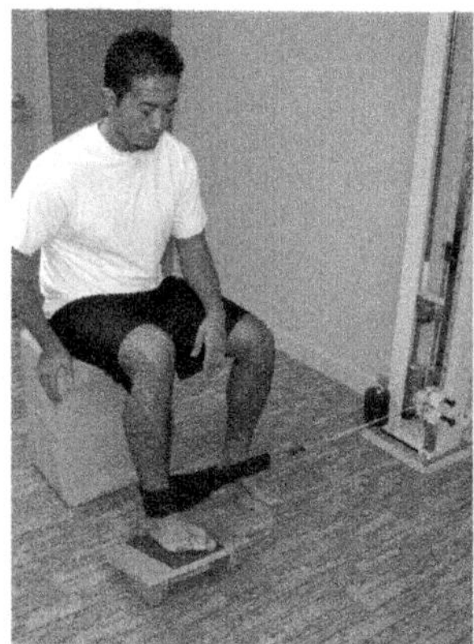 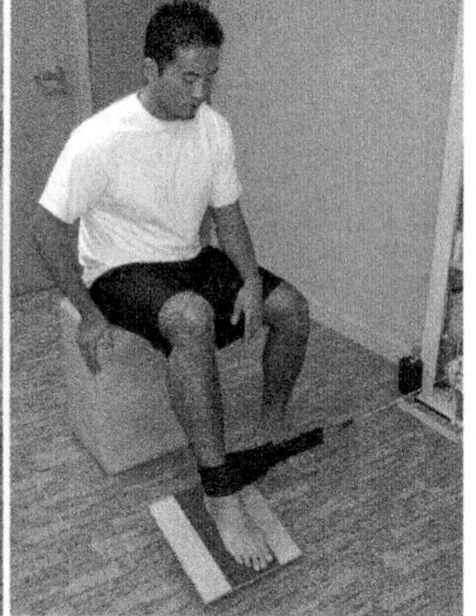

Figure 1.45a,b: Partial weight bearing plantarflexion and dorsiflexion with a secondary valgus force from a pulley or band medial to the ankle. The patient is instructed to maintain a neutral ankle from inversion/eversion during the motion. Initially visual feedback will assist with the coordination but should eventually be removed from the exercise. From a tissue training approach, the valgus force may initially be used to prevent inversion from occurring, straining tissues injured in an inversion injury. Alternatively, for an eversion sprain, a valgus force moment provides a greater challenge to the muscles of inversion to maintain a neutral ankle and protect the medial structures. The fixed axis of the wobble board makes this initial approach very safe from creating excessive inversion to prevent reinjury of sprained lateral structures.

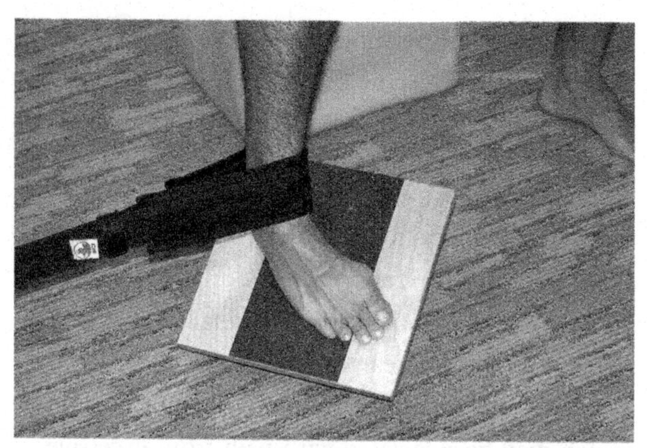

Figure 1.46: Changing the angle of the wobble board allows for a combined plantarflexion and inversion moment with an opposite dorsiflexion and eversion moment (Shown with a varus force).

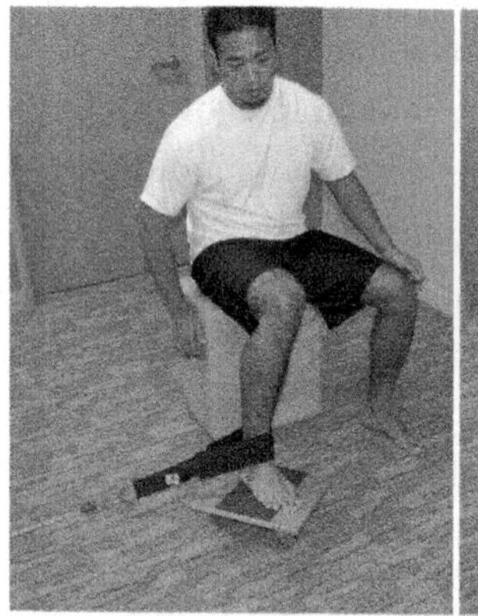 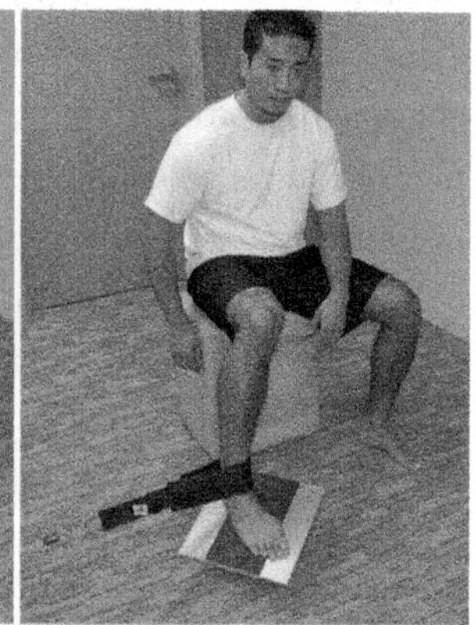

Figure 1.47a,b: A progression of partial weight bearing plantarflexion and dorsiflexion with a secondary varus force from a pulley or band lateral to the ankle. The wobble board is rotated to allow for combined plantarflexion and inversion toward the unstable range of a lateral ankle sprain. The muscles of inversion must work against the vector of the pulley as well as the motion of the wobble board.

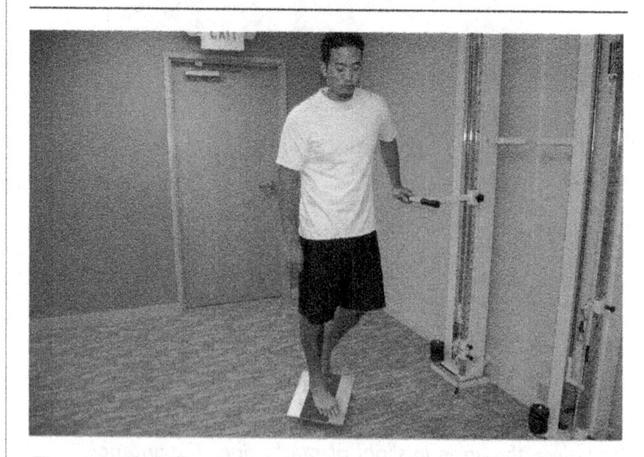

Figure 1.48: Full weight bearing plantarflexion/dorsiflexion. Here the wobble board is again rotated 45° to allow for combined plantarflexion and inversion. This is a more significant challenge to the lateral ankle sprain to maintain stability of the ankle while the wobble board moves toward the pathological range.

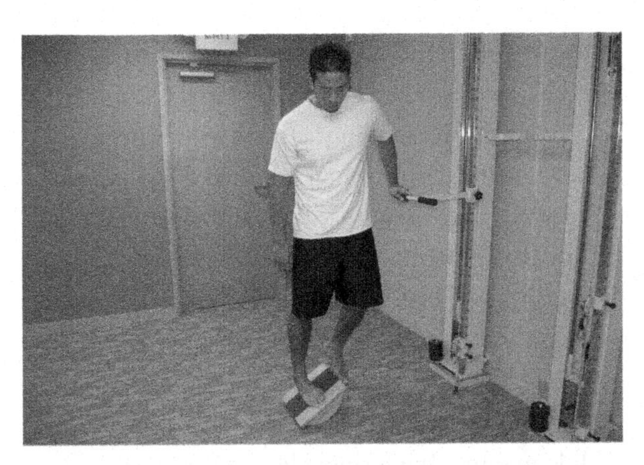

Figure 1.49: Full weight bearing eversion and inversion on a wobble board. The patient begins with slow motions in a limited pain free range. The range of motion is gradually increased but should remain pain free. For an inversion sprain the patient is instructed to begin in full eversion. The ankle is slowly moved into inversion and quickly moved back to the start position. The patient is not to stop or pause in the inverted position to avoid further tissue injury.

Figure 1.50: As a progression from above, a varus force moment is added from a lateral pulley while performing active eversion. This is a final progression to load the muscles of eversion with an emphasis on eccentric training. As with the previous example of an inversion sprain, the patient is instructed to begin in full eversion, move slowly into inversion and quickly moved back to the start position. This exercise may be performed with the wobble board in the plane of plantarflexion and dorsiflexion only. The wobble board can be gradually rotated with each set or training session until this exercise of pure eversion/inversion can be performed safely. Any combination of vectors from the foot position on the wobble board, to the force moment direction from the pulley, can be adjusted to match the tissue injury and/or impairment.

Balance Testing / Training

Balance control is a combination of sensory input from the vestibular system, somatosensory systems and the visual system. Integration of information from these three systems allows the body to recognize position in space, control the body's orientation in space, and interaction with the center of gravity. Disruption of somatosensory input from injury or disease to the foot and ankle can lead to both local and central deficits in motor performance related to balance and proprioception. Reduction of ankle range of motion, due to capsular and ligamentous tightness but not muscle tightness, has been associated with reduced balance (Mecagni et al. 2000). Manual intervention for improving joint play and range of motion should be a precursor to beginning dynamic balance training exercises.

Muscle weakness is not a major contributing factor in ankle instability (Soderberg et al. 1991). As maximum strength is not required to maintain balance, appropriate motor patterns and timing is more the issue. Ankle instability has been associated with a delay in onset of the peroneal muscles with sudden ankle inversion (Konradsen and Ravn 1991/1990), and a delay in the tibialis anterior with linear perturbations (Brunt et al. 1992). Balance testing identifies alterations in stability and performance, assisting in the design of specific exercises related to tested impairments.

Single Leg Balance Testing

A single leg balance test is a basic starting point to identifying impairments in lower limb balance. The subject stands on one leg with the eyes open and the contralateral hip in neutral with the knee at 90 degrees flexion. The test is performed for 30 seconds per limb with measurements including: 1) the number of times the elevated foot contacts the ground and 2) the relative degree of trunk and upper limb motion needed to maintain balance (Forkin et al. 1996). If appropriate, the test is repeated with the eyes closed. Trojian et al. (2006) found a failed single leg balance test (SLB) to be a reliable predictor of future ankle sprain, testing American male football players and women's volleyball and soccer players. Athletes were expected to hold a single leg stance position, with eyes open, for 10 seconds. A positive test was defined as the athlete failing to maintain balance for the 10

1. Exercise for the Ankle and Foot

seconds or describing a sense of imbalance. Athletes with two or more previous ankle sprains were also more likely to have a positive SLB test. More complex computerized systems measuring increased postural sway have also been successful in predicting ankle injury and measuring proprioceptive deficits in athletes (Tropp et al. 1984, Leanderson et al. 1993, McGuine et al. 2000). Once injury has occurred, retraining balance and coordination has a significant impact on retraining proprioception in the unstable ankle (Bernier and Perrin 1998).

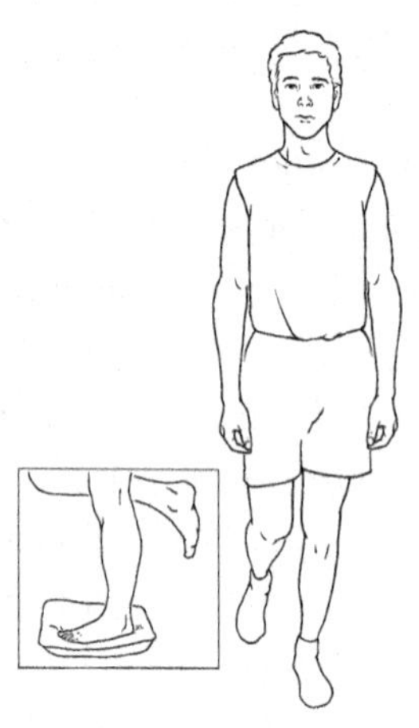

Figure 1.51: Single leg balance test/treatment. The patient performs single leg stance as a timed test or exercise. The patient is instructed to gaze forward to avoid visual compensation for somatosensory deficits in balance reflexes. Alternative surfaces can be used to challenges the balance systems. Foam will reduce somatosensory input, emphasizing contributions of the vestibular system.

Star Excursion Balance Test (SEBT)

The Star Excursion Balance Test (SEBT) may offer a simple, reliable, low-cost alternative to more sophisticated instrumented methods for testing dynamic balance. The test requires the patient to maintain balance on a single limb, while manipulating the other limb. Dynamic postural stability has been defined as the extent to which a person can lean or reach without moving the feet and still maintain balance (Goldie et al. 1989). Kinzey and Armstrong (1998) first published the test with reliability estimates ranged from 0.67 to 0.87, increasing to above 0.86 with duplicate practice sessions. The goal of the SEBT is to reach with one leg as far as possible in each of eight prescribed directions, while maintaining balance on the contralateral leg (Olmsted et al. 2000). A star pattern is placed on the floor with four lines of tape: vertical, horizontal and two diagonals at 45 degrees angle. Alternatively, one line of tape can be placed on the floor with the subject altering the body position relative to the line of tape for each test, replicating all eight test directions.

Directions of reach are relative to the stance leg. While maintaining a single-leg stance, the subject reaches with the contralateral leg (reach leg) as far as possible along the appropriate vector. The subject lightly touches the furthest point possible on the line with the most distal part of the reach foot, and returns to the center (Olmsted et al. 2000). Measurement of the distance along the tape is recorded in centimeters. For testing purposes, each direction is repeated three times, with a 15-second rest between each reach. A trial is deemed invalid and repeated if the subject (1) does not touch the line with the reach foot while maintaining weight bearing on the stance leg, (2) lifts the stance foot from the center grid, (3) loses balance at any point in the trial, or (4) does not maintain the start and the return positions for one full second (Olmsted et al. 2000).

Areas identified as having balance deficits can easily be turned into clinical and home exercises. Testing should not only include measurement of the reach distance, but also to identify abnormal compensatory movement patterns that need to be corrected for training. The most common compensation is to lean the trunk in the opposite direction of the lower limb reach, as a counter balance. The patient should be instructed to maintain the head and trunk directly above the stance foot. With a vertical trunk, the pelvis may compensate by shifting in the opposite direction of the reach as a counter balance. The patient should be instructed to maintain alignment of the hip, knee and ankle in a vertical line. Knee valgus and excessive ankle pronation are also common compensations that should be avoided. The distance of the balance reach exercise should be only to the range that can be coordinated with proper alignment of the trunk and lower limb.

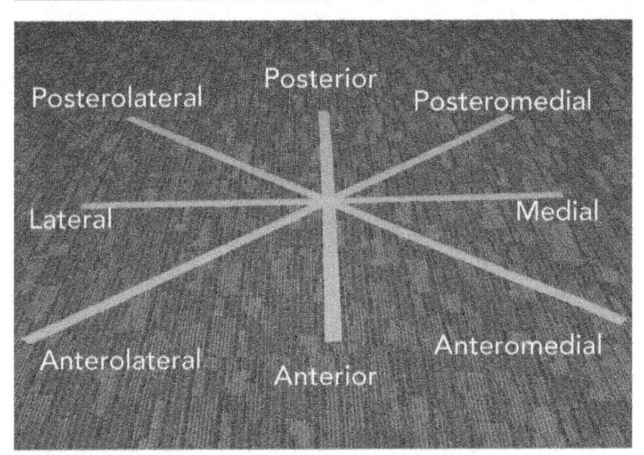

Figure 1.52: Star pattern for balance reach, stance on the right foot with left lower limb reaches. The eight lines positioned on the grid are labeled according to the direction of excursion relative to the stance leg: anterolateral (AL), anterior (A), anteromedial (AM), medial (M), posteromedial (PM), posterior (P), posterolateral (PL) and lateral (L).

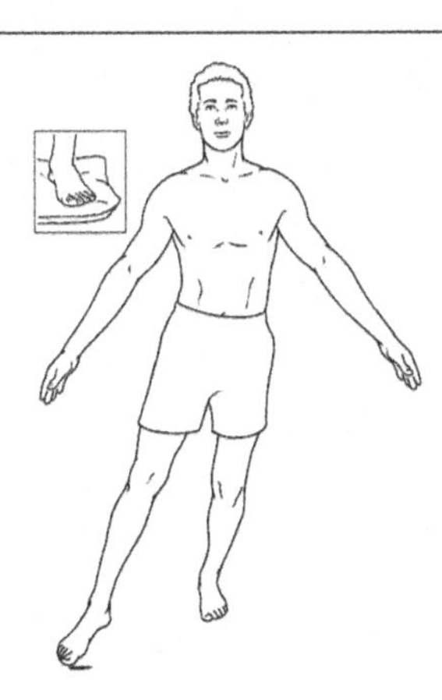

Figure 1.53: Single Leg Stance-Lower quarter reach (Patient performs a SLS and slowly hovers the opposite foot away from the center of gravity). This is an example where decreasing the speed increases the required coordination.

The STAR pattern of lower limb reach can easily be turned into an exercise program, and combined with a complete balance rehabilitation program. The entire star pattern can be trained, or just the directions showing deficiency. Chaiwanichsiri et al. (2005) compared a general star pattern of balance training in acute ankle sprains, after the subject could walk pain free, without medication. The control group continued with traditional physical therapy utilizing heat, ultrasound and exercises for range of motion, stretching and strengthening. All subjects improved in single limb balance testing, but the training group improved twice as fast (eyes open and closed) with fewer incidents of a second sprain. Training on labile surfaces is still recommended for complete balance rehabilitation, but the STAR pattern is a simple and effective approach for both the clinic and home setting.

Figure 1.54: Single Leg Stance-Lower quarter reach (The exercise can be more easily performed by reaching both arms laterally across the body while performing this exercise). This will shift the center of gravity over the stance leg and reduce lever arm needed for the muscles.

Star Upper Limb Reach

Similar to the Star Excursion Balance Tests for the lower limb reach, upper limb reaching can also be performed in the star pattern. Duncan et al. (1990) described functional reach as a measure of dynamic balance. Functional reach was defined as the difference between arm's length and maximal forward reach, using a fixed base of support. Upper limb reaches are described in the STAR pattern as previously described. Reaching may also include a vertical component of reaching down toward the floor (inferior), horizontal reach and superior reach or as diagonal patterns. Testing is similar to the lower limb reach with measurement being the distance from fingertip to stance foot. When testing, dropping an object straight down from the hand to the line of tape on the floor can provide a more objective measure for distance. Reaching measurements horizontally or with an upward vector can be measured with a floor measurement from the toe to the base of the object that is being touched. A vertical component may also be measured if the emphasis is on the height of the reach, rather than the distance.

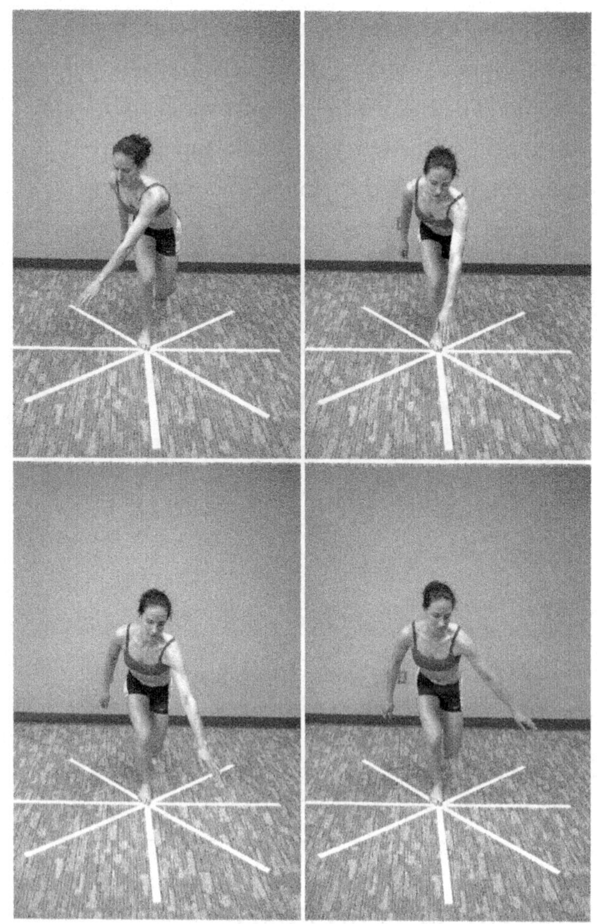

Figure 1.55a,b,c,d: Balance training with upper limb reaching is performed with a similar STAR pattern used in lower quarter reaches. Upper limb reaching, however, allows for a superior, horizontal and inferior component, to the direction in the transverse plane. The exercise is named for the direction the hand is reaching relative to the stance leg: a) right, inferior, lateral reach, b) right, superior, medial reach.

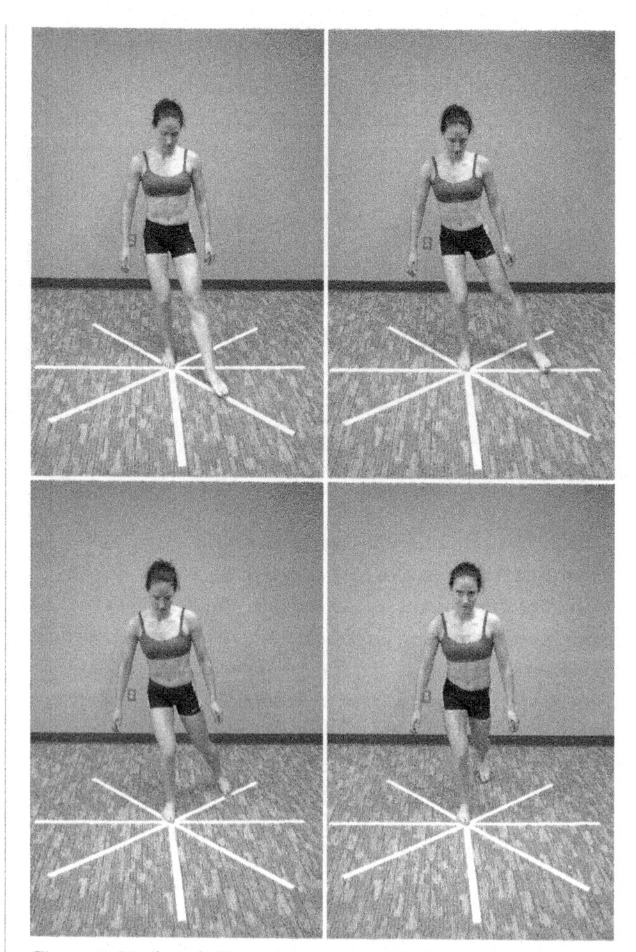

Figure 1.56a,b,c,d: Single leg stance with lower quarter reaches in multiple directions—STAR pattern. The patient is instructed to maintain balance for at least one second before moving the limb. The toe teaches out one inch off the ground without making contact and brought back to the start position. Balance is to be maintained for at least one second at the end of each reach.

	Direction of Training	Dosage and Frequency
Balance Training for Unilateral Lower Limb Stance	• Stable surface, eyes open • Stable surface, eyes closed • Stable surface, eyes open, cervical motion • Stable surface, eyes closed, cervical motion • Stable surface, eyes closed, trunk external challenge • Stable surface, eyes open, lower limb reach • Stable surface, eyes closed, lower limb reach • Stable surface, eyes open, upper limb reach • Stable surface, eyes closed, upper limb reach	• 15 seconds at each progression • 45 second rest between • Daily 5–15 minutes.
Options	• As above with labile surface: Tilt board—single plane challenge Tilt board with foot at 45 degrees diagonal Wobble board—3-dimensional challenge • Foam—(vestibular challenge)	Progress: • 30 seconds per exercise • 15 second rest

Table 2.2: General guidelines for balance training.

Restoring normal muscle strength of the lower quarter is a component of normalizing postural reflexes for balance (Carter et al. 2002, Messier 2002). Pain free weight bearing status for lower quarter and lumbar patients is necessary to have safe and effective balance training. Proprioceptive loss at the knee and ankle from trauma to the somatosensory system has been shown to significantly delay ankle muscle responses while upper leg and trunk responses are not delayed (Bloem 2002). Without retraining of the lower leg somatosensory system the majority of lower leg balance correcting responses will be initiated by hip and trunk proprioceptive inputs (Bloem 2002). Chong et al. (2001) assessed balance improvements with training on the BAPS board (biomechanical ankle platform system) to determine if balance improvements could be attributed to retraining deficits in the ankle and foot. Improvements observed during training were attributed from diffuse enhancement of proprioception in other body segments such as the knees, hips, spine and upper extremities rather than targeting proprioception specific to the ankle.

Waddington et al. (2000) demonstrated that eight weeks of wobble board training in healthy soccer players improved discrimination of discrete ankle inversion movements, compared to jump-landing training and control. This improvement enables greater accuracy in the making of inversion movements of the foot and ankle in preparation prior to ground contact.

Utilization of balance board training as a preventative measure for ankle sprains in healthy individuals has conflicting evidence. Studies have demonstrated non-preventative influence on ankle sprains (Soderman et al. 2000), while other studies have shown a preventative influence (Verhagen et al. 2004, Bahr et al. 1997, Troop et al. 1985, Wedderkopp et al. 1999). The studies of Bahr et al. (1997) and Tropp et al. (1985) found the preventative influence of balance board training was the greatest for players with a history of previous ankle sprains.

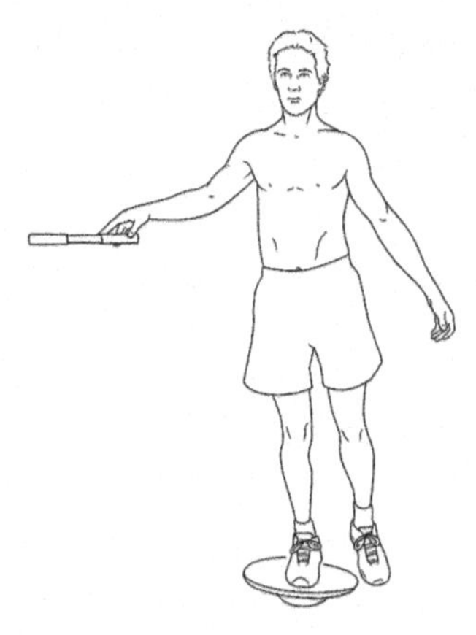

Figure 1.57: Single leg stance on wobble board. Standing balance on a wobble board can provide further challenges to the somatosensory system and challenge coordination.

Figure 1.58: Wobble board proprioception training with reduced body weight in standing. Sitting wobble board training, as shown in Stage 1, may be the initial starting position to emphasize isolated ankle proprioceptive and range of motion training, but integrating patterns that include the knee, hip and lumbar spine require an upright posture. This may require unloading of body weight for a pain free challenge.

The rehabilitation program should include multifaceted concepts to specifically address the individual needs of the patient with specific reference as to what is the goal of the exercise. If you are attempting to restore coordinated movement from injured mechanoreceptors, as in a lateral ankle sprain, balance training should be performed on a stable surface to engage the ankle. The use of foam pads and other soft surfaces should be reserved for vestibular training or emphasizing more proximal joints, as this type of surface would remove the normal ground reaction force at the ankle, in sorts "blinding" the ankle mechanoreceptors.

1. Exercise for the Ankle and Foot

The following is a sample progression of a logical sequence that can be used to increase weight bearing/proprioception when treating a lateral ankle sprain with squatting activities.

- *Unloaded squats using lateral heel wedge and/or medially directed pulley to keep lateral ligaments out of end range tension.*
- *Unloaded squats without heel wedge/pulley assistance in pain free range of motion.*
- *Unloaded squats throughout full ROM while decreasing assistance.*
- *Full body weight squats with medially directed pulley to decrease tension on lateral collateral ligaments.*
- *Full body weight squats working towards full range of motion.*
- *Full body weight squats using laterally directed pulley to increase challenge/tension on lateral ligaments and muscles.*

Tissue Support or Challenge with Balance Training

Locking techniques in Stage 1 described utilizing a wedge during standing exercises to support the ankle on the side of pathology. A lateral wedge was used to create a slight eversion of the subtalar joint, protecting lateral joint structures. Similar approaches were presented with a medial or posterior wedge. An alternative to this approach is to apply an external resistance at the joint line that serves to assist the muscles in dynamically stabilizing the joint on a flat surface.

In the example of a lateral ankle sprain, a force moment from a pulley system or elastic resistance can be applied driving the ankle medially. This medial vector will reduce the force necessary in the muscles of the lower limb to protect the ligaments. In later stages the vector may be reversed, pulling the ankle toward the direction of pathology, to increase the challenge to stabilizing muscles. These types of challenges can be utilized in all weight bearing activities such as unilateral standing balance, squats and step training.

Figure 1.59a,b: Unilateral standing balance with medial force assist. A medial force creates a pronation force moment to assist muscles for stabilization of an inversion sprain.

Figure 1.60a,b: Unilateral standing balance with lateral force challenge. A lateral force creates a supination force moment to challenge muscles for stabilization of an inversion sprain. This type of challenge should be used with caution, and may be more safely applied in later stages of training.

Motions and Directions

In Stage 1 motion was directed away from the pathology to safely provide motion for tissue repair, motor facilitation, pain inhibition etc. Muscles responsible for dynamic joint stabilization are trained concentrically away from the pathological range, with a return to neutral. During Stage 2, emphasis is placed on sensitizing the muscle spindles in these muscles to improve recruitment of motor patterns. Isometric work allows for greater resistance than what could be used concentrically, which will more quickly improve sensitization of afferent input from muscle spindles. The Manual

Therapy Lesion concept, discussed in Chapter 8, describes the possible effects of tissue damage and mechanoreceptor damage leading to reduced afferent input, resulting in reduced efferent drive for motor performance. Dynamic instability results, with both tissue damage and reduced motor control. The clinical challenge is to replace the afferent input from damaged mechanoreceptors with increased sensitivity of muscle spindles in the local muscle. By increasing the muscle spindle's intrafusal muscle fibers sensitivity to stretch reflexive activity can increase extrafusal motor facilitation.

Evidence supports the proposition that activation of large muscle afferents, from muscle spindles and tendon organs, are capable of shaping the activation of extensor muscles and the timing of locomotor patterns by signalling loading and extension of the leg (Conway et al. 1987, Forssberg et al. 1980, Grillner and Rossignol 1978, Hiebert and Pearson 1999, Hiebert et al. 1996). Yang et al. (1991) demonstrated that afferent feedback from leg muscle spindles contributes to control of the plantarflexor muscles activity during stance. Sorensen et al. (2002) found strong evidence that ankle muscle spindles play a significant role in the control of posture and balance during the swing phase of locomotion.

Muscle spindle feedback provides information describing the movement of the body's center of mass with respect to the support foot. Progressive training on the coordination-endurance-strength continuum is thought to aide in establishing dynamic stability with muscle spindle afferent input improving motor recruitment and motor control.

Similar exercises are performed from Stage 1, but one of the sets is changed to an isometric hold of 10–30 seconds with increased resistance. The joint, or tissue is maintained in a neutral position. Several holds can be performed, but emphasis however, is on neurological adaptation for motor recruitment, as opposed to making structural changes in muscle tissue. The tissue is never positioned in a range of pain, or one that cannot be stabilized. Stage 3 will continue to train these same muscles with emphasis on eccentric work toward the pathological range. Stage 2 is typically a combination of concentric and isometric work.

Generalized Progression of the Direction of Training for Stabilization

- *Stage 1: Concentric work with muscles moving away from the pathological ranges.*
- *Stage 2: Isometric of the same muscles in mid range and continued concentric work away.*
- *Stage 3: Train the same muscles with an eccentric emphasis, starting away from the pathology, performing eccentrically toward the pathological range with a concentric return to the start position. Next the antagonist group works with a concentric emphasis toward the pathological range with a one second pause up to, but not into, the pathological range.*
- *Stage 4: Diagonal patterns through full range of motion around a normal physiological axis of motion.*

Progression Concepts Applied to Dynamic Balance Training Following a Right Lateral Ankle Sprain

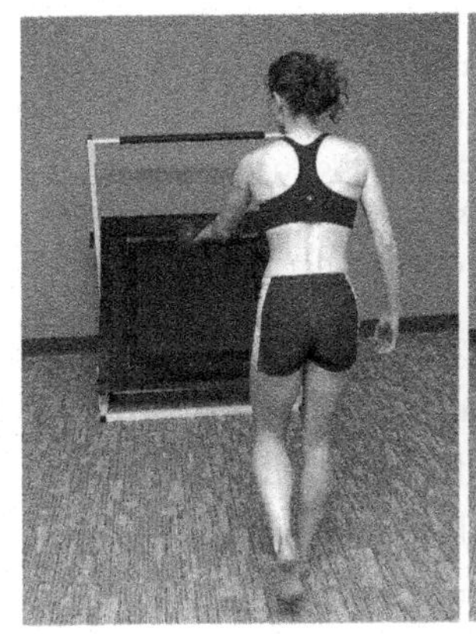

Figure 1.61a,b: Concentric Away—Single leg stance with ball catch (Stage 1). Single leg stance for balance is performed while catching a ball on the left side of the body. When beginning a dynamic balance program, it might first be necessary to protect the healing tissues. Start by throwing the ball towards the opposite side of the patient's body (left side). When the ball is caught the body will decelerate into turning to the left driving the right ankle into pronation. Pronation is motion away from the supination associated with an inversion ankle sprain.

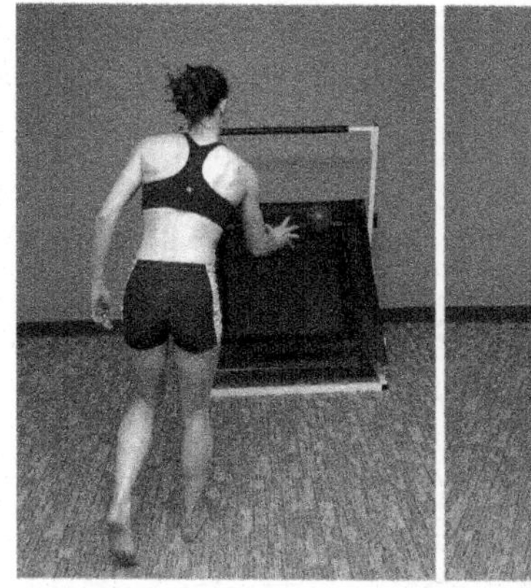
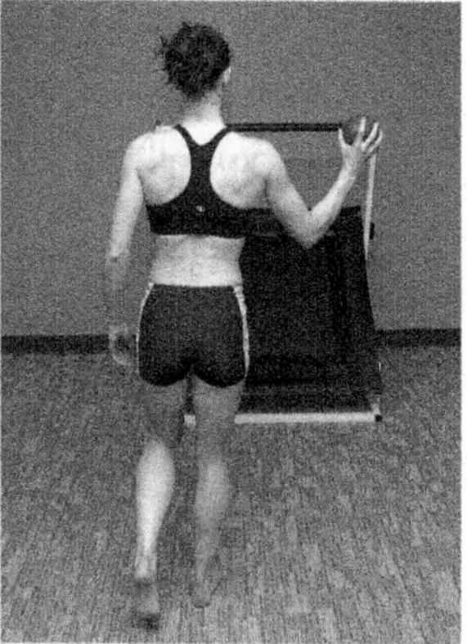

Figure 1.62a,b: Isometric—Single leg stance with ball catch (Stage 2). The first progression is to toss the ball to the center or to the right side of the patient. Catching to the right requires the body to decelerate to the right driving the foot into supination, or toward the pathology. The patient is instructed to catch the ball and isometrically stabilize without allowing the body to turn.

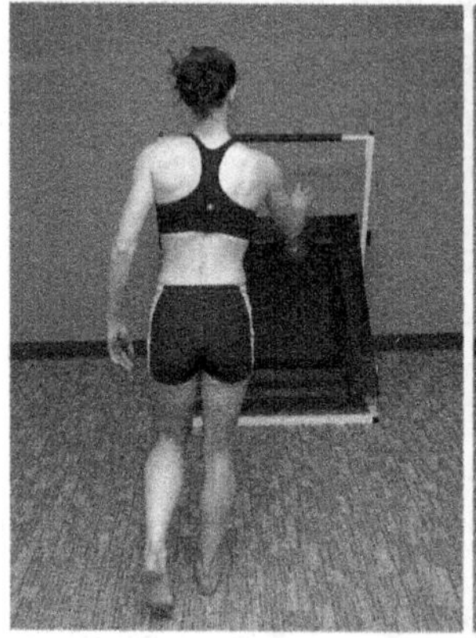

Figure 1.63a,b: Eccentric toward—Single leg stance with ball catch (Stage 3). The final progression is to catch the ball on the left side and allow the body to turn to the right. The muscles of the lower limb must decelerate the lower limb and ankle from supination requiring eccentric control, followed by a strong concentric return. This would protect the instability and emphasize eccentric control into supination.

Balance testing and training identifies directions of motion that are not stable. These directions can be directly trained through the addition of weight bearing exercises such as lunges, step training and gait training. The concept of moving away from pathology first still applies. Movement away will provide a safe tissue challenge as work is performed to unload weight bearing tissue or shorten tissues under tensile load. Exercise can then be progressed in a direction that will further challenge the tissue in lesion, as well as the muscles that stabilize the movement toward the pathological range.

Lunge / Step Progression Examples with a Right Lateral Collateral Ligament Strain

Motion is performed initially in a direction to shorten the lateral tissues of the ankle that have been strained. A progressive increase in the range of inversion allowed places increasing tension on the collateral ligaments, increasing the challenge to the muscles of eversion to stabilize the ankle from the pathological range of plantarflexion and inversion.

Figure 1.64a,b: Right foot medial lunge. The right foot steps across the body, medial the left, landing in a position that drives the foot into eversion. Minimal demand is placed on the muscles of eversion, with the body weight landing on the lateral side of the foot driving the foot into eversion. The ligament undergoes repetitive tension from its mid to shortened position, providing a stress for repair but not excessive strain at the end range of tension (inversion).

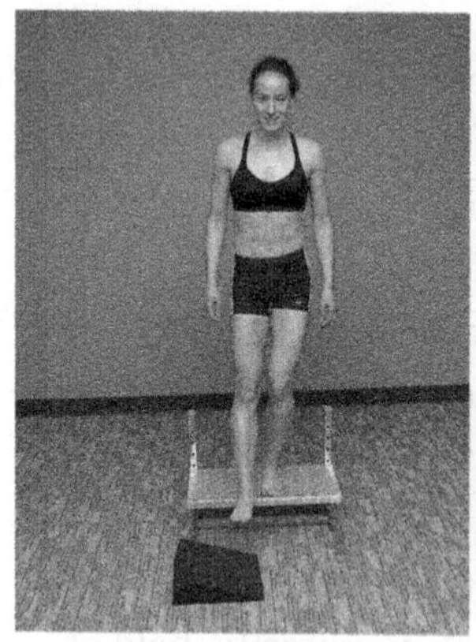

Figure 1.65a,b: Right foot medial step down onto a wedge creating an eversion moment. The right foot steps down anteriorly with the foot landing on a foam wedge to create an eversion moment of the rear foot. Minimal demand is placed on the muscles of eversion with an increased impact load from stepping down. A significant sprain, or a chronic sprain with alterations in normal motor patterns, may hang in a position of plantarflexion and inversion during the step down motion, bias for inversion strain. The wedge prevents this from occurring until the patient can maintain neutral.

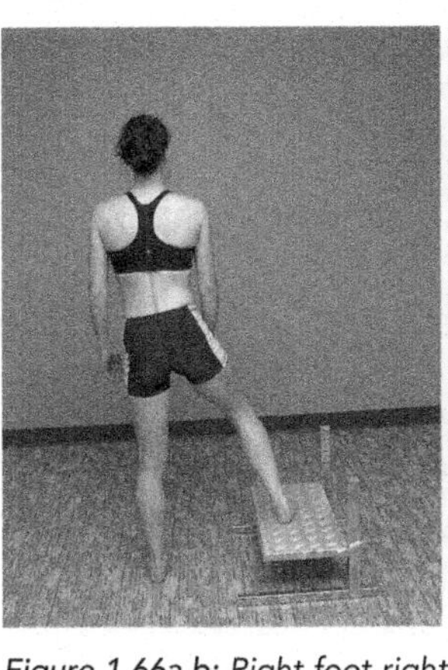
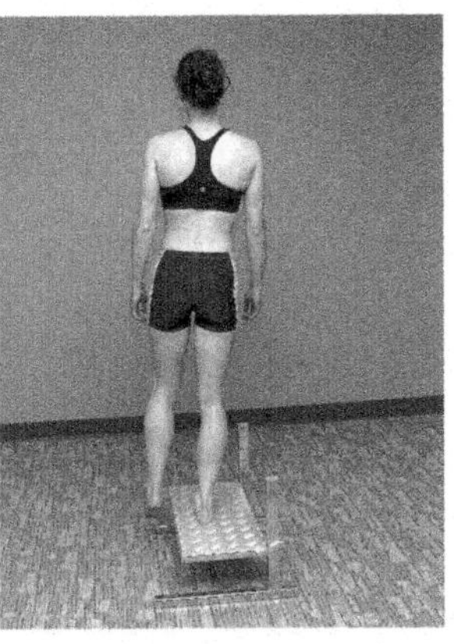

Figure 1.66a,b: Right foot right lateral step-up. The right foot steps lateral and up onto a step, beginning with a short height and progressing. The muscles of eversion work isometrically to stabilize the ankle during the motion while creating minimal tension on the lateral collagenous structures.

Figure 1.67: Right foot right lateral lunge onto a wedge. The right foot lunges lateral onto an angled surface. With the lateral lunge the foot will hit the ground on the medial side driving the ankle into inversion, requiring significant demand on the muscles of eversion. To initially reduce the load on these muscles the foot lands on a slant, rather than a flat surface, to lock the ankle for full inversion, while training the muscles of the lower limb to stabilize lateral movement.

Figure 1.68: Right foot right lateral lunge onto flat surface. The right foot lunges lateral onto a flat surface, driving the ankle into inversion and requiring significant demand on the muscle of eversion.

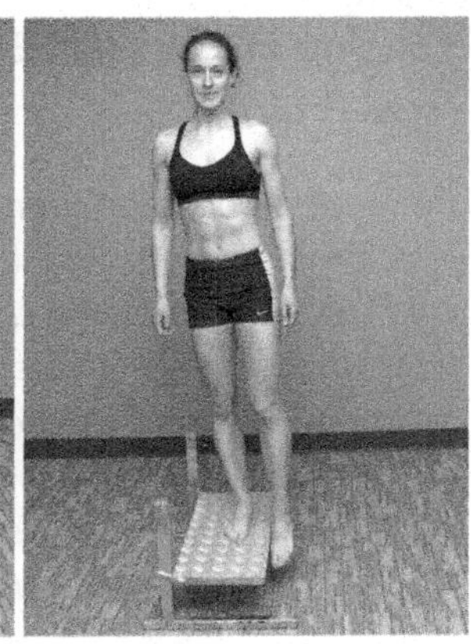

Figure 1.69a,b: Right foot right lateral lunge step-up. The right foot lunges lateral and up onto a step, driving the ankle into end range of inversion end range. Significant demand is placed on the muscles of eversion to stabilize on impact and pull the tibia back in a neutral position from inversion.

The above example demonstrates the concept of performing early weight bearing functional movements with a progression from unloading the primary tissue in lesion, toward a challenge at end range. Muscle coordination, endurance and strength training can be dosed in conjunction with these progressions in motion. This concept can be applied in any plane of motion, related to the individual's tissue in lesion. The previous example demonstrated one plane of training only. A progression should include different angles of lunging and stepping for stabilization training through a greater range. Rotary components may also be necessary, with the lower limb landing in ranges from neutral to lateral rotation. Sport activities may dictate specific directions of performance that must be retrained prior to return to athletic performance.

For tissues under weight bearing load (cartilage / bone), rather then a program of progressive tension of injured collagenous structures, a progressive program focuses on incremental increases in loading. Initial exercises include unloading body weight for pain free and coordinated loading to the involved joint. Range of motion is progressed first, then loading increases to full body weight. Further progression may includ loaded body weight and, if necessary, more ballistic loading or plyometric jump training programs may be appropriate. The progressive loading may still haves a component of direction as it relates to motor control.

Loading Progression Example for a Right Heel Bruise

For a heel bruise, motions are performed initially in a direction to reduce the loading impact on the heel, while training the muscles of the lower limb to absorb load. The activities performed progressively load the heel until full heel impact is possible with gait and running. A progression from Stage 1 through 4 should involve modified loading, to normal loading through over loading.

Figure 1.70a,b: Right foot posterior lunge. The involved foot lunges back landing on the forefoot, with a slow deceleration of the heel to the floor. The concept here is to move in the opposite direction of that causing the primary pain (i.e., anterior lunge landing on the heel). The initial impact is made in the forefoot with the load of the body absorbed by the muscles of the lower limb. Eccentric performance of plantarflexors of the ankle, with extension at the hip and knee, decelerating and the body and lowering the heel to the floor gently to avoid compressional pain.

Figure 1.71: Right foot lateral lunge. The involved foot lunges back and lateral, continuing to land on the forefoot with a slow deceleration of the heel to the floor. The concept here is to move in a direction perpendicular, or in another plane of motion, then that of the primary plane of pain. The anterior lunge would be the most painful, with a lateral lunge being perpendicular. The patient continues to land on the forefoot, reducing the impact to the heel during training.

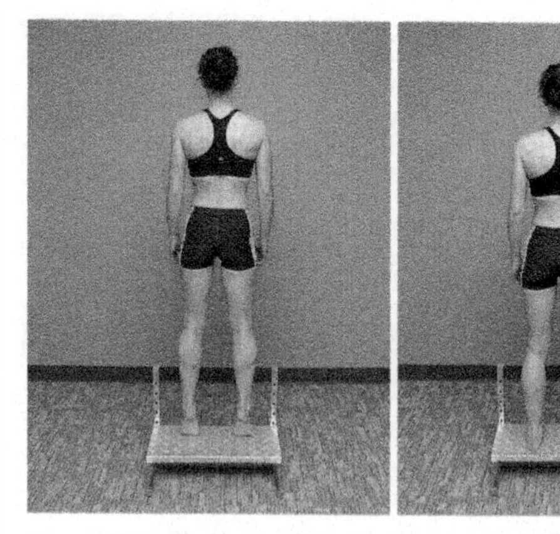

Figure 1.72a,b: Right foot plantarflexion. A heel raise is performed with emphasis on slow increase in loading of the heel during the eccentric phase placing the heel on the floor. The exercise begins bilaterally, for support with the opposite limb and progresses to unilateral heel raise.

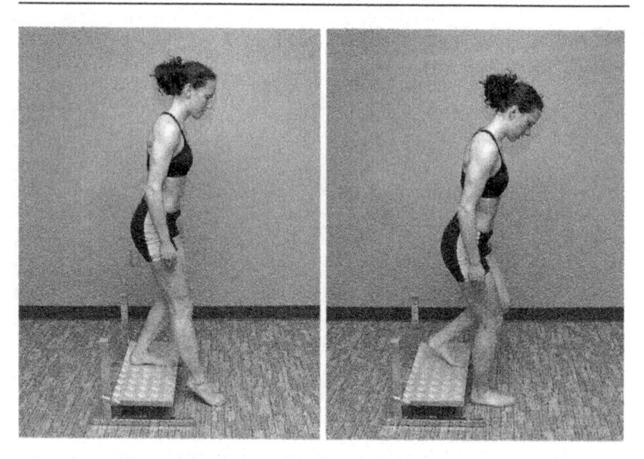

Figure 1.73a,b: Right foot anterior step down. The anterior step down is performed contacting the ground with the forefoot, and slowly lowering the heel to the ground. This is only a modest increase in loading compared to the calf raise, but the exercise begins the process of coordinating muscles for anterior motion.

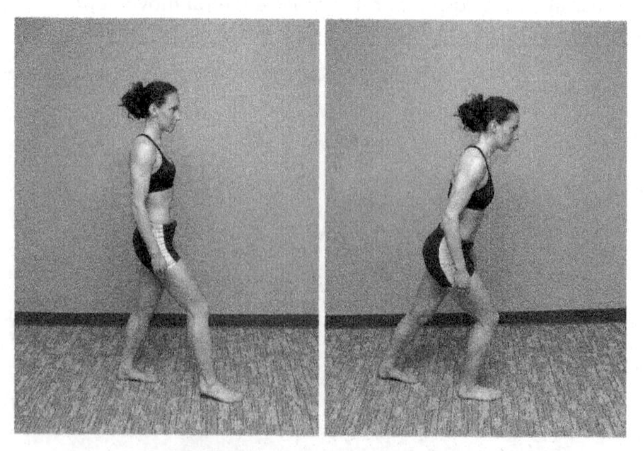

Figure 1.74a,b: Right foot anterior step-lunge. An anterior step-lunge is performed by placing the right foot forward with the body weight maintained relatively posteriorly on the left foot and then gradually moving body weight anterior onto the involved foot loading the heel.

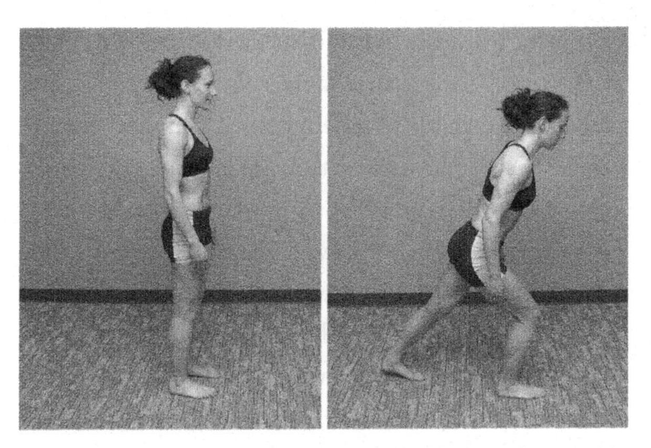

Figure 1.75a,b: Right foot anterior lunge. Anterior lunge with normal impact on loading of the heel.

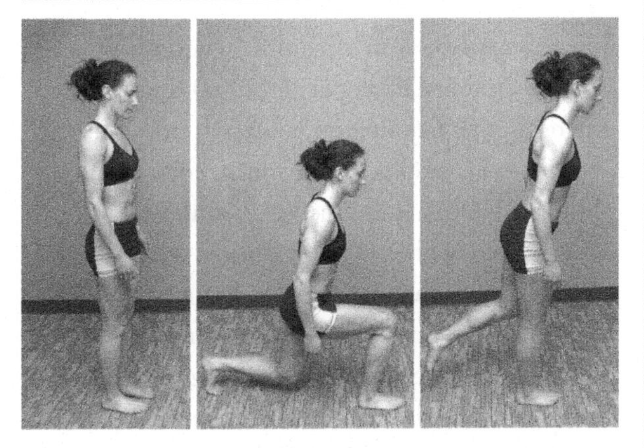

Figure 1.76a,b: Right foot anterior walk-lunge. Anterior walk-lunge will increase the loading on the heel, with emphasis on a functional progression toward running.

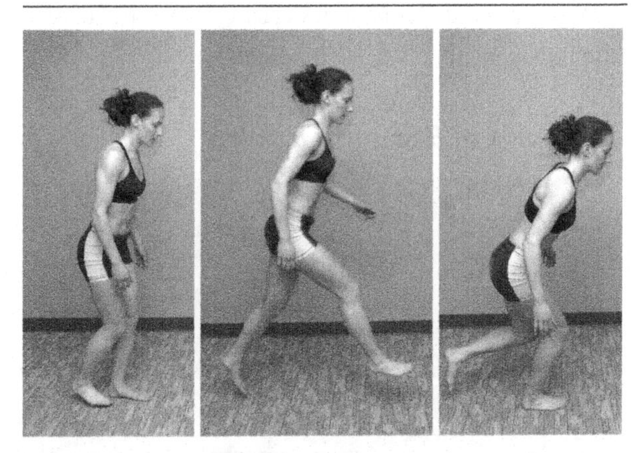

Figure 1.77a,b,c: Right foot anterior hop-and-stop. An anterior hop is performed with full body weight landing on the heel, as a functional progression toward running. Landing may begin on a softer surface, with a gradual progression to a surface consistent with functional demands. Returning to jogging and running is now possible, as the tissue in lesion has been progressively loaded to tolerate normal weight bearing impact. At the same time, the muscles from the trunk to the heel have been trained to coordinate the movement pattern to decelerate weight bearing forces, reducing the overall load placed on the tissue in lesion.

Isometric Training

Stage 1 emphasized tissue training, mobilization, pain inhibition and resolution of muscle guarding. Isometric training was performed below 20% of 1RM, as higher percentages reduce circulation with elevation in intramuscular pressure. Isometric work above 20% 1RM creates a constant intramuscular pressure that can lead to additional ischemia in muscles already in guarding. This is why isometric endurance is trained by strength training utilizing repetitive motion, not sustained isometric work. Isometric endurance is the only functional quality that is obtained by training a different quality. Stage 1 training for vascularity resolves the ischemia and guarding allowing isometric work at a heavier resistance to improve the muscle spindles sensitivity to tension. Increased sensitivity of muscle spindles, and other mechanoreceptors, are necessary to provide dynamic stability to newly gained range of motion. Isometrics allow for heavier resistance that could not be used for dynamic motion.

The heavier the resistance the greater the afferent input. Fixation of strength by isometric exercise is specific to joint angle (Kitai 1989 and Weir 1994). An example of training with isometrics is to perform multiple angle isometrics every 15–20 degrees and held for six to ten seconds (Hall 2002, Houglum 2001 and Kisner 2002). When dealing with hypermobile joints, the ability to use greater loads with isometric work can more quickly improve muscle strength through neurological adaptation, to achieve dynamic stabilization.

The addition of isometric work is added by first increasing the resistance to 75–85% of 1RM. One isometric hold can be done as a second set between two isotonic sets. The resistance should be increased at least 20–30% of 1RM higher than the resistance used for the dynamic sets. The joint angle initially is kept in the inner range of the motion for stabilization and at the outer range for mobilization. Over the course of progressing the program, and especially in Stage 3, multiple isometric holds can be performed at joint angles closer to the pathological range.

1. Exercise for the Ankle and Foot

Eccentric Training for Tendinopathy or Collagen Injury

In the previous example of a gastrocnemius muscle lesion in the vascularization section of Stage 1, training directions avoided plantarflexion directly with a focus on dorsiflexion, eversion and inversion. Exercise was dosed to for vascularization and to provide passive tension for tissue repair to the muscles of plantarflexion. This initial treatment avoided concentric work of the gastroc/soleus complex, as the overuse tendinopathy was already oxygen deprived.

Concentric training focuses on improving vascularity and coordination of muscle. The tendon requires modified tension the line of stress for repair. Pure eccentric training, with the concentric phase removed or assisted, can be performed to any tendon, whether in suffering from an overuse tendinopathy, chronic pain or after surgical repair. The main guidelines are to form coordinated motions around a normal joint axis, limit repetitions below 15 to prevent repetitive strain, and to stay in the pain free range. Training can be one set only working up to multiple sets daily. Rest between sets for oxygen recovery is irrelevant, but time between training sessions allows for protein synthesis and collagen repair.

Eccentric training to chronic mid-portion Achilles tendinosis has been shown to reduce tendon volume, intratendinous signal (Shalabi et al. 2004) and promotes normalized tendon structure (Ohberg et al. 2004). Eccentric training has also been shown to have superior results to concentric training in reduction of pain, clinical outcomes and return to preinjury levels (Fahlstrom et al. 2003, Mafi et el. 2001, Ohberg et al. 2004, Roos et al. 2004). Alfredson et al. found the use of heavy-load eccentric training successfully returned athletes with long standing chronic Achilles tendinosis to pre-injury levels, including running (Alfredson et al. 1998). These results were after only 12 weeks of training and to patients who have not responded to conventional non surgical treatment.

Eccentric work for tendinopathies are performed in the mid to shortened range of collagen tension, working up to, but not into pain. Begin eccentric exercise at one to three sets, eight to ten repetitions per set and 50% of 1RM. The weight should be progressed to as heavy as will allow one set with no more than mild soreness for 10–12 reps. Pure eccentric training is initially performed slowly, for coordination, one to two times per day until the tendon is able to tolerate pain-free loading. The number of training sessions per day can be increased as tissue tolerance improves.

The exact mechanism of chronic tendon pain is not known, but Ohberg et al. (2002) found that sclerosing the pathological neovessels in patients appears to be an effective treatment for painful chronic Achilles tendinosis, suggesting that neovessels may play a role in chronic tendon pain. Like sclerosing therapy, eccentric training has also been shown to normalize tendon structure and eliminate neovascularization (Ohberg et al. 2004). Eccentric work also requires 70–75 percent less oxygen than comparable work load with concentric work, as well as producing a greater peak torque in the muscles, causing a positive strain for repair in both thee parallel and series elastic components of the collagen fibers of the tendons.

Examples of Eccentric Training

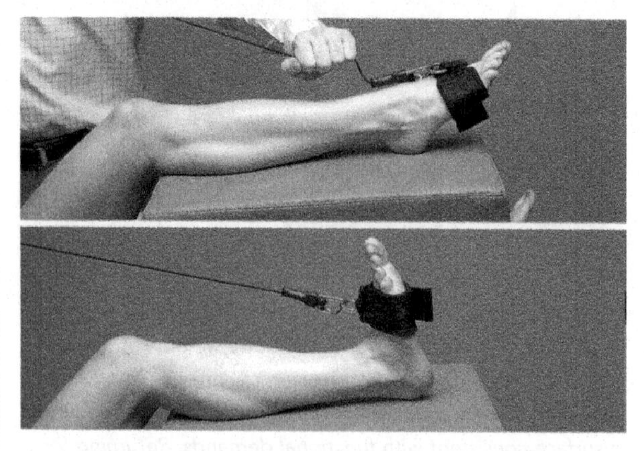

Figure 1.78a,b: Eccentric plantarflexion in supine with pulley resistance. The operator assists concentric phase using a weight that exceeds the concentric ability of the muscle, with the patient lowering eccentrically. Pulley resistance is recommended as elastic resistance reduces resistance toward the lengthened range of the muscle, thus reducing the positive strain to the collagen.

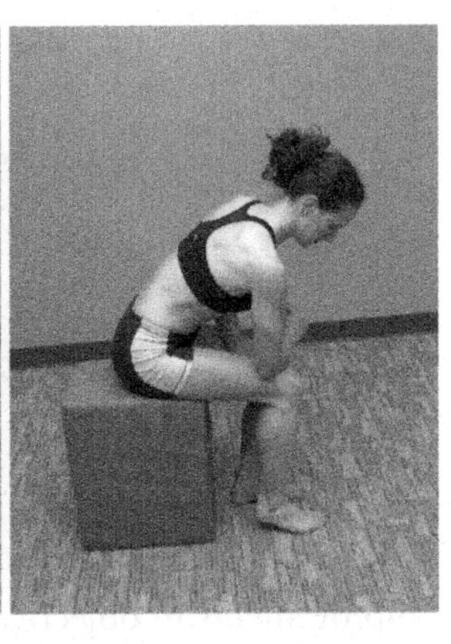

Figure 1.79a,b: Eccentric plantarflexion in sitting, patient assists concentric phase. Heel is positioned forward to limit dorsiflexion to the pain free range. Progression would first include sliding the heel back to increase the range of motion. Resistance can be added by placing a weight on the knee or by resting the elbow on the knee.

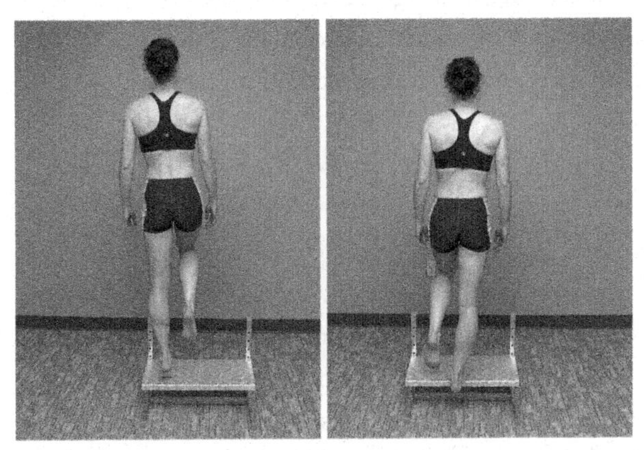

Figure 1.80a,b: Standing eccentric unilateral calf raise ("heel drops" or "drop and stop"). The concentric lift is performed unilaterally by the uninvolved limb followed by eccentric lowering with the involved limb. Initially the uninvolved limb may assist the eccentric phase down until it can safely be performed unilaterally. Progression may also include adding weight held in one hand, with the other used for balance support, or use a backpack with added weight.

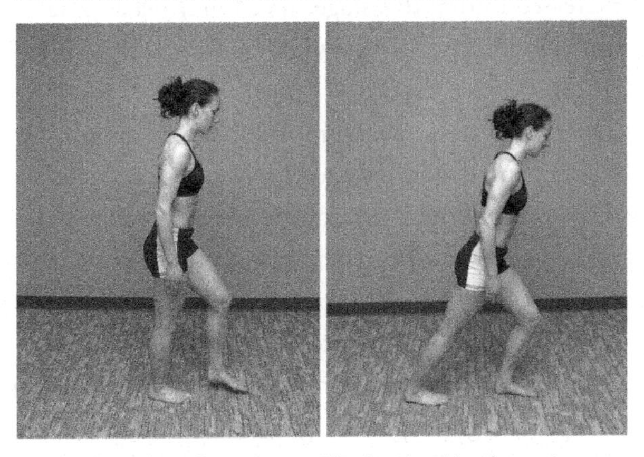

Figure 1.81a,b: Fall out lunge. The involved heel remains on the ground as the uninvolved limb performs an anterior lunge. The plantarflexors work eccentrically controlling the anterior movement of the tibia over the talus. This is a more functional progression that requires a higher level of balance and coordination. With walking and running, dorsiflexion of the ankle is controlled eccentrically by the posterior plantarflexors, with the peak activity at mid stance (Reber et al. 1993).

Progressive training for collagen repair with eccentric training is not the entire picture for rehabilitation. The goal for a thicker and more elastic tendon is a precursor to returning to function, but is not the only variable in avoiding reinjury. Ying et al. (2003) measured Achilles tendon cross sectional area (CSA) in healthy subjects that exercised, compared to a group that did not, finding an increase in those that exercise but no difference with dominant and nondominant sides in either group.

Kongsgaard et al. (2005) measured the CSA in the Achilles tendon bilaterally in elite runners and volleyball players, comparing them to a control group of kayakers. All athletes sustained a unilateral Achilles tendon rupture. The authors reported this study to be the first to show larger CSA in tendons that are subjected to intermittent high loads. Despite larger CSA, Achilles tendon rupture patients did not display differences in structural or loading properties of the tendons. CSA is not the only variable in determining risk for rupture.

Summary

Rehabilitation must also include assessment of joint mechanics of the foot and lower limb to assess abnormal mechanics placed on the ankle. Assessment must also include myokinematic coordination of lower limb performance. Going one step further, assessment of the lumbar spine is necessary. A facilitated segment in the lower lumbar spine can result in increased resting tone of the Triceps surae, leading to increased muscle stiffness and increased risk for tensile injury to collagen in the muscle. Finally, training must involve a progressive return to the functional activities or sports in which the tendon will be subjected.

1. Exercise for the Ankle and Foot

The Following is a Sample Progression for Eccentric Training of Achilles Tendinopathy.

- *Sitting Eccentric only (patient lifts knee with arms leaving only toes in contact with ground, then eccentrically lowers into zero degrees of dorsiflexion).*
- *Sitting Eccentric only (same technique as above, now patient can add resistance through body weight by leaning onto knees).*
- *Non-weight bearing unilateral concentric (uninvolved Achilles and arms) with bilateral eccentric lowering (hip sled or pulley setup providing eccentric plantarflexion and return to plantarflexed position with assist of the other foot).*
- *Non-weight bearing unilateral concentric (uninvolved Achilles and arms) with unilateral eccentric lowering (involved Achilles).*
- *Weight-bearing on flat surface with unilateral concentric (uninvolved Achilles and arms) with bilateral eccentric lowering.*
- *Weight-bearing on flat surface with unilateral concentric (uninvolved Achilles and arms) with unilateral eccentric lowering (involved Achilles).*
- *Weight-bearing with heels off of a step, unilateral concentric (uninvolved Achilles and arms) with bilateral eccentric lowering.*
- *Weight-bearing with heels off of a step, unilateral concentric (uninvolved Achilles and arms) with unilateral eccentric lowering (involved Achilles).*
- *As above with increased weight and/or speed.*
- *Walking up hill progressing to jogging and then running.*
- *Plyometric jumping on level surface.*
- *Plyometric jumping off of box height to level surface.*
- *Plyometric jumping off of box with increased load (ie weight vest).*

Section 3: Stage 3 Exercise Progression Concepts for the Ankle and Foot

Training Goals

Progression to Stage 3 concepts is defined by improvements in objective status, such as achieving full range of motion around a physiological axis, coordinated motor patterns through simple planes of motion and some level of endurance. Tissue tolerance is greatly increased for compression and/or tension. Muscle guarding has resolved and motor patterns are nearing normal but may lack significant levels of performance in terms of endurance, strength and functional activities.

Concentric and eccentric exercises are performed and dosed up to 75–90% 1RM for sets of 16–7 repetitions respectively, improving muscle strength and hypertrophy. Increased resistance is also necessary for continued improvement in neurological adaptation for improved muscle spindles sensitivity to tension as a part of dynamic stability. Exercises should be advanced out of straight plane movements into more functional combined planes of motion. Hypermobilities progress similarly with the addition of 70–80% 1RM isometric contractions being added throughout the full range of motion, except at the pathological end range. Since the load is being increased during this stage, speed may need to be initially decreased to allow for coordination.

This stage represents the introduction of more global pattern training exercise concepts such PNF diagonal patterns. These motions are typically performed below functional speeds as the patient is still developing a level of coordination for these more complex motor patterns. PNF concepts are certainly introduced prior to Stage 3 but performing complex diagonal pattern three dimensionally is not possible with the impairments associated with Stage 1 and 2.

Weight Bearing / Proprioception

Progressing to Stage 3 is defined by achieving full weight bearing status and should now include more complex multi-planar movement patterns that are consistent with retraining a specific motor deficit or directed toward a specific functional goal. Coordination is emphasized first on these more complex patterns with dosage progressing up the functional quality continuum to include endurance and strength (75–85% RM). Each exercise may be dosed differently to emphasize the specific functional quality being trained.

In Stage 2 we were increasing weight bearing tolerance while protecting the injured tissues, now we can start to challenge the healing tissues and muscles related to stabilization of the area. The same exercise as mentioned previously to protect the lateral collateral ligaments of the ankle can now be arranged to challenge the peroneal muscles and ligaments towards the instability. Most of the activities by Stage 3 should be looking more functional and incorporate more global movements and patterns than muscle specific work. Resistance might also be added to further challenge the tissues.

Figure 1.82a,b: Squats with supination challenge to lateral ankle. A pulley or elastic band at the distal leg creates a supination force moment at the ankle, challenging the lateral ankle stability, and selectively facilitate the peroneal muscles, increasing the reactivity and proprioception to protect the instability. The is a more simplistic challenge to previous exercises shown on the wobble board with similar lateral resistance. Squatting provides a functional motion of the entire lower quarter in which this secondary lateral force can increase the need for stability of the ankle evertors. The squat with a lateral pulley force can also be performed on two labile surfaces to allow for a greater need of stability.

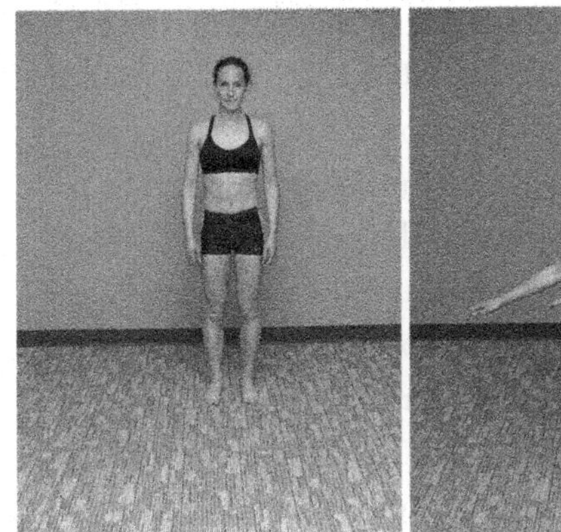

Figure 1.83a,b: Lunging anterior to lateral, with a lateral upper limb reaches, creates an inversion moment of the ankle by shifting more weight lateral to the axis of motion. This increases the challenge to the lateral stabilizing muscles of the ankle.

Eccentric Training

One of the key concepts with Stage 3 is eccentric training for stabilization of the new range gained. Muscles are trained to decelerate motion toward an instability or pathological range and then concentrically return to a more neutral position. Exercises may continue to have both concentric and eccentric components. This is not pure eccentric training with removal of concentric work, as with collagen training in tendinopathy rehabilitation. To emphasize the eccentric component during the exercise two variables are changed other than increased resistance, the work order and the length tension curve.

The work order is the key change to emphasizing eccentric work. The typical exercise is a performed by lifting a weight concentrically first and then lowering the weight eccentrically. This is reversed. The motion is performed eccentrically first toward the pathological range and then concentrically returned. This is done slowly to become coordinated with the motion and the change in work order. As the new movement pattern is coordinated speed is increased gradually to a more functional level while maintaining stability.

The second variable to change is the length tension curve. When emphasizing eccentric work the length

1. Exercise for the Ankle and Foot

tension curve is set to the maximum resistance where the muscle is 20–30 percent into shortened range. The range of motion for the joint to be exercised is assessed. From the mid position of this range, the muscle is shortened roughly 20–30 percent. The pulley rope is set to be perpendicular to the lever arm in this position. Starting eccentrics in this position is safer for end stage eccentric training to a hypermobile joint. As the muscle reaches its lengthened range eccentrically, the angle of the rope moves more toward parallel with the lever arm. This decreases the force moment and lessens the chance of taking the joint into a pathological range.

There may be more complex exercises in Stage 3 but the exercises from Stage 2 may be used with emphasis on the work order and the length tension curve for eccentric training. In Stage 3, resistance can be increased as necessary to build strength and continue to sensitize the afferent receptor systems. The exercises may begin slowly to allow the patient to adjust to the increased weight and the change in work order. As coordination is achieved with this new movement speed is increased toward a more ballistic nature. The transition from eccentric deceleration to concentric acceleration away from the pathological range must be as short as possible. As with jump training, the transition from eccentric deceleration to concentric jumping must be short to transfer as much energy in the collagen structures back into the concentric phase. If this transition is performed slowly then energy is wasted as heat. Functionally the body must learn to make quick reactions to external forces and be able to dynamically control motion from entering the pathological range.

Eccentric training can also be applied while performing dynamic balance activities. Single leg stance ball toss is an example where you can change the variables to emphasize eccentric, isometric or concentric work. Performing reactive exercises, rather than repeated planned motions, is a greater challenge to the somatosensory system for stabilization training to the unstable ankle. Catch and throw activities provide a reactive activity in which the patient must focus on the ball and not the ankle motion. Previously, throwing plyoballs at a rebounder was demonstrated in a safe progression for training ankle stability. This same pattern can be progressed performing on a labile surface.

Ball toss and catch stabilization progression from beginning away from the direction of pathology, to isometrics toward and finally eccentric stabilization toward. Example of right inversion ankle sprain.

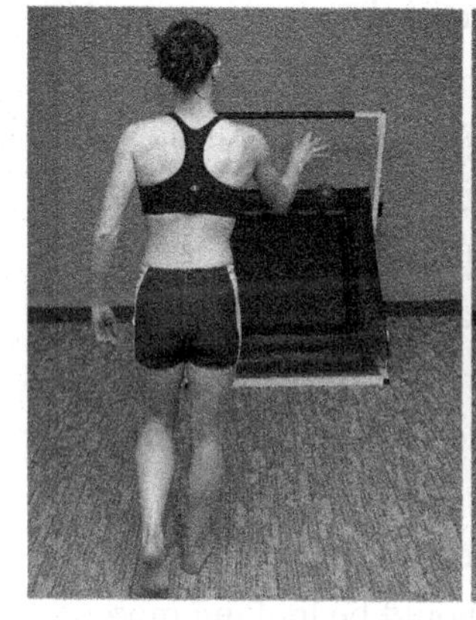

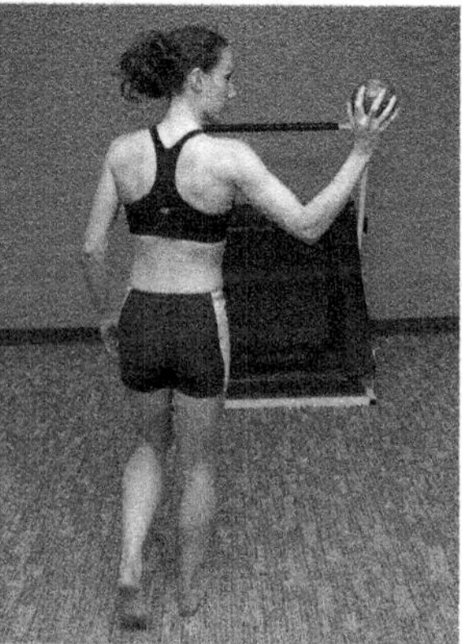

Figure 1.84a,b: Concentric work away/labile surface. Single leg stance right (involved ankle with eversion sprain), with ball catch on the right side (motion away). Catching the ball on the right creates a right rotation moment of the trunk and relative external rotation of the lower limb, causing an inversion moment at the ankle. The ankle is driven into inversion, away from the injured medial structures. The labile surface provides a greater challenge to the somatosensory system to provide stability to the ankle.

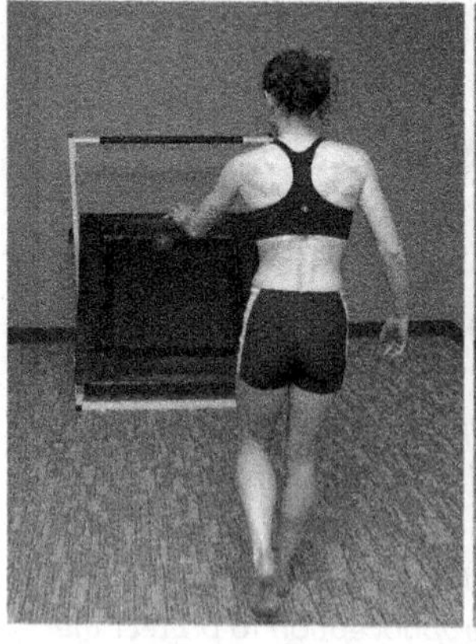

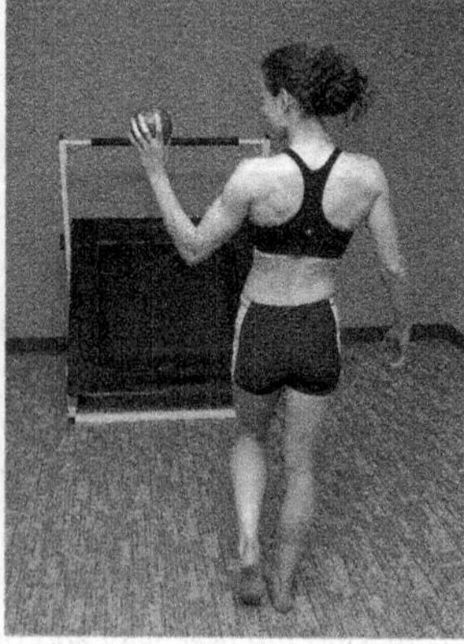

Figure 1.85a,b: Isometric work/labile surface. Single leg stance right, with ball catch on the left side, isometric stabilization. Catching the ball on the left creates a left rotation moment of the trunk, which would create an inversion moment at the ankle. The patient catches the ball with isometric stabilization of the trunk, not allowing left rotation or inversion of the lower leg or ankle.

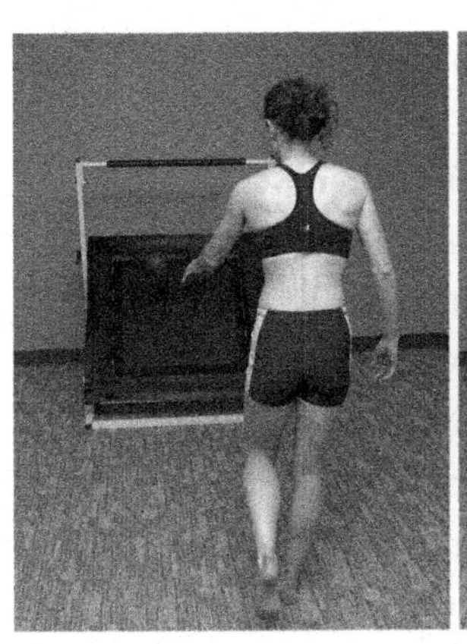

Figure 1.86a,b: Eccentric work toward the instability/labile surface. Single leg stance right, with ball catch on the left side, eccentric stabilization to the left. The ball is caught on the left with the trunk allowed to rotate to the left. The ankle is driven into inversion requiring eccentric stabilization with the muscles of inversion.

Isometric Training for Strength Fixation

There is continued use of isometrics in Stage 3 and the resistance will increase as the training level improves. Isometric work, which began in mid range or away from the pathological range, is performed closer to the end range being stabilized, or toward the pathological range. Isometric work can be done separately from the dynamic sets, or can be substituted for one of the dynamic sets of the exercise. A simple approach is to substitute the second set of a three set series with one set of isometric work. The weight is increased for the isometric set, as the peak torque for isometrics is greater than with concentric work. Isometric training can also be emphasized with more functional activities, as with the previous examples of single leg stance ball toss.

Balance Training

Stage 3 implies a higher level of tissue tolerance, coordination and normalized range of motion. Balance training is progressed from single leg stance and upper/lower limb reaches. Static characteristics of single-leg balance may not adequately challenge the postural control system resulting in normal tests, when deficits are noted with higher levels of function (Colby et al. 1999). Progressing functional tests to include hopping, jumping, agility running or sports replication may be necessary to identify more subtle deficits, leading to functionally specific exercise for injury prevention. As static balance improves, more advanced, dynamic, balance challenges may be functionally necessary, though not for every patient. Frequency of balance training may also reduce, as the difficulty is progressed, in combination with other lower limb endurance and strengthening exercises. In a pain free population with ankle instability, Eils and Rosenbaum (2001) made significant improvement in ankle stability training one time a week for six weeks. A twelve-station program was completed two times in each session with 45 seconds at each station, 30 seconds rest between. Different surfaces, ankle positions and motion challenges where used to vary the balance challenge to the ankle and lower limb.

Functional Training and Directions

Symptoms of the chronic ankle instability do not necessarily affect higher levels of functional performance. Demeritt et al. (2002) found no difference in performance of shuttle run tests and agility hop tests in subjects with chronic ankle instability. These subjects are no longer painful during functional activities, but perceive a level of instability in their ankles. Over time the knee and ankle are thought to compensate for proprioceptive deficits in the ankle, despite remaining passive instability. Other authors have reported no differences in more functional testing in the chronically unstable ankle (Munn et al. 2002, Worrell et al. 1994). Acute ankle injuries, and post surgical patients, will likely have more significant functional deficits, requiring a more gradual progression on the continuum of balance, functional training and agility training.

Step-lunge. The step lunge is a beginning stage lunge in which the foot is placed forward with the body weight remaining on the back foot. After the foot has made contact with the ground the weight is slowly transferred to the lunging limb. The anterior lunge is performed with the back heel on the ground and back knee straight. The trunk is in a

1. Exercise for the Ankle and Foot

forward position, inline with the lower limb. This alignment, contrary to a vertical trunk and flexed back knee, increases the load at the hip and trunk, emphasizing proximal stabilization rather than emphasizing the knee only.

Lunge Agility Progression

Step-lunge

Step-lunge, upper limb reaches

Fall-out lunge

Fall-out lunge, upper limb reaches

Ballistic Lunges

Lunge step-up

Lunge step-down

Lunge onto labile surface

Lunge and land

Hop and stop

Step-lunge, upper limb reaches. The step-lunge is performed as above, but include resisted arm movements. Reaching in the direction of the lunge will increase the weight over the lunging limb, to increase the overall load and create and additional balance challenge. Reaching the opposite direction of the lunge will serve as a counter weight to reduce the load on the lunging limb.

Fall-out lunge. The fall-out lunge increases the impact load and balance challenge to the lunging limb. Rather than setting the foot down first and then transferring the weight on to it, the body falls forward in a straight line with the foot hitting the ground with the body in the full lunge position. A quick isometric stabilization is required in the lower limb and trunk to quickly stabilize the body.

Fall-out lunge, upper limb reaches. The fall-out lunge can have a further challenge of upper limb movements. Reaching in the direction of the lunge will increase the challenge, while reaching in the opposite direction will reduce the load on the lunging limb. Reach directions may also be determined as it relates to a work or sport activity.

Ballistic Lunges. As performance improves, with increased range of motion and alternate directions, speed, rather than weight, is used to progress all lunges. Ballistic lunges involve a fast transition from landing to return, as in plyometric jumping. Speed will add forces of inertia that act as an increase in resistance, thus increasing the load to performing muscles. Speed will also add a further balance and coordination challenge.

Lunge-and-land. When significant balance deficits are present, the lunge-and-land is a good progression option. The subject lunges away from the stance leg, and then transfers all the body weight to the lunging foot for a five-second single leg balance.

Lunge step-up. Changing the emphasis of the lunge from decelerating the body weight, to a concentric performance of lifting the body, is achieved with a lunge step-up. Distance from the step, height of the step and weights in the hands, can all be used as increased challenges.

Lunge step-down. Stepping down increases the balance and loading for eccentric deceleration and stabilization of the landing foot. Distance from the step during the lunge, height of the step and weights in the hands, will all increase the challenge.

Hop and stop. The hop and stop is a progression of the lunge-and-land listed above. Rather than lunging, the subject hops with both feet off the ground, and then landing for single leg stabilization and a five-second hold. The hop-and-stop can be performed in any direction in the "star" pattern or performed linearly with alternating feet.

Lunge onto labile surface. When retraining balance deficits, lunging onto a labile surface can serve as a further challenge to the central and peripheral balance mechanisms. Each set of lunges may be challenged with different labile and fixed surfaces to vary the challenge.

Examples of Lunge Patterns Designed to Increase or Decrease the Stress / Strain to Collagenous or Contractile Tissues.

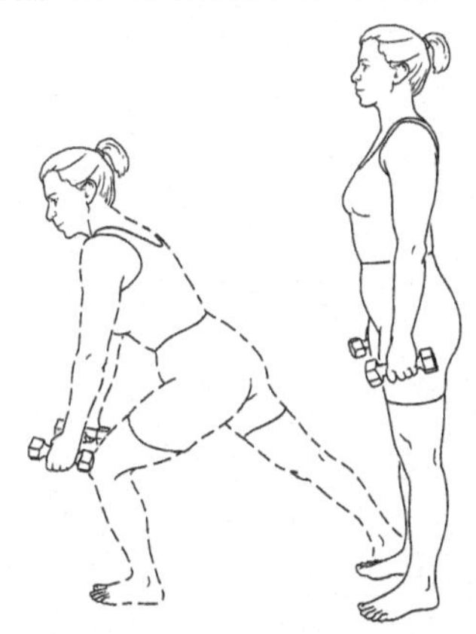

Figure 1.87: Left anterior fallout lunge. The anterior lunge has minimal tissue stress to the collaterals, while maximizing Achilles tension. The trunk remaining in line will emphasize the gluteal muscles and back, while maintaining a vertical trunk with flexion of the back knee will emphasize the quadriceps of the anterior knee.

Figure 1.88: Left anterolateral lunge. The anterolateral lunge increases the stress to the lateral structures of the ankle, but not to the extent of a straight lateral lunge. The trunk position in line with the back leg will again emphasize the gluteal and back muscles, and de-emphasize the quadriceps.

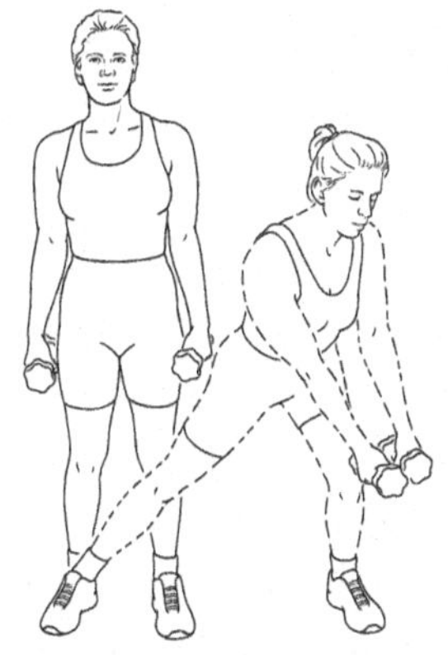

Figure 1.89: Left lateral lunge. The lateral lunge maximizes the stress to the lateral structures of the ankle, producing the greatest stabilization challenge to a lateral collateral ligament sprain. The lateral lunge also places the least stress to the Achilles, being a preferred initial starting direction, with a gradual progression to a pure anterior or posterior lunge. A lateral upper limb reach will increase the deep external rotators of the hip by creating a rotational moment in the transverse plane. Muscles of pure abduction will have less load, as weight is transferred lateral to the axis of the hip joint. Conversely, a medial reach will increase the challenge to the muscles of abduction by increasing the load and lever arm medial to the axis.

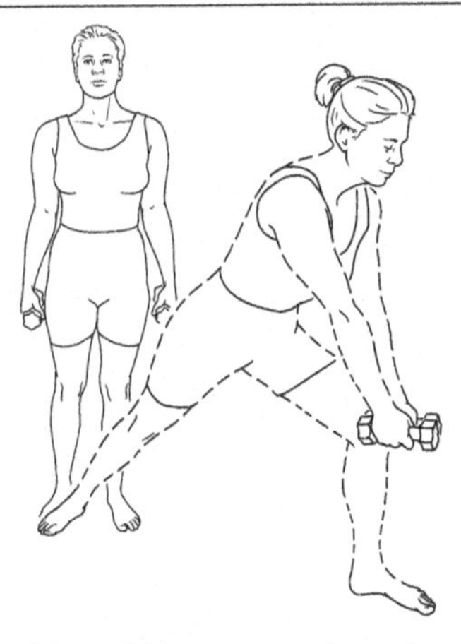

Figure 1.90: Left lateral lunge, external rotation. The addition of rotational moments at the ankle may mimic more functional or athletic movements, increase the rotatory stress at the knee, or minimize the lateral (inversion) ankle strain during lateral lunging.

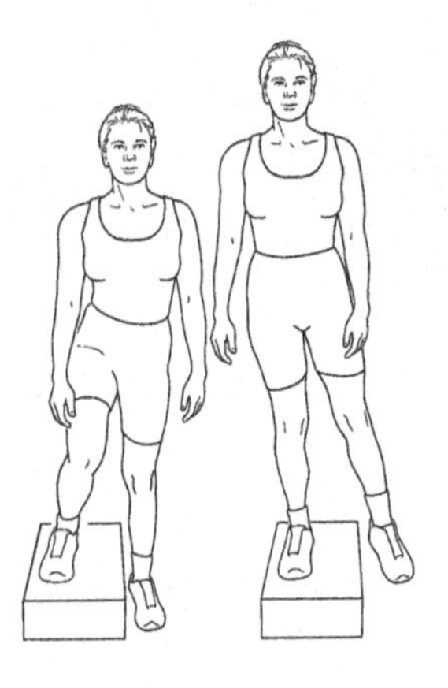

Figure 1.91a,b: Left lateral step-up lunge. The lateral step-up lunge places the ankle in greater eversion prior to loading and stepping up. This would be a significant challenge to the collagenous tissues of the lateral ankle, as well as increasing the torque moment of the muscles of ankle eversion.

Agility Hop Test / Exercise

The agility hop is performed unilaterally, stopping to regain balance and hold for five seconds in each

1. Exercise for the Ankle and Foot

position of the pattern shown below (Bernier and Perrin 1998). An eyes-closed version is performed with the eyes open during hoping, but immediately closed for the five-second sustained balance portion of the test. Other challenges during the test or exercise may include 1) immediately after each hop, bring the arms to the sides and fully extend the involved hip and knee (forcing the subject to make postural corrections at the ankle), and 2) keep the uninvolved leg directly to the side of the test leg upon landing (Demeritt et al. 2002). For testing, an error is scored for each of the following: 1) the test foot is moved or the subject does not "stick" the landing, 2) position was not held for the full 5 seconds on each spot, 3) subject moved arm(s) for balance, 4) contralateral leg moved away from the test leg, 5) the contralateral leg touched down or 6) body sways excessively in any direction (Demeritt et al. 2002).

Figure 1.92a,b: Jumping on a soft surface, such as a mat surface or rebounder, will minimize the loading forces to the joint structures, which may allow an early addition of coordinative jumping to a rehabilitation program. Progressions in later stages would involve surfaces more consistent with functional requirements.

Improvements in tissue tolerance, coordination, endurance and strength training, as they relate to balance reaching and lunging, serve as tests for progressing to more hopping or agility running activities. When lunging can be performed without tissue pain, coordination of lower limb muscles and proper alignment of the lower limb than the patient is ready to progress to more aggressive hopping activities. Attempting hopping activities without proximal stability will increase the tissue strain to ankle structures and increase the chance of a negative outcome with training. The direction of hopping can coincide with the progression of lunging that was initiated by moving in directions that reduced loading to the primary tissue in lesion. A gradual progression toward the direction of lesion achieves the greatest challenge to involved contractile and noncontractile tissues.

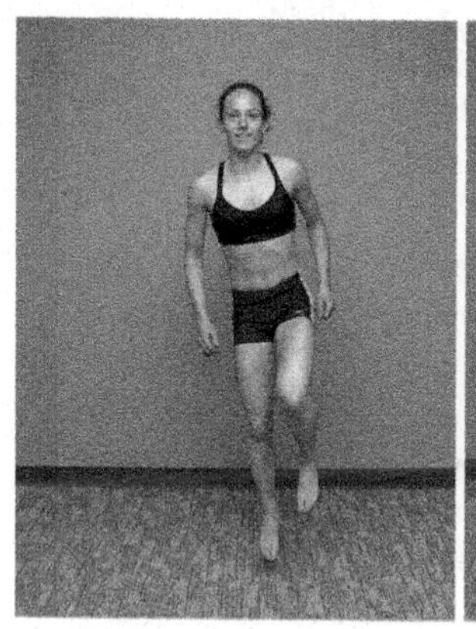

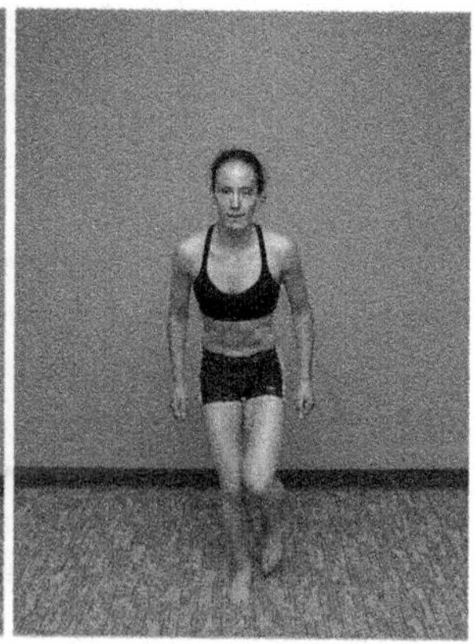

Figure 1.93a,b: Unilateral hopping lateral onto the involved limb. Hopping can begin in one plane, with continued progressive motion. Variations may include hopping back and forth or a tape line on the floor in the desired direction. Hopping may begin bilateral, to reduce the amount of tissue loading and reduce the level of coordination required.

Knowing when to return to running activities can be a challenge for the clinician. Difficult questions like these are best answered by the patient's performance. The hop-and-stop provides a good transitional test and exercise for the ability of the lower limb to land solidly on the ground with good pelvic and lower limb stability. The patient is required to hop forward, land and hold the position for several seconds. The amount of adjustments required at the ankle, knee, hip or lumbar spine serves as a measure of overall stability. Less stable patients will incorporate excessive trunk and upper limb motion. Classic signs of instability include loss of balance requiring toe touch of the opposite limb, collapse of the foot into pronation, collapse of the knee into valgus and excessive adduction of the hip. Patients that demonstrate these deficiencies are likely not ready to return to running or agility activities as they lack the necessary stability to prevent excessive tissue loading and with possible re-injury during higher end activity.

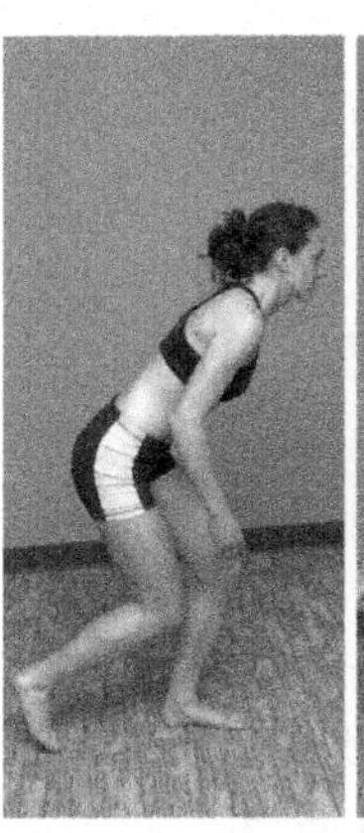

Figure 1.94a,b,c: Anterior hop-and-stop. The hop-and-stop emphasizes proximal stability during the landing phase of hopping. The patient is required to hop, landing on one foot and hold the position for three to five seconds to emphasize balance of the lower limb. The exercise can be progressed by shortening the amount of holding time that is spent on the limb, provided the patient is stable. The direction of each successive hop can be altered in a random pattern to challenge the client. As with the plyometric lunge, the hop-and-stop can also be progressed off of a raised platform to increase the load through the landing limb.

Single Leg Jump-Landing Test

A single-leg hop test has been suggested as a higher level dynamic balance test that serves to strain the postural control system (Ross and Guskiewicz 2004). Subjects were first measured for maximum double-legged vertical jump, as measured by overhead reach. The test involved a double-legged jump at 50–55% of the maximum jump height, landing on one leg and sustaining this position for 20 seconds. Subjects with chronic ankle instability where found to take longer to stabilize in frontal and sagittal planes on landing, as measured on a force plate (Ross and Guskiewicz 2004, Ross and Guskiewicz 2005). In the absence of a force plate, visual assessment for time to stabilization can be used, as well as noting strategies such as touching legs together to gain stability or touching the ground with the non stance limb.

Lack of stability when landing is not primarily attributed to peripheral deafferentation or motor weakness, but alterations in central processing. Caulfield and Garrett (2002) were able to demonstrate increases in ankle and knee joint angular displacement in the immediate pre and post impact period during landing from a jump. Despite the delayed motor responses in subjects with ankle instability, recruitment still occurred faster than the time needed for a peripheral reflex. The absence of external influences suggests that differences must be related to differences in the feed-forward motor program. Pre-activation of ankle musculature does occur prior to ground contact during landing from a jump (Duncan and McDonagh 1997, Dyhre-Poulsen et al. 1991), suggestive of feedforward programming for presetting of muscle.

For acute ankle sprains and chronic ankle instability, the position of the ankle just prior to landing may be altered, increasing the risk of injury. Delahunt et al. (2006) observed a delay in peroneal function just prior to landing, resulting in a slightly inverted position of the ankle on impact. Reduced joint mobility, decreased flexibility and delayed recruitment of the muscles of eversion may all play a role in an increase in the relative inversion and plantar flexed position of the ankle prior to impact. Manual intervention may be necessary prior to exercise training to address joint and/or myofascial restrictions contributing to altered range of motion and motor recruitment.

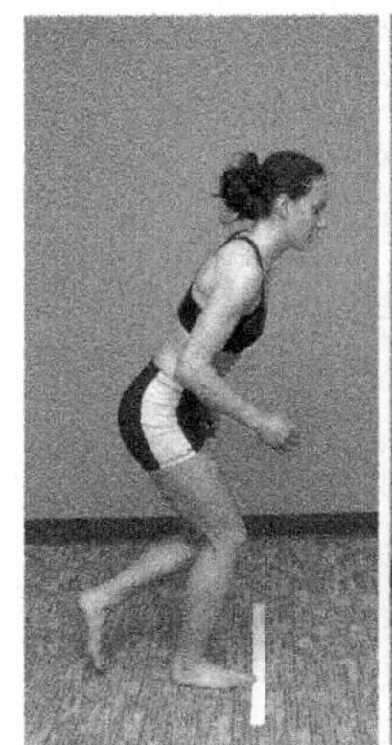
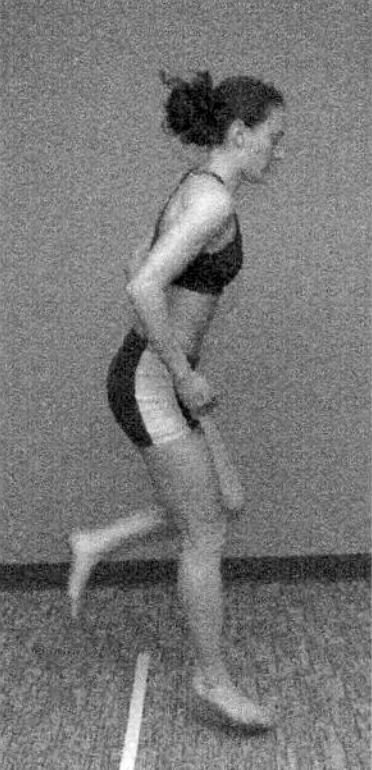

Figure 1.95a,b,c: Single leg jump-landing test.

Gait

Normal gait is a prerequisite for advancement into jogging, running and sport specific activities. Exercises for coordination and balance should no longer involve any assistive devices, upper quarter support or verbal cues. Exercises may involve

1. Exercise for the Ankle and Foot

many combinations of diagonal and rotational movements, working towards functional patterns. Training exercises may be towards specific repetitive motions or may involve reacting quickly to specific instruction to alter direction, such as drill exercises.

Challenges in gait may initially include more simple tasks of walking backward, lateral with a shuffle step, lateral with a crossover step, forward with eyes closed and forward with cervical motions in rotation or flexion/extension. Quick reversals in direction of motion provide an additional challenge. This may be a planned movement change or in response to a verbal command. Gait challenges may also include stepping over objects to increase dorsiflexion for toe clearance or work on balance. Uphill walking will also provide a challenge to end range dorsiflexion.

Progressive Gait Challenge Examples for Tissue Stress

Achilles Tendon

Lateral, medial, anterior, posterior

Lateral Joint structures

Anterior, posterior, medial, lateral

Medial Joint structures

Anterior, posterior, lateral, medial

Progressive Gait Challenge Examples for Coordination Training

Forward

Lateral shuffle step

Lateral crossover step

Backward

As above, with eyes closed

Stepping over or around objects

Uphill walking

Fast walking

Section 4: Stage 4 Exercise Progression Concepts for the Ankle and Foot

Training Goals

As the functional qualities of Stage 3 are achieved, the patient is ready to progress to Stage 4. Indications for Stage 4 are decreased exercise time and coordination of each exercise at multiple speeds. The goals for Stage 4 are to restore full functional stability and coordinate tonic and phasic muscles throughout the physiological range of motion. The exercises should be focused towards the functional requirements of daily living, sports or work related activities. The more sport/activity specific the strengthening exercises are, the more functional the joint complex will be. Therefore, it is imperative that the training regime be of multiple weight bearing (closed chain) exercises that incorporate the lower limb, trunk and upper body.

In cases of instability, many forms of bracing or taping may be used in conjunction with stabilization training. Athletes may be concerned about reduced performance with continued use of bracing during training or athletic competition. Paris (1992) demonstrated no significant differences in speed, balance and agility with and without the use of tape, nonelastic adhesive tape, Swede-O-brace, the New Cross-brace and the McDavid-brace. Bracing has more lasting structural stabilization than tape, but all of these methods increase afferent input from cutaneous receptors, improving proprioception and motor recruitment. Gloria et al. (2001) confirmed that stability of knees with knee anterior cruciate ligament repairs was not due to structural stability of bracing but from receptor feedback (Wu et al. 2001).

Directions of Training

As previously described, the direction of training should take into account the primary tissue in lesion, the motor performance impairments and functional requirements of the task. Whether walking, jogging, running, lunging, hopping, jumping or performing step activities, the exercise design must be customized to the patient. Stage 4 is defined by symptoms being minimal allowing for more aggressive combinations of training directions and activities.

Functional Training

In Stage 3 plyometric and running types of activities were introduced. Stage 4 activities should more closely mimic the sport, work or recreational activities to which the client will be returning. All planes of motion and activities may be utilized in any rehabilitation program, but different sports and activities require different performance emphasis. For example runners, cyclists and swimmers primarily need forward motion training. Lateral and rotary impairments are addressed, but the level of training may not be as intense as for sports that require balance, agility and strength in all planes such as tennis, basketball and gymnastics. Volleyball players may require emphasis on vertical movement for jumping, while a long jumper may require a more horizontal training emphasis. Identifying performance needs is essential for designing an end stage rehabilitation program that mimics a specific sport or task. These aspects of performance can initially be addressed in Stage 3, and then primarily emphasized at higher intensities with Stage 4 training. Often the final stage becomes a long-term home program that the patient is to continue independently, with a trainer or a coach. The patient should have a thorough understanding of the performance and goals of the specific exercises, as well as how they relate to their sport requirements and how to progress them in the future. Communication with trainers or coaches may also be necessary to ensure a smooth and safe transition back to athletic activity, while continuing the rehabilitation process. Remember that the complete healing time for collagen, cartilage and bone can far exceed the time line of formal supervised rehabilitation. Pain free functional return to sport often occurs prior to complete repair of the injured tissues and restoration of neural pathways to mechanoreceptors for afferent feedback. Patients should be encouraged to train along a time line that is consistent with the type and severity of their specific tissue injury.

Strength Training

The dosage of exercises in Stage 4 should be according to the desired functional quality, typically within the strength/endurance range with sets of 10–15 repetitions. Additions of agility training and plyometrics might also be added to allow return to pre morbid activity levels. If fast ballistic training is required, it should added in Stage 4 after the patient has demonstrated an adequate performance level at slower speeds through functional ranges and directions. Muscles involved in fast ballistic movements tend to favor multijoint phasic muscles with a reduced activation of single joint tonic muscles (Mackenzie et al. 1995, Thorstensson et al. 1985). Proper arthrokinematic endurance, coordination and strength around a physiological axis are a prerequisite for safe ballistic training.

Agility / Balance and Agility

Exercises to challenge the weight bearing capacity can be further progressed at this time to include tri-planar motion through full range of motion and loaded weight bearing. The muscular capacity can be progressed by either increasing the depth of the squat or lunge as well as changing the plane that the exercises are performed. By Stage 4, static balance and performance should be relatively good, even for the patient with functionally unstable ankles. Ross et al. (2005) demonstrated that patients with functionally unstable ankles take significantly longer to recover after a single-leg jump-landing test compared to those with stable ankles (Ross et al. 2005), while single leg stance was comparable in both groups. By Stage 4, emphasis needs to be on dynamic balance, plyometric work and jump training. If the patient is an athlete, technical training that is sport specific should be emphasized along with balance and proprioception work.

1. Exercise for the Ankle and Foot

Figure 1.96a,b: Unilateral hopping or bilateral jumping for agility. Emphasis may be placed on a specific direction or a change in direction, height and/or distance. A four square pattern of tape on the floor can provide a pattern for hopping back and forth, in circles from square to square, crisscross patterns and/or star patterns.

Figure 1.97: Hopping over objects (foam roll pictured) with a lateral force moment from a pulley system. The relative medial pulley line increases the resistance on the push off foot, as well as providing a greater balance challenge on landing phase.

Figure 1.98a,b: Catch and throw with unilateral balance on labile surface. Functional reactive activities, such as throwing and catching, distract the patient from focusing on the ankle, requiring postural reflexes to take over for stability. The use of a labile surface further increases the challenge of the somatosensory system working with postural reflexes.

Figure 1.99a,b,c: Lateral hopping with ball catching and throwing. Performing different directions of lunging, step-ups or jumping in conjunction with activities requiring reaction, rather than planned motor tasks, further increases the coordination challenge. Exercises can be designed to match functional tasks or athletic performance.

Progressing to running is necessary for return to most athletic activities. Stage 3 outlined progressions for walking, including coordination and tissue challenges. These same progressions are used for a return to jogging, light running and sprinting. Additional challenges may include running with quick directional changes, as in cutting. Lateral shuttle running can be used to emphasize lateral ankle stability via eccentric deceleration during changes of direction. Lateral scissor step running can also challenge balance, as well as dynamic stabilization of the ankle. The different types of training exercises chosen should relate to the tissue injury, initial direction of balance and/or strength impairments as well as the functional requirements for return to activity.

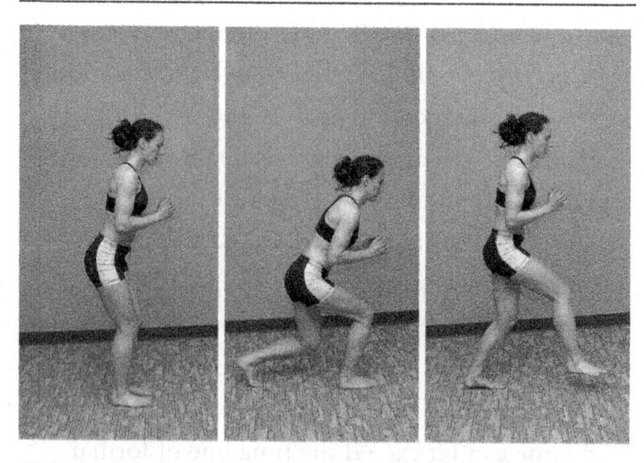

Figure 1.100a,b,c: The ballistic lunge. The ballistic lunge refers to increasing the speed of the activity, with emphasis on the transition from eccentric deceleration to an explosive

concentric return to the center. This is similar to plyometric jump training in which the emphasis is placed on the quick transition from eccentric to concentric work.

Figure 1.101a,b: Ball catch and throw with unilateral standing on a labile surface (mini-trampoline).

Shuttle Run Test

Functional testing and exercise may include agility running, such as the shuttle run. This test involves running back and forth between two lines. A distance of 6.1 meters is marked on the floor with two separate pieces of tape. Subjects start behind the first piece of tape, then run and touch the opposite tape, completing four consecutive 6.1 m lengths for a total of 24.4 m Demeritt et al. (2002). Subjects are instructed to change directions by pushing off the involved limb. This timed test may begin as a walking test, progressing to jogging and, if appropriate, to running.

Agility running may include lateral, backward and shuttle running among many variations of patterns. For instance, figure 8 and zigzag running combine quick directional changes that require significant ankle stability. Success with these types of drills may be a good indication of readiness to return to more competitive sport activities.

Plyometrics

Jump training began in Europe; with interest increasing in the 1970's after Eastern European athletes began excelling in the world of sport competition (Chu 1998). An American track and field coach coined the term plyometrics (Wilt 1975). The terms jump training and plyometrics are often used synonymously, but for the purposes of this chapter jump training is used in Stage 3 to represent the beginning stage of training during the pathological state. Plyometric training is more associated with end stage training with emphasis on improving athletic performance.

By definition, plyometric exercises involve a muscle reaching its maximum strength in as short a time as possible (Chu 1998). This is a natural progression from strength training to power training. Power is associated with the ability to over come a heavy resistance in a short time period. Jump training was added in Stage 3 as a progression after coordination had been established with lower level activities, such as the lunge. Lunging required a solid landing while maintaining single limb alignment of the lower kinetic chain. Ballistic lunging incorporates more of the plyometric concepts of the stretch-shortening cycle, with a fast transition between eccentric deceleration and concentric propulsion.

The eccentric phase of jump training has been associated with delayed onset muscle soreness (DOMS) (Byrne and Eston 2002). But these types of studies commonly reflect healthy subjects training at excessively high levels. Jump training is added later in a rehabilitation program following progressive training to improve tissue tolerance, coordination, endurance and strength so the system is prepared for the higher intensity joint loading forces. Early low-level jump training should not represent a level of eccentric performance that is at risk of creating any significant DOMS if properly performed and dosed.

Plyometric jumping exercises emphasize quick explosions so that as little time as possible is spent on the ground. A quick transition is performed between the eccentric deceleration and concentric propulsion to maximize elastic recoil and the myotatic stretch reflex. Early jump training may involve a slower performance that focuses on coordination of the landing and avoiding pain. With a slower transition between the eccentric and concentric phases, energy is lost as heat. Plyometrics

1. Exercise for the Ankle and Foot

emphasizes a ballistic, or explosive transition, with the goal of maximum energy transfer into propelling the body. The process has also been referred to as the stretch-shortening cycle, associated with collagen elongation during the eccentric phase, followed by a recoil in the concentric propulsion phase. Tension in mechanoreceptors of the collagen, and the muscle spindle in the muscle (Chu 1998), are believed to provide afferent feedback to increase motor output. Assessment of motor responses in the ankle during jump performance suggests that the muscle response for an immediate jump after landing needs to occur much faster in order for afferent signals to reach the spinal cord and influence the motor response. A feedforward response refers to a preprogrammed motor response that anticipates the impact of landing with subsequent stabilization for landing or propulsion. Duncan and McDonagh (1997) demonstrated an alteration in this feedforward programming with reduced and delayed pre-activation of ankle musculature prior to ground contact during landing from a jump.

Early Jumping Training Options

Jumps-in-place

Standing Jumps—for distance or height

Bilateral jumps forward and back

Bilateral jumps side to side

Hop and stop

Unilateral hops forward and back

Unilateral hops side to side

4x4 grid

Diagonals on grid

Jumping over a barrier—anterior or lateral

Jumping focuses on increasing speed to achieve greater performance. Dosage may be slightly reduced in the clinical setting, as a part of end stage rehabilitation, than in the normal athletic population. Reduced load, fewer repetitions and longer rest times between training sessions may be necessary to avoid tissue irritation and muscle fatigue. Plyometric training at 30% of 1RM has been found to be superior in increasing vertical jump than training at 80% of 1RM (McBride et al. 2002). Optimal loading for ballistic jump squat training has been reported at 30% of 1RM (Wilson et al. 1993). Loading at 45% of 1RM has been reported as optimal for jump squats (Baker et al. 2001A), while ballistic bench press training is best achieved at 60% of 1RM (Baker et al. 2001B).

Depth jumps involve jumping off a box, landing and immediately jumping again for height. Depth jump training has been shown to be more effective than weight training, jump and reach training and horizontal hops for improving speed and strength capabilities (Verhoshanski and Tatyan 1983). Determining the height for depth jumps has been studied, but is too individual to have a predetermined protocol. Chu (1998) recommends a testing procedure for each athlete. First vertical jump is measured with the jump and reach. The depth jump is then performed with the initial test on an 18-inch box height. If the jump after landing is equal to the initial vertical jump test, than the box height is raised in six-inch increments until the athlete can no longer reach the original test height. If the first test on the 18-inch box is lower then the vertical jump test, than the box height is lower.

Summary for Final Stage

To restore function the initial stages focused on treatment of pain inhibition using soft tissue and joint mobilization techniques. Goals then focsed on addressing joint hypomobilities to restore normal arthrokinematics and exercise for vascularity when muscle guarding is present. Combined elevation with active pain-free exercises were used to activate the venous pump and lymphatic drainage systems when significant swelling is present. For overuse tendinopathy exercise the surrounding muscles for vascularity was used, except for the muscle of the involved tendon. Pure eccentric exercises were used for the involved tendon. Begin closed chain exercises according to weight bearing precautions and pain tolerance to gradually restore proprioceptive function. Progress exercises towards the goals of restoring dynamic stability and normal gait mechanics.

No rehabilitation program can be considered complete with out a home program designed to direct self-care. The findings of the study by Jones et al. (2005) suggested that interventions to ensure realistic expectations might increase success and prevent potential negative effects of failure. Patient education should be emphasized during the rehabilitation process so that the patient is aware of realistic tissue healing time frames and outcomes before discharge.

1. Exercise for the Ankle and Foot

Bibliography

Adams CA, Hamblen DL. *Outline of Orthopedics.* Eleventh Edition, Churchill Livingstone, New York, 1990.

Alfredson H, Pietila T, Jonsson P, Lorentzon R. Heavy-load eccentric calf muscle training for the treatment of chronic Achilles tendinosis. Am J Sports Med 26(3):360–6, May–June, 1998.

Aquino A, Payne C. Function of the plantar fascia. Foot 9:73–78, 1999.

Arnason A, Gudmundsson A, Dahl HA, Johannsson E. Soccer injuries in Iceland. Scand J Med Sci Sports 6(1):40–45, Feb, 1996.

Astrom M, Arvidson T. Alignment and joint motion in the normal foot. J Orthop Sports Phys Ther 22(5):216–222, Nov, 1995.

Bahr R, Bahr IA. Incidence of acute volleyball injuries: a prospective cohort study of injury mechanisms and risk factors. Scand J Med Sci Sports 7(3):166–171, Jun, 1997.

Bahr R, Karlsen R, Lian O, Ovrebo RV. Incidence and mechanisms of acute ankle inversion injuries in volleyball: a retrospective cohort study. Am J Sports Med 22(5):595–600, Sep–Oct, 1994.

Bahr R, Lian O, Bahr IA. A twofold reduction in the incidence of acute ankle sprains in volleyball after the introduction of an injury prevention program: a prospective cohort study. Scand J Med Sci Sports 7(3):172–177, Jun, 1997.

Baker D, Nance S, Moore M. The load that maximizes the average mechanical power output during explosive bench press throws in highly trained athletes. J Strength Cond Res 15(1):20–24, Feb, 2001a.

Baker D, Nance S, Moore M. The load that maximizes the average mechanical power output during explosive jump squats in power-trained athletes. J Strength Cond Res 15:92–97, Feb, 2001b.

Bandy WB, Irion JM, Briggler M. The effect of static stretch and dynamic range of motion training on the flexibility of the hamstring muscles. J Orthop Sports Phys Ther 27(4):295–300, Apr, 1998.

Barrett SJ, O'Malley R. Plantar fasciitis and other causes of heel pain. Am Fam Physician 59(8):2200–6, Apr 15, 1999.

Baumhauer J, Alosa D, Renstron F, Trevino S, Beynnon B. A prospective study of ankle injury risk factors. Am J Sports Med 23(5):564–570, Sep–Oct, 1995.

Bernier J, Perrin D, Rijke A. Effect of unilateral functional instability of the ankle on postural sway and inversion and eversion strength. J Athl Train 32(3):226–232, Jul, 1997.

Bernier JN, Perrin DH. Effect of coordination training on proprioception of the functionally unstable ankle. J Orthop Sports Phys Ther 27(4):264–75, Apr, 1998.

Beynnon BD, Murphy DF, Alosa DM. Predictive factors for lateral ankle sprains: A literature review. J Athl Train 37(4):376–380, Dec, 2002.

Beynnon BD, Vacek PM, Murphy D, Alosa D, Paller D. First-time inversion ankle ligament trauma: the effects of sex, level of competition, and sport on the incidence of injury. Am J Sports Med 33(10):1485–91, Oct, 2005.

Bolgla LA, Malone TR. Plantar fasciitis and the windlass mechanism: A biomechanical link to clinical practice. J Athl Train 39(1):77–82, Jan, 2004.

Bomar RE, Cordova ML, Demchak TJ, Storsved JR. External ankle stabilization: male high school basketball players' perceptions regarding specific use and functional performance. J Athl Train 39:S103, 2004.

Bonnin JG. *Injuries to the Ankle. Darien,* CT: Hafner Publishing Co. 118, 1950.

Bosien WR, Staples S, Russell SW. Residual disability following acute ankle sprains. J Bone Joint Surg Am 37–A:1237–1243, Dec, 1955.

Braun BL. Effects of ankle sprain in a general clinic population 6 to 18 months after medical evaluation. Arch Fam Med 8(2):143–8, Mar–Apr, 1999.

Brooks SC, Potter BT, Rainey JB. Treatment for partial tears of the lateral ligament of the ankle: a prospective trial. Br Med J (Clin Res Ed). 282(6264):606–7, Feb 21, 1981.

Browne K, Lee J, Ring PA. The sensation of passive movement at the metatarso-phalangeal joint of the great toe in man. man. J Physiol 126(3):448–58, Dec 10, 1954.

Brunt D, Andersen JC, Huntsman B, Reinhert LB, Thorell AC, Sterling JC. Postural responses to lateral perturbation in healthy subjects and ankle sprain patients. Med Sci Sports Exerc 24(2):171–6, Feb, 1992.

Bullock-Saxton JE, Janda V, Bullock MI. The influence of ankle sprain injury on muscle activation during hip extension. Int J Sports Med 15(6):330–334, Aug, 1994.

Burwell HN, Charnley AD. The treatment of displaced fractures at the ankle by rigid internal fixation and early joint movement. J Bone Joint Surg 47(4):634–660, Nov, 1965.

Byrne C, Eston R. The effect of exercise-induced muscle damage on isometric and dynamic knee extensor strength and vertical jump performance. J Sports Sci 20(5):417–25, May, 2002.

Caulfield BM, Garrett M. Functional instability of the ankle: Differences in patterns of ankle and knee movement prior to and post landing in a single leg jump. Int J Sports Med 23(1):64–68, Jan, 2002.

Chaiwanichsiri D, Lorprayoon E, Noomanoch L. Star excursion balance training: effects on ankle functional stability after ankle sprain. J Med Assoc Thai 88 Suppl 4:S90–4, Sep, 2005.

Chandler TJ, Kibler WB. A biomechanical approach to

the prevention, treatment and rehabilitation of plantar fasciitis. Sports Med 15:344–352, 1993.

Chong RK, Ambrose A, Carzoli J, Hardison L, Jacobson B. Source of improvement in balance control after a training program for ankle proprioception. Percept Mot Skills 92(1):265–72, Feb, 2001.

Chu DA. *Jumping Into Plyometrics*, 2nd Edition. Human Kinetics. Champaigne, IL, 1998.

Clancy WG, Neidhart D, Brand DL. Achilles tendinitis in runners. A report of five cases. Am J Sports Med 4(2):46–57, Mar–Apr, 1976.

Colby SM, Hintermeister RA, Torry MR, Steadman JR. Lower limb stability with ACL impairment. J Orthop Sports Phys Ther 29(8):444–51, Aug, 1999.

Collins N, Teys P, Vicenzino B. The initial effects of a Mulligan's mobilization with movement technique on dorsiflexion and pain in subacute ankle sprains. Man Ther 9(2):77–82, May, 2004.

Conway BA, Hultborn H, Kiehn O. Proprioceptive input resets central locomotor rhythm in the spinal cat. Exp Brain Res 68(3):643–656, 1987.

Cordova ML, Cardona CV, Ingersoll CD, Sandrey MA. Long-term ankle brace use does not affect peroneus longus muscle latency during sudden inversion in normal subjects. J Athl Train. 2000 Oct;35(4):407–411, Oct, 2000.

Cordova ML, Dorrough JL, Kious K, Ingersoll CD, Merrick MA. Prophylactic ankle bracing reduces rearfoot motion during sudden inversion. Scand J Med Sci Sports 17(3):216–22, Jun, 2007.

Cordova ML, Ingersoll CD, Leblanc JM. Influence of ankle support on joint range of motion before and after exercise: a meta-analysis. J Orthop Sports Phys Ther 30(4):170–177, Apr, 2000.

Cordova ML, Ingersoll CD, Palmieri RM. Efficacy of prophylactic ankle support: An experimental perspective. J Athl Traning 37(4):446–457, Dec, 2002.

Cordova ML, Ingersoll CD. Peroneus longus stretch reflex amplitude increases after ankle brace application. Br J Sports Med 37(3):258–62, Jun, 2003.

Cordova ML, Scott BD, Ingersoll CD, LeBlanc MJ. Effects of ankle support on lower-extremity functional performance: a meta-analysis. Med Sci Sports Exerc 37(4):635–41, Apr, 2005.

Cornwall MW, McPoil TG. Plantar fasciitis: etiology and treatment. J Orthop Sports Phys Ther 29(12):756, Dec, 1999.

Cornwall MW. Common pathomechanics of the foot. Athl Ther Today 5(1):10–16, 2005.

Crosby W, Humble RN. Rehabilitation of plantar fasciitis. Clin Podiatr Med Surg 18(2):225–31, Apr, 2001.

Davis PF, Severud E, Baxter DE. Painful heel syndrome: results of nonoperative treatment. Foot Ankle Int 15(10):531–5, Oct, 1994.

Den Hartog BD. Flexor digitorum longus transfer with medial displacement calcaneal osteotomy. Biomechanical rationale. Foot Ankle Clin 6(1):67–76, Mar, 2001.

Denegar CR, Miller SJ. Can chronic ankle instability be prevented? Rethinking management of lateral ankle sprains. J Athle Training 37(4):430–435, Dec, 2002.

Demeritt KM, Shultz SJ, Docherty CL, Gansneder BM, Perrin DH. Chronic ankle instability does not affect lower extremity functional performance. J Athle Train 37(4):507–511, Dec, 2002.

DiGiovanni BF, Nawoczenski DA, Lintal ME, Moore EA, Murray JC, Wilding GE, Baumhauer JF. Tissue-specific plantar fascia-stretching exercise enhances outcomes in patients with chronic heel pain. A prospective, randomized study. J Bone Joint Surg Am 85–A(7):1270–7, Jul, 2003.

Digiovanni BF, Nawoczenski DA, Malay DP, Graci PA, Williams TT, Wilding GE, Baumhauer JF. Plantar fascia-specific stretching exercise improves outcomes in patients with chronic plantar fasciitis. A prospective clinical trial with two-year follow-up. J Bone Joint Surg Am 88(8):1775–81, Aug, 2006.

Donatelli R. *The Biomechanics of the Foot And Ankle*. FA Davis Company, Philadelphia, 1990.

Donatelli RA, Wooden M, Ekedahl SR, Wilkes JS, Cooper J, Bush AJ. Relationship between static and dynamic foot postures in professional baseball players. J Orthop Sports Phys Ther 29(6):316–330, Jun, 1999.

Duncan A, McDonagh MJN. The role of short latency spinal stretch reflexes in human lower leg muscles when landing from a jump. J Physiol (Lond) 501P:42P, 1997.

Duncan PW, Weiner DK, Chandler J, Studenski S. Functional reach: a new clinical measure of balance. J Gerontol 45(6):M192–7, Nov, 1990.

Dyhre-Poulsen P, Simonsen E, Voight M. Dynamic control of muscle stiffness and H-reflex modulation during hopping and jumping in man. J Physiol (Lond) 437:287–304, Jun, 1991.

Ebig M, Lephart SM, Burdett RG, Miller M, Pincivero DM. The effect of sudden inversion stress on EMG activity of the peroneal and tibialis anterior muscles in the chronically unstable ankle. J Orthop Sports Phys Ther 26(2):73–77, Aug, 1997.

Eiff MP, Smith AT, Smith GE. Early mobilization versus immobilization in the treatment of lateral ankle sprains. Am J Sports Med 22(1):83–8, Jan–Feb, 1994.

Eils E, Rosenbaum D. A multi-station proprioceptive exercise program in patients with ankle instability. Med Sci Sports Exerc 33(12):1991–8, Dec, 2001.

Eisenhart AW, Gaeta TJ, Yens DP. Osteopathic manipulative treatment in the emergency department for patients with acute ankle injuries. J Am Osteopath Assoc 103(9):417–21, Sep, 2003.

Ekstrand J, Tropp H. The incidence of ankle sprains in soccer. Foot Ankle 11(1):41–44, Aug, 1990.

Faugli HP. *Medical Exercise Therapy*. Laerergruppen for Medisnsk Treningsterapi AS, Norway, 1996.

Fahlstrom M, Jonsson P, Lorentzon R, Alfredson H. Chronic Achilles tendon pain treated with eccentric calf-muscle training. Knee Surg Sports Traumatol Arthrosc 11(5):327–22, Sept, 2003.

Fallat L, Grimm DJ, Saracco JA. Sprained ankle syndrome: prevalence and analysis of 639 acute injuries. J Foot Ankle Surg 37(4):280–285, Jul–Aug, 1998.

Finsen V, Saetermo R, Kibsgaard L, Farran K, Engebretsen L, Bolz KD, Benum P. Early postoperative weightbearing and muscle activity in patients who have a fracture of the ankle. J Bone Joint Surg Am 71(1):23–7, Jan, 1989.

Forkin DM, Koczur C, Battle R, Newton RA. Evaluation of kinesthetic deficits indicative of balance control in gymnasts with unilateral chronic ankle sprains. J Orthop Sports Phys Ther 23(4):245–50, Apr, 1996.

Forssberg H, Grillner S, Halbertsma J, Rossignol S. The locomotion of the low spinal cat. II. Interlimb coordination. Acta Physiol Scand 108(3):283–295, Mar, 1980.

Freeman E, Appel D. *The Wisdom and Ideas of Plato.* Fawcett Publications, Incs., Greenwhich, Conn, 1966.

Freeman MA, Dean MR, Hanham IW. The etiology and prevention of functional instability of the foot. J Bone Joint Surg Br 47(4):678–85, Nov, 1965.

Gardiner MD. T*he Principles of Exercise Therapy*. G. Bell and Sons, LTD, 1954.

Gatev P, Thomas S, Kepple T, Hallett M. Feedforward ankle strategy of balance during quiet stance in adults. J Physiol 514(Pt 3):915–28, Feb 1, 1999.

Gerber JP, Williams GN, Scoville CR, Arciero RA, Taylor DC. Persistent disability associated with ankle sprains: a prospective examination of an athletic population. Foot Ankle Int 19(10):653–660, Oct, 1998.

Glasoe WM, Allen MK, Awtry BF, Yack HJ. Weight-bearing immobilization and early exercise treatment following a grade II lateral ankle sprain. J Orthop Sports Phys Ther 29(7):394–9, Jul, 1999.

Glencross D, Thornton E. Position sense following joint injury. J Sports Med Phys Fitness 21(1):23–27, Mar, 1981.

Godbout C, Ang O, Frenette J. Early voluntary exercise does not promote healing in a rat model of Achilles tendon injury. J Appl Physiol 101(6):1720–6, Dec, 2006.

Goldie P, Evans O, Bach T. Postural control following inversion injuries of the ankle. Arch Phys Med Rehabil 75(9):969–975, Sep, 1994.

Goldie PA, Bach TM, Evans OM. Force platform measures for evaluating postural control: reliability and validity. Arch Phys Med Rehabil 70(7):510–517, Jul, 1989.

Grampp J, Willson J, Kernozek T. The plantar loading variations to uphill and downhill gradients during treadmill walking. Foot Ankle Int 21(3):227–31, Mar, 2000.

Green T, Refshauge K, Crosbie J, Adams R. A randomized controlled trial of a passive accessory joint mobilization on acute ankle inversion sprains. Phys Ther 81(4):984–94, Apr, 2001.

Grillner S, Rossignol S. On the initiation of the swing phase of locomotion in chronic spinal cats. Brain Res 146(2):269–277, Mar 12, 1978.

Grimsby O. Manual Therapy of the Extremities-MT 6 workbook. The Ola Grimsby Institute, San Diego, CA, 1996.

Grimsby O. Neurophysiological view points on hypermobilities. J Manual Therapy. The Nordic Group of Specialists. Manual Therapy 2:2–9, 1988.

Grimsby O. Post-Graduate Manual Therapy Residency Part I course outline. The Ola Grimsby Institute, San Diego, CA, 1991/1996.

Gross MT. Effects of recurrent lateral ankle sprains on active and passive judgments of joint position. Phys Ther 67(10):1505–1509, Oct, 1987.

Hall CM, Brody LT. *Therapeutic Exercise Moving Toward Function.* Philadelphia, Pa: Lippincott Williams & Wilkins; 1999.

Hannukainen J, Kalliokoski KK, Nuutila P, Fujimoto T, Kemppainen J, Viljanen T, Laaksonen MS, Parkkola R, Knuuti J, Kjaer M. In vivo measurements of glucose uptake in human Achilles tendon during different exercise intensities. Int J Sports Med 26(9):727–31, Nov, 2005.

Harty J, Soffe K, O'Toole G, Stephens MM. The role of hamstring tightness in plantar fasciitis. Foot Ankle Int 26(12):1089–92, Dec, 2005.

Hather BM, Tesch PA, Buchanan P, Dudley GA. Influence of eccentric actions on skeletal muscle adaptations to resistance training. Acta Physiol Scand 143(2):177–85, Oct, 1991.

Hedstrom M, Ahl T, Dalen N. Early postoperative ankle exercise. A study of postoperative lateral malleolar fractures. Clin Orthop Relat Res (300):193–6, Mar, 1994.

Hertel J, Braham RA, Hale SA, Olmsted-Kramer LC. Simplifying the star excursion balance test: analyses of subjects with and without chronic ankle instability. J Orthop Sports Phys Ther 36(3):131–7, Mar, 2006.

Hertel J, Miller SJ, Denegar CR. Intratester and intertester reliability during the Star Excursion Balance Tests. J Sport Rehabil 9:104–116, 2000.

Hertel J. Functional anatomy, pathomechanics, and pathophysiology of lateral ankle instability. J Athl Train 37(4):364–375, Dec, 2002.

Hiebert GW, Pearson KG. Contribution of sensory feedback to the generation of extensor activity during walking in the decerebrate cat. J Neurophysiol 81(2):758–770, Feb, 1999.

Hiebert GW, Whelan PJ, Prochazka A, Pearson KG. Contribution of hind limb flexor muscle afferents to the timing of phase transitions in the cat step cycle. J Neurophysiol 75(3):1126–1137, Mar, 1996.

Hoiness P, Glott T, Ingjer F. High-intensity training with a bi-directional bicycle pedal improves performance in mechanically unstable ankles—a prospective randomized study of 19 subjects. Scand J Med Sci Sports 13(4):266–71, Aug, 2003.

Holten O, Faugli HP. Medical Exercise Therapy. The Norwegian M.E.T. Institute. Laerergruppen for Medisinsk, 1996.

Holten O, Grimsby O. *Medical Exercise Therapy*: A Course Workbook. The Holten Institutt for Medisinsk Treningsterapi, First edition, 1986.

Hopkins WG, Hawley JA, Burke LM. Design and analysis of research on sport performance enhancement. Med Sci Sports Exerc 31(3):472–485, Mar, 1999.

Horstmann T, Mayer F, Maschmann J, Niess A, Roecker K, Dickhuth HH. Metabolic reaction after concentric and eccentric endurance-exercise of the knee and ankle. Med Sci Sports Exerc 33(5):791–5, May, 2001.

Houglum PA. *Therapeutic Exercise for Athletic Injuries.* Champaign, Ill: Human Kinetics; 2001.

Hubbard TJ, Kaminski TW. Kinesthesia is not affected by functional ankle instability status. J Athl Train 37(4):481–486, Dec, 2002.

Hunt G. Functional biomechanics of the subtalar joint. Orthopedic Physical Therapy, Home Study Course 92–1. Orthopedic Section of the APTA, Mar, 1992.

Hyland MR, Webber-Gaffney A, Cohen L, Lichtman PT. Randomized controlled trial of calcaneal taping, sham taping, and plantar fascia stretching for the short-term management of plantar heel pain. J Orthop Sports Phys Ther 36(6):364–71, Jun, 2006.

Inman V, Ralston H, Todd F. *Human Walking*. Williams and Wilkins, Baltimore, 1981.

Inman V. The influence of the foot-ankle complex on the proximal skeletal structures. Artificial Limbs 13(1):59, Spring, 1969.

Inman V. *The Joints of the Ankle*. Williams and Wilkins, Baltimore, 1978.

Isakov E, Mizrahi J. Is balance impaired by recurrent sprained ankle? Br J Sports Med 31(1):65–7, Mar, 1997.

James SL, Brubaker CE. Biomechanics of running. Orthop Clinc North Am 4(3):605–615, Jul, 1973.

Jenkins W, Bronner S, Mangine R. Functional evaluation and treatment of the lower extremity. In: Brownstein B, Bronner S. *Functional Movement in Orthopaedic and Sports Physical Therapy*, Churchill Livingstone, New York, 1997.

Johnston RB, Howard ME, Cawley PW, Losse GM. Effect of lower extremity muscular fatigue on motor control performance. Med Sci Sport Ex 30(12):1703–1707, Dec, 1998.

Jonsson P, Alfredson H. Superior results with eccentric compared to concentric quadriceps training in patients with jumper's knee: a prospective randomised study. Br J Sports Med 39(11):847–50, Nov, 2005.

Kaminski TW, Buckley BD, Powers ME, Hubbard TJ, Ortiz C. Effect of strength and proprioception training on eversion to inversion strength ratios in subjects with unilateral functional ankle instability. Br J Sports Med 37(5):410–415, 2003.

Kaminski TW, Hartsell HD. Factors contributing to chronic ankle instability: A strength perspective. J Athl Train 37(4):394–405, Dec, 2002.

Kaminski TW, Perrin DH, Gansneder BM. Eversion strength analysis of uninjured and functionally unstable ankles. J Athl Train 34(3):239–245, Jul, 1999.

Karlsson J, Andreasson G. The effect of external ankle support in chronic lateral ankle joint instability. Am J Sports Med 20(3):257–261, May–Jun, 1992.

Karlsson J, Wiger P. Longitudinal split of the peroneus brevis tendon and lateral ankle instability: Treatment of concomitant lesions. J Athl Train 37(4):463–466, Dec, 2002.

Kelsey DD, Tyson E. A new method of training for the lower extremity using unloading. J Orthop Sports Phys Ther 19(4):218–223, Apr, 1994.

Kern-Steiner R, Washecheck HS, Kelsey DD. Strategy of exercise prescription using an unloading technique for functional rehabilitation of an athlete with an inversion ankle sprain. J Orthop Sports Phys Ther 29(5):282–7, May, 1999.

Khan KM, Cook JL, Bonar F, Harcourt P, Astrom M. Histopathology of common tendinopathies. Update and implications for clinical management. Sports Med 27(6):393–408, Jun, 1999.

Khan KM, Cook JL, Kannus P, Maffulli N, Bonar SF. Time to abandon the "tendinitis" myth. BMJ 324(7338):626–7, Mar 16, 2002.

Khan KM, Cook JL, Maffulli N, Kannus P. Where is the pain coming from in tendinopathy? It may be biochemical, not only structural, in origin. Br J Sports Med 34(4):318, Aug, 2000.

Kinzey SJ, Armstrong CW. The reliability of the star-excursion test in assessing dynamic balance. J Orthop Sports Phys Ther 27(5):356–360, May, 1998.

Kisner C, Colby LA. *Therapeutic Exercise Foundations and Techniques*. 4th ed. Philadelphia, Pa: FA Davis; 2002.

Kitai TA, Sale DG. Specificity of joint angle in isometric training. Eur J Appl Physiol Occup Physiol 58(7):744–748, 1989.

Klein J, Hoher J, Tiling T. Comparative study of therapies for fibular ligament rupture of the lateral ankle joint in competitive basketball players. Foot Ankle 14(6):320–4, Jul–Aug, 1993.

Kleinrensink GJ, Stoeckart R, Meulstee J, Kaulesar Sukul DM, Vleeming A, Snijders CJ, van Noort A. Lowered motor conduction velocity of the peroneal nerve after inversion trauma. Med Sci Sports Exerc 26(7):877–83, Jul, 1994.

Knight KL, Londeree BR. Comparison of blood flow in the ankle of uninjured subjects during therapeutic applications of heat, cold, and exercise. Med Sci Sports

Exerc 12(1):76–80, Spring, 1980.
Knuttegen HG, Klausen K. Oxygen debt in short-term exercise with concentric and eccentric muscle contractions. J Appl Physiol 30(5):632635, May, 1971.
Knuttgen HG, Petersen FB, Klausen K. Exercise with concentric and eccentric muscle contractions. Acta Paediatr Scand (Suppl)217:42–6, 1971.
Kokkonen J, Nelson AG, Cornwell A. Acute muscle stretching inhibits maximal strength performance. Res Q Exerc Sport 69(4):411–5, Dec, 1998.
Kongsgaard M, Aagaard P, Kjaer M, Magnusson SP. Structural Achilles tendon properties in athletes subjected to different exercise modes and in Achilles tendon rupture patients. J Appl Physiol 99(5):1965–71, Nov, 2005.
Konradsen L, Magnusson P. Increased inversion angle replication error in functional ankle instability. Knee Surg Sports Traumatol Arthrosc 8(4):246–251, 2000.
Konradsen L, Ravn JB, Sorensen AI. Proprioception at the ankle: the effect of anaesthetic blockade of ligament receptors. J Bone Joint Surg Br 75(3):433–6, May, 1993.
Konradsen L, Ravn JB. Ankle instability caused by prolonged peroneal reaction time. Acta Orthop Scand 61(5):388–90, Oct, 1990.
Konradsen L, Ravn JB. Prolonged peroneal reaction time in ankle instability. Int J Sports Med 12(3):290–292, Jun, 1991.
Konradsen L. Factors contributing to chronic ankle instability: Kinesthesia and joint position sense. J Athl Trainnig 37(4):381–382, Dec, 2002.
Kulig K, Burnfield J, Requejo SM, Terk M. Selective activation of tibialis posterior: evaluation by magnetic resonance imaging. Med Sci Sports Exerc 36(5):862–867, May, 2004.
Kvist M. Achilles tendon injuries in athletes. Sports Med 18(3):173–201, Sept, 1994.
Kwong PK, Kay D, Voner PT, White MW. Plantar fasciitis: mechanics and pathomechanics of treatment. Clin Sports Med 7(1):119–126, Jan, 1988.
Langberg H, Bulow J, Kjaer M. Blood flow in the peritendinous space of the human Achilles tendon during exercise. Acta Physiol Scand 163(2):149–53, Jun, 1998.
Langberg H, Ellingsgaard H, Madsen T, Jansson J, Magnusson P, Aagaard P, Kjaer M. Eccentric rehabilitation exercise increases peritendinous type I collagen synthesis in humans with Achilles tendinosis. Scand J Med Sci Sports 17(1):61–6, Feb, 2007.
Leanderson J, Wykman A, Eriksson E. Ankle sprain and postural sway in basketball players. Knee Surg Sports Traumatol Arthrosc 1(3–4):203–5, 1993.
Lentell GB, Bass B, Lopez D, McGuire L, Sarrel M, Synder P. The contributions of proprioceptive deficits, muscle function, and anatomic laxity to functional instability of the ankle. J Orthop Sports Phys Ther 21(4):206–215, Apr, 1995.
Lentell GL, Katzman L, Walters M. The relationship between muscle function and ankle stability. J Orthop Sport Phys Ther 11:605–611, 1990.
Leonard CT. *The Neuroscience of Human Movement.* St Louis: Mosby, 1998.
Lofvenberg R, Karrholm J, Sundelin G, Ahlgren O. Prolonged reaction time in patients with chronic lateral instability of the ankle. Am J Sports Med 23(4):414–417, Jul–Aug, 1995.
Lynch SA, Eklund U, Gottlieb D, Renstrom PA, Beynnon B. Electromyographic latency changes in the ankle musculature during inversion moments. Am J Sports Med 24(3):362–369, May–Jun, 1996.
Mafi N, Lorentzon R, Alfredson H. Superior short-term results with eccentric calf muscle training compared to concentric training in a randomized prospective multicenter study on patients with chronic Achilles tendinosis. Knee Surg Sports Traumatol Arthrosc 9(1):42–7, 2001.
Manfroy PP, Ashton-Miller JA, Wojtys EM. The effect of exercise, prewrap, and athletic tape on the maximal active and passive ankle resistance of ankle inversion. Am J Sports Med 25(2):156–63, Mar–Apr, 1997.
Mantar JT. Movements of the subtalar and transverse tarsal joints. Anat Record 80:397–410, 1941.
Marek SM, Cramer JT, Fincher AL, Massey LL, Dangelmaier SM, Purkayastha S, Fitz KA, Culbertson JY. Acute effects of static and proprioceptive neuromuscular facilitation stretching on muscle strength and power output. J Athl Train 40(2):94–103, Jun, 2005.
Matsusaka N, Yokoyama S, Tsurusaki T, Inokuchi S, Okita M. Effect of ankle disk training combined with tactile stimulation to the leg and foot on functional instability of the ankle. Am J Sports Med 29(1):25–30, Jan–Feb, 2001.
Mattacola CG, Dwyer MK. Rehabilitation of the ankle after acute sprain or chronic instability. J Athletic Training 37(4):413–429, Dec, 2002.
McBride JM, Triplett-McBride T, Davie A, Newton RU. The effect of heavy vs light-load jump squats on the development of strength, power, and speed. J Strength Cond Res 16(1):75–82, Feb, 2002.
McGuine TA, Greene JJ, Best T, Leverson G. Balance as a predictor of ankle injuries in high school basketball players. Clin J Sport Med 10(4):239–44, Oct, 2000.
McGuine TA, Keene JS. The effect of a balance training program on the risk of ankle sprains in high school athletes. Am J Sports Med 34(7):1103–11, Jul, 2006.
McKay GD, Goldie PA, Payne WR, Oakes BW. Ankle injuries in basketball: injury rate and risk factors. Br J Sports Med 35(2):103–108, Apr, 2001.
McKenzie RT. *Exercise in Education and Medicine.* Philadelphia, 1910.
Mecagni C, Smith JP, Roberts KE, O'Sullivan SB. Balance and ankle range of motion in community-dwelling women aged 64 to 87 years: a correlational

study. Phys Ther 80(10):1004–11, Oct, 2000.

Milgrom C, Shlamkovitch N, Finestone A, Eldad A, Laor A, Danon YL, Lavie O, Wosk J, Simkin A. Risk factors for lateral ankle sprain: a prospective study among military recruits. Foot Ankle 12:26–30, 1991.

Morasso PG, Sanguineti V. Ankle muscle stiffness alone cannot stabilize balance during quiet standing. J Neurophysiol 88(4):2157–62, Oct, 2002.

Muir IW, Chesworth BM, Vandervoort AA. Effect of a static calf-stretching exercise on the resistive torque during passive ankle dorsiflexion in healthy subjects. J Orthop Sports Phys Ther 29(7):421–4, Jul, 1999.

Mulligan EP. Lower leg, ankle, and foot rehabilitation. In: Andrews JR, Harrelson GL, Wilk KE. Physical Rehabilitation of the Injured Athlete, WB Saunders, Philadelphia, 1998.

Munn J, Beard D, Refshauge K, Lee RJ. Do functional-performance tests detect impairment in subjects with ankle instability? J Sport Rehabil 11:40–50, 2002.

Murrell GA, Jang D, Deng XH, Hannafin JA, Warren RF. Effects of exercise on Achilles tendon healing in a rat model. Foot Ankle Int 19(9):598–603, Sep, 1998.

Nawoczenski DA, Owen ML, Ecker B, Altman B, Epler M. Objective evaluation of peroneal response to sudden inversion stress. J Orthop Sports Phys Ther 7:107–109, 1985.

Nelson AG, Kokkonen J, Arnall DA. Acute muscle stretching inhibits muscle strength endurance performance. J Strength Cond Res 19(2):338–43, May, 2005.

Nelson AG, Kokkonen J. Acute ballistic muscle stretching inhibits maximal strength performance. Res Q Exerc Sport 72(4):415–9, Dec, 2001.

Nelson RT, Bandy WB. Eccentric training and static stretching improve hamstring flexibility of high school males. J Athl Train 39(3):254–258, Sep, 2004.

Nilsson G, Ageberg E, Ekdahl C, Eneroth M. Balance in single-limb stance after surgically treated ankle fractures: a 14-month follow-up. BMC Musculoskelet Disord 7:35, Apr 6, 2006.

Nishikawa T, Grabiner MD. Peroneal motoneuron excitability increases immediately following application of a semirigid ankle brace. J Orthop Sports Phys Ther 29:168–73, 1999.

Nitz A, Dobner J, Kersey D. Nerve injury and grades II and III ankle sprains. Am. J. Sports Med 13(3):177–182, May–Jun, 1985.

Ohberb L, Lorentzon R, Alfredson H. Eccentric training in patients with chronic Achilles tendinosis: normalized tendon structure and decreased thickness at follow up. Br J Sports Med 38(1):8–11, Feb, 2004.

Ohberg L, Alfredson H. Effects on neovascularisation behind the good results with eccentric training in chronic mid-portion Achilles tendinosis? Knee Surg Sports Traumatol Arthrosc 12(5):465–70, Sept, 2004.

Ohberg L, Alfredson H. Ultrasound guided sclerosis of neovessels in painful chronic Achilles tendinosis: pilot study of a new treatment. Br J Sports Med 36(3):173–5, Jun, 2002.

Olmsted LC, Carcia CR, Hertel J, Shultz SJ. Efficacy of the star excursion balance tests in detecting reach deficits in subjects with chronic ankle instability. J Athl Train 37(4):501–506, Dec, 2002.

Palmieri RM, Ingersoll CD, Hoffman MA, Cordova ML, Porter DA, Edwards JE, Babington JP, Krause BA, Stone MB. Arthrogenic muscle response to a simulated ankle joint effusion. Br J Sports Med 38(1):26–30, Feb, 2004.

Paris DL. The effects of the Swede-O, New Cross, and McDavid ankle braces and adhesive ankle taping on speed, balance, agility, and vertical Jump. J Athl Train 27(3):253–256, 1992.

Pellow JE, Brantingham JW. The efficacy of adjusting the ankle in the treatment of subacute and chronic grade I and grade II ankle inversion sprains. J Manipulative Physiol Ther 24(1):17–24, Jan, 2001.

Perry J. *Gait Analysis. Normal and Pathological Function.* McGraw-Hill, Inc, New York, 1992.

Pope R, Herbert R, Kirwan J. Effects of ankle dorsiflexion range and pre-exercise calf muscle stretching on injury risk in Army recruits. Aust J Physiother 44(3):165–172, 1998.

Porter D, Barrill E, Oneacre K, May BD. The effects of duration and frequency of Achilles tendon stretching on dorsiflexion and outcome in painful heel syndrome: a randomized, blinded, control study. Foot Ankle Int 23(7):619–24, Jul, 2002.

Porter GK Jr, Kaminski TW, Hatzel B, Powers ME, Horodyski M. An examination of the stretch-shortening cycle of the dorsiflexors and evertors in uninjured and functionally unstable ankles. J Athl Train. 2002 Dec;37(4):494–500, Dec, 2002.

Powers CM, Chen PY, Reischl SF, Perry J. Comparison of foot pronation and lower extremity rotation in persons with and without patellofemoral pain. Foot Ankle Int 23(7):634–640, Jul, 2002.

Provins KA. The effect of peripheral nerve block on the appreciation and execution of finger movements. J Physiol 143(1):55–67, Aug 29, 1958.

Quaschnick MS. The diagnosis and management of plantar fasciitis. Nurse Pract 21(4):50–4, 60–3, quiz 64–5, Apr, 1996.

Reber L, Perry J, Pink M. Muscular control of the ankle in running. Am J Sports Med 21(6):805–10, Nov–Dec, 1993.

Refshauge KM, Kilbreath SL, Raymond J. The effect of recurrent ankle inversion sprain and taping on proprioception at the ankle. Med Sci Sports Exerc 32(1):10–15, Jan, 2000.

Reischl SF, Powers CM, Rao S, Perry J. Relationship between foot pronation and rotation of the tibia and femur during walking. Foot Ankle Int 20(8):513–520, Aug, 1999.

Reitdyk A, Patla A, Winter D, Ishac M, Little C. Balance

1. Exercise for the Ankle and Foot

recovery from medio-lateral perturbations of the upper body during standing. J. Biomech. 32(11):1149–1151, Nov, 1999.

Ricard MD, Sherwood SM, Schulthies SS, Knight KL. Effects of Tape and Exercise on Dynamic Ankle Inversion. J Athl Train 35(1):31–37, Jan, 2000.

Riddle DL, Pulisic M, Pidcoe P, Johnson RE. Risk factors for plantar fasciitis: a matched case-control study. J Bone Joint Surg Am 85–A(5):872–7, May, 2003.

Riemann BL. Is there a link between chronic ankle instability and postural instability? J Athl Train 37(4):386–393, Dec, 2002.

Robbins S, Waked E, Rappel R. Ankle taping improves proprioception before and after exercise in young men. Br J Sports Med 29(4):242–7, Dec, 1995.

Rome K, Howe T, Haslock I. Risk factors associated with the development of plantar heel pain in athletes. Foot 11:119–125, 2001.

Roos EM, Engstrom M, Lagerquist A, Soderberg B. Clinical improvement after 6 weeks of eccentric exercise in patients with mid-portion Achilles tendinopathy—a randomized trial with 1-year follow-up. Scand J Med Sci Sports 14(5):286–95, Oct, 2004.

Roos SE, Guskiewicz KM, Yu B. Single-leg jump-landing stabilization times in subjects with functionally unstable ankles. J Athl Train 40(4):298–304, Oct–Dec, 2005.

Ross S, Guskiewicz K. Single-leg jump-landing stabilization times in subjects with functionally unstable ankles. J Athl Train 40(4):298–304, Oct–Dec, 2005.

Ross SE, Guskiewicz KM. Examination of Static and Dynamic Postural Stability in individuals with functionally stable and unstable ankles. Clin J Sport Med 14(6):332, Nov, 2004.

Rovere GD, Clarke TJ, Yates CS, Burley K. Retrospective comparison of taping and ankle stabilizers in preventing ankle injuries. Am J Sports Med 16(3):228–233, May–Jun, 1988.

Rozzi S, Lephart S, Sterner R, et al. Balance training for persons with functionally unstable ankles. J Orthop Sports Phys Ther 29(8):478–486, Aug, 1999.

Ryan L. Mechanical stability, muscle strength, and proprioception in the functionally unstable ankle. Aust J Physiother 40:41–47, 1994.

Sahrmann, SA. *Diagnosis and Treatment of Movement Impairment Syndromes.* St. Louis, MO: Mosby; 2002.

Sammarco GJ, Burnstein AH, Frankel VH. Biomechanics of the ankle: a kinematic study. Orthop Clin North Am 4(1):75–96, Jan, 1973.

Segal D, Wiss, D, Whitelaw G. Functional bracing and rehabilitation of ankle fractures. Clin Orthop 199:39–45, Oct, 1985.

Shalabi A, Kristoffersen-Wilberg M, Svensson L, Aspelin P, Movin T. Eccenric training of the gastrocnemius-soleus complex in chronic Achilles tendinopathy results in decreased tendon volume and intratendinous signal as evaluated by MRI. Am J Sports Med 32(5):1286–96, Jul–Aug, 2004.

Silbernagel KG, Thomee R, Thomee P, Karlsson J. Eccentric overload training for patients with chronic Achilles tendon pain—a randomised controlled study with reliability testing of the evaluation methods. Scand J Med Sci Sports 11(4):197–206, Aug, 2001.

Sitler MR, Horodyski M. Effectiveness of prophylactic ankle stabilisers for prevention of ankle injuries. Sports Med 20(1):53–57, Jul, 1995.

Smith RW, Reischl SF. Treatment of ankle sprains in young athletes. Am J Sports Med 14(6):465–471, Nov–Dec, 1986.

Soderberg GL, Cook TM, Rider SC, Stephenitch BL. Electromyographic activity of selected leg musculature in subjects with normal and chronically sprained ankles performing on a BAPS board. Phys Ther 71(7):514–22, Jul, 1991.

Sondernaa K, Holgard. U, Smith D, Alho. A. Immobilization of operated ankle fractures. Acta Orthop Scand 57(1):59–61, Feb, 1986.

Sorensen KL, Hollands MA, Patla E. The effects of human ankle muscle vibration on posture and balance during adaptive locomotion. Exp Brain Res 143(1):24–34, Mar, 2002.

Staples OS. Result study of ruptures of lateral ligaments of the ankle. Clin Orthop 85:50–58, 1972.

Surve I, Schwellnus MP, Noakes T, Lombard C. A fivefold reduction in the incidence of recurrent ankle sprains in soccer players using the sport-stirrup orthosis. Am J Sports Med 22(5):601–6, Sep–Oct, 1994.

Tibero D, Gary G. Kinematics and kinetics during gait. In: Donatelli R, Wooden M. *Orthopaedic Physical Therapy*, Churchill Livingstone, New York, 1989.

Trojian TH, McKeag DB. Single leg balance test to identify risk of ankle sprains. Br J Sports Med 40(7):610–3, Jul, 2006.

Tropp H, Askling C, Gillquist J. Prevention of ankle sprains. Am J Sports Med 13(4):259–62, Jul–Aug, 1985.

Tropp H, Ekstrand J, Gillquist J. Stabilometry in functional instability of the ankle and its value in predicting injury. Med Sci Sports Exerc 16(1):64–6, 1984.

Tropp H. Commentary: Functional Ankle Instability Revisited. J Athl Train 37(4):512–515, Dec, 2002.

Tropp H. Pronator muscle weakness in functional instability of the ankle joint. Int J Sports Med 7(5):291–294, Oct, 1986.

Vaes P, Duquet W, Van Gheluwe B. Peroneal reaction times and eversion motor response in healthy and unstable ankles. J Athl Train 37(4):475–480, Dec, 2002.

Vaes P, Van Gheluwe B, Duquet W. Control of acceleration during sudden ankle supination in people with unstable ankles. J Orthop Sports Phys Ther

31(12):741–52, Dec, 2001.

van der Wees PJ, Lenssen AF, Hendriks EJ, Stomp DJ, Dekker J, de Bie RA. Effectiveness of exercise therapy and manual mobilisation in ankle sprain and functional instability: a systematic review. Aust J Physiother 52(1):27–37, 2006.

van Mechelen W, Hlobil H, Zijlstra WP, de Ridder M, Kemper HC. Is range of motion of the hip and ankle joint related to running injuries? A case control study. Int J Sports Med 13(8):605–10, Nov, 1992.

Vasli S. The operative treatment of ankle fractures. Acta Chir Scand 114(3):242, Mar 8, 1958.

Vasli S. Operative treatment of ankle fractures.Acta Chir Scand (Suppl)226:1–74, 1957.

Verhagen E, van der Beek A, Twisk J, Bouter L, Bahr R, van Mechelen W. The effect of a proprioceptive balance board training program for the prevention of ankle sprains: a prospective controlled trial. Am J Sports Med 32(6):1385–93, Sep, 2004.

Verhagen RA, de Keizer G, van Dijk CN. Long-term follow-up of inversion trauma of the ankle. Arch Orthop Trauma Surg 114(2):92–96, 1995.

Verhoshanski V, Tatyan V. Speed-strength preparation of future champions. Sove Sports Review 18(4):166–170, 1983.

Viel E, Esnault M. The effect of increased tension in the plantar fascia: a biomechanical analysis. Physiother Pract 5:69–73, 1989.

Vitale TD, Fallat LM. Distal tibiofibular synostosis and late sequelae of an ankle sprain. J Foot Surg 2(1):33–36, Jan–Feb, 1990.

Waddington G, Seward H, Wrigley T, Lacey N, Adams R. Comparing wobble board and jump-landing training effects on knee and ankle movement discrimination. J Sci Med Sport 3(4):449–59, Dec, 2000.

Warren C, Lehman J, Koblanski J. Heat and stretch procedures: An evaluation using rat tail tendon. Arch Phys Med Rehabil 57(3):122–126, Mar, 1976.

Wearing SC, Smeathers JE, Urry SR, Hennig EM, Hills AP. The pathomechanics of plantar fasciitis. Sports Med 36(7):585–611, 2006.

Wearing SC, Smeathers JE, Yates B, Sullivan PM, Urry SR, Dubois P. Sagittal movement of the medial longitudinal arch is unchanged in plantar fasciitis. Med Sci Sports Exerc 36(10):1761–7, Oct, 2004.

Wedderkopp N, Kaltoft M, Lundgaard B, Rosendahl M, Froberg K. Prevention of injuries in young female players in European team handball: a prospective intervention study. Scand J Med Sci Sports 9(1):41–47, Feb, 1999.

Weinstein ML. An ankle protocol for second-degree ankle sprains. Mil Med 158(12):771–4, Dec, 1993.

Weir JP, Housh TJ, Weir LL. Electromyographic evaluation of joint angle specificity and cross-training after isometric training. J Appl Physiol 77(1):197–201, Jul, 1994.

Wilkerson G, Nitz A. Dynamic ankle stability: mechanical and neuromuscular interrelationships. J Sport Rehabil 3:43–57, 1994.

Wilkerson G. Biomechanical neuromuscular effects of ankle taping and bracing. J Athl Train 37(4):436–445, Dec, 2002.

Wilkerson GB, Pinerola JJ, Caturano RW. Invertor vs. evertor peak torque and power deficiencies associated with lateral ankle ligament injury. J Orthop Sports Phys Ther 26(2):78–86, Aug, 1997.

Willems T, Witvrouw E, Verstuyft J, Vaes P, De Clercq D. Proprioception and muscle strength in subjects with a history of ankle sprain and chronic instability. J Athl Train 37(4):487–493, Dec, 2002.

Williams P, Wawick R, Dyson M, Bannister L. *Gray's Anatomy*, 37th ed. Churchill Livingstone, Edinburgh, London, Melbourne and New York, 1989.

Wilson GJ, Newton RU, Murphy AJ, Humphries BJ. The optimal training load for the development of dynamic athletic performance. Med Sci Sports Exerc 25(11):1279–1286, Nov, 1993.

Wooden MJ. Overuse syndromes of the foot and ankle. Orthopedic Physical Therapy Home Study Course 95–1. Topic: The foot and ankle. Orthopedic Section of the APTA, Mar, 1995.

Worrell T, Booher LD, Hench KM. Closed kinetic chain assessment following inversion ankle sprain. J Sport Rehabil 3:197–203, 1994.

Wu GK, Ng GY, Mak AF. Effects of knee bracing on the sensorimotor function of subjects with anterior cruciate ligament reconstruction. Am J Sports Med 29(5):641–5, Sep–Oct, 2001.

Xu D, Hong Y, Li J, Chan K. Effect of tai chi exercise on proprioception of ankle and knee joints in old people. Br J Sports Med 38(1):50–4, Feb, 2004.

Yaggie JA, McGregor SJ. Effects of isokinetic ankle fatigue on the maintenance of balance and postural limits. Arch Phys Med Rehabil 83(2):224–8, Feb, 2002.

Yanagisawa O, Niitsu M, Yoshioka H, Goto K, Itai Y. MRI determination of muscle recruitment variations in dynamic ankle plantar flexion exercise. Am J Phys Med Rehabil 82(10):760–5, Oct, 2003.

Yang JF, Stein RB, James KB. Contribution of peripheral afferents to the activation of the soleus muscle during walking in humans. Exp Brain Res 87(3):679–687, 1991.

Ying M, Yeung E, Li B, Li W, Lui M, Tsoi CW. Sonographic evaluation of the size of Achilles tendon: the effect of exercise and dominance of the ankle. Ultrasound Med Biol 29(5):637–42, May, 2003.

Young B, Walker MJ, Strunce J, Boyles R. A combined treatment approach emphasizing impairment-based manual physical therapy for plantar heel pain: a case series. J Orthop Sports Phys Ther 34(11):725–33, Nov, 2004.

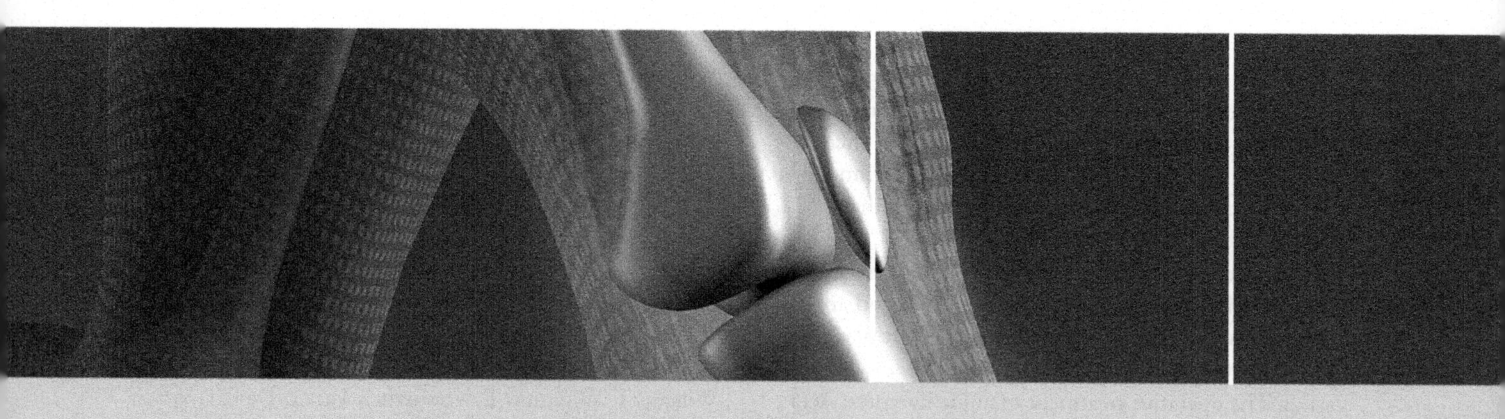

Per Henrik Ling (1776–1839) recognized that exercise was necessary for all persons. He maintained that exercise programs should be devised based on individual needs and that physical educators must possess knowledge of the effects of exercise on the human body. He used science and physiology to better understand the importance of fitness.

Authors: *Brian Power, Jim Rivard, Bill Hayner, Ola Grimsby*

CHAPTER 2

Exercise Rehabilitation of the Knee

Introduction

In this chapter:

The knee is one of the most researched joints in the human body. From anatomy and biomechanics to exercise design, studies have attempted to learn more about this complex region. Discussions have focused on open chain, closed chain, concentric, eccentric, isotonic and isokinetic exercise programs, as well as combinations of these. Emphasis on evidence-based practice can guide treatment, but should not be limited to these techniques alone, as they may not be adequate for a comprehensive rehabilitation program of the individual patient.

From a biomechanical perspective, the knee joint has the contrary requirements of providing stability in weight bearing and mobility for dynamic function. Stability derives from the powerful ligamentous system, the menisci, a complex capsular design and the supporting musculature. The neurological system, via mechanoreceptor afferent feedback in ligaments such as the anterior cruciate ligament (ACL) and joint capsule, also significantly contribute to the regulation of muscular stiffness and dynamic stability around the knee (Johansson et al. 1990). The knee requires proper biomechanical alignment and adequate mobility to allow for the necessary re-orientations of the foot in response to irregularities of the ground during walking and running. The myokinetic portion that governs knee motion is a combination of synergistic motor patterns and muscle performance that encompasses the entire lower quarter kinematic chain including the hip and ankle. More proximal, the lumbar spine has implications on knee stability, both biomechanical attachment to the lower quarter chain and neurological influence for complex movement patterns. Passive treatment to normalize soft tissue and joint mobility cannot be over looked prior to exercise rehabilitation, but is beyond the scope of this text. This chapter will outline the stages of knee rehabilitation exercise including treatment design guidelines, dosage and progressions with reference to specific pathologies. A thoughtful application of the best evidence for treatment will be applied using a foundation of the knowledge of anatomy, biomechanics, neurophysiology, exercise physiology, pathology and clinical experience.

Section 1: Stage 1 Exercise Progression Concepts for the Knee

Training Goals

Exercise prescription is one of the keys to a successful rehabilitation program. It is not always the specific exercise chosen, but the dosage and design that will determine the outcome. Intervention not only focuses on the knee, but on any joint that has an adverse effect on the function of the lower limb. Training for the initial functional qualities of range of motion, pain inhibition, tissue repair and resolution of edema are all addressed with high numbers of repetitions and minimal resistance. The high number of repetitions will also serve to improve coordination of basic movement patterns. The lighter resistance allows for proper motor unit recruitment and sequencing that results in the restoration of functional movement patterns.

Minimal resistance in the beginning phases is also important to increase tissue tolerance to stress and strain. When attempting to train tissue tolerance, over exertion is avoided to allow energy for protein synthesis after training. Properly dosing resistance, repetitions and rest breaks are also necessary to avoid abnormal tissue stress and further collagen breakdown. Pain and compensatory movement patterns are hallmarks of abnormal joint loading and must be avoided. Proper rest breaks can be determined based on the respiration rate. If it exceeds the steady state due to oxygen debt, a longer rest is needed after the set to allow for recovery.

Delayed onset tissue soreness (DOTS) and delayed onset muscle soreness (DOMS) can also be indications of over training and are associated with higher level muscle training, specifically eccentric work. Exceeding tissue tolerance should be avoided, as it can produce inflammatory responses leading to increased tissue soreness, stiffness and reflexive muscle guarding. DOTS should be avoided in initial training programs focused on tissue repair. The utilization of thermal agents and medication can reduce the pain associated with DOTS, but the tissue has still been over trained and the overall recovery will be delayed. Ice after exercise should not be necessary, as pain and inflammation should be reduced with training, not aggravated. Thermal agents may be more appropriate prior to exercise, along with soft tissue work and joint mobilization, to reduce symptoms and muscle guarding allowing for a higher level of function and training.

Stage 1—Dosed Functional Qualities

- *Optimal stimulus for tissue repair and tolerance to activities of daily living: collagen, cartilage, bone, muscle, disc and nerve.*
- *Edema reduction: venous pump*
- *Range of motion: osteokinematic motion*
- *Joint mobility: arthrokinematic motion*
- *Vascularity to guarded muscles*
- *Facilitation: recruitment patterns*
- *Coordination: neurological adaptation*

Joint / Myofascial Mobilization

Normalizing joint mobility (arthrokinematic motion) and range of motion (osteokinematic motion) are a necessary first step prior to training for muscular qualities such as endurance, strength and power. The instantaneous axis of motion in a hypomobile joint shifts toward the restriction creating abnormal tissue stress that can lead to additional trauma and compensatory movement patterns. Achieving full passive motion ensures proper force distribution through the knee to avoid these consequences. For example, limited end range extension during weight bearing activity creates abnormal compression through the patellofemoral and tibiofemoral joints. Passive articulations may be necessary to restore arthrokinematic motion prior to an active exercise approach.

Mobilization of the knee can be addressed through translatoric glides or distraction forces. The techniques are best performed manually by a trained physical therapist, but basic techniques can be performed as mobilization exercises in the clinic and at home. As arthrokinematic joint motion improves, high repetition and low load active exercises can be safely employed to improve osteokinematic joint motion. Arthrokinematic mobilization forces can be achieved in non-weight bearing positions utilizing a pulley and straps to replace the mobilizing hands of a therapist. Mobilizations should begin with distraction, as it is typically the safest and best tolerated direction, deforms the greatest amount of collagen, and fires the greatest number of mechanoreceptors for pain inhibition and muscle facilitation. The closed packed position (full extension) should be avoided with all mobilization techniques and may occur much earlier in the range of motion under pathological conditions. A common clinical error is to attempt to mobilize a translatoric glide at the end of the available range, which is the new relative closed packed position, resulting in increasing joint compression and not improving gliding movements.

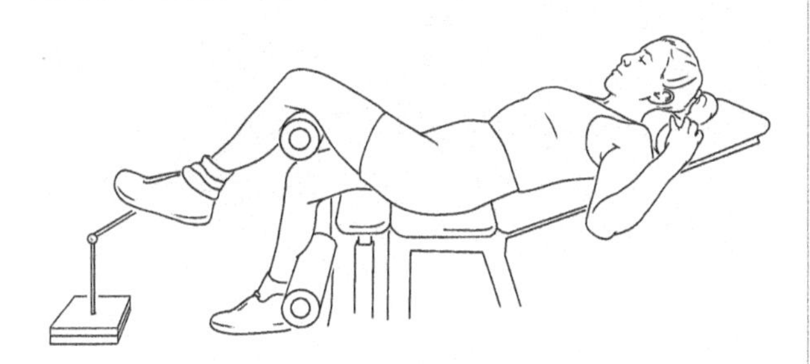

Figure 2.1: Knee distraction mobilization. The patient lies supine on angle bench with knee over the bolster in the resting position . A strap is placed around the ankle and attached to a wall pulley to produce a distraction force perpendicular to the joint surface for ten seconds. Active knee flexion will compress the joint, unloading the capsule from the mobilizing force, held for up to five second pause between each mobilization.

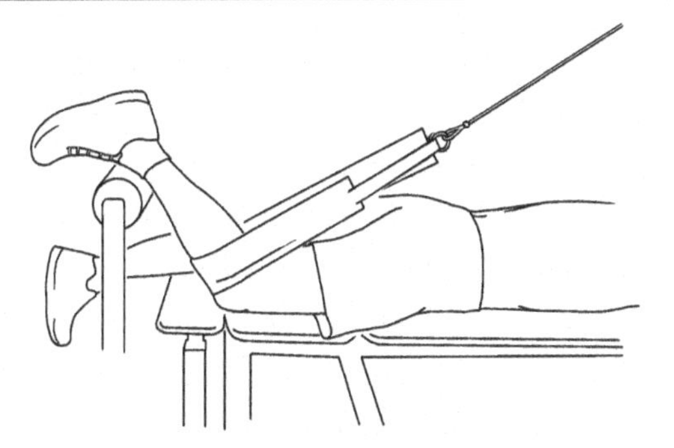

Figure 2.2: Passive knee flexion mobilization. The patient is prone with the knee in the resting position. A strap is placed parallel to the joint plane as close to the tibial plateau as possible and attached to a wall pulley. The patient relaxes the knee allowing the pulley to produce a posterior glide moment for ten seconds. Active extension takes tension off the knee capsule and serves as a rest break.

Figure 2.3: Passive knee extension mobilization. The mobilization is performed as above with an anterior force moment for an anterior glide of the tibia (extension). The pulley line remains parallel to the joint plane. The knee should be positioned in its resting position.

These mobilization examples follow the normal arthrokinematic motions of the knee complex. The force moment from the strap should always be parallel to the joint surface to achieve a translatoric glide. More classic exercises attempt to mobilize via osteokinematic torque, rather than arthrokinematic gliding translations. The former technique may initially improve range of motion, but can lead to intracapsular swelling and increased muscle guarding that ultimately increases joint stiffness. For example, the prone knee hang is a popular technique used to help improve knee extension. A weight is placed around the ankle and the lower leg is dangled over the edge of a table. The combination of the long moment arm, represented by the tibia, and resistance from ankle weight create a rotary force at the tibiofemoral joint that places abnormal compression on the cartilage and tension on the joint capsule and ligaments. This may be an effective technique for a total knee arthroplasty that is devoid of these structures, but it will be problematic for a normal joint with traumatised tissue or surgically repaired tissues (i.e., post ACL or PCL repair).

Supine knee hanging is another common approach to early mobilization to post-surgical knees.

The above supine knee hang has been used for improving knee extension. The patient lies supine, with the heel placed on a pad. A weight is placed on the knee to push it into extension and providing tension on the posterior structures. This approach is not recommended, as a torque moment is created at the knee, rather than an anterior arthrokinematic glide of the tibia on the femur. Even for joint replacement, when no cartilage is present to be irritated from an abnormal torque, a sustained stretch is not recommended for the capsule. Mobilization should focus on previous examples of distraction or translatoric gliding. If a sustained stretch for an athrokinematic glide is performed, the hold time is 10–15 seconds to deform collagen, consistent with previous discussions on collagen mobilization. This approach avoid pain from abnormal tissue stress and the potential for an increase in intracapsular edema from abnormal joint compression forces.

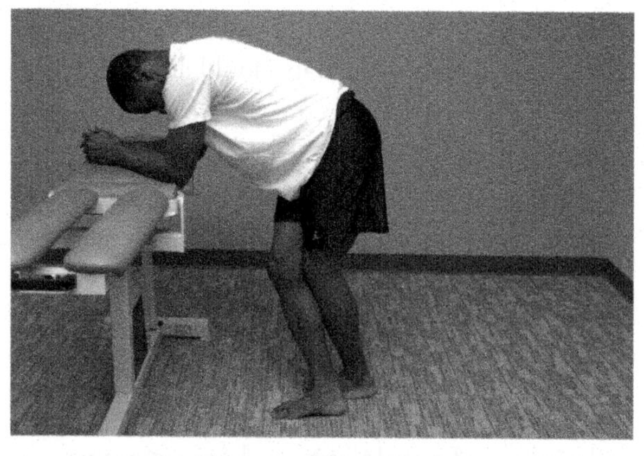

Figure 2.4: Crossed leg passive knee extension mobilization—osteokinematic mobilization for total knee arthroplasty only. The uninvolved leg is crossed over the involved knee to perform an extension

Rather than placing a torque on the joint to improve osteokinematic motion, mobilizing exercises should focus on improving arthrokinematic motion. The external mobilizing force (the strap or roll) should be placed as close to the joint line as possible to produce a pure translation, rather than a rotatory torque moment. Muscle activity can be used to help influence the mechanics and coordinate motion.

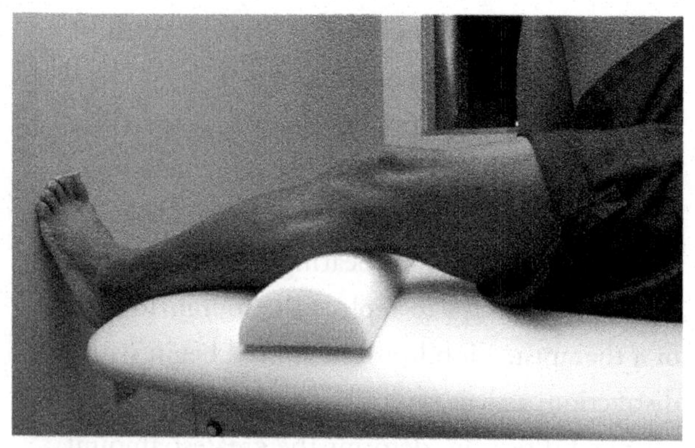

Figure 2.5: Anterior glide of the tibia (extension) with partial weight bearing terminal knee extension. Patient is supine with a bolster or towel roll placed under the proximal end of the tibia. Partial weight bearing is achieved by placing the ball of the foot against a wall during contraction of the quadriceps. An extension pattern of plantarflexion, knee extension and hip extension is performed by pushing the foot into the wall. The knee is pushed down onto the bolster with the tibia blocked by the roll creating a relative posterior glide of the femur on a fixed tibia.

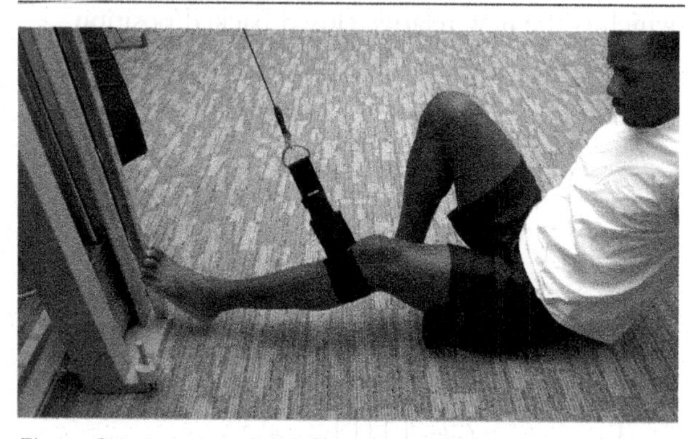

Figure 2.6: Anterior glide of the tibia (extension) with partial weight bearing active knee extension. Patient lies supine with the foot against a wall for partial weight bearing. A anterior force moment is created by placing a wall pulley strap posteriorly on the proximal. The patient is instructed to push the foot into the wall extending the entire lower limb. A common mistake is to place the strap directly behind the knee against the tibia and femur, resisting the osteokinematic extension, and not producing the arthrokinematic anterior glide of the tibia.

Knee flexion is improved by performing a posterior glide of the tibia on the femur. As in the previous examples, care is taken not to perform the mobilization at the end range of available flexion where the joint will be maximally compressed. It should be performed with the knee close to the resting position, or at least roughly 20 degrees from the end of available flexion range.

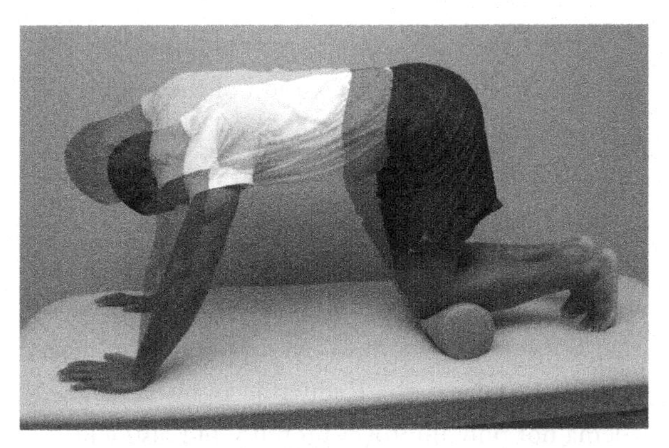

Figure 2.7: Knee flexion mobilization in quadruped. The patient kneels on a bolster or folded towel roll placed just below the tibial tubercle and joint line. Simply kneeling in this position will create a relative posterior glide of the tibia as the femur glides anterior toward the floor. A stretch mobilization is performed for ten seconds then weight is shifted to the uninvolved side. As range of motion improves the patient can slowly shift the body backward to increase the flexion angle. Care is taken to avoid creating a rotatory torquing moment into flexion (osteokinematic motion). Adding slightly internal rotation of the tibia will follow normal arthrokinematic motion for knee flexion.

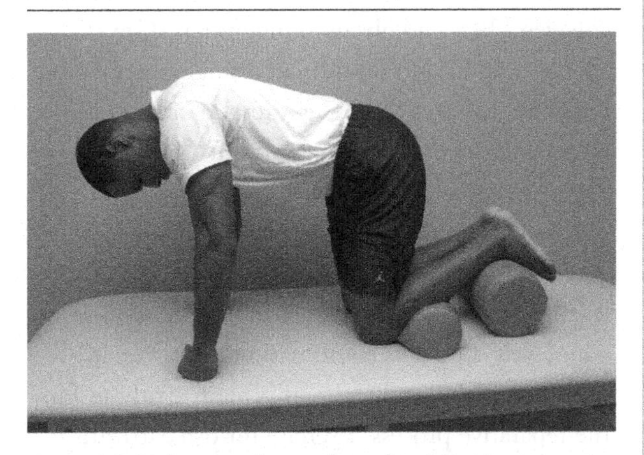

Figure 2.8: Performing the quadruped mobilization beyond 90° of flexion requires that a bolster is placed under the ankle. The femur must remain vertical, or leaning toward the shoulders, to avoid a flexion torque moment the knee. A torque would abnormally compress the cartilage of the posterior compartment. The compression may be pain free at the time, as normal cartilage is not innervated, but may lead to inflammation causing secondary pain after training. Placing a bolster under the ankle increases the amount of knee flexion allowing for a mobilizing force with the knee beyond 90° of flexion while maintaining a vertical femur for a more pure tranlatoric mobilization. The patient should be educated to avoid dropping the hips back too far, keeping the femur vertical. As the knee reaches the end of available flexion an abnormal closed packed position is achieved resulting in maximum joint compression. With this level of compression a translatoric glide cannot be performed and the mobilization becomes ineffective.

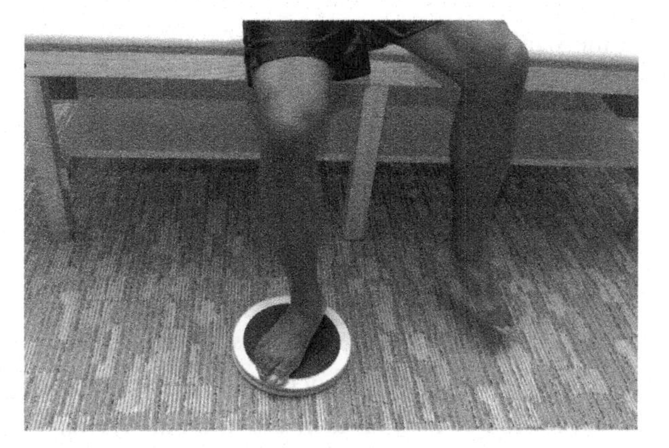

Figure 2.9: Sitting active mobilization for tibial internal and external rotation. Tibial rotation is required for knee motion. Internal rotation occurs with flexion, while external rotation occurs with extension. These motions can simply be performed as isolated rotation in sitting with the weight on the heel to improve arthrokinematic spin of the tibia for osteokinematic flexion and extension motions.

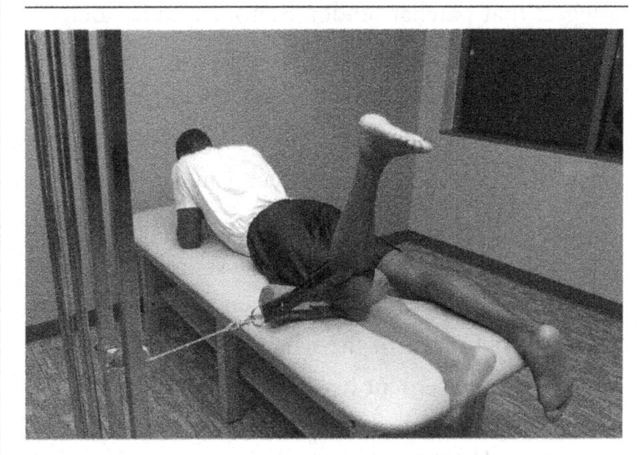

Figure 2.10: Lateral glide mobilization with active knee flexion and extension. A lateral force moment is placed as close to the joint line as possible and within the plane of the joint. The patient performs a slow, pain free flexion and extension motion concurrently with the lateral glide mobilization. This technique is an active version of Mulligan's (2004) passive mobilization with movement concept.

Selective Tissue Training (STT)

SST addresses issues of tissue repair, edema resolution and pain inhibition. Repetitive motion in the pain free range can assist in improving range of motion by reducing muscle tension and increasing collagen elasticity of the joint capsule. Muscle guarding of the quadriceps due to joint pain and/or inflammation can limit knee flexion, while extension may be limited by guarding in the hamstrings, adductors and gastrocnemius-soleus complex. Repetitive motion also provides a mechanical stimulus for repair of collagen, cartilage and bone. A thorough evaluation for differential diagnosis is an important step in exercise prescription so that the most effective stimulus can be applied to the dysfunctional tissue. For example, training collagen in a tendinopathy is different than addressing cartilage issues in patellofemoral syndrome. Progressive tensile forces can address collagen dysfunction, while progressive compression or loading exercises stimulate cartilage and bone repair. Surgical procedures, such as ACL reconstruction, can dictate the focus of a rehabilitation program. For instance, an exercise prescription for the donor site in a patellar tendon autograft will need to be considered along with protection of the graft fixation. Data from studies suggest that patellar tendon donor sites remodel over an 18–24 month period (Kiss et al. 1998). It is also common to have meniscal, articular cartilage and surrounding ligamentous damage that needs to be addressed concurrently, which can act to delay and complicate functional progression in the clinic.

Resolution of inflammation, when present, is a part of the initial phase of the rehabilitation program. Effusion management involves modifying tissue stress with manual techniques and exercise to avoid pain, which is of paramount importance in rehabilitating traumatized or post surgical tissues (Iles et al. 1990). Edema can be a cause for decreased range of motion, increased pain and altered motor function. It must be noted that inflammation is a complex interaction of cellular signals and responses and not all conditions are related to inflammation and respond to anti-inflammatory medication (Scott et al. 2004). Muscle inhibition due to arthrogenic pathologies has been attributed to presynaptic neuronal reflex activity in which altered afferent input originating from the injured joint results in a diminished efferent motor drive to the quadriceps muscles (Palmieri et al. 2005). The effects of long-term effusions contribute to the loss of proprioception observed in some clinical conditions (McNair et al. 1995). Edema is better treated with active muscle pumping approaches rather than more passive modalities such as cryotherapy or compression (Wester et al. 1996). Stimulation of the venous pump via muscle contraction allows for active transport of fluid and exudate through the venous system and lymphatics. Elevation of the limb during training can further assist in decreasing edema by reducing the mean fluid pressure of the blood that is forcing fluid out of the capillaries and into the extracellular space. Reduction in intracapsular edema not only improves mobility but also will assist in normalizing the afferent feedback from the joint capsule, and increase motor facilitation. McNair et al. (1996) demonstrated a reduction in quadriceps facilitation after a 60 mL saline and dextrose injection into the knee joint, with torque measurements returning to normal after three to four minutes of submaximal exercise.

Repetitive mechanical loading of collagen in the line of stress stimulates fibroblasts to produce new collagen fibers and glycosaminoglycans (GAG). Miller et al. (2005) found that a single acute bout of strenuous, non-damaging exercise increased the rates of synthesis of tendon and muscle collagen, as well as muscle myofibrillar and sarcoplasmic protein that peaked after 24 hours and slowly decreased toward resting values by 72 hours. The level of strain to healing collagen is increased, as pain levels drop and coordination improves, in order to maximize the rate of protein synthesis for the reparative process. Exercise for osteoarthritis of the knee has been extensively researched. Bautch et al. (2000) disputed concerns that exercise for osteoarthritis (OA) would contribute to cartilage degradation. In a systematic review of randomized clinical trials (RCTs) of exercise therapy for patients with osteoarthritis of the hip and knee, van Baar (1999) found sufficient evidence for including exercise as a major component of rehabilitation. Roddy et al. (2005a) performed a systematic review of RCTs comparing aerobic walking to quadriceps strengthening for OA that showed evidence for improving disability and reducing pain in both cases. In an evidence-based review, recommendations included utilizing an individualized exercise approach consisting of proprioceptive training and compliance with the program (Roddy et al. 2005b).

Prior to initiating exercise in a degenerative joint, arthrokinematic motion needs to be restored to a sufficient level (Grimsby 1991). Passive joint mobilization may be necessary to improve joint motion, reduce pain, improve afferent input for motor facilitation and proprioception as well as lubricate avascular tissues. Deyle et al. (2000) found a combination of passive manual physical therapy treatments and supervised exercise to be beneficial for patients with OA of the knee, possibly delaying or preventing the need for surgical intervention. A comparison of patients receiving home exercise alone to those receiving both manual therapy (MT) and supervised exercise found the home exercise group only had about half the improvement compared to the MT and supervised exercise group after one month (Deyle et al. 2005). Long-term outcomes after one year were similar, as subjects continued similar home exercise programs through this period. Factors other than joint restriction that may impede outcome include quadriceps inhibition, obesity, passive knee laxity, poor alignment, fear of physical activity and a lack of self-efficacy (Fitzgerald 2005). In the presence of any of these variables, exercise design may need to be modified to a lower level with a slower progression to higher level training.

Vad et al. (2002) presented a five-stage program recommended for rehabilitation of athletes with early knee OA. Stage 1 related to the use of protected mobilization and pain control modalities. Medical intervention focused on pain medications, nonsteroidal anti-inflammatory drugs, chondroprotective agents such as glucosamine in some cases, and injection therapy including intra-articular injections of corticosteroids or viscosupplementation. Stages 2–5 involved progressive exercises consistent with the four stages outlined in this chapter. The initial stage focused on open kinetic chain exercises for resolution of edema, increasing range of motion, controlling pain and improving coordination. Progressive closed kinetic chain exercises were added in the next stage with an emphasis on tissue repair, and some functional retraining. This was followed by an increased focus on functional activities and training related to specific athletic performance. The final stage emphasized patient education for long term training to reduce the risk of re-injury, improve the overall training state of the knee joint and provide long-term activity to delay the progression of the disease.

Moderate exercise in patients at high risk of developing OA has been associated with symptom reduction and improved function, as well as an improvement in the GAG content of the knee cartilage. After 12 weeks of training no significant deleterious effects of the osteoarthritic joints were found, as reflected by chondroitin sulphate synovial fluid markers. Pain reduction is a consistent finding with exercise programs for OA. van Baar et al. (1998) reported positive outcomes for improvements in pain and range of motion after 12 weeks of exercise. Aquatic exercise has been shown to significantly improve knee and hip flexibility, as well as strength and aerobic fitness in subjects with hip or knee OA, but it did not improve self-reported physical functioning and pain (Wang et al. 2007). Both aquatic and land-based training have been shown to be effective in improving knee range of motion, thigh girth, subjective pain ratings, and time for a one-mile walk (Wyatt et al. 2001). Training can be effective in improving motor performance and providing a positive stimulus for cartilage repair. In the obese patient, benefits of exercise may be limited or slow due to constant overloading of cartilage. Weight loss through dietary intervention is beneficial in over weight patients with OA (Focht et al. 2005). Martin et al. (2001) were successful in reducing pain levels in obese women with OA of the knee with weekly nutrition classes and an exercise-walking program.

Active motion with minimal to no resistance allows for hundreds of repetitions to facilitate a muscle pumping action for fluid transport without associated fatigue or pain. The tension that is developed during the contraction with low resistance (<25% of 1RM) is not great enough to increase intramuscular pressure high enough to collapse the capillaries, which would decrease blood flow. Slings, powder boards, gantries and slant boards can reduce the resistance to allow for early range of motion with a high number of repetitions. Slings and powder boards allow for horizontal movement perpendicular to gravity with zero resistance. Slant boards and gantries serve to reduce the body weight and compressive forces through the knee joint. As tissue tolerance improves, higher levels of tension and compression can be used through an increased range of motion, by increasing resistance and/or weight bearing forces. More passive approaches, such as the use of a continuous passive motion (CPM), does not have the same effect as active exercise, related to early training of coordination, circulation and tissue repair.

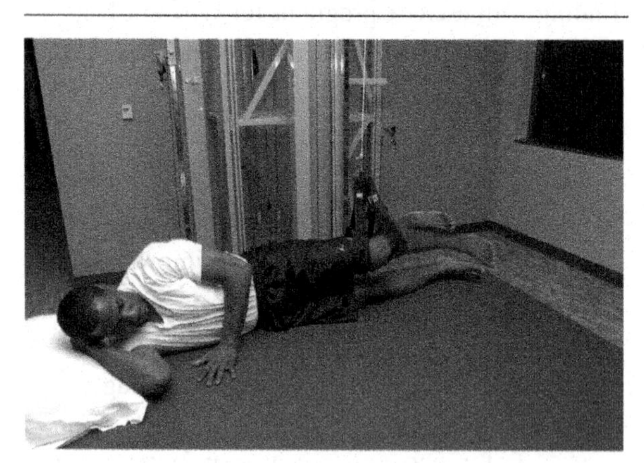

Figure 2.11: Sling training for knee flexion and extension. Two straps are placed on either side of the joint and attached to a fixed line. The axis of the knee joint is placed directly underneath the fixed line. Zero resistance in the sling suspension allows for pure concentric work of alternating knee flexion and extension. Hundreds of repetitions can be performed in the pain free range to positively influence pain, inflammation, tissue repair and early coordination. This is an excellent approach for acute post surgical knees that require a pain free safe environment for early high repetition training.

Beaupre et al. (2001) demonstrated that the addition of a CPM or slide board to a standard exercise program in the rehabilitation of total knee arthroplasty did not improve quality of life or outcome measures at both three and six months. Use of a CPM should be based on the individual situation and not a specific research study. Other variables such as pain levels, psychological motivation and general health may warrant the addition of a passive component to early rehabilitation. But these procedures can be costly and time consuming and may not significantly alter the end stage functional outcomes.

Figure 2.12a,b: Supine heel slides. Repetitive motion can be performed actively with heel slides in supine. Work emphasis is on the hamstrings and knee extension range. To reduce friction a sock can be worn with the foot on a sheet or vinyl. The knee is slowly flexed and extended in the pain free range. Forcing motion in to the painful range will only reduce range by increasing muscle tension, as well as increase tissue irritation and inflammation.

Figure 2.13: Wall slides are performed as with the above heel slide, but work emphasis is on the quadriceps and knee flexion range.

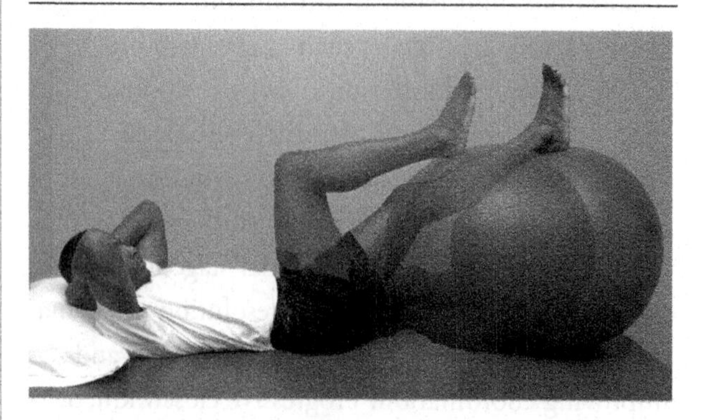

Figure 2.14: Fitness ball knee flexion-extension. The patient is in supine with the calf resting on top of a fitness ball. The

knee is flexed rolling the ball toward the body, with contact transferring to the heel, then returned to an extended position. Work emphasis is on the hamstrings, range emphasis on knee extension.

Open chain low resistance knee flexion and extension exercises create a greater tissue stress and a higher motor demand for early coordination training than sling suspension training. A pulley assist can be used to reduce the weight of the lower limb to allow for 40–50 repetitions with resistance below 50% of 1RM. Pulleys provide an objective and consistent method of decreasing load.

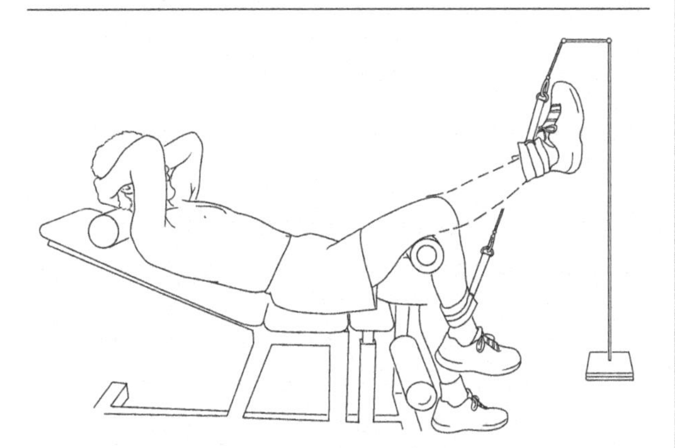

Figure 2.15: Assisted terminal knee extension. The knee is supported to allow for the desired range of motion. A vertical assist is provided to reduce the weight of the lower leg. The patient is instructed to perform knee extension in the available range with a one to two-second hold in full extension. Long isometric holds are initially avoided to prevent compromising circulation through increased intramuscular pressure.

Range of motion limitations from muscle guarding can be trained with repetitive concentric contraction of the involved muscles. Passive stretch is avoided in the acute stage when pain is present, as reduced flexibility is a function of neurological reflexive muscle guarding and not collagen restrictions. Repetitive motion at 60% of 1RM helps to increase circulation providing oxygen to tonic muscles in guarding. A slow eccentric phase can be added to allow for lengthening of the muscle to the end range for dynamic stretching. A sustained stretch on the last repetition of up to 30 seconds to improve flexibility can also be used. Pain is avoided at all times during the exercises, as a reflexive muscle guarding response will again limit range.

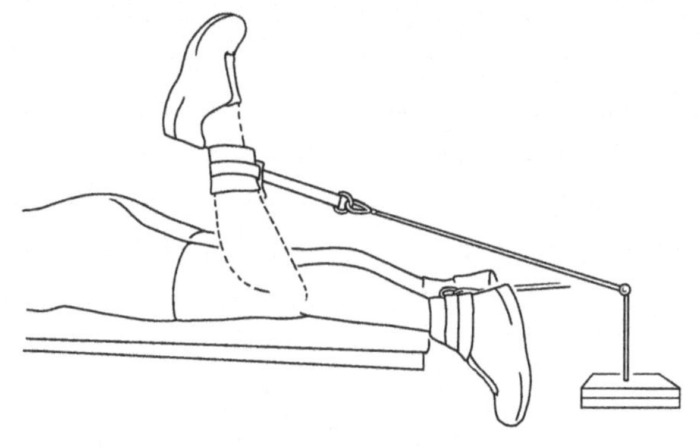

Figure 2.16: Prone knee flexion with eccentric emphasis for knee extension. Training for extension can be achieved through active hamstring work. An active hamstring contraction is followed by eccentric work to the end range. In the presence of muscle guarding three sets of 30 to 50 repetitions at 40–60% of 1RM is used to increase local circulation and provide oxygen to tonic muscles. The last repetition may involve a relaxation at the end range to stretch passive structures. Using pulleys or free weights with eccentric emphasis to gain mobility are preferred over elastic resistance, as minimal resistance is present toward the end range of motion.

Protocols for open kinetic chain (OKC) quadriceps strengthening have been recommended in the range of 20 degrees to full extension, as this is deemed to be more effective due to the muscle effort of the quadriceps being the highest in this range (Wild et al. 1982, Escamilla et al. 1998). But muscle recruitment can be difficult toward the end range due to active insufficiency and patients with lateral subluxation or dislocation may need to avoid terminal extension to maintain stability of the patella in the trochlear groove. Also with end range extension the patella is superior to the femoral groove, so tissue repair may be reduced with lack of normal cartilage compression for a cellular stimulus. But, terminal knee extension has been recommended with PFPS to avoid cartilage contact in the trochlear groove (Alaca et al. 2002).

Lesions on the proximal aspect of patellar surface may be painful with closed kinetic chain (CKC) exercises with the knee flexed in the 60–90 degrees range due to compression of the lesion with the femoral trochlea (Ahmed et al. 1983, Huberti et al. 1984, Huberti et al. 1998). Training may be better tolerated in a 20–60 degrees range, in order to modify the amount of cartilage compression to a pain free level. This mid range position also serves to stimulate both type I and II mechanoreceptors for maximizing quadriceps facilitation. Ultimately pain and coordination, along with the specific pathology, will determine the range and level of training that is most effective for each patient.

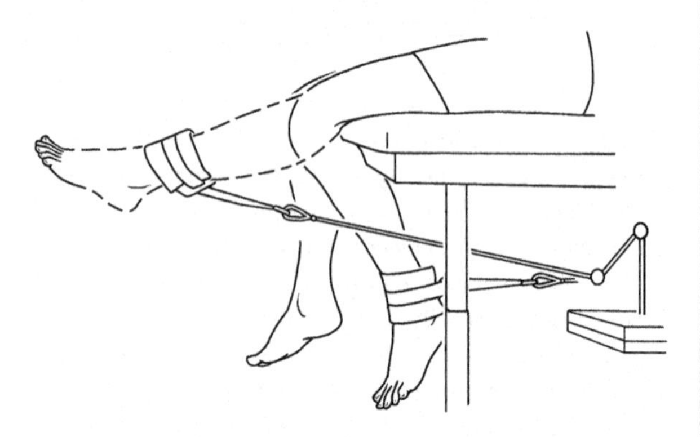

Figure 2.17: Sitting knee extension. Training for knee flexion with active quadriceps work emphasizing the eccentric phase into flexion. The patient actively contracts the quadriceps muscles to extend the knee, then performs a slow eccentric knee flexion to end range (three sets of 30–50 repetitions at 40–60% of 1RM). The last repetition is held as a passive stretch for 15–30 seconds. This exercise would be appropriate for dynamically stretching contractures of the quadriceps with TKAs and/or increasing vascularization to the quadriceps to help resolve muscle guarding and improve local muscle endurance.

Knee extension training in the presence of patellofemoral subluxation initially avoids terminal knee extension to maintain the patella in the femoral groove. Full extension can lead to abnormal lateral motion of the patella and remove the cartilage stimulus for repair. These poor mechanics may also lead to abnormal forces during weight bearing, causing pain and limiting tolerance to weight bearing training. For weight bearing training the tibia can be placed in slight internal rotation during, creating a relative external rotation of the femur, serving to unload the lateral patellar facet enough to allow for pain free training. Unloading the body weight of the patient can also serve to decrease compression forces.

Figure 2.18: Standing open chain hip extension, eccentric knee flexion. This exercise can address hip extension endurance training, while at the same time emphasizing knee flexion mobilization with a slow eccentric phase of hip and knee flexion in the pain free range.

Dosage for Stage 1 Functional Qualities

Tissue Repair, Pain Inhibition, Edema Reduction and Joint Mobilization

- *Sets*: 1–5 sets or training sessions daily
- *Repetitions*: 2–10 hours during the day
- *Resistance*: Assisted exercise up to 25% 1RM
- *Frequency*: Multiple times daily

Coordination

- *Sets*: 1–5 sets or training sessions daily
- *Repetitions*: 20–50 during the day
- *Resistance*: Assisted exercise up to 50% 1RM
- *Frequency*: 1–3 times daily

Vascularity / Local Endurance

- *Sets*: 2–3 sets or training sessions daily
- *Repetitions*: 25
- *Resistance*: Assisted to 50–60% 1RM
- *Frequency*: One time daily

Muscle in Atrophy

- *Sets*: 2–3 sets or training sessions daily
- *Repetitions*: 40 or more
- *Resistance*: Assisted to 30–40% 1RM
- *Frequency*: One time daily

Direction of Exercise

The initial direction of training can be influenced by the mobility restriction, tissue injury and/or functional deficits. Exercises may be selected simply to facilitate the muscles that govern basic tasks or motions or directly influence an injured muscle or tendon. Alterations in normal joint mobility have specific implications in exercise design, as it relates to direction and range of training. Initial training of hypomobile joints emphasizes restoring arthrokinematic joint play and osteokinematic range of motion. Repetitive eccentric work toward the restriction improves elasticity of collagen and muscle extensibility. The opposing pattern, or concentric work toward the restriction, ensures dynamic stabilization of the newly gained range.

Concentric Emphasis

Initial training for hypermobile joints involves a concentric emphasis within the beginning and mid ranges of motion away from the injury. Motion is performed in the same plane of tissue injury to provide modified tension in the line of stress for tissue repair. The end range, or pathological range, is avoided to prevent further tissue strain or injury. Concentric work away from the pathological range is performed to begin the process of gaining motor control of the muscles that provide stability. These same muscles are also eventually trained eccentrically toward the hypermobility to increase muscle spindle sensitivity and improve stability toward the hypermobile range.

Eccentric Training for Tendinopathy and Collagen Repair

Patellar tendinopathies, patellar tendon repairs, hamstring and patellar grafts for ACL reconstruction, Osgood Schlatter's disease, as well as hamstring and quadriceps strains represent impaired collagen states associated with a lowered ability to elongate and absorb tensile forces. All of these conditions can benefit from a specific tissue training program that focuses on repetitive modified tension in the line of stress for collagen repair.

Tendon is a metabolically active tissue that imposes an increased energy demand during exercise. Blood flow increases in the peritendinous tissue during exercise in humans (Boushel et al. 2000, Langberg et al. 1998, Langberg et al. 1999). Oxidative enzymes exist within fibroblasts and tenocytes of tendon (Joza et al. 1979, Kvist et al. 1987) and oxygen consumption increases during exercise (Boushel et al. 2000). A coupling exists between the exercise-induced drop in tissue oxygenation and increase in blood flow (Boushel et al. 2000). Glucose uptake has been shown to increase during exercise in the quadriceps tendon (Kalliokoski et al.

Stage 1 Direction of Exercise Concepts

Joint Mobility/ Grade	*Direction of Exercise*	*Hold Time*	*Histological Effect*	*Neurological Effect*
Hypermobile *Grades 4, 5 and 6*	*Concentric work away from the hypermobility – beginning to mid range only.*	*None*	• *Optimal stimulus for repair* • *Vascular effect on edema*	• *Sensitize spindles to stretch* • *Pain inhibition*
Hypomobile *Grades 1 and 2*	*Eccentric work toward the hypomobility –end range training .*	*Slow speed with 10 second hold*	• *Improve elasticity and plasticity deformity* • *Optimal stimulus for repair*	• *Pain inhibition*

2005). The increase in tendon glucose uptake is less pronounced and not correlated with that in muscle, as there is an unknown independent regulatory mechanism. Glucose uptake is more enhanced in the quadriceps tendon compared with the patella tendon in response to knee-extension exercise (Kalliokoski et al. 2005).

The pathology specifies the tissue in lesion while histology dictates the intervention and approach for rehabilitation. For example, overuse tendinopathies are aggravated and become chronic conditions if additional exercise and activity further contribute to oxygen and energy depletion. Typically it leads to the hallmark histopathological signs of tendinosis including collagen degeneration with fiber separation, increased mucoid ground substance and an absence of inflammatory cells often found in the Achilles, patellar, rotator cuff and extensor carpi radialis brevis tendons (Khan et al. 1999). The lack of inflammatory markers in these injured tendons is an essential point that alters the rehabilitation approach, as these conditions are often misdiagnosed as inflammatory issues rather than degenerative collagen disorders. Studies performed on chronic painful tendons, classically defined as tendonitis, have provided contrary evidence to assumed inflammatory responses following tendon injury. The evidence available suggests degenerative changes in type I collagen are the result of increased levels of matrix turnover affecting tendons that are exposed to higher levels of strain. Changes in cellular activity lead to a tendon that is mechanically weaker and more susceptible to damage (Riley 2004). Appropriate training in these cases, involves pure concentric work to the surrounding muscles in the area, except the one in lesion (Holten 1996). This indirect approach provides needed oxygen to the region, without placing an increased metabolic demand on the injured muscle/tendon complex. The tissue in lesion is then trained directly via pure eccentric work to stimulate collagen repair with little oxygen consumption, as the energy demand is only a third of that required for concentric work (Knuttgen et al. 1971). Eccentric exercise has been shown to lead to less fatigue, as well as lower lactate and ammonia reaction than concentric exercise at comparable work levels (Horstmann et al. 2001). The eccentric work phase also provides the necessary tension for collagen hypertrophy, where the concentric phase is more associated with vascularization via increased capillary density (Hather et al. 1991).

Application of eccentric training based on these histological concepts for patellar tendonitis is not new, as one of the first trials was performed by Cannell (1982). Several additional studies also assessed pure eccentric training for patellar tendonopathy (Jensen K, DiFabio 1989, Cannell et al. 2001, Young 2002, Stasinopoulos and Stasinopoulos 2004, Purdam 2004, Jonsson and Alfredson 2005, Young et al. 2005, Taunton 2006, Bahr et al. 2006,). Visnes et al. (2005) did not find benefit with eccentric training for jumper's knee in volleyball players, although subjects continued to train and compete during the trial treatment.

Eccentric Training Options for Patellar Tendinopathy

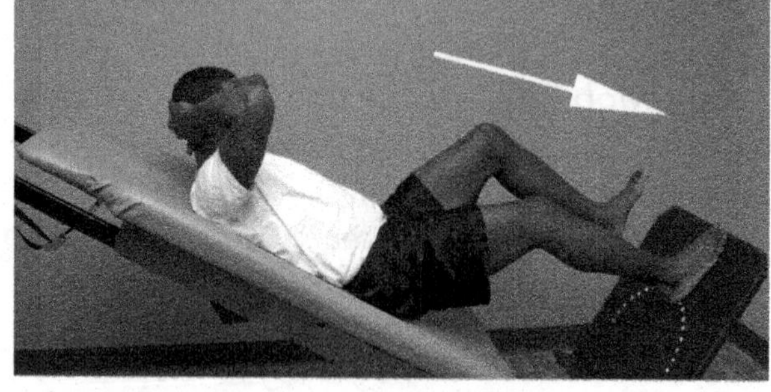

Figure 2.19a,b: Pure eccentric unilateral squat on an unloading slide board. The uninvolved left knee performs the concentric knee extension from a flexed position (top picture). The involved right knee performs the eccentric unilateral squat back to the start position. If the body weight cannot be reduced enough for pain free eccentric lowering unilaterally, then the uninvolved knee can assist with the eccentric performance as well.

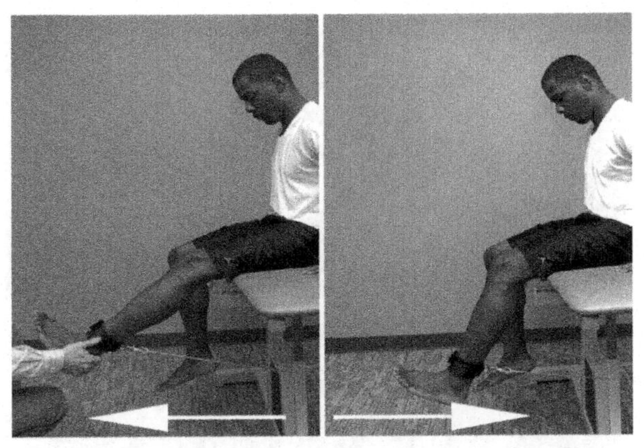

Figure 2.20a,b: Seated open chain eccentric quadriceps knee extension using a wall pulley. The therapist raises the weight (a) and the patient fixes the joint position with a brief isometric contraction of the quadriceps. The knee is then slowly lowered eccentrically in the pain free range (b). Progression may include lying back in a supine position placing the hip in neutral to increase muscle lengthening of the proximal quadriceps. After increasing length, speed and finally resistance are increased.

Figure 2.21a,b: Door jam squats with upper limb support for eccentric phases. Weight is shifted to the involved knee for eccentric lowering, then shifted onto the uninvolved side for the concentric return.

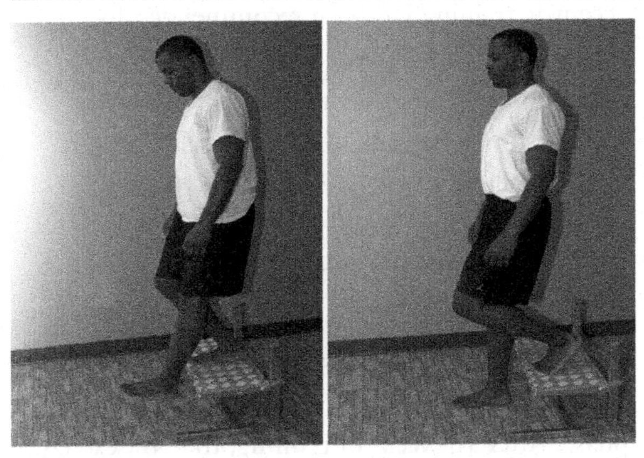

Figure 2.22a,b: Step-downs with pure eccentric emphasis of the involved limb. Eccentric step-downs are performed by the involved limb while the uninvolved limb performs a concentric step-up, back up onto the step.

Dosage for pure eccentrics is not as specific as concentric training for specific muscle performance qualities. Vines and Bahr (2007) attempted to summarize the previous clinical trials for patellar tendinopathy training, but could not recommend one specific protocol. They did recommend including a decline board for squats and allowing some level of discomfort with training. These studies included training anywhere from three times per week to two times per day, with most exercises dosed at three sets of fifteen repetitions.

General guidelines can be used with pure eccentric training, but the patient will determine the starting level and speed of progression. The amount of resistance used is dictated by the integrity of the collagen. The general rule is to use as heavy a weight as possible without causing pain or losing coordination. The number of repetitions is initially limited to a range of eight to 15, with fewer repetitions in more significant collagen injuries. Warren et al. (1993) presented a material fatigue model for collagen assessing load versus repetitions for collagen breakdown. Findings included a significant increase in damage after eight repetitions. This study was performed in the laboratory on rattails, and does not exactly translate into the clinical setting, but the concept of initially limiting the repetitions and using a heavy load is otherwise utilized. The need for additional sets is another consideration as strain is important to stimulate cellular activity, but the added repetitions may contribute to material breakdown in the collagen. Generally only one or two sets is recommended, and progressed to multiple training sessions per day as tolerated. Rest, for oxygen recovery, between sets is not necessary when training collagen, but time for collagen production and bonding is necessary.

Range of motion should be progressed prior to increasing the resistance. Training through the full length of the tendon, the typical range of injury, is more important than increasing load tolerance in the shortened range. Speed is also increased before resistance is increased. Kubo et al. (2005) compared ballistic drop jumps with isometric leg presses involving slow contractions lasting up to ten seconds. Ballistic drop jumps were more associated with muscle strengthening and delayed onset muscle soreness, while the slow prolonged work resulted in greater effects on the tendon. Initially, slower eccentric training will be more effective for stimulating tissue reparative responses in the tendon. Adding speed later is more related to simulating functional requirements. Once eccentrics can be performed at functional speeds through the entire range, resistance can be increased, which may require a slight and temporary drop in speed. Progression then continues by alternating increases in speed and weight.

Basic Patellar Tendinosis Rehab Concepts

- *Initial concentric training emphasis for vascularity for all motions in the region of the tendon, except the primary motion of the tendon (three sets of 24 at 60% of 1RM)*
- *Eccentric work one to two sets of eight to 15 repetitions with resistance as heavy as possible to allow for coordinated motion. Mild discomfort during the exercise is tolerated if coordination is not negatively affected and pain does not linger after completion of the exercise.*
- *Train through the full range of motion first, then progress by adding speed, followed by increasing resistance. Finally alternate progressions in speed and weight as appropriate.*
- *Avoid interventions that weaken collagen or slow repair: rest or immobility, stretching, NSAIDs, corticosteroids and heating the tendon with ultrasound.*

In more acute injuries or post surgical situations pure eccentrics are performed open chain, in non-weight bearing positions. Pulley resistance is preferred as consistent tension can be applied throughout the range. Free weights lose the force moment when parallel to gravity and elastic resistance decreases toward the lengthened range of muscle where tension is needed.

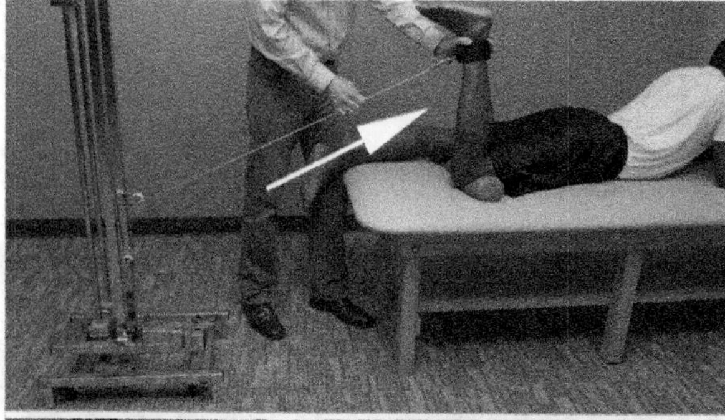

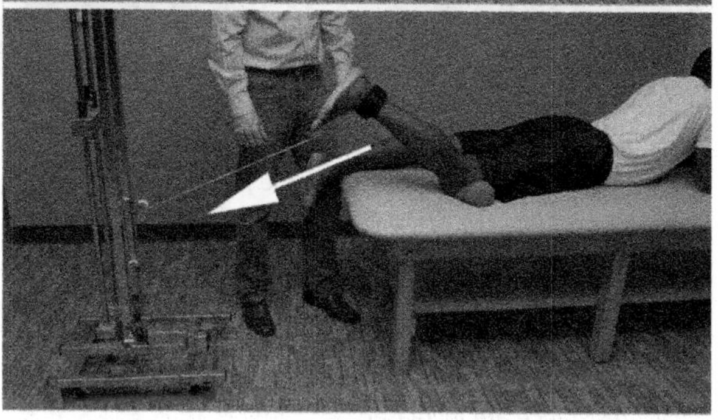

Figure 2.23a,b: Prone eccentric hamstrings knee flexion with a wall pulley or free weight. The therapist raises the weight (a) with the patient fixing the joint position via a brief isometric contraction followed by eccentrically lowering the knee into extension within the pain free range (b).

The hamstrings are frequently subjected to injuries due to eccentric forces developed when the leg is decelerating. The greatest tension develops immediately following toe off when there is an increase in the force moment during the transition between hip extension and flexion. This range is where the hamstrings need to have the greatest tensile strength so they have the ability to tolerate forces placed upon the collagen matrix. Pure eccentric training has been recommended for hamstring injury (Proske et al. 2004). Mjolsnes et al. (2004) compared the effects of a 10-week training program on hamstring muscle strength among male soccer players with two different exercises. The first was a traditional hamstring curl (HC) that included both CW and EW. The second was a Nordic hamstrings (NH) exercise that involved a partner assisting the CW phase with emphasis on heavier resistance for the EW phase. After 10 weeks of training, the NH exercise was found to be more effective than the HC exercise in developing maximal eccentric hamstring strength. Improved training effects may be at least partially contributed to remodeling of endomysial type IV collagen after a bout of eccentric muscle contractions as suggested by Mackey et al. (2004).

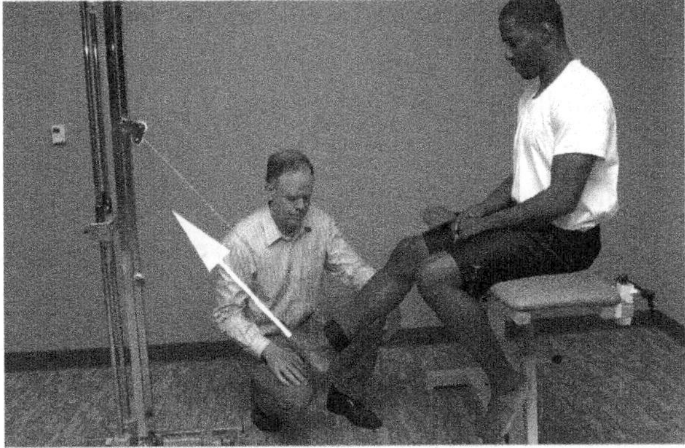

Figure 2.24a,b: Seated eccentric quadriceps knee flexion using a wall pulley. The therapist pulls the weight down (a) and the patient fixes the joint position with a brief isometric contraction of the hamstrings. The knee is then slowly extended eccentrically (b).

Nautilus type machines can be effective for pure eccentric training with an emphasis on early tissue repair. As one bolster is typically set for bilateral performance, the uninvolved knee can easily provide the concentric lift with a relatively heavier weight than the involved side can perform without pain. The involved knee can then perform the eccentric phase lowering a heavier resistance. Both quadriceps and hamstrings can be trained in this manner for rehabilitation of tendinopathy. The fixed axis of these machines reduces the coordination requirement for a more controlled performance that can focus on positive tissue strain. Beyond the tendinopathy, any collagen issue these muscles can be trained. This may include muscle tears, surgical repair or grafting. In the example of a hamstring graft for ACL repair, faster collagen repair and strength recovery can be achieved.

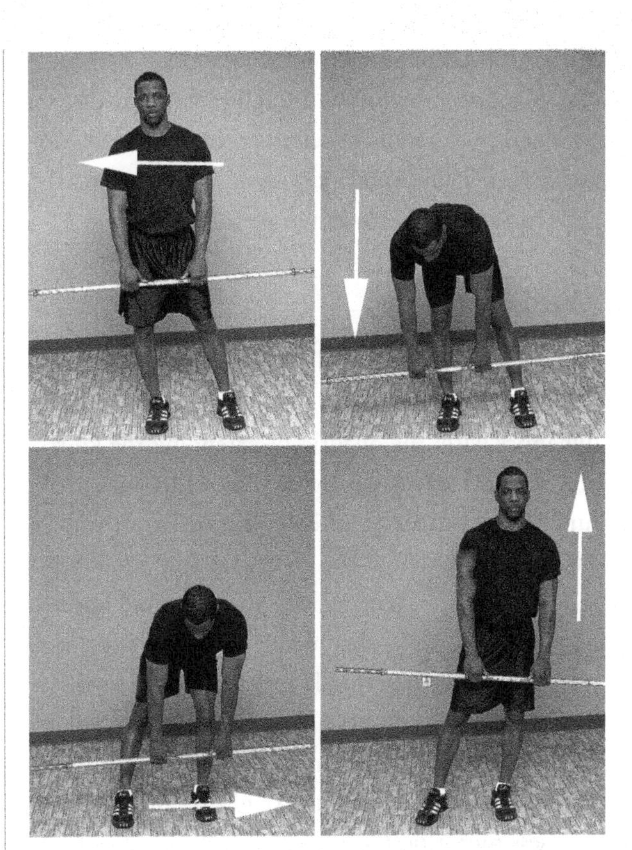

Figure 2.25a,b,c,d: "The Hamstring Shuffle"—Unilateral dead lift for eccentric training emphasizing right proximal hamstrings (advanced eccentric training for hamstring strain). With the knees slightly flexed, weight is shifted to the involved side (a) and a slow eccentric lowering into hip flexion is performed (b). Weight is shifted toward the uninvolved side (c) for the concentric return (d).

Decline Squat

A decline squat protocol offers greater clinical gains during a rehabilitation program for patellar tendinopathy in athletes who continue to train and play with pain, as based on a study by Young et al. (2005). In this study a decline squat group was compared to a step-up group. The decline group was to perform single leg squats on a 25 degree decline board during a 12 week period. The step group performed single leg squats on a 10 centimeter step, exercising without pain and progressing by increasing speed then load. Similar findings were concluded in the study by Purdam et al. (2004). In this study eccentric training on a 25 degree slant board was performed twice a day at three sets of 15 repetitions for 12 weeks and compared to flat foot eccentrics. The group that trained with the decline board exhibited greater pain reduction and returned more quickly to previous levels of activity. Jonsson and Alfredson (2005) stopped half way through their study due to pain and poor results concluding that eccentric, not concentric, quadriceps training on a decline board seemed to reduce pain in jumper's knee.

Myokinematics

Basic motor function at the knee is commonly described in reference to open kinetic chain (OKC) dynamics with, the quadriceps extending the knee and the hamstrings flexing the knee. Dynamic function is actually more complicated than this simple description, as the knee primarily functions in closed kinetic chain (CKC) scenarios and also depends on muscles that act on the ankle, hip, pelvis and lumbar spine. This requires assessment of the entire kinetic chain from the lumbar spine to the foot including both non-weight bearing muscle testing and more functional weight bearing testing to assess motor performance.

In weight bearing, muscles of the lower limb work in synergy to resist the force of gravity. The hamstrings do not work antagonistic to the quadriceps, but in concert to support the lower limb. Isometric stabilization and eccentric deceleration help to prevent excessive valgus moments from occurring at the hip and knee as well as abnormal pronation at the foot and ankle. Pronation at the foot is classically described as a combination of dorsiflexion, abduction and eversion of the rear foot. The knee motion includes a combination of flexion, tibial internal rotation and abduction. The hip collapses into flexion, internal rotation and adduction. This collapse of the lower limbs is associated with lumbar flexion with bilateral limb support. Unilateral stance involves lumbar flexion, ipsilateral side bending, depending upon the coupled forces of the individual, ipsilateral rotation. With unilateral stance propelling against gravity involves the opposite patterns: 1) ankle plantar flexion, inversion and adduction, 2) knee extension with tibial external rotation and adduction, 3) hip extension, external rotation and abduction and 4) lumbar extension, contralateral side bending and ipsilateral rotation.

The hamstring muscles contribute to concentric knee flexion and hip extension, as well as eccentric deceleration of knee extension and hip flexion in the swing phase of gait. The lateral hamstrings generate an external rotation moment at the tibia, that act to decelerate internal rotation during flexion of the lower limb. Tibial rotation in weight bearing is controlled primarily by passive structures of the tibia on the talus, but motor control is necessary for coordination of the movement pattern and absorption of ground reaction forces. The medial hamstrings generate an internal rotational moment on the tibia, and act to decelerate external rotation of the lower leg. The hamstrings, as a group, contribute to concentric hip extension and adduction as well. With a lateral lunge they act to eccentrically decelerate hip flexion and abduction.

Within the context of closed chain kinetics, the quadriceps can be described as knee extensors that also contribute to external rotation and adduction of the tibia, with opposing eccentric function. The rectus femoris portion of the quadriceps also contributes to eccentric deceleration of hip extension in the open chain swing through phase of gait. The quadriceps, as a group, also helps to dynamically control the axis of motion and tracking of the patellofemoral joint. The vastus medialis obliquus (VMO) imparts a medial vector proximally on the patella that is opposed by the lateral vector of the vastus lateralis and iliotibial band with a direct superior vector imparted by the rectus femoris and vastus lateralis muscles. Short arc extension exercises are commonly used for specific VMO strengthening as this muscle is often solely implicated in terminal knee extension deficiencies (Tria et al. 1992). This is a simplified view of motor performance, as muscles do not function in isolation. Lieb and Perry (1968) demonstrated that terminal knee extension involves the entire quadriceps complex, not only the VMO. Joint angles can play an important role for training of functional movement patterns. Mid range positions of the knee are more effective for muscle recruitment and may be favorable in early training sessions. For instance, VMO activation and contraction intensity has been measured to be greatest at 60 degrees of knee flexion (Tang et al. 2001). This information would suggest terminal knee extension quad sets for VMO training would be best performed closure to 60 degrees of flexion, rather than at full extension.

Quadriceps strengthening is an important component of a rehabilitation program for treatment of knee conditions ranging from jumper's knee to post-operative ligament repair and total knee arthroplasty. The introduction of quadriceps exercises into rehab programs, particularly following ACL repairs, has been tentative secondary to the belief that contraction of the quadriceps would increase the anterior translation of the tibia on the femur and place unwanted tension on the ACL. Tagesson et al. (2005) showed that with the performance of cycling, heel raises and knee extension exercises did not increase tibial translation in healthy knees. Sagittal tibial translation was measured during Lachman's testing, maximal isometric quadriceps contractions, one-legged squats and gait using the CA-4000 electrogoniometer. Electromyographic activity of the vastus medialis, vastus lateralis, gastrocnemius and hamstring muscle group was recorded. The results showed that none of the exercises influenced the amount of translation in healthy individuals. Beutler et al. (2002) concluded from the electromyographic levels of quadriceps activation in their study that one-legged squats and step-ups would be effective in muscle rehabilitation and may also be protective of anterior cruciate ligament grafts.

The force developed from a quadriceps contraction is not able to generate enough tension to disrupt the anterior cruciate ligament and is the primary protector of the ACL from injury during anterior tibial translation (Aune et al. 1997). According to Bodor (2001) the quadriceps is the primary restraint of anterior tibial translation during closed kinetic chain activities such as running, jumping, walking and standing. It creates an inferior/ posteriorly directed force vector during closed chain knee extension and an opposite superior/ anteriorly directed force vector during open chain knee extension. The inferiorly directed vector has an anterior femoral-tibial or posterior tibial-femoral component, which protects the anterior cruciate ligament (ACL) from anterior tibial-femoral shear during weight bearing activities. The quadriceps complex protects the ACL regardless of the activity of the hamstrings and findings by Bodor (2001) suggest that: (1) weak quadriceps are a risk factor for non-contact ACL injuries, (2) strong quadriceps are important for ACL injury prevention and rehabilitation and (3) preservation of quadriceps strength is an important surgical goal.

Emphasis in the past has been placed on the quadriceps/hamstring strength ratio for stabilization training of ACL deficient and reconstructed knees. Studies have shown that although these ratios are important factors, lower limb alignment and dynamic stability also rely on contributions from the soleus, gastrocnemius and popliteus muscles, as well as the hip external rotators and extensors.

The posterior hip muscles must also be considered in CKC knee function. The gluteus maximus contributes to hip extension and the gluteus medius, as a primary abductor of the hip, stabilizes the leg during the unilateral stance phase of gait. The deep hip external rotators help to decelerate pronation of the hip into flexion, internal rotation and adduction. Weakness in these muscles can lead to excessive valgus stress during weight bearing eccentric deceleration. Concentric propulsion into the lower quarter extension pattern may then occur with improper alignment and an incorrect instantaneous axis of motion for any of the joints in the kinetic chain.

The soleus primarily decelerates subtalar pronation while contributing to propulsion into supination. It also acts to directly affect the knee by contributing to eccentric deceleration of the tibia on the step-through phase of gait, although it does not cross the joint. The gastrocnemius has similar affects on the ankle, but also assists the hip external rotators by contributing to eccentric deceleration of the femur into internal rotation. Tightness and/or weakness of the gastrocnemius may also contribute to genu recurvatum of the knee.

The soleus and gastrocnemius also play key roles in reducing ACL strain by eccentrically controlling the anterior motion of the tibia during closed chain knee flexion. The soleus is the primary muscle controlling tibial motion for low-level activity and gait. During higher-level functional activities both muscles are emphasized for control of joint translation. Coactivation of the quadriceps and gastrocnemius muscles is important for knee stability during squatting motions, whereas hamstring muscle co-activation is relatively insignificant (Kvist and Gillquist 2001). More emphasis should thus be placed on training the gastrocnemius muscle due to its contribution in providing functional stability to the ACL-deficient knee (Lass et al. 1991).

Nyland et al. (2005) described the popliteus as a dynamic guidance system for monitoring and controlling subtle transverse and frontal plane knee joint movements. This included guiding anterior, posterior and lateral meniscal translations, unlocking and internally rotating the knee joint during flexion initiation, assisting with 3-dimensional dynamic lower extremity postural stability during single-leg stance, preventing forward femoral dislocation on the tibia during flexed-knee stance, and providing for postural equilibrium adjustments during standing.

The majority, if not all of these functions, are most important during mid-range knee flexion when the supportive structures are the least able to tolerate stress. The popliteus has attachments close to the borders of both collateral ligaments providing the potential for instantaneous kinesthetic feedback of tibial movement. It also acts in synergy with the hip musculature for dynamic control of femoral internal rotation and adduction, as well as subtalar control of tibial abduction-external rotation or adduction-internal rotation (Nyland et al. 2005). Afferent feedback from this muscle may help to prevent athletic knee joint injuries and assist the quadriceps femoris, hamstrings and gastrocnemius in providing stability to the knee within the sagittal plane. Muscle spasm, altered timing or inhibition of the popliteus may contribute to an alteration in normal lower quarter mechanics.

Abnormal Myokinematics

Abnormal motor patterns result from a myriad of causes and may be either a primary or secondary phenomenon. If motor dysfunction stems from local joint or nerve pathology, then the latter impairments are addressed prior to muscle performance training. Restoration of proper joint mechanics can assist in the normalization of afferent mechanoreceptor feedback improving motor performance. These motor patterns should be retested after resolving more acute impairments, associated with the primary pathology, to assess their level of contribution to the existing abnormal motor dysfunction.

Both down training of abnormally facilitated muscle and uptraining of inhibited muscle may be necessary to normalize dynamic function. Excessive strain on compensatory muscles may lead to pathologies such as tendinopathy. The tendinopathy may be the initial painful experience, but it is actually a secondary impairment to the primary motor weakness. For example, an Achilles tendinopathy may develop with excessive activity of the gastrocnemius-soleus complex when attempting to compensate for weakness in the extensors of the knee and/or hip to eccentrically stabilize knee flexion during weight bearing activity.

Specific motor dysfunction leads to compensatory muscle recruitment patterns that also place excessive stress and strain on passive joint structures. Improper lower limb alignment during functional activities may alter the axis of motion creating abnormal cartilage compression and collagen tension in the ligaments and joint capsules of the lower quarter. Inhibition of the VMO may lead to lateral tracking of the patella and secondary patellofemoral pain syndrome (PFPS). An altered recruitment ratio between the vastus lateralis (VL) and the VMO is commonly referred to as the reason for this abnormal lateral tracking of the patella. In fact, studies have highlighted reduced VMO/VL EMG activity ratios in individuals symptomatic for PFPS when compared with asymptomatic individuals (Souza and Gross 1991, Cemy 1995, Boucher et al. 1992, Miller et al. 1997). Alternatively, Powers (2000) described no significant difference in VL:VMO and VL:VML ratios between subjects with and without patella femoral pain syndrome (PFPS) across knee flexion angles ranging from 45º–0º. The author refutes the theory that symptomatic individuals have reduced VMO motor unit activity resulting in excessive lateral glide of the patella causing PFPS.

These altered mechanics could be related to insufficiency of passive structures. Conlan et al. (1993) found that the medial patellofemoral ligament provided 53 percent of the static restraining force to patellar lateral displacement and an average of 22 percent was offered by the patellomeniscal ligament and associated deep retinacular fibers of the capsule. Also the Q-angle, which describes the direction of force placed on the knee by the quadriceps tendon, causes the patella to track laterally. Increased anatomic valgus at the knee and dysplasias of the trochlea where the groove is shallow laterally can also contribute to lateral displacement of the patella (Grelsamer et al. 2001).

Hip extensor and/or external rotator weakness may lead to excessive internal rotation of the femur creating increased compression of the lateral patella within the lateral trochlear groove. Tibial torsions can have the same affect with internal rotation increasing the medial contact area and external rotation increasing the lateral contact area. Chondromalacia from chronic, abnormal compressive forces placed across the patella can then lead to a decrease in the moment arm for the quadriceps tendon. The quadriceps muscles then have to contract more forcefully to extend the knee and increase contact forces on the patellofemoral joint (Ho and Jaureguito 1994).

Potential for injury to capsules, ligaments and menisci are also increased. Jonsson and Karrholm (1994) measured joint motion on ACL-deficient knees during step training, finding little change in tibial adduction and abduction. But they did note a shift of the femur anteriorly as well as a distal and anterior shift of the rotary axis in sagittal plane motions. The abnormal adduction component of the femur during weight bearing training, as seen with a valgus moment at the knee, may not be as significant as deficient external rotation support at the hip. Hamstring weakness is also associated with poor outcome in ACL-deficient knees (Tsepis et al. 2004). The clinical emphasis is not to identify one specific muscle that is weak, but to assess the entire lower limb for motor pattern dysfunction. Individual muscles may be emphasized in a specific exercise design, but are facilitated in synergy with the supporting muscles.

Muscle weakness or inhibition associated with joint damage can be highly selective. Knee disorders commonly produce quadriceps atrophy with limited affect on the hamstrings predisposing the joint to a position of flexion (Young et al. 1987). This phenomenon can be attributed to altered mechanoreceptor activity selectively inhibiting the quadriceps muscles. By initially focusing on reducing pain and effusion to normalize mechanoreceptor function, the recruitment patterns of the quadriceps can be restored. This approach may prove to be more effective than simply stretching tissue and potentially facilitating a flexion pattern of recruitment.

Tightness of the iliotibial band (ITB) is often associated with knee pathology, including lateral tracking of the patella. Classically this dysfunction is described as inhibition of the gluteus medius with compensatory over activity of the tensor fascia lata and iliacus muscles leading to ITB tightness. Secondary symptoms may include hip bursitis or lateral knee pain. Iliotibial band syndrome (ITBS) may not really exist as a freestanding pathology, but rather as a symptom within a cascade of mechanical breakdowns throughout the lower limb in closed chain activities that frequently correlates with gluteus medius weakness. EMG studies have shown that both athletic and arthritic hips with symptoms of ITBS usually have poor gluteus medius facilitation and an inability to isolate gluteus medius recruitment from the tensor facia lata (TFL) in standing (Fredericson 2000, Kasman 2002). In long distance runners with ITBS, symptom improvement and return to preinjury training correlates with increased hip abductor strength (Fredericson 2000). Rather than a stretching program for the ITB, emphasis is placed on uptraining of the gluteus medius and down training of the TFL/ITB.

The Manual Therapy Lesion concept attempts to provide a pathological model that incorporates joint biomechanics, myokinematics and neurophysiology with tissue pathology (Grimsby 1988). The basic concept is that damage to mechanoreceptors from primary tissue pathology leads to tonic muscle inhibition, altered joint motion and further abnormal tissue strain. The tissue and receptor trauma is not exclusive to the primary joint of dysfunction as Bullock-Saxton et al. (1994) demonstrated that subjects sustaining ankle sprains, with associated collagen and receptor damage, had reduced hip extension strength. Lower lumbar spine pathology can lead to inhibition of hip muscles, contributing to abnormal stress on the entire lower quarter. Hart et al. (2006) identified quadriceps weakness after fatiguing exercises to the lumbar spine paraspinals that was even more significant in subjects with a history of back pain.

In more chronic states of pathology, motor dysfunction is not so much associated with reduced feedback from peripheral receptors, as it is related to altered cortical feedforward control of specific motor patterns. Feedforward mechanisms have been identified in tests where motor performance requires such a fast response that there is not sufficient time for an afferent feedback loop to influence motor performance. The drop jump test suggests a centrally pre-programmed activity and the associated elastic behavior of the series elastic component in the knee extensor muscle, in conjunction with the muscle contractile properties, playing a major role in regulating motor performance (Horita et al. 2002). The implications of the manual therapy lesion concept requiring assessment and treatment from the lumbar spine to the foot for all lower limb patients goes beyond the scope of this text.

Alterations in afferent input affecting motor patterns may occur prior to tissue trauma, as a result of muscle fatigue or edema. Tissue injury may then be a secondary consequence of the altered movement patterns, as described in the manual therapy lesion concept. Joint position sense is believed by some authors to be primarily signaled by the muscle spindle (McCloskey 1978, Gandevia et al. 1995, Proske et al. 2000, Forestier et al. 2002, Proske 2005). Local muscular fatigue modifies the peripheral proprioceptive system by increasing the threshold for muscle spindle discharge and affecting alpha– gamma motor neuron co-activation (Pedersen et al. 1999). When local muscles are fatigued metabolic products of muscular contraction, including bradykinin, arachidonic acid, prostaglandin E2, potassium and lactic acid activate nociceptors. These inflammatory substances and metabolites have a direct impact on the discharge pattern of muscle spindles (Pedersen et al. 1999).

Simple muscle fatigue can thus play a significant role in the breakdown of normal myokinematics. Tonic muscles, responsible for arthrokinematic control, can be more significantly affected in cases of pathology due to inhibition or excessive fatigue from pain and muscle guarding. Previous studies in the lower extremity of healthy young adults have shown that exercise-induced local muscle fatigue adversely alters joint position sense, impairing neuromuscular control (Skinner et al. 1986, Marks 1994, Lattanzio et al. 1997). Other studies of the knee (Marks and Quinney 1993), elbow (Sharpe and Miles 1993) and shoulder joint (Pedersen et al. 1999) have revealed no such effects.

Basic Stages of the Manual Therapy Lesion (Grimsby 1988)

1. Collagen/tissue Trauma:

Acute injury, surgery, degenerative joint disease, repetitive strain, postural strain, hypo/hypermobile joint. There may be an acute painful beginning to a pathological state, or a slow progression with an insidious onset related to gradual degeneration.

2. Receptor Damage:

Structural damage to receptor, disruption in a neural pathway and/or imbedded in non-mobile capsules. Type I receptors are more easily damaged than type II, as they are located in more superficial layers of the joint capsule. Simple muscle fatigue also alters neuromuscular spindle feedback impairing normal motor recruitment patterns.

3. Reduced Muscle Fiber Recruitment:

Reduction in afferent signals reduces the normal feedback mechanisms for proprioception and motor facilitation. Alteration in central processing of afferent signals from cutaneous, articular and muscle/tendon receptors can lead to changes in feedforward mechanisms to alpha motor neurons (extrafusal muscle fibers) and gamma motor neurons (Intrafusal muscle fibers). The end result includes altered reflex responses for proprioception and kinesthesia, motor weakness, motor delay and/or poor timing. Central spinal segmental dysfunction, both with pain and without, alters central programming for distal performance.

4. Tonic Fiber Atrophy:

Reduced motor recruitment eventually leads to atrophy over time. Tonic muscle fiber is primarily affected from type I mechanoreceptor loss and inhibition from the type IV mechanoreceptor (pain) system. Phasic muscle fiber atrophy occurs with more significant tissue trauma that affects type II mechanoreceptor function, and leads to an overall reduction in higher level activity.

5. Reduced Anti-Gravity Stability:

Reduced recruitment of the tonic system results in loss of dynamic arthrokinematic control of joint motion, static postural stability/alignment and central balance mechanisms. Central stability at the lumbar spine and pelvis is reduced, which may alter the static and dynamic alignment of the hip, knee and ankle.

6. Motion Around Non-physiologic Axis:

Loss of dynamic control increases the neutral zone of function for the joint, abnormally altering the instantaneous axis of motion that can affect the entire lower quarter biomechanical chain. Reduced tonic function creates an abnormal relationship between tonic (arthrokinematic) and phasic (osteokinematic) muscles causing compensatory motor recruitment patterns and abnormal mechanics.

7. Trauma/Acute Locking/Degeneration

Altered axis of motion leads to abnormal tissue stress/strain, resulting in further tissue and receptor damage. Hypermobile joints are prone to acute locking and joint degenerative changes occur over time.

8. Pain/Guarding/Fear-anxiety of movement

The type IV mechanoreceptor responds to tissue trauma with nociceptive signals and tonic reflexogenic muscle guarding. Psychological influences of pain may lead to altered motor patterns and reduced effort. The overall effect is reduced tissue tolerance, abnormal joint/tissue loading, reduced afferent feedback, altered feedforward efferent drive, reduced motor recruitment and function.

Exercise-induced local muscle fatigue has been shown to alter knee joint position sense in older adults (Riberior et al. 2007). Early training focuses on coordination and endurance training of the tonic system attempting to normalize movement patterns and reduce the effects of fatigue. Contributions of the Golgi tendon organ (GTO) to modulations in motor performance are not fully understood. Proske and Gregory (2002) found that the GTO signals a rise in tension only when the motor unit is contracting with direct action on the GTO. It did not fire with passive tension placed on the tendon. thorough understanding of the GTO mechanism of function may not be completely clear, but enhanced motor performances following exercise is due to both improved mechanical properties of the muscles and better kinesthetic sensibility (Bouet and Gahery 2000). Closed kinetic chain training as well as knee extension exercises in supine have both been shown to improve knee proprioception (Lin et al. 2007).

The basic approach for treatment within the manual therapy lesion concept is to initially focus on the primary tissue in lesion with the optimal stimulus for repair. Repetitive motion for repair results in improved tissue tolerance, as well as reductions in pain and abnormal muscle guarding patterns. Motor recruitment patterns are improved through joint capsule mobilizations to increase collagen mobility and allow for normal receptor feedback. Intracapsular edema and muscle guarding are resolved by increasing local vascularity with repetitive motion that also helps to normalize neuromuscular spindle feedback. Clinically it is impossible to directly assess the level of receptor damage or recovery.

These receptors may take months to repair, or may never fully recover, depending upon the level of damage. Other receptors in the system must be sensitized through training to replace the afferent information to the joint system. The muscle spindle then becomes the primary focus through direct training of tonic muscles to coordinate joint motion around a physiological axis. As noted above, these receptors provide joint position sense and contribute to alpha motor neuron firing for motor recruitment. Other receptors that can contribute to motor facilitation during early coordination training include cutaneous receptors (taping/bracing), adjacent mechanoreceptors in other joints of the kinetic chain (repetitive motion), receptors in the vestibular system (weight bearing balance training), receptors of the ocular system (watching motion or EMG computer screen) and auditory receptors (EMG signals turned to auditory feedback).

Emphasis on the muscle spindle is first addressed by concentrically training the tonic muscles associated with arthrokinematic control of joint motion. Resistance below 50% of 1RM allows for high repetitions to train coordination, avoid fatigue and compensatory movement patterns, as well as improve circulation to resolve muscle guarding. Increasing resistance to 60% of 1RM will serve to improve endurance qualities by stimulating tissue capillarization. Isometric work is then used to help stabilize joint motion in the newly gained range for hypomobilities or toward the range of pathology or instability for hypermobilities. Isometric work allows for training at greater resistances than with concentric work, which helps to efficiently increase muscle spindle sensitivity. Eccentric motion is then performed toward, but not into the pathological range to further increase muscle spindle sensitivity with dynamic training.

The use of cutaneous receptors to augment afferent feedback from functioning receptors, and to replace lost signals from damaged receptors, can be achieved through the use of taping and bracing. A study testing the ability of the knee to replicate joint angles after application of a neoprene sleeve (sleeve effect) was found to be significantly less during the supine closed kinetic chain test (0.3 degrees, +/- 1.4 degrees) than during the sitting open kinetic chain test (Birmingham et al. 1998). Sleeve effects were small in this study, particularly during the supine closed kinetic chain test, but 72 percent of subjects felt that the sleeve improved their test performance. Taping techniques to the skin have also been used to improve knee proprioception. Callaghan et al. (2007) studied the effect of taping in joint position sense, active angle reproduction and passive angle reproduction on patients with PFPS. As a whole, taping did not improve these three variables, but a subgroup of those with significant deficits in proprioception did show significant improvements. Similar results were found in ankle taping, as only those subjects with deficits in proprioception demonstrated improvement (Callaghan et al. 2002). Taping and bracing can be used early to improve proprioception, but the patient should be gradually weaned from dependency on cutaneous input for normal motor function. Clark et al. (2000) did demonstrate that exercise, not taping, is responsible for earlier discharge from therapy in patients with anterior knee pain.

Many techniques can be used to increase the firing pattern of the quadriceps. One of the most common is to use the afferent receptors of the skin to facilitate quadriceps activity. There are many uncontrolled studies stating that the McConnell taping regimen has given 92–96 percent pain reduction after five to eight sessions (McConnell 1986) This technique calls for an immediate pain reduction of 50 percent with the corrective taping in order for its use to be deemed appropriate. Ebourne and Bannister (1996) compared McConnell taping with isometric quadriceps training, and found them to be equally effective at controlling patellofemoral pain. The Cochrane Review (2002) for patellofemoral pain found moderate evidence for McConnell taping and strong evidence for therapeutic exercise in managing patients with patellofemoral pain syndrome.

Vascularity / Circulation

Dosage for improving vascularity and circulation requires low resistance, high repetition dynamic motion. Isometric training is avoided in this stage, as it creates intramuscular pressures that lead to circulatorystasis within the particular muscle(s) involved. Pain is also avoided with training to prevent onset of muscle guarding that can lead to abnormal movement patterns. In the presence of significant tonic fiber atrophy resistance set at <40% of 1RM is enough to facilitate strength gains, prevent quick fatigue and compensatory motor recruitment patterns, as well as minimize intramuscular pressures that reduce local circulation at higher RM percentages. Emphasis on local muscle circulation and endurance is most efficiently achieved at around 60% of 1RM. Specificity of repetitions, resistance and rest breaks ensures enough energy is left in the system for protein synthesis of repairing tissues. Performing a high number of repetitions also provides both a mechanical as well as a neurological influence for tissue repair and pain inhibition.

Coordination / Motor Learning

Coordination is the contraction and relaxation of muscles in a specific sequence and magnitude that produces movement around a physiological axis of motion for proper joint function. It is a function of neuromuscular control or neurological adaptation that improves with repetitive training. Coordination relates to the execution of a particular skill, while motor learning refers to the process of acquisition of the skill. Coordination training with motor learning techniques should emphasize natural patterned movements when possible, rather than attempting to isolate a specific muscle. For example, the vastus medialis obliquus is often exercised in isolation to help normalize patellar mechanics and knee function. Attempting to train in this manner is both nonfunctional, and inconsistent with the neurological drive that governs patterned movements. Cerny (1995) showed that the VMO could not be isolated or preferentially recruited with various open kinetic chain or closed kinetic chain exercises, which was also confirmed by Laprade et al. (1998) and Vaatainen et al. (1995). Exercises attempting to isolate the VMO are actually training the knee extensor pattern which includes all the muscles of the quadriceps complex. Although it cannot be selectively isolated, the VMO can be emphasized with specific exercises. Muscle isolation programs may have some benefit in improving kinesthetic function, however they lack functional relevance according to Cowan et al. (2002). As previously described, normal myokinematics for the knee are not achieved by a single muscle, but involves muscles of the entire lower limb, including the lumbar spine, pelvis, hip, knee and ankle. Initial exercise programs should progress to more functional designs as soon as tolerated by the injured tissue. Weight bearing exercises have been shown to be more effective in rehabilitation of injured limbs, and is partly attributed to the greater recruitment of the afferent neural receptors in the inert and contractile tissues of the knee. Exercise design should include groupings of muscles working in synergy to complete a movement pattern, as this is more consistent with normal motor patterns.

The literature contains studies outlining various approaches of motor recruitment for coordination training. Many of these studies emphasize afferent mechanoreceptor function in the joint structures and their influence on normal motor recruitment. Freeman and Wyke (1967) stated that the joint receptors likely contribute to the "coordination of muscle tone in posture and movement" via the gamma motor neuronloop. Sojka et al. (1989) concluded that the cruciate ligaments play a sensory role acting on the central nervous system via reflex actions of the gamma motor spindle system that regulates muscular stiffness of the knee joint, and improve dynamic stability. Elmqvist et al. (1988) suggests that the reason for decreased maximum and total knee extensor performance is due to a change in knee joint receptor afferent inflow. Passive manual techniques, as well as mobilization exercises, to improve capsular mobility will assist in normalizing afferent input for proper motor recruitment. Resolution of spinal segmental dysfunction or facilitation can assist in reducing pain levels at the knee and improving lower quarter motor recruitment. Similar improvements in afferent responses and lower quarter patterns will also be noted with restoration of normal mobility in the joints of the hip, ankle and foot. Mobilizing exercises for these regions are covered in their respective chapters.

Quadriceps training can and should be incorporated early in rehabilitation programs, including acute stage ACL reconstruction (Morrissey et al. 2004). Weight bearing exercises should be added as soon as indicated for early synchronization of the vasti and hamstring muscles to facilitate co-contraction and provide stability to the knee.

Open Kinetic Chain and Close Kinetic Chain Exercise

The lower quarter kinetic chain is comprised of the hip, knee and ankle joints (Palmitier et al. 1991). Closed kinetic chain (CKC) exercise refers to having the foot fixed on a surface with free movement of the proximal segments. They employ specificity of training principles through recruitment of all hip, knee and ankle extensors in synchrony. The spine is typically left out of the discussion in relation to CKC training, but is an important part of the kinetic chain. Lumbopelvic stability will affect the alignment of the pelvis on the fixed femur, and have direct implications on function of the lower limb joints. Hip weakness, commonly associated with knee pathology, may have a lumbar origin that requires direct intervention in conjunction with a knee rehabilitation program.

For testing performance, CKC movements are unable to effectively evaluate specific muscle function and instead are more geared toward assessing functional performance. Conversely, open kinetic chain (OKC) testing is better able to isolate and detect specific muscle impairment. Augustsson and Thomee (2000) found a significant correlation between tests of functional performance and closed and open kinetic chain tests of muscular strength. Petschnig et al. (1998) found moderately strong correlation between isokinetic quadriceps strength tests and four different functional performance tests in healthy subjects and post-ACL repair patients. Blackburn and Morrissey (1998) found weak correlation between open kinetic chain knee extensor strength and the low vertical and standing long jump tests.

Östenberg et al. (1998) recommended not using functional performance and isokinetic testing interchangeably. For clinical purposes, OKC testing may assist in identifying specific muscles to emphasize during training that may even translate to CKC training. Functional CKC testing may serve as an indicator for when to progress with functional exercises and return to athletic activities. The effectiveness of training or rehabilitation programs should be based on changes in performance rather than tests of muscular function (Murphy and Wilson 1997). Carter and Edinger (1999) found only half of the competitive athletes in their study had achieved at least 80 percent of their leg strength six months after ACL surgery. Despite these strength deficits, it is not uncommon to allow patients to return to full activities at six months, with some surgeons advocating return to sports as early as four months after ACL reconstruction.

Both OKC and CKC exercises are necessary when rehabilitating the knee. CKC training includes unloading exercises in which body weight is reduced to allow for early return to simulated functional weight bearing tasks. Rehabilitation programs for the injured or post surgical knee typically begin with non-weight bearing OKC exercises progressing to CKC exercises. OKC exercises have been shown to be safe and effective in increasing tissue response to exercise (Perry et al. 2005) and provides a safe and low level stress necessary for restoring joint motion, resolving edema, inhibiting pain and normalizing muscle tone. CKC exercises are increasingly added as an integral part of accelerated rehabilitation programs (Kibler 2000). In subacute or chronic conditions, it may be appropriate to start with a combination of open and closed chain exercises. The end stage progression remains focused on returning to functional weight bearing activities, but a gradual transition from OKC to CKC exercises is made with improving tissue tolerance.

Even though most knee function occurs in closed chain positions, several studies have demonstrated no significant increase in functional gains during the early stages of rehabilitation between open and closed chain exercises (Perry et al. 2005, Witvrouw et al. 2004, Hooper et al. 2002, Ross et al. 2001, Cohen et al. 2001, Hooper et al. 2001, Morrisey et al. 2000). Functional gains may be minimal in the initial stages of training, but early modified weight bearing training using unloading devices allows for coordination training of the entire lower limb, increased compression loading for cartilage repair and reduced time for return to full weight bearing training. For post-surgical patients the clinician should be aware of any weight bearing restrictions from the referring surgeon and communicate with the physician to ensure a safe, but aggressive treatment design.

CKC training not only provides a more functional approach for coordination training, but also benefits early motor facilitation of inhibited muscles through cross-education training. Recruiting muscles of the uninvolved limb will have a neurological cross-education effect and improve facilitation of the same muscles on the involved limb. Seger and Thorstensson (2005) demonstrated a specific cross-education effect with both concentric and eccentric training. Work type and velocity specific increases in strength occur in the contralateral, untrained leg and are accompanied by a specific increase in eccentric to concentric EMG ratio after eccentric training.

Tissue Stress / Strain with OKC and CKC

Support in the literature exists for both OKC and CKC exercises for all knee pathology, including ACL deficient knees and those with patellofemoral pain syndrome (PFPS). Proper exercise design and dosage will determine the degree of safety and quality of the functional outcomes. Both OKC and CKC exercises can be modified and implemented for quadriceps training after ACL reconstruction without causing excessive ACL strain or patellofemoral joint stress (Fitzgerald 1997, Ross et al. 2001).

Co-contraction and synergistic muscle recruitment during closed chain activities may be safer in the presence of knee laxity, as less load is placed on the capsuloligamentous structures. OKC exercises are commonly restricted to seated knee extension for quadriceps training and prone hamstrings curls. But OKC training can also includes sling exercises that allow for essentially zero resistance flexion and extension to stimulate tissue repair, reduce pain and resolve edema. Some authors report an increase in co-contraction of the hamstrings and quadriceps with CKC exercises, believed to increase dynamic stability to ACL deficient knees (Beynnon and Fleming 1998, Fleming et al. 2001b). Other authors have noted little hamstring recruitment during squatting motions, but rather a synergistic action between the quadriceps and gastrocnemius muscles (Kvis and Gillquist 2001).

Benefits of CKC Exercises

- *Reduce anterior tibial forces on the tibia relative to the femur (Graham et al. 1993, Bynum et al. 1995, Beynnon and Fleming 1998, Escamilla et al. 1998, Fleming et al. 2001, Kvist and Gillquist 2001, Fleming et al. 2003, Heijne et al. 2004).*
- *Increase compressive forces of the tibiofemoral joint (Bynum et al. 1995, Escamilla et al. 1998, Fleming et al. 2001b, Fleming et al. 2003).*
- *Increase co-contraction of the quadriceps and hamstrings (Beynnon and Fleming 1998, Fleming et al. 2001b).*
- *More consistent with functional activities than OKC exercises (Escamilla et al. 1998, Hooper et al. 2001).*
- *Reduces the incidence of patellofemoral complications (Bynum et al. 1995, Escamilla et al. 1998, Hooper et al. 2001).*

Anterior tibial translation is associated with open chain knee extension exercise, but not with closed chain weight bearing exercise (Yack et al. 1993). In fact, anterior tibial translation progressively increases as loading increases with open chain exercises (Yack et al. 1994, Beynnon et al. 1997). Excessive compression to the patellofemoral joint has also been associated with OKC isokinetic training (Kaufman et al. 1991). In a review article, Fleming (2005) makes the case that the strain to the ACL ligament during the Lochman's test is greater than or equal to many OKC exercises.

Morrisey et al. (2000) compared ACL laxity after six weeks of either OKC or CKC exercises, finding a nine percent increase in laxity for the OKC group. In a follow up study this group of authors found no difference in laxity with OKC or CKC exercises, but exercise was not initiated until eight weeks after surgery (Perry et al. 2005). Minimal ACL stress may be noted with a particular test or exercise, but repetitive training over a prolonged period with progressive resistance can magnify the tissue strain. Kirkley et al. (2001) found that repetitive loading exercises contributed to an increase in anterior translation in normal, ACL-deficient and ACL-reconstructed knees. A notable finding was that the ACL reconstructed knee using a bone-tendon-bone autograft demonstrated a significantly smaller increase in translation. Reduced laxity in the reconstructed knee may be related to a stiffer central third of the patellar tendon graft versus a normal ACL, increased stiffness from early degenerative arthritis or a minimal generalized fibrosis that may occur following major reconstructive surgery.

Choosing and designing exercises that minimize the potential for anterior shear forces, and ensuring proper technique during the exercise, can minimize the cumulative effect on the ACL deficient or repaired knee during rehabilitation. Restoration in the integrity of the ligamentous structures is difficult to predict. Nawata et al. (1999) did measure about a one third increase in ACL laxity in normal knees after 20 minutes of running at seven km/hr., but integrity was restored within one hour after running. Sumen et al. (1999) assessed post-exercise translation in ACL reconstructed knees with a hamstring graft 15 months after surgery, finding increased laxity in the control group, but no increase in the reconstructed group.

Weight bearing motion in the pathological knee typically does not demonstrate abnormal adduction or abduction of the tibia on a fixed foot, but rather abnormal femoral motion. Jonsson and Karrholm (1994) measured joint motion on ACL deficient knees during step training and found little change in tibial adduction/adduction, but noted an anterior shift in the femur (PCL strain) and a distal and anterior shift of the joint axis for sagittal plane rotation. The abnormal adduction component of the femur during weight bearing training, as seen with valgus moments at the knee, may not be as significant as the deficient external rotation support at the hip.

It has been widely accepted that female athletes sustain disproportionately more ACL injuries than do male athletes who compete in similar sports (Arendt and Dick 1995, Ferretti et al. 1992, Gray

ACL Strain During Commonly Prescribed Rehabilitation Tests and Exercises

Rehabilitation Exercise/Test	*Peak Strain Percent (mean standard error)*
Isometric quadriceps contraction at 15 degrees (30 Nm extension torque) (OKC)	4.4 (0.6)
Squats with sport cord (CKC)	4.0 (0.6)
Active flexion–extension of the knee with 45 N weight boot (OKC)	3.8 (0.5)
Lachman test (150 N of anterior shear load; 30-degree flexion)	3.7 (0.8)
Squats (CKC)	3.6 (0.5)
Active flexion–extension (no weight boot) of the knee (OKC)	2.8 (0.8)
Simultaneous quadriceps and hamstring contraction at 15 degrees (OKC)	2.8 (0.9)
Isometric quadriceps contraction at 30 degrees (30 Nm extension torque) (OKC)	2.7 (0.5)
Stair climbing (CKC)	2.7 (1.2)
Leg press at 20-degree flexion (40% body weight) (CKC)	2.1 (0.5)
Anterior Lunge (CKC)	1.9 (0.5)
Stationary cycling (CKC)	1.7 (0.7)
Isometric hamstring contraction at 15 degrees (to 10 Nm flexion torque) (OKC)	0.6 (0.9)
Simultaneous quadriceps and hamstring contraction at 30 degrees (OKC)	0.4 (0.5)
Isometric quadriceps contraction at 60 degrees (30 Nm extension torque) (OKC)	0.0
Isometric quadriceps contraction at 90 degrees (30 Nm extension torque) (OKC)	0.0
Simultaneous quadriceps and hamstring contraction at 60 degrees, 90 degrees (OKC)	0.0
Isometric hamstring contraction at 30, 60 and 90 degrees (10 Nm flexion torque) (OKC)	0.0
** The failure strains of the normal ACL are approximately 15% (Butler et al. 1992).	

Table 2.1: Peak ACL strains comparison measured during commonly prescribed rehabilitation tests and exercises (adapted from Fleming et al. 2005).

et al. 1985, Hutchinson and Ireland 1995, Nielsen AB, Yde 1991, Zelisko et al. 1982). It has also been widely thought that material properties of the ACL are affected by the normal fluctuations of the hormones associated with the menstrual cycle, possibly leading to increased risk of injury (Liu et al. 1996, Liu et al. 1997, Myklebust et al. 2003, Slauterbeck et al. 1999, Slauterbeck et al. 2003, Wojtys et al. 2002). Contrary to these previous authors, Pollard et al. (2006) did find an increase in ACL laxity in females, as opposed to males, after training, but the stage of menstruation or estrogen levels did not correlate with increased laxity. Belanger et al. (2004) also demonstrated that ACL laxity is not significantly different during the follicular, ovulatory and luteal phases of the menstrual cycle. The authors did list intrinsic factors that influence the anatomy and physiology of the knee directly for injury risk, including generalized ligamentous laxity, ACL size, femoral notch dimensions, limb alignment and ligamentous physiology. Extrinsic factors were considered more remote, but nevertheless influence the development of loads in the joint. These included the level of strength and conditioning, body mechanics, neuromuscular performance and footwear.

During weight bearing squats the tibial plateau tips forward, sloping anterior. Body weight causes an anterior translation of the femur on a fixed tibia, placing tension on the PCL and unloading the ACL. Lutz et al. (1993) demonstrated significantly greater compression forces and increased muscular co-contraction at the same angles at which the open-kinetic-chain exercises produced maximum shear forces and minimal muscular co-contraction.

Open chain extension also involves anterior movement of the tibia on a fixed femur, placing excessive strain on the ACL. Jenkins et al. (1997) found an increase in anterior tibial displacement in open chain, compared to closed chain isometric work at 30 degrees and 60 degrees. This issue is most relevant to the ACL deficient, ruptured or repaired knee, and is dependent upon the amount of distal resistance.

A generalized ACL rehabilitation program has been previously described in detail (DeMaio et al. 1992a, DeMaio et al. 1992b, Mangine et al. 1992, Noyes and Barber-Westin 1997) and divided into four phases consistent with the four stages in this chapter (Barber-Westin and Noyes 1993, Barber-Westin et al. 1999). This program was assessed for long term ACL laxity, and was found that the rehabilitation program was not itself injurious and resulted in an acceptable failure rate of 5 percent (Barber-Westin et al. 1999).

- ***Phase 1:** The assisted ambulatory phase (approximately lasting up to the fourth to eighth week after surgery). Patient using crutch or cane support with exercises for range of motion: straight leg raises (extension, flexion, abduction and adduction), quadriceps muscle isometrics, electrical muscle stimulation, and closed-chain exercises (minisquats, toe raises).*
- ***Phase 2:** The early strength training phase (from 4th to 8th postoperative week to the 12th to 16th). Exercises incorporate balance, proprioception and gait-training.*
- ***Phase 3:** The intensive strength training phase (varies for each patient lasting from between the 12th to16th postoperative week to between the 24th and 52nd). More strenuous training consists of progressive resistive exercises, swimming, bicycling, ski machines, stair climbing machines and running programs.*
- ***Phase 4:** The return to sports phase (begins after successful completion of phase 3). Formal rehabilitation is discontinued with return to sport.*

Studies of knee laxity after heavy resistance squat training are consistent with increased PCL rather than ACL tension due to anterior femoral translation during closed chain activities. Increased PCL tension throughout the range has been demonstrated for a closed chain squat (Escamilla 2001), while open chain knee extension increased ACL tension from 40 degrees to full extension (Wilk et al. 1996). Stuart et al. (1996) had similar findings of a net posterior tibial shear force (PCL strain) with power squats, front squats and lunges. Assessment of both normal subjects and post ACL reconstruction patients, support safer and more beneficial outcomes with a closed chain emphasis when compared to open chain activities. Chandler et al. (1989) found no increased laxity with power lifters and weight trainers after eight weeks of heavy resistance squat training. Panariello et al. (1994) found no increase in laxity with squat training in football players. Yack et al. (1993) demonstrated significantly less stress to the anterior cruciate ligament using closed chain parallel squats, compared to the relative anterior tibial displacement during knee extension. Heijne et al. (2004) showed that ACL strain produced during step-up, step-down, lunge and one-legged sit to stand exercises was similar to that produced during other rehabilitation exercises (i.e., squatting, active extension of the knee) previously tested.

Athletic activity has been attributed to increased laxity of the knee. Steiner et al. (1986) suggests that repetitive physiologic stresses at a high strain rate produce significant ligamentous laxity, where this is avoided with fewer repetitions at higher loads. Increased ligamentous laxity was found in college runners and basketball players up to 90 minutes after participation, but not in weight trainers performing squats. Kvist et al. (2006) measured increased laxity in both male and female volleyball players, comparing measurements before and after performance, while swimmers were found to have no increased laxity. Grana and Muse (1988) measured a significant increase in ACL laxity in controls and in ACL deficient knees after 20 minutes on a bicycle ergometer, though Belanger et al. (2004) found no significant ACL laxity in a 10-week cycling program in healthy female college athletes, suggesting over time these temporary changes in laxity in training may not of concern.

Muscle Performance with OKC/CKC

Muscle fatigue is a relevant issue pertaining to excessive ACL strain and injury. Increased injury rates during the later portion of games in a variety of sports have been identified including rugby (Gabbett 2000 and 2002), in the arms of baseball pitchers (Lyman et al. 2001), soccer (Rahnama et al. 2002) and with rapid run and stop testing (Nyland et al. 1994). More specifically Chappell et al. (2005) compared anterior tibial shear with fatigue related to stop-jump tasks. Subjects performed vertical, anterior and posterior stop-jumps with assessment of jump height and anterior tibial motion. Fatigue significantly increased the peak anterior shear force on the proximal tibia for both male and female subjects. Fatigue also increased the valgus moment at the knee in both male and female subjects. Russell et al. (2006) also found that women land from a single leg drop jump in more knee valgus than men, both before and upon impact. The authors concluded that excessive valgus knee angles displayed in women might be one explanation for the sex disparity in anterior cruciate ligament injury. Adequate rest breaks during early training to avoid fatigue can assist in reducing the overall anterior shear forces during rehabilitation, and may be more important in females. Ensuring ample overall training time to reestablish endurance and strength will also reduce fatigue upon returning to sport, and thus reduce risk of re-injury.

The issue of recruitment ratios within the different portions of the quadriceps complex is irrelevant in closed chain training. Synergistic function between the different portions of the quadriceps musculature is necessary for all knees, but may be of particular importance in designing training programs for establishing control of the patellofemoral joint in subjects with patellofemoral pain syndrome (PFPS). The importance of the recruitment ratio between vastus medialis obliquus (VMO) and vastus lateralis (VL) for controlling patellar tracking is debatable. Mellor and Hodges (2005) demonstrated a neuromuscular motor unit synchronization of the VMO and the VL in closed chain exercises, not found with open chain extension. Stensdotter et al. (2003) also found more balanced initial quadriceps activation in closed chain versus open chain exercises. Tang et al. (2001) did find lower VMO/VL ratios of PFPS subjects compared to unimpaired subjects, though not statistically significant, during knee isokinetic CKC exercises. Slightly greater coupling between the VMO and VL has been shown with CKC versus OKC (Mellor et al. 2005). Stensdotter et al. (2003) assessed the onset of EMG activity of the four different muscle portions of the quadriceps with isometric knee extension, findings more simultaneous contractions in CKC than in OKC isometrics. In OKC isometrics the rectus femoris had the earliest EMG onset, and the vastus medialis obliquus was activated last with smaller amplitude. CKC training seems to promote more balanced initial quadriceps activation and may be more efficient for coordination training of quadriceps coupling in the PFPS patient. Knee and trunk positions as well as Q-angle and femoral/tibial torsion must be considered with PFPS patients as all these factors all affect the patellofemoral joint. For instance deeper squats and lunges increase patellofemoral contact forces especially with knee flexion between 60–90 degrees (Huberti et al. 1984, Ho et al. 1994, Grelsamer et al. 2001), but trunk flexion reduces quadriceps tension and joint compression at any given angle (Ohkoshi et al. 1991). Tibial and femoral torsions, as mentioned earlier, affect patellar contact locations and increased Q-angles can magnify loads placed on the lateral aspect of the patellofemoral joint. Also patella alta can reduce contact between the quadriceps tendon and femoral groove, which affects the mechanical ability to relieve pressure on the patellofemoral joint at high flexion angles (Huberti et al. 1984). Therefore it is important to monitor positioning of the entire leg from the hip to the foot, the range of the exercise and trunk position so that forces across the knee are properly managed during rehabilitation.

OKC extension training in normal knees is also associated with a relative shift in the instantaneous axis of motion. Distal resistance on the tibia shifts the axis from the femoral condyles toward the attachment of the quadriceps tendon at the tibial tubercle. This causes the tibia to translate posteriorly during extension, rather than anteriorly, increasing tension on the posterior cruciate ligament (PCL) while relaxing the ACL. Reduced recruitment of the quadriceps and increased facilitation of the hamstrings ensues, with some eccentric participation of the quadriceps (Grimsby 1980). Applying a pulley strap proximal to the tibial insertion of the quadriceps for extension training and starting in a range greater than roughly 30–40 degrees of flexion, will produce a normal anterior tibial translation.

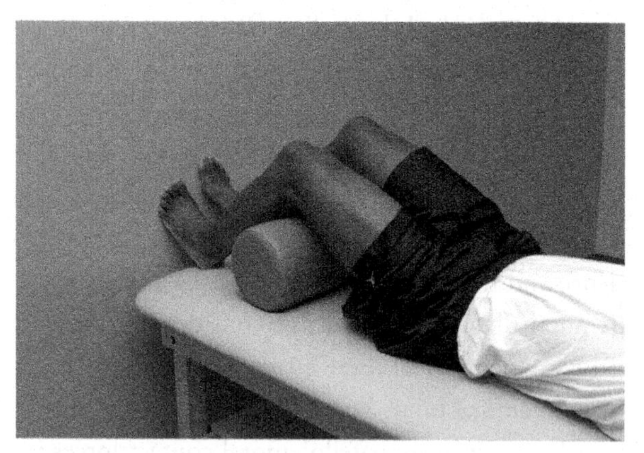

Figure 2.26: Quad set training for patellofemoral pain syndrome. The knee is placed in flexion to ensure the patella is in the grove of the femoral condyles, avoiding full extension. More selective VMO training is noted around 60° of knee flexion (Tang et al. 2001). Pushing the foot into a wall creates a simulated weight bearing environment to engage extensor muscles of the entire lower limb. Having a block on the tibial surface, rather than directly behind the knee, will create a relative anterior glide of the tibia for an extension mobilizing force. This added stress to the joint capsule may increase the amount of recruitment via increased mechanoreceptor firing from tension to the joint capsule.

Early training emphasizes coordination and range of motion with progression to functional tasks. CKC exercises are more appropriate than OKC exercises for restoring coordination and tissue tolerance to weight bearing tasks of the lower limb. Graham et al. (1993) compared range of motion (ROM) and muscle performance during six exercises: unilateral one quarter squats, leg extensions on an N-K table (lifting 25 percent of body weight), lateral step-ups (20.3 cm step) and movements on the Fitter (two cords if body weight was less than 73 kg; three cords if greater than 73 kg), Stairmaster 4000 (manual setting, level seven, steady climb) and slide board. Lateral step-ups, sitting leg extension and the Stairmaster were most effective for ROM training. The slide board and Fitter were associated with greater hamstring recruitment and proprioceptive enhancement, which would be good choices for post ACL construction using a hamstring autograft.

Exercises that are more closely related to a required task will more efficiently improve muscle performance qualities for that task. Blackburn and Morrissey (1998) demonstrated that CKC muscle strengthening was more highly related to jump performance than knee extensor OKC strengthening. Isokinetic quadriceps training for patients with PFPS has shown positive effects on passive position sense, as well as increasing strength and work capacity (Hazneci et al. 2005). But isokinetic training has also shown poor correlation with function hopping tests (Alaca et al. 2002), suggesting the need for inclusion, or an eventual progression, to more functional exercises.

Clinical Comparisons of OKC/CKC

Previous research identified specific traits of OKC and CKC training as it relates to motor recruitment, forces across the joint, tissue stress and isolated training variables. Bynum et al. (1996) compared two rehabilitation protocols for ACL reconstruction, which included both an open chain and closed chain program. The CKC group was found to have lower mean KT-1000 arthrometer side-to-side differences for anterior shear, less patellofemoral pain and more often returned to normal daily activities and sports sooner than expected. Hooper et al. (2001) also compared OKC and CKC programs in patients with ACL reconstruction by measuring walking, stair ascent and stair descent. Seventeen knee variables were studied and only the angle of knee flexion during stair ascent, favoring earlier increases in ROM with the OKC group, was noted as a statistically significant difference between the groups.

Stiene et al. (1996) compared OKC and CKC exercises for patients with patellofemoral dysfunction to controls. Both groups improved in peak torque measures of the quadriceps at different speeds, but only the CKC group demonstrated significant improvement in CKC testing and perceived functional status. Witvrouw et al. (2000) compared an OKC to a CKC five-week program in subjects with patellofemoral pain. No significant difference was noted in muscle characteristics, subjective symptoms and functional performance at the end of the treatment period, and three months later. The CKC group tested better for the triple-jump test, the frequency of locking in the knee joint, clicking sensations in the knee, night pain and pain during isokinetic testing, suggesting that CKC training was little more effective than the OKC program. A five-year follow up again found little statistical significance between the two training groups (Witvrouw et al. 2004).

Figure 2.27a,b: Partial squats using the pull down machine with the gravity assist cage. Body weight is reduced using the weight stack as a counter balance. This will allow the person to perform a coordinated movement pattern with more pain free repetitions.

Adjunct Treatment

Treating mobility issues in the knee, whether hypomobile or hypermobile, requires an assessment of both proximal and distal structures. A basic tenet of stabilizing hypermobile joints is to initially restore mobility to any adjacent joint or tissue restrictions. Mobilization exercises for the foot, hip and lumbopelvic region may be a necessary precursor to functional stabilization training for the knee. A hypomobile knee may be placing excessive stress on surrounding tissues and any accompanying restrictions in adjacent joints of the lower limb need to be identified and addressed in Stage 1. A comprehensive treatment plan consisting of both active and passive interventions must consider the joint and muscle function from the lumbar spine to the ankle, yet a complete description of the biomechanical relationships between these joints goes beyond the scope of this chapter.

The Biomechanical Chain

Relative to a fixed foot on the ground, and a less mobile hip and pelvis, the knee is by nature unstable. Though it primarily acts as a hinge joint with one degree of freedom, accessory motions of the knee include spin and lateral gliding between the tibia and femur. The lateral gliding movements are a part of normal valgus and varus moments at the knee. Weight bearing internal rotation of the tibia occurs with ankle dorsiflexion and knee flexion, as plantarflexion and knee extension are accompanied by tibial external rotation. In open chain movements the screw home mechanism involves an external rotation of the tibia. Inert structures, such as ligaments and joint capsules, assist in guiding motion in the lower limb.

Basic descriptions of biomechanical breakdown include excessive supination creating an abnormal varus force moment at the knee, while excessive pronation would cause a valgus moment and collapse of the medial arch. Addressing the passive inert structures may require an alteration in footwear or an orthotic lift to support the arch. Limited dorsiflexion at the talocrural joint will reduce the amount of internal rotation of the tibia. A compensatory increase in external rotation of the femur can result creating the potential for tissue damage, such as meniscal and cruciate ligament injury. Conversely, reduced hip internal rotation from a restricted capsule may lead to compensations at the knee or foot. Reduced extension of the first metatarsophalangeal joint can lead to excessive external rotation of the entire lower limb transmitting abnormal forces across the medial knee joint. Numerous restrictions from joint capsules, and myofascial structures can all contribute to mechanical changes in mobility and force transmission in the lower quarter.

Not only mobility, but also alignment of the joints in the biomechanical chain can affect motion and motor recruitment. Abnormal movement patterns, compensations and weak muscles may all normalize, or at least improve, when normal biomechanical relationships are restored. Joint restrictions should be addressed first followed by a reassessment of motor function to help direct specific exercises for improved motor performance. Treatment at the hip for reduction of knee pain has been shown to be effective by Cliborne et al. (2004). After just a single session of hip mobilizations, patients experienced increases in knee range of motion as well as decreases in pain complaints, and painful test findings. These improvements may be from inhibition of muscle guarding, improved lower quarter biomechanical function, increased muscle facilitation and/or resolution of referred pain originating from the hip. This study illustrates the need to assess the mechanics and function of all surrounding joints in order to determine their potential effects on the knee.

The Myokinematic Chain

The myokinematic chain refers to synergistic muscle function of the pelvis and lower limb. Functionally all muscles are synergistic working together around the joint axis to properly coordinate motion of the joint system. The brain acts to fire muscles within global functional patterns, rather than in isolation. Tepperman et al. (1986) found that the facilitation of either ankle plantarflexors or dorsiflexors during isometric knee extension significantly increased quadriceps peak torque. Alterations in function of any muscle or muscles in the chain of a movement pattern, can have deleterious effects on function.

Knee dysfunction, including patellofemoral dysfunction, is often blamed on weakness or poor timing of the vastus medialis obliquus (McConnell 1986, Boucher et al.1992), while weak hamstrings have been implicated in anterior tibial shearing with anterior cruciate ligament dysfunction (Tsepis et al. 2004). While these specific muscle tests constitute important findings, evaluating more complex muscle recruitment patterns can serve as better indicators of abnormal function.

The gluteus medius, gluteus maximus and deep external rotators are the main stabilizers at the hip during lower quarter weight bearing activity. Eccentric function of these muscles in weight bearing stabilizes the pelvis on the hip, as well as preventing adduction collapse of the femur relative to the knee. Leetun et al. (2004) found that basketball and track athletes who did not sustain lower extremity injuries had stronger hip abduction and external rotation strength. Impairments such as lateral patellar tracking problems, joint degeneration, patellar tendinopathy or ITBS (Iliotibial Band Syndrome) often present with motor weakness proximally at the hip and less often at the ankle. Predominant proximal weakness at the hip, pelvis and lumbar spine may contribute to unopposed hip adduction, flexion and internal rotation leading to abnormal knee valgus and compression of the patellofemoral joint. Weakness of the gluteus medius is compensated for by excessive tensor fascia lata (TFL) facilitation, placing increased tension through the iliotibial band (ITB). The ITB is then determined to be tight and treated as a primary cause of dysfunction, rather than as a secondary compensation for posterior hip weakness. Stretching the ITB will not improve proximal hip stability, reduce the knee valgus moment or the tissue irritation at the lateral knee.

Proximal dynamic stability includes not only the hip, but the lumbar spine as well. The hip abductors and external rotators along with the ipsilateral spinal muscles all act together to provide weight bearing stabilization to the pelvis on the femur. Lumbar dysfunction resulting in motor inhibition will negatively affect stability of the lower kinetic chain. Hart et al. (2006) measured reduced quadriceps strength after performance of spinal extension exercises to fatigue, in both normal subjects and those with a history of low back pain.

Motor function at the knee may also be influenced by mobilization of adjacent joints in the kinetic chain. Normalizing hip capsular mobility will assist in normalizing motor function of the hip musculature and improving lower quarter dynamic support. Uptraining the gluteal muscles will reduce the abnormal alignment and forces transmitted to the knee joint. Acute low back pain or chronic segmental facilitation, can result in reduced motor recruitment of the hip and knee musculature. Treating the lower lumbar segments may serve to normalize tone, as well as improve facilitation and coordination of these muscles. For example, segmental dysfunctions at L4/5 or L5/S1 can increase the resting tone of the hamstring muscle group leading to increased patellar compression in flexion. Rather than stretching the tight hamstrings, treatment to the lumbar spine may be more effective in restoring hamstring length.

Distal function at the ankle also effects knee and hip motor function. Plantarflexor weakness reduces extension stability of the knee from mid stance to step off and may lead to overuse pathologies of the patellar tendon. Bulluck-Saxton et al. (1994) demonstrated how previous ankle ligamentous and receptor injury decreased the activation of the hip extensors. Passive viscoelastic ligaments and dynamic viscoelastic muscles provide stability to joint systems. The viscoelastic effects of the ligaments are activated and applied strictly upon the geometric and kinematic configuration of the joint moving through its range of motion. The musculature can apply both passive as well as variable dynamic viscoelastic effects and are the primary structures for absorbing and transmitting forces up the biomechanical chain. Ligaments act to primarily guide the joint around its axis of motion and provide afferent feedback for proprioception and recruitment via mechanoreceptor activation.

Just as ligamentous strain (hypermobility) or capsular stiffness (hypomobility) can alter normal mechanics and function of the lower limb, abnormal motor responses result in reduced performance and excessive tissue stress and strain from the lumbar spine to the foot. Preservation of joint stability must be considered as a synergistic function in which bones, joint capsules, ligaments, muscles, tendons and sensory receptors and their spinal and cortical neural projects and connections function in harmony. The role of the musculature in maintaining stability while controlling joint motion is considered to be one of the most important factors for proper joint stability Solomonow and Krogsgaard (2001).

The Neurological Chain

The neurological system plays an obvious role in function. Nerve pathology proximal in the lumbar spine, or associated with peripheral nerve entrapment, will impact motor function and pain levels. Adverse neural tension may even have an impact on gross mobility, and has been suggested as a component of hamstring strain injury (Turl and George 1998). Any dysfunction of the nervous system should take precedence, or be treated in conjunction with, the primary knee pathology. But even with normal nerve conduction and mobility, the impact of the central nervous system and peripheral afferent system on motor function should be addressed. Issues of central balance will be addressed later in this chapter.

The osteopathic lesion, or segmental facilitation, was first described by Korr (1947) as an abnormal afferent bombardment of a segment resulting in an altered efferent response with sensitization of the gamma motor system, as well as increased sympathetic and even glandular output. Facilitation in the L3 lumbar segment can increase pain sensitivity and alter normal motor patterns. Addressing segmental sensitization, when present, prior to direct training to the knee may reduce pain and improve motor performance allowing for a higher level of training. Mechanoreceptors embedded in collagen provide afferent feedback for muscle activation, pain modulation and balance reflexes. Spinal treatment at the level that correlates with the symptomatic region may have a profound result in decreasing tone or improving motor recruitment to the involved joint.

Slow adapting mechanoreceptors act to monitor joint position and slow-adapting receptors offer feedback for speed and acceleration variables with joint motion (Schutte et al. 1987). Small changes in the tension of the ligaments can influence the fusimotor system of the muscles acting at the knee joint, participating in the regulation of stiffness, and ultimately to dynamic stability (Sjolander et al. 1989, Johansson et al. 1991). The receptors of the collateral ligaments in response to tension act to regulate muscle stiffness and increase stability at the knee as those found in the ACL (Sojka et al. 1991). The neurophysiological system that monitors motion, speed, chemical changes, pain and pressure within joint systems must be taken into consideration with any rehabilitative program. If the environment of the joint structure is not conducive to proper receptor function, the desired outcome of the exercise program cannot be achieved.

Summary

Numerous combinations of joint restriction, joint hypermobility, muscle stiffness, muscle weakness, poor motor timing and poor motor conditioning can lead to a multitude of pathological presentations. Thorough evaluation of the biomechanics, motor performance and neurological function will assist in a complete treatment plan and more specific exercise design and dosage.

Section 2: Stage 2 Exercise Progression for the Knee

Training Goals

Stage 1 exercises resolved the acute issues of muscle guarding, joint restriction, inflammation and pain. Emphasis may be placed on one or all of these depending on the patient's pathology, stage of healing, type of surgical intervention, training level and general health. Each exercise is progressed as the specific functional quality being trained is realized. Progression to Stage 2 implies that proper joint mechanics have been restored, alignment of the lower limb is normalized, and a level of tissue healing has occurred to allow for more challenging exercises. The focus shifts more toward an emphasis on motor performance and stabilization within the newly gained range of motion.

Training Goals Stage 2

- *Increase repetitions (endurance)*
- *Increase number of sets*
- *Increase number of exercises*
- *Increase speed (strength/endurance)*
- *Histologically—increase lubrication of tissue*
- *Increase range of motion (if hypomobile or pain free)*
- *Body/limb change from recumbent to dependent*
- *Add isometrics in the range that strength/ stability is wanted*
- *Locking: change to coordinative locking or remove altogether*
- *Planar motions with exercise*
- *Add upper extremity closed chain exercise*

Basic Stage 2 concepts include increasing the range and speed of training as well as increasing the overall number of repetitions in the program. The focus is on continuing to improve tissue tolerance and coordination while shifting from predominantly open chain to closed chain weight bearing activities and working toward achieving full 3-dimensional range of motion. A closed chain emphasis should address the lower limb as a functional unit, rather than focusing only on the knee. Specific hip or ankle exercises should be employed if deficits exist (refer to the hip and ankle chapters of this text).

Increase Speed not Weight

Increasing weight or resistance is often the first choice made by therapists when progressing exercises. Speed, rather than weight, should be the initial change as it acts to further challenge coordination and actually increases resistance due to the forces of inertia. Coordination is speed specific. Increasing speed to the level of functional requirement is necessary for a return to activity or sport. Recruitment strategies require slow controlled movements initially to establish neurological pathways for properly timed patterns of motor unit activity. This phenomenon of neurological adaptation leads to improvements in coordination with more efficient recruitment patterns that allow for increases in the speed of training. Clinically this often occurs naturally with the patient performing the exercise in less time, which may signal the need to progress the activity.

As mentioned above speed increases the inertial forces of acceleration and deceleration, which adds resistance to the performance. Weight is not increased as these forces provide an adequate increase in tissue stress and demands on muscle performance. Once the speed of the task has reached the functional requirement, then the weight can be increased with a slight reduction in speed to ensure proper coordination. Alternating increases in speed and resistance will continue through Stage 3 and Stage 4.

Speed may also be adjusted to assist or challenge motor performance. Speed changes have different impacts on concentric work (CW) compared to eccentric work (EW). Slow CW allows for a greater load to be overcome, as time allows for additional motor units to be recruited. For example, performing step-ups slowly allows for improved coordination and greater recruitment of the lower limb musculature to assist in overcoming body weight. For eccentric training, speed is increased to increase peak muscle force that acts to reduce the workload. A slow eccentric step-down may be painful and difficult to coordinate. Increasing the speed where the patient quickly "falls" off the step landing on the uninvolved lower limb allows the exercise to be performed early to reestablish coordination for descending stairs. As coordination improves, speed is gradually reduced with eccentric work to better match more normal functional patterns of movement.

Increase Repetitions

A basic progression of Stage 2 is to increase the number of repetitions in the overall training session, or in a daily program. This is accomplished by adding additional sets to the existing exercises, and/or adding exercises to the program. Initial exercises consisting of one or two sets may, if appropriate, increase to three or more sets. Additional exercises may not only address the knee but focus on the lumbar spine, pelvis, hip or ankle. These areas are typically not directly addressed in Stage 1, as the focus is on more acute issues within the knee. The additional repetitions act to continue the process of selective tissue training for repair, and further enhance coordination.

Endurance Training

Tonic muscle performance is emphasized early, as it relates to arthrokinematic and postural support for the lower limb. Tonic muscle fiber is designed for lower load, high endurance performance and is initially targeted with exercise dosage. Endurance training is a function of improved vascularization, cardiovascular conditioning, and tissue capacity. It can be improved through increased repetitions, sets, and time of training as well as reduced rest breaks. Initial coordination training in non-weight bearing from Stage 1 should progress to coordinating lower limb alignment with functional weight bearing activities and exercises.

Isometrics

Isometric work (IW) was avoided in Stage 1 due to the presence of muscle guarding, as circulation is reduced with significant increases in intramuscular pressure. It also fails to provide the repetitive motion necessary for early tissue repair, pain modulation and coordination training. With resolution of muscle guarding and increasing joint mobility, isometrics are added to assist in fixing strength within the newly gained range of motion.

The main emphasis of IW early in rehabilitation is to more quickly establish joint stability. IW can generate greater peak torques than CW and allows for the use of greater resistances to improve strength qualities. The primary use of isometric work in Stage 2 is to sensitize the muscle spindle to stretch. As the manual therapy lesion described earlier, mechanoreceptor damage in the joint reduces the normal afferent reflex responses for motion. Sensitizing the spindle increases the afferent input to assist in normalizing movement patterns, replacing some of the afferent input lost with the mechanoreceptor damage. As the muscle lengthens toward the hypermobile range, the spindle responds to the tension from the changing length of the muscle. Afferent input from the spindle will assist in the recruitment of alpha motor neuron pools for normal motor recruitment. The primary muscles that would eccentrically stabilize the hypermobile range are trained isometrically in their shortened range. The shortened range is a joint position opposite of the hypermobility. In this range isometric work can be safely performed to begin the process of sensitizing the spindles. In later stages, as a progression, the isometric work can be performed in a range closer to the hypermobile range.

A basic clinical example of this concept is a knee with an ACL injury or surgical repair. The hamstrings would function eccentrically to decelerate the lower leg toward knee extension in open chain activity. Damage to the ligament and joint capsule would damage mechanoreceptors and reduce muscle recruitment for lower limb motor patterns. Stage 1 began concentric work from a mid to shortened range of the muscle, or mid to full flexion of the knee. End range extension is avoided. Stage 2 begins isometric training of the hamstrings with higher resistance, but with the knee in full flexion. As the spindle becomes more sensitive to stretch into the lengthened position, it will fire more easily when the moving toward full extension. The spindle will fire recruiting alpha motor neuron pools for improved recruitment of the hamstrings to decelerate the leg from end range extension.

Isometric Training Goals

- *Sensitize the muscle spindles in the shortened range of the muscle*
- *To fix strength*

The concept of fixing strength with IW should be performed in the ranges requiring stability and are started with the joint in the opposite range as the instability and the muscles in their shortened position. Stage 2 emphasizes the shortened position of the muscle to sensitize the muscle spindle. Later stages, a progression, may add IW in different joint positions closure to the pathological range of motion. In 1953 Hettinger and Muller published their pioneering work on isometric strength training demonstrating that a six-second isometric hold at 75% of 1RM was sufficient to increase strength. Isometric holds for six seconds at 2/3 of an isometric maximum (IM) were found to lead to the best increase in strength (Muller 1957). Lower levels of resistance can be used with longer hold durations. One isometric hold, at a heavier resistance, can replace one dynamic set in a three set exercise. Isometric peak torque is greater than that of concentric peak torque allowing for a greater resistance with isometric work than used during the dynamic set.

Joint Locking Training Techniques

Joint locking concepts are more related to the spine, but the concept is simply to protect tissues from a particular range of motion. For the knee, basic locking may relate to limiting the degree of knee flexion and/or extension. This can be accomplished by having the weight stack come to rest with open chain exercises. An artificial block of the pelvis may be used in weight bearing or partial weight bearing closed chain exercise to create a safe environment for initial squat training.

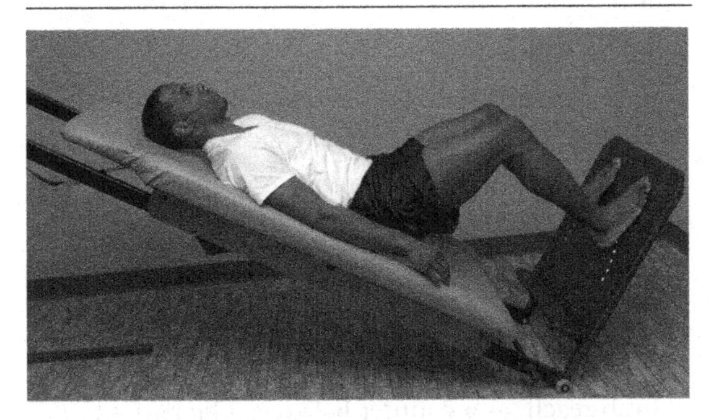

Figure 2.28: Decline sleds, such as the Vigor Gym® pictured, allow for a reduction in body weight for closed chain exercise, but a block for flexion can be set at any range. With the sled in resting position the pelvis is slid down toward the feet until the desired degree of flexion is attained. Repetitive unloaded squatting exercises can then be performed with a mechanical block of end range flexion.

Locking techniques may also be used to influence muscle performance. For example, placing a heel wedge under the medial calcaneous creates slight plantarflexion and inversion of the hindfoot. With squat training this acts to slightly inhibit the soleus and place a greater demand on the quadriceps.

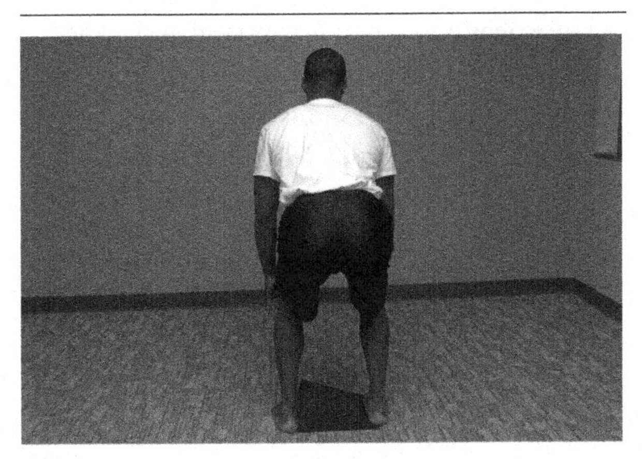

Figure 2.29: Soleus inhibition, with quadriceps emphasis during squatting. A medial heel wedge creates slight inversion and plantarflexion leading to partial inhibition of the soleus function of decelerating the tibia anteriorly during the squat, placing more demand on the quadriceps.

Functional Testing

As Stage 2 shifts toward a more closed chain weight bearing emphasis, functional testing can provide an objective baseline for establishing deficits and guide exercise design. This testing may simply relate to daily weight bearing activities and at times they can be incorporated into the exercise program (de Vreede et al. 2004). For example, the patient recovering from a total joint replacement may have difficulty transferring out of a chair or climbing stairs. Duplicating the functional activity is necessary to establish coordination, and unloading may be necessary to reduce the body weight to allow for a coordinated and safe movement pattern.

Examples of Functional Tests / Measures

- *Chair Raise: The coordination of rising from a high chair and the variability of rising from various chair heights indicate a functional recovery after TKA implantation (Boonstra et al. 2006).*
- *Single Leg Raise: The single leg rising performance demonstrates a positive relationship with knee extensor strength (Aasa et al. 2003).*
- *6-Minute Walk: The six-minute walk (6mw) is a well-established measure of aerobic capacity in elders with cardiorespiratory and peripheral vascular disease and may be an accurate measure of functional performance in healthy elders (Bean et al. 2002).*
- *Self-Timed Walking: Self timed walking test for knee osteoarthritis. Measurements of self-paced walking time can provide both reliable and valid data for evaluating functional performance (OA) of the knee (Marks 1994).*
- *For more athletic patients there are: Single hop, triple hop, cross-over hop and timed 6-meter hop tests (Noyes et al. 1991).*

Functional testing of the lower extremity is an effective and necessary predictor for successful return to pre-injury activity. Most reliable functional tests can be too strenuous for a patient in the acute stage of recovery, and are typically performed in Stage 2. Functional testing provides a baseline of objective measures to mark improvement, predict capacity for return to work/ sport, identify movement dysfunction and motivate the patient. Exercise programs built around functional tasks are feasible and show promise of being more effective for improving functional performance than a resistance exercise program in elderly women (de Vreede et al. 2004).

Balance Testing / Training

Star Excursion Balance Test (SEBT)

The Star Excursion Balance Test (SEBT) may offer a simple, reliable, low-cost alternative to more sophisticated instrumented methods for testing dynamic balance. The test requires the patient to maintain balance on a single limb, while manipulating the other limb. The SEBT is more extensively outlined in the Ankle Chapter, but will briefly be covered as it relates to the knee.

The SEBT involve single leg stance while reaching in different directions with the opposite lower limb. The reach is performed as far as possible in each of eight prescribed directions (the star pattern), while maintaining balance on the contralateral leg (Olmsted et al. 2000). Directions of reach are listed relative to motion away from the stance leg. A star pattern can be placed on the floor with four lines of tape: vertical, horizontal and two diagonals at 45 degrees angle. Alternatively, one line of tape can be placed on the floor with the subject altering the body position relative to the line for each test, and replicate all eight test directions. The subject lightly touches the furthest point possible on the line with the most distal part of the reach foot, and returns to the center (Olmsted et al. 2000). Measurement of the distance along the tape is recorded in centimeters. For testing purposes, each direction is repeated three times, with a 15 second rest between each reach. A trial is deemed invalid and repeated if the subject 1) does not touch the line with the reach foot while maintaining weight bearing on the stance leg, 2) lifts the stance foot from the center grid, 3) loses balance at any point in the trial or 4) does not maintain start and return positions for one full second (Olmsted et al. 2000). Directions of reach identified with balance deficits can easily be used as clinical and home exercises. Chaiwanichsiri et al. (2005) demonstrated effective outcomes with STAR pattern exercise for ankle instability. This approach can be used for any functional balance deficit. Testing should not only include measurement of the reach distance, but also identification of abnormal compensatory movement patterns that need to be corrected for training. The distance of the balance reach exercise is limited to the coordinated range with proper alignment of the trunk and lower limb. Common compensations include:

- Trunk lean in the opposite direction of the lower limb reach, as a counter balance. The patient should be instructed to maintain the head and trunk directly above the stance foot.
- With a vertical trunk, the pelvis may compensate by shifting in the opposite direction of the reach as a counter balance. The patient should be instructed to maintain alignment of the hip, knee and ankle in a vertical line.
- Knee valgus is common with hip external rotation and/or extensor weakness. This valgus moment can lead to numerous pathologies at the knee and ankle.
- Hip extensor weakness may present as the knee flexing forward beyond the toe with the hip and pelvis remaining vertical without posterior motion. Instruction should be to initiate the motion by bringing the pelvis posterior, as if sitting back in a chair, and then flexing at the knee.
- Weakness at the hip and/or knee may lead to excessive ankle pronation, as forces are focused in the lower leg. Passive arch supports may be necessary for training, but the patient can also be instructed to actively externally rotate through the lower limb to dynamically stabilize the medial arch of the foot.

Hertel et al. (2006) looked at whether all directions of the SEBT needed to be performed to identify functional deficits in individuals with chronic ankle instability (CAI). The complete test was found to be somewhat redundant with reaching in the antero-medial direction, where the medial and postero-medial directions were found to be the most sensitive to testing functional deficits. Future studies or clinical experience may identify specific directions of reaches being more important to test and train in specific diagnoses, but training should focus on the deficient directions.

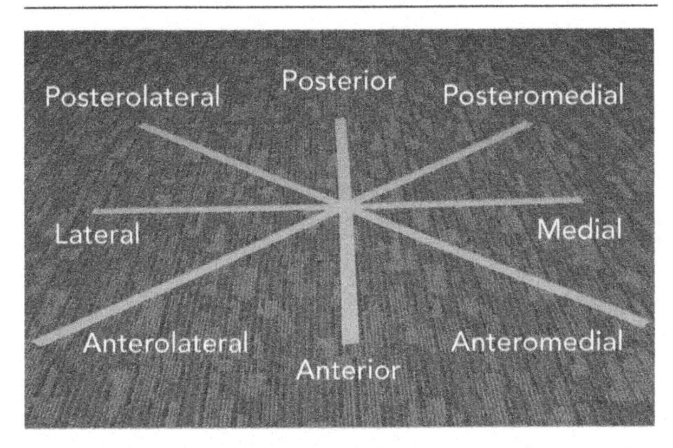

Figure 2.30: Star pattern for balance reach. The eight lines positioned on the grid are labeled according to the direction of excursion relative to the stance leg (relative right stance leg facing forward is shown): anterolateral (AL), anterior (A), anteromedial (AM), medial (M), posteromedial (PM), posterior (P), posterolateral (PL) and lateral (L) (Hertel et al. 2006).

Functional Balance Training

Balance retraining is a fundamental rehabilitation goal for lower limb patients involving cooperation of the vestibular and visual systems as well as the peripheral and axial mechanoreceptors or proprioceptive system. Loss of balance has also been shown to be a predictor of injury demonstrating the need for balance training even in the healthy population (McGuine et al. 2000). Strength training alone does not appear to enhance standing balance (Schlicht et al. 2001, Topp et al. 1993). The speed of torque production, and not just the peak torque generated, is also a factor in motor control for balance (Robinovitch et al. 2002). Combined programs of strength and balance training have shown benefits (Judge et al. 1993). Balance training designed to improve intersensory interaction from the visual, vestibular and somatosensory systems can effectively improve balance performance (Hu and Woollacott 1994). Identification of the primary system causing the balance deficits can allow for a more focused and effective rehabilitation program.

Loss of any proprioceptive input due to joint mobility restrictions or adaptively shortened muscles can slow the rehabilitation process and increase the prevalence of injury. Ankle motion can be limited from a tight gastrocnemius muscle, and play a significant role in the stability of the knee (Lass et al. 1991). Lower limb exercises must address hip, knee and ankle function as a unit, and not solely focus on the knee. Elias et al. (2003) described the agonistic behavior of the soleus muscle to the anterior cruciate ligament, as it decelerates the anterior motion of the tibia beyond mid stance. Soleus strength and function should be included in rehabilitation strategies to reduce the incidence of ACL injury, and to improve function in ACL deficient knees or postsurgical repairs.

Pain free weight bearing status for lower quarter and lumbar patients is necessary for a safe and effective balance training program. Pain reduction alone is not enough to restore weight bearing balance function (Bennell and Hinman 2005). Decreased proprioception and balance is noted in knee osteoarthritis (Hinman et al. 2002, Wegener 1997). Proprioceptive loss at the knee and ankle from trauma to the somatosensory system has been shown to significantly delay ankle muscle responses while upper leg and trunk responses are not delayed (Bloem 2002). Without retraining of the lower leg somatosensory system, the majority of lower leg balance correcting responses will be initiated by hip and trunk proprioceptive inputs (Bloem 2002). Isakov and Mizrahi (1997) found foot-ground reaction forces were the same between normal and sprained ankles while standing with either eyes open or closed. But standing with eyes closed, irrespective of the ankle status, produced significantly higher reaction forces than standing with eyes open. Primary endings of ankle muscle spindles play a significant role in the control of posture and balance during the swing phase of locomotion by providing information describing the movement of the body's center of mass with respect to the support foot (Sorensen et al. 2002).

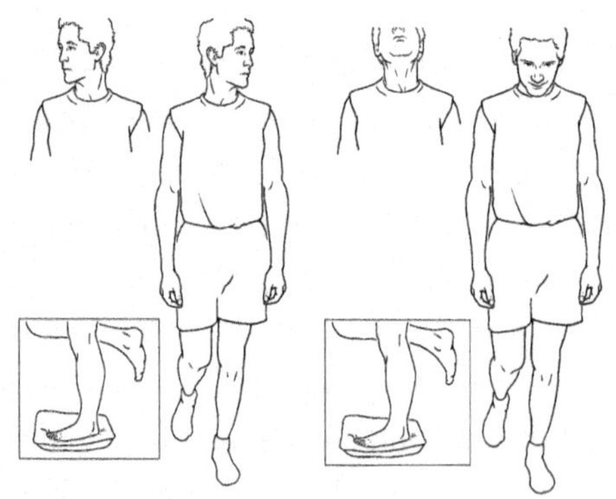

Figure 2.31a,b: Balance testing/training with bilateral stance and cervical motions. With the eyes open, the head and neck move in all three planes of motion. If performance is at an accepted level, the same testing can be performed with the eyes closed to remove compensation with ocular input. Directions of poorest performance can be modified for safe daily exercises. Labile surfaces can be added as a progression as balance improves.

Adding balance and proprioceptive training to rehabilitation does improve outcomes in individuals with ACL deficient knees. Cooper et al. (2005) studied improvements in joint position sense, muscle strength, perceived knee joint function, and hop testing after proprioceptive and balance exercises. The study showed some benefits in the proprioceptive group for measures of strength and proprioception, however no improvements were noted for any functional activities. Strength training had a more significant impact on functional tests. Balance training should be included as soon as weight bearing tolerance allows, as early proprioceptive training is needed to effectively progress to a functional strengthening program.

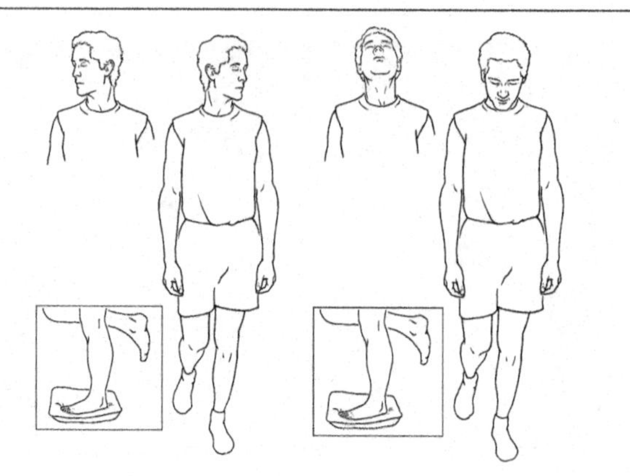

Figure 2.32a,b: Balance testing/training with eyes closed, bilateral close stance position and cervical motion. Feet are positioned next to each other to reduce the base of support. With the eyes open, the head and neck move in all three planes of motion. If performance is at an accepted level, the same testing can be performed with the eyes closed to remove compensation with ocular reflexes. Directions of poorest performance are given as daily exercises. Labile surfaces can be added as a progression as balance improves.

	Direction of Training	*Dosage and Frequency*
Balance Training for Unilateral Lower Limb Stance	• *Stable surface, eyes open* • *Stable surface, eyes closed* • *Stable surface, eyes open, cervical motion* • *Stable surface, eyes closed, cervical motion* • *Stable surface, eyes closed, trunk external challenge* • *Stable surface, eyes open, lower limb reach* • *Stable surface, eyes closed, lower limb reach* • *Stable surface, eyes open, upper limb reach* • *Stable surface, eyes closed, upper limb reach*	• *15 seconds at each progression* • *45 second rest between* • *Daily 5–15 minutes.*
Options	• *As above with labile surface:* *Tilt board—single plane challenge* *Tilt board with foot at 45 degrees diagonal* *Wobble board—3-dimensional challenge* • *Foam—(vestibular challenge)*	*Progress:* • *30 seconds per exercise* • *15 second rest*

Table 2.2: General guidelines for balance training.

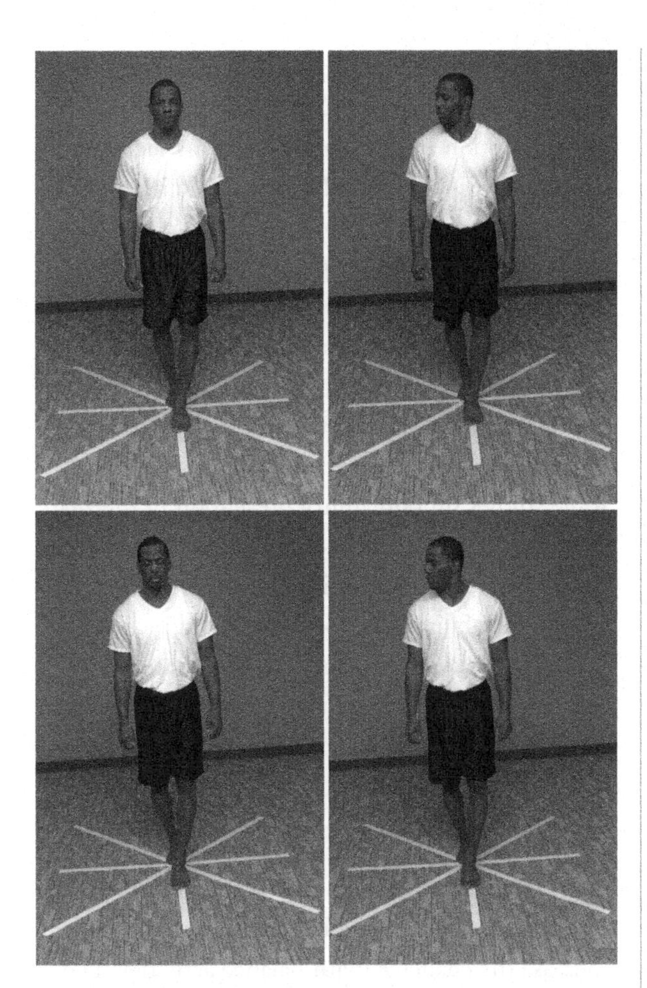

Figure 2.33a,b,c,d: Balance testing/training with bilateral stance feet positioned in tandem. A narrow base of support is a basic balance challenge. The addition of cervical motion, closing the eyes, and even performing upper extremity motions can further increase the challenge. Each set may involve a different option to vary the challenge.

Figure 2.34: The Leaning Tower. In bilateral standing, the body is actively swayed in different directions and returned back to neutral: anterior/ posterior, medial lateral and/or diagonals.

Lower limb reaching requires a combination of balance, coordination and strength. The strategies of compensation can provide insight into which joints or muscle groups are impaired. Trunk lean in the opposite direction of the reaching limb reduces the load on hip muscles but is an abnormal pattern. The patient should be instructed to maintain a vertical trunk or perform the activity next to a wall to prevent the lean. Hip extensor weakness can create an abnormal pattern of reduced hip flexion and increased anterior translation of the knee over the distal end of the foot with single leg squatting.

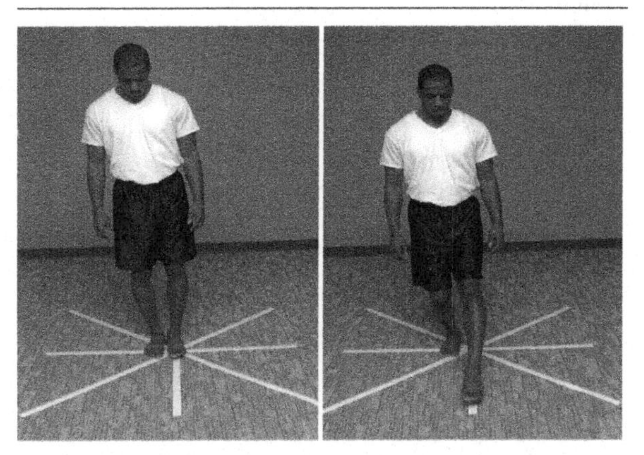

Figure 2.35a,b: Star Excursion Balance Test. Star pattern for lower limb balance reach involves standing one foot with the opposite lower limb reaches in different directions. The eight lines positioned on the grid are labeled according to the direction of excursion relative to the stance leg: anterolateral (AL), anterior (A), anteromedial (AM), medial (M), posteromedial (PM), posterior (P), posterolateral (PL), and lateral (L).

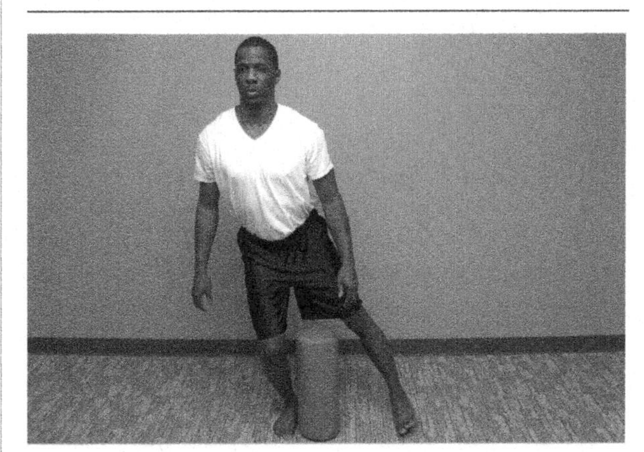

Figure 2.36: Single leg stance-lower quarter reach. Patient performs single leg balance while reaching the opposite foot away from the center of gravity. External feedback is provided by placing a bolster on the medial knee to avoid excessive valgus from occurring at the knee. This exercise is an example where decreasing the speed can increase the required coordination.

A block in front of the knees will limit the anterior

moment, requiring posterior motion of the pelvis and increased hip flexion. Upper limb reaching may also be used as a counter weight to balance body weight on either side of the axis of motion of the hip. The patient can be instructed to maintain the arms at the side for an increased challenge. Weakness at the hip may also result in a valgus moment at the knee. A bolster placed on the medial side of the knee can provide external feedback to engage hip muscles and prevent occurrence of the valgus moment.

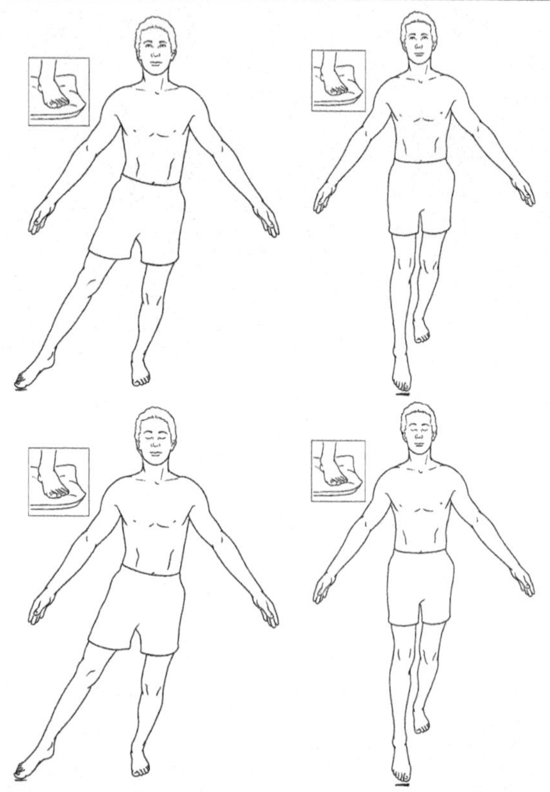

Figure 2.37a,b,c,d: Lower quarter reach balance training. The exercises start with standing on a firm surface with eyes open. The foot reaches forward staying roughly one inch off the floor and then returns to neutral. The patient reaches only as far as can be coordinated to return without touching the floor to regain balance, or without abnormal mechanics of the stance limb. Progressions include closing the eyes, using a labile surface to challenge lower quarter somatosensory input, use of foam to challenge the vestibular system, and to perform all of these options with lower quarter reaches in multiple directions.

As discussed in the exercise prescription section, Chong et al. (2001) assessed balance improvements with training on the BAPS® board (biomechanical ankle platform system) to determine if they could be attributed to retraining deficits in the ankle and foot. Improvements observed during training were attributed from diffuse enhancement of proprioception in other body segments such as the knees, hips, spine and upper extremities rather than targeting proprioception specific to the ankle. Improvements in the proprioception of the entire body, as well as local gains in coordination and strength may be of clinical benefit, but may not be optimal for balance training. Utilization of balance board training as a preventative measure for ankle sprains in healthy individuals is controversial as some studies support injury prevention in the ankle (Verhagen et al. 2004) and others refute a positive impact for lower quarter injuries (Soderman et al. 2000). Rehabilitation programs should include multifaceted concepts to specifically address the individual needs of patients with specific reference to the goal of the exercise.

Upper Limb Reaches

Upper limb reaching can also be used with SEBT testing and exercise to either assist the motion or increase the challenge with balance activities. The subject stands on the involved limb while reaching either extremity away from the center of gravity and then returns while maintaining single leg balance. Reaching can take place in any direction deemed functionally necessary. A hand can reach toward a specific target as a measurement of function or as an external cue for training. Simple distance away from the center point can be measured by having the subject drop a sand bag at the end point, with measurement being from the tip of the first toe to the sand bag (Gray 2000).

Figure 2.38a,b: Example of right anterior upper limb horizontal reach. The subject stands on one leg while

reaching forward as far as possible, and then dropping an object for measurement. The distance from the 1st toe to the bag is measured to score the test. The drop counts only if the subject can return to neutral without losing control of single leg balance.

Performing upper limb reaching is a simple way to train more functional applications of single leg balance. Targets can be set for the hand to repetitively and/or sequentially hit while maintaining balance. Combinations of using either extremity in different directions can replicate any necessary functional requirements. Speed should be slow and consistent through the performance initially to establish coordination. It can then be increased to meet the functional needs of the task being trained. The opposing lower limb may be used as a counter balance for the upper limb reach to establish coordination with the motion. The non-weight bearing limb can then be held in neutral, rather than used as a counter balance, to place more significant challenge to the stance limb.

Figure 2.39a,b: Example of right lower limb balance with left upper limb anterolateral and inferior reach. A bolster is used as a target to touch with the finger. Lower limb alignment is maintained, avoiding abnormal compensation. The speed is fixed through the performance to improve coordination.

Multidirectional exercise concepts can be added to balance reach training. The main force line is gravity, which can be increased by the use of a free weight in the reaching hand. Increasing gravity will emphasize the extensor muscles of the hamstrings, gluteus maximus and long trunk extensors. Often the trunk instability accompanied with lower quarter weakness includes rotary patterns as well.

Providing a horizontal line of resistance from a pulley or elastic band will recruit a rotational vector of muscle for core facilitation. The lumbar multifidi, transverse abdominus and deep hip rotators are facilitated improving the proximal stabilization for balance and reach training.

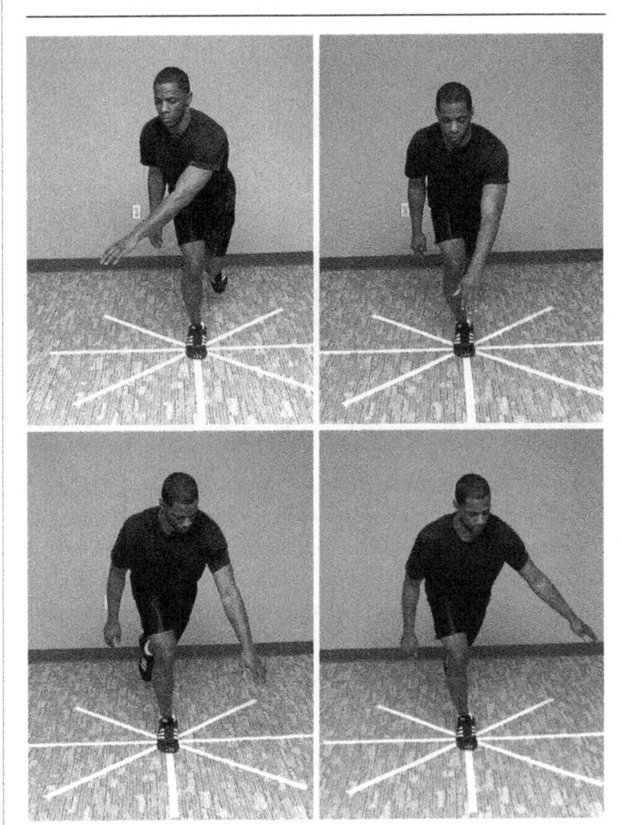

Figure 2.40a,b,c,d: Upper quarter reach balance training. Reaches can be performed in any degree within the star pattern. Alternatives including the addition of an inferior or superior vector to the reach. A dumbbell weight can be added in the hand to increase the extension vector against gravity. A lateral pulley line can be added to the non reaching hand to increase the rotational vector to the trunk. Closing the eyes or standing on a labile surface will further challenge central balance.

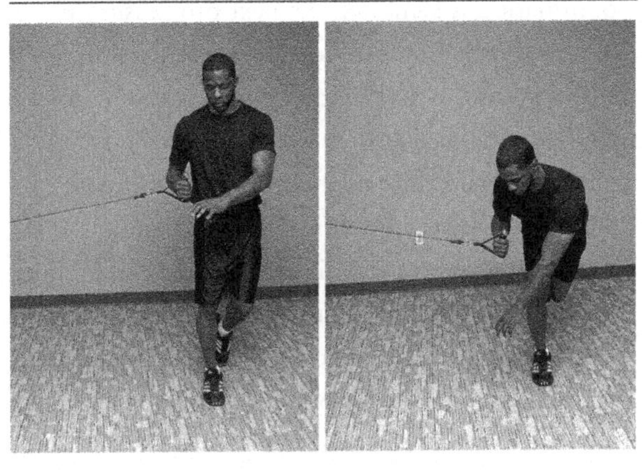

Figure 2.41a,b: Multidirectional training of upper quarter

reaches in a diagonal pattern with a right lateral force from a pulley, held in the right hand. The pulley creates a right rotational trunk moment. The right lumbar multfidi, deep hip external rotators and the transverse abdominus are recruited through the upper extremity to resist the force. Increasing central recruitment for stabilization will improve the control of the weight bearing leg for balance reach training, whether for upper quarter or lower quarter reach training.

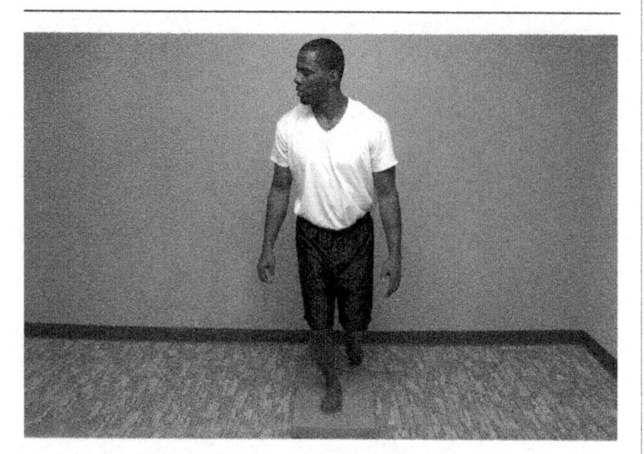

Figure 2.42a,b: Balance foam will reduce afferent receptor input from the cutaneous receptors on the plantar surface of the foot as well as joint mechanoreceptors of the foot and ankle. This may emphasize the vestibular system more and be less functional than a regular wobble board or balance reach training on a level surface.

Balance Training with Upper Limb and Trunk Motion

Ball toss and catch requires a combination of coordination, balance and stabilization or strength. An unplanned pattern of motion reacting to a ball provides a greater challenge on coordination, with focus on the upper limb performance rather than the stance leg. For an unstable joint, the catch is first performed on the side that will move the body away from the primary direction of instability. As coordination is achieved, the next progression is to catch the ball on the side that will move the body toward the instability but with an isometric stabilizing catch that does not allow the body to move toward the range of instability or tissue trauma. Finally the catch is made with an eccentric deceleration toward the direction of instability, but not into the pathological range, followed by a quick concentric return and throw. A plyoball rebounder can be used, or the therapist can manually toss and catch a ball with the patient.

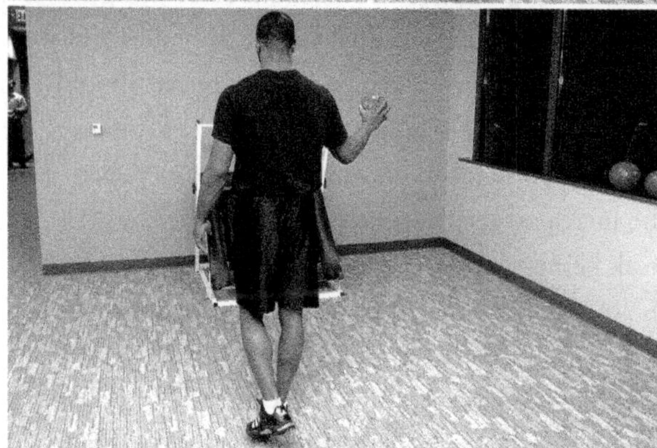

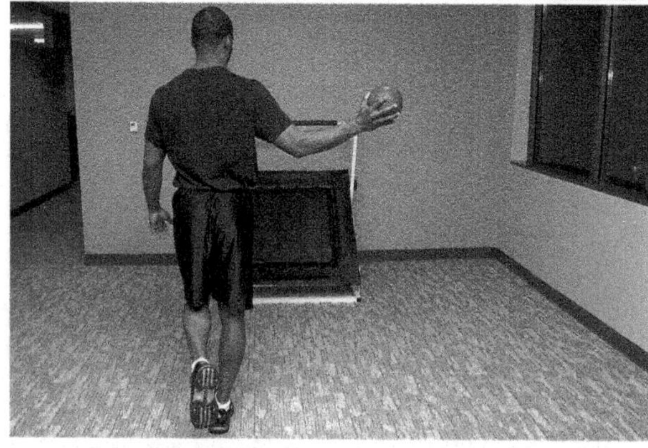

Figure 2.43a,b,c: Example of a ball toss progression for right ACL internal rotary instability of the tibia. The first stage is to catch the ball on the left side, rotating the body to the left and the femur into internal rotation, with relative external rotation of the tibia. The next progression is to catch the ball isometrically on the right side, not allowing any rotation of the trunk or lower limb. The final progression is to catch the ball on the right decelerating with eccentric work into right rotation of the body, external rotation of the femur with relative internal rotation of the tibia. Muscles of the hip must control the eccentric deceleration to prevent rotary forces from translating down to the ACL.

Motions and Directions

Improved coordination is coupled with more efficient neurophysiological synchronization in firing of the motor unit, which increases stability.

Stage 1 exercises moved away from pain or instability, but motion is now directed toward the pathological range. Additional planes of motion are added to increase the challenge to the tissue in lesion and to coordination.

Closed Kinetic Chain Training

Closed kinetic chain (CKC) functional training may provide a method for more effectively rehabilitating an injured or reconstructed knee. Once the muscular system has proper neurological recruitment, the rehabilitation program can progress toward endurance and strength training (Nyland and Brosky 1994). Weight bearing can be progressed through the addition of weights or by utilizing the effects of gravity to increase the work. As the functional patterns being trained become better coordinated, the need for external support or cueing decreases.

The high correlation between quadriceps strength and functional performance suggests that improved postoperative quadriceps strengthening could be more important to enhance the potential benefits of TKA. Postoperatively, quadriceps strength as measured by functional activities declined early after TKA but recovered more rapidly when quadriceps strength improved (Mizner et al. 2005).

Unloading

Progression of closed chain exercise is important to increase the stimulus to the repairing tissue, further challenge coordination of the lower limb, and provide a more functional type of training. The decline board option—such as a Total Gym®, Vigor Gym or horizontal Nautilus squat machine—provides an easy way to adjust the angle of the body and amount of weight borne through the extremity. Weight bearing is progressed as tolerated with both bilateral and unilateral exercises.

Adjunct Popliteus Training

Nyland et al. (2005) described the popliteus as a dynamic guidance system for monitoring and controlling subtle transverse and frontal plane knee joint movements. The affect of the popliteus included controlling anterior, posterior and lateral meniscus movement, unlocking and internally rotating the knee joint (tibia) during flexion initiation, assisting with 3-dimensional dynamic lower extremity postural stability during single-leg stance, preventing forward femoral dislocation on the tibia during flexed-knee stance and providing for postural equilibrium adjustments during standing. Early training may involve direct training of the popliteus with resisted internal rotation training of the tibia in non-weight bearing.

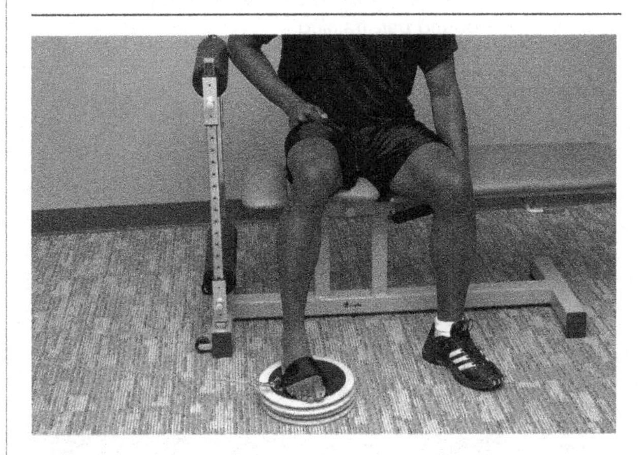

Figure 2.44. Sitting internal rotation of the tibia for mobilization and/or training of the popliteus muscle. The patient is sitting with the foot on a rotation plate. Resistance to internal rotation is applied with pulley or elastic band.

Gait Training

Early gait training may emphasize balance, coordination and range of knee flexion to clear the toe with step through. Initial emphasis may simply be on walking with the eyes looking forward, rather than down at the ground. Central balance deficits may be more provoked when attempting to walk forward but looking right to left or up and down. Lateral walking, cross over stepping and turning are all basic applications to early gait training. Stepping over objects creates an external cue to increase hip and knee flexion during walking and also places a greater challenge on coordination and balance strategies. Further progression to uneven surfaces may provide functional training for challenges with outdoor walking.

Figure 2.45a,b: Cup walking. Short cups are spaced apart on the floor with the patient having to step over each cup while walking forward. A taller cup is used to increase the challenge. This exercise encourages and coordinates hip flexion range during gait training.

Figure 2.46a,b: Hurdle Walk—Step over hurdle and hold position for one to two seconds then step over the next hurdle and repeat with the opposite leg. To increase difficulty close the eye of the leg that is stepping over the hurdle.

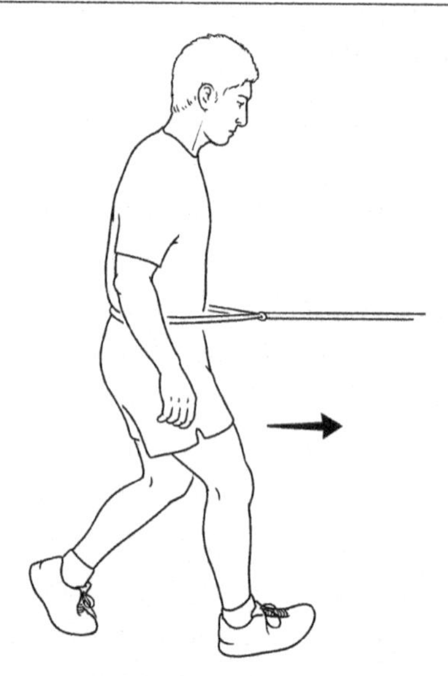

Figure 2.47: Eccentric gait. An anterior resistance from a pulley or sport cord is attached to the waist. The patient walks slowly toward the resistance, attempting control and decelerate the resistance.

Basic Squatting

Closed chain squatting can place excessive stress on healing or repaired tissues. The angle of the lower limb joints plays an important role in distributing forces safely for replication of functional movement patterns. Tibial translation increases with squatting except when the center of gravity is behind the feet. Proper activation of the soleus and gastrocnemius is necessary to control the anterior tibial moment and co-activation of the quadriceps and gastrocnemius muscles is important for knee stability. Kvist and Gillquist (2001) found hamstring muscle co-activation relatively insignificant for control of anterior shearing in ACL deficient knees, as it was more related to gastrocnemius eccentric function and displacement of the center of body mass posterior to the feet during the squat motion.

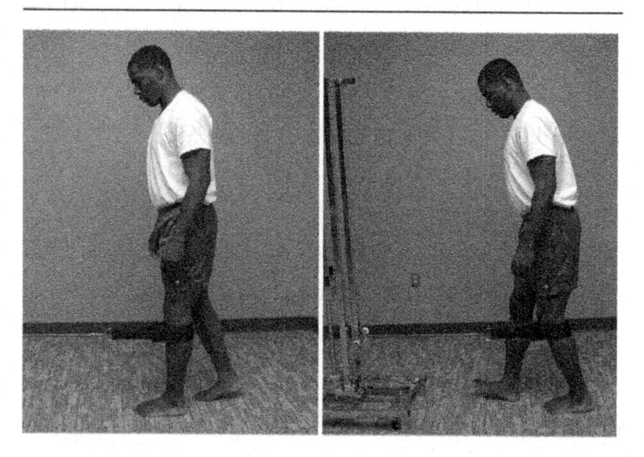

Figure 2.48a,b. Soleus training with step through for eccentric training of deceleration of the tibia on the stance leg during step through. The patient stands on the involved limb. The uninvolved limb starts from behind, with a slow movement from toe off to heel strike. Emphasis is placed on slow control of the stance leg as the tibia moves forward toward heel strike. An anterior force moment from a wall pulley attached to the waist can increase the eccentric challenge of the soleus (pictured). For the unstable knee, such as ACL laxity or repair, this exercise improves stability of the knee in the presence of weakened quadriceps by increasing the function of the soleus to control the anterior moment of the tibia during gait.

Posterior displacement of the body mass behind the feet is not only important for ACL deficient knees, but should be emphasized with all knee dysfunction. It acts to increase recruitment of the hip muscles reducing the forces transmitted to the knee, which can cause abnormal tissue strain.

External cues are more effective than internal cues for motor learning when teaching the proper squat technique. Rather than cueing patients to move their pelvis posterior, a chair placed behind them to sit back on can serve as an external cue for more efficient motor learning. Another common compensation is to shift weight toward the uninvolved limb. Neitzel et al. (2002) described a significant load shift toward the uninvolved lower extremity for up to 12–15 months post-anterior cruciate ligament reconstruction. Even loading with bilateral training should be emphasized, as load shifting can place unwanted stress on the uninvolved limb and create compensatory pathologies and repetitive strain injuries.

Tibial rotation during squat training may also be of benefit. Miller et al. (1997) demonstrated that leg rotation does not affect the VMO/VL EMG activity ratio in subjects with PFPS, though a general decrease in VMO/VL EMG activity ratio was found when the leg was externally rotated in the asymptomatic control group. Leg rotation may not have a an effect within the quadriceps for motor recruitment, but can effect the firing pattern of the entire lower limb and change the tissue strain moment of the patella. Internal rotation of the tibia moves the insertion of the quadriceps medially, reducing the lateral moment on the patella. This subtle change in foot position may be enough to reduce pain and allow for training. As tissue tolerance and motor performance improves the foot position is gradually returned to normal alignment.

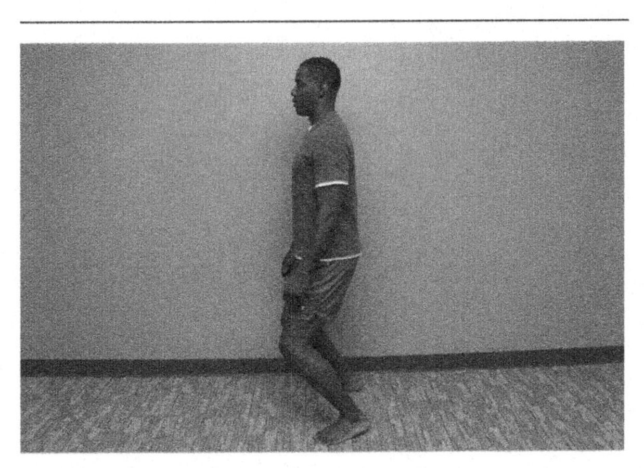

Figure 2.49: Incorrect squatting technique with excessive anterior tibial translation. This is a classic compensatory squatting technique in the presence of weakened hip and trunk muscles. The body remains vertical without any hip flexion moment, placing excessive stress on the quadriceps, increasing patellofemoral compression and creating a less functional movement pattern.

Figure 2.50a,b: Squats: unloaded with a pull down machine and gantry. The patient supports the body weight in a gantry or with a pull down bar. Weight is placed on the pull down machine to assist the lower quarter with the squat motion.

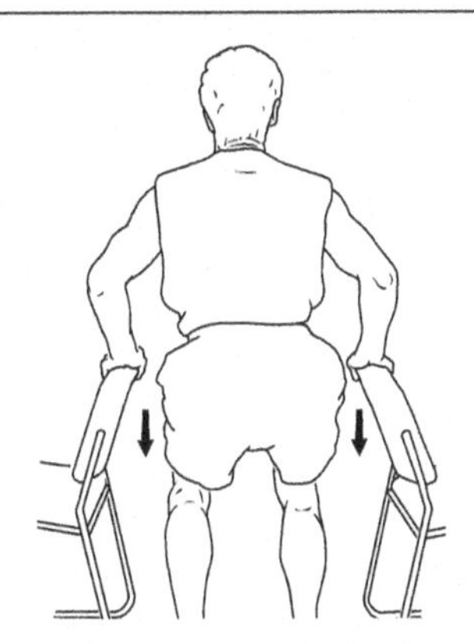

Figure 2.51: Unloaded squats may be performed in the home setting with support through the arms on two chairs.

Figure 2.52: Door jam squats. The patient stands with the hands holding on to a door jam, feet close to the door jam. A squat motion is performed, keeping the knees above the toes and dropping the hip backward. The arms provide as much resistance as necessary to avoid pain, allow for the desired repetitions and/or achieve the desire range.

The use of external cues as targets can more quickly enhance motor learning than internal cues where the patient focuses on a particular muscle or movement such as preventing the knees from moving beyond the toes when squatting. Placing a block in front of the knees provides an external cue that instantly creates a posterior pelvic movement. Having a posterior target, such as a chair, to sit back on is also an effective external cue to help with motor learning.

Figure 2.53: Correct squat technique with the center of gravity behind the feet and the knees not extending beyond the toes. This technique encourages the larger muscles of the hip to be involved with the motion, as well as the trunk extensors. This approach reduces abnormal forces to injured tissues in the knee for earlier squat training. Sitting back when performing a squat was been shown to decrease amount of translation at the knee joint (Kvist and Gillquist 2001), which is important for ACL deficient knees that have difficulty dynamically controlling this anterior shear moment.

Abnormal valgus moments at the knee result from weakness of the hip external rotators and abductors leading to femoral adduction and internal rotation Hip, pelvis and lumbar dysfunction can be potential contributors to weakness of these muscles and must be screened during the initial evaluation. Closed chain training of knee flexion may require improved facilitation of these muscles to ensure proper alignment and reduce transmission of forces to distal muscles and joint structures of the lower quarter. Gravity provides the primary force for closed chain knee flexion training. Providing a secondary medially directed line of force at the knee creates a stimulus for recruitment of the hip abductors and external rotators. This can be achieved using a wall pulley or elastic resistance for bilateral squats, unilateral squats, lunging, step-up and step-down training. In more acute cases with higher levels of dysfunction a varus force can be used to assist weak muscles at the hip to allow for earlier pain free weight bearing training. These approaches have been used since the 1970's in the Medical Exercise Therapy approach from Norway (Holten/Faugli 1996).

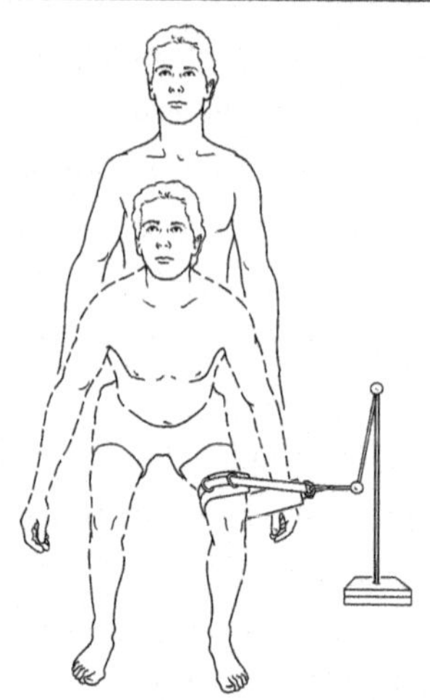

Figure 2.54: Squatting with a varus force moment to assist muscles of hip abduction and external rotation. This technique may be used to support/protect tissues at the knee or ankle that would be further traumatized with valgus at the knee (i.e., MCL strain or repair). A light varus force is placed at the knee joint line from an elastic band or wall pulley to help maintain the knee in neutral alignment. Resistance is increased until knee flexion can be performed without pain and with proper alignment of the hip, knee and ankle.

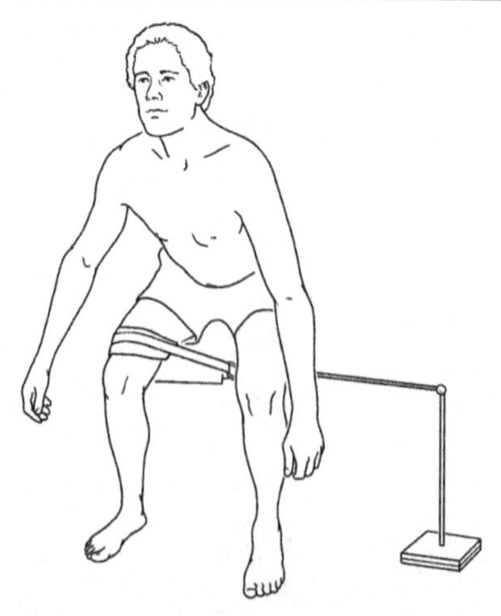

Figure 2.55: Squatting with a valgus force moment to facilitate muscles of hip abduction and external rotation. Resistance from a wall pulley or elastic band is placed at the knee joint line creating a valgus moment. This multidirectional training technique allows for earlier pain free training of the quadriceps for sagittal plane flexion by increasing the motor facilitation in the frontal and transverse planes at the

hip joint. Improved facilitation at the hip will again improve dynamic control of the entire lower extremity, reducing tissue strain at the knee. A heavy resistance for the valgus moment is not necessary to improve muscle facilitation and may be less safe in early training.

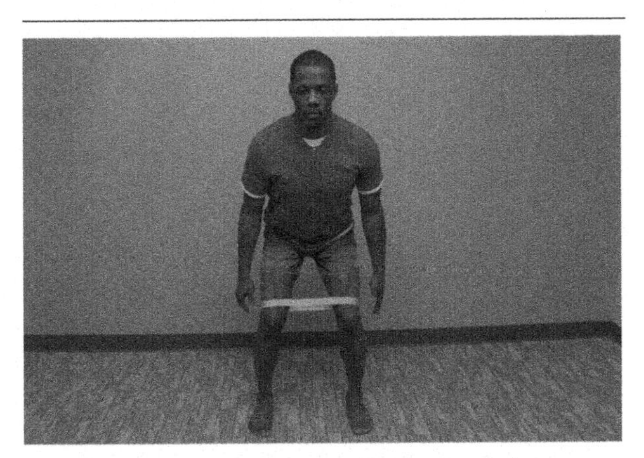

Figure 2.56: Bilateral knee valgus moment for multidirectional squat training. An elastic band may be looped around both knees creating a valgus moment at both knees improving bilateral facilitation of hip muscles. An elastic band provides an easy home application, as well as creates bilateral facilitation at the hip, which may improve the crossover effect of recruitment.

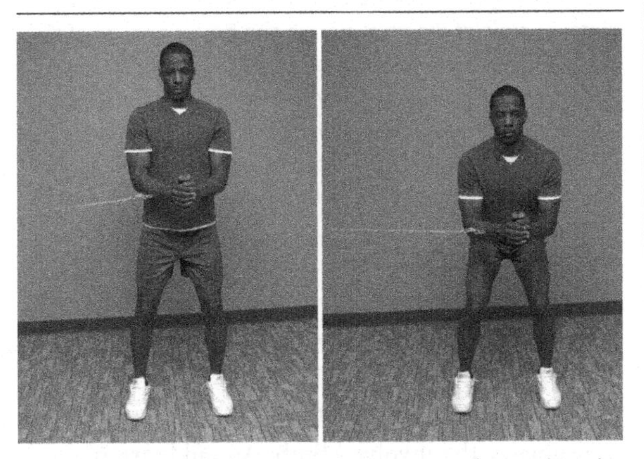

Figure 2.57a,b: Multidirectional squat training for combined lumbar and hip stabilization training. Proximal weakness in the knee patient may extend beyond the hip, requiring improved stability of the pelvis and lumbar spine. In these cases, a lateral force moment from a wall pulley is held anterior away from the body. Here both hands are shown holding the line for bilateral trunk recruitment. The force moment from the pulley naturally recruits the fiber direction of the transverse abdominus, the oblique abdominal muscles, the rotational vector for the lumbar multifidi of the lumbar spine and the deep rotators of the hips. The pulley is stabilized anteriorly while the squat motion is being performed. Further progression may involve rotational motions of the trunk during a normal squatting motion. As a motor assessment tool, if the patient is able to squat further with less pain when holding a horizontal resistance, the exercise may help in diagnosing or identifying proximal weakness as a component of the overall motor impairment.

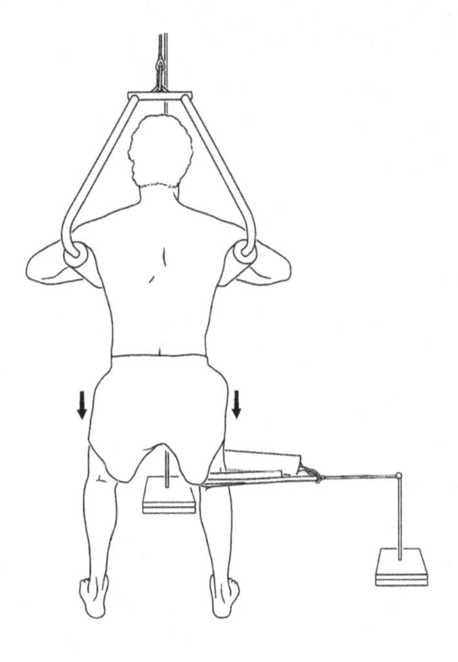

Figure 2.58: Unweighted squat with unilateral valgus force moment (multiple resistance exercise). Regardless of the planes of weakness, and what type of proximal facilitation that is required, the body weight may still be too high to allow for weight bearing training without some type of unloading device. The use of a varus or valgus force moment can be added to any unloading device, whether a pull down machine with an unloading gantry, horizontal Nautilus® type machine or Total Gym®. If a portable pulley is not available, or there is no proper attachment sites for an elastic band, the therapist may be required to manually hold a strap around the knee to create the required force moment.

The stance phase of gait is associated with knee flexion and ankle dorsiflexion with facilitation of both the hamstrings and quadriceps influenced by mechanoreceptors in the cruciate ligaments. Body weight is transferred to the femur, which glides anteriorly on the tibial platform applying tension to the PCL, not on the ACL (Grimsby 1980). Mechanoreceptors in the ACL assist in facilitation of concentric quadriceps and eccentric hamstrings, whereas the PCL receptors assist in facilitation of concentric hamstring work and eccentric quadriceps. A preloading of the anteromedial and posterolateral fascicles in the ACL and the anterior and posterior fascicles in the PCL has an effect on facilitation of the hamstrings and quadriceps (Raunest et al. 1996). Static and dynamic loading of the anteromedial fibers of the ACL, create increased EMG activity in the hamstrings and a simultaneous decrease of the quadriceps. Loading of the posterolateral fibers of the ACL create a significant excitation of the quadriceps. When mechanical shear is applied both to the anterior and posterior fascicles of the PCL an activation of the ipsilateral quadriceps muscles is induced with a simultaneous inhibition of the hamstrings. The quality of mechanical loading (the static or dynamic shear) determines the amount of muscular activity, which demonstrates a proprioceptive mechanism of the cruciate ligaments (Raunest et al. 1996).

During the deceleration of knee flexion during the stance phase of gait, the quadriceps work eccentrically. As the knee approaches extension during "toe off" the quadriceps work concentrically, increasing the compressive tension on the patella and preventing further anterior glide of the femur. Knee extension involves anterior roll of the femoral condyles increasing tension on the PCL, which is reduced with concurrent posterior glide of the femur on the tibial plateau. The ACL, in turn, is less exposed to tension during normal gait than the PCL, which explains the fact that the PCL collagen fibers are thicker and stronger.

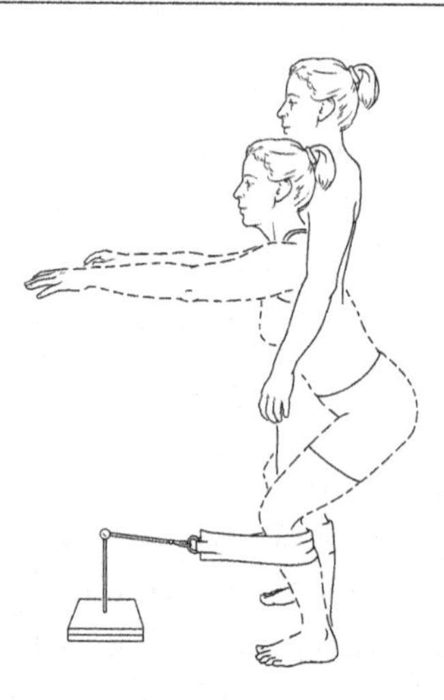

Figure 2.59: Squatting with an anterior glide of the tibia from a wall pulley. The strap is placed at the most proximal portion of the tibia creating an anterior moment of the tibia consistent with arthrokinematic motion for extension, and increasing collagen tension to stimulate greater afferent feedback. The result is increased motor recruitment for the extensor pattern of the lower limb. The concept of using a anterior force moment for improving quadriceps facilitation can also be applied to step-ups, step-downs, side steps and even lunges.

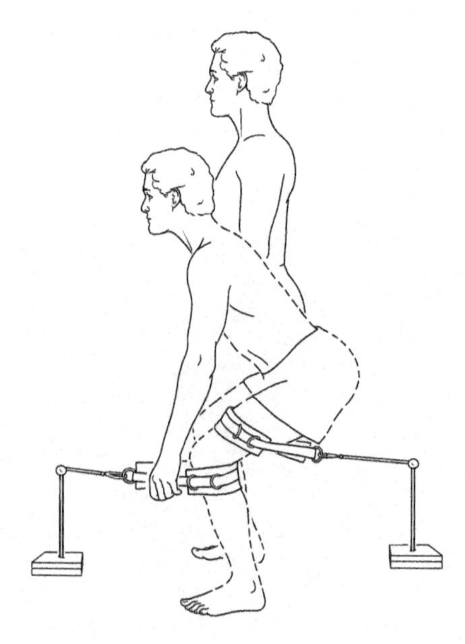

Figure 2.60: Squat with an anterior force moment on the tibia and posterior moment on the femur. As in the previous exercise example, it leads to an increase in recruitment of the extensor pattern musculature by promoting normal arthrokinematic motion at the knee joint and increasing collagen tension for improved afferent feedback. The opposing forces in this version have a greater biomechanical impact for mobilization into end range extension (relative anterior glide of the tibial surface and posterior glide of the femur). The patient is instructed to perform a squat. The lower strap creates an anterior force moment on collagen structures increasing mechanoreceptor afferent input for increased quadriceps and co-contraction of hamstrings. The double pulley set up can also be used with step-ups and lateral steps.

Basic Step / Pull Through Training

Step through training begins with mid stance gait training on the involved limb. An ankle strap around the swing leg is attached to a sport cord or wall pulley. The front-pull begins with the swing leg positioned posteriorly in a toe-off position and swings through to heel contact with body weight remaining on the stance leg. The back-pull exercise is reversed with the uninvolved limb swinging posteriorly from heel contact to a toe-off position. Hopkins et al. (1999) compared these exercises with sport cord on healthy subjects, comparing EMG activity between a lateral step-up and a unilateral squat. The front-pull and back-pull exercises with elastic resistance produced higher levels of biceps femoris activity than the lateral step-up during the knee extension phase. During knee flexion both the front-pull and back-pull exercises produced higher levels of biceps femoris activity than unilateral one-quarter squats and the lateral step-ups. The front pull was also associated with higher levels of recruitment for vastus medialis activity than the unilateral one-quarter squat, lateral step-up and back pull during knee extension.

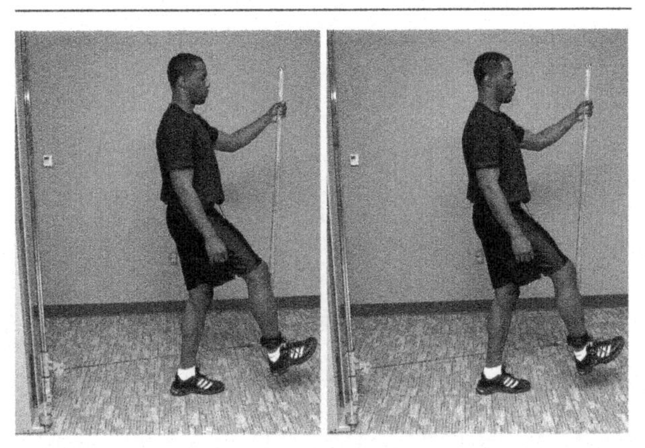

Figure 2.61a,b: The front-pull step through with wall pulley (optional sport cord) for knee stabilization training. An ankle strap on the uninvolved limb is attached to a wall pulley. The patient stands on the involved limb, while the uninvolved limb swings from toe-off to heel strike. Weight is maintained on the involved limb. The emphasis is placed on weight bearing stability of the involved leg, the non-weight bearing swing phase can also be performed on the involved side, with emphasis only on proximal hip and trunk strengthening.

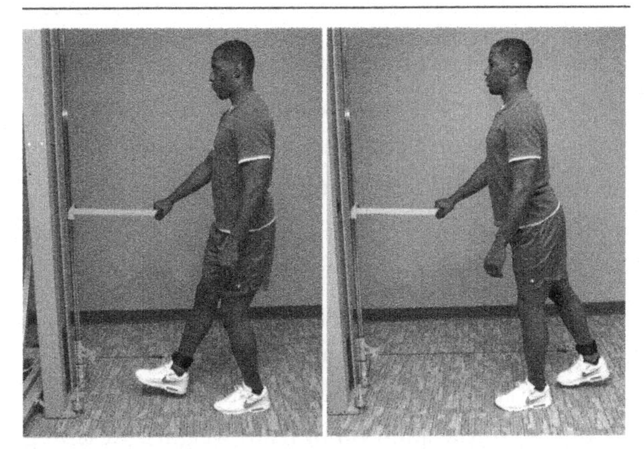

Figure 2.62a,b: The back-pull step through with wall pulley (optional sport cord) for knee stabilization training. An ankle strap on the uninvolved limb is attached to a wall pulley. The patient stands on the involved limb, while the uninvolved limb swings from a heel strike position back to a toe-off position. Weight is maintained on the involved limb.

Schulthies et al. (1998) analyzed the front-pull, back-pull, crossover-pull and reverse crossover-pull using elastic tubing. The front-pull had significantly greater VMO and vastus lateralis (VL) contributions, the back-pull was greatest for the VL, the front cross-over pull was greatest for semitendinosus and semimembranosus, and the back crossover pull was greatest for the VL and biceps femoris. During these tests, the hamstring to quadriceps ratio was greatest for the front-pull and crossover pull exercises.

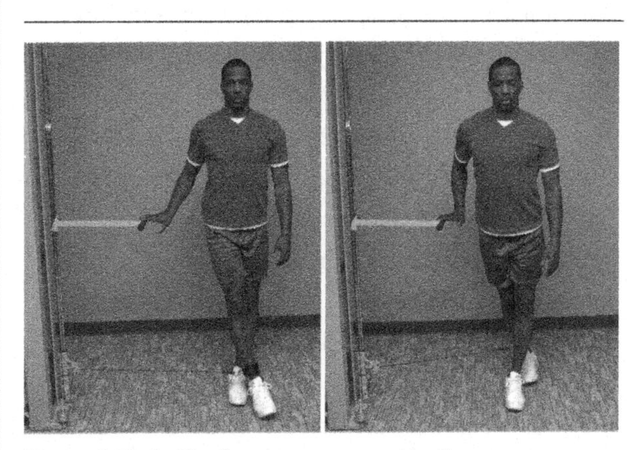

Figure 2.63a,b: The front crossover and back crossover.

Basic Lunging

Lunges can be performed in many ways with emphasis placed on the needs of the patient including range of motion, tissue repair, balance and/or stability. The basic anterior lunge recruits the vastus lateralis, vastus medialis and biceps femoris muscles as a unit during both the concentric and eccentric phases (Pincivero and Aldworth 1999). The lunge has been shown to increase quadriceps muscle activity and decrease hamstring muscle activity, compared with the power and front squats (Stuart et al. 1996).

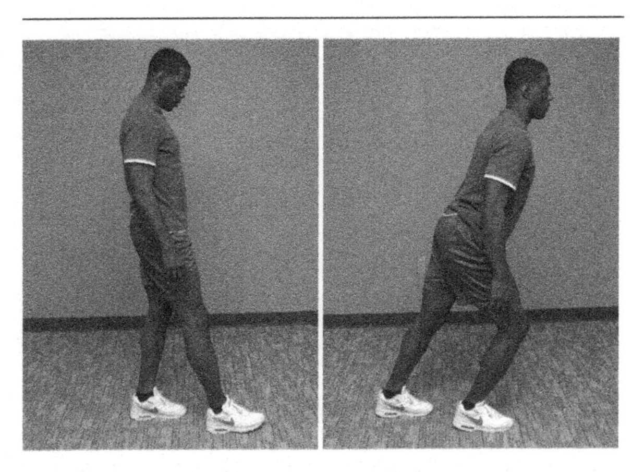

Figure 2.64a,b: Step-and-lunge. With pain or poor coordination lunging begins with the step-and-lunge. Weight

is maintained on the posteriorly placed uninvolved limb, while the involved limb is placed a short distance anteriorly on the ground. Weight is then slowly shifted onto the involved limb. As coordination and weight bearing tolerance improve, the amount of weight shifted anteriorly is increased prior to increasing the distance of the lunge.

Lunging can be performed in many ways to emphasize different aspects of joint motion and muscles recruitment. Early in knee rehabilitation the emphasis is often on restoring hip coordination and reducing the excessive forces transmitted to the knee. The fall-out lunge technique increases recruitment of the lumbar and hip muscles while reducing the load on the quadriceps and compression forces at the patella. The trunk remains in line with the posterior leg, the knee remains extended and the heel is fixed to the ground. Weight is transferred over the anterior leg with the nose over the front toes. Load should be felt in the front foot, without excessive force through the quadriceps and knee. The patient is instructed to avoid an abnormal valgus force moments at the knee. When the uninvolved leg steps forward, the involved limb is also trained for eccentric deceleration of the tibia from the soleus. More traditional lunges involve the trunk remaining vertical, the back knee flexing with the majority of the load placed on the quadriceps with excessive patellar compression. This may be a progression for later stages, but early on the emphasis is on activating the back and hip for more efficient force distribution and unloading injured tissues distal to the hip.

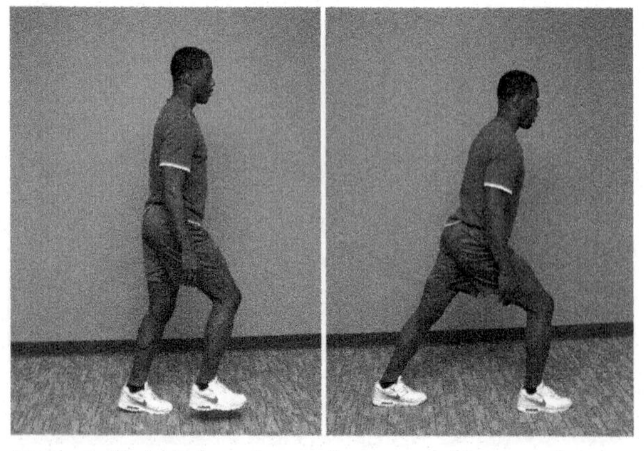

Figure 2.65a,b: Fall-out lunges for increased hip muscle recruitment, with decreased quadriceps emphasis and reduced patellofemoral compression. The patient steps forward with a slow weight transfer to the front foot. The back leg is kept straight in line with the spine, with the knee extended and the heel on the ground. The trunk is in line with the back leg, with the nose over the toe of the lead foot. As the center of body mass moves anterior, increased gluteal recruitment occurs with reduced quadriceps activation. If the weight remains more posterior, as occurs with a vertical trunk position, excessive force is placed on the knee with reduced hip muscle recruitment.

Figure 2.66: Anterior lunge with posterior upper quarter resistance. An anterior lunge is performed onto the involved limb. A posterior resistance attached to the waist, or bilaterally through the upper limbs (pictured), decreases the weight transfer anterior onto the involved limb. A twin wall pulley or sport cord can be used as resistance. This exercise may also be used to emphasize the stabilization challenge to the anterior trunk muscles or to push off on the back leg.

Initial coordination training can involve the arms as an assist. Reaching lateral during the anterior lunge will create a relative external rotation moment in the trunk and lower limb, preventing a valgus moment at the knee. The rotary moment can also improve facilitation of the deep hip rotator muscles during the exercise for improved dynamic stability. Progressions include lunging and reaching with different combinations and in different directions and/or with the addition of hand weights. Reaching lateral also shift more weight lateral to the hip axis, unloading the gluteus medius for hip and pelvis stabilization. With significant lateral hip weakness the lateral reach may serve as an assist. Later progression may include reaching forward and then medial to gradually increase the relative body weight medial to the axis of hip motion for abduction. Creative thinking during exercise design can unload or load specific joint tissues and/or muscles.

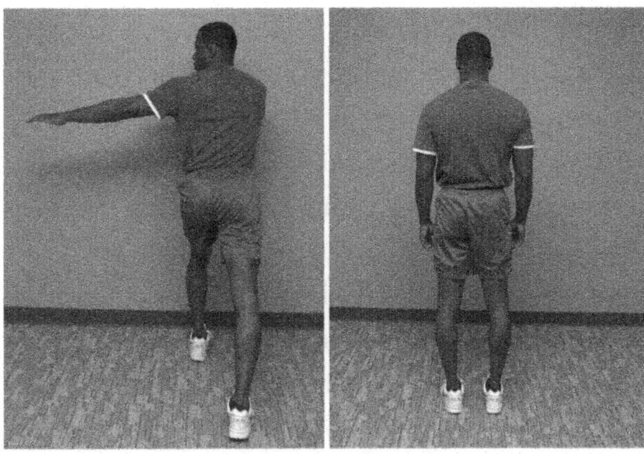

Figure 2.67a,b: Example of left anterior lunge with left lateral arm reach. The lateral reach will assist weakened hip abductors by shifting weight from medial to the joint axis to lateral. Anterior lunge with a lateral upper limb reach will reduce body weight medial to the sagittal axis of abduction for the hip. The rotational vector also increases muscle recruitment of the left hip external rotators.

A similar rotary moment at the hip can be created with a medially directed force at the lower thigh. A pulley line or elastic resistance is placed just proximal to the knee joint, which acts to facilitate the hip abductors and external rotators.

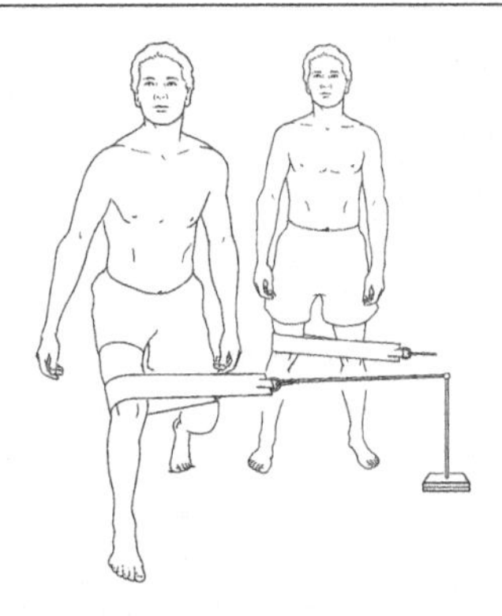

Figure 2.68: Step with unilateral hip abduction challenge—sagittal plane vector. The patient steps forward with appropriate weight added to the pulley for active recruitment of the hip abductors, maintaining the hip and knee in proper alignment. Elastic resistance can be used in the home for the valgus force moment.

The same concept can be applied to improve recruitment of the gastrocnemius and soleus complex for eccentric deceleration the tibia on the talus during gait. The line of resistance is placed to create an anterior tibial force moment that must be countered by the gastrocnemius-soleus complex while the uninvolved limb performs an anterior lunge, as previously shown.

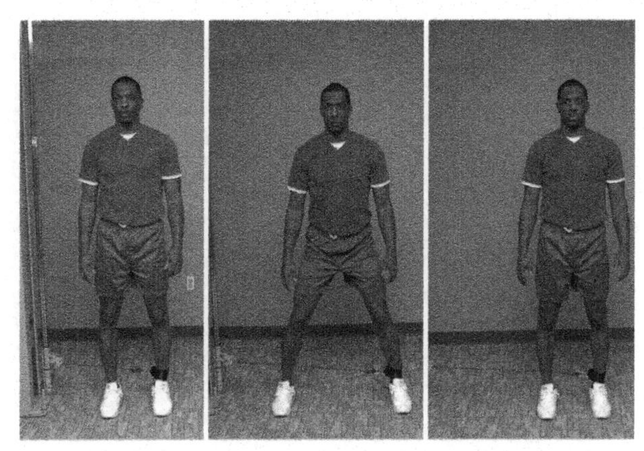

Figure 2.69a,b,c: Resisted left lateral walking with a pulley attached at the ankle. Emphasis is placed on the open chain abduction to the hip abductors on the left. The gluteus medius will be targeted with the abduction motion, but the tensor fascia lata will all have significant recruitment. This is not recommended early hip bursitis or "TFL syndrome."

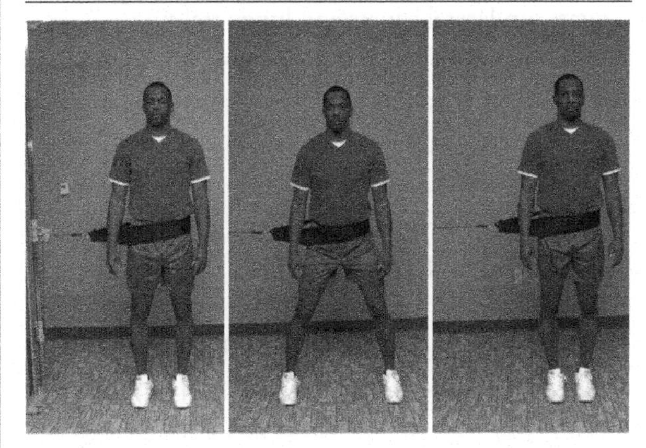

Figure 2.70a,b,c: Resisted left lateral walking with a pulley force moment attached at the pelvis, proximal to the hip. Emphasis is placed on the closed chain abduction of the right hip, including the hip abductors and external rotators.

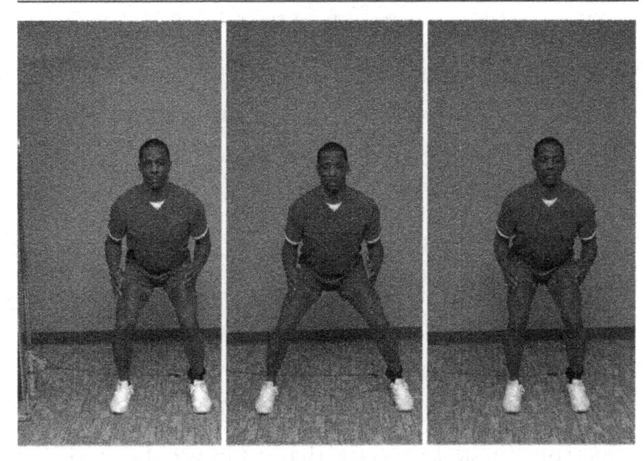

Figure 2.71a,b,c: Resisted left lateral crab walking with a pulley force moment attached at the ankle. Emphasis is still placed on the closed chain abduction of the right hip. The flexed position of the hip reduces the relative contribution of the tensor fascia lata, with increased emphasis on the gluteus medius as the primary abductor.

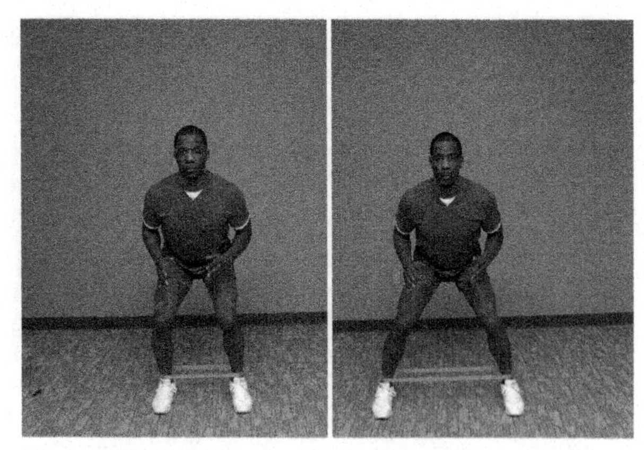

Figure 2.72a,b: Crab walk with elastic band for resistance.

Basic Step-Up / Step-Down Training

Step training is a basic progression from bilateral squats, unilateral squats and lunge training. Coordination of hip and knee muscles is necessary to avoid a valgus knee moment or hip adduction during the step-up motion. Step-downs require eccentric stabilization of the entire kinetic chain, from the lumbar spine to the foot, to avoid collapse of knee during the landing phase. Step-ups and the single leg squats demonstrate high and sustained levels of quadriceps activation, making them effective in rehabilitation (Beutler et al. 2002).

The step-up can be performed as a pure concentric activity, with the uninvolved limb performing the step-down eccentric phase. On the other hand, the step-down can be performed as a pure eccentric activity, with the uninvolved limb performing the step-up concentric phase. Progression includes performing both the concentric and eccentric phases in the step-up and step-down exercises.

The step height should initially be set at a functional level that can be managed by the available range of motion of the involved knee. It is more functional to train at a normal step height and provide assistance to allow for pain free coordinated motion, than to lower the step height below a typical level of six to eight inches. Many types of unloading devices can be used to reduce the body weight to allow for a pain free performance. Utilizing work physiology concepts associated with speed of training precludes the need for unloading equipment. Concentric peak torque is increased at slower speeds, which may provide time for additional motor units to be recruited when performing a slower step-up. Conversely, eccentric peak torque is greater with increased speed that may allow for pain free completion of the step-down at faster speeds. The involved limb acts to eccentrically lower the body with the uninvolved limb accepting the body weight on landing. Speed can be gradually decreased with the step-down to increase the challenge and train a more functional movement.

- *Decrease speed to assist concentric step-up*
- *Increase speed to assist eccentric step-down*

Figure 2.73a,b: Unloaded step-ups. Body weight is reduced for step-ups by holding on to a gantry from a pull down machine. The patient places the foot of the involved limb on the step, shifts weight onto the foot and then steps up. Enough weight is added to the pull down machine to allow for the motion to be pain free, coordinated with proper lower limb alignment and achieve the desire number of repetitions.

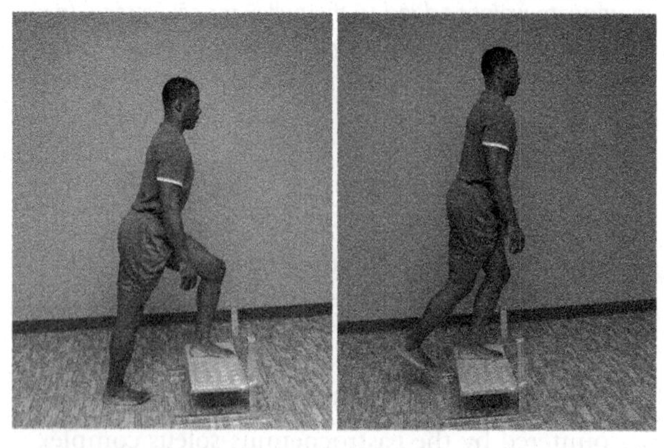

Figure 2.74a,b: Anterior step-up. The patient places foot on step of appropriate height. Weight is transferred to the anterior leg prior to extending the hip and knee. The back

foot can be prevented from plantar flexing for assistance with dorsiflexion of the toes prior to the opposite leg stepping up.

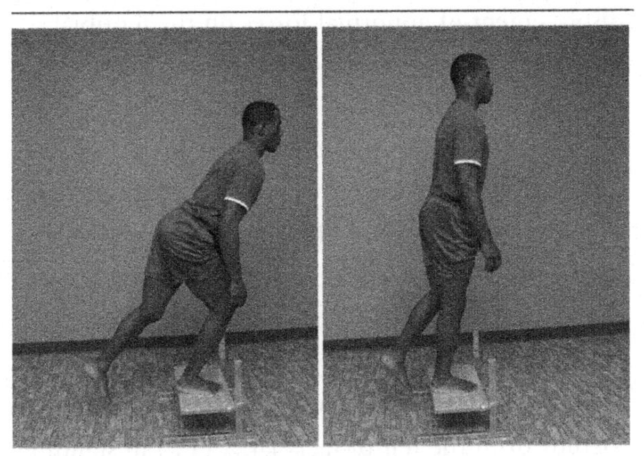

Figure 2.75a,b: Step-up with increased hip musculature emphasis. The hip can be emphasized by placing the foot on the step, flexing at the hip to lean the trunk over the foot, stepping up and then straightening the trunk.

Figure 2.76a,b: Step-downs—unloaded. The patient is placed in the unloading cage with the desired amount of weight added to the pulley and standing on an elevated surface of an appropriate height. The patient steps off the surface in a controlled manner while keeping the weight bearing knee in alignment with the toes and avoiding knee valgus.

Anderson and Herrington (2003) established a relationship between patellofemoral pain syndrome and poor eccentric control of weight bearing knee flexion by the quadriceps. A break in eccentric performance was noted in a higher percentage of patellofemoral pain patients than controls, assessed with eccentric isokinetic testing and functionally descending stairs. The break, or giving-way reflex, was theorized as a response, not just to the degree of pain but as a way to prevent further stress on joint and soft tissue structures. These findings stress the neurological retraining aspects of dysfunction.

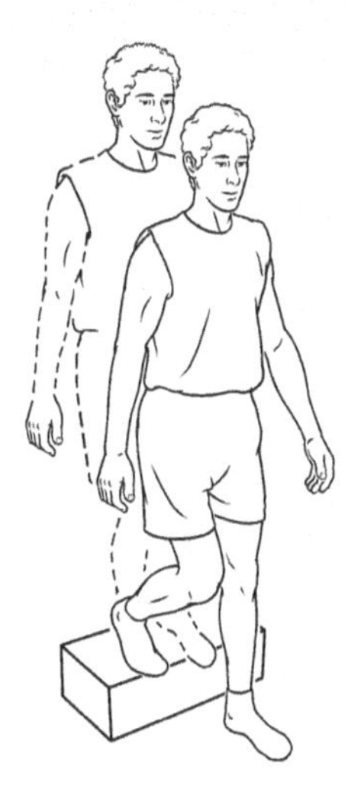

Figure 2.77: Step-down (Eccentric emphasis). From the edge of the platform, the patient steps forward and down onto the uninvolved foot. The uninvolved foot then step back and up onto the step. In this way the involved side limb performs the eccentric phase only. As peak torque for eccentric work increases with speed, the patient can be instructed to allow the body weight to fall down more quickly on the eccentric phase to allow for a pain free and coordinated movement for the desired number of repetitions. Lowering the step height can also reduce the challenge, removing pain, but training at a higher speed eccentrically at a functional step height is more desirable for early functional training.

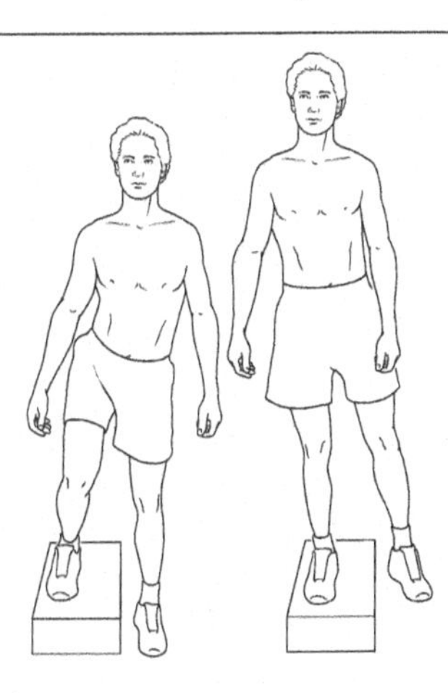

Figure 2.78a,b: Lateral step-ups (concentric emphasis). The patient stands next to the lateral side of a step. Lateral steps increase emphasize lateral hip and medial thigh musculature. The foot of the involved limb is placed up onto the step. The patient lifts the body weight up until both feet are on the step. To increase contribution from the hip, the trunk can be slightly flexed forward at the hip prior to the step-up. To avoid plantarflexion from the uninvolved limb assisting the motion the patient can be instructed to dorsiflex the digits of the uninvolved foot. Swinging the arms up and forward will assist the motion. To minimize the eccentric phase, the patient can be instructed to fall off the step, catching the body weight with the uninvolved limb, rather than eccentrically lowering with the involved limb.

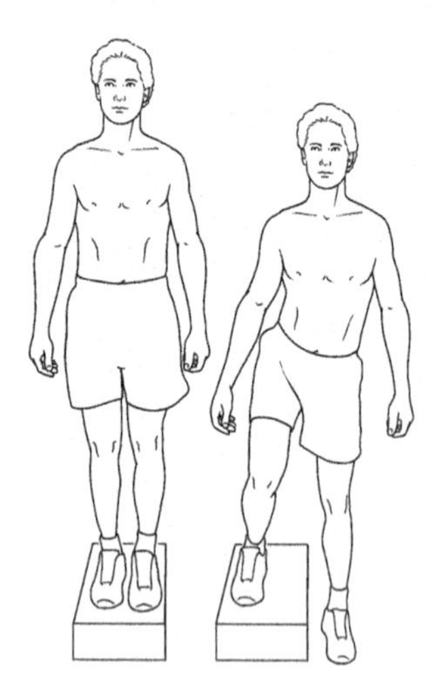

Figure 2.79a,b: Lateral step-downs (eccentric emphasis). The patient stands with the foot of the uninvolved limb close to the edge of the step. The foot steps down with the involved knee eccentrically controlling the movement. The lateral step down places more emphasis on the hip abductors, rather than the quadriceps, during the step-down maneuver. This approach places less tissue stress on the cartilage of the patellofemoral joint, compared to the anterior step-down, as well as reducing the potential for an abnormal valgus moment at the knee. Again, slightly flexing the trunk forward at the hip will increase the contribution of hip musculature to the movement, unloading the quadriceps and knee joint.

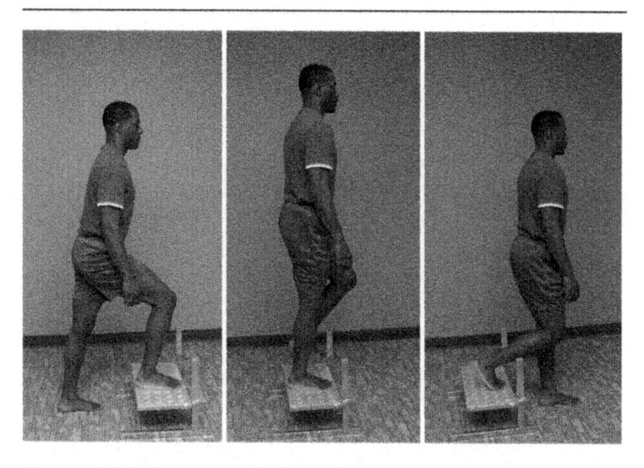

Figure 2.80a,b,c: Step-up and over. A combination of concentric and eccentric work is achieved with the step-up and over exercise. Subjects with reconstructed ACLs have demonstrated reduced force generation and slower performance times than controls when performing this exercise (Mattacola et al. 2004).

Adjunct Hip Training

Anterior fibers of the iliacus contribute to hip flexion, with the posterior fibers contributing to hip abduction and internal rotation (Pare et al. 1981). Weakness or inhibition of the gluteus medius can cause compensation by the anterior fibers of the iliacus to aide hip abduction leading to overuse and displacement of abnormal forces on the iliotibial band (ITB). Iliotibial band syndrome (ITBS) should not be considered a freestanding pathology, but rather a symptom of mechanical breakdown of forces throughout the lower limb with closed chain activities. Symptoms of ITBS frequently correlate with gluteus medius weakness. EMG studies have shown that both athletic and arthritic hips with symptoms of ITBS usually have poor gluteus medius facilitation and an inability to isolate gluteus medius from the tensor facia lata (TFL) in standing (Fredericson 2000, Kasman 2002). In long distance runners with ITBS, symptom improvement and return to preinjury training correlates with increased hip abductor strength (Fredericson 2000).

Retraining coordinated function of the gluteal muscles requires establishing a normal neurophysiological movement pattern. Starting in a partial or non-weight bearing position and, in some cases, adding an external line of force may be necessary to facilitate the gluteus medius. Progressing to full weight bearing training in functional triplanar patterns, as well as incorporating jump training for eccentric and plyometric capacity, are required to achieve a higher level of dynamic stability.

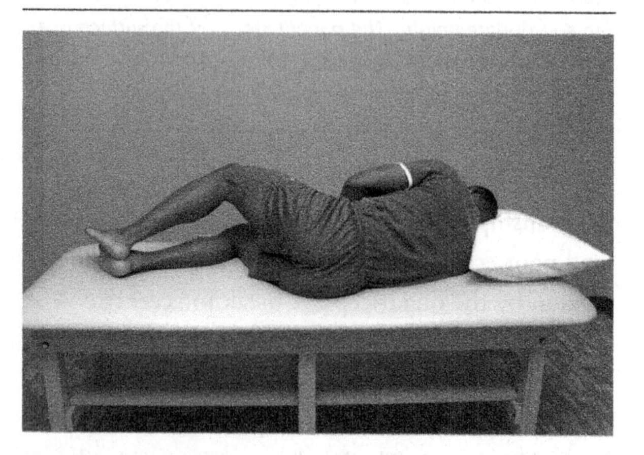

Figure 2.81: Hip Clam Exercise: sidelying horizontal hip abduction to attempt to isolate the gluteus medius and reduce recruitment of the tensor fascia lata. The knee is lifted keeping the feet together and without rotation of the pelvis and lumbar spine. Performing muscle isolation exercise in non-weight bearing is of limited functional value, but may be a necessary starting point in painful or deconditioned training

states. A cuff weight at the knee may allow for more specific exercise dosage, but this exercise typically emphasizes the functional quality of coordination with only the weight of the leg. As coordination of the exercise is achieved, a functional progression into weight bearing may be more appropriate than the addition of weight.

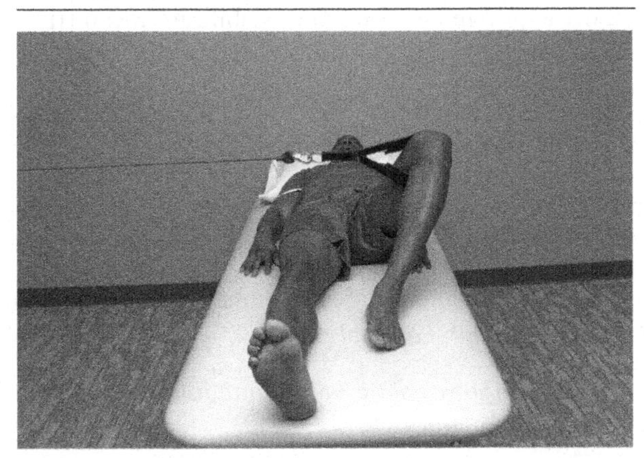

Figure 2.82. Horizontal hip abduction in supine with pulley resistance. Gluteus medius training is emphasized with down training of the tensor fascia lata. As a non-weight bearing exercise for the hip, there may be limited carry over to weight bearing function. Progression to a weight bearing option is more appropriate than progressive weight training.

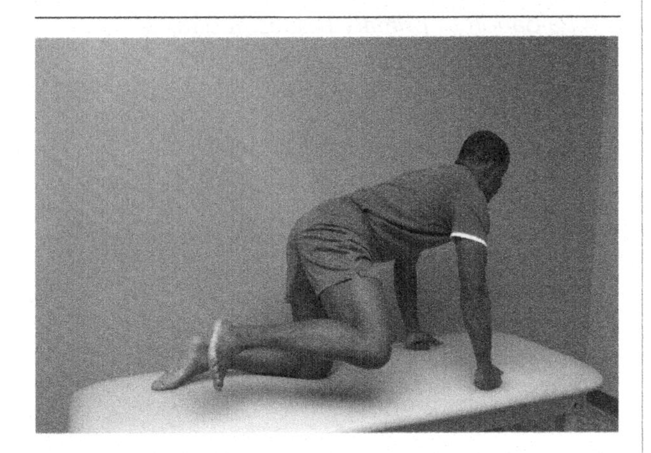

Figure 2.83: Hip Fire Hydrant: quadruped horizontal hip abduction to attempt to facilitate the gluteus medius and reduce recruitment of the tensor fascia lata. The knee is lifted laterally in isolation, followed by rotation of the pelvis and finally abduction of the weight bearing hip. The gluteus medius is facilitated bilaterally with increased activity associated with increased afferent input with partial weight bearing and the cross over effect.

Non-weight bearing training of the hip abductors may be a necessary step for limitations in tissue tolerance and overall strength. Muscle strength needed for non-weight bearing does not meet the torque levels needed for weight bearing training. Functional training in weight bearing is a necessary progression for normalizing hip strength.

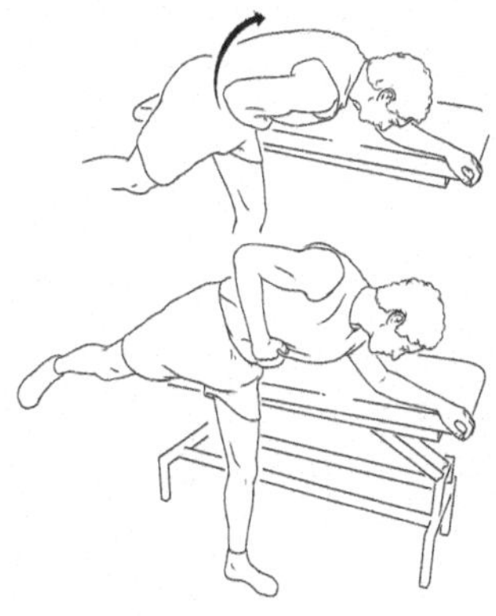

Figure 2.84a,b: Weight bearing hip horizontal abduction. Horizontal left hip abduction, with the hip in a flexed position, will increase the demand on the gluteus medius while reducing recruitment of the tensor fascia lata by placing it in a relatively shortened position with hip flexion.

Squatting can also be modified to emphasize function at the hip. Combining upper limb reaches with squats, lunges and step training can either assist or increase the challenge of balance, lower limb alignment and loading on specific muscles. Reaching with the upper extremity shifts the relative amount of body weight in the direction of the reach. Weight shifted lateral to the hip joint moves more weight lateral to the hip axis of hip abduction, thus reducing the amount of weight the hip abductors need to stabilize. The direction of arm reach can also increase the recruitment of specific muscles, as in an anterior reaching increasing hamstrings and gluteus maximus recruitment.

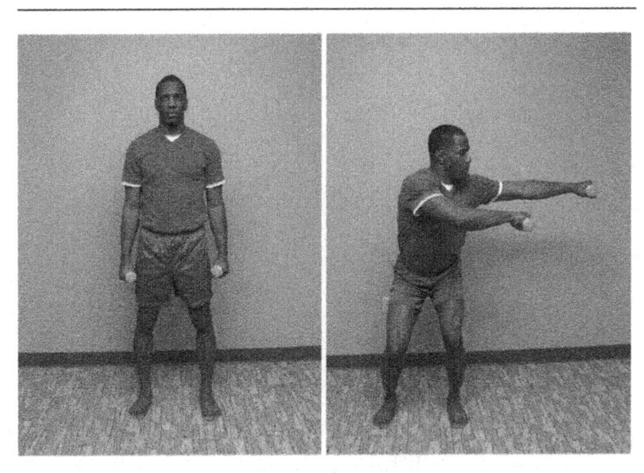

Figure 2.85a,b: Squat with medial reach (relative to involved limb)—away from involved knee (right). Reaching across the body shifts the center of gravity away from the involved knee, reducing the level of torque required at the hip and knee. This may allow for squatting without a valgus compensation at the knee and/or pain. As coordination improves the arms can be held at the side, progressing to a lateral reach.

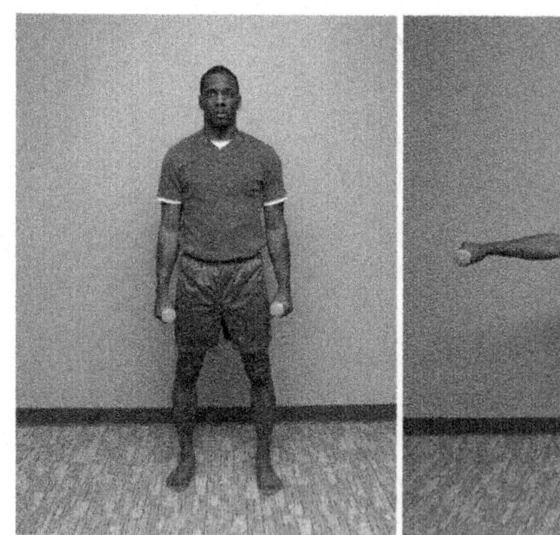

Figure 2.86a,b: Squat with lateral arm reach (relative to involved right). The lateral reach will shift the center of gravity toward the involved lower limb increasing the torque moment at the hip and knee. This may present a greater challenge for avoiding a valgus compensation at the knee. During the eccentric phase of the squat the arms reach lateral to the involved knee that acts to also increase the rotational components of the gluteus medius in the sagittal plane.

Summary

Stage 2 focuses on normalizing knee range of motion and establishing coordination in partial to full weight bearing functional exercises including squats, lunges and steps. Pain is no longer an issue, other than with excessive loading or training. Progression to more complex movement patterns, as well as strength and power training are addressed in Stages 3 and 4.

Section 3: Stage 3 Exercise Progression for the Knee

Training Goals

Progression to Stage 3 concepts comes with resolution of impairments from Stage 2. Pain is resolved with general activity, although the healing tissues may be still be reactive with higher loads or longer duration of activity. Range of motion is normalized due to restoration of capsular mobility and/or elimination of muscle guarding patterns. Tissue tolerance has improved for normal joint loading and mobility, but requires increased loading for continued improvement in stress/strain tolerance. Coordination of basic movement patterns has been established, but dynamic stabilization is still necessary with an eccentric emphasis shifting toward more functional activities and sport requirements. Isometric stabilization can still be utilized to improve dynamic stabilization via sensitization of muscle spindle receptors at higher loads in specific ranges of instability.

Tissue and Functional States—Stage 3

- *Full arthrokinematic and osteokinematic motion with cardinal planes.*
- *Full weight bearing with limitation in loaded weight bearing.*
- *Joint may be painful with excessive repetitions.*
- *Edema has resolved pre and post activity.*
- *Muscle guarding has resolved.*
- *Palpation of primary tissues negative, provocation tests negative, trigger points may be positive to deep palpation.*
- *Fair to Good coordination in planar motions—limitations in tri-planar motion.*
- *Fair to Good balance/functional status—limitation in functional challenges.*
- *May have reduced fast coordination, endurance, strength and power.*

A level of endurance has also been achieved with previous training, though strength is not fully returned. Quadriceps strengthening is an effective approach regardless of the patient's age, sex, body composition, athletic level, duration of symptoms, or biomechanical malalignment in the lower extremities (Kannus and Niittymäki 1994). Most protocols speak of strength training much earlier, but joint motion and coordination must be established along with a level of muscle endurance prior to beginning true strength training. For athletes, exercise design should attempt to incorporate more sport-specific exercises to improve neuromuscular coordination and timing to protect against future injury. Educating the patient on how to reduce the risk of re-injury is necessary for any patient. Some conditions, such as osteoarthritis, may result in some level of permanent tissue damage and impairment. A progressive rehabilitation program that includes active therapeutic exercise may help delay the progression of disorders such as osteoarthritis and give patients more years of pain-free activity and improved quality of life (Vad et al. 2002).

Basic Training Goals Stage 3: Dynamic Stability

- *Increase resistance*
- *Resistance increased to 70–80% of 1RM*
- *Strength and endurance*
- *Add isometrics toward pathological range*
- *Weight bearing functional exercises*
- *Avoid compensations/malalignment*
- *Concentric and eccentric exercises*
- *Functional patterns*

Increase Resistance / Speed

Weight bearing exercises have many variables that can be modified to increase the tissue challenge. In Stage 3 resistance is progressively increased along the coordination-endurance-strength spectrum ranging between 70–80% of 1RM. The increased resistance will provide greater tissue stress/strain and hypertrophy of the targeted muscles for long-term strength gains.

Speed can also be used as a variable to increase resistance to a muscle group instead of increasing the load (weight). In some cases, increasing speed may be of more importance than resistance, as in examples of sports that require a greater speed element than a slow strength performance. Speed not only affects local muscle performance but can also have an impact on the cardiovascular system. For example, running provides a more significant increase in cardiovascular challenge than walking.

Eccentric Stabilization

Basic stabilization concepts involve progressive training of tonic muscles that contribute to dynamic control of arthrokinematic motion. Concentric work emphasis was employed first to emphasize coordination, reduce muscle guarding and/or facilitate inhibited muscle. Isometric work in Stage 2 was then incorporated to improve neurological adaptation for motor facilitation. Stage 3 concepts now emphasize eccentric work for greater muscle spindle sensitivity and dynamic control of the joint system Eccentric emphasis or "lengthening contractions" were used in the 1960's for meniscectomy rehabilitation (Kressley 1963). The progression to eccentric stabilization is more deliberate for upper limb and axial skeletal training. To emphasize EW, the work order is changed to EW first with CW return. The force moment from the resistance is adjusted for eccentric length tension in the shortened range of the muscle. Eccentric emphasis for the lower limb is already built into the functional demand of closed chain training, and requires less of a deliberate adaptation of training unlike the upper limb and axial skeleton. Open chain training requires more specific set up of the length tension curve and work order, whereas closed chain training naturally adjusts these variables for the lower limb.

The squat motion begins with the knee in full extension. The entire lower quarter is involved in eccentric deceleration of the body weight against gravity with a concentric return to the start position. Emphasis is placed on a quick transition between the eccentric and concentric phases. A slow transition results in a loss of energy during the stretch-shortening cycle, and is dissipated as heat. A progression to jump training emphasizes a quicker, more ballistic, transition between the work phases.

Endurance to Strength Training

Dynamic stabilization, through coordination training and sensitization of afferent receptors, requires a progression into strength training. Resistance is elevated 10–20 RM percentage points from previous levels to approximately 75–85% of 1RM. Patients starting progressive resistive training at lower training states may take longer to reach these levels. Improved vascularization and hypertrophy are both desired outcomes with progressive resistive training. Oxygen debt and depletion of glucose with training will create a response for more capillaries to bring blood to the area improving perfusion to the tissue. The rest periods between sets and training sessions allow the body to replenish the system and progressively decrease as endurance improves within the muscle group. But increasing resistance will actually require greater restitution times due to tissue breakdown from increased forces placed on the joint complex. At the end of a strength training session, the muscle system should be exhausted and may take up to 24 hours to fully recuperate from the challenge. The body responds by creating more sarcomeres and capillaries leading to hypertrophy and increased strength. Higher end strength training at 90% of 1RM may require several days between training sessions for muscle recovery, but this is more associated with training in normal subjects, not in rehabilitation of pathology.

Motions and Directions

Full range of motion for the knee is initially measured and trained in a non-weight bearing position, as there is less demand on the joint. Full range in weight bearing requires significant coordination and strength as well as tissue tolerance. As passive joint motion improves motor function must also improve within the new available range for better dynamic control. Since the knee primarily functions eccentrically, the basic concept is to train the quadriceps to eccentrically decelerate knee flexion and concentrically contract to reestablish an erect standing posture. However, training concepts may be predicated upon treatment needs for specific pathologies. For example, prone hamstring eccentric training is necessary for ACL deficient knees to control excessive anterior translation of the tibia. Pure eccentric training for hamstring strains is another strategy utilized for collagen repair as outlined in section 1.

Direction of training also needs to reflect gross functional requirements that can be assessed with gait, balance, squatting, lunging, step and jumping tests. Stage 3 should reflect a more specific match between the impairments and functional needs of the patient. Generalized protocols may not be appropriate for all situations. Neeter et al. (2006) found that testing power for knee extension, knee flexion and leg-press had a high ability to determine deficits in leg power six months after ACL injury and reconstruction. They were able to discriminate between the leg power performance on the injured and uninjured side, both in patients with an ACL injury and those who have undergone an ACL reconstruction surgery.

These types of tests may help in deciding when and whether patients can safely return to strenuous physical activities after an ACL injury or reconstruction, but they should be used in conjunction with functional performance testing. Functional tests that measure increases in strength alone may not be the most sensitive in determining the success of a rehabilitation program. Augustsson and Thomee (2000) showed that a moderately strong correlation exists between the test of functional performance and muscular strength tests for both closed and open chain positions. They suggested that the effect of training or rehabilitation interventions should not be based exclusively on tests of muscular strength but rather on various forms of dynamometry including functional performance tests.

Weight Bearing Closed Chain Progression

Bracing or assistive devices are discarded during training exercises in Stage 3 as full weight bearing tolerance has been achieved. Bracing may still be used as an adjunct to athletic performance or competition outside the rehabilitation setting. Stairs are ascended and descended in alternating steps and step-ups can be progressed by increasing the height of the step and/or emphasizing the eccentric component at different speeds. Resistance can be increased with aerobic equipment, by inclining a treadmill or using a stair stepper. Basic weight bearing exercises are progressed in range, direction, resistance and speed as well as with balance and surface challenges.

Squat Progressions

The basic progression for squats is to increase the resistance. A weight bar can be used posteriorly across the shoulders or held anteriorly with crossed arms to unload body weight. Dumbbells or weighted boxes can also be used for more functional squat-lift training to increase resistance. Single-limb squats can be incorporated to increase the overall torque on the lower limb, challenge balance and replicate any necessary functional requirements. Single-limb squats at increased levels of resistance have been shown to increase the co-activation of the hamstring muscles along with the quadriceps, which is essential to optimize neuromuscular control of the knee (Sheilds et al. 2005). Single-limb squats can initially be assisted by unweighting the body, or providing upper limb assist and progressed to full body weight and an eventual addition of resistance.

Figure 2.87a,b: Unilateral partial squats unloading with a pull down machine with gravity assist cage. Enough weight is used to decrease the body weight to allow for a partial squat with proper coordination and alignment. If a gantry is not available a regular lat bar or double grip handle can be used. The entire weight stack is used to create a fixed line. The patient pulls up on the handle during the squat motion with enough assist to allow for a pain free squat performance. The similar approach can be used for step-up training.

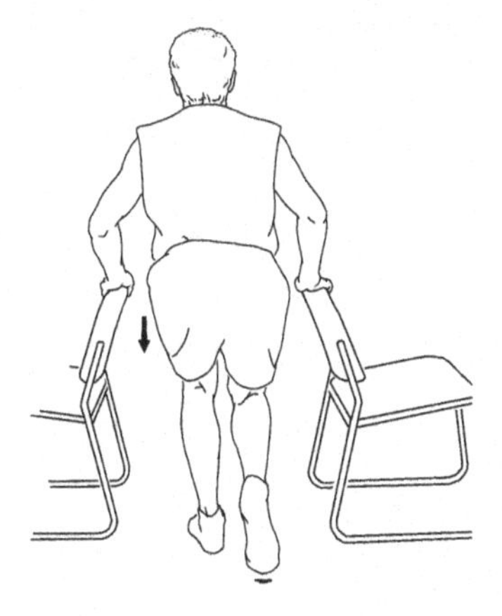

Figure 2.88: Home exercise option: unilateral partial squats unloading through the arms using two chairs.

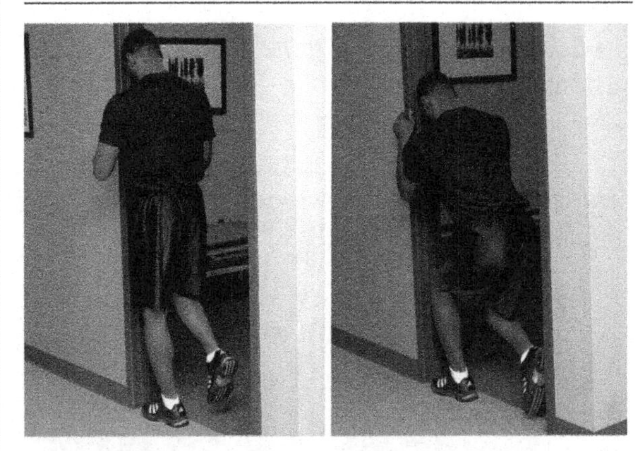

Figure 2.89a,b: The door jam squat can be also be performed as a single leg squat. Upper limb support reduces the load on the knee and allows for increased knee flexion to coordinate motion through a greater range of motion.

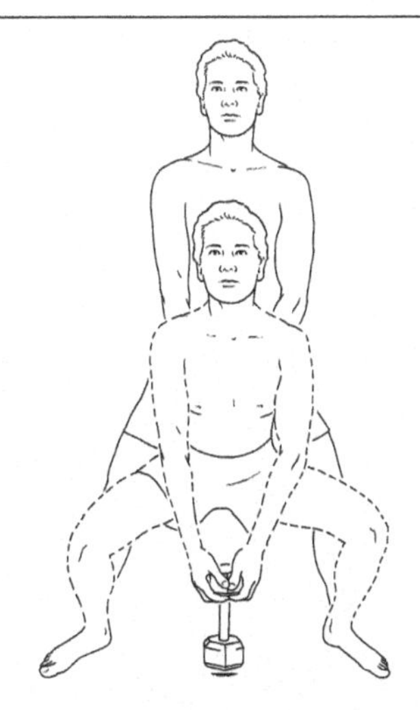

Figure 2.90: The Plie squat, or Sumo squat, is performed with a wide stance, abducting and externally rotating the hips. Emphasis is placed on posterior motion of the pelvis while maintaining the knees over the feet. This stance prevents a valgus moment at the knee and may unload patellofemoral tissues from lateral tracking issues. Progression is gradually made back to a more traditional foot position as muscle coordination and tissue tolerance improve.

Squat Variations / Challenges

- *Bilateral squat with full body weight*
- *Resisted squats—dumbbells, weight bar, sport cord or weight vest*
- *Squat lift training with box weights*
- *Unilateral squats—body weight and resisted*
- *Plie squat—wide foot stance*
- *Narrow foot stance*
- *Squat with upper limb reaches*
- *Squatting on labile surfaces*
- *Squatting with pulley trunk diagonal lifting patterns*

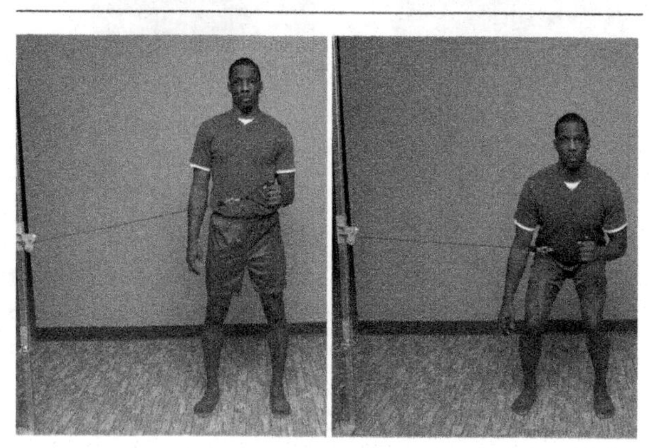

Figure 2.91a,b: Multidirectional resistance for squat training with left arm resistance from a right lateral force moment. The pulley recruits a diagonal line of muscle from the left arm, crossing the lumbodorsal fascia, into the right hip. Posterior musculature is emphasized, including the lumbar multifidi and the deep hip external rotators.

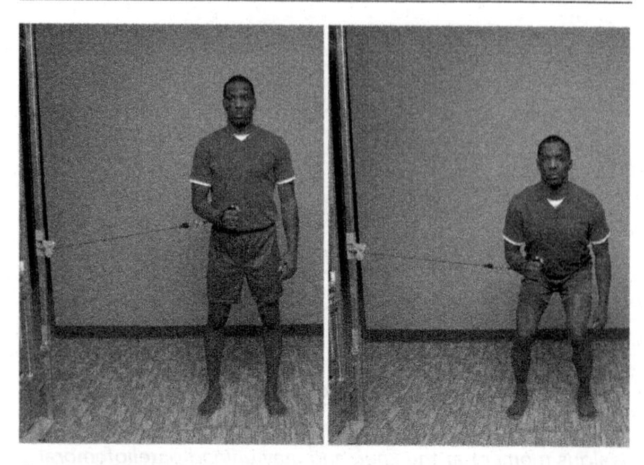

Figure 2.92a,b: Multidirectional squat training with right arm resistance from a right lateral force moment. The pulley recruits a diagonal line of muscle from the right arm, crossing anteriorly to the left hip. Anterior musculature of the hip and trunk are emphasized, including the transverse abdominus, the obliquus' and left hip flexors.

Step Training Progression

Basic step training in Stage 2 focused on improving coordination and endurance with ascending and descending stairs. Progression of step training can focus on different variables, depending upon the functional demand. Increasing weight will impart strength qualities and increasing the step height will help to improve dynamic control through a greater range of motion. Balance, coordination and agility may be addressed by adding lower limb rotations to the task. Any one of these options or a combination may be emphasized to vary the challenge based on functional needs of the patient.

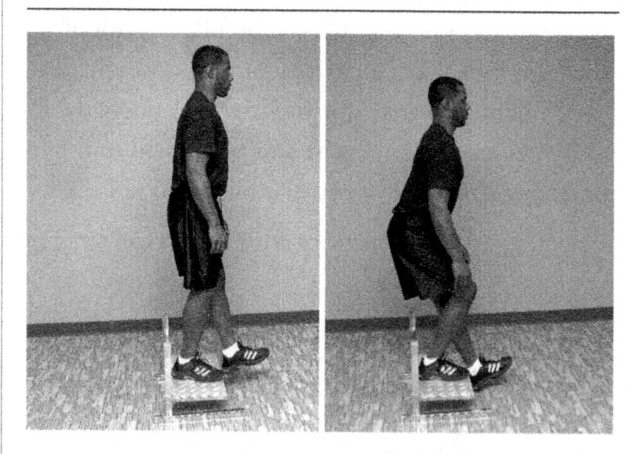

Figure 2.93a,b: Step-down with return. The patient steps down off the platform, keeping the knee centered over the foot, lightly touching the heel and then returning to the start position. Eccentric deceleration is emphasized to prevent transfer of weight onto the forward foot.

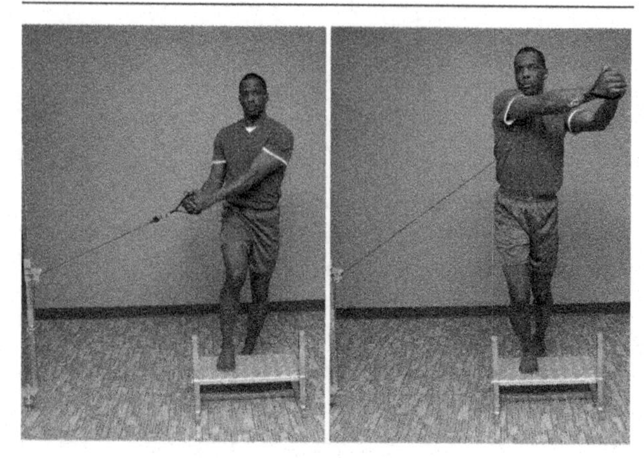

Figure 2.94a,b: Multidirectional right step-up with a right inferiolateral force moment. An extension moment for the hip

and trunk is created with gravity during the step-up motion. The right lateral force increases the extension moment from the inferior vector, but also a right trunk rotational moment. The arms swing through to the left, creating an left rotational trunk pattern, increasing activity of the right lumbar multifidi, right hip deep external rotators, the oblique abdominals and the transverse abdominus.

Step Progression / Challenges

- *Step-up with increase speed (CW) for motor challenge*
- *Step down with decrease speed (EW) for motor challenge*
- *Increase step height*
- *Step with hand weights/weight bar*
- *Step-up/down in different directions (i.e., anterior, anterolateral, lateral or posterior etc.)*
- *Step-up/down with external rotation of foot*
- *Step-up with external rotation of pelvis*
- *Step-up/down with arm reaches*
- *Step-up and hop*
- *Step-down and jump—unilateral and bilateral jump*

Lunge Progressions

Lunges in Stage 2 emphasized coordination of the lower kinetic chain for maintaining alignment during the loading phase. In the frontal plane, the hip, knee and ankle remain in alignment with each other and in relation to the trunk. For the fallout lunge technique the back knee is kept straight with the trunk leaning forward in line with the back leg to emphasize hip and lumbar musculature recruitment. For a quadriceps emphasis the trunk is kept vertical, the back knee flexed and the majority of load is placed on the knee. The latter option is not typically recommended for patellofemoral or tendinopathy issues, as excessive stress and strain is placed on tissue of the anterior aspect of the knee. Many option on lunging exist, there is not "one way" to do them. The style of squat used should match the tissue injury and the functional need.

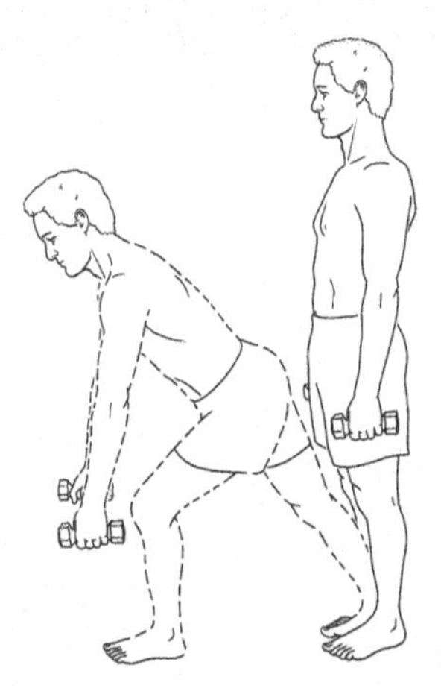

Figure 2.95: Fallout lunge. The patient stands straight and leans forward keeping the trunk and legs in a straight line. The foot makes contact with the floor catching the body in the end position with quick isometric stabilization. The back leg is kept straight in line with the trunk. The patient pushes off the front leg through the heels to return to the starting position. Options for resistance include a weight bar placed on the shoulders or dumbbells held in the hands.

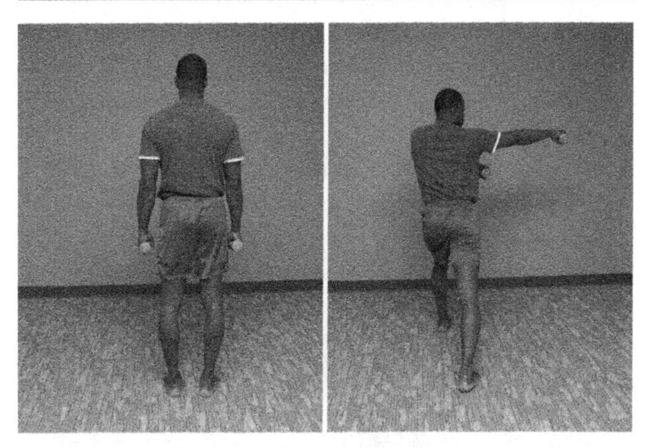

Figure 2.96a,b: Left anterior lunge with medial upper extremity reach. The medical reach will increase the load on left hip abductor muscles and increase the balance challenge

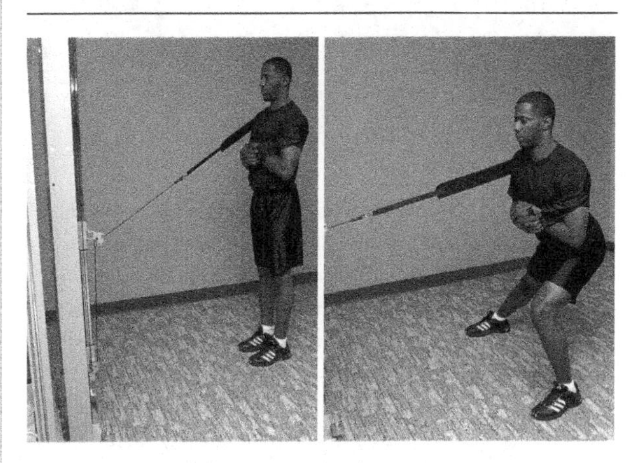

Figure 2.97a,b: Multidirectional left lateral lunge training with an anterior force moment at the opposite shoulder (for lumbar weakness). The lateral lunge emphasizes the hip and lower extremity, with an extension moment against

gravity. The anterior pulley line further increases the extensor moment from the trunk, but the line is also attached to the right shoulder and wrapped around the body. Attachment to the right shoulder causes a relative left rotational trunk moment. The lumbar multifidi on the left are recruited, along with the deep external rotators of the left hip, to resist this force moment.

Figure 2.98a,b: Multidirectional right lateral lunge training with left arm resistance from a right lateral force moment. The pulley recruits a diagonal line of muscle from the left arm, crossing the lumbodorsal fascia, into the right hip. Posterior musculature is emphasized, including the lumbar multifidi and the deep hip rotators. The primary resistance for the right hip is the landing of the lunge, but further stabilization is required to control the rotary vector caused by the pulley. This approach is used primarily for the knee patient that has associated lumbar instability due to poor motor control of posterior muscles, such as the multifidi.

Figure 2.99a,b: Multidirectional lateral lunge training with right arm resistance from a right lateral force moment. The pulley recruits a diagonal line of muscle from the right arm, crossing anteriorly to the left hip. Anterior musculature is emphasized, including the transverse abdominus and hip flexors. Greater recruitment is achieved through the left side internal obliques, right external obliques, transverse abdominus and hip flexors to control the rotary vector from the pulley. This approach is used primarily for the knee patient that has associated lumbar instability due to poor motor control of anterior muscles, such as the transverse abdominus muscle.

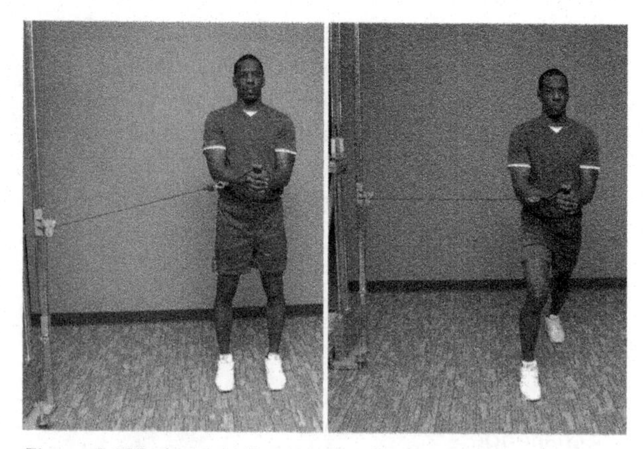

Figure 2.100a,b: Multidirectional right anterior lunge training with a secondary right lateral force moment. A lateral force moment from a wall pulley is held anterior away from the body during a anterior fallout lunge. The force moment from the pulley is stabilized with both hands, requiring motor recruitment of both anterior and posterior muscles of the trunk and pelvis. The trunk is pulled into right rotation, which is resisted by the deep rotators of the right hip and the right lumbar multifidi. More indirect challenges can be added to central muscles of stability by closing the eyes, requiring improved processing of somatosensory input. Lunging onto a labile surface, such as a wobble board or foam, will also increase the balance challenge through the entire lower extremity. Head motions, as shown previously, can further challenge cervical contributions to posture and balance. Head motions may provide a more functional athletic challenge than a labile surface.

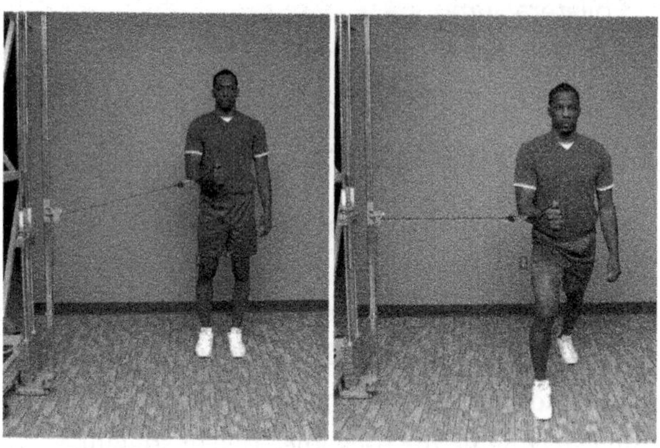

Figure 2.101a,b: Multidirectional anterior lunge training with right arm resistance from a right lateral force moment (for anterior abdominal weakness). Body weight against gravity provides the extension moment for the exercise. The pulley recruits transverse line of muscle fibers from the right arm, crossing anteriorly in the trunk. The abdominals, including the transverse abdominus, are naturally recruited to resist the force moment. The deep external rotators of the right hip are also recruited to resists the right rotational trunk moment created by the pulley. As noted with the previous lateral lunge with a similar external resistance, this approach is used primarily for the knee patient that has associated lumbar instability due to poor motor control of anterior muscles, such as the transverse abdominus muscle.

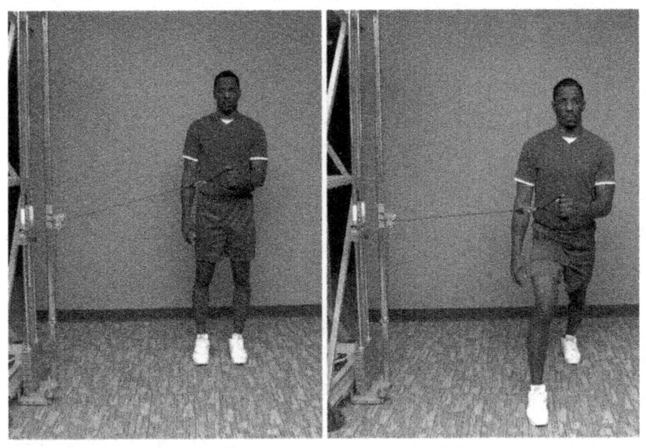

Figure 2.102a,b: Multidirectional anterior lunge training with left arm resistance from a right lateral force moment (for posterior hip and/or back weakness). The pulley recruits transverse line of muscle fibers from the left arm, crossing posteriorly. The deep rotators of the right hip are still recruited to resist the right rotational trunk moment, but the force line is transmitted posteriorly through the back muscles to the right hip.

Lunge Progressions / Challenges

- *Traditional vertical trunk lunge*
- *Fall-out lunge*
- *Lunging in different directions*
- *Lunging with external rotation of the leg*
- *Lunging with resistance to the trunk*
- *Lunging with weight in the hands*
- *Lunge with upper limb reaches in any vector*
- *Lunge up onto a step—anterior to lateral*
- *Lunge down off of a step—different directions*
- *Walk-lunge*
- *Lunge onto labile surface*
- *Lunge off labile surface to a solid surface*
- *Lunge with upper limb arm swings—with arm weights*
- *Resisted trunk lunge—trunk attached to sport cord or pulley*
- *Ballistic lunge—high speed with a fast transition from the eccentric to concentric work*
- *Lunge with reactive activities, such as catch and throw activities*

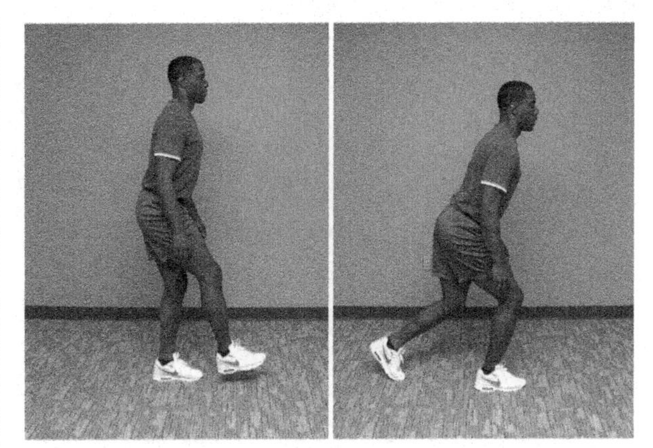

Figure 2.103a,b: The ballistic lunge involves a fast plyometric transition between the eccentric deceleration and concentric propulsion. Ballistic applications can be performed in any lunge direction.

Figure 2.104a,b,c: Resisted left lateral walking with a pulley held in both hands. Emphasis is placed on training central stabilization of the trunk during lateral walking. Both hands will recruit both anterior and posterior musculature, with greater overall recruitment.

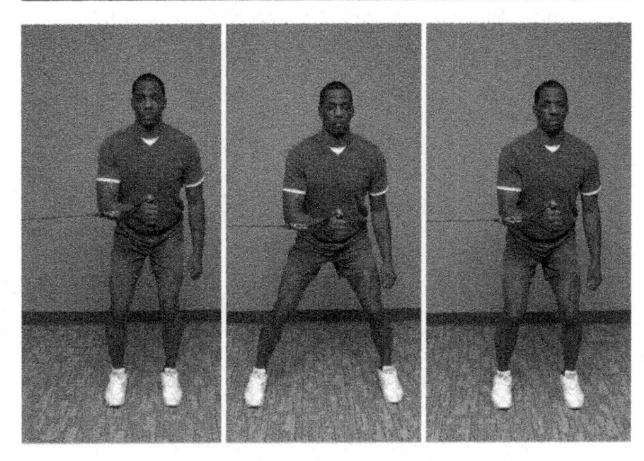

Figure 2.105a,b,c: Resisted left lateral walking with a pulley held in the right hand. Emphasis is on the anterior muscles, with a line of recruitment through the right arm, to the pectorals, the right external obliques, the left internal obliques, transverse abdominus and the left hip flexors. The right hip also has a significant increase in the hip rotators to resist the right rotation trunk moment, as well as abduction.

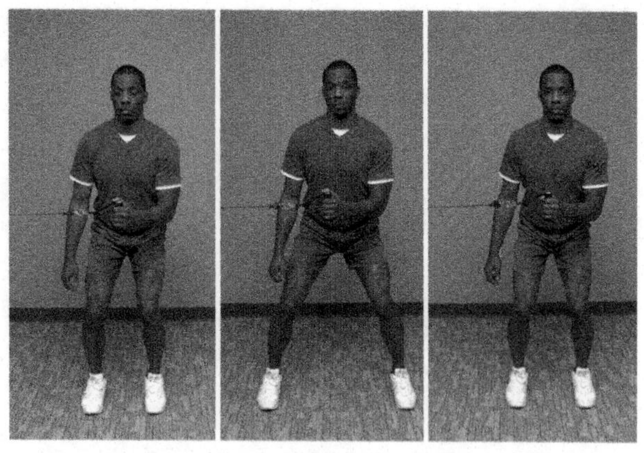

Figure 2.106a,b,c: Resisted left lateral walking with a pulley held in the left hand. Emphasis is placed on pushing off with the right hip and the posterior muscles of the trunk, including the right multifidi.

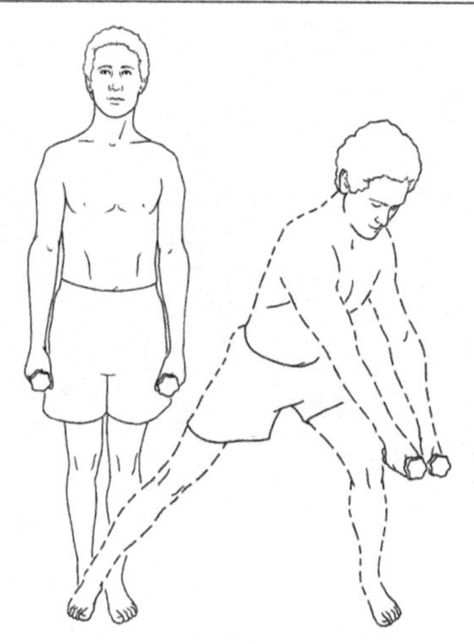

Figure 2.107: Lateral lunge with dumbbells. The patient steps lateral holding two dumbbells at knee level to increase the work load. The foot is maintained pointing forward. The initial progression is to add speed to increase the overall resistance. As weight is added, speed is decreased to allow for coordination at the new level of resistance.

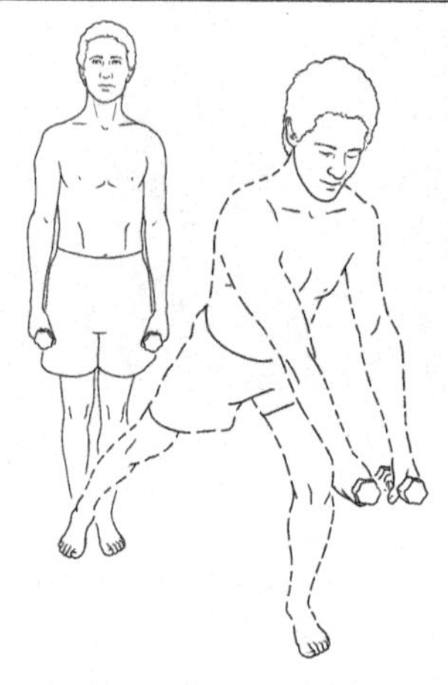

Figure 2.108: Anterolateral lunge with dumbbells. The patient steps in an anterolateral direction holding two dumbbells at knee level. An noted above, the initial progression is to add speed to increase the overall resistance.

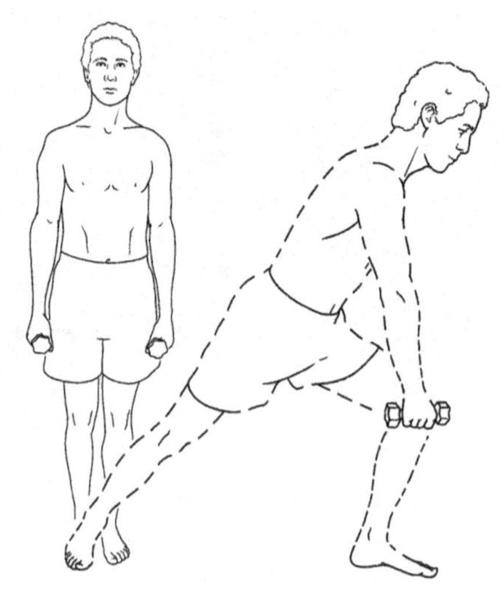

Figure 2.109: An additional challenge with lunging is to add a rotational component. The toe is pointed in the direction of the lunge to add a rotational component at the hip.

One Leg Hop Test / Exercise

Hopping is the next progression from lunging. The hop-and-stop is a task that requires concentric forward propulsion by the back stance leg followed by eccentric stabilization for the landing on the opposite leg. Both the quadriceps and hamstring muscles contribute to the forward propulsion phase, but the hamstring seems to play a more important role during the single leg hop enabling subjects to jump further (Pincivero et al. 1997). Landing, or "stopping," on the forward leg requires eccentric deceleration and stabilization from the lumbar spine through the entire lower limb. Distance and speed of hopping are reduced to allow for proper coordination of landing avoiding a valgus collapse of the lower limb. An indicator for a safe return to running involves first establishing a solid hop and stop performance. Training then can be progressed to bounding, or successive hopping, with a solid landing followed by an immediate propulsion phase forward. Once these exercises can be performed with proper coordination and an absence of secondary tissue irritation, a gradual return to running can be safely attempted.

Hopping for distance and stability is a necessary functional requirement on the progression from walking to running. Landing from a hop with a stable limb involves maintaining proper alignment of the lower kinetic chain as well as avoiding excessive ankle, arm and/or trunk compensations to maintain balance. The knee should land solidly, maintaining position without significant movement. Testing for hopping stability is a good indicator of functional improvement to determine when an athlete is ready to progress to jumping, jogging and running activities. Louw et al. (2006) assessed the front foot landing phase of adolescent pitchers finding that the uninjured players had significantly greater hip and knee flexion angles, as well as eccentric activity, on landing than that of the injured players.

Mattacola et al. (2002) assessed 20 subjects from 10 to 18 months post anterior cruciate ligament reconstruction (ACLR) with a patellar tendon autograft. Significantly shorter distances were identified with the single-leg hop for distance on the involved side versus the uninvolved side. This group of young athletes had already returned to athletic activity, but their performance levels with functional testing did not meet the requirements for a safe return to athletic performance. Weakness and/or instability with hopping and landing can lead to abnormal tissue strain and degeneration over time.

The single leg hop for distance test is a commonly employed functional test in the evaluation of patients with ACL deficient or reconstructed knees. It is scored by hop distance as well as the hop index, which is the ratio or percentage of hop distance of the involved leg relative to the opposite leg. Norms for the hop index were empirically established by Barber et al. (1990), who found that greater than 90 percent of subjects without a history of ACL injury had a hop index equal to or greater than 85 percent, while Daniel et al. (1982 and 1988) found the number to be equal to or greater than 90 percent. Reliability of the hop test score has been well established (Bolgla and Keskula 1997, Kramer et al. 1992, Bandy et al. 1994, Greenberger and Paterno 1994). Validity studies have revealed low sensitivity rates in detecting abnormal limb symmetry in ACL deficient subjects (Barber et al. 1990, Noyes et al. 1991).

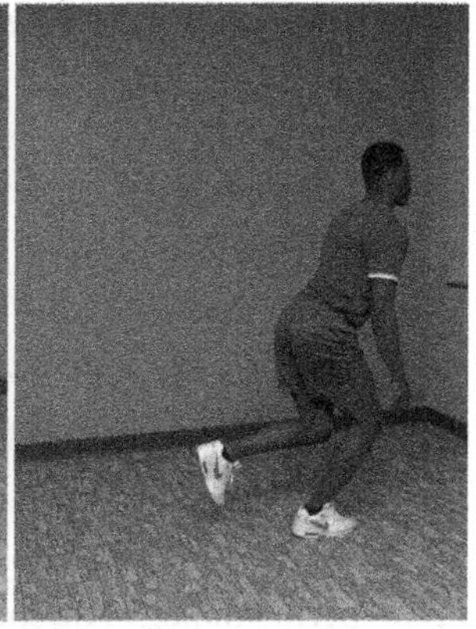

Figure 2.110a,b: The hop-and-stop. The patient hops forward onto the involved knee. Emphasis is placed on a stable landing without lower limb collapse. The patient is instructed to maintain the single stance position until the unilateral stance position is completely controlled, then hop to the other foot. The hop-and-stop can be performed in any direction. Emphasis may be placed on the height of the jump or the distance, depending upon the athletic demand.

Progressions Options Hopping—All of the Options Below Can Be Progressed Form a Bilateral to a Unilateral Leg Exercise

- *Lateral line hops—hop back and forth over a line. Can be effective training for improving lateral plyometric agility for such sports as skiing, basketball and tennis. Single leg hop for distance.*
- *Hopping back and forth in different directions—anterior, anterolateral, lateral, posterolateral and posterior. Can be effective training for plyometric agility in multiple planes for such athletes as football running backs, defensive backs, basketball and volleyball players. Hop in four-square pattern.*
- *Alternate line hops along a 10-foot line—performed by jumping and landing on either side of a line. This can be performed in any direction and is effective training for skiers.*
- *Lunge hops—performed by starting in a lunge position, exploding upward and switching legs in the air and landing in the a lunge position with the other leg forward. This can be effective training for telemark skiers hopping with 45 degrees rotation.*
- *Bounding—successive hopping without stopping.*

Agility Training and Early Stage Plyometrics

Return to athletic activities may require a higher degree of functional coordination training, rather than just simple strength training with basic cardinal plane movements. Performing agility walking, lunging, hopping and jumping tasks in different directions can provide significant neuromuscular challenges that better mimic sport performance. The specific exercises and directions chosen for training should reflect the functional demands of the sport or activity to be performed. Houck et al. (2006) demonstrated that unanticipated straight walking and side-stepping were associated with increased levels of hip abduction, with abnormal foot placement causing excessive trunk motion to compensate.

The slideboard can be incorporated to add challenges to the proprioceptive system and improve balance with weight shifting. EMG analysis by Heller and Pincivero (2003) found that the VMO, vastus lateralis and anterior tibialis act as the prime movers during the push off phase in normal and ACL deficient knees. Blanpied et al. (2000) also found the slideboard to be an effective strengthening exercise for post ACL reconstruction. Concentric firing of the gluteus medius to push off the board with speed requires coordination, strength and power. Greater degrees of knee flexion during the slide impart an increased challenge to the quadriceps and gluteus medius muscles, increasing their capacity to tolerate more stress. Throwing and catching a weighted ball while sliding will add more complexity to the exercise, as it shifts the center of gravity and challenges the base of support requiring a higher level of recruitment and coordination. This exercise can be further progressed by throwing the ball to different spots, challenging reactive reach and balance responses.

Jump Training

Jump training is added once coordination has been established with performance of lower level activities such as lunges. Lunging requires a solid single leg landing while maintaining proper alignment of the lower kinetic chain. Ballistic lunging incorporates a fast transition between eccentric deceleration and concentric propulsion, which then progresses to low level jumping. Landing from the jump requires a fast and synchronous recruitment of the lower limb muscles to stabilize the hip, knee and ankle. Even a low degree of valgus collapse at the knee can lead to abnormal tissue strain, particularly with high repetition rehabilitation activities. In a cadaveric study Withrow et al. (2006) demonstrated a 30 percent increase in ACL ligament strain with improper muscle support allowing a valgus moment in knee flexion with simulated landing. Chappell et al. (2005) found fatigue with stop-jump tasks led to increased peak anterior tibial shear, as well as an increased valgus moment at the knee. Caution should be taken to avoid fatigue, with observation of proper knee alignment during the exercise.

The eccentric phase of jump training has been associated with delayed onset muscle soreness (DOMS) (Byrne and Eston 2002). But these types of studies commonly reflect healthy subjects training at excessively high levels. Jump training is added later in a rehabilitation program once progressive training has improved tissue tolerance, coordination, endurance and strength to levels that are appropriate and safe for performance of higher level activities such as jumping. Properly performed and dosed, early low-level jump training should not represent a level of eccentric performance that is at risk of creating any significant DOMS.

The beginning of light plyometrics can be introduced on a rebounder, to reduce impact forces on healing joint systems. Mats on the floor can also be used to reduce these forces and eventually removed with improved tissue tolerance. Light jumping in place can also be employed for an initial plyometric training program as it limits the range of training and loading of the lower quarter.

More aggressive jump training for agility can be achieved by incorporating greater speeds, changes of directions and higher jumps. Initially a taped line or cross on the floor can be used as a marker for cardinal planes of training. Multiple combinations of jumping with both legs in different directions can be trained and progressed to single leg training. Boxes of varied heights can then replace the tape to offer greater ballistic challenges by jumping on and/ or over them with both legs, or a single leg.

Examples of Early Jumping Training Options and Progressions

- *Jumps-in-place*
- *Standing Jumps—for distance or height*
- *Bilateral jumps forward and back*
- *Bilateral jumps side to side*
- *Hop and stop*
- *Unilateral hops forward and back*
- *Unilateral hops side to side*
- *4x4 grid*
- *Diagonals on grid*
- *Jumping over a barrier—anterior or lateral*

Circuit Training

Circuit training involves performing a group of different exercises in a series without a rest break. Exercises are chosen to train the entire body but can also include specific exercises related to the primary pathology. Measurements for improvement include reduced heart rate at the end of the circuit, reduced time to complete the circuit, as well as decreased time to recover to resting respiration rate and resting heart rate. Endurance can also be improved through the performance of circuit training. Performance is measured on the amount of sets/reps completed in a given time period. As the time improves, the challenge is increased so they are performing increased sets/reps in the same period of time. This type of training can be performed in the clinic and with aerobic cross training as well. Adding more repetitions or sets via increasing the amount of exercises will affect the endurance capacity of the involved muscle groups. For the knee, a circuit training program might include squats, step-ups, step-downs, lunges, lateral steps with resistance, hopping/jumping drills and more.

Section 4: Stage 4 Exercise Progression for the Knee

Training Goals

Stage 3 training marks the end of all primary impairments and symptoms focusing on improving the overall training state of healing tissues as well as incorporating more specific training for functional demands. Exercises are modified to train deficiencies with functional tasks as they relate to endurance, coordination, speed, strength and directions of dysfunction. Balance and proprioceptive retraining is also continued and further challenged during activity simulation. All exercises are performed through the full physiological range with a focus on strengthening around 80% of 1RM. Resistances, however, may be reduced to lower percentage RMs when initially performing more complex activities at higher speeds, as they require a greater degree of coordination. With improved motor control, the resistance can then be elevated to train for strength and power. The latter functional quality is addressed with explosive training at near maximal resistance and may include such activities as pushing blocking sleds, bungee running and resisted block starts.

Tissue and Functional States—Stage 4

- *Full active and passive range of motion*
- *Pain free joint motion at a significant level of exercise*
- *Good coordination for functional motions*
- *Limited endurance and/or strength with functional performance*
- *Limitations in athletic performance*

Training Goals Stage 4—Return to Functional Activity

- *Explosive exercise training*
- *Resistance increase to 80% of 1RM*
- *Low repetitions*
- *Speed added*
- *Full physiological movement*
- *Plyometric jump training*
- *Aggressive agility drills*

Motions and Directions

Tissue that is unaccustomed to eccentric exercise can sustain muscle fiber damage and delayed-onset muscle soreness (DOMS) within a few days after the exercise. A second bout of eccentric exercise, less than a week after the first bout, can typically be tolerated with much less damage and soreness (Brockett et al. 2001). Increases in leg girth in some subjects that reported muscle soreness suggested inflammation from muscle fiber damage. As the tissue matures and tensile strength increases, activities and exercises should be progressed accordingly to maximize its hypertrophy and functional capacity. Aggressive agility tasks including plyometrics and other triplanar exercises can be employed. For rehabilitation of ACL repairs, typically the tissue has healed sufficiently by six months to allow for the initiation of aggressive agility exercises including figure-8's, lateral cuts, ball toss on the slide board, scissor running and quick start/stops.

Prevention programs have long been thought to help in reduce the incidence of ACL ruptures and other ankle, knee and hip injuries. Recently more evidence has been compiling in favor of the development and application of these prevention programs. Olsen et al. (2005) concluded that a structured program of warm-up exercises prevent knee and ankle injuries in young people playing sports. Preventative training should be introduced as an integral part of sports programs and include running, cutting, balance, proprioceptive and

Power Training General Guidelines	*Resistance*	*Volume*	*Rest Intervals*	*Velocity*	*Frequency*
Beginning Rehab/ Unhealthy/Non-trained					
Lower Body Multiple Joint Exercise	*35–45% of 1RM*	*1–3 sets, 10–15 repetitions*	*2–3 minutes*	*Moderate*	*2–3x/week*
End Stage Rehab/ Healthy/Trained					
Lower Body	*45–70% of 1RM*	*2–3 sets, 6–12 repetitions*	*2–3 minutes*	*Moderate to fast*	*1–2x/week*
Jump Training	*35–45% of 1RM*	*2–3 sets, 6–10 repetitions*	*3–5 minutes*	*Fast to ballistic*	*1–2x/week*
Olympic Lifts	*80–90% of 1RM*	*6–8 sets, 1–5 repetitions*	*5 minutes*	*Fast to ballistic*	*2x/week*

Table 2.?: Dosage guidelines for lower body power training/jump training.

strength training as well as instruction in various landing techniques (eccentric and plyometric), which require an extremely high level of stability.

Weight Bearing Progression

Exercises to challenge weight bearing capacity can be further progressed at this time to include tri-planar motion through the full ROM. Step-ups and eccentric step-downs can be challenged by adding a rotation component. Increasing the depth of squats or lunges as well as changing the direction of motion are other examples of increasing weight bearing challenges.

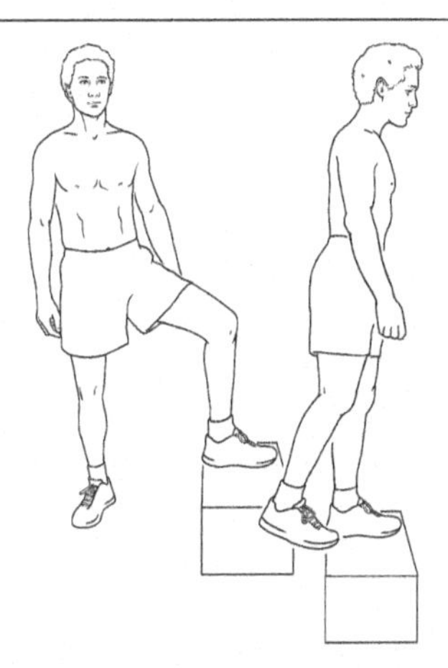

Figure 2.111: Step-up with external rotation. The patient stands to the side of the platform, depending on how much rotation is desired. One foot is placed on the platform as the patient remains facing straight ahead. As the body weight is shifted to the stance leg, the trunk is rotated toward the platform with the knee kept centered over the foot.

Figure 2.112: Resisted squats with weight bar. The weight bar is rested posteriorly on the shoulders, off the cervical spine. The gaze is forward or up to assist in maintaining a neutral thoracic and lumbar spine. The patient is instructed to initiate the motion by shifting the hips posteriorly, rather than the knees forward. The depth of the squat should be no lower than is pain free, with the thigh going no lower then a horizontal position relative to the floor.

Figure 2.113: Front squat is a variation of the standard squat with the barbell resting on the front shoulders rather than on the back. The more upright position of the front squat creates less hip involvement with less gluteal contribution, places more emphasis on quadriceps, reduces spinal flexion as the patient cannot lean forward and increases abdominal contribution to stability.

Advanced Lunge and Step Training

Figure 2.114: Right anterior fallout lunge with posterior weight bar. The fallout lunge technique places more emphasis on the hip musculature compared to the knee. A weight bar on the shoulders will increase the demand on the spinal extensors, as well as increase torque at the hip.

Figure 2.115: Right anterior fallout lunge with right rotation:

weight bar. The fallout lunge is performed with a weight bar on the shoulders. At the end position of the forward lunge the trunk is rotated toward the front foot (right rotation shown). The addition of rotation will significantly increase the contribution of the deep hip external rotators, working eccentrically to decelerate the motion, as well as the rotational vectors of the lumbar multifdi and abdominals.

Figure 2.116: The traditional anterior lunge with shoulder flexion (anterior reach): free weights. The anterior lunge is performed with a vertical trunk and shoulder flexion performed at the end range. A more vertical trunk will reduce hip and back contribution, emphasizing the quadriceps and hamstrings. Initial emphasis is on coordination, flexibility and balance. As performance improves free weights can be added, held at the side to increase overall load to the lower quarter muscles. The anterior reach will increase load on the spinal extensors and hip extensors. The load on the quadriceps increases as the center of gravity moves anterior.

Figure 2.117: Fallout anterior lunge with shoulder flexion (anterior reach): free weights. The fallout anterior lunge is performed with the trunk and posterior leg in alignment in a forward lean or "fallout" position. The fallout position will increase contribution of the hip, reducing the torque moment at the knee. Shoulder flexion is performed at the end lunge position. Free weights can be added to increase resistance. The anterior reach will create a significant increase in load of the spinal and hip extensors due to the long lever arms. Changes in speed, direction of the lunge and direction of the reach can also be modified to alter the challenge of the lunge or duplicate functional and/or sport performance.

Figure 2.118: Lateral lunge with shoulder flexion. The patient steps laterally maintaining alignment of the hip, knee and foot of the lunging foot. The knee of the anchor foot remains in extension. Shoulder flexion is performed at the end of the lunge position. The lateral lunge emphasizes lateral hip musculature, with shoulder flexion increasing torque on the lumbar and hip extensors. For push off back to the start position, emphasis is placed on slight pronation of the foot and slight valgus at the knee to recruit the entire lower quarter extension pattern. Allowing the foot to supinate at push off reduces recruitment of extensor muscles of the hip and knee. A common compensation with lower quarter weakness is to shift the pelvis more lateral, outside the functional axis of the lower limb, maintaining the trunk more medial rather than outside of the foot. The patient should be instructed to keep the shoulders and trunk over the foot with the hip, knee and ankle in alignment.

Figure 2.119a,b: Anterior lunge onto a labile surface. To emphasize balance, stability and proprioception, rather then strength training, having the patient lunge onto a labile surface will increase the demand on the somatosensory system to coordinate the movement. Different types of surfaces can be used to vary the challenge. The patient can be allowed to look at the surface to assist with coordinating the motion, but may later be progressed by being asked to look forward during the movement. A progression may also be to have the patient move the cervical spine, looking lateral, up or down, increasing the challenge to postural and balance reflexes. These types of training challenges may be more indicated when proximal weakness at the hip and knee is identified, or significant central balance deficits are present.

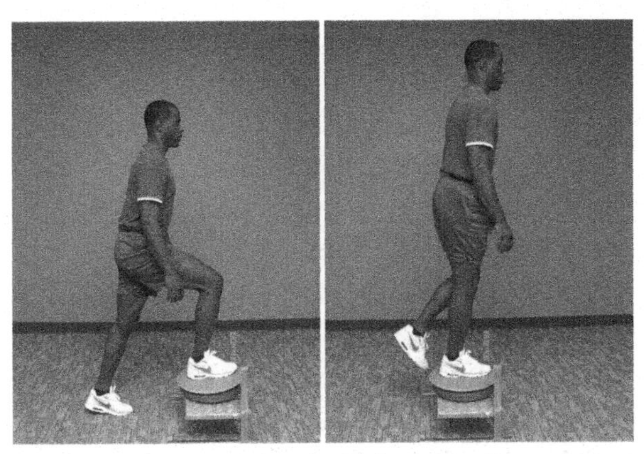

Figure 2.120a,b: Step-up onto a labile surface. To emphasize balance, stability and proprioception, rather then strength training, having the patient perform a step-up (pictured), step-down or lateral step onto a labile surface. The labile surface will increase the demand on the somatosensory system to coordinate the movement. The patient can be allowed to look at the surface to assist with coordinating the motion, but may later be progressed by being asked to look forward during the movement. This type of training may be more indicated when proximal weakness at the hip and knee is identified, or significant central balance deficits are present.

Figure 2.121a,b: Step and jump. Duplication of athletic performance is necessary prior to returning to sport. Reactive, unplanned movements, are necessary to train lower quarter patterns with distraction from upper quarter or visual activities. The patient may be asked to lunge by reacting to a verbal or visual command for the direction. Lunging can also be performed in any direction with a catch-and-throw activity (pictured), creating a distraction form lower quarter function, challenging hand-eye coordination.

Strength Training

The functional quality of strength is addressed in the later stages of rehabilitation. When all the lower level training qualities including joint mobility, range of motion, tissue tolerance, coordination, speed and endurance have been established strength can be safely trained. However it can be initiated earlier in the rehabilitation progression for the uninvolved joints. For instance, strength training for the hip may begin prior to the knee in order to help protect it from further physical trauma during weight bearing activity. Dosage for strength training is typically set between 80–90% of 1RM with a range of roughly eleven to four repetitions respectively. There are a variety of weight training concepts available to challenge the neuromuscular system including super sets, tri-sets, giant sets, pyramid training and negative training.

Power is a function of adding speed to an exercise dosed for strengthening and is typically associated with athletic performance. Initial strength training levels are usually dosed at 70–75% of 1RM with pure strengthening emphasized at approximately 80–90% of 1RM with eight to 12 repetitions per set. Rest periods between sets are increased with high end strength training due to increased energy consumption. The amount of time is dependent on cardio-vascular conditioning ranging from two to five minutes for a return to steady state. General weight training and body building concepts do have some application in the clinical setting. Options for alternating different types of exercise, loads and/or directions of training can be applied to vary the motor challenge and prevent boredom.

Super Set

Super sets refer to training two separate, often opposing muscle groups, for two to three sets with a two to three-minute rest between each set. Dosage for super set training is typically in the strength training range for healthy weight trainers, but dosage can reflect training for coordination and endurance as well. Body builders and power lifters may employ this method to improve strength gains. The knee may include opposing quadriceps and hamstring exercises. For rehabilitation purposes a modification in the super set concept may include the addition of rest breaks between the sets of each exercise. Super sets may also involve exercise groupings that include one trunk and one lower limb exercise, or one upper and one lower limb exercise, rather than antagonistic muscle groups.

Figure 2.122a,b,c,d: Superset with squats with 4 sets performed with different balance challenges during endurance/strength training: a) sumo/plie' squats with the feet wide apart b) balance foam, c) squats while standing on two labile surfaces and d) squats on one labile surface.

Tri-Set

A Tri-set includes three or more similar exercises in a series that have different influences on the same muscle group with a rest break between each series. For example squats can be trained in a multitude of ways with various influences on recruitment of the lower quarter musculature. Specific variations include altering stance width, performing on labile surfaces, altering the trunk angle, and adding or subtracting resistance using dumbbells, barbells, pulleys, elastic resistance and/or squat rack machines. Lunges can also be varied with directional changes, arm reaches, steps and/or labile surfaces. These variations alter the coordination demands, emphasize recruitment of different muscle fibers and help break up monotonous exercise routines of performing the same exercise.

Giant Set

A Giant Set refers to three different exercises performed consecutively per set on the same group of muscles without a rest break. This is basically the same as the tri-set example, but without a rest break until all bouts are completed.

Drop Sets/ Stripping/Pyramid Training

Drop sets involve multiple sets beginning with a high weight that allows for only one to three repetitions. Resistance is gradually decreased for the subsequent two to five sets, but repetitions may vary based on fatigue. These types of techniques are typically reserved for strength or power training for healthy athletes. Lower RM percentages can be used as a modification for patients at lower tissue training states. Stripping involves a specific number of repetitions for each set with gradually decreasing weight. Pyramid training begins with light weights and high repetitions then progressively increases weight and deceases repetitions over two to three sets followed by a reversal of the training pattern. Rest breaks are taken between each set based on the dosage. These types of techniques are effective in adding variety to the exercise regime, preventing stagnation in training and providing a different neurological challenge.

Negatives

Negatives are eccentric repetitions that are usually set at a resistance higher than the one lift concentric maximum (1RM) and performed for strength training. A spotter is required to assist with the concentric phase of the lift. This type of heavy lifting is not a common approach to clinical rehabilitation, but similar concepts are applied to pure eccentric training for tendinopathy.

Oscillation Repetitions

Oscillation repetitions involve performing a lift with back and forth concentric-eccentric contractions through the entire range of the exercise (Garfield 2003). Alternating acceleration and deceleration is performed five to ten times through one full repetition of the motion. Overcoming inertia during the transition between the deceleration and acceleration phases of the oscillation adds to the overall workload. This is a result of the greater demand placed on the neurological system to coordinate motor performance throughout the range of the exercise. The total number of repetitions may need to be reduced, as fatigue will occur with fewer repetitions. As an example, squatting activities can be performed where the motion is stopped at several ranges for both eccentric and concentric phases.

Functional Training for Jumping, Running and Agility

Agility training and plyometrics might also be added to allow return to pre injury activity levels. If fast ballistic training is required, it should be performed when the patient has attained an adequate level of functional coordination, eccentric stability and tissue tolerance that allows for safe, pain free performance of these activities. This can be achieved through progressive resistance and anaerobic training in the earlier stages of rehabilitation It must be remembered that specificity of training is important for all rehabilitation stages, including plyometrics, to ensure that the exercises chosen are applicable to the sport or skill being trained. This is achieved by determining specific needs during the initial evaluation and periodic reassessments, as the patient progresses through the physical therapy regimen.

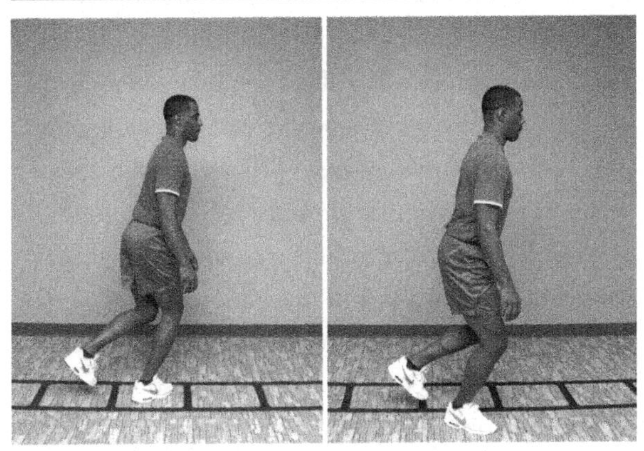

Figure 2.123a,b: Agility ladder hopping with 45° lower extremity internal and external rotation. Unilateral hopping is performed with targets (ladder squares). The patient performs alternating rotational motions of the lower quarter while maintaining the shoulders and trunk facing forward. Small rotational motions are progressed as coordination is established. This would represent a torsional challenge for rotary dysfunction retraining in the knee, as in ACL pathology.

Figure 2.124a,b: Agility stepping/running with ladder. The patient is asked to perform quick stepping within the agility ladder. Many variations are possible with alternating the direction (forward, backward and laterally) or by changing the target sequence of the squares in the ladder.

Figure 2.125a,b: Bounding is a progression of the hop-and-stop. Rather than the emphasis being on a stable landing, emphasis is on a quick, plyometric explosion forward to the alternate foot. The stride distance may start short to establish coordination but is progressed to exceed that of a normal running stride length. This is an exaggeration of a normal running stride with emphasis on improving stride length and stability on landing.

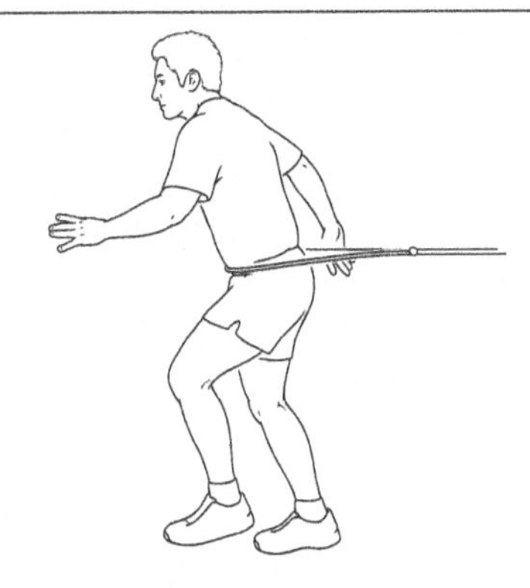

Figure 2.126: Resisted running or lunging. A sport cord or speed pulley is attached to the waist. The patient performs resisted motions of walking, jogging, lunging and/or running. Resisted motions can also include lateral, backward or in combined vectors.

Plyometrics / Jump Training

Jump training has its roots in Europe, with interest increasing in the 1970's after Eastern European athletes began excelling in the world of sport competition (Chu 1988). Jump training was not referred to as plyometrics until the term was coined by an American track and field coach, Fred Wilt (1975). The terms jump training and plyometrics can be used synonymously, but for the purposes of this chapter, jump training is used in Stage 3 to represent the less aggressive initial stage of training for healing tissue states. Plyometric training, on the other hand, is more associated with end stage training emphasizing improvement of athletic performance. Plyometric exercises help a muscle reach its maximum strength in as short a time as possible (Chu 1988). They combine speed with strength, otherwise known as power training, where the muscles are trained to overcome heavy resistance through a short burst of activity.

Figure 2.127a,b: Step, jump and catch. More complex exercises include athletic simulation with combinations of stepping, jumping and reaction to a catch and throw activity.

Motor activation involved in fast ballistic movements tends to favor multi-joint phasic muscles, with reduced activation of single joint tonic muscles (Mackenzie et al. 1995, Thorstensson et al. 1985). Plyometrics rely on anaerobic capacity that requires rest periods for adequate recovery between repetitions and sets to maintain optimal training conditions (Chu 1988). It must be noted that bouts lasting four to 15 seconds deplete the creatine phosphate energy stores requiring a significant amount of rest for recovery, whereas bouts that last 30–90 seconds help to train lactic acid tolerance which may be preferable for some patients (Chu 1988). Recovery periods between sessions typically range from 48–72 hours depending upon the intensity of training to allow for tissue repair (Chu 1988).

Figure 2.128a,b: Rebounder jumping with catch and throw. An additional challenge to a reactive exercise of catch and throw while jumping is to use a more labile surface for jumping such as a rebounder.

In the example of jumping, emphasis is placed on spending as little time on the ground as possible. A quick transition is performed between the eccentric deceleration and concentric propulsion. Early jump training may involve a slower performance, focusing on coordination with landing to maintain proper lower quarter alignment and avoid pain. Plyometrics emphasize a ballistic, or explosive transition, to maximize energy transfer for propelling the joint system. The process has also been referred to as the stretch-shortening cycle, where collagen elongation during the eccentric phase is followed by an elastic recoil in the concentric propulsion phase. It also is governed by preset muscle tension via feedback from muscle spindles and mechanoreceptors to the spinal cord that can act to increase motor output. A delay in this process or transition, also known as the amortization phase, leads to a loss of energy via friction-generated heat and can be shortened by applying learning and skill training (Chu 1988). Assessment of motor responses in the ankle during jump performance suggests that the muscle response for an immediate jump after landing occurs too quickly for afferent signals to reach the spinal cord and influence the motor response. Instead it seems to involve a feedforward mechanism, which refers to a preprogrammed motor response that anticipates the impact of landing with subsequent activation of muscles necessary for the landing and propulsion phases. Duncan and McDonagh (1997) demonstrated an alteration in this feedforward programming with reduced and delayed pre-activation of ankle musculature prior to landing from a jump. Jumping emphasizes speed to achieve greater performance. Dosage for jump training may be slightly reduced within the clinic, as a part of end stage rehabilitation, than in the normal healthy athletic population. Reduced load, fewer repetitions and longer rest times between training sessions may be necessary to avoid tissue irritation and muscle fatigue. Plyometric training at 30% of 1RM has been found to be superior in increasing vertical jump than training at 80% of 1RM (McBride et al. 2002). Optimal loading for ballistic jump squat training has been reported at 30% of 1RM by Wilson et al. 1993 and 45% of 1RM by Baker et al. 2001A, while ballistic bench press training is best achieved at 60% of 1RM (Baker et al. 2001B).

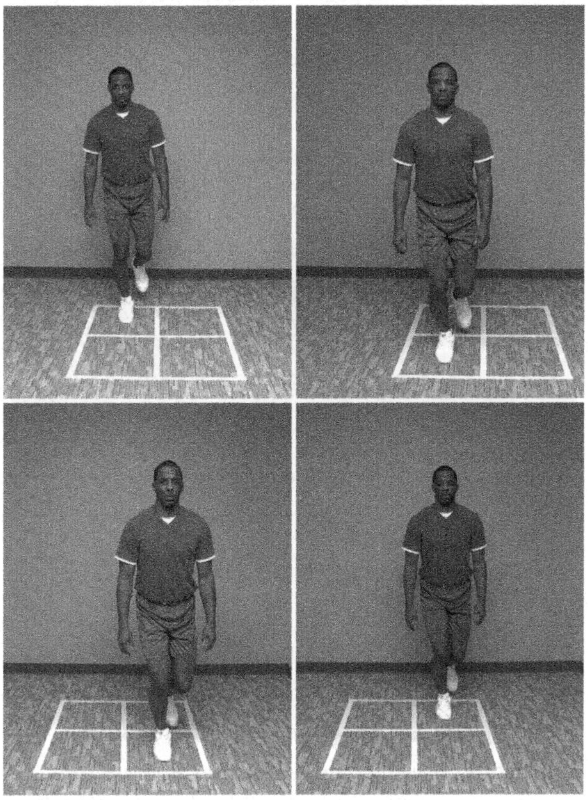

Figure 2.129a,b,c,d: Box Drills. The patient stands lateral to the box with the foot of the involved leg on the box and opposite arm lifted up. A ballistic step-up is performed, pumping the arms to jump up off of the box. The foot lands back on the box with the opposite foot landing back on the ground. They can be performed bilaterally or unilaterally with direction changes including anterior, posterior and lateral.

Depth jumps involve jumping off a box, landing and immediately jumping again for height. Depth jump training has been shown to be more effective than weight training, jump and reach, and horizontal hops for improving speed and strength capabilities (Verhoshanski and Tatyan 1983). Determining the height for depth jumps has been studied, but is too individualized for a predetermined protocol, but Chu (1988) recommends a testing procedure for each athlete. First vertical jump is measured with the jump and reach and then the depth jump is tested using an 18-inch box. If the jump after landing is equal to the initial vertical jump test, than the box height is raised in six-inch increments until the athlete can no longer reach the original test height. If the first test on the 18-inch box is lower then the vertical jump test, than the box height is lowered.

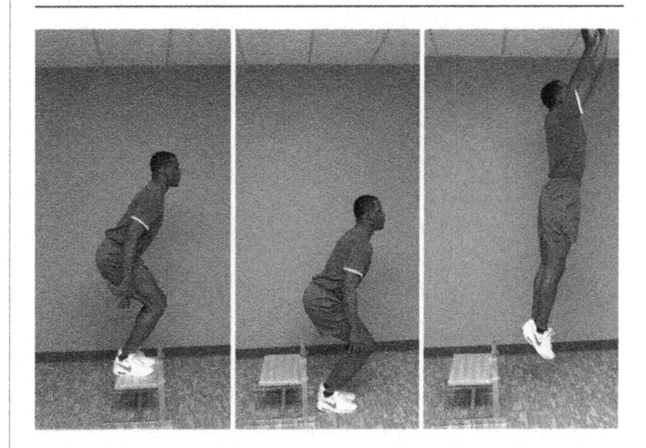

Figure 2.130a,b,c: Depth Jumps. The patient stands on top of a step or platform (the lower the height the greater the speed component, the greater the height, the greater the strength component). The patient steps slightly forward off the platform, landing toward the forefeet. They are then instructed to react as quickly as possible to the ground and spring immediately back up into the air. The arms are used to add speed by drawing them back prior to stepping off the platform and swinging them vigorously upward as the feet hit the ground. The back remains in neutral alignment with the gaze forward.

Plyometric Jump Options

- *Jumps-in-Place—Jump and land from same spot. Repeated in succession with a progressive increase in the rate of jumping to improve transition between eccentric landing and concentric propulsion phases. They can also be progressed by jumping higher.*
- *Standing Jumps—a single maximal effort in a horizontal or vertical direction that may be repeated following a complete recovery from the previous effort.*
- *Multiple Hops and Jumps—Maximal effort jumps performed one after another over a distance of less than 30 meters that can be performed with or without barriers (boxes, cones, hurdles, etc.). They act as a precursor to box drills described earlier.*
- *Ninety Degree Jumps—Jump and turn 90 degrees in the air. Land and jump and turn 90 degrees opposite back to the starting position. Jump and turn 180 degrees then 270 degrees and finally 360 degrees.*
- *Anterior barrier hops—unilateral, bilateral, or alternating feet*
- *Lateral barrier hops—unilateral, bilateral, or alternating feet*
- *Stair (stadium) hops—bilateral step jumping*
- *Step-up and jump*
- *Lateral step-up and jump*
- *Agility running*
- *Scramble-Up—Start from a prone position and scramble up to a balance standing position as quickly as possible. To increase the difficulty, perform the exercise with the eyes closed.*
- *Standing long jump with sprint: sprint in any direction (i.e., anterior, lateral and posterior).*

Running Options

- *Bounding.*
- *Jogging/running different directions: anterior, posterior, lateral.*
- *Cutting—jogging/running with direction change (i.e., anterior and cut lateral).*

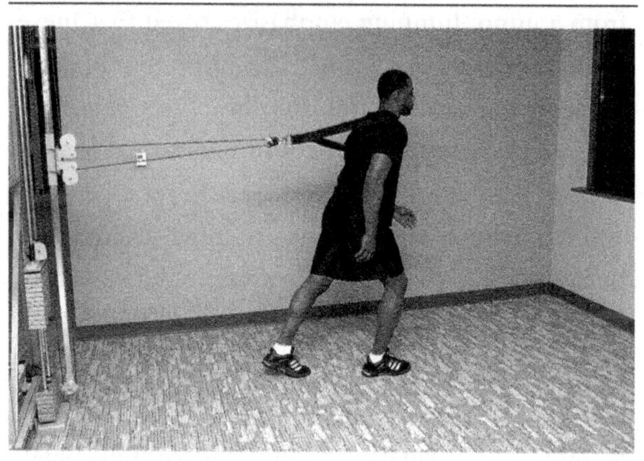

Figure 2.131: Resisted walk/run. A chest harness is placed around the trunk or a strap is used around the plevis. Appropriate weight is added to challenge balance and recruitment. The patient starts from a standing or crouched position (start position in sprinting), then pushes through the stance limb to propel the body forward. If starting from crouched position, the patient "explodes" to a standing position. In both cases, the patient takes a few steps to challenge balance and weights can be placed in the hands to further increase the challenge.

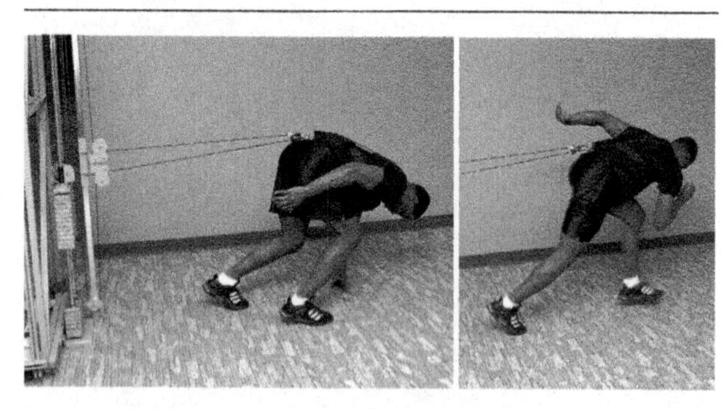

Figure 2.131a,b: 3-Point resisted track start with speed pulley.

Summary

It has long been thought that therapeutic exercise plays an important role in the rehabilitation of joint pathology. A review of the literature found that numerous clinical trials used either exercise or a patellofemoral physical therapy treatment plan that included therapeutic exercise as a control (Antich 1986, Thomme 1997). These trials all showed significant improvement related to one or more of the following:

- *pain reduction*
- *strength gains*
- *improved EMG activity*
- *improved patellofemoral congruence*

The Philadelphia Panel (2001) found good evidence to include strengthening, stretching and functional exercises together as an intervention for osteoarthritic knees. Numerous studies have also shown correlations between obesity and prevalence of knee OA (Coggon et al. 2001, Felson et al. 1997, Cicuttini et al. 1997). Bynum (1995) performed a randomized clinical trial (RCT) comparing open-chain and closed chain kinetic exercise following ACL reconstruction, demonstrating that subjects that performed closed kinetic exercises experienced a significant reduction of preoperative patellar pain and significantly less patellar femoral pain following surgery and subsequent rehabilitation

Nyland et al. (1994) stated that an understanding of the afferent neural system of the knee is imperative to properly plan a rehabilitation program since there is an intricate relationship that exists between the afferent receptors and the contractile tissues of the knee. They believe that closed kinetic chain functional training (CKCFT) may provide a method for more effectively rehabilitating an injured or reconstructed knee by incorporating sensorimotor integration through motor learning.

This section has outlined an integration of the sensory motor afferent system beginning with initial coordination exercises aimed at developing proper recruitment patterns. Once these neuromuscular recruitment patterns are established, the rehabilitation program is progressed along the endurance-strength continuum by gradually increasing load and lowering repetitions to meet the functional demand. Generally, the average patient population requires strength capacity of approximately 85% of 1RM, which should be sufficient for performance of a majority of activities of daily living. A home program is also necessary to direct self-care. Findings by Chamberlain et al. (1982) suggest that if patients are not held accountable for their home exercise program however, the majority will not remain compliant. The effects of home-based exercise, even for more chronic issues such as osteoarthritis, have been shown to be beneficial (Thomas et al. 2002), and act to serve the greatest benefit to those who adhere to the program (Hart et al. 2003). Post rehab exercise need not involve continued progressive resistive training, but may include a simple return to normal activity levels and/or sport.

Bibliography

Abernethy P, Wilson G, Logan P. Strength and power assessment. Issues, controversies and challenges. Sports Med 19(6):401–17, Jun 1995.

Ahmed AM, Burke DL, Yu A. Invitro measurement of static pressure distribution in synovial joints. Part II: Retropatellar surface. J Biomech Eng 105:226–236, 1983.

Alaca R, Yilmaz B, Goktepe AS, Mohur H, Kalyon TA. Efficacy of isokinetic exercise on functional capacity and pain in patellofemoral pain syndrome. Am J Phys Med Rehabil 81(11):807–13, Nov, 2002.

Anderson G, Herrington L. A comparison of eccentric isokinetic torque production and velocity of knee flexion angle during step down in patellofemoral pain syndrome patients and unaffected subjects. Clin Biomech (Bristol, Avon) 18(6):500–4, Jul, 2003.

Arendt E, Dick R. Knee injury patterns among men and women in collegiate basketball and soccer: NCAA data and review of literature. Am J Sports Med 23:694–701, 1995.

Augustsson J, Thomee R. Ability of closed and open kinetic chain tests of muscular strength to assess functional performance. Scand J Med Sci Sports 10(3):164–8, Jun, 2000.

Aune AK, Cawley PW, Ekeland A. Quadriceps muscle contraction protects the anterior cruciate ligament during anterior tibial translation. Am J Sports Med. 25(2):187–9, Mar–Apr, 1997.

Bahr R, Fossan B, Loken S, Engebretsen L. Surgical treatment compared with eccentric training for patellar tendinopathy (Jumper's Knee). A randomized, controlled trial. J Bone Joint Surg Am 88(8):1689–98, Aug, 2006.

Baker D, Nance S, Moore M. The load that maximizes the average mechanical power output during explosive bench press throws in highly trained athletes. J Strength Cond Res 15:20–24, 2001a.

Baker D, Nance S, Moore M. The load that maximizes the average mechanical power output during explosive jump squats in power-trained athletes. J Strength Cond Res 15:92–97, 2001b.

Bandy W, Rusche K, Tekulve F. Reliability and limb symmetry for five unilateral functional tests of the lower extremities. Isokinetics and Exercise Science 4(3):108–111, 1994.

Barber SD, Noyes FR, Mangine RE, McCloskey JW, Hartman W. Quantitative assessment of functional limitations in normal and anterior cruciate ligament-deficient knees. Clin Orthop 255:204–214, 1990.

Barber-Westin SD, Noyes FR, Heckmann TP, Shaffer BL. The effect of exercise and rehabilitation on anterior-posterior knee displacements after anterior cruciate ligament autograft reconstruction. Am J Sports Med 27(1):84–93, Jan–Feb, 1999.

Barber-Westin SD, Noyes FR. The effect of rehabilitation and return to activity on anterior-posterior knee displacements after anterior cruciate ligament reconstruction. Am J Sports Med 21:264–270, 1993.

Bautch JC, Clayton MK, Chu Q, Johnson KA. Synovial fluid chondroitin sulphate epitopes 3B3 and 7D4, and glycosaminoglycan in human knee osteoarthritis after exercise. Ann Rheum Dis 59(11):887–91, Nov, 2000.

Beaupre LA, Davies DM, Jones CA, Cinats JG. Exercise combined with continuous passive motion or slider board therapy compared with exercise only: a randomized controlled trial of patients following total knee arthroplasty. Phys Ther 81(4):1029–37, Apr, 2001.

Belanger MJ, Moore DC, Crisco JJ 3rd, Fadale PD, Hulstyn MJ, Ehrlich MG. Knee laxity does not vary with the menstrual cycle, before or after exercise. Am J Sports Med 32(5):1150–7, Jul–Aug, 2004.

Bencke J, Naesborg H, Simonsen EB, Klausen K. Motor pattern of the knee joint muscles during side-step cutting in European team handball. Influence on muscular co-ordination after an intervention study. Scand J Med Sci Sports 10(2):68–77, Apr, 2000.

Bennell KL, Hinman RS. Effect of experimentally induced knee pain on standing balance in healthy older individuals. Rheumatology (Oxford) 44(3):378–81, Mar, 2005.

Beutler AI, Cooper LW, Kirkendall DT, Garrett WE Jr. Electromyographic analysis of single-leg, closed chain exercises: Implications for rehabilitation after anterior cruciate ligament reconstruction. J Athl Train 37(1):13–18, Mar, 2002.

Beynnon BD, Fleming BC, Labovitch R, Parsons B. Chronic anterior cruciate ligament deficiency is associated with increased anterior translation of the tibia during the transition from non-weightbearing to weightbearing. J Orthop Res 20:332–337, 2002.

Beynnon BD, Fleming BC. Anterior cruciate ligament strain in-vivo: A review of previous work. J Biomech 31:519–525, 1998.

Beynnon BD, Johnson RJ, Fleming BC, Stankewich CJ, Renstrom PA, Nichols CE. The strain behavior of the anterior cruciate ligament during squatting and active flexion-extension. A comparison of an open and a closed kinetic chain exercise. Am J Sports Med 25(6):823–9, Nov–Dec, 1997.

Birmingham TB, Kramer JF, Inglis JT, Mooney CA, Murray LJ, Fowler PJ, Kirkley S. Effect of a neoprene sleeve on knee joint position sense during sitting open kinetic chain and supine closed kinetic chain tests. Am J Sports Med 26(4):562–6, Jul–Aug, 1998.

Bizzini M, Childs JD, Piva SR, Delitto A. Systematic review of the quality of randomized controlled trials for patellofemoral pain syndrome. J Orthop Sports Phys Ther 33(1):4–20, Jan, 2003.

Blackburn JR, Morrissey MC. The relationship between open and closed kinetic chain strength of the lower limb and jumping performance. J Orthop Sports Phys

Ther 27(6):430–5, Jun, 1998.

Blanpied P, Carroll R, Douglas T, Lyons M, Macalisang R, Pires L. Effectiveness of lateral slide exercise in an anterior cruciate ligament reconstruction rehabilitation home exercise program. J Orthop Sports Phys Ther 30(10):602–8, Oct, 2000.

Bloem BR, Allum JH, Carpenter MG, Verschuuren JJ, Honegger F. Triggering of balance corrections and compensatory strategies in a patient with total leg proprioceptive loss. Exp Brain Res 142(1):91–107, Jan, 2002.

Bodor M. Quadriceps protects the anterior cruciate ligament. J Orthop Res 19(4):629–33, Jul, 2001.

Bolgla L, Keskula D. Reliability of lower extremity functional performance tests. J Orthop Sports Phys Ther 26(3):138–142, 1997.

Boucher JP, King MA, Lefebvre R, Pépin A. Quadriceps femoris muscle activity in patellofemoral pain syndrome. Am J Sports Med. 20(5):527–32, Sep–Oct, 1992.

Bouet V, Gahery Y. Muscular exercise improves knee position sense in humans. Neurosci Lett 289(2):143–6, Aug 4, 2000.

Boushel R, Langberg H, Green S, Skovgaard D, Bulow J, and Kjaer M. Blood flow and oxygenation in peritendinous tissue and calf muscle during dynamic exercise in humans. J Physiol 524:305–313, 2000.

Brindle TJ, Nyland J, Ford K, Coppola A, Shapiro R. Electromyographic comparison of standard and modified closed-chain isometric knee extension exercises. J Strength Cond Res 16(1):129–34, Feb, 2002.

Brockett CL, Morgan DL, Proske U. Human hamstring muscles adapt to eccentric exercise by changing optimum length. Med Sci Sports Exerc 33(5):783–90, May, 2001.

Bullock-Saxton JE, Janda V, Bullock MI. The influence of ankle sprain injury on muscle activation during hip extension. Int J Sports Med 15:330–334, 1994.

Butler DL, Guan Y, Kay MD, Cummings JF, Feder SM, Levy MS. Location-dependent variations in the material properties of the anterior cruciate ligament. J Biomech 25:511–518, 1992.

Bynum EB, Barrack RL, Alexander AH. Open versus closed chain kinetic exercises after anterior cruciate ligament reconstruction. A prospective randomized study. Am J Sports Med 23(4):401–6, Jul–Aug, 1995.

Byrne C, Eston R. The effect of exercise-induced muscle damage on isometric and dynamic knee extensor strength and vertical jump performance. J Sports Sci 20(5):417–25, May, 2002.

Callaghan MJ, Selfe J, Bagley PJ, Oldham JA. The effects of patellar taping on knee joint proprioception. J Athl Train 37(1):19–24, Mar, 2002.

Callaghan MJ, Selfe J, McHenry A, Oldham JA. Effects of patellar taping on knee joint proprioception in patients with patellofemoral pain syndrome. Man Ther, Feb 10, 2007.

Cannell LJ, Taunton JE, Clement DB, Smith C, Khan KM. A randomised clinical trial of the efficacy of drop squats or leg extension/leg curl exercises to treat clinically diagnosed jumper's knee in athletes: pilot study. Br J Sports Med 35(1):60–4, Feb, 2001.

Cannell LJ. The effects of an eccentric-type exercise versus concentric-type exercise in the management of chronic patellar tendinitis [master's thesis], Richmand: University of British Columbia; 1982.

Carter TR, Edinger S. Isokinetic evaluation of anterior cruciate ligament reconstruction: hamstring versus patellar tendon. Arthroscopy 15(2):169–72, Mar, 1999.

Cemy K. Vastus medialis oblique/vastus lateralis muscle activity ratios for selected exercises in persons with and without patellofemoral pain syndrome. Phys Ther 75:672–683, 1995.

Chaiwanichsiri D, Lorprayoon E, Noomanoch L. Star excursion balance training: effects on ankle functional stability after ankle sprain. J Med Assoc Thai 88 Suppl 4:S90–4, Sep, 2005.

Chandler TJ, Wilson GD, Stone MH. The effect of the squat exercise on knee stability. Med Sci Sports Exerc 21(3):299–303, Jun, 1989.

Chappell JD, Herman DC, Knight BS, Kirkendall DT, Garrett WE, Yu B. Effect of fatigue on knee kinetics and kinematics in stop-jump tasks. Am J Sports Med 33(7):1022–9, Jul, 2005.

Chu DA. *Jumping Into Plyometrics*, 2nd Edition. Human Kinetics. Champaigne, IL, 1998.

Cicuttini FM, Spector T, Baker J. Risk factors for osteoarthritis in the tibiofemoral and patellofemoral joints of the knee. J Rheum 24(6):1164–7, June, 1997.

Clark DI, Downing N, Mitchell J, Coulson L, Syzpryt EP, Doherty M. Physiotherapy for anterior knee pain: a randomised controlled trial. Ann Rheum Dis 59(9):700–4, Sep, 2000.

Coggon D, Reading I, Croft P, McLaren M, Barrett D, Cooper C. Knee osteoarthritis and obesity. Int J Obes Relat Metab Disord 25(5):622–7, May, 2001.

Cohen ZA, Roglic H, Grelsamer RP, Henry JH, Levine WN, Mow VC, Ateshian GA. Patellofemoral stresses during open and closed kinetic chain exercises. An analysis using computer simulation. Am J Sports Med 29(4):480–7, Jul–Aug, 2001.

Conlan T, Garth WP, Lemons JE. Evaluation of the medial soft tissue restraints of the extensor mechanism of the knee. J Bone Joint Surg Am 75(5):682–693, May, 1993.

Cooper RL, Taylor NF, Feller JA. A randomized controlled trial of proprioceptive and balance training after surgical reconstruction of the anterior cruciate ligament. Res Sports Med 13(3):217–30, Jul–Sept, 2005.

Cooper RL, Taylor NF, Feller JA. A systematic review of the effect of proprioceptive and balance exercises on people with an injured or reconstructed anterior

cruciate ligament. Res Sports Med 13(2):163–78, Apr–Jun, 2005.
Cowan SM, Bennell KL, Crossley KM, Hodges PW, McConnell J. Physical therapy alters recruitment of the vasti in patellofemoral pain syndrome. Med Sci Sports Exerc 34(12):1879–85, Dec, 2002.
Daniel D, Malcom L, Stone ML, Perth H, Morgan J, Riehl B. Quantification of knee stability and function. Contemp Orthop 5(1):83–91, 1982.
Daniel DM, Stone ML, Riehl B, Moore MR. A measurement of lower limb function: The one-leg hop for distance. Am J Knee Surg 1(4):212–214, 1988.
Dean E. Physiology and therapeutic implications of negative work. A review. Phys Ther 68(2):233–7, Feb, 1988.
DeMaio M, Mangine RE, Noyes FR, Barber SD. Advanced muscle training after ACL reconstruction: Weeks 6 to 52. Orthopedics 15:757–767, 1992a.
DeMaio M, Noyes FR, Mangine RE. Principles for aggressive rehabilitation after reconstruction of the anterior cruciate ligament. Orthopedics 15:385–392, 1992b.
Deyle GD, Allison SC, Matekel RL, Ryder MG, Stang JM, Gohdes DD, Hutton JP, Henderson NE, Garber MB. Physical therapy treatment effectiveness for osteoarthritis of the knee: a randomized comparison of supervised clinical exercise and manual therapy procedures versus a home exercise program. Phys Ther 85(12):1301–17, Dec, 2005.
Deyle GD, Henderson NE, Matekel RL, Ryder MG, Garber MB, Allison SC. Effectiveness of manual physical therapy and exercise in osteoarthritis of the knee. A randomized, controlled trial. Ann Intern Med 132(3):173–81, Feb 1, 2000.
Doucette SA, Child DD. The effect of open and closed chain exercise and knee joint position on patellar tracking in lateral patellar compression syndrome. J Orthop Sports Phys Ther 23(2):104–10, Feb, 1996.
Duncan A, McDonagh MJN. The role of short latency spinal stretch reflexes in human lower leg muscles when landing from a jump. J Physiol (Lond) 501P:42P, 1997.
Elias JJ, Faust AF, Chu YH, Chao EY, Cosgarea AJ. The soleus muscle acts as an agonist for the anterior cruciate ligament. An in vitro experimental study. Am J Sports Med 31(2):241–6, Mar–Apr, 2003.
Elmqvist LG, Lorentzon R, Johansson C, Fugl-Meyer AR. Does a torn anterior cruciate ligament lead to change in the central nervous drive of the knee extensors? Eur J Appl Physiol Occup Physiol 58(1–2):203–7, 1988.
Escamilla RF, Fleisig GS, Zheng N, Barrentine SW, Wilk KE, Andrews JR. Biomechanics of the knee during closed kinetic chain and open kinetic chain exercises. Med Sci Sports Exerc 30:556–569, 1998.
Escamilla RF. Knee biomechanics of the dynamic squat exercise. Med Sci Sports Exerc 33(1):127–41, Jan, 2001.
Evcik D, Sonel B. Effectiveness of a home-based exercise therapy and walking program on osteoarthritis of the knee. Rheumatol Int 22(3):103–6, Jul, 2002.
Eyigor S, Hepguler S, Capaci K. A comparison of muscle training methods in patients with knee osteoarthritis. Clin Rheumatol 23(2):109–15, Apr, 2004.
Faugli HP. *Medical Exercise Therapy*. Laerergruppen for Medisnsk Treningsterapi AS, Norway, 1996.
Felson DT, Chaisson CE. Understanding the relationship between body weight and osteoarthritis. Baillieres Clin Rheum. 11(4):671–81, Nov 1997.
Ferretti A, Papandrea P, Conteduca F, Mariani PP. Knee ligament injuries in volleyball players. Am J Sports Med 20:203–207, 1992.
Fitzgerald GK, Childs JD, Ridge TM, Irrgang JJ. Agility and perturbation training for a physically active individual with knee osteoarthritis. Phys Ther 82(4):372–82, Apr, 2002.
Fitzgerald GK. Open versus closed kinetic chain exercise: issues in rehabilitation after anterior cruciate ligament reconstructive surgery. Phys Ther 77(12):1747–54, Dec, 1997.
Fitzgerald GK. Therapeutic exercise for knee osteoarthritis: considering factors that may influence outcome. Eura Medicophys 41(2):163–71, Jun 2005.
Flanagan S, Salem GJ, Wang MY, Sanker SE, Greendale GA. Squatting exercises in older adults: kinematic and kinetic comparisons. Med Sci Sports Exerc. 35(4):635–43, Apr, 2003.
Fleming BC, Ohlen G, Renstrom PA, Peura GD, Beynnon BD, Badger GJ. The effects of compressive load and knee joint torque on peak anterior cruciate ligament (ACL) strains. Am J Sports Med 31:701–707, 2003.
Fleming BC, Oksendahl H, Beynnon BD. Open- or closed-kinetic chain exercises after anterior cruciate ligament reconstruction? Exerc Sport Sci Rev 33(3):134–40, Jul, 2005.
Fleming BC, P. Renstrom P, Beynnon BD, Engstrom B, Peura GD, Badger GJ. The effect of weightbearing and external loading on anterior cruciate ligament strain. J Biomech 34:163–170, 2001b.
Fleming BC, Renstrom PA, Ohlen G, Johnson RJ, Peura GD, Beynnon BD, Badger GJ. The gastrocnemius muscle is an antagonist of the anterior cruciate ligament. J Orthop Res 19:1178–1184, 2001a.
Focht BC, Rejeski WJ, Ambrosius WT, Katula JA, Messier SP. Exercise, self-efficacy, and mobility performance in overweight and obese older adults with knee osteoarthritis. Arthritis Rheum 53(5):659–65, Oct 15, 2005.
Forestier N, Teasdale N, Nougier V. Alteration of the position sense at the ankle induced by muscular fatigue in humans. Med Sci Sports Exerc 34(1):117–122, 2002.
Fredericson M, Cookingham CL, Chaudhari AM,

Dowdell DC, Oestreicher N, Sahrmann SA. Hip abductor weakness in distance runners with iliotibial band syndrome. Clin J Sport Med 10(3):169–75, Jul, 2000.

Freeman, Wyke B. The innervation of the knee joint. An anatomical and histological study in the cat. J Anat 101(Pt 3):505–32, Jun, 1967.

Fyfe I, Stanish WD. The use of eccentric training and stretching in the treatment and prevention of tendon injuries. Clin Sports Med 11(3):601–24, Jul, 1992.

Gabbett TJ. Incidence of injury in amateur rugby league sevens. Br J Sports Med 36:23–26, 2002.

Gabbett TJ. Incidence, site, and nature of injuries in amateur rugby league over three consecutive seasons. Br J Sports Med 34:98–103, 2000.

Gandevia SC, Enoka RM, McComas AJ, Stuart DG, Thomas CK. Neurobiology of muscle fatigue. Advances and issues. Adv Exp Med Biol 384:515–525, 1995.

Goldie PA, Bach TM, Evans OM. Force platform measures for evaluating postural control: reliability and validity. Arch Phys Med Rehabil 70(7):510–517, Jul, 1989.

Graham VL, Gehlsen GM, Edwards JA. Electromyographic evaluation of closed and open kinetic chain knee rehabilitation exercises. J Athl Train 28(1):23–30, Spring, 1993.

Grana WA, Muse G. The effect of exercise on laxity in the anterior cruciate ligament deficient knee. Am J Sports Med 16(6):586–8, Nov–Dec, 1988.

Gray G. Functional balance. Course workbook. Wynn Marketing, Inc., Nov, 2000.

Gray J, Taunton JE, McKenzie DC, Clement DB, McConkey JP, Davidson RG. A survey of injuries to the anterior cruciate ligament of the knee in female basketball players. Int J Sports Med 6:314–316, 1985.

Greenberger HB, Paterno MV. Relationship of knee extensor strength and hopping test performance in the assessment of lower extremity function. J Orthop Sports Phys Ther 22(5):202–6, Nov, 1995.

Grelsamer RP, Weinstein CH. Applied biomechanics of the patella. Clinical Orthopaedics Related research. 389:9–14, Aug, 2001.

Grimsby O. Manual Therapy of the Extremities-MT 6 workbook. The Ola Grimsby Institute, San Diego, CA, 1996.

Grimsby O. Neurophysiological view points on hypermobilities. J Manual Therapy. The Nordic Group of Specialists. Manual Therapy 2:2–9, 1988.

Grimsby O. Personal communication regarding the biomechanics and influences of neuro-receptors on the stability of the knee. The Ola Grimsby Institute, San Diego, CA, 2006.

Grimsby O. Post-Graduate Manual Therapy Residency Part I course outline. The Ola Grimsby Institute, San Diego, CA, 1991/1996.

Hakkinen K, Komi PV. Training-induced changes in neuromuscular performance under voluntary and reflex conditions. Eur J Appl Physiol Occup Physiol 55(2):147–55, 1986.

Hart JM, Fritz JM, Kerrigan DC, Saliba EN, Gansneder BM, Ingersoll CD. Reduced quadriceps activation after lumbar paraspinal fatiguing exercise. J Athl Train 41(1):79–86, Jan–Mar, 2006.

Hart LE. Home exercise for knee pain and knee osteoarthritis. Clin J Sport Med 13(6):388–9, Nov, 2003.

Hather BM, Tesch PA, Buchanan P, Dudley GA. Influence of eccentric actions on skeletal muscle adaptations to resistance training. Acta Physiol Scand 143(2):177–85, Oct, 1991.

Hazneci B, Yildiz Y, Sekir U, Aydin T, Kalyon TA. Efficacy of isokinetic exercise on joint position sense and muscle strength in patellofemoral pain syndrome. Am J Phys Med Rehabil 84(7):521–7, Jul, 2005.

Heijne A, Fleming BC, Renstrom PA, Peura GD, Beynnon BD, Werner S. Strain on the anterior cruciate ligament during closed kinetic chain exercises. Med Sci Sports Exerc 36(6):935–41, Jun, 2004.

Heintjes E, Berger MY, Bierma-Zeinstra SM, Bernsen RM, Verhaar JA, Koes BW. Exercise therapy for patellofemoral pain syndrome. Cochrane Database Syst Rev (4):CD003472, 2003.

Heller BM, Pincivero DM. The effects of ACL injury on lower extremity activation during closed kinetic chain exercise. J Sports Med Phys Fitness 43(2):180–8, Jun, 2003.

Hendry M, Williams NH, Markland D, Wilkinson C, Maddison P. Why should we exercise when our knees hurt? A qualitative study of primary care patients with osteoarthritis of the knee. Fam Pract 23(5):558–67, Oct, 2006.

Herbert R. Exercise, not taping, improves outcomes for patients with anterior knee pain. Aust J Physiother 47(1):66, 2001.

Hertel J, Braham RA, Hale SA, Olmsted-Kramer LC. Simplifying the Star Excursion Balance Test: Analyses of subjects with and without chronic ankle instability. J Orthop Sports Phys Ther 36(3): 131–137, Mar, 2006.

Hettinger T, Muller EA. Muscle capacity and muscle training. Arbeitsphysiologie 15(2):111–26, 1953.

Higuchi H, Terauchi M, Kimura M, Shirakura K, Katayama M, Kobayashi F, Takagishi K. Characteristics of anterior tibial translation with active and isokinetic knee extension exercise before and after ACL reconstruction. J Orthop Sci 7(3):341–7, 2002.

Hinman RS, Bennell KL, Metcalf BR, Crossley KM. Temporal activity of vastus medialis obliquus and vastus lateralis in symptomatic knee osteoarthritis. Am J Phys Med Rehabil 81(9):684–90, Sep, 2002.

Ho S, Jaureguito JW. Functional anatomy and biomechanics of the patellofemoral joint. Operative techniques Sports Med 2(4):238–247, Oct, 1994.

Holten O, Faugli HP. *Medisinsk Treningsterapi* (Medical Exercise Therapy), Ilniversitetsforlaaet. 0608 Oslo,

Norway, 1996.
Hooper DM, Hill H, Drechsler WI, Morrissey MC. Range of motion specificity resulting from closed and open kinetic chain resistance training after anterior cruciate ligament reconstruction. J Strength Cond Res 16(3):409–15, Aug, 2002.
Hooper DM, Morrissey MC, Drechsler W, Morrissey D, King J. Open and closed kinetic chain exercises in the early period after anterior cruciate ligament reconstruction. Improvements in level walking, stair ascent, and stair descent. Am J Sports Med 29(2):167–74, Mar–Apr, 2001.
Hopkins JT, Ingersoll CD, Sandrey MA, Bleggi SD. An Electromyographic Comparison of 4 Closed Chain Exercises. J Athl Train 34(4):353–357, Oct, 1999.
Horita T, Komi PV, Nicol C, Kyrolainen H. Interaction between pre-landing activities and stiffness regulation of the knee joint musculoskeletal system in the drop jump: implications to performance. Eur J Appl Physiol 88(1–2):76–84, Nov, 2002.
Horstmann T, Mayer F, Maschmann J, Niess A, Roecker K, Dickhuth HH. Metabolic reaction after concentric and eccentric endurance-exercise of the knee and ankle. Med Sci Sports Exerc 33(5):791–5, May, 2001.
Houck JR, Duncan A, De Haven KE. Comparison of frontal plane trunk kinematics and hip and knee moments during anticipated and unanticipated walking and side step cutting tasks. Gait Posture 24(3):314–22, Nov, 2006.
Houk J. Muscle activation patterns of selected lower extremity muscles during stepping and cutting tasks. J Electromyogr Kinesiol 13(6):545–54, Dec, 2003.
Howell SM. Comparison of closed and open kinetic chain exercise in the anterior cruciate ligament-deficient knee. Am J Sports Med 21(4):632; author reply 632–3, Jul–Aug, 1993.
Hu MH, Woollacott MH. Multisensory training of standing balance in older adults: I. Postural stability and one-leg stance balance. J Gerontol 49(2):M52–61, Mar, 1994.
Huang MH, Lin YS, Lee CL, Yang RC. Use of ultrasound to increase effectiveness of isokinetic exercise for knee osteoarthritis. Arch Phys Med Rehabil 86(8):1545–51, Aug, 2005.
Huberti HH, Hayes WC. Contact pressures in chondromalacia patellae and the effects of capsular reconstructive procedures. J Orthop Res 6:499–508, 1988.
Huberti HH, Hayes WC. Patellofemoral contact pressures. The influence of Q-angle and tendofemoral contact. J Bone Joint Surg Am 66(5):715–24, Jun, 1984.
Hurley MV. The effects of joint damage on muscle function, proprioception and rehabilitation. Manual Therapy 2(1):11–17, 1997.
Hutchinson MR, Ireland ML. Knee injuries in female athletes. Sports Med 19:288–302, 1995.
Iles JF, Stokes M, Young A. Reflex actions of knee joint afferents during contraction of the human quadriceps. Clin Physiol 10(5):489–500, Sep, 1990.
Isakov E, Mizrahi J. Is balance impaired by recurrent sprained ankle? Br J Sports Med 31(1):65–7, Mar, 1997.
Jenkins WL, Munns SW, Jayaraman G, Wertzberger KL, Neely K. A measurement of anterior tibial displacement in the closed and open kinetic chain. J Orthop Sports Phys Ther 25(1):49–56, Jan, 1997.
Jensen K, DiFabio RP. Evaluation of eccentric exercise in treatment of patellar tendinitis. Phys Ther 69(3):211–216, 1989.
Johansson H, Sjolander P, Sojka P. A sensory role for the cruciate ligaments. Clin Orthop Relat Res 268:161–78, Jul, 1991.
Johansson H, Sjolander P, Sojka P. Actions on gamma-motoneurones elicited by electrical stimulation of joint afferent fibres in the hind limb of the cat. J Physiol 375:137–52, Jun, 1986.
Johansson H, Sjolander P, Sojka P. Activity in receptor afferents from the anterior cruciate ligament evokes reflex effects on fusimotor neurones. Neurosci Res Apr;8(1):54–9, 1990.
Johansson H, Sjolander P, Sojka P. Receptors in the knee joint ligaments and their role in the biomechanics of the joint. Crit Rev Biomed Eng 18(5):341–68, 1991.
Jonsson H, Karrholm J. Three-dimensional knee joint movements during a step-up: evaluation after anterior cruciate ligament rupture. J Orthop Res 12(6):769–79, Nov, 1994.
Jonsson P, Alfredson H. Superior results with eccentric compared to concentric quadriceps training in patients with jumper's knee: a prospective randomized study. Br J Sports Med 39(11):847–50, Nov, 2005.
Jozsa L, Balint JB, Reffy A, and Demel Z. Histochemical and ultrastructural study of adult human tendon. Acta Histochem 65:250–257, 1979.
Judge JO, Lindsey C, Underwood M, Winsemius D. Balance improvements in older women: effects of exercise training. Phys Ther 73(4):254–62; discussion 263–5, Apr, 1993.
Kalliokoski KK, Langberg H, Ryberg AK, Scheede-Bergdahl C, Doessing S, Kjaer A, Boushel R, Kjaer M. The effect of dynamic knee-extension exercise on patellar tendon and quadriceps femoris muscle glucose uptake in humans studied by positron emission tomography. J Appl Physiol 99(3):1189–92, Sep, 2005.
Kapandji IA. The knee ligaments as determinants of trochleo-condylar profile. Med Biol Illus 17(1):26–32, Jan, 1967.
Karlsson J, Kalebo P, Goksor LA, Thomee R, Sward L. Partial rupture of the patellar ligament. Am J Sports Med 20(4):390–395, 1992.
Kasman GS. Surface EMG Evaluation and Feedback Training: Course Manual, 2002.
Kaufman KR, An KN, Litchy WJ, Morrey BF, Chao EY.

Dynamic joint forces during knee isokinetic exercise. Am J Sports Med 19(3):305–16, May–Jun, 1991.

Khan KM, Cook JL, Bonar F, Harcourt P, Astrom M. Histopathology of common tendonopathies-update and implications for clinical management. Sports Med 27(6):393–408, Jun, 1999.

Kibler WB, Livingston B. Closed-chain rehabilitation for upper and lower extremities. J Am Acad Orthop Surg 9(6):412–21, Nov–Dec, 2001.

Kibler WB. Closed kinetic chain rehabilitation for sports injuries. Phys Med Rehabil Clin N Am 11(2):369–84, May, 2000.

Kirkley A, Mohtadi N, Ogilvie R. The effect of exercise on anterior-posterior translation of the normal knee and knees with deficient or reconstructed anterior cruciate ligaments. Am J Sports Med 29(3):311–4, May–Jun, 2001.

Kiss ZS, Kellaway DP, Cook JL, Khan KM. Postoperative patellar tendon healing: an ultrasound study. VIS Tendon Study Group. Australas Radiol 42(1):28–32, Feb, 1998.

Klein KK. Specific progressive resistive exercise for increasing medial-lateral collateral ligament stability and the use of the knee ligament testing instrument for test-retest measurement. J Assoc Phys Ment Rehabil 18:135–7, Sep–Oct, 1964.

Klein KK. The maintenance of bilateral strength balance following progressive exercise for post operative and post injury knee conditions. J Assoc Phys Ment Rehabil 18:47–8, Mar–Apr, 1964.

Knuttgen HG, Petersen FB, Klausen K. Exercise with concentric and eccentric muscle contractions. Acta Paediatr Scand, suppl. 42:S217, 1971.

Korr IM. The neural basis of the osteopathic lesion. JAOA 47:191–198, 1947.

Kramer JF, Nusca D, Fowler P, Webster-Bogaert S. Test-retest reliability of the one-leg hop test following ACL reconstruction. Clin J Sport Med 2:240–243, 1992.

Kressley NW. Progressive exercise program following meniscectomy: utilization of lengthening contractions. J Am Phys Ther Assoc 43:263–4, Apr, 1963.

Kubo K, Kanehisa H, Fukunaga T. Influences of repetitive drop jump and isometric leg press exercises on tendon properties in knee extensors. J Strength Cond Res 19(4):864–7, Nov, 2005.

Kvist J, Cunningham D, Tigerstrand-Wejlemark H. Gender differences in post-exercise sagittal knee translation: a comparison between elite volleyball players and swimmers. Knee 13(2):132–6, Mar, 2006.

Kvist J, Gillquist J. Sagittal plane knee translation and electromyographic activity during closed and open kinetic chain exercises in anterior cruciate ligament-deficient patients and control subjects. Am J Sports Med 29(1):72–82, Jan–Feb, 2001.

Kvist M, Jozsa L, Jarvinen MJ, and Kvist H. Chronic Achilles paratendonitis in athletes: a histological and histochemical study. Pathology 19:1–11, 1987.

Langberg H, Bulow J, and Kjaer M. Blood flow in the peritendinous space of the human Achilles tendon during exercise. Acta Physiol Scand 163:149–153, 1998.

Langberg H, Skovgaard D, Karamouzis M, Bulow J, and Kjaer M. Metabolism and inflammatory mediators in the peritendinous space measured by microdialysis during intermittent isometric exercise in humans. J Physiol 515:919–927, 1999.

Lass P, Kaalund S, leFevre S, Arendt-Nielson L, Sinkjaer L, Sinkjaer T, Simonsen O. Muscle Coordination following rupture of the anterior cruciate ligament. Electromyographic studies of 14 patients. Acta Orthop Scand 62(1):9–14, Feb, 1991.

Lattanzio PJ, Petrella RJ, Sproule JR, Fowler PJ. Effects of fatigue on knee proprioception. Clin J Sport Med 7(1):22–27, 1997.

Leetun DT, Ireland ML, Willson JD, Ballantyne BT, Davis IM. Core stability measures as risk factors for lower extremity injury in athletes. Med Sci Sports Exerc 36:926–934, 2004.

Lieb FJ, Perry J. Quadriceps function: An anatomical and mechanical study using amputated limbs. J Bone Joint Surg (Am) 50:1535–48, 1968.

Lin DH, Lin YF, Chai HM, Han YC, Jan MH. Comparison of proprioceptive functions between computerized proprioception facilitation exercise and closed kinetic chain exercise in patients with knee osteoarthritis. Clin Rheumatol 26(4):520–8, Apr, 2007.

Liu SH, al-Shaikh R, Panossian V, Yang RS, Nelson SD, Soleiman N, Finerman GA, Lane JM. Primary immunolocalization of estrogen and progesterone target cells in the human anterior cruciate ligament. J Orthop Res 14:526–533, 1996.

Liu SH, Al-Shaikh RA, Panossian V, Finerman GA, Lane JM. Estrogen affects the cellular metabolism of the anterior cruciate ligament: a potential explanation for female athletic injury. Am J Sports Med 25:704–709, 1997.

Louw Q, Grimmer K, Vaughan C. Knee movement patterns of injured and uninjured adolescent basketball players when landing from a jump: a case-control study. BMC Musculoskelet Disord 7:22, Mar 7, 2006.

Lutz GE, Palmitier RA, An KN, Chao EY. Comparison of tibiofemoral joint forces during open-kinetic-chain and closed-kinetic-chain exercises. J Bone Joint Surg Am 75(5):732–9, May, 1993.

Lyman S, Fleisig GS, Waterbor JW, Funkhouser EM, Pulley L, Andrews JR, Osinski ED, Roseman JM. Longitudinal study of elbow and shoulder pain in youth baseball pitchers. Med Sci Sports Exerc 33:1803–1810, 2001.

Macintyre JG. 5-year follow-up of open or closed kinetic chain exercises for patellofemoral pain. Clin J Sport Med 15(3):199–200, May, 2005.

Mackey AL, Donnelly AE, Turpeenniemi-Hujanen T,

Roper HP. Skeletal muscle collagen content in humans after high-force eccentric contractions. J Appl Physiol 97(1):197–203, Jul, 2004.
Mafi N, Lorentzon R, Alfredson H. Superior short-term results with eccentric calf muscle training compared to concentric training in a randomized prospective multicenter study on patients with chronic Achilles tendinosis. Knee Surg Sports Traumatol Arthrosc 9(1):42–7, 2001.
Mangine RE, Noyes FR, DeMaio M. Minimal protection program: Advanced weightbearing and range of motion after ACL reconstruction—Weeks 1 to 5. Orthopedics 15:504–515, 1992.
Marks R, Quinney HA. Effect of fatiguing maximal isokinetic quadriceps contractions on ability to estimate knee-positon. Percept Mot Skills 77:1195–1202, 1993.
Marks R. Effect of exercise-induced fatigue on position sense of the knee. Aust J Physiother 40(3):175–181, 1994.
Martin K, Fontaine KR, Nicklas BJ, Dennis KE, Goldberg AP, Hochberg MC. Weight loss and exercise walking reduce pain and improve physical functioning in overweight postmenopausal women with knee osteoarthritis. J Clin Rheumatol 7(4):219–223, Aug, 2001.
Mattacola CG, Jacobs CA, Rund MA, Johnson DL. Functional assessment using the step-up-and-over test and forward lunge following ACL reconstruction. Orthopedics 27(6):602–8, Jun, 2004.
Mattacola CG, Perrin DH, Gansneder BM, Gieck JH, Saliba EN, McCue FC 3rd. Strength, functional outcome, and postural stability after anterior cruciate ligament reconstruction. J Athl Train 37(3):262–268, Sep, 2002.
McBride JM, Triplett-McBride T, Davie A, Newton RU. The effect of heavy vs light-load jump squats on the development of strength, power, and speed. J Strength Cond Res 16:75–82, 2002.
McCloskey DI. Kinesthetic sensibility. Physiol Rev 58(4):763–820, 1978.
McConnell J. The management of chondromalacia patellae: A long term solution. Aus J Physio 32(4):215–223, 1986.
McGuine TA, Greene JJ, Best T, Leverson G. Balance as a predictor of ankle injuries in high school basketball players. Clin J Sport Med 10(4):239–44, Oct, 2000.
McNair PJ, Marshall RN, Maguire K, Brown C. Knee joint effusion and proprioception. Arch Phys Med Rehabil 76(6):566–8, Jun, 1995.
McNair PJ, Marshall RN, Maguire K. Swelling of the knee joint: effects of exercise on quadriceps muscle strength. Arch Phys Med Rehabil 77(9):896–9, Sep, 1996.
Mellor R, Hodges PW. Motor unit synchronization of the vasti muscles in closed and open chain tasks. Arch Phys Med Rehabil 86(4):716–21, Apr, 2005.
Meyers MC, Sterling JC, Marley RR. Efficacy of stairclimber versus cycle ergometry in postoperative anterior cruciate ligament rehabilitation. Clin J Sport Med 12(2):85–94, Mar, 2002.
Miller BF, Olesen JL, Hansen M, Dossing S, Crameri RM, Welling RJ, Langberg H, Flyvbjerg A, Kjaer M, Babraj JA, Smith K, Rennie MJ. Coordinated collagen and muscle protein synthesis in human patella tendon and quadriceps muscle after exercise. J Physiol 567(Pt 3):1021–33, Sep 15, 2005.
Miller JP, Sedory D, Croce RV. Leg rotation and vastus medialis oblique/vastus lateralis electromyogram activity ratio during closed chain kinetic exercises prescribed for patellofemoral pain. J Athl Train 32(3):216–220, Jul, 1997.
Mine T, Kimura M, Sakka A, Kawai S. Mine T, Kimura M, Sakka A, Kawai S. Innervation of nociceptors in the menisci of the knee joint: an immunohistochemical study. Arch Orthop Trauma Surg 120(3–4):201–4, 2000.
Mizner RL, Petterson SC, Snyder-Mackler L. Quadriceps strength and the time course of functional recovery after total knee arthroplasty. J Orthop Sports Phys Ther 35(7):424–36, Jul, 2005.
Mizner RL, Stevens JE, Snyder-Mackler L. Voluntary activation and decreased force production of the quadriceps femoris muscle after total knee arthroplasty. Phys Ther 83(4):359–65, Apr, 2003.
Mjolsnes R, Arnason A, Osthagen T, Raastad T, Bahr R. A 10-week randomized trial comparing eccentric vs. concentric hamstring strength training in well-trained soccer players. Scand J Med Sci Sports 14(5):311–7, Oct, 2004.
Morrissey MC, Hooper DM, Drechsler WI, Hill HJ. Relationship of leg muscle strength and knee function in the early period after anterior cruciate ligament reconstruction. Scand J Med Sci Sports 14(6):360–6, Dec, 2004.
Morrissey MC, Hudson ZL, Drechsler WI, Coutts FJ, Knight PR, King JB. Effects of open versus closed kinetic chain training on knee laxity in the early period after anterior cruciate ligament reconstruction. Knee Surg Sports Traumatol Arthrosc 8(6):343–8, 2000.
Mueller, K. *Statische und Dynamische Muskelkraft*. (Static and Dynamic Muscle Strength) Deutsch, Fankfurt/M. Thun, 1987.
Muller EA. The regulation of muscular strength. J Assoc Ment Rehab 11:41, 1957.
Mulligan BR. *Manual Therapy NAGS SNAGS MWM*. Wellington, New Zealand: Plane View Services, 2004.
Myklebust G, Engebretsen L, Braekken IH, Skjolberg A, Olsen OE, Bahr R. Prevention of anterior cruciate ligament injuries in female team handball players: a prospective intervention study over three seasons. Clin J Sport Med 13:71–78, 2003.
Nawata K, Teshima R, Morio Y, Hagino H, Enokida M, Yamamoto K. Anterior-posterior knee laxity increased by exercise. Quantitative evaluation of physiologic

changes. Acta Orthop Scand 70(3):261–4, Jun, 1999.

Neeter C, Gustavsson A, Thomee P, Augustsson J, Thomee R, Karlsson J. Development of a strength test battery for evaluating leg muscle power after anterior cruciate ligament injury and reconstruction. Knee Surg Sports Traumatol Arthrosc 14:1–10, Feb, 2006.

Neitzel JA, Kernozek TW, Davies GJ. Loading responses following anterior cruciate ligament reconstruction during the parallel squat exercise. Clin Biomech (Bristol-Avon) 17(7):551–4, Aug, 2002.

Nielsen AB, Yde J. Epidemiology of acute knee injuries: a prospective hospital investigation. J Trauma 31:1644–1648, 1991.

Noel G, Verbruggen LA, Barbaix E, Duquet W. Adding compression to mobilization in a rehabilitation program after knee surgery. A preliminary clinical observational study. Man Ther 5(2):102–7, May, 2000.

Noyes FR, Barber SD, Mangine RE. Abnormal lower limb symmetry determined by function hop tests after anterior cruciate ligament rupture. Am J Sports Med 19(5):513–8, Sep–Oct, 1991.

Noyes FR, Barber-Westin SD. A comparison of results in acute and chronic anterior cruciate ligament ruptures of arthroscopically assisted autogenous patellar tendon reconstruction. Am J Sports Med 25:460–471, 1997.

Nyland J, Brosky T, Currier D, Nitz A, Caborn D. Review of the afferent neural system of the knee and its contribution to motor learning. J Orthop Sports Phys Ther 19(1):2–11, Jan, 1994.

Nyland J, Lachman N, Kocabey Y, Brosky J, Altun R, Caborn D. Anatomy, function, and rehabilitation of the popliteus musculotendinous complex. J Orthop Sports Phys Ther 35(3):165–79, Mar, 2005.

Nyland JA, Shapiro R, Stine RL, Horn TS, Ireland ML. Relationship of fatigued run and rapid stop to ground reaction forces, lower extremity kinematics, and muscle activation. J Orthop Sports Phys Ther 20:132–137, 1994.

O'Donnell S, Thomas SG, Marks P. Improving the sensitivity of the hop index in patients with an ACL deficient knee by transforming the hop distance scores. BMC Musculoskelet Disord 7:9, Feb 1, 2006.

Ohkoshi Y, Yasuda K, Kaneda K. In vivo analysis of the compression force on the patellofemoral joint (abstr). Transactions of the 37th Annual Meeting, Orthopaedic Research Society, 572, 1991.

Olmsted LC, Carcia CR, Hertel J, Shultz SJ. Efficacy of the star excursion balance tests in detecting reach deficits in subjects with chronic ankle instability. J Athl Train 37(4):501–506, Dec, 2002.

Olsen OE, Myklebust G, Engebretsen L, Holme I, Bahr R. Exercises to prevent lower limb injuries in youth sports: cluster randomized controlled trial. BMJ 26;330(7489):449, Feb, 2005.

Ostenberg A, Roos E, Ekdahl C, Roos H. Isokinetic knee extensor strength and functional performance in healthy female soccer players. Scand J Med Sci Sports 8(5 Pt 1):257–64, Oct, 1998.

Palmieri RM, Weltman A, Edwards JE, Tom JA, Saliba EN, Mistry DJ, Ingersoll CD. Pre-synaptic modulation of quadriceps arthrogenic muscle inhibition. Knee Surg Sports Traumatol Arthrosc 13(5):370–6, Jul, 2005.

Palmitier RA, An KN, Scott SG, Chao EY. Kinetic chain exercise in knee rehabilitation. Sports Med 11(6):402–13, Jun, 1991.

Panariello RA, Backus SI, Parker JW. The effect of the squat exercise on anterior-posterior knee translation in professional football players. Am J Sports Med 22(6):768–73, Nov–Dec, 1994.

Pare EB, Stern JT, Schwartz JM. Functional differentiation within the tensor fasciae latae. J Bone Joint Surg Am 63:1457–1471, 1981.

Paterno MV, Myer GD, Ford KR, Hewett TE. Neuromuscular training improves single-limb stability in young female athletes. J Orthop Sports Phys Ther 34(6):305–16, Jun, 2004.

Pedersen J, Lonn J, Hellstrom F, Djupsjobacka M, Johansson H. Localized muscle fatigue decreases the acuity of the movement sense in the human shoulder. Med Sci Sports Exerc 31(7):1047–1052, 1999.

Peers KH, Lysens RJ. Patellar tendinopathy in athletes: current diagnostic and therapeutic recommendations. Sports Med 35(1):71–87, 2005.

Perry MC, Morrissey MC, King JB, Morrissey D, Earnshaw P. Effects of closed versus open kinetic chain knee extensor resistance training on knee laxity and leg function in patients during the 8- to 14-week post-operative period after anterior cruciate ligament reconstruction. Knee Surg Sports Traumatol Arthrosc 13(5):357–69, Jul, 2005.

Perry MC, Morrissey MC, Morrissey D, Knight PR, McAuliffe TB, King JB. Knee extensors kinetic chain training in anterior cruciate ligament deficiency. Knee Surg Sports Traumatol Arthrosc. 13(8):638–48, Nov, 2005.

Petschnig R, Baron R, Albrecht M. The relationship between isokinetic quadriceps strength test and hop tests for distance and one-legged vertical jump test following anterior cruciate ligament reconstruction. J Orthop Sports Phys Ther 28(1):23–31, Jul, 1998.

Pincivero DM, Aldworth C, Dickerson T, Petry C, Shultz T. Quadriceps-hamstring EMG activity during functional, closed kinetic chain exercise to fatigue. Eur J Appl Physiol 81(6):504–9, Apr, 2000.

Pincivero DM, Lephart SM, Karunakara RG. Relation between open and closed kinematic chain assessment of knee strength and functional performance. Clin J Sport Med 7(1):11–6, Jan, 1997.

Pollard CD, Braun B, Hamill J. Influence of gender, estrogen and exercise on anterior knee laxity. Clin Biomech (Bristol, Avon) 21(10):1060–6, Dec, 2006.

Proske U, Gregory JE. Signalling properties of muscle spindles and tendon organs. Adv Exp Med Biol 508:5–12, 2002.

Proske U, Morgan DL, Brockett CL, Percival P. Identifying athletes at risk of hamstring strains and how to protect them. Clin Exp Pharmacol Physiol 31(8):546–50, Aug, 2004.
Proske U, Wise AK, Gregory JE. The role of muscle receptors in the detection of movements. Prog Neurobiol 60(1):85–96, 2000.
Proske U. Kinesthesia: the role of muscle receptors. Muscle Nerve 34(5):545–58, Nov, 2006.
Proske U. What is the role of muscle receptors in proprioception? Muscle Nerve 31(6):780–7, Jun, 2005.
Purdam CR, Jonsson P, Alfredson H, Lorentzon R, Cook JL, Khan KM. A pilot study of the eccentric decline squat in the management of painful chronic patellar tendinopathy. Br J Sports Med 38(4):395–7, Aug, 2004.
Rabin A. Is there evidence to support the use of eccentric strengthening exercises to decrease pain and increase function in patients with patellar tendinopathy? Phys Ther 86(3):450–6, Mar, 2006.
Rahnama N, Reilly T, Lees A. Injury risk associated with playing actions during competitive soccer. Br J Sports Med 36:354–359, 2002.
Rannou F, Poiraudeau S, Revel M. Cartilage: from biomechanics to physical therapy. Ann Readapt Med Phys 44(5):259–67, 2001.
Raunest J, Sager M, Burgener E. Proprioceptive mechanisms in the cruciate ligaments: an electromyographic study on reflex activity in the thigh muscles. J Trauma 41(3):488–93, Sep, 1996.
Ribeiro F, Mota J, Oliveira J. Effect of exercise-induced fatigue on position sense of the knee in the elderly. Eur J Appl Physiol 99(4):379–85, Mar, 2007.
Riley G. The pathogenesis of tendinopathy. A molecular perspective. Rheum (Oxford) 43(2):131–142, Jul 16, 2004.
Robinovitch SN, Heller B, Lui A, Cortez J. Effect of strength and speed of torque development on balance recovery with the ankle strategy. J Neurophysiol 88(2):613–20, Aug, 2002.
Roddy E, Zhang W, Doherty M, Arden NK, Barlow J, Birrell F, Carr A, Chakravarty K, Dickson J, Hay E, Hosie G, Hurley M, Jordan KM, McCarthy C, McMurdo M, Mockett S, O'Reilly S, Peat G, Pendleton A, Richards S. Evidence-based recommendations for the role of exercise in the management of osteoarthritis of the hip or knee—the MOVE consensus. Rheumatology (Oxford) 44(1):67–73, Jan, 2005b.
Roddy E, Zhang W, Doherty M. Aerobic walking or strengthening exercise for osteoarthritis of the knee? A systematic review. Ann Rheum Dis 64(4):544–8, Apr, 2005a.
Rohmert W. Ermittung von Erholung-spausen fur statische Arbeit des Menschen. Physologist 18:123–120, 1960.
Rooks DS, Huang J, Bierbaum BE, Bolus SA, Rubano J, Connolly CE, Alpert S, Iversen MD, Katz JN. Effect of preoperative exercise on measures of functional status in men and women undergoing total hip and knee arthroplasty. Arthritis Rheum 55(5):700–8, Oct 15, 2006.
Roos EM, Dahlberg L. Positive effects of moderate exercise on glycosaminoglycan content in knee cartilage: a four-month, randomized, controlled trial in patients at risk of osteoarthritis. Arthritis Rheum 52(11):3507–14, Nov, 2005.
Roos EM, Engstrom M, Lagerquist A, Soderberg B. Clinical improvement after 6 weeks of eccentric exercise in patients with mid-portion Achilles tendinopathy—a randomized trial with 1-year follow-up. Scand J Med Sci Sports 14(5):286–95, Oct, 2004.
Ross MD, Denegar CR, Winzenried JA. Implementation of open and closed kinetic chain quadriceps strengthening exercises after anterior cruciate ligament reconstruction. J Strength Cond Res 15(4):466–73, Nov, 2001.
Russell KA, Palmieri RM, Zinder SM, Ingersoll CD. Sex differences in valgus knee angle during a single-leg drop jump. J Athl Train 41(2):166–71, Apr–Jun, 2006.
Schlicht J, Camaione DN, Owen SV. Effect of intense strength training on standing balance, walking speed, and sit-to-stand performance in older adults. J Gerontol A Biol Sci Med Sci 56(5):M281–6, May, 2001.
Schulthies SS, Ricard MD, Alexander KJ, Myrer JW. An electromyographic investigation of 4 elastic-tubing closed kinetic chain exercises after anterior cruciate ligament reconstruction. J Athl Train 33(4):328–335, Oct, 1998.
Schutte MJ, Dabezies EJ, Zimny ML, Happel LT. Neural anatomy of the human anterior cruciate ligament. J Bone Joint Surg Am 69(2):243–7, Feb,1987.
Scott A, Khan KM, Roberts CR, Cook JL, Duronio V. What do we mean by the term "inflammation"? A contemporary basic science update for sports medicine. Br J Sports Med 38(3):372–380, Jun, 2004.
Seger JY, Thorstensson A. Effects of eccentric versus concentric training on thigh muscle strength and EMG. Int J Sports Med 26(1):45–52, Jan–Feb, 2005.
Sernert N, Kartus J, Kohler K, Stener S, Larsson J, Eriksson BI, Karlsson J. Analysis of subjective, objective and functional examination tests after anterior cruciate ligament reconstruction. A follow-up of 527 patients. Knee Surg Sports Traumatol Arthrosc 7(3):160–5, 1999.
Sharpe MH, Miles TS. Position sense at the elbow after fatiguing contractions. Exp Brain Res 94(1):179–182, 1993.
Shelbourne KD, Davis TJ. Evaluation of knee stability before and after participation in a functional sports agility program during rehabilitation after anterior cruciate ligament reconstruction. Am J Sports Med 27(2):156–61, Mar–Apr, 1999.

Shields RK, Madhaven S, Gregg E, Leitch J, Peterson B, Salata S, Wallerich S. Neuromuscular control of the knee during a resisted single-limb squat exercise. Am J Sports Med 33(10):1520–6, Oct, 2005.

Silbernagel KG, Thomee R, Thomee P, Karlsson J. Eccentric overload training for patients with chronic Achilles tendon pain—a randomised controlled study with reliability testing of the evaluation methods. Scand J Med Sci Sports 11(4):197–206, Aug, 2001.

Sjolander P, Johansson H, Sojka P, Rehnholm A. Sensory nerve endings in the cat cruciate ligaments: a morphological investigation. Neurosci Lett 17;102(1):33–8, Jul, 1989.

Skinner HB, Wyatt MP, Hodgdon JA, Conard DW, Barrack RL. Effect of fatigue on joint position sense of the knee. J Orthop Res 4(1):112–118, 1986.

Slauterbeck J, Clevenger C, Lundberg W, Burchfield DM. Estrogen level alters the failure load of the rabbit anterior cruciate ligament. J Orthop Res 17:405–408, 1999.

Slauterbeck JR, Fuzie SF, Smith MP, lark RJ, Xu K, Starch DW, Hardy DM. The menstrual cycle, sex hormones, and anterior cruciate ligament injury. J Athletic Train 37:275–280, 2002.

Sojka P, Johansson H, Sjolander P, Lorentzon R, Djupsjobacka M. Fusimotor neurones can be reflexly influenced by activity in receptor afferents from the posterior cruciate ligament. Brain Res 27;483(1):177–83, Mar, 1989.

Sojka P, Sjolander P, Johansson H, Djupsjobacka M. Influence from stretch-sensitive receptors in the collateral ligaments of the knee joint on the gamma-muscle-spindle systems of flexor and extensor muscles. Neurosci Res 11(1):55–62, Jun, 1991.

Solomonow M, Krogsgaard M. Sensorimotor control of knee stability. A review. Scand J Med Sci Sports 11(2):64–80, Apr, 2001.

Sorensen KL, Hollands MA, Patla E. The effects of human ankle muscle vibration on posture and balance during adaptive locomotion. Exp Brain Res 143(1):24–34, Mar, 2002.

Souza DR, Gross MT. Comparison of vastus medialis obliquus: vastus lateralis muscle integrated electromyographic ratios between healthy subjects and patients with patellofemoral pain. Phys Ther 71:310–320, 1991.

Stasinopoulos D, Stasinopoulos I. Comparison of effects of exercise programme, pulsed ultrasound and transverse friction in the treatment of chronic patellar tendinopathy. Clin Rehabil 18(4):347–52, Jun, 2004.

Steiner ME, Grana WA, Chillag K, Schelberg-Karnes E. The effect of exercise on anterior-posterior knee laxity. Am J Sports Med 14(1):24–9, Jan–Feb, 1986.

Stensdotter AK, Hodges PW, Mellor R, Sundelin G, Hager-Ross C. Quadriceps activation in closed and in open kinetic chain exercise. Med Sci Sports Exerc 35(12):2043–7, Dec, 2003.

Stiene HA, Brosky T, Reinking MF, Nyland J, Mason MB. A comparison of closed kinetic chain and isokinetic joint isolation exercise in patients with patellofemoral dysfunction. J Orthop Sports Phys Ther 24(3):136–41, Sep, 1996.

Stuart MJ, Meglan DA, Lutz GE, Growney ES, An KN. Comparison of intersegmental tibiofemoral joint forces and muscle activity during various closed kinetic chain exercises. Am J Sports Med 24(6):792–9, Nov–Dec, 1996.

Sumen Y, Ochi M, Adachi N, Urabe Y, Ikuta Y. Anterior laxity and MR signals of the knee after exercise. A comparison of 9 normal knees and 6 anterior cruciate ligament reconstructed knees. Acta Orthop Scand 70(3):256–60, Jun, 1999.

Tagesson S, Oberg B, Kvist J. Passive and dynamic translation in the knee is not influenced by knee exercises in healthy individuals. Scand J Med Sci Sports 15(3):137–47, Jun, 2005.

Tang SF, Chen CK, Hsu R, Chou SW, Hong WH, Lew HL. Vastus medialis obliquus and vastus lateralis activity in open and closed kinetic chain exercises in patients with patellofemoral pain syndrome: an electromyographic study. Arch Phys Med Rehabil 82(10):1441–5, Oct, 2001.

Taunton J. Comparison of 2 eccentric exercise protocols for patellar tendinopathy in volleyball players. Clin J Sport Med 16(1):90–1, Jan, 2006.

Tepperman PS, Mazliah J, Naumann S, Delmore T. Effect of ankle position on isometric quadriceps strengthening. Am J Phys Med 65(2):69–74, Apr, 1986.

Thomas KS, Muir KR, Doherty M, Jones AC, O'Reilly SC, Bassey EJ. Home based exercise programme for knee pain and knee osteoarthritis: randomised controlled trial. BMJ 325(7367):752, Oct 5, 2002.

Thorstensson CA, Roos EM, Petersson IF, Ekdahl C. Six-week high-intensity exercise program for middle-aged patients with knee osteoarthritis: a randomized controlled trial. BMC Musculoskelet Disord 6:27, May 30, 2005.

Tibone JE, Antich TJ. Electromyographic analysis of the anterior cruciate ligament-deficient knee. Clin Orthop Relat Res (288):35–9, Mar, 1993.

Topp R, Mikesky A, Wigglesworth J, Holt W Jr, Edwards JE. The effect of a 12-week dynamic resistance strength training program on gait velocity and balance of older adults. Gerontologist 33(4):501–6, Aug, 1993.

Tria AJ, Palumbo RC, Alicea JA. Conservative care for the patellofemoral pain. Orthop Clin North Am 23:545–53, 1992.

Tsepis E, Vagenas G, Giakas G, Georgoulis A. Hamstring weakness as an indicator of poor knee function in ACL-deficient patients. Knee Surg Sports Traumatol Arthroscl2(1):22–9, Jan, 2004.

Turl SE, George KP. Adverse neural tension: a factor in repetitive hamstring strain? J Orthop Sports Phys Ther

27(1):16–21, Jan, 1998.
Vad V, Hong HM, Zazzali M, Agi N, Basrai D. Exercise recommendations in athletes with early osteoarthritis of the knee. Sports Med 32(11):729–39, 2002.
van Baar ME, Assendelft WJ, Dekker J, Oostendorp RA, Bijlsma JW. Effectiveness of exercise therapy in patients with osteoarthritis of the hip or knee: a systematic review of randomized clinical trials. Arthritis Rheum 42(7):1361–9, Jul, 1999.
van Baar ME, Dekker J, Oostendorp RA, Bijl D, Voorn TB, Bijlsma JW. Effectiveness of exercise in patients with osteoarthritis of hip or knee: nine months' follow up. Ann Rheum Dis 60(12):1123–30, Dec, 2001.
van Baar ME, Dekker J, Oostendorp RA, Bijl D, Voorn TB, Lemmens JA, Bijlsma JW. The effectiveness of exercise therapy in patients with osteoarthritis of the hip or knee: a randomized clinical trial. J Rheumatol 25(12):2432–9, Dec, 1998.
van Gool CH, Penninx BW, Kempen GI, Rejeski WJ, Miller GD, van Eijk JT, Pahor M, Messier SP. Effects of exercise adherence on physical function among overweight older adults with knee osteoarthritis. Arthritis Rheum 53(1):24–32, Feb 15, 2005.
Verhoshanski V, Tatyan V. Speed-strength preparation of future champions. Sove Sports Review 18(4):166–170, 1983.
Visnes H, Bahr R. The evolution of eccentric training as treatment for patellar tendinopathy (jumper's knee): a critical review of exercise programmes. Br J Sports Med 41(4):217–23, Apr, 2007.
Visnes H, Hoksrud A, Cook J, Bahr R. No effect of eccentric training on jumper's knee in volleyball players during the competitive season: a randomized clinical trial. Clin J Sport Med 15(4):227–34, Jul, 2005.
Waddington G, Seward H, Wrigley T, Lacey N, Adams R. Comparing wobble board and jump-landing training effects on knee and ankle movement discrimination. J Sci Med Sport 3(4):449–59, Dec, 2000.
Wang TJ, Belza B, Elaine Thompson F, Whitney JD, Bennett K. Effects of aquatic exercise on flexibility, strength and aerobic fitness in adults with osteoarthritis of the hip or knee. J Adv Nurs 57(2):141–52, Jan, 2007.
Warren GL, Hayes DA, Lowe DA, Prior BM, Armstrong RB. Materials fatigue initiates eccentric contraction-induced injury in rat soleus muscle. J Physiol 464:477–89, May, 1993.
Wegener L, Kisner C, Nichols D. Static and dynamic balance responses in persons with bilateral knee osteoarthritis. JOSPT 25(1):13–8, Jan, 1997.
Wester JU, Jespersen SM, Nielsen KD, Neumann L. Wobble board training after partial sprains of the lateral ligaments of the ankle: a prospective randomized study. J Orthop Sports Phys Ther 23(5):332–6, May, 1996
White DM. Comparison of closed and open kinetic chain exercise in the anterior cruciate ligament-deficient knee. Am J Sports Med 21(4):633, Jul–Aug, 1993.
Wild JJ Jr, Franklin TD, Woods Gw. Patellar pain and quadriceps rehabilitation. An EMG study. Am J Sports Med 10(1):12–5, Jan–Feb, 1982.
Wilk KE, Escamilla RF, Fleisig GS, Barrentine SW, Andrews JR, Boyd ML. A comparison of tibiofemoral joint forces and electromyographic activity during open and closed kinetic chain exercises. Am J Sports Med 24(4):518–27, Jul–Aug, 1996.
Wilk KE, Reinold MM, Hooks TR. Recent advances in the rehabilitation of isolated and combined anterior cruciate ligament injuries. Orthop Clin North Am 34(1):107–37, Jan, 2003.
Williams GN, Chmielwski T, Rudolph K, Buchanan TS, Snyder-Mackler L. Dynamic knee stability: current theory and implications for clinicians and scientists. J Orthop Sports Phys Ther 31(10):546– 66, Oct, 2001.
Wilson GJ, Newton RU, Murphy AJ, Humphries BJ. The optimal training load for the development of dynamic athletic performance. Med Sci Sports Exerc 25:1279–1286, 1993.
Withrow TJ, Huston LJ, Wojtys EM, Ashton-Miller JA. The effect of an impulsive knee valgus moment on in vitro relative ACL strain during a simulated jump landing. Clin Biomech (Bristol, Avon) 21(9):977–83, Nov, 2006.
Withrow TJ, Huston LJ, Wojtys EM, Ashton-Miller JA. The relationship between quadriceps muscle force, knee flexion, and anterior cruciate ligament strain in an in vitro simulated jump landing. Am J Sports Med 34(2):269–74, Feb, 2006.
Witvrouw E, Cambier D, Danneels L, Bellemans J, Werner S, Almqvist F, Verdonk R. The effect of exercise regimens on reflex response time of the vasti muscles in patients with anterior knee pain: a prospective randomized intervention study. Scand J Med Sci Sports 13(4):251–8, Aug, 2003.
Witvrouw E, Danneels L, Van Tiggelen D, Willems TM, Cambier D. Open versus closed kinetic chain exercises in patellofemoral pain: a 5-year prospective randomized study. Am J Sports Med 32(5):1122–30, Jul–Aug, 2004.
Witvrouw E, Lysens R, Bellemans J, Peers K, Vanderstraeten G. Open versus closed kinetic chain exercises for patellofemoral pain. A prospective, randomized study. Am J Sports Med 28(5):687–94, Sep–Oct, 2000.
Witvrouw E, Lysens R, Bellemans J, Peers K, Vanderstraeten G. Open versus closed kinetic chain exercises for patellofemoral pain. A prospective, randomized study. Am J Sports Med 28(5):687–94, Sep–Oct, 2000.
Wojtys EM, Huston LJ, Boynton MD, Spindler KP, Lindenfeld TN. The effect of the menstrual cycle on anterior cruciate ligament injuries in women as determined by hormone levels. Am J Sports Med

30:182–188, 2002.
Wyatt FB, Milam S, Manske RC, Deere R. The effects of aquatic and traditional exercise programs on persons with knee osteoarthritis. J Strength Cond Res 15(3):337–40, Aug, 2001.
Yack HJ, Collins CE, Whieldon TJ. Comparison of closed and open kinetic chain exercise in the anterior cruciate ligament-deficient knee. Am J Sports Med 21(1):49–54, Jan–Feb, 1993.
Yack HJ, Collins CE, Whieldon TJ. Comparison of closed and open kinetic chain exercise in the anterior cruciate ligament-deficient knee. Am J Sports Med 21(1):49–54, Jan–Feb, 1993.
Yack HJ, Riley LM, Whieldon TR. Anterior tibial translation during progressive loading of the ACL-deficient knee during weightbearing and nonweightbearing isometric exercise. J Orthop Sports Phys Ther 20(5):247–53, Nov 1994.
Young A, Stokes M, Iles JF. Effects of joint pathology on muscle. Clin Orthop Relat Res. (219):21–7, Jun, 1987.
Young J. The effectiveness of eccentric quadriceps muscle training in decreasing tendon pain and improving lower extremity function in patients with chronic patellar tendinitis. Doctoral Dissertation: The Ola Grimsby Institute, 2002.
Young MA, Cook JL, Purdam CR, Kiss ZS, Alfredson H. Eccentric decline squat protocol offers superior results at 12 months compared with traditional eccentric protocol for patellar tendinopathy in volleyball players. Br J Sports Med 39(2):102–5, Feb, 2005.
Zelisko JA, Noble HB, Porter M. A comparison of men's and women's professional basketball injuries. Am J Sports Med 10:297–299, 1982.

Authors: Dan Washeck, Jim Rivard, Ola Grimsby

Exercise Rehabilitation of the Hip Joint

Introduction

In this chapter:

Review of published physical therapy literature presents little information on treatment progressions of the hip using manual therapy techniques for joint mobilization and exercise. Manual therapy involves passive stretching and manipulation techniques aimed at restoring joint function (Cyriax 1996). The current definition of manual therapy practice includes exercise as a component and is not limited to passive joint treatments. Orthopaedic Manual Therapy is a specialized area of physiotherapy/physical therapy for the management of neuromusculoskeletal conditions, based on clinical reasoning, using highly specific treatment approaches including manual techniques and therapeutic exercises (IFOMT). Long axis distraction is one of the most commonly used manual therapy techniques for hip pain (Kisner et al. 1985, Woerman et al. 1989), and has shown to be beneficial for both pain inhibition and increasing hip range of motion (Kaltenborn 1980).

Exercise therapy administered by a physical therapist, described as active and passive exercises aimed at improvement of pain, range of motion, muscle function and ambulation (Minor 1994, Hofmann 1993) is also reported to be effective in the treatment of patients with osteoarthritis of the hip (Van Baar 1998, Smidt et al. 2005). While both exercise and manual therapy have been shown to be effective on painful hips, Hoeksma et al. (2004) compared the effectiveness of manual therapy, consisting of stretching and manipulation techniques, versus exercise alone on hip osteoarthritis. They found manual therapy to be more effective than exercise alone in the treatment of hip osteoarthritis. Exercise is a part of manual therapy and a combined approach of both passive and active treatment is recommended, as neither can be fully replaced by the other.

Passive joint mobilization and soft tissue work may be necessary prior to initiating exercise. Passive manual therapy techniques are aimed at restoring arthrokinematic mobility, improving range of motion, reducing pain, resolving muscle guarding and assisting mechanoreceptors to facilitate normal motor patterns (Grimsby

1991). If functional qualities and specific tissues are not addressed in the exercise routine, then outcomes may be suboptimal. For example, if the femoral head does not have normal arthrokinematic mobility, then applying torque with an exercise consisting of long lever movements may abnormally load the cartilage and lead to further degeneration. The dosage and exercise selection will also vary depending on the functional quality being trained. Early training to stimulate articular cartilage repair, restore joint mobility and reduce tone in guarded muscles will have a better treatment outcome than strength training a restricted joint.

Joint mobilization techniques should be focused on restoring proper joint mechanics and arthrokinematic mobility. While both exercise therapy and joint mobilization have been shown to be effective in the short term treatment of hip osteoarthritis, the benefits of each one taken individually decline over time and finally disappear at six to nine month follow-ups (Hoeksma et al. 2004, Van Baar et al. 2001). Strictly using a passive approach such as joint mobilization may lead to resolution of muscle guarding and restoration of joint mobility, however, adding exercise is necessary to provide the stimulus to repair tissue and to coordinate movement in the newly gained ranges of motion.

Providing only exercise therapy may not address specific soft tissue and joint restrictions that impede function, while providing only passive joint treatments does not improved tissue tolerance or training qualities leading to better function. Assigning random exercises with random dosage without attention to the tissue in lesion will be ineffective. The exercises and dosage of treating articular cartilage of the hip as a primary tissue will look much different than exercises focused on mobilizing collagen of a restricted joint capsule. The following sections provide options for addressing different functional qualities from Stage 1 to Stage 4. These exercise options incorporate concepts of manual therapy and exercise benefits that have been shown individually with hip pathology.

Section 1: Stage 1 Exercise Progression Concepts for the Hip Joint

Training Goals

Treatment for acute any hip pathology begins with an emphasis on reducing pain, decreasing muscle spasm, improving range of motion and normalizing soft tissue mobility. Stage 1 goals also focus on normalizing joint mobility, improving coordination, increasing circulation to the tonic muscles, preventing atrophy, resolving inflammation and stimulating protein synthesis for repairing tissues (Grimsby 1995). Initially, soft tissue and joint mobilization techniques should be employed to resolve muscle guarding and reduce pain through mechanoreceptor activation for afferent inhibition. Basic training concepts for Stage 1 exercises include performing many repetitions with minimal resistance at slower speeds for an emphasis on pain inhibition, tissue repair, coordination and improving range of motion.

In a simplified description, hypomobile joints are trained in the outer range of available motion to improve elasticity, while hypermobile joints are trained within the inner range for dynamic stabilization, avoiding the pathological outer range. Variations of tissue presentation within the individual will dictate the exact range of training and progression. Significant individual impairment and/or pain may require the use of suspensions slings and eccentric stop wheels to attain the desired number of repetitions. This section will provide examples of these type of techniques as part of the Stage 1 exercises concepts and dosage parameters for specific functional qualities. The sections covering Stage 1 through 4 concepts may employ similar exercise, but it is not the exercise itself that determines the outcome of training. The outcome is dependent upon the design, dosage and progression models that are presented to emphasize different tissues, impairments and functional qualities.

Basic Outline for Stage 1

- *3–5 Exercises*
- *Many repetitions minimal resistance (coordination)*
  - *< 50% of 1RM for joint mobilization*
  - *< 60% of 1RM (vascularity)*
  - *< 50% of 1RM—Tissue repair, edema reduction and range of motion*
  - *< 40% of 1RM tonic fiber atrophy (hypertrophy)*
- *Low speed (coordination)*
  - *Deformity of collagen*
  - *Concentric facilitation*
- *Hypomobile: train in the outer range of available normal physiologic motion*
- *Hypermobile: train in the mid to inner range of available normal physiologic motion*
- *Selective tissue training (STT): for bone, collagen, muscle and cartilage*
- *Neurophysiologic influences: pain inhibition and coordination*

Options

- *Joint in resting position, start contraction from length tension position*
- *Concentric work only for vascularity of muscles in guarding*
  - *Use a pulley with an eccentric stop wheel to remove the eccentric phase*
  - *Weight stack rests between repetitions*
- *Artificial locking may be required to avoid motion around a non-physiologic axis*

Joint / Myofascial Mobilization

Before training for coordination, joint mobility must be normalized so that arthrokinematic motion occurs around a physiological axis. If the joint does not glide properly then exercise directed toward muscle performance may actually increase dysfunction. Evaluation procedures should distinguish whether the limited motion is due to collagen restriction (hypomobile model) or muscle guarding (hypomobile and hypermobile models), as exercise will be dosed differently for these two different tissue states (Grimsby 1991). A collagen restriction requires many repetitions for improving elasticity, as well as sustained stretch of 10–15 seconds for plastic deformity. Muscle guarding, on the other hand, can be reduced with joint mobilization exercises that increase local circulation and reduce metabolic rigor as well as stimulate type I and II mechanoreceptors for neurophysiological tone reduction (Grimsby 1991). The high number of repetitions will also influence tissue repair and improve coordination. Flexibility in normal joints has been shown to increase equally with either multiple end range repetitions or static stretches (Webright et al. 1997).

Potential Stage 1 Tissue States and Functional Status:

- *Reduced arthrokinematic motion*
- *Decrease in active and passive range of motion*
- *Painful joint at rest and/or with motion*
- *Abnormal respiration patterns*
- *Edema*
- *Muscle guarding*
- *Reduced coordination, timing and recruitment*
- *Poor balance/functional status*
- *Positive palpation to involved tissues*

Abnormal mobility of the hip can be a component of dysfunction from neighboring joints in the biomechanical and neurophysiological chain. Determining a biomechanical or neurological origin of pathology affecting the hip is an important first step in treatment. Spinal segmental facilitation may be a component of muscle guarding patterns that restrict hip mobility. Leg length discrepancy is an example of a biomechanical deficit that has been associated with hip pain and hip arthrosis

(Friberg 1983). Leg length issues may relate to true structural asymmetry, muscle guarding, or pelvic torsions. Sacroiliac dysfunction and pelvic obliquity are predisposing factors of hip arthritis (Bjerkeim 1974). Sacroiliac manipulation has been found effective in treating hip pain in runners, which helps to confirm the potential for pelvic influence on hip pain (Cibulka et al. 1993). Additionally, foot mechanics such as pes planus leading to a lower quarter internal rotation moment during weight bearing function can lead to abnormal loading of the hip cartilage and muscle imbalances.

General Guidelines for Mobilization

Restriction from Capsular Tightness: Treat for Deformity of Collagen

- *Direction of Force*: 1) Distraction 2) Glide—arthrokinematic motion
- *Dosage*: 1 set of 3–5 reps, 0–50% 1RM, one to multiple times daily
- *Speed*: Slow, hold for 10–15 seconds

Restriction from Capsular Tightness: Treat for Elasticity of Collagen

- *Direction of Force*: Modified tension in the line of stress, toward end range joint play
- *Dosage*: 2-3 sets of 30–50 repetitions, 30–50% of 1RM, one time daily
- *Speed*: Slow for coordination, increase speed to increase lubrication of collagen/cartilage

Restriction from Muscle Tension: Treat for Vascularity of Tonic Muscle

- *Direction of Force*: Exercise into guarded pattern, least restricted direction first
- *Dosage*: 2–3 sets of 24 repetitions, 40–60% 1RM, one time daily
- *Speed*: Slow for coordination, rest breaks of one minute for local oxygen recovery

Passive mobilization of the hip joint can be performed in the clinic with a pulley system and strap. True mobilization exercises emphasize arthrokinematic motions of the hip joint, including distraction and glides. Using pulleys and belts to replace the hands of a manual therapist may not be as effective as a manual mobilization but may be necessary in larger limbs. If there is apprehension of the patient in allowing the therapist to manually mobilize. This is also useful when hold times and repetitions can not be maintained by the therapist. Initial plastic stretches are performed three to five times holding for 10–15 seconds each. Increase in repetitions and/or time is dependent upon the tissue responses to the mobilization. The patient and strap should be positioned comfortably without resistance. The therapist then slowly releases the resistance from the pulley line to ensure no pain is produced. The patient is instructed to relax and let the pulley stretch the hip, avoiding active muscle tension to resist the mobilizing force.

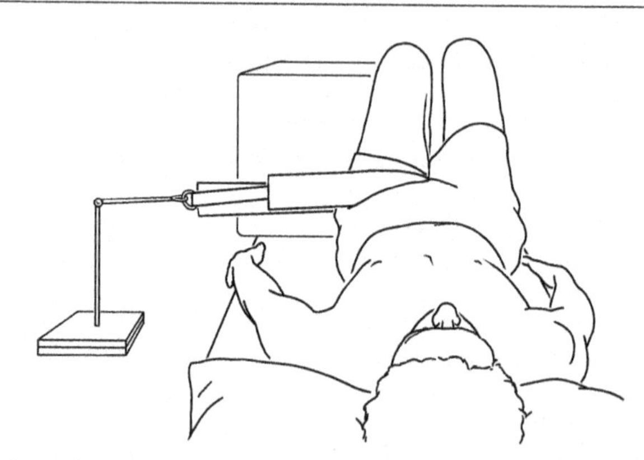

Figure 3.1: Lateral distraction mobilization of the hip with a pulley system for plastic stretch of the joint capsule to improve general motion.

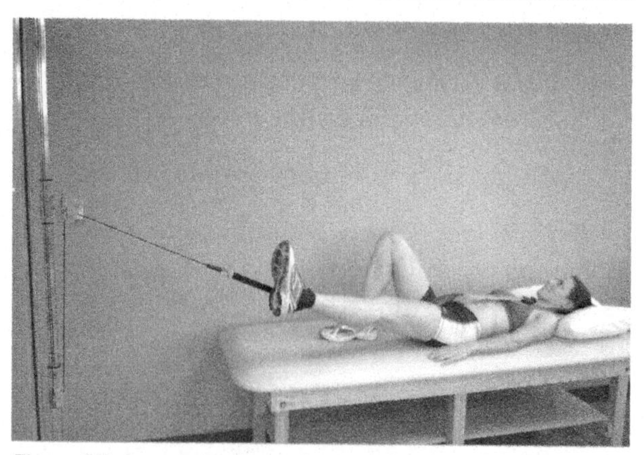

Figure 3.2: Long axis distraction mobilization of the hip joint with a pulley for plastic stretch of the joint capsule to improve general motion. The opposite lower extremity is flexed with a foot support to stabilize the pelvis. This is more commonly used when the patient is significantly larger than the therapist or prolonged hold times are desired.

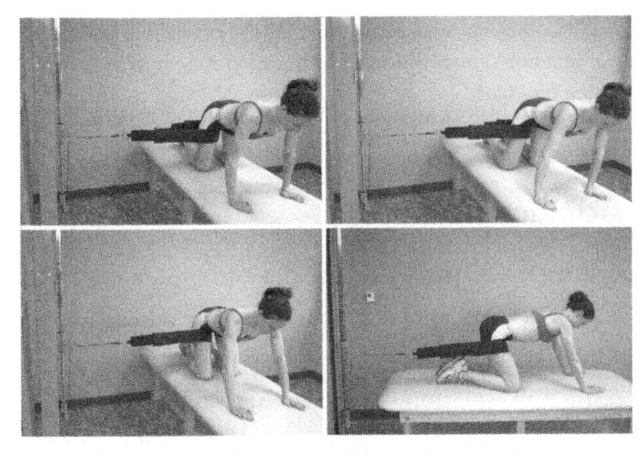

Figure 3.3a,b,c,d: Lateral distraction mobilization in quadruped. The vector from the pulley can be set for lateral distraction or a distal distraction. The patient can rock back and forth for inhibition of pain and muscle tension.

Mobilization exercises may also involve general motions emphasizing osteokinematic motion to improve elasticity of the joint capsule. Non weight bearing and partial weight bearing motions can be performed in the pain free range with emphasis on eccentric motion toward the restriction and then returning. Repetitive movements allow muscles to lengthen and increases lubrication of the joint capsule to improve elasticity. If muscle guarding is reduced, a long hold may also be effective in creating plasticity in the collagen of the fascia, muscle and capsules.

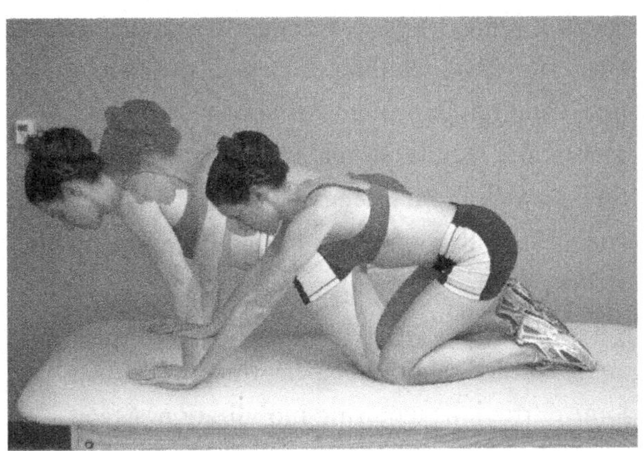

Figure 3.4: Partial weight bearing motion through the hip in quadruped can be performed as an exercise to stimulate joint mechanoreceptor inhibition of pain, reduce muscle spasm and provide repetitive motion to the capsule for lubrication. Motions may include flexion/extension, horizontal abduction/ adduction and circumduction. If effective, this exercise can be performed multiple times per day with hundreds of repetitions. Modifications include gradually separating the knees further apart to place more strain to the anterior inferior medial joint capsule or placing the lower legs in an external or internal rotated position during the mobilization.

Tissue Repair/ Edema Resolution / Pain Inhibition

Exercise for tissue repair and reduction of acute symptoms requires many repetitions in a pain free range of motion to allow for fluid exchange, vascularization, protein synthesis, inhibition of pain and resolution of muscle guarding. Inhibitory techniques must take into account the predominate mechanoreceptor in the area to determine whether static holds (stretch for type I) or repetitions (oscillation or type II) are more effective (Grimsby 1991). Response to a particular stimulus, however, will ultimately determine which technique should be utilized regardless of the mechanoreceptor predominance.

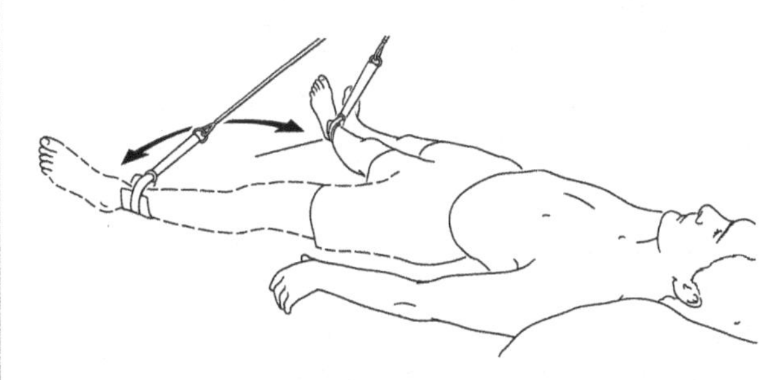

Figure 3.5: Sling suspension, with the fixed line directly above the axis of the hip joint, allows for zero resistance for hip abduction and adduction. Hundreds of repetitions can be performed without fatigue to influence pain, lubrication of cartilage, elasticity of joint capsules and to begin low level coordination training. Alternating pure concentric work will also influence circulation in local muscles.

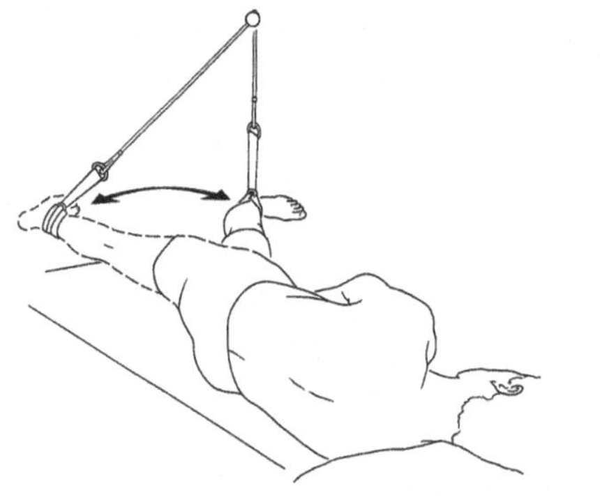

Figure 3.6: Sling suspension for hip flexion and extension. To emphasize extension mobilization the bottom hip and back can be placed in flexion to lock motion from occurring in the spine as the top hip moves into extension.

Edema reduction might benefit from elevation in the acute state to assist with venous return, although repetitive active motion is superior for normalizing fluid mechanics. Tissue repair for low metabolic tissues, such as collagen and cartilage, will require thousands of repetitions and the dosage must be low enough to prevent pain. The cartilage of the hip, for example, requires multiple hours of weight bearing per day for optimal health requiring multiple sets throughout the day.

Tissue Repair, Pain Inhibition, Edema Reduction and Joint Mobilization

- *Sets*: 1–5 sets or training sessions daily
- *Repetitions*: 2–10 hours during the day
- *Resistance*: Assisted exercise up to 25% 1RM
- *Frequency*: Multiple times daily

Coordination

- *Sets*: 1–5 sets or training sessions daily
- *Repetitions*: 20–50 during the day
- *Resistance*: Assisted exercise up to 50% 1RM
- *Frequency*: 1–3 times daily

Vascularity / Local Endurance

- *Sets*: 2–3 sets or training sessions daily
- *Repetitions*: 25
- *Resistance*: Assisted to 50–60% 1RM
- *Frequency*: One time daily

Muscle in Atrophy

- *Sets*: 2–3 sets or training sessions daily
- *Repetitions*: 40 or more
- *Resistance*: Assisted to 30–40% 1RM
- *Frequency*: One time daily

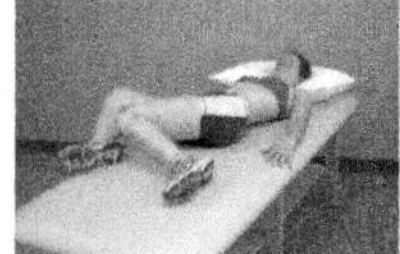
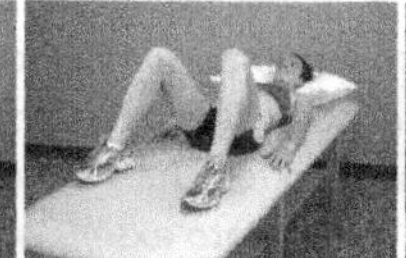
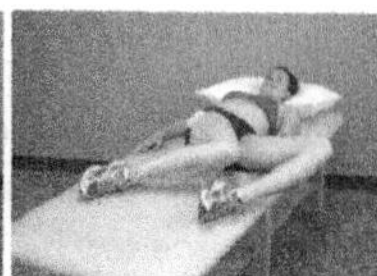

Figure 3.7a,b,c: The supine hip roll exercise can be modified to emphasize hip motion rather than back motion. With the feet apart about the distance of the length of the patient's tibia, the hips are internally and externally rotated attempting to place the knee inside the arch of the opposite side foot. Artificial locking, with a belt holding the pelvis to a bench, will prevent lumbar motion and isolating the mobilizing force to the hip joint.

Adjunct Treatment

It must be emphasized that the entire biomechanical and neurophysiological chain must be addressed when training function. The hip joint cannot be trained in isolation. Biomechanical restrictions in the foot, ankle, knee, pelvis and lumbar spine all play a role in forces transmitted through the hip. Lumbar segmental dysfunction, facilitation and nerve root pathologies also play a role in normal muscle recruitment and motor patterns. True bony leg length asymmetry, abnormal foot alignment or mobility, hallux rigidus of the 1st toe, excessive pronation and loss of full knee extension are just a few common biomechanical chain considerations for weight bearing and gait. Assessment of assistive devices may be appropriate, short-term, to avoid additional stress to the tissues and to discourage abnormal movement patterns (Norkin 1992). Reviewing chapters in these other joint systems will provide additional ideas in designing a complete exercise program for patients with hip dysfunction.

Biomechanical Chain

The hip joint is a part of a larger biomechanical chain of joints including the rest of the lower limb, pelvis and lumbar spine. Functional activities, such as walking, require normal mobility on the opposite limb as well. From a biomechanical standpoint, the goal of mobilization is to normalize alignment and achieve full range of motion of the associated joints. Clearing for normal range of motion of the knee, ankle and foot should be performed to identify joint restrictions requiring mobilization. Many possibilities exist for associated joint restriction that may be asymptomatic but alter the normal force transmission and mobility of the lower limb. Loss of terminal knee extension, for example, will result in a shorter stride length and reduced posterior hip muscle facilitation on heel strike during gait. If the knee is flexed the hip will be flexed as well. The majority of abnormal tissue stress will surely be at the knee but the hip, pelvis and back are also affected by the altered force lines and reduced muscular support.

A restricted subtalar joint has reduced shock absorbing capabilities which lead to increased forces through the hip and knee that may contribute to osteoarthritic changes. Loss of talocrural dorsiflexion or reduced extension of the first metatarsophalangeal joint will reduce stride length. This will reduce push off force of the ankle plantarflexors requiring additional force of the hip flexors. An anterior tipped pelvis, with associated internal rotation of the femur and valgus moment at the knee, will create an abnormal contact of the anterior femoral head and the acetabulum leading to abnormal tissue stress of the articular surfaces. Similar conditions may result with an anterior torsion of the innominate but also involve a functional leg length change. A change in mobility of the lumbar spine altering the normal lordosis for stance, as well as mobility for gait, will be compensated for in the hip joint.

Neurological Chain

The neurological chain for the spine, pelvis and lower limb must also be considered. The L4–S1 nerve roots all have an impact on the hip and lower quarter motor patterns. Acute or residual nerve damage from spinal pathology will have a significant impact on motor recruitment for stability in the hip. Segmental facilitation at these levels does not involve nerve conduction issues reduction of efferent motor signals, but segmental sensitivity will increase pain levels in innervated tissues. Healing time and progress with exercise training will be slower. A thorough evaluation of these structures must be performed prior to initiating a treatment program.

Myokinematic Chain

All muscles in the lower quarter must be assessed, preferably in weight bearing functional testing, to identify patterns of weakness that place excessive stress to ankle and foot structures. When specific muscle weakness is identified, this should be assessed in conjunction with lumbar spine segmental dysfunction or facilitation. For example, weakness or poor endurance of the gluteus medius, gluteus minimus and tensor fascia latae reduces the eccentric deceleration of internal rotation. The resulting overall internal rotation moment of the lower extremity in weight bearing may occur through a range which places excessive demand on the tissues. The collapse into internal rotation leads to poor shock absorption and ineffective proximal control of pronation (Sahrmann 2002). Hip muscle weakness may be a symptom of concurrent proximal or distal impairments. This has been demonstrated with measurements of reduced hip extension strength in subjects with ankle sprains (Bullock-Saxton et al. 1994). The lower quarter motor patterns should be assessed as a whole to identify global motor dysfunction.

Direction of Exercise

The initial direction of training is influenced by the restriction in mobility, tissue injury and/or functional deficits. Exercises may be selected simply to facilitate the muscles that influence basic tasks of rolling, transfers, standing or walking. Exercises may also be chosen to directly influence injured muscle and tendon, as with a collagen injury to the proximal hamstring tendon.

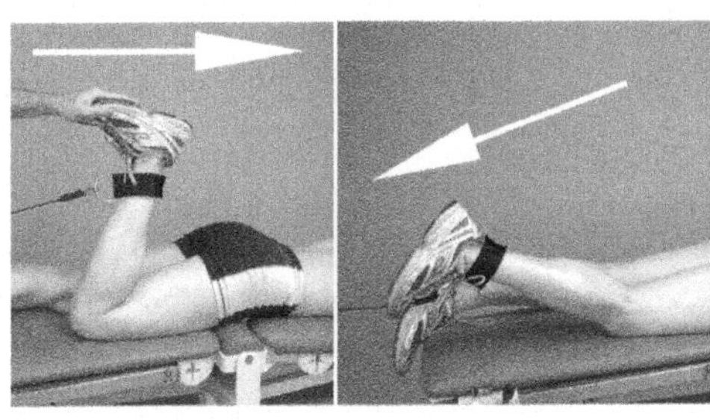

Figure 3.8a,b: Prone pure eccentric hamstring curls for tissue repair of the proximal hamstring strain. The concentric lifting phase is performed by the opposite leg, or by the therapist. The exercise is therefore pure eccentric lowering with the involved muscle. Resistance is set as high as is pain free and can be coordinated for one to two sets of 10–15 repetitions daily to influence collagen repair and tensile strength. Stretching is contraindicated with a musculotendinous strain, as plastic stretching of injured collagen will only further lower its tensile strength.

When the emphasis is first on gaining hip mobility, the initial concept of choosing direction of motion are the same as in the shoulder joint. If the restriction is believed to be from the hip capsule then exercises are chosen to influence the collagen of the capsule for elasticity with exercises in all directions. If the limitation is from muscle guarding then exercise is performed into the guarded pattern for local vascularity to reduce muscle tension and allow for full eccentric lengthening. A capsular pattern of restriction, with a loss of extension and internal rotation, may be mimicked by the primary tonic muscles that guard the hip. Increased tone in the piriformis, and other deep external rotators, will limit gross internal rotation, while tone in the psoas and iliacus limit extension. Antagonists to these guarded muscles will be inhibited creating weakness during testing and function. Exercise of the guarded muscles at 60% of 1RM for 3 sets of 24 repetitions will maximally increase circulation and reduce abnormal tone by providing oxygen to tonic muscle. Emphasis is on training into the guarded pattern, or away from pain, and increasing the eccentric range of the muscles. The initial pattern of training for reducing guarding will be external rotation, flexion and adduction.

Hip Capsular Pattern

- *Internal Rotation*
- *Extension*
- *Abduction*
- *External Rotation*
- *Flexion*

The body position is typically in non-weight bearing postures for early training to reduce the level of tissue stress and coordination required for training. The joint itself is positioned in a more open packed position to allow for the greatest mobility and potentially avoid pain. The close packed position of maximum joint congruity and collagen tension is avoided, as no further motion can occur and external forces are transmitted into compressive forces that may irritate the joint cartilage. Cartilage is a non-innervated tissue and is pain free, except in severe degenerative joint disease where blood vessels grow in from the subchondral bone bringing mechanoreceptors. Exercise that overloads damaged cartilage may cause deep aching pain and increased muscle tension within hours after training, caused by increase inflammation within the joint. These are signs of over training.

When training with restricted motion the hip is positioned in a more open packed position, as in prone with slight flexion, with a gradual change in position closer to, but not into, the close packed position as mobility improves. The close packed position is stated as full extension, internal rotation and abduction. But this must be remembered as referring to a normal joint. The relative close packed position for a pathological hip may occur much earlier than in these end range positions. Manual assessment of mobility aids in locating the true range of the patient's close packed position.

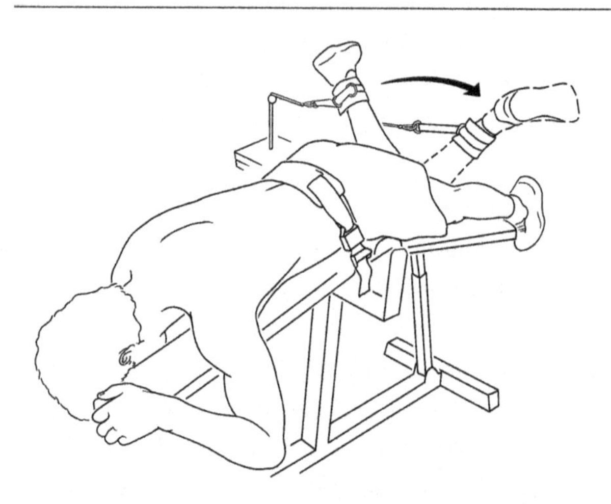

Figure 3.9: Training the guarded pattern: Hip external rotation for reducing tonic guarding of the hip external rotators. Belting the pelvis will increase the emphasis on joint mobility and muscle elongation into full internal rotation. The final repetition of each set can also be held as a stretch for 15–30 seconds at end range internal rotation to create plastic deformity of the hip joint capsule. The joint is positioned in slight flexion to start the exercise in a more open packed position and is progressed by gradually positioning the hip toward a more neutral or extended position while performing the same rotational exercises.

Close Packed Position for a Healthy Hip

- *Limited Extension*
- *Limited Internal Rotation*
- *Limited Abduction*

Piriformis syndrome is a description of symptoms from increased tone or volume of the piriformis causing irritation locally and referred symptoms from the sciatic nerve but is not itself a diagnosis. Increased guarding of the piriformis has been related to lumbosacral and sacroiliac dysfunction along with segmental facilitation (Grimsby 1991). Besides acute symptoms from guarding, non-discogenic sciatica has been documented from local tissue changes in the piriformis and obturator internus muscles (Meknas 2003). External rotation training to increase local circulation (60% of 1RM) can assist in normalizing the resting tone, by increasing oxygen to the muscle and improving the dynamic length of the muscle. On the other hand, exercising the antagonists with internal rotation training can lead to a neurological inhibition of the external rotators (i.e., piriformis).

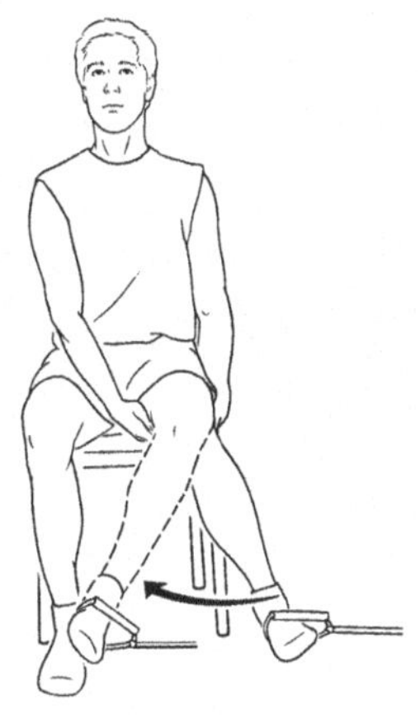

Figure 3.10: Sitting option for external rotation training. Emphasis is placed on maintaining the thigh in neutral, avoiding compensatory abduction with the exercise. Weight shifting off of the involved hip should also be avoided to keep mobilizing forces focused at the hip joint.

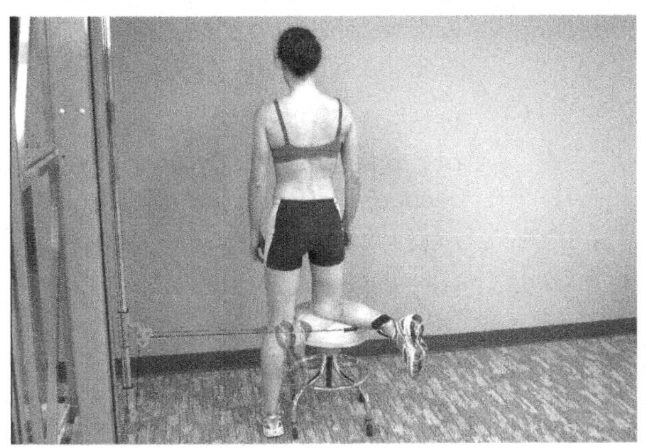

Figure 3.11: Partial weight bearing option for external rotation training using the stool. The trunk is supported with slight flexion with partial weight bearing through the involved knee and hip. The rotation training ensures a fixed axis of motion but the patient must avoid medial and lateral shifting of the pelvis to compensate for hip mobility limitations.

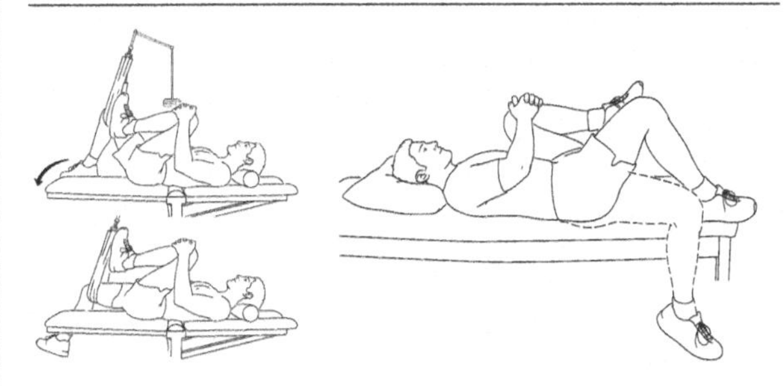

Figure 3.12a,b: Training the guarded pattern: Hip flexion for reducing in tonic guarding of the hip flexors allowing for full extension. Depending upon the training level the hip flexors may need an assist, as pictured above, or added resistance with cuff weights to provide for 3 x 24 at 60% of 1RM. The opposite hip and lumbar spine are locked into flexion to localize the influence to the hip joint prevent the lumbar spine from participating. The final repetition of each set can also be held as a stretch for 15–30 seconds at end range extension to create plastic deformity of the hip joint capsule.

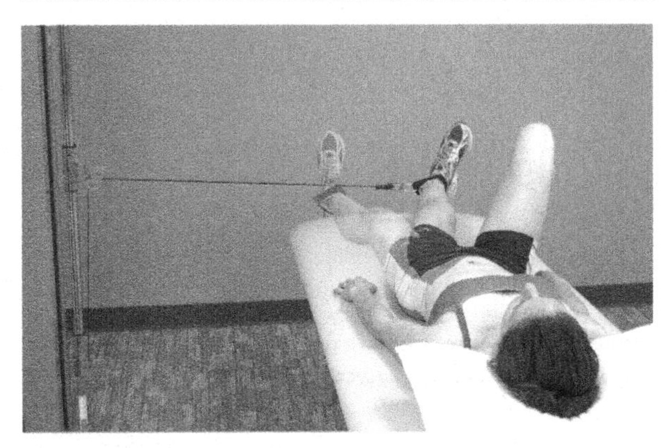

Figure 3.13: Training the guarded pattern: Hip adduction for reducing in tonic guarding of the hip adductors allowing for full abduction.

Partial weight bearing training should begin as soon as possible for coordination training and improved tissue tolerance. Many options exist for reducing body weight to allow for squatting activities or transfer training. Using a pull down machine bar to reduce body weight is a simple approach. Horizontal nautilus machines allow for horizontal squat training with adjusted weights. Hip sleds with adjustable angles also allow for changing the amount of body weight through the limb. Water therapy may also be an effective approach to reduce joint load, given caution to avoid increased speed of performance, and muscular resistance causing additional joint compression. Matching the patient's coordination level and functional goals can aid in determining the best starting point.

Figure 3.14a,b: Assisted squats with a pull down machine and gantry. Holding a regular lat bar in front of the body is also an option if the gantry is not available. Emphasis is placed on even weight bearing, avoiding compensatory shifting toward the non involved limb.

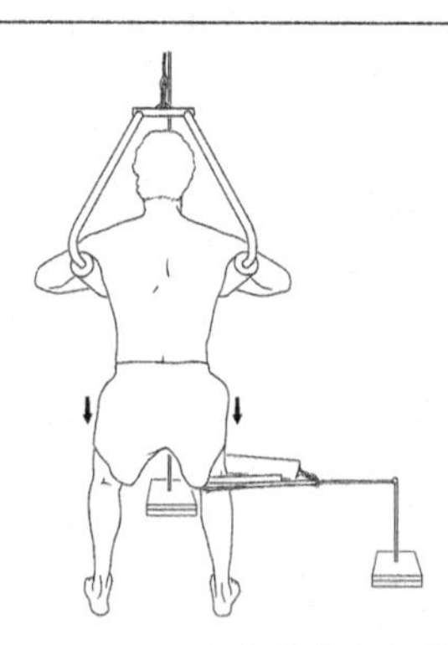

Figure 3.15: In cases were weight is being shifted to the non involved limb, secondary force moment from a pulley, elastic resistance or manually can increase the weight bearing on the involved limb. Resistance can be applied at the waist or knee.

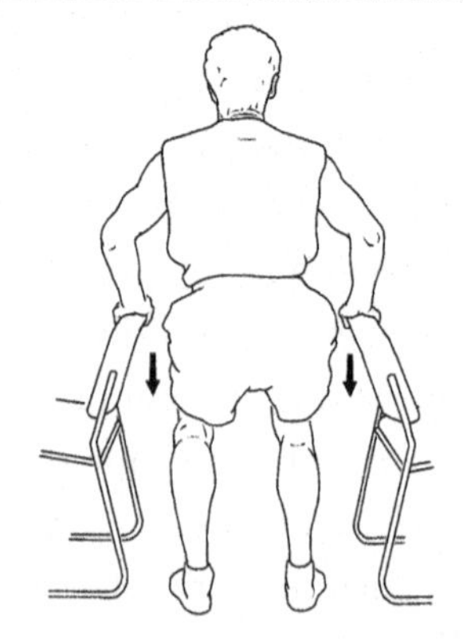

Figure 3.16: Upper quarter assisted squats with chairs. Emphasis is placed on even weight bearing, rather than shifting toward the non involved limb.

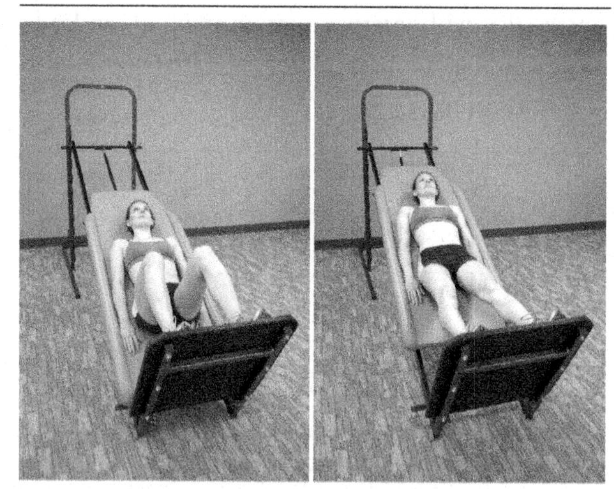

Figure 3.17a,b: Unloaded squats on the Vigor Gym.

Common Abnormal Motor Patterns of the Hip

Pathological Motor Change	*Possible Compensations*	*Possible Clinical Presentations*
Inhibited gluteus medius from lumbar or SIJ dysfunction.	*Internal rotation/adduction of hip joint with weight bearing activity.*	*Tensor Facia Lata compensation with secondary ITB "tightness" or "friction syndrome".* *Valgus moment at knee and/ or pronation at the ankle with secondary tissue breakdown.*
Hip articular pathology leading to tonic muscle guarding pattern in hip of deep external rotators (Piriformis, superior/inferior gemelli, obturator internus/ externus) and in the hip flexors (iliacus and psoas).	*Secondary inhibition of the gluteus maximus and medius from flexion and internal rotation guarding pattern.*	*Loss of range of motion with a false capsular pattern of restriction with loss of extension and internal rotation.* *Increased hamstring activity for hip extension.*

Coordination / Motor Learning

The previous exercise options have dealt with tissue stimulation, joint biomechanics and pain reduction. Coordination and motor learning exercises are the first functional qualities needed to improve muscle performance and are required before progression into endurance, strength and power training. Training for coordinated movement patterns can take as many as 5,000 to 6,000 repetitions or more (Tropp 1985).

As noted earlier, the hip must be viewed as a portion of a biomechanical and neurological chain involving the lumbar spine, pelvis and entire lower quarter. Abnormal movement patterns in the hip may be only a symptom of underlying dysfunction in the spine or pelvis. Manual treatment to these areas should be attempted first if dysfunction is found. Correcting these dysfunctions may result in a resolution of abnormal movement patterns without specific motor training. If the patterns do not improve with manual intervention then exercise is initiated for coordination and motor learning. As in all joints, afferent receptor systems play a role in the modulation of motor patterns for normal function. Joint mechanoreceptors provide information for joint position sense for posture, coordination and measuring tension in the collagen of the joint capsule. Type I receptors have demonstrated a function of "limiting detectors" signaling proximity of the hip joint to its limitation in rotation (Rossi and Grigg 1982). The function of receptors has been demonstrated in their loss following surgery. Joint position sense after total hip replacement has been partially attributed to afferent systems being disrupted with joint and capsular damage after surgery (Grigg et al. 1973, Karanjia and Ferguson 1983). As with all joints, common abnormal motor patterns are seen with a myriad of diagnoses. Tonic muscle of the hip will go into segmentally driven patterns of guarding resulting in arthrokinematic loss. These guarded patterns can significantly limit joint motion by increasing joint compression. Muscles may be inhibited or altered recruitment may be established to compensate for altered function. Treatment first addresses arthrokinematic dysfunction, and then attempts to reduce actively guarded muscles with down training to normalize motor patterns. Chronic conditions may result in altered patterns in which over worked muscles

Example Of Close Chain Compensation For Right Hip Abductor Dysfunction

Joint	*Planes Of Compensation*	*Potential Medial Joint Axis Dysfunction*	*Potential Lateral Joint Axis Dysfunction*
Lumbar spine	*Right side bending, extension and left rotation*	*Left lumbar tension pathology*	*Right lumbar compression pathology*
Sacroiliac Joint	*Right relative posterior torsion/ left anterior*	*Right posterior torsion tension in the right SIJ ligaments: sacrotuberous, sacrospinous and iliolumbar*	*Left SIJ anterior torsion with left long dorsal ligament tension*
Right Hip	*Hip adduction, internal rotation and flexion*	*Anteromedial joint compression*	*Right hip joint lateral tension pathologies*
Right Knee	*Valgus Moment*	*Right medial knee tension pathologies*	*Lateral right knee compression pathologies*
Ankle/Foot	*Pronation*	*Medial arch tension pathologies*	*First MTP compression pathologies*

require a strategy of down training, while up training the inhibited muscles.

Gluteus Medius Inhibition

The most classic example of the altered motor patterns in the hip is inhibition of the gluteus medius with compensatory overactivity of the tensor fascia lata (Kasman 2002). The anterior fibers of the tensor facia lata (TFL) contribute to hip flexion, while the posterior fibers contribute to hip abduction and internal rotation (Pare et al. 1981). With weakness or inhibition of the gluteus medius the anterior fibers of the TFL attempt to compensate leading to overuse with local muscle soreness and abnormal forces through the iliotibial band (ITB). Secondary symptoms associated with hip bursitis or lateral knee pain may result but the treatment emphasis must be on restoring the normal motor pattern. Iliotibial band syndrome (ITBS) may not be considered a freestanding pathology but merely a symptom of an underlying mechanical breakdown of forces through the lower limb in close chain activities. Symptoms of ITBS frequently correlate with gluteus medius weakness. EMG studies have shown that both athletic and arthritic hips with symptoms of ITBS usually have poor gluteus medius facilitation and an inability to isolate gluteus medius from the tensor facia lata (TFL) in standing (Fredericson 2000, Kasman 2002). In long distance runners with ITBS, symptom improvement and return to pre-injury training levels were found to correlate with increases in strength of the hip abductors (Fredericson 2000).

Example of Close Chain Compensation for Right Hip Abductor Dysfunction

Weakness in hip abductors leads to a chain of events through the entire biomechanical chain. Muscular dysfunction can lead to motion occurring too far or too fast resulting in compression injuries on one side of a joint axis of motion and tension on the opposite side. Abnormal levels of tension may result in tendinopathies, bursitis or ligament/capsular strain. Abnormal compression may result in impingement pathologies, arthritis, labral damage or stress fracture. Conceptually abnormal biomechanics can occur in any plane of muscle weakness with a resulting 3-dimensional abnormality in force attenuation. Treatment starts with a focus on the tissue injury but must look for and address any underlying motor dysfunction proximal and distal, to the tissues in lesion. When the normal motor pattern shifts to reduced gluteus medius activity and excessive TFL activity weight bearing training may augment the abnormal pattern. The resistance of the body weight is too high for the gluteus medius to stabilize the hip laterally. Resistance must be reduced to allow for the gluteus medius to control the activity. This may require non-weight bearing or partial weight bearing exercise. Attempting to isolate the gluteus medius from the TFL can be performed clinically but may lack functional carry over. Attempt should be made to remain as functional as possible for the goal of coordination. Coordination should relate to performing at least a part of a functional activity, and not to performing an isolated rehabilitation exercise that does not relate to daily activity. The following exercise examples include more non functional training of the gluteus medius with ideas for more functional applications. One exercise does not solve the coordination problem for every patient. The clinician should be armed with the concepts of what the normal pattern of muscle recruitment is, how to gain coordination, and be able to create an exercise to serve this purpose. Early non-weight bearing approaches may be necessary given the patient's pathology or training level. A progression to more functional weight bearing can take place when tissue tolerance improves and/or overall coordination allows for safe training. Unloading, or partial weight bearing, may also provide a transition to full weight bearing.

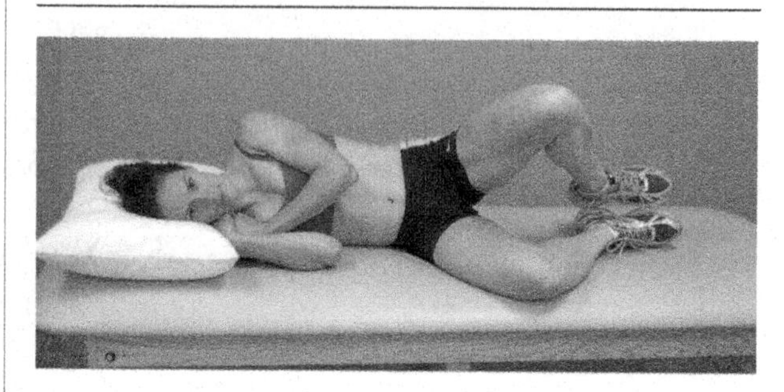

Figure 3.18: Hip Clam Exercise: side lying horizontal hip

abduction to attempt to isolate the gluteus medius and reduce recruitment of the tensor fascia lata. The knee is lifted keeping the feet together without rotation of the pelvis and lumbar spine. Performing muscle isolation exercise in non-weight bearing is a limited functional value, but may be a necessary starting point in painful or deconditioned training states. A cuff weight at the knee may allow for more specific exercise dosage, but this exercise typically emphasizes the quality of coordination with only the weight of the leg. As the exercise becomes easy, a progression into weight bearing is more appropriate than the addition of weight.

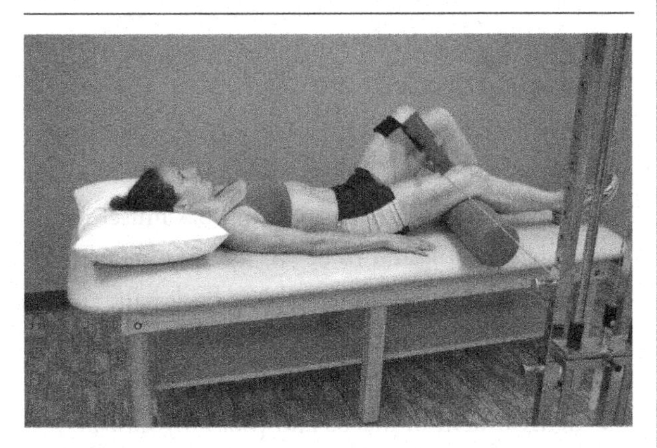

Figure 3.19: Horizontal hip abduction in supine with pulley resistance can also be used to train with specific resistance. Gluteus medius training is emphasized with down training of the tensor fascia lata. As a non-weight bearing exercise for the hip, there may be limited carry over to weight bearing function. Progression to a weight bearing option is more appropriate than progressive resistance training in supine for carry over to more functional motions.

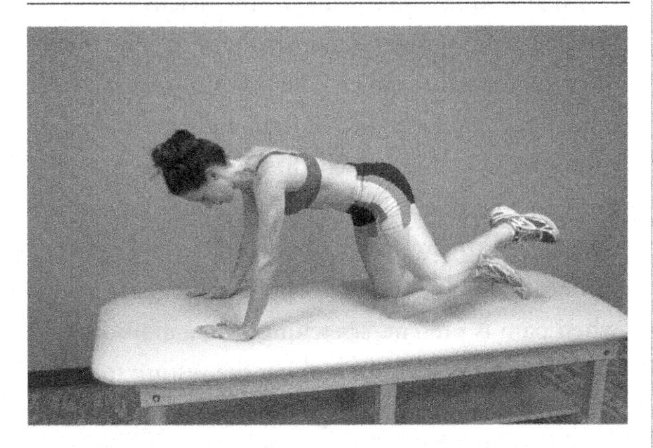

Figure 3.20: Hip Fire Hydrant: quadruped horizontal hip abduction. This exercise attempts to emphasize facilitation the gluteus medius while reducing recruitment of the tensor fascia lata. The patient is in a quadruped position with weight on the uninvolved hip. The involved side knee is lifted laterally in isolation without pelvic or lumbar motion. As a progression the addition of pelvic rotation, lumbar rotation and finally abduction of the weight bearing hip increases the challenge to the overall pattern. The gluteus medius is facilitated bilaterally with increased activity associated with increased afferent input with partial weight bearing and the cross over effect. The exercise can also be performed with the involved hip performing in weight bearing while abducting the uninvolved side hip.

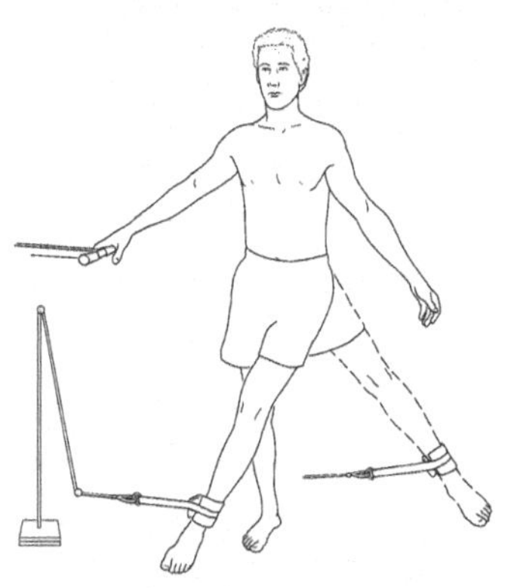

Figure 3.21: Standing hip abduction with pulley resistance. Open chain training of the left hip abductors with closed chain training of the right. Arm support reduces the need for balance and coordination to focus on muscle training for endurance or strength. Arm support can be removed to increase the balance challenge and level of difficulty.

Gluteus Medius Assist in Weight Bearing

The gluteus medius is described by basic anatomy and kinesiology texts as performing hip extension, abduction and external rotation. This description is certainly true, but from a rehabilitation standpoint refers to non-weight bearing (open chain) training. Closed chain training in weight bearing emphasizes the eccentric functions of decelerating hip adduction, flexion and internal rotation, as in the stance phase of gait from heel strike to mid stance (Norkin and Levangie 1992). An increase in the excursion or speed of these motions in weight bearing function is a sign of hip abductor weakness defined as the Trendelenburg sign (Grimsby 1991).

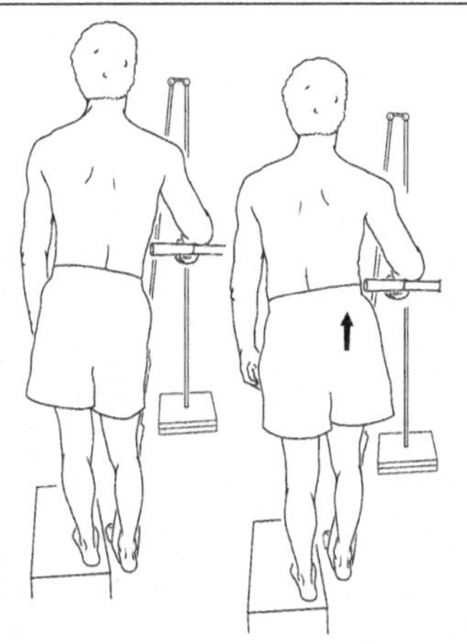

Figure 3.22a,b: Pulley assisted left hip abduction training in weight bearing. Body weight is reduced with a pulley force cranially on the uninvolved hip. The assist allows for dosing the exercise for coordination at 25–30 repetitions or more. A slight external rotation of the hip (pelvis facing right) at the end of the motion will aid in emphasizing the gluteus medius rather than the tensor fascia lata for hip stability.

Motor dysfunction is often related to loads across a muscle being higher than can be controlled. Body weight is often too high of a load to use to begin gluteus medius training. Reducing the relative amount of weight the hip abductors need to control may allow for earlier, more functional, weight bearing training. The hip abductors lie lateral to the hip axis for abduction and adduction with a relative majority of the body weight medial to the axis. The body weight creating a downward force on the medial side of the axis is counter balanced by the downward pull of the hip abductors on the lateral side. This is a classic type one lever, like the see-saw ride for children. To assist the gluteus medius either the body weight is reduced with a pulley assist on the opposite side, or additional weight is applied lateral to the joint axis creating a downward force.

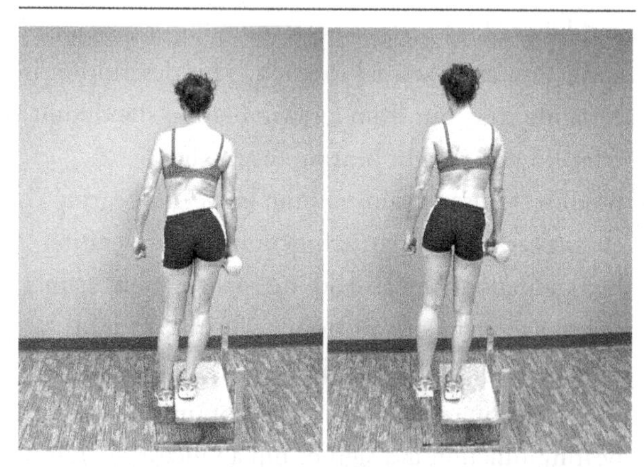

Figure 3.23a,b: Free weight assisted right hip abduction training in weight bearing. The relative body weight to the right of the axis of motion in the left hip joint is reduced with a free weight force caudally on the involved side. The weight pulls the body down on the left side assisting the hip abductors with the hip hike motion.

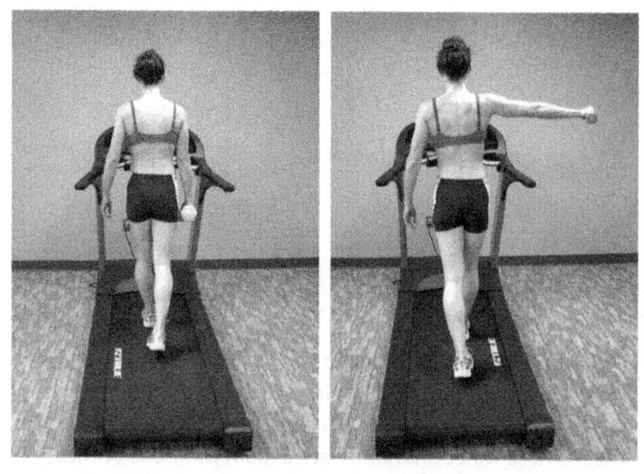

Figure 3.24a,b: A positive Trendelenburg during gait, from a weak gluteus medius, can be assisted by carrying a weight, up to 10 percent of the body weight (Norkin and Levangie 1992) on the involved side. This can be performed on a treadmill while holding a heavy weight at the involved side (A). With gait training, squatting or lunging activities weight can be held on the same side to assist or the arms can reach out lateral to the involved hip (right hip pictured). Using light weights with the arms out laterally (B) creates a longer lever arm and greater force on the lever for counter balance. Functionally the patient should be instructed to carry objects such as purses or grocery bags on the involved side to counter balance the weight medial to the hip joint axis.

The Hip Flexors

The underlying cause of muscle guarding, and/or inhibition, must also be considered when looking at hip pathology. The iliacus and psoas muscle group has been found to be weak and decreased in size in patients with unilateral low back pain and unilateral disc herniation (Dangaria et al. 1998, Barker et al. 2004). Ingber (1989) reported success in the treatment of iliacus and psoas myofascial restrictions in "failed" chronic low back pain patients with results of decreased pain reports, increases in spine and hip extension and return to normal activity. Whether the lumbar pathology was first leading to guarding of the iliacus, psoas and thus hip dysfunction, or the hip pathology was first leading to guarding of the iliacus and psoas altering spinal mechanics, both regions must often be treated together. The order of treatment, however, is typically first central, addressing the spinal segmental issues which may have an impact on reducing muscle guarding in the hip. Direct hip treatment is then typically less irritable and exercise can begin at a higher level. In the acute patient, when pain is present, stretching of the iliopsoas is avoided to prevent provoking the resting tone of the muscle. Training a guarded psoas at a level to increase local circulation will assist in breaking up the guarded pattern by providing oxygen to tonic muscle fibers and allowing the muscle to lengthen with the exercise. There is no such muscle as the iliopsoas muscle but two separate muscles of the psoas and iliacus (McGill 2002). The psoas originates from T12–L5, however its activation is not consistent with spinal movement but primarily with hip flexion (Juker et al. 1998, Andersson et al. 1995). The iliacus also produces hip flexion but if left alone to do so would create an anterior pelvic tilt forcing the spine into lumbar extension. The psoas counteracts this anterior tilting moment (McGill 2002). The iliacus eccentrically stabilizes the pelvis during contralateral hip extension (Andersson et al. 1995).

Local Endurance (Vascularity) / Up Training / Down Training

Reduced hip flexor extensibility, for any reason, has a significant negative impact on the biomechanics of the hip, pelvis and lumbar spine. Many different causes of "tight" hip flexors provide different conceptual options on how to correct the issue. The psoas may be working excessively as a tonic muscle in guarding for any articular pathology of the hip, sacroiliac joint or associated levels of innervation of the spinal segments. Chronic increased tone of the tonic motor units in the psoas leads to ischemia and the "vicious cycle" of muscle guarding. Direct exercise to the psoas at a level to increase local circulation can provide oxygen to assist in normalization of the muscles tone when performed at 60% of 1RM for three sets of 24 repetitions. Emphasizing pure concentric training or allowing resistance to come to rest between each repetition can improve the intramuscular circulation during the exercise.

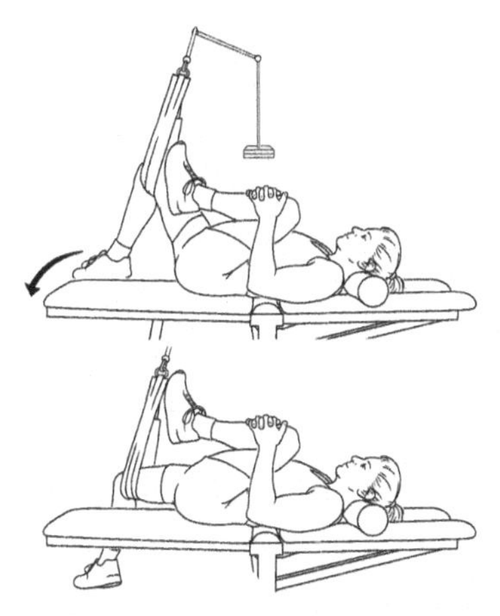

Figure 3.25a,b: Supine active hip flexion training. The pulley is used as a vertical assist to reduce the weight of the leg and allow for 3 sets of 24 repetitions at 60% of 1RM. The weight of the leg, or adding cuff weights at the knee, may be necessary in higher trained individuals to create fatigue in the 24 repetition range.

Increased hip flexor activity may relate to compensatory firing due to weakness in other muscles in the pattern of movement. Lumbar pathology may cause tonic muscle spasm of the multifidi and inhibition of the lower abdominals, leading to an anterior tipping of the pelvis. With weak lower abdominals the psoas will assist in controlling anterior tilting of the pelvis (McGill 2002). Direct stretching or exercise to the psoas in this case is treating a symptom of the overall motor pattern and may not be successful in restoring normal motion. Initiating treatment at the lumbar segments, or training the multifidi for vascularity, would better deal with the inhibition of the lower abdominals even without abdominal training exercises. If training of the lower abdominals is necessary it can be performed while inhibiting the psoas at the same time.

The debate over removing the stimulus for the hip flexors during an abdominal curl up has focused on whether the hips are flexed to shorten the hip flexors making them actively insufficient in the shortened position. This, however, is not correct as the hip flexors will continue to fire whether the hips are flexed or extended as a global flexor pattern is being recruited with the sit up motion. The key is to create an extensor pattern in the lower extremity to facilitate the hip extensors, and inhibit the hip flexors, while performing the sit up exercise (Janda 1999). Emphasis for the hip patient may not be actually increasing lower abdominal strength, but to inhibit chronic muscle guarding in the psoas and iliacus muscles. The severely deconditioned hip patient may require improved lower abdominal strength for more central stabilization.

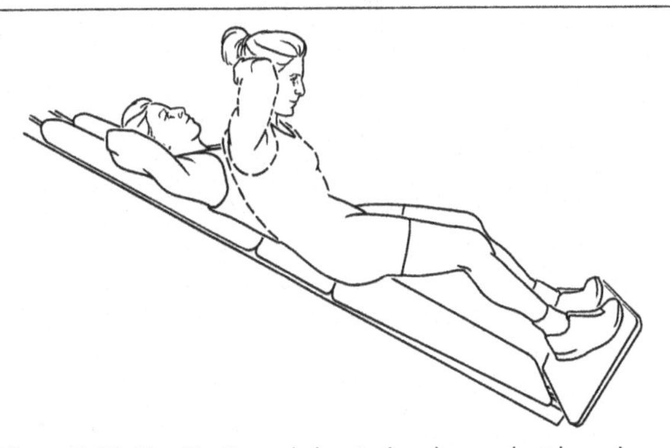

Figure 3.26: The Sit-Over abdominal curl on a slant board.

Down training the iliacus and psoas muscles, while up training the abdominals, can be performed with a curl-up motion on an incline. The hips and knees are slightly flexed with the ankle in slight plantar flexion, facilitating an extensor pattern of the lower extremities, which inhibits the flexor pattern. As the curl up motion is performed the lower abdominals have greater contribution to the motion while the hip flexors are inhibited. Holding a lateral resistance, in the direction of the fibers of the transverse abdominus, will also increase the facilitation of the tonic fibers of the lower abomdinals for initial stabilization. The lateral force can come from holding a pulley or elastic band in front of the stomach with both hands. A free weight can also be held in one hand, positioned out to the side of the body, adding a rotary force moment to the trunk during the flexion motion.

Figure 3.27: The Ball Sit Over- abdominal curl inclined on a ball. The same extensor pattern can be performed on a ball with the feet against a wall, hips and knees are slightly flexed with the ankle in slight plantar flexion facilitating a lower limb extensor pattern. The body is positioned to be inclined, rather than horizontal. As the curl up motion is performed the lower abdominals perform the motion while the hip flexors are inhibited. The ball is more limited in terms of the angle that can be effectively trained, while the slant board options allows an almost vertical position with a gradual progression to supine.

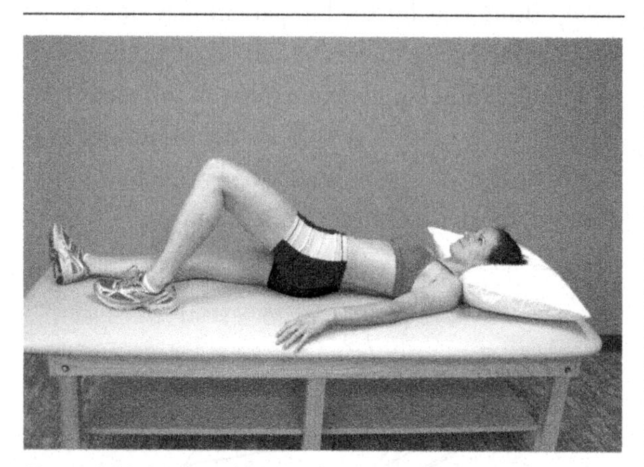

Figure 3.28: Unilateral bridge for down training the hip flexors. Unilateral hip extension is performed rotating the pelvis onto the opposite side. The knee is maintained vertically toward the ceiling, requiring the psoas muscle to relax and lengthen with reciprocal inhibition. The exercise is also a self mobilization for the sacroiliac joint and the lower lumbar spine.

An additional option to inhibit tone in the psoas is to simply train the antagonistic hip extensors to passively lengthen the psoas muscle. Non weight bearing options may be more beneficial in inhibiting the psoas, as weight bearing postures will naturally require some level of psoas firing to maintain an erect posture. Training prone active hip extension held for up to 30 seconds or until fatigue has been shown to effectively length the psoas (Winters et al. 2004).

Figure 3.29: Weight bearing open chain training of the right hip extensors to down training the hip flexors. As the hip extensors are facilitated the hip flexors are inhibited and allowed to lengthen with the exercise. Standing hip extension with a pulley is only one option in a list of many exercises for training hip extension.

Figure 3.30a,b: More direct training of the hip flexors in a flexion, adduction and external rotation pattern can emphasize eccentric lengthening to gain full extension of the joint capsule and/or reduce tone in guarded iliacus and psoas muscles. Pulley attachment can be at the ankle or thigh, with changing the body orientation to change the angle of force.

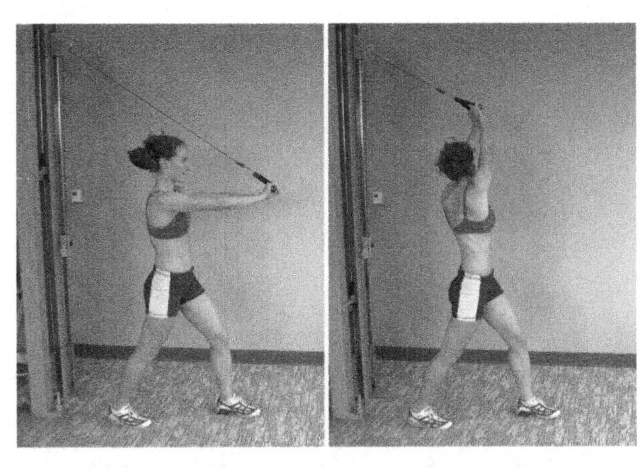

Figure 3.31a,b: Dynamic lengthening of the hip flexors with proximal extension and opposite rotation. The patient stands with the involved hip in extension and the back to a pulley. The patient works eccentrically into trunk extension and opposite side rotation to dynamically lengthen the hip flexors, lower abdominals and fascia.

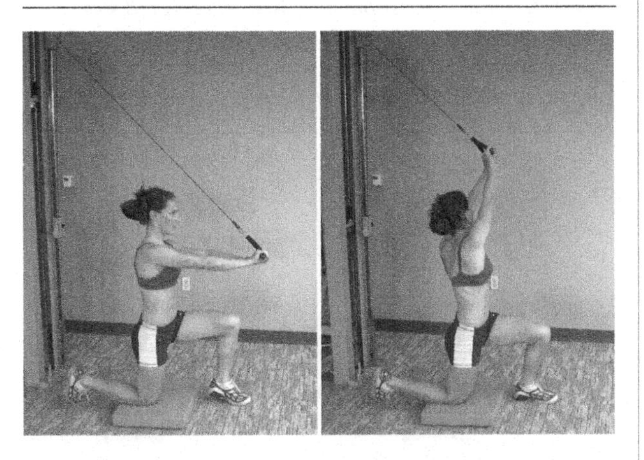

Figure 3.32a,b: Dynamic lengthening of the hip flexors in kneeling with proximal extension and opposite rotation.

Balance Training

Balance training is difficult to begin prior to normalizing joint mobility and range of motion allowing for straight posture. Tissue tolerance must also reach a level to perform bilateral and unilateral positions to initiate balance training. Restoring normal muscle strength of the lower quarter is a component of normalizing postural reflexes for balance (Carter et al. 2002, Messier 2002). Muscle strength must increase enough to support the lower limb to allow for weight bearing training. Improving strength for balance training is not a component of muscle training with heavy resistance and low repetitions, but more related to neurological adaptation for improved weight bearing motor patterns to allow for a higher level of function. Balance training must be initiated long before the time it would take to strength train for structural changes in muscle tissue. With this said, improving strength is especially important in the elderly population as strength can decline as much as 30–50 percent between the ages of 30 and 80 (Frischknecht 1998). Strength training alone, however, does not appear to enhance standing balance (Schlicht et al. 2001, Topp et al. 1993). Combined programs of strength training and balance training have shown benefits (Judge et al. 1993). Balance training designed to improve integration from the visual, vestibular, and somatosensory systems can effectively improve balance performance (Hu and Woollacott 1994). Not only muscle strength and coordination need to be assessed but joint mobility in all joints from the lumbar spine to the foot in the biomechanical chain. Restricted joint motion will reduce normal afferent feedback systems altering normal motor patterns. Loss of ankle range of motion in the elderly has been correlated with balance deficits (Mecagni et al. 2000). Any loss of mechanoreceptor input from joint restriction, trauma or surgery alters afferent input, reducing function. Bullock-Saxton et al. (1994) demonstrated how previous ankle ligament and receptor injury decreased the activation of the hip extensors.

With regard to the elderly hip, approximately 90 percent of hip fractures result from falls (Parkkari et al. 1999), and out of these 12–20 percent are fatal (Peck et al. 1988). The available data is mixed on whether evidence-based exercises can help prevent falls in elderly. Some of the more recent studies, since 1996, incorporating various aspects of strength, endurance, flexibility and balance training support the effects of exercise for fall prevention in the elderly (Khan et al. 2001). Exercise prescriptions for hip patients should incorporate aspects of balance training focused on the primary deficit, whether it be somatosensory, vestibular, visual or a combination. Even a total hip replacement from a fracture due to a fall may have underlying vestibular, ocular or cervical issues that contribute to reduced balance and weight bearing performance.

A balance progression should address the region of primary deficit in the patient as well as the functional long-term goals. Basic balance testing should begin with bilateral stance with assessment of more central impacts of cervical motions. Closing the eyes will reduce the compensation of ocular reflexes on postural reflexes affected by altered afferent input from mechanoreceptors in the lower limbs. Initially the feet are placed shoulder width apart to increase the base of support. Performing similar tests with the feet together or parallel will reduce the base of support and serve as a greater challenge to the lower limbs. Balance training is dosed with similar concepts of dosing coordination. Many repetitions are required on a daily basis to improve neurological adaptation. Shorter training sessions of three to five minutes multiple times daily will be more effective than attempting long training bouts in which muscle fatigue is the dominating factor in poor performance. Balance should also be emphasized during daily activity and initially should emphasize shifting weight from the uninvolved limb to the involved limb for standing activities. Maintaining bilateral stance but simply shifting weight from side to side may be an easy initial starting point to reinforce dynamic balance responses throughout the day.

Basic Bilateral Stance Balance Tests / Exercises

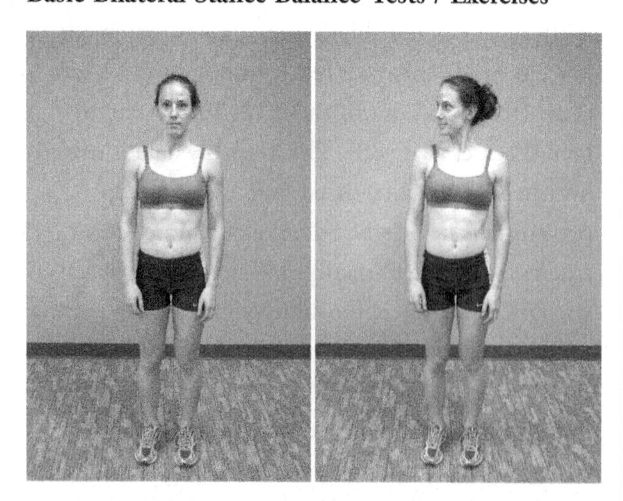

Figure 3.33a,b: Balance testing/training with bilateral stance and cervical motions. With the eyes open, the head and neck move in all three planes of motion. If performance is at an accepted level, the same testing can be performed with the eyes closed to remove compensation with ocular input. Directions of poorest performance can be modified for safe daily exercises.

Figure 3.34a,b: Balance testing/training with eyes closed, bilateral close stance position and cervical motion. Feet are positioned next to each other to reduce the base of support. With the eyes open, the head and neck move in all three planes of motion. If performance is at an accepted level, the same testing can be performed with the eyes closed to remove compensation with ocular reflexes. Directions of poorest performance are given as daily exercises.

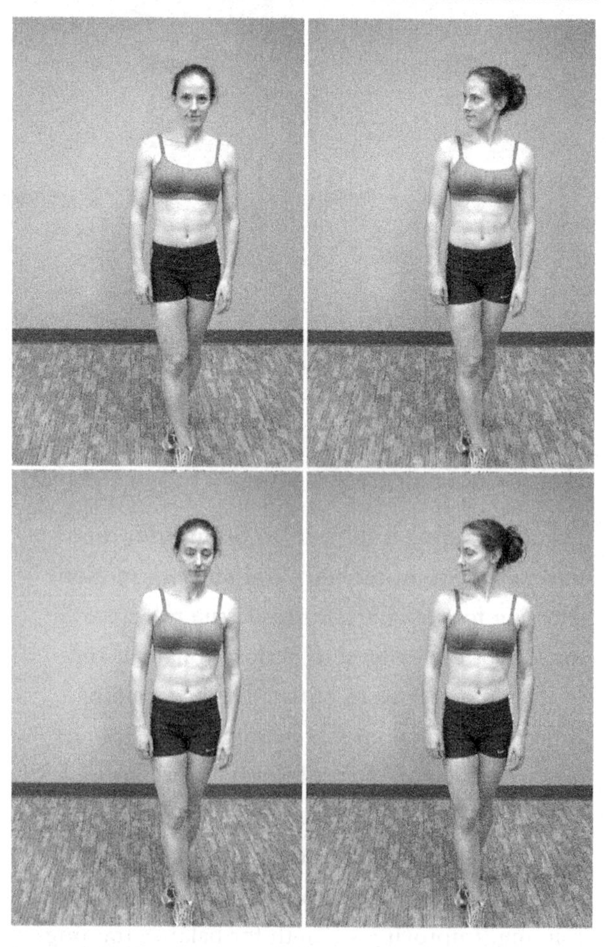

Figure 3.35a,b,c,d: Balance testing/training with bilateral stance feet positioned in tandem (a). A narrow base of support is a basic balance challenge. The addition of cervical motion (b), closing the eyes (c), and even performing cervical or arm motions (d) can further increase the challenge. Each set may involve a different option to vary the challenge.

Respiration

Respiration patterns should be addressed with all early exercises. Basic concepts of inspiration during eccentric work and expiration with concentric work should be established. Breath holding is a common abnormal response when performing new exercises or trying to gain coordination. Breathing patterns do not directly affect hip mechanics but can have an indirect contribution as deep muscles of the hip joint are related to the sacrum and pelvic floor. The pelvic floor muscles are antagonistic to the respiratory diaphragm and abnormal elevation in pelvic floor muscle tone can occur with reduced diaphragm function. Establishing normal breathing mechanics will not only improve oxygen uptake, but have a significant influence on maintaining proper motor patterns of the hip and pelvis.

Section 2: Stage 2 Exercise Progression Concepts for the Hip Joint

Training Goals

Progression to Stage 2 concepts of training are indicated as muscle guarding, edema and pain have significantly resolved. The patient will also have demonstrated an increased speed of training coordinated around a physiological axis, improved tissue tolerance and range of motion along with a decrease in perception of fatigue. The goals for Stage 2 are to improve muscle endurance and fast coordination by increasing repetitions and speed of the movement patterns. Repetitions are increased by adding sets and exercises to the program. During Stage 2 the goal may be to work up to 5 to 10 exercises of two to three sets each. Speed is also increased to emphasize fast coordination and to increase the relative resistance of the exercises. The exercises from Stage 1 may be carried over, but the desired functional qualities will shift with these changes. Isometric contractions within the mid to inner range of motion are also introduced to help fixate strength (Grimsby 1991). The following are examples of Stage 2 exercises dosed for specific functional qualities.

Basic Stage 2 Training Concepts

- *Increased repetitions with increased exercises (five to ten exercises)*
- *Increase the number of repetitions with additional sets (endurance)*
- *Isometric contractions (strength)*

Dosage: Increase Speed Not Weight

Dosage of training continues to be set for each exercise depending upon the functional quality desired from the exercise. Too often the progression focus is on increasing resistance of training when emphasis should remain on achieving coordination in the full range of motion, in more challenging postures and in more functional patterns. Simple exercises in one plane of motion need to be progressed into more challenging motions that reflect a functional goal for the patient.

Initially, Stage 1 exercises were performed slowly to allow for maximal concentric motor recruitment, slow coordination and to avoid the potential for abnormal tissue stress. As tissue tolerance and coordination improve the patient may naturally increases the speed of the exercise. If speed does not increase, the patient may be instructed to increase the speed of the exercise as a form of progression. The addition of speed is adding the additional forces of inertia, acceleration and deceleration, increasing the demand on the motor system. When performing concentric activities, increasing the speed decreases the overall amount of time to recruit motor units resulting in more selective motor units performing the motion. As a skill is learned (coordination), the ability to control all degrees of freedom increases, leading to more selective motor function (Bernstein 1967.) Increased speed can also place a greater demand on the level of coordination required.

Initial exercises may have involved a slower training speed to emphasize coordination. As coordination improves fewer motor units are required to perform the activity and the patient perceives increased ease of performance. Here the patient may ask for an increase in resistance for the exercise. Instead, the patient is asked to perform the same exercise with a slightly increased speed increasing the work demand. Exercises previously performed at slower than normal speed are increased to a more functional level to improve fast coordination. Other exercises may exceed functional speeds to reach more athletic speeds of performance.

Increase Repetitions

The most basic progression to an exercise program is the addition of more repetitions to the program to further increase tissue training qualities and to further enhance coordination. Tissue repair and coordination are a function of high repetitions in training. Increasing the number of repetitions is achieved by performing more sets of the current exercises and/or adding new exercises. Adding more exercises may involve the opposing, or antagonistic patterns, to the initial exercises. However, by adding new exercises recalculation of the dosage of all previous exercises in the same synergy may be required. Training the guarded pattern in Stage 1 of external rotation, flexion and adduction is followed up with adding the opposite pattern of internal rotation, extension and abduction. These motions may have been left out in the initial stages until the muscle guarding of the external rotators and hip flexors could be normalized to allow for normal recruitment of the antagonist pattern. As the overall time of training increases appropriate dosage and rest breaks are necessary to avoid excessive fatigue and to allow enough reserve energy for recovery and tissue repair.

Functional Testing

Basic testing during assessment should identify motor deficits for specific retraining. Stage 1 may have addressed individual muscle dysfunction. A progression in training is to incorporate more functional testing for a progression in training. The ankle and knee chapters cover basic balance testing, including balance training with upper and lower limb reaches, which can be applied to hip testing and exercise design. The reader is referred to the ankle chapter for details on the Star Excursion Balance Tests (SEBTs). The test requires the patient to maintain balance on a single limb, while manipulating the opposite lower limb or reaching with the upper limbs. Dynamic postural stability has been defined as the extent to which a person can lean or reach without moving the feet and still maintain balance (Goldie et al. 1989). The goal of the SEBTs is to reach with one leg as far as possible in each of eight prescribed directions, while maintaining balance on the contralateral leg (Olmsted et al. 2002). Similar patterns can be performed with the upper limbs. A star pattern is placed on the floor with four lines of tape: vertical, horizontal and two diagonals at 45 degrees angle. Alternatively, one line of tape can be placed on the floor with the subject altering the body position relative to the line of tape for each test, replicating all eight test directions. Hip patients, particularly elderly hip patients, may require an even more basic balance assessment than the SEBT to set a baseline for a safe level of training and to assess risk for falls.

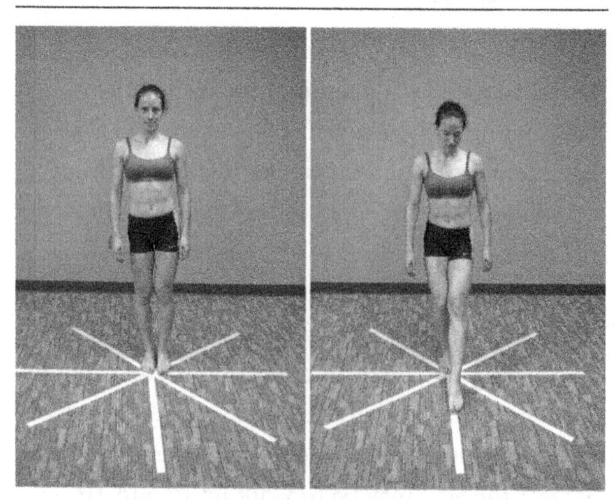

Figure 3.36a,b: Star Excursion Balance Test. Star pattern for lower limb balance reach involves standing one foot with the opposite lower limb reaches in different directions. The eight lines positioned on the grid are labeled according to the direction of excursion relative to the stance leg: anterolateral (AL), anterior (A), anteromedial (AM), medial (M), posteromedial (PM), posterior (P), posterolateral (PL), and lateral (L).

Functional Reach Test/Multidirectional Reach Test
The Functional Reach Test (FRT) was designed to measure how far a subject can reach forward without a loss of balancd. FRT has been found to be reliable, precise, portable, inexpensive and a reasonable clinical measure of the margin of limits of stability (Duncan et al. 1990). The test takes generally one to two minutes to complete. The original measurement consisted of a horizontally mounted yardstick set to the end of the fingertips with the arm flexed forward. The subject reaches forward down the yardstick maintaining balance on both feet, with the distance measured. FRT is a useful tool for detecting balance impairment and changes in balance performance over time. As body weight is shifted forward the hip must stabilize the center of gravity over the foot to maintain balance, incorporating motor performance of the trunk and entire lower limb. This simple test does correlate with more extensive questionnaires and functional testing. Weiner et al. (1992) assessed the FRT finding it to be a practical instrument that correlates with the Physical and Instrumental Activities of Daily Living tests (PADL, IADL).

The Multidirectional Reach Test (MDRT)
The MDRT is a modification of the FRT test in different directions. Newton (2001) performed a similar yardstick assessment of reach forward, backward, to the right and left. Instructions given were "without moving your feet or taking a step, reach as far (direction given) as you can, and try to keep your hand along the yardstick." For the backward direction the subject was asked to "lean as far back as you can." Individuals used their typical strategy to accomplish the task. Participants used their arm of choice for the forward and backward tasks, and used the respective arm for the right and left reaches. Feet are to be maintained flat on the floor with the test discarded if the feet move.

The MDRT can be further modified for reaching in different directions. Activities of daily living and sport performance may not only require balance with reaching activity 360 degrees in the horizontal plane of the forward reach, but may also require a vector of reaching up or down. Testing can be individualized based on functional demands and can then be turned in to active balance reach exercises in the directions of impairment. Alternatively, measurements may include reaching for specific targets anywhere from the floor to above the head. Another form of measurement is to have the subject hold onto an object, such as a sand bag, dropping the object at the final reach position. The distance on the ground from the toe to the sand bag provide an objective measurement.

Measurement of distance for upper quarter reach testing is only one form of assessment. The strategy used can also provide clues for areas of restriction, weakness or compensation. Limited range of joint motion in the arm, trunk or hip may be the primary limiting factor. Vestibular balance issues may be more noticeable in one direction or one plane of movement. Unilateral stance during the tests may reveal more significant balance deficits in one limb. Weakness may be identified as primarily in the trunk, hip, knee or ankle. Exercises can then be designed to more specifically address the functional deficits identified with the testing.

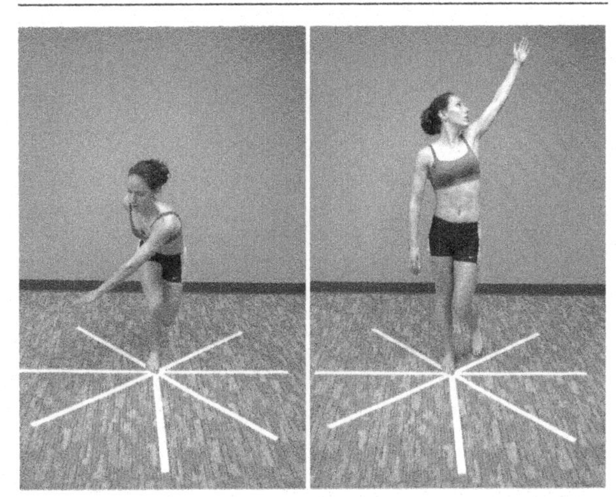

Figure 3.37a,b: Functional Reach test/exercise. The subject stands on both feet, reaching forward along a fixed yardstick. Measurement is taken of the distance of the reach in which the subject can return to the start position without loss of balance. The test can be modified with bilateral or unilateral stance (pictured), or unilateral arm reach and reaching in any direction. Measurement can also be taken by having the subject reach in any direction, drop a sand bag at the end position and then measure the distance from the toe to the sand bag. Vectors of impairment can be turned into exercises to train balance and stability.

The Berg Balance Scale (BBS)

The BBS provides and effective assessment tool for balance with sit to stand activities and gait. Non-fallers, compared to those patients prone to falling, have demonstrated faster reaction times and higher scores on the Berg Balance scale (Lajoie et al. 2004). A score of lower than 45 out of 56 has been correlated with an increase risk for falls (Bogle Thorbahn and Newton 1996). Tinetti Gait and Balance Assessment Tool is interpreted as follows the maximum score for the gait component is 12 points. The maximum score for the balance component is 16 points. The maximum total score is 28 points. Patients who score below 19 are at a high risk for falls. Patients who score in the range of 19–24 indicate that the patient has a risk for falls.

Balance Section (Tinetti et al. 1986)

Patient is seated in hard, armless chair.

Sitting Balance	*Leans or slides in chair* *Steady, safe*	*= 0* *= 1*
Rises from chair	*Unable to without help* *Able, uses arms to help* *Able without use of arms*	*= 0* *= 1* *= 2*
Attempts to rise	*Unable to without help* *Able, requires > 1 attempt* *Able to rise, 1 attempt*	*= 0* *= 1* *= 2*
Immediate standing Balance (first 5 seconds)	*Unsteady (staggers, moves feet, trunk sway)* *Steady but uses walker or other support* *Steady without walker or other support*	*= 0* *= 1* *= 2*
Standing balance	*Unsteady* *Steady but wide stance and uses support* *Narrow stance without support*	*= 0* *= 1* *= 2*
Nudged	*Begins to fall* *Staggers, grabs, catches self* *Steady*	*= 0* *= 1* *= 2*
Eyes closed	*Unsteady* *Steady*	*= 0* *= 1*
Turning 360 degrees	*Discontinuous steps* *Continuous* *Unsteady (grabs, staggers)* *Steady*	*= 0* *= 1* *= 0* *= 1*
Sitting down	*Unsafe (misjudged distance, falls into chair)* *Uses arms or not a smooth motion* *Safe, smooth motion*	*= 0* *= 1* *= 2*
	Balance scores	*/16*

Gait Section (Modified from Tinetti et al. 1986)

Patient stands with therapist, walks across room (+/- aids), first at usual pace, then at rapid pace.

Indication of Gait
(Immediately After Told to 'Go'.)

Any hesitancy or multiple attempts = 0
No hesitancy = 1

Step Length And Height	Step to Step through Right Step through Left	= 0 = 1 = 1
Foot Clearance	Foot drop Left foot clears floor Right foot clears floor	= 0 = 1 = 1
Step Symmetry	Right and left step length not equal Right and left step length appear equal	= 0 = 1
Step Continuity	Stopping or discontinuity between steps Steps appear continuous	= 0 = 1
Path	Marked deviation Mild/moderate deviation or uses walking aid Straight without walking aid	= 0 = 1 = 2
Trunk	Marked sway or uses walking aid No sway but flexion knees or back or uses arms for stability No sway, no flexion, use of arms or walking aid	= 0 = 1 = 2
Walking Time	Heels apart Heels almost touching while walking	= 0 = 1
	Gait score Balance score carried forward Total Score = Balance + Gait score	/12 /16 /28
	Risk of Falls	≤18 High 19–23 Moderate ≥24 Low

Sit-to-Stand Test

Csuka and McCarty (1985) first described the Sit-to-Stand Test (SST) as a simple, rapid, reproducible method for quantification of lower extremity muscle strength. Netz et al. (2004) found the test to be a better assessment of general endurance, rather than specific lower limb muscle strength or endurance. The timed test involves standing from a chair 10 times. The authors found their results correlated well with published data on the strength of knee flexor and extensor muscles in groups of men and women of various ages. The SST is commonly used to assess lower extremity strength and balance (Lord et al. 2002, Bohannon 1998, Bohannon et al. 1995, Cheng et al. 1998, Huges 1998, Newcomer 1993), as a crucial factor in measuring independence in older adults living in the community (Nevitt 1989, Campbell et al. 1989), to assess subjects with arthritis (Newcomer 1993), renal disease (Bohannon 1995), stroke (Brunt et al. 2002), and, as an outcome measure of intervention (Drabsch et al. 1998, Headley et al. 2002, Skelton et al. 1995). Kerr et al. (2007) assessed the sit-to-stand motion as a measure of risk for falling. Normal elderly subjects tested with an average total time of 1.80 seconds, while those elderly subjects at risk for falling averaged 4.15 seconds. Whitney et al. (2005) used a 5-times SST to discriminate between subjects with and subjects without balance disorders. The test did display discriminative and concurrent validity properties making the test potentially useful in clinical decision making.

Timed Up and Go Test (TUG)

The TUG is a measure of the time it takes to stand up from an armchair, walk three meters, turn, walk back to the chair, and sit down (Steffen et al. 2002). The original test was a clinical measure of balance in elderly people. Scoring on an ordinal scale of one to five was based on an observer's perception of the performer's risk of falling during the test (Mathias et al. 1986). Podsiadlo and Richardson (1991) modified the test as a timed task to assess basic mobility skills for frail community-dwelling elderly. This test may not provide specific information for exercise, but more provide a general assessment of mobility that can be used as a retest for performance improvement.

Expanded Timed Get Up and Go Test

Wall et al. (2000) created a sensitive and objective assessment of function in the "Expanded Timed Get-Up-and-Go" test (ETGUG). It isolates functional deficits, helping to guide treatment and re-evaluation. This test requires the subject to repeat the above test, beginning in an unsupported chair with stop watch timing at the following intervals during the task: when the subject is standing upright, as the subject passed the 2-meter mark, as the subject passed the 8-meter mark, as the subject passed the 8-meter mark when returning; and finally as the subject passed the 2-meter mark when returning. These measurements correlate with the tasks of sit-to-stand, gait initiation, walk one, turn around, walk two, slow down, stop, turn around, and sit down.

Comfortable and Fast Gait Speeds (CGS and FGS)

Gait speeds can be measured over a relatively short distance to assess a subject's ability to increase or decrease walking speed above or below a "comfortable" pace. Performance at different speed suggests a potential to adapt to varying environments and task demands (e.g., crossing streets; avoiding obstacles). Steffen et al. (2002) used the test by assessing two consecutive trials of gait speed as subjects walked on a marked 10-meter walkway in a gym area at a "normal, comfortable speed," followed by two trials "as fast as you can safely walk." The examiner walks to the side of and behind each subject to ensure safety. Gait speed can be measured in hundredths of a second with a stopwatch. Several trials at each speed can averaged for final speed analysis. Van Loo et al. (2004) found the 10-meter walking test for speed and distance to have excellent test-retest reliability. Longer walking distances, such as with the 6-minute walk test (described in Stage 3), may be better predictors of walking in real-life environments than the 10-meter walk test, as they include a more significant component of endurance.

For subjects over 60 years of age without known impairments have average gait speeds ranging from 0.60 to 1.45 m/s for comfortable walking speeds (Bohannon 1997, Murray 1969, Oberg et al. 1993, Ferrandez et al. 1990) and from 0.84 to 2.1 m/s for fast walking speeds (Bohannon 1997, Murray 1969, Oberg et al. 1993, Ferrandez et al. 1990, Elble et al. 1991). Depending on the study, the comfortable walking speed of the older adults was an average of 71–97 percent slower than that of the young adults and the fast walking speed of the older adults averaged from 71–95 percent slower than that of the young adults (Bohannon 1997, Murray 1969, Oberg et al. 1993, Ferrandez et al. 1990). Older adults have been shown to be able to increase their walking speed from 21–56 percent above a comfortable pace when instructed to "walk as fast as possible" or "very fast" (Bohannon 1997, Murray et al. 1969, Ferrandez et al. 1990, Elble et al. 1991, Himann et al. 1988).

Side Step Test

Fujisawa and Takeda (2006) confirmed the reliability and validity of the side step test. Their procedure involved performing the side-step test barefoot without support, measuring both the affected and unaffected side. A starting line is set with a 10-meter perpendicular line to measure the distance. Subjects begin with the legs and feet together on the starting line performing five repetitions of side-steps, attempting to step as wide as possible. The body is not supported with the arms, though a physical therapist can stand behind the patient for safety, and subjects are instructed not to jump. The maximum sidestep length is calculated by dividing the total distance moved by the number of steps. The measurement is performed three times with the maximum value taken. A sufficient interval between measurements was taken to avoid fatigue of subjects. Hip and knee motion was important for short distance lateral steps, where adequate ankle mobility is necessary to reach longer step lengths (Fujisawa and Takeda 2006).

Viton et al. (1999) identified changes in equilibrium and lateral movement control strategies in patients with arthritis of the knee. The duration of the bilateral stance phase was increased, with reduction in the unilateral stance phase on the involved leg. The arthritis patients were found to develop a modified postural strategy, mainly aimed at shortening the monopedal phase when on the affected leg. A follow up study testing pre and post operative performance in subjects undergoing a total knee arthroplasty did show improvements in the lateral step test, though still not reaching normal values, and continued compensatory upper body strategies were employed (Viton et al. 2002).

Motions and Directions

Establishing full range of motion and increasing weight bearing tolerance are the key goals of Stage 2. Improving motor performance and strength output must follow the achievement of full range of motion around a physiological axis. Full hip extension and internal rotation mobility is crucial to upright standing and normalization of gait patterns. Exercises should involve all functional planes of movement that are limited, painful, poorly coordinated or lacking in endurance and strength. Motor challenges are much more significant in terms of coordination and torque in weight bearing training versus non-weight bearing training. Emphasis should be shifting toward full weight bearing exercises.

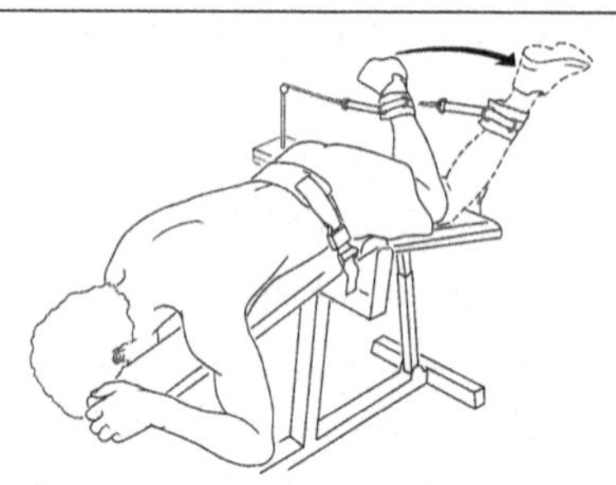

Figure 3.38: Internal rotation training of the hip in open packed position with belted stabilization. If joint limitation was due to a true capsular restriction, rotation training may have already been added.

Training Directions

- *Increase ROM of previous Stage 1 exercises*
- *Add the antagonistic muscle pattern to those trained in Stage 1*
- *Train internal and external rotation closer to full extension*
- *Increase weight bearing challenge*

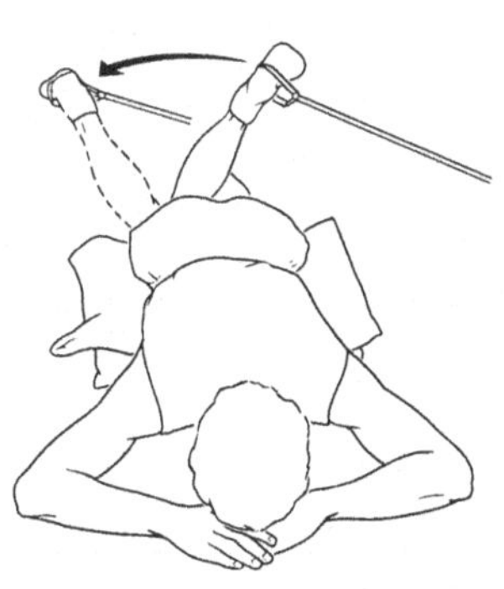

Figure 3.39: Home version Internal rotation training of the hip in open packed position with elastic resistance.

Unloading Concepts for Progressive Close Chain Exercise

Reducing body weight is an important progression to more quickly begin training in weight bearing postures. Many options are available to emphasize tissue loading and/or motor recruitment. The type of unloading used should match the needs of the patient, rather than the preference of the clinician. Pool therapy, as previously noted, is an example of a technique for reducing body weight for training tissue. Though not entirely functional, pool therapy can emphasize tissue stimulus with 30 to 60 minutes of modified compression/ decompression with gliding for cartilage repair. Horizontal Nautilus weight machines and vertically adjusting sliding boards are available for close chain training in recumbent postures. Lying on the back and pushing with the legs does not simulate the exact motor pattern of weight bearing function for the hip. Emphasis is on knee flexion with slightly abnormal hip, pelvis and back motion. Training in these positions may be adequate for pain inhibition, tissue stimulus, edema and motor performance.

Coordination training of muscle performance, as well as training functional patterns, requires normal standing postures. Vertical unloading of the body weight can provide for erect postures with training. More elaborate unloading machines can reduce body weight while walking on a treadmill. Using a pull down machine with an overhead gantry, or simply holding the bar under the arms, can reduce body weight for squats, lunging and step training. Emphasis should be placed on coordination of the pattern of motion. Often hip patients will compensate with excessive knee and back motion with little hip flexion. The squat motion should involve moving the pelvis and hips posterior first, rather than the knees forward or the trunk forward first. For motor learning, an external cue is more effective than in internal cue. An internal cue would be asking the patient to focus on the flexing the hips and moving the pelvis backward. This may not be as effective as standing the patient with back to the wall and asking them to feel their tailbone hit the wall before the knees move forward. Another external cue can include keeping a chair in front of the knees to provide an external block for anterior knee movement, requiring posterior pelvic motion with hip flexion.

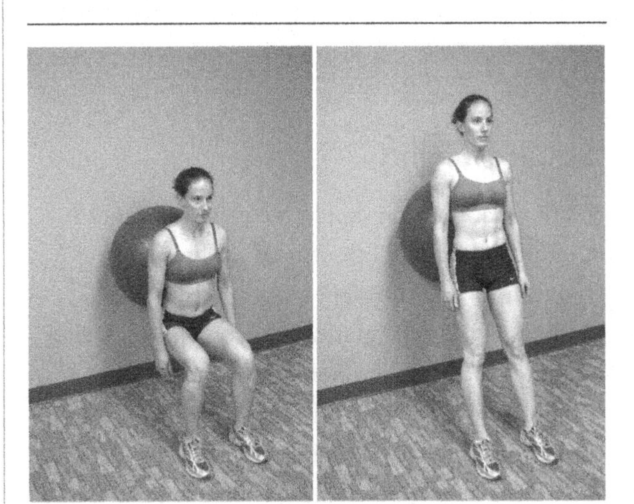

Figure 3.40a,b: Performing ball squats against a wall can reduce body weight for early tissue training. The pattern of motion emphasizes the knees moving anterior without the hips, pelvis and back moving posterior. This is not an optimal choice when emphasizing coordination, endurance or strength training.

Figure 3.41a,b: Unloaded squats with a lat machine or with chairs for home version. Coordination and endurance training

3. Exercise for the Hip Joint

can occur with normal functional patterns in an unloaded environment. Home versions include arm support with two chairs or using arm support in a door jam (figure 3.44).

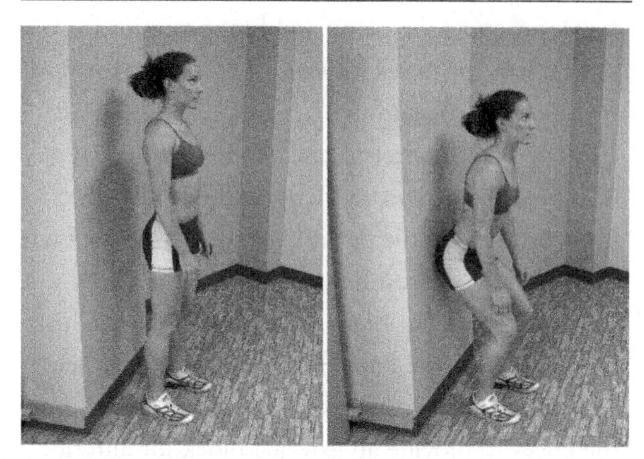

Figure 3.42a,b: Squatting with an external cue of making contact with the sacrum against the wall. Emphasis is placed initiating the motion with hip flexion with posterior pelvic motion, rather than lumbar flexion or anterior motion of the knees past the toes.

Squat Training

Early squat training should focus on technique, coordination and range of motion. Emphasis is on improving the degree of hip flexion with participation of the gluteal muscles with the pelvis and hips moving posterior. Hip weakness can lead to compensation with excessive quadriceps work with the knees moving anterior, but the hips remaining over the feet. The arms may reach forward to counter balance posterior movement of the center of gravity, reducing the load on the posterior hip muscles. With reduced hip flexion mobility, or weakness of the hip extensors, the lumbar spine may also compensate with excessive flexion. Limited ankle dorsiflexion, from joint restriction or muscle tightness, may cause the heels to rise, or limited depth, during a deep squat. Joint mobilization of the ankle may be necessary prior to squat training. If dorsiflexion is limited by muscle tightness in the Achilles, repetitive squats can provide an eccentric load to improve flexibility. A lift of one to two centimeters under the heels, placing the ankle in slight plantar flexion, can increase the depth of squatting before reaching the end of available dorsiflexion. Using a heel lift does not represent a functional platform for normal squatting activities, and should be avoided, or trained out of, as soon as possible. Use of a heel lift is more appropriate when joint restriction is present, but the heel lift should be avoided if muscle tension is the limitation to allow for full lengthening of the Achilles tendon.

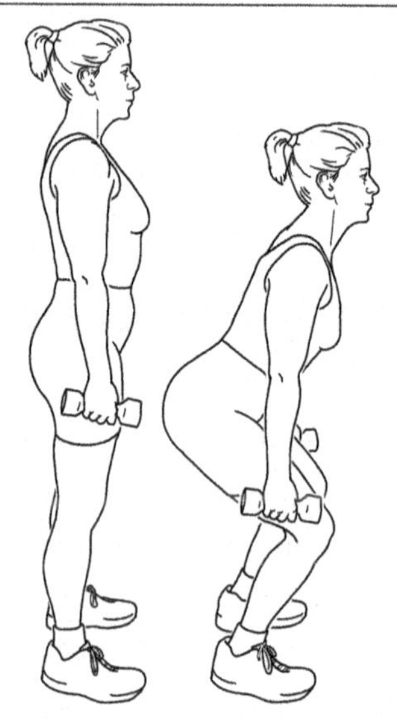

Figure 3.43a,b: Proper squat technique should involve a flat lumbar spine, posterior movement of the pelvis with hip flexion and the knees remaining over the toes. The weight of the trunk and upper limbs should remain over the feet.

Figure 3.44: Door Jam Squats. Upper limb support provides support to reduce load on the lower limbs and allows for proper technique of the hips moving posterior.

Weight Bearing Gluteus Medius Training Concepts

Weight bearing training for hip stability is preferred

but may be limited by pain, coordination and/or strength. Unloading approaches attempt to reduce the load of body weight via an external force against gravity. Unloading may allow for earlier training than would be otherwise tolerated. Movement of the pelvis on the femur is generally more consistent with functional performance than motion of the femur on the pelvis in non-weight bearing, or open chain approaches.

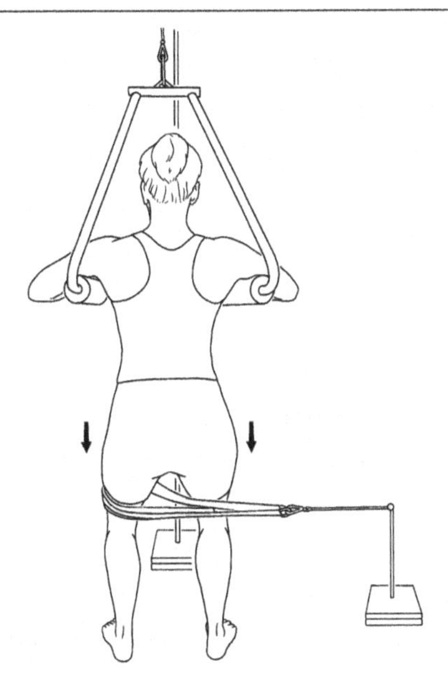

Figure 3.45: Increased gluteus medius recruitment during squat training. A valgus moment on the involved lower limb from a secondary force moment will increase the demand on the hip abductors to avoid adduction of the femur.

Stage 1 training involved addressing local neurological drive for excessive tone in the hip flexors. As articular pain resolves, the neurological drive for muscle guarding is eliminated. Chronic guarding requires resolving tissue ischemia by training for local circulation. As guarding resolves, the focus of training can now shift to training coordination. The hip flexors and lower abdominals can now work together in a smooth pattern, as in performing caudal hip flexion with an abdominal curl. With weakness in the lower abdominals there will be a slight stop, or hitch, in the middle of the motion as the primary movers shifts from the hip flexors to the trunk flexors (abdominals). Resistance (body weight) must be reduced to allow for the motion to be performed smoothly, and throughout the desired number of repetitions (> 25 when emphasizing coordination).

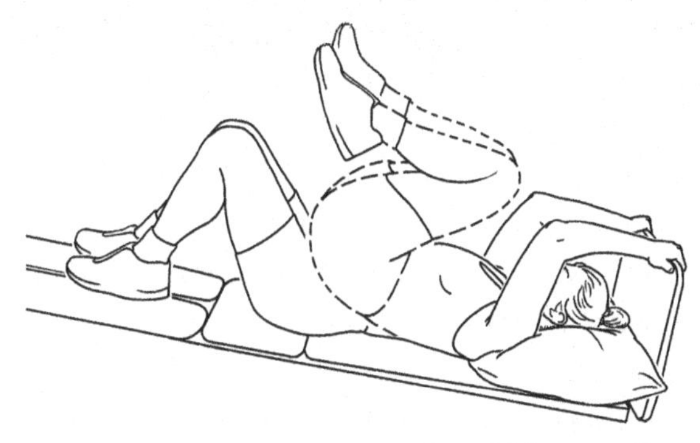

Figure 3.46: Coordination training of caudal flexion with smooth transition from hip flexors to trunk flexors. The slant board allows the angle to be adjusted from that of a decline position, to horizontal or to an incline to match the patient's performance level at the desired number of repetitions.

Stretching / Mobilization

Earlier exercises from Stage 1 emphasized mobilization of arthrokinematic motions to improve joint play and range of motion. As arthrokinematic movements are restored, these types of self-mobilization exercises are discontinued in Stage 2 and replaced with more functional movements into patterns of restriction. Passive muscle stretching was avoided in Stage 1, as most of the limitations in muscle extensibility were associated with tonic muscle guarding from pain. Treatment attempted should address the cause of pain leading to abnormal muscle tone and to exercise for local circulation to supply oxygen to tonic muscles in guarding. Passive stretching is still not an intervention of choice in relation to muscle performance as the plastic deformity of collagen from the prolonged stretch will reduce the tensile properties of tendons, reduce normal motor recruitment and functionally weaken the muscle.

As pain decreases and the neurological drive for tonic guarding and secondary shortened muscles reduces, treatment focuses more on active motions, rather than passive stretching, to obtain muscle lengthening while also improving coordination of functional patterns (Grimsby 1991). Dynamic Range of Motion (DROM) techniques involve exercising the antagonist for reciprocal inhibition to improve functional lengthening of a muscle (Brandy et al. 1988). Training the agonist directly with an emphasis on eccentric contractions to the end range is the preferred method to address neurological aspects of motor function as well as to influence structural changes in collagen within the muscle. This type of active stretching was referred to as ARM (Active Range of Motion stretching). Actively training a muscle eccentrically to the end range is just as effective as passive stretching (Nelson and Bandy 2004, Doll 1999) but is also associated with improved coordination, muscle strength and more functional movements.

Figure 3.47a,b: Self mobilization of the anterior medial hip capsule. The involved hip is placed on a step stool in a lateral lunge position with external rotation. The patient can repetitively stretch into the lunge or push the pelvis and hip anterior to stretch the capsule. Upper limb support is used to reduce the need for balance, allowing for improved control and range of motion.

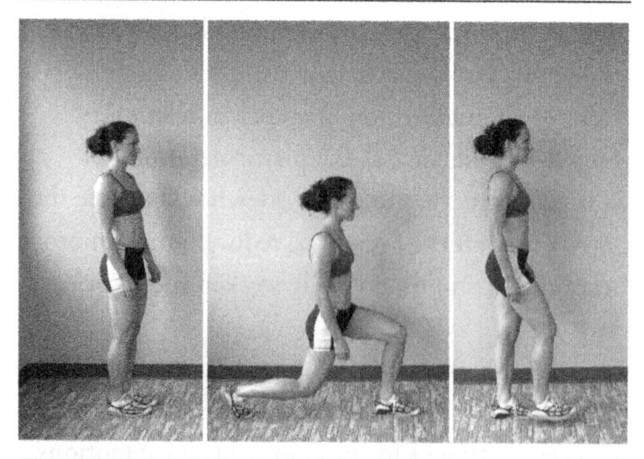

Figure 3.48a,b,c: The anterior hip capsule, as well as the hip flexors, can be actively lengthened by performing a walk – lunge anteriorly. The forward leg creates hip flexion and lumbar flexion allowing the full extension of the hip and eccentric lengthening of the contralateral hip flexors. Functional goals of coordination and endurance can simultaneously be achieved during this exercise. If balance is poor the lunge can be performed next to a wall for upper limb support. Emphasis is placed on a deeper and longer stride to increase range.

Closed Chain Training

As range of motion and tissue tolerance increase, exercises need to progress from dependent postures of lying or sitting to more functional positions of standing with partial, full or loaded weight bearing. Weight bearing provides increased compressive forces through the joint to further stimulate cells for bone or cartilage repair requiring thousands of overall repetitions. Signals from afferent receptors are increased in weight bearing improving the neural drive and normalizing motor patterns. Coordination training involves many repetitions in functional patterns of weight bearing. Endurance training requires a more specific repetition to resistance ratio to achieve specific changes in muscle fiber and vascularity. Closed chain, weight bearing training, is preferred when tolerated. Below are examples of weight bearing exercises focused on gluteus medius/hip stabilization progressions.

Exercise dosage for Stage 2 emphasizes fast coordination with increases in speed, not increased weight. Establishing coordination and a level of endurance with squats, lunges and step training is the initial goal with gradual increases in speed to simulate eventual functional or sport specific tasks. If initial weight bearing training requires some type of an assist with reduction in body weight through unloading techniques, then a load progression might involve a reduction in the level of assist until full body weight is achieved. No additional loading with weights are used until coordination is established through a full functional range of motion in different directions and planes of motion. Earlier functional testing identified directions of motion lacking balance and/or coordination. Normalizing these deficits must occur prior to adding additional load.

Gluteus Medius Training

Closed chain training of the hip muscles, including the gluteus medius, is necessary to coordinate functional patterns. Non-weight bearing hip abduction isolates the hip abductors only and does not train the entire biomechanical chain and motor pattern. Initial open chain training may have been necessary in Stage 1 due to joint pain, tissue repair or severe weakness but close chain training should be initiated as soon as possible to establish coordination in more functional movements.

Figure 3.49: Standing forward bent horizontal hip abduction is a progression of the quadruped version presented in Stage 1. Longer lever arms of the extended leg and unsupported trunk significantly increase the resistance for the gluteus medius, lumbar multifidi and lower abdominals. Afferent feedback from the entire lower limb increases the functional carry over to normal activities. A progression may include using a dumbbell in the right hand, performing horizontal shoulder abduction during the motion. The trunk remains in neutral, with the pelvis abducting on the femur.

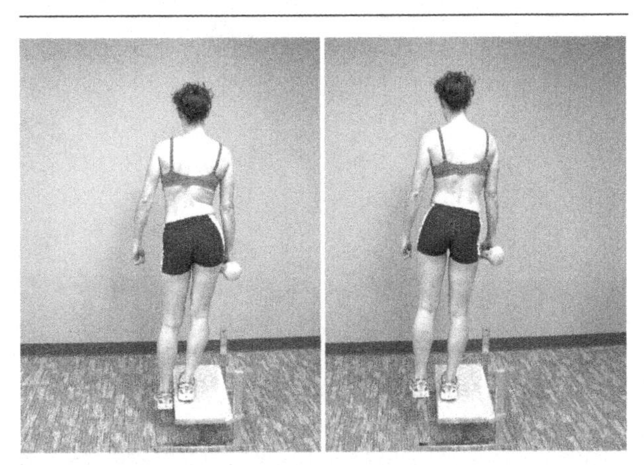

Figure 3.50a,b: Assisting the gluteus medius in Stage 1 may progress to performing a hip hike exercise (closed chain right hip abduction) with full body weight. To add resistance to allow for dosage of specific functional qualities a free weight is held on the involved side. The relative body weight is on the left side of the axis of motion in the right hip joint. By adding weight in the right hand, weight on the right side of the joint axis is increased. This increase in weight helps to counter balance the body weight on the left side of the hip joint axis. Functionally, patients can be instructed to carry bags and purses ipsilateral to assist the muscles stabilizing the lateral hip.

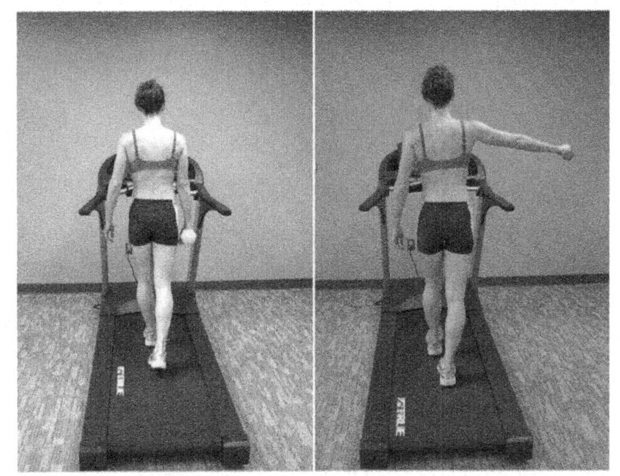

Figure 3.51a,b: As gluteus medius strength improves the positive Trendelenburg sign during gait resolves. To increase the load on the left hip abductors during gait weight is now held on the opposite side. A heavy weight can be held at the side during walking. A lighter weight can be used with active shoulder abduction during heel strike and foot flat on the involved side.

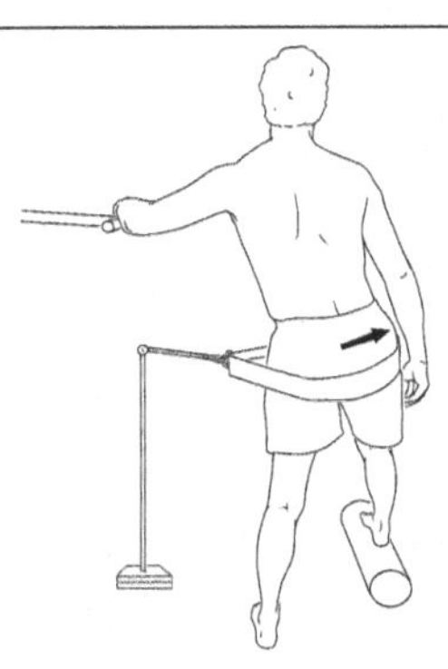

Figure 3.52: Left gluteus medius training with unilateral balance challenge. Hip abduction is performed by shifting weight from lateral to the axis of the right hip to medial without the shoulders moving. The right foot is on tiptoe or lightly resting on a roll to prevent the right hip adductors from assisting the left hip abductors with the weight shift.

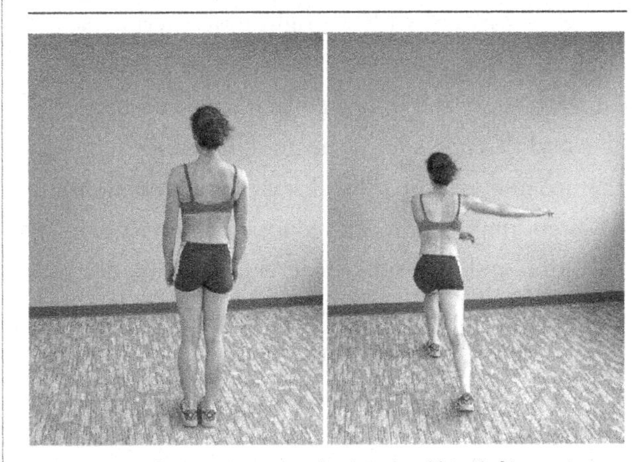

Figure 3.53a,b: Lunging onto the involved hip (left) requires eccentric function of the gluteus medius to decelerate hip flexion, adduction and internal rotation. Holding weight in the right hand, or reaching right, will increase the weight on the medial side of the left hip joint axis increasing the challenge to the hip abductors.

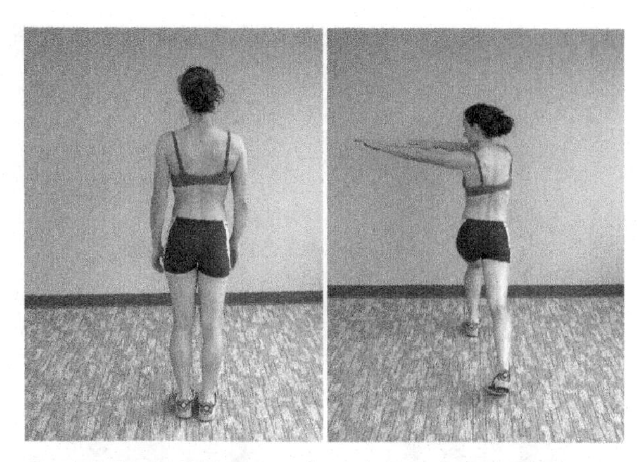

Figure 3.54a,b: Performing a lateral upper limb reach on the involved side hip does shift weight from medial to lateral in relation to the axis reducing the abduction moment for the gluteus medius. Reaching lateral from above creates a rotation of the trunk that requires an increase in activity of the gluteus medius in relation to the transverse plane. One plane of motion is assisted (frontal) while another plane of motion has an increased challenge (transverse).

Lunge Training

Lunging in specific directions provides an easy weight bearing approach to stress different portions of the joint, influence specific muscles in the lower quarter, and train normal functional motions. When first initiating lunges emphasis is placed on moving away from the direction of pain. As coordination and tissue tolerance improves the lunging direction can change to more aggressively challenge the joint and motor system. Initial lunges are performed in a 180 degrees half circle and are named for the leg doing the work, or lunging. Lunging forward with the right foot is termed a right anterior lunge. Options would also include anterolateral, lateral, posterolateral and posterior. Beyond this any specific angle of lunging could be named and measured for distance and direction. A "step-lunge" approach reduces the impact on the joint and decreases the level of coordination required to more easily establish the motor pattern. The step-lunge is performed by placing the lunging foot forward on the ground while the body weight remains on the stance leg. After the foot hits the ground body weight is transferred to forward foot.

How the lunge is performed will also determine which muscles are emphasized. Repetitive lunges provide an eccentric deceleration to lengthen muscle and soft tissue functionally in a normal pattern of movement. Lateral lunges will increase the demand on the hip adductors and anterior inferior hip joint capsule, while forward lunges will increase the eccentric demand on the hip extensors, knee flexors and posterior hip capsule. The posterior lunge provides an eccentric lengthening of the hip flexors and anterior hip capsule, often restricted in early hip training.

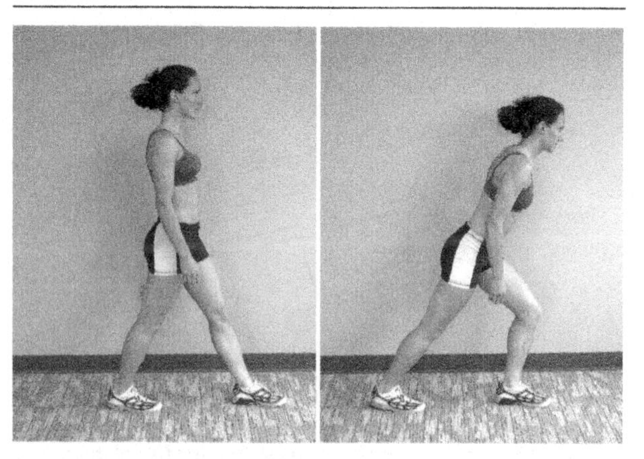

Figure 3.55a,b: The step lunge involves first placing the lunging foot on the ground and then shifting the weight onto the lower limb. Step lunge provides a reduction in impact force, and need for balance, coordination and strength.

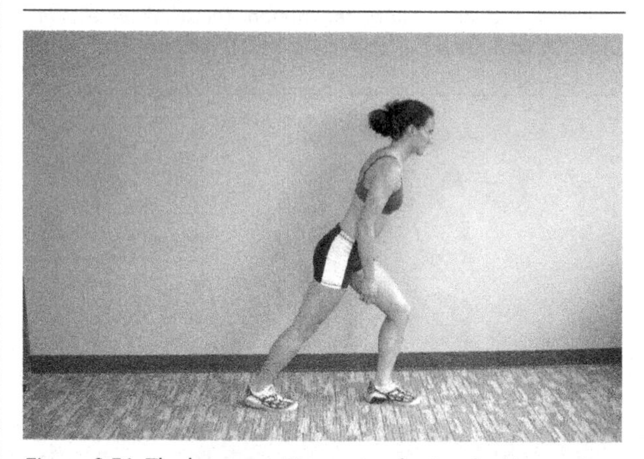

Figure 3.56: The lunge position to emphasize the hip requires the back knee in extension, heel on the ground, trunk parallel to the back leg, weight shifted to the front foot with the nose over the toes. The arms can hang at the side, hold weights or perform reaching motions in different directions to challenge specific muscles or balance.

The technique of the lunge can also emphasize the knee or the hip. To emphasize the hip in the anterior lunge the back knee remains in full extension with the heel on the ground. The trunk remains neutral in the lumbar spine but leans forward at the hip, remaining parallel to the back leg with the nose over the forward toes. The patient should feel more work in the hip than the knee. With a weak hip the patient will compensate by keeping the trunk vertical, not transferring enough weight to the lunging foot and having excessive anterior thrust of the knee.

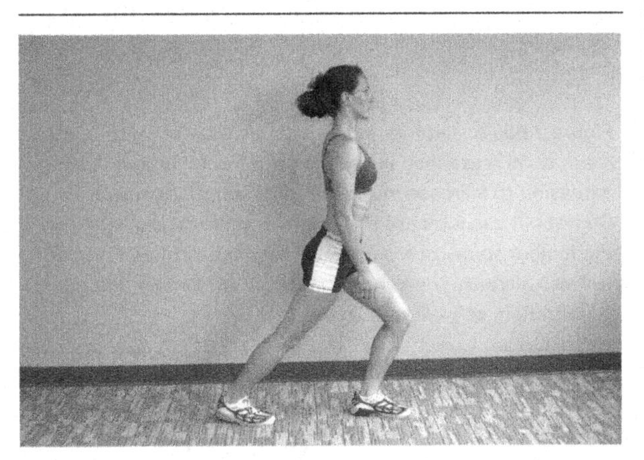

Figure 3.57: The incorrect lunge position for hip emphasize involves compensation with excessive quadriceps creating additional stress to the articular structures of the knee. The back knee flexes with the heel coming off the ground, or remains vertical with weight remaining on the back foot.

Multidirectional Lunge Training

Weakness in the hip extensors and deep external rotators often leads to a collapse in adduction and internal rotation at the hip with secondary valgus at the knee. The valgus collapse into the lower limb can result in secondary pathologies such as chondromalacia patella, tendinopathy of the patellar or Achilles tendons, or pronation foot pathologies such as plantar fasciitis. Anterior lunges facing a mirror may provide visual feedback for the patient to actively correct lower limb alignment. Providing a secondary lateral varus force moment from a pulley or elastic band can assist weakened hip muscles to train coordination of the lunge pattern. A medial valgus force moment will increase the facilitation and coordination of the hip abductors and external rotators. An original technique from the Medical Exercise Therapy curriculum is to use a pulley to provide a medial or lateral force moment to the knee when performing squat training, lunges or step exercises (Holten and Faugli 1996). Lateral force moments are used to increase weight bearing on the limb and to assist the hip muscles in stabilizing the limb preventing a valgus collapse at the knee. A medial force moment is used to increase facilitation of the hip muscles to dynamically stabilize the knee.

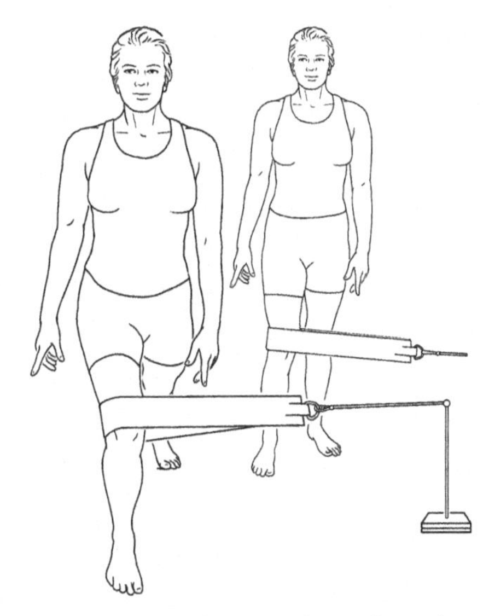

Figure 3.58: The anterior lunge can be performed as a multidirectional exercise with a secondary valgus force moment at the knee to facilitate the hip abductors and external rotators. The secondary valgus force may also be used to unload specific joint structures (i.e., lateral collateral ligament or lateral joint compartment of the knee).

The use of pulleys or elastic bands at the knee provides a caudal force moment acting at the hip to assist or resist posterior hip muscles. A caudal force moment can also be created by reaching with the arms, and to a greater extent by using hand weights. As the arms reach lateral to the lunging hip, upper body weight is shifted lateral to the sagittal axis of hip adduction and abduction, reducing the load on the hip abductors. Reaching lateral also creates a rotational motion of the trunk in the transverse plane requiring increased muscle recruitment at the hip to eccentrically control this rotation during the lunge. The deep hip external rotators, as well as the more horizontal fibers of the gluteal muscles, will increase activity to control the force moment created by the upper extremities. Reaching medial to the involved hip during the anterior lunge can create the opposite effect. Body weight is shifted medial to the sagittal axis of the hip increasing the load on the hip abductors. An external rotation force moment is created at the hip, which may reduce the transverse plane recruitment of the deep hip external rotators placing a greater demand on the lower limb to stabilize the valgus moment.

A potential reach progression for weight transfer would begin with a medial reach to reduce load lateral to the axis, followed by an anterior reach and finally to lateral reach to increase load lateral to the axis. A progression for motor performance is more dependent upon which muscles, or fiber directions, that need to be emphasized. A lateral hip reach will increase the transverse plane force moment requiring an increase in muscles resisting external rotation (deep rotators and gluteus medius). An anterior lunge while reaching anterior increases the load anterior to the frontal axis of flexion and extension, requiring recruitment of the hip extensors. An anterior lunge with an overhead reach will shift weight posterior reducing recruitment of the hip extensors resulting in an increase demand on the knee extensors. Many combinations of of lunging exist, along with upper limb reaches and trunk motion, to influence specific muscles, fiber directions or functional performance.

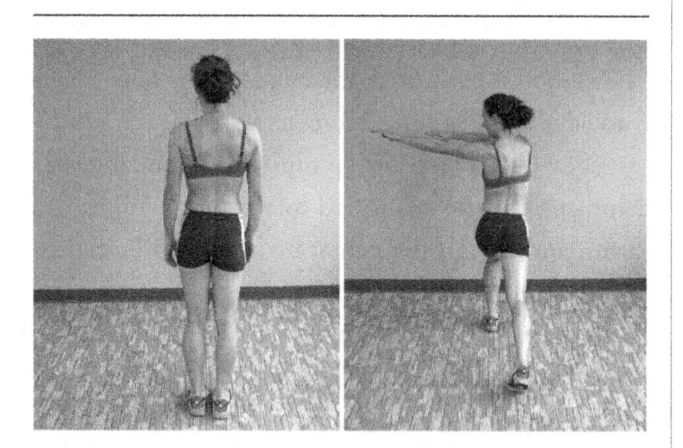

Figure 3.59a,b: Anterior lunge with a lateral upper limb reach will reduce body weight medial to the sagittal axis of abduction, while increasing muscle recruitment of the hip external rotators.

Reaching activities with lunging can be performed with the added challenge of dumbbell weights in the hands. But reaching and free weights still emphasize an extensor pattern primarily, as gravity is always the line of resistance. Asymmetrical arm motions, with or without dumbbells, can further enhance the challenge for central stabilization during active lower extremity exercises.

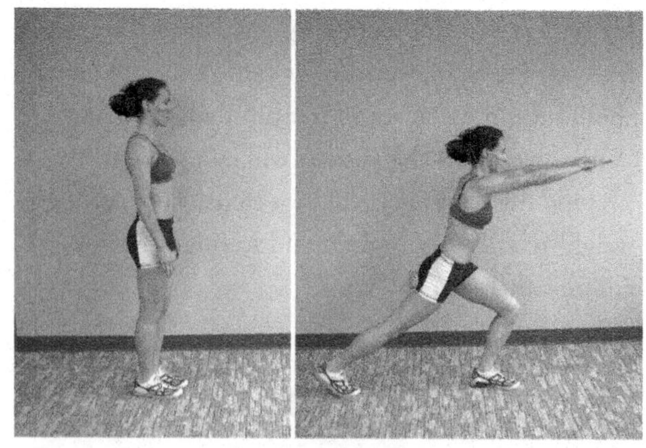

Figure 3.60a,b: Anterior lunge with an anterior upper limb reach to increase load anterior to the frontal axis of flexion/ extension to increase muscle recruitment of extension. Dumbbells can increase the extensor moment challenge for the lumbar spine, pelvis and hip. Asymmetrical arm motions will also increase the central challenge for lumbar and pelvic stabilization, as well as balance.

True "Core" Training Examples

Core muscle training is often thought to be emphasized with either lying down postures with cognitive recruitment of trunk muscles, or when working on a labile surface such as a fitness ball or wobble board. These approaches do provide some level of deep motor recruitment, but these central stabilizing core muscles work with every motion as a preset for stability. A functional progression for core training employs a second line of resistance that specifically recruits muscle fiber lines as a secondary action during a primary functional task. The following exercises demonstrate the value of this concept applied to simple squatting, lunging and step exercises. A lateral force moment from a wall pulley, or elastic resistance, is held anteriorly away from the body. The force moment from the pulley is in line with the transverse abdominus, the rotational vector of the lumbar multifidi and the deep hip rotators. These muscles are naturally recruited to resist the force from the pulley while a larger motion is being performed. Unilateral arm holds can further emphasize the anterior or posterior core muscles, depending upon which arm is used.

Manual therapy evaluation and treatment for lower quarter pathology involves the biomechanical chain from the lumbar spine to the foot. Soft tissue work and joint mobilization in the lower limb, and even at the spinal level of innervation, can positively influence motor performance at the hip. This exercise approach takes this concept further by combining spinal and lower quarter exercises together, when necessary.

The pulley ropes provides an external cue for muscle facilitation and motor learning. External cues from a resistance outside the body is a more efficient and effective learning style than internal cueing, which emphasizes the patients ability to concentrate on specific muscles contracting during an exercise or activity. The external cue allows the patient to focus on a functional motor performance while the desired muscles are naturally recruited by the direction of the force moment from the pulley. Improving central stabilization for distal performance is a basic technique from the original PNF concept. The concept of "core" training has been popularized in the fitness arena, and is not always appropriately applied to the rehabilitation arena. The concepts themselves can be applied to many different exercises for the lower and upper extremities. The following exercises are just a few examples of applying these concepts with basic lower quarter exercises.

Figure 3.61a,b: Multidirectional squat training for combined lumbar and hip stabilization training. A lateral force moment from a wall pulley is held anterior away from the body. The force moment from the pulley naturally recruits the fiber direction of the transverse abdominus, the rotational vector for the lumbar multifidi and the deep rotators of the hips. The pulley is stabilized anteriorly while the squat motion is being performed. Further progression may involve rotational motions of the trunk during a normal squatting motion.

Figure 3.62a,b: Multidirectional squat training with left arm resistance from a right lateral force moment (posterior trunk muscle emphasis). The pulley recruits a diagonal line of muscle from the left arm, crossing the lumbodorsal fascia, into the right hip. Posterior musculature is emphasized, including the left dorsal scapular muscles, lumbar multifidi and the deep hip external rotators.

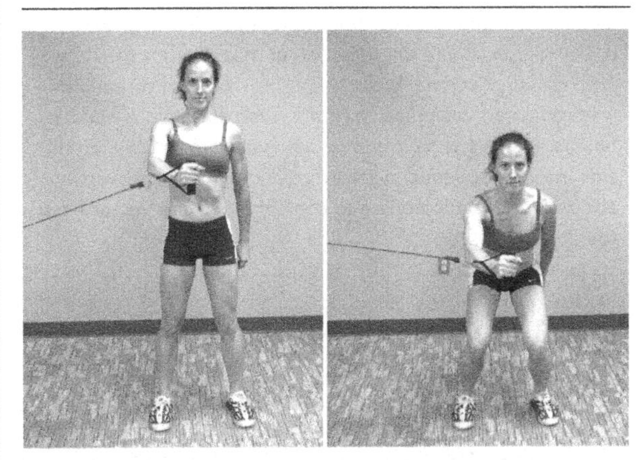

Figure 3.63a,b: Multidirectional squat training with right arm resistance from a right lateral force moment (anterior trunk muscle emphasis). The pulley recruits a line diagonal line of muscle from the right arm, crossing anteriorly to the left hip. Anterior musculature is emphasized, including the transverse abdominus, the obliquus' and left hip flexors.

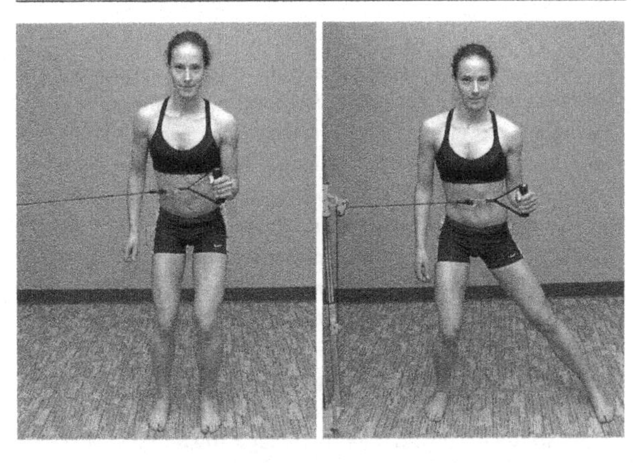

Figure 3.64a,b: Multidirectional right lateral lunge training with left arm resistance from a right lateral force moment (posterior trunk muscle emphasis). The pulley recruits a line diagonal line of muscle from the left arm, crossing the lumbodorsal fascia, into the right hip. Posterior musculature

is emphasized, including the lumbar multifidi and the deep hip rotators. The primary resistance for the right hip is to push off for the lunge, but further stabilization is required to control the rotary vector caused by the pulley.

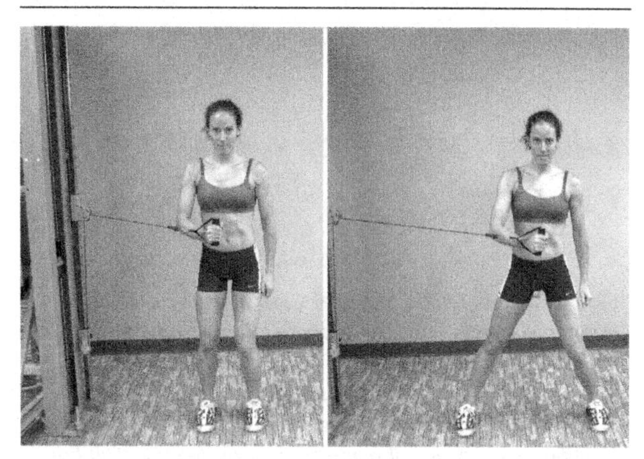

Figure 3.65a,b: Multidirectional left lateral lunge training with right arm resistance from a right lateral force moment held in the right hand (anterior trunk muscle emphasis). The pulley recruits a line diagonal line of muscle from the right arm, crossing anteriorly to the left hip. Anterior musculature is emphasized, including the transverse abdominus and hip flexors. Greater recruitment is achieved through the left side internal obliques, right external obliques, transverse abdominus and hip flexors to control the rotary vector from the pulley.

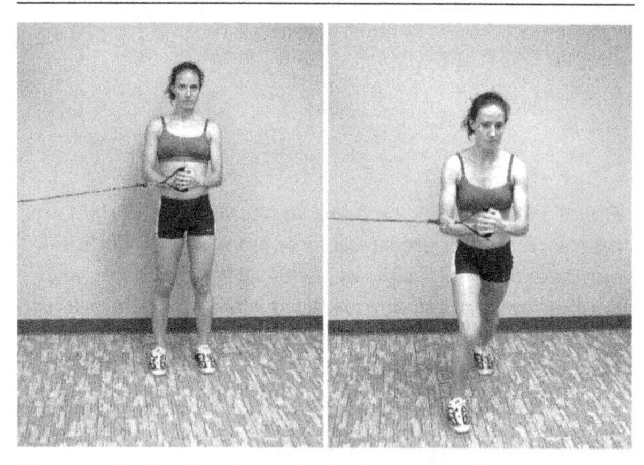

Figure 3.66a,b: Multidirectional right anterior lunge training with a secondary right lateral force moment. A lateral force moment from a wall pulley is held anterior away from the body during a anterior fallout lunge. The force moment from the pulley is stabilized with both hands, requiring motor recruitment of both anterior and posterior muscles of the trunk and pelvis. The trunk is pulled into right rotation, which is resisted by the deep rotators of the right hip and the right lumbar multifidi. More indirect challenges can be added to central muscles of stability by closing the eyes, requiring improved processing of somatosensory input. Lunging onto a labile surface, such as a wobble board or foam, will also increase the balance challenge through the entire lower extremity. Head motions, as shown previously, can further challenge cervical contributions to posture and balance. Head motions may provide a more functional athletic challenge than a labile surface.

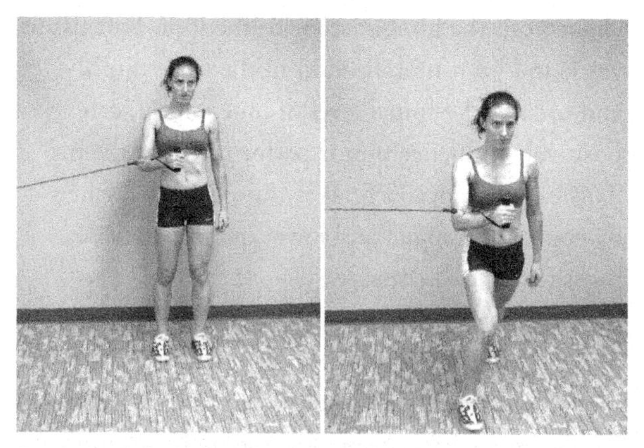

Figure 3.67a,b: Multidirectional anterior lunge training with right arm resistance from a right lateral force moment. Body weight against gravity provides the extension moment for the exercise. The pulley recruits transverse line of muscle fibers from the right arm, crossing anteriorly. The abdominals, including the transverse abdominus, are naturally recruited to resist the force moment. The deep external rotators of the right hip are also recruited to resists the right rotational trunk moment created by the pulley.

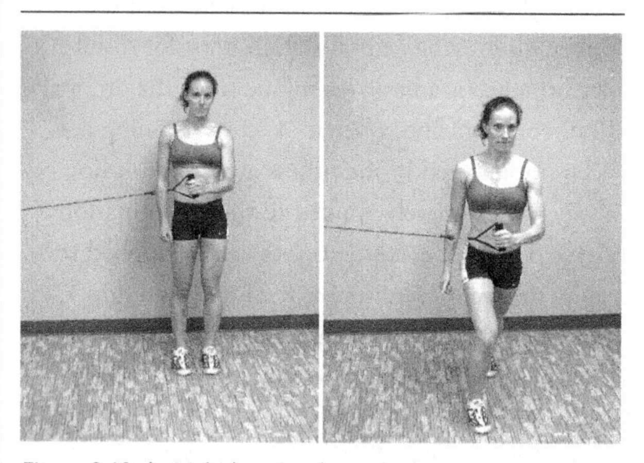

Figure 3.68a,b: Multidirectional anterior lunge training with left arm resistance from a right lateral force moment. The pulley recruits transverse line of muscle fibers from the right arm, crossing posteriorly. The deep rotators of the right hip are still recruited to resist the right rotational trunk moment, but the force line is transmitted posteriorly through the back muscles to the right hip.

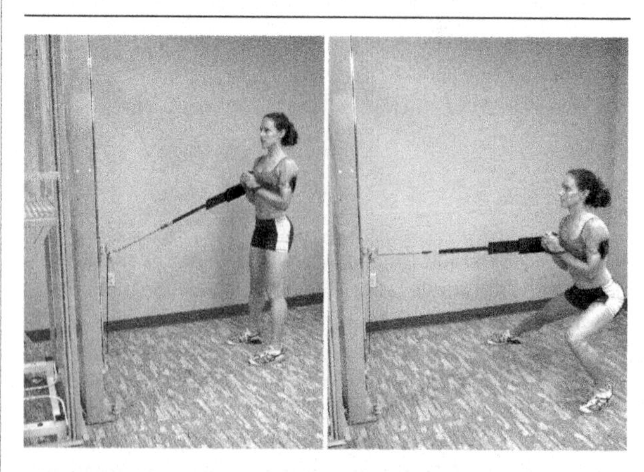

Figure 3.69a,b: Multidirectional left lateral lunge training with an anterior force moment. The lateral lunge emphasizes the

hip and lower extremity, with an extension moment against gravity. The anterior pulley line further increases the extensor moment from the trunk, but the line is also attached to the right shoulder and wrapped around the body. Attachment to the right shoulder causes a relative left rotational trunk moment. The lumbar multifidi on the left are recruited, along with the deep external rotators of the left hip, to resist this force moment.

The above examples are only a few applications with the emphasis on a rotary line of recruitment. The force moment can be changed in any vector to emphasize stabilization in any plane, or combined planes, of motion. The concept must be understood, beyond the application of a few of these simple exercises, to allow the clinician to create unique exercises to match the motor deficits of the patient.

Step Training

Step height is best kept at functional step levels. Starting on a 2-inch box with full body weight may not be as effective as using a normal step height and providing assistance or unloading to coordinate ascending or descending stairs. Assistance is reduced until the patient can perform normally with full body weight. Emphasis remains on training the functional quality of coordination so the resistance must be light enough to allow for the proper motor pattern while performing a high number of repetitions. Coordination training should focus on resistance being less than 50% of 1RM with repetitions from 20–30. Excessive fatigue may still occur with early weight bearing training as excessive muscle recruitment and compensation occurs in trying to learn the motor pattern. Adequate rest breaks, or performing active rests training a different muscle group, can ensure a more successful training session.

The step up motion is primarily a concentric activity that is assisted with slow performance. With the step up the two joint muscles of the quadriceps work concentrically at the knee but eccentrically at the hip, while the hamstrings work concentrically to extend the hip and eccentrically to control flexion at the knee. Slower speeds with the predominant concentric work during the step up will increase the peak torque. A slower speed will reduce the level of coordination required.

A correct step should involve the foot being placed fully on the step, weight transferred to the foot and the hip with the knee extending to lift the body. Typical compensations include excessive activity of the back ankle plantar flexing to vault the body upward. Performing a slow weight transfer to the front foot, and stepping up reducing the mechanical advantage of the plantar flexors of the back foot, can reduce plantar flexion compensation. An alternative is to contract the antagonists to the plantar flexors with dorsiflexion of the digits of the stance foot. This method is effective but is also an abnormal performance and may slow the overall motor learning process.

Figure 3.70: Anterior step-up with unloading from a pull down machine and gantry. Coordination and endurance training can occur with normal functional patterns in an unloaded environment. Performing step-ups slowly will increase concentric peak torque.

Step down training can also be performed earlier if the body weight can be reduced. As in early step up training the functional quality emphasis is coordination for stepping down from a normal step height. The step down is named for the leg performing the work. A right anterior step down involves standing on the right leg while the left leg is lowered down to the floor. The uninvolved hip performs stepping back onto the step.

Figure 3.71: Anterior step down with unloading from a pull down machine and gantry to allow for early coordination training on a standard stair height.

The step down is an eccentric activity that is aided by increasing the speed of the task as peak torque with eccentric performance is increased with speed. Rather than using an unloading device it may be possible to simply increase the speed of the activity to be able to coordinate a step down motion without pain. It may be possible to perform the motion as a controlled fall in which the back leg shifts body weight off of the step and the forward leg rapidly catches the body weight, requiring minimal eccentric muscle activity from the back leg. As coordination improves the speed of the performance is slowed down to increase the eccentric work of the back leg with reduced work of the landing leg.

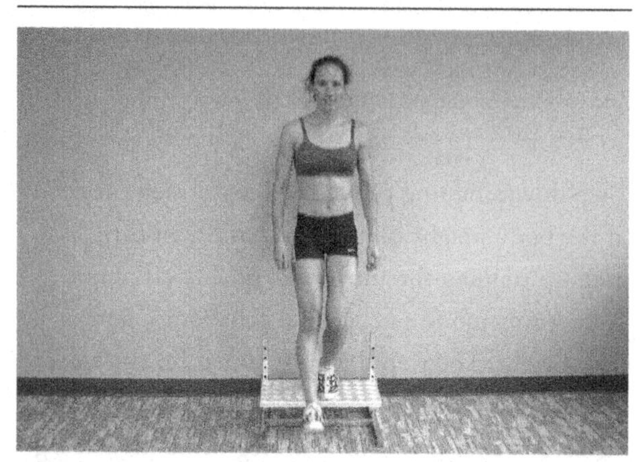

Figure 3.72: Anterior step-down with high speed performed as a controlled fall onto the front foot. Increased speed will increase the eccentric peak torque of the extensor muscles of the back leg.

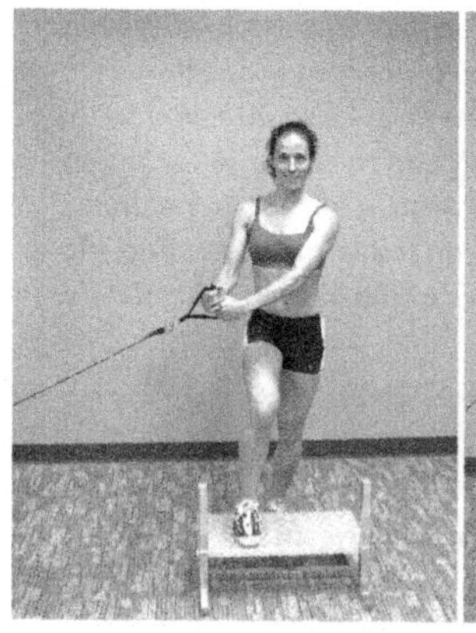

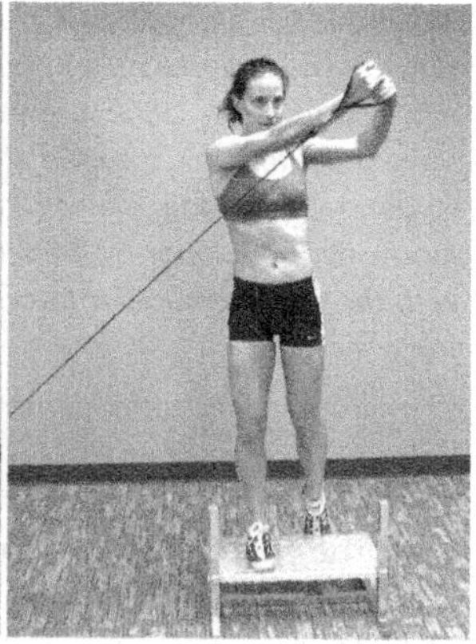

Figure 3.73a,b: Multidirectional right step-up with a right inferiolateral force moment. An extension moment for the hip and trunk is created with gravity during the step-up motion. The right lateral force increases the extension moment from the inferior vector, but also a right trunk rotational moment. The arms swing through to the left, creating an left rotational trunk pattern, increasing activity of the right lumbar multifidi, right hip deep external rotators, the oblique abdominals and the transverse abdominus.

Balance Training

Initial balance training was performed with bilateral stance emphasizing central challenges of moving the cervical spine and closing the eyes. The next progression is to unilateral stance balance training. As noted earlier, restoring normal muscle strength is a component of normalizing reflexes for balance (Carter et al. 2002, Messier 2002), but strength training alone has been shown not to enhance standing balance (Schlicht et al. 2001, Topp et al. 1993). The speed at which torque can be produced, as opposed to peak torque, is also a factor in motor control for balance (Robinovitch et al. 2002). If hip weakness contributes to balance deficits, the pelvis may drop into a Trendelenburg position of adduction, internal rotation and flexion. The patient should be instructed to maintain a level pelvis during the unilateral balance exercises. Balance training is dosed with similar concepts of dosing coordination. Many repetitions are required on a daily basis to improve neurological adaptation, requiring shorter training sessions of three to five minutes multiple times daily. This will be more effective than attempting long training bouts in which muscle fatigue is the dominating factor in poor performance. Balance exercises may be alternated between other exercises using different muscle groups to avoid excessive fatigue limiting performance. Balance should also be emphasized during daily activity such as encouraging unilateral stance for brushing the teeth or working at a table.

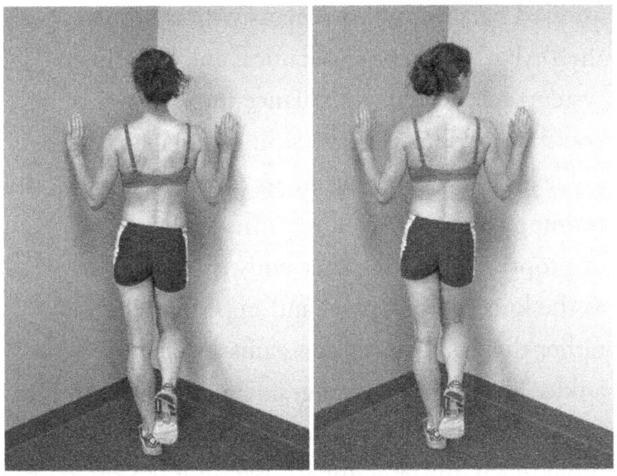

Figure 3.74a,b: Unilateral stance can be initiated with bilateral upper limb support as needed. Using a wall corner or door jam provides support on both sides for safety during training. Progression of cervical motions or closing the eyes can be added as performance improves.

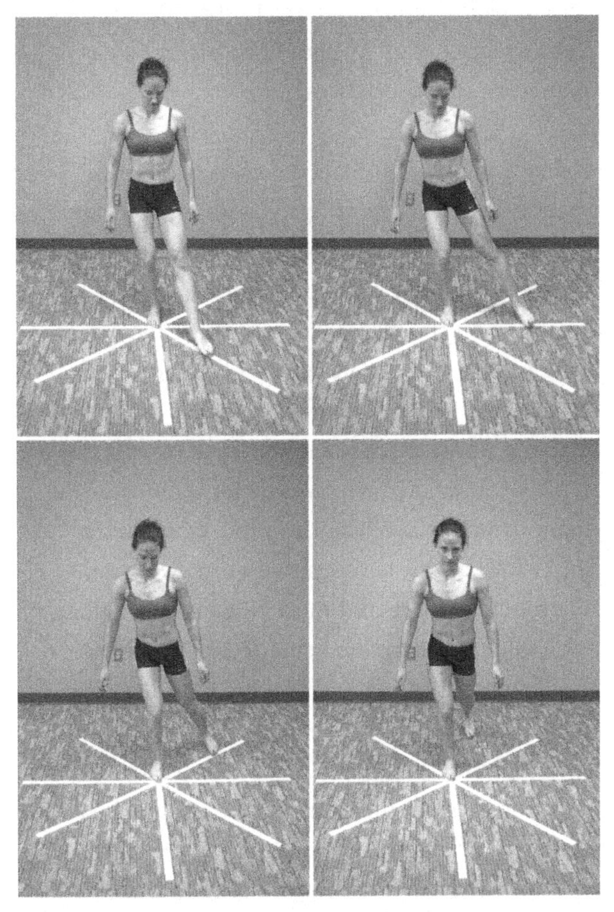

Figure 3.75a,b,c,d: Star Excursion Balance Training. Star pattern for lower limb balance reach involves standing one foot with the opposite lower limb reaches in different directions. The eight lines positioned on the grid are labeled according to the direction of excursion relative to the stance leg: anterolateral (AL), anterior (A), anteromedial (AM), medial (M), posteromedial (PM), posterior (P), posterolateral (PL), and lateral (L).

Primary emphasis with balance reach training is to maintain a straight line between the axes of motion for the hip, knee, ankle and foot. Trunk and hip weakness often leads to a valgus moment at the knee, as well as foot pronation. Early training should involve only the range in which normal alignment can be maintained.

Initial reach directions can take a tissue training or directional approach. This involves reaching in directions that move the joints away from the direction of pain or pathology. Later progression would involve reaching in vectors that increase the strain to the involved tissues.

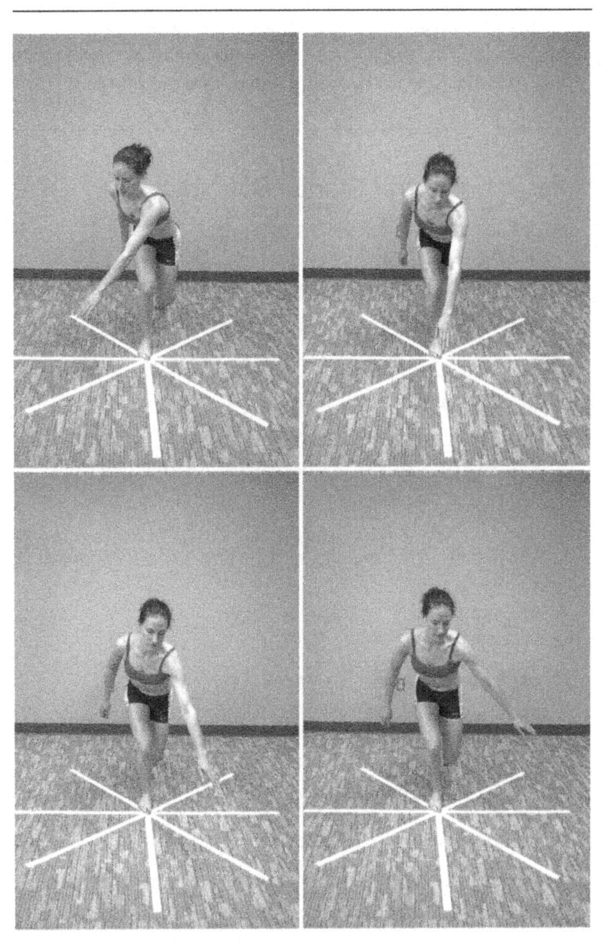

Figure 3.76a,b,c,d: Upper quarter reach balance training. Reaches can be performed in any degree within the star pattern. Alternatives including the addition of an inferior or superior vector to the reach. A dumbbell weight can be added in the hand to increase the extension vector against

gravity. A lateral pulley line can be added to the non reaching hand to increase the rotational vector to the trunk. Closing the eyes or standing on a labile surface will further challenge central balance.

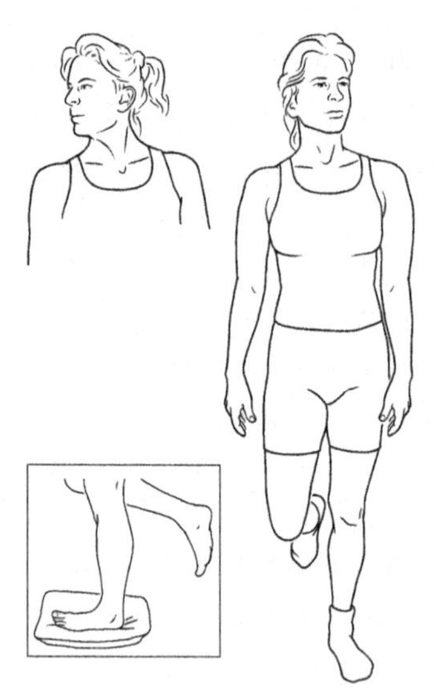

Figure 3.77: Balance training with single leg stance and cervical motions. Challenges to the vestibular system can be made with cervical AROM. Even greater challenges can be obtained by closing the eyes (removing vision) or standing on foam (removing somatosensory input from the feet).

Basic Unilateral Stance Balance Assessment / Exercises

- *Unilateral stance on a stable surface with eyes open*
- *Unilateral stance on a stable surface, eyes open with cervical motions*
- *Unilateral stance on a stable surface, eyes closed with cervical motions*
- *As above with labile surfaces as appropriate*

Balance with Labile Surfaces

The use of labile surfaces such as wobble boards and foam provide a greater challenge to balance training. Not all patients may require this progression as their functional requirements may not involve uneven or mobile surfaces. Emphasis may remain on flat surfaces consistent with functional demands, such as challenges related to narrowing the base of support, closing the eyes, upper limb and lower limb reaching. Retraining on balance boards may have a functional aspect for some patients, however it is more common functionally to have the foot on a fixed surface while the body moves over it. The functional needs of the patient should be assessed and the identified deficits should be addressed in the rehabilitation program. Chong et al. (2001) assessed balance improvements with training on the BAPS board (biomechanical ankle platform system) to determine if balance improvements could be attributed to retraining deficits in the ankle and foot. Improvements observed during training were attributed to diffuse enhancement of proprioception in other body segments such as the knees, hips, spine, and upper extremities rather than proprioception gains specific to the ankle. These diffuse changes may be helpful in retraining lower limb function in hip rehabilitation programs with less emphasis placed upon the foot and ankle mechanics. Many athletic activities do require balance and functional stability on moveable surfaces. The progression of balance training should attempt to mimic these functional demands to retrain global motor patterns for balance responses. Labile surface training should be prescribed as it matches a functional demand.

Balance training on foam is another common practice within the orthopaedic setting, but not always for the correct reasons. Foam reduces much of the normal afferent input from cutaneous receptors on the plantar surface of the foot as well as joint mechanoreceptors of the foot and ankle. This may be an effective tool to increase the demand of more proximal joint systems but is more typically associated with vestibular training as the foam dampens much of the lower limb afferent input for balance and postural reflexes. Again, the functional demand must be considered along with which systems are being reduced and which are being challenged. Closing the eyes increases the demand on the vestibular system and receptor systems in the periphery. Foam reduces lower limb receptor input to increase demand on proximal system, particularly the vestibular system. If the desired outcome is to improve somatosensory processing from the lower limb, then a labile surface may be a more efficient surface on which to train.

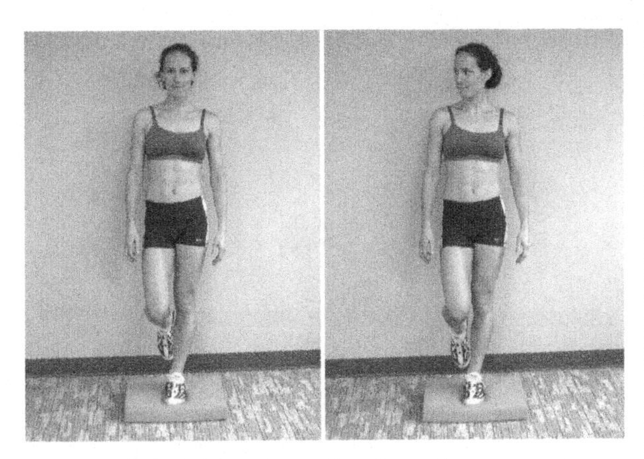

Figure 3.78a,b: Balance foam will reduce afferent receptor input from the cutaneous receptors on the plantar surface of the foot as well as joint mechanoreceptors of the foot and ankle. This may emphasize the vestibular system more and be less functional than a regular wobble board or balance reach training on a level surface.

Balance Reach for Deficits in Weight Shifting

Initial balance exercise addressed the ability to maintain the center of gravity with bilateral and unilateral stance while moving the head and closing the eyes. The conceptual progression is to now move body segments away from the center of gravity, while maintaining balance, and being able to return under control to the starting position. Balance reach exercises first involve reaching with the upper limbs, unilateral and bilateral, in all three planes of motion.

The initial direction of balance reach training may be chosen for different reasons. Testing may simply reveal a direction or plane of dysfunction with both upper and lower limb reaches that can be turned into a training exercise. Another basic option for upper quarter reaches is to affect weight bearing. Reaching laterally away from the involved side will increase body mass lateral to the hip axis for abduction, reducing the relative resistance to the muscles of hip abduction.

Balance Reach to Unload Tissue

With specific tissue injury, directions may initially move away from the primary tissue to reduce the level of stress or strain. For example, a left anterior hip impingement is most symptomatic with internal rotation of the hip. To stand on the left limb performing a lateral reach across the left hip will cause an internal rotation moment that may cause further tissue irritation. Performing a medial reach, however, will create an external rotation moment at the hip to reduce the potential for tissue irritation. Gradual progression in balance reach training would involve anterior reaches and then lateral reaches. Stage 3 training concepts will emphasize this eccentric control toward the direction of tissue lesion with the ability to concentrically return.

Choosing Balance Reach Directions

- *Perform in direction of deficits identified in testing (Stage 2-4)*
- *To decrease weight bearing toward the involved side (Stage 2)*
- *To increase weight bearing toward the involved side (Stage 2 or 3)*
- *To initially reduce stress on injured tissue (Stage 2)*
- *Increase stress and strain to involved tissue under control (Stage 3)*
- *To selectively facilitate specific muscles (Stage 2 or 3)*
- *Reproduce a functional activity (Stage 3 or 4)*

Balance Reach to Recruit Specific Muscles

Direction of upper and lower limb reaching not only challenges general balance with functional motions but may be directed toward facilitating specific muscles within the pattern of motion. An anterior lower limb reach will increase the force moment for the hip extensors. Reaching laterally will add an increased moment for more lateral hip fibers for abduction and external rotation to control the adduction and internal rotation moment either for isometric stability or eccentric control into the pattern. Reaching in a direction opposite to the axis of motion in the plane the muscle functions can facilitate specific fiber direction. Reaching in a different plane of motion challenges the 3-diminsional functions of the muscles, as in the above example facilitating the rotational moment of the gluteus medius in the transverse plane.

All balance reach activities should be initially performed slowly and under control. The limb should be extended out, held for one second in the end position, and then returned to the start position without losing control. Reaching away from the body requires predominantly eccentric motor control, which is aided with faster speeds. Patients will attempt to perform the motion quickly and return to the more stable starting position as quickly as possible. Forcing the patient to perform slowly will be safer, emphasize early coordination and prohibit the assistance of speed on eccentric peak torque. Performing at slower than normal speeds will actually make the activity more difficult.

Advanced Unilateral Stance Balance Assessment / Exercises

- *Unilateral stance on a stable surface with eyes open, upper limb reaching in planes to assist a motor pattern or unload pathological tissue.*
- *As above with eyes closed.*
- *Unilateral stance on a stable surface with eyes open, lower limb reaching in planes to assist a motor pattern or unload pathological tissue.*
- *As above with eyes closed.*

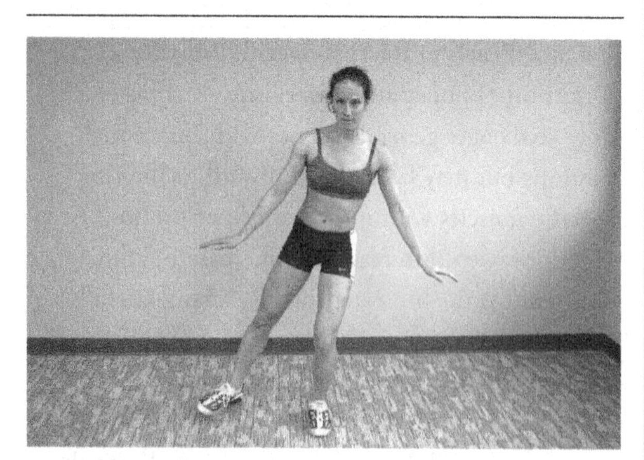

Figure 3.79: Left lower limb medial reach. Balance on the left lower limb with the right limb reaching medially relative to the working stance leg. Lateral reach will unload the anterior hip structures by creating an abduction/external rotation moment at the hip. This may be the first direction of choice for an anterior hip impingement patient, where the reach direction is chosen based on tissue pathology. As coordination and tissue tolerance are improved, an anterolateral reach with the right leg will maximally challenge the anterior hip joint structures.

Figure 3.80: Left lower limb b1alance with upper limb lateral reach. A lateral reach will increase the weight lateral to the sagittal axis for hip abduction/adduction, reducing the load resisted by the hip abductors. Reaching with both arms across will cause a rotational moment from above that will increase the fiber recruitment of gluteal fibers oriented more transversely for rotation.

Figure 3.81: Multidirectional training of upper quarter reaches in a diagonal pattern with a right lateral force moment from a pulley, held in the right hand. The pulley creates a right rotational trunk moment. The right lumbar multfidi, deep hip external rotators and the transverse abdominus are recruited through the upper extremity to resist the force moment. Increasing central recruitment for stabilization will improve the control of the weight bearing leg for balance reach training, whether for upper quarter or lower quarter reach training.

Gait Training

As tissue tolerance improves to weight bearing activity, walking is one of the most basic functions to restore. Hip pathology will tend to cause a flexion moment at the hip with the trunk forward bent and the head and eyes oriented toward the floor. Visual input will compensate for reduced afferent input from the hip and lower limb. Gait training may involve a simple progression of keeping the eyes forward and the trunk more erect. Other compensations usually involve reduced stride length from reduced balance or limited hip extension. Challenges may include backward walking with arm support on a wall to encourage hip extension motion and to further challenge balance. Straight lateral walking or with a cross over step can also be performed initially with the hands on a wall for support.

Section 3: Stage 3 Exercise Progression Concepts for the Hip Joint

Training Goals

Progression on the functional quality continuum continues in Stage 3. Concepts for providing the optimal stimulus for tissue repair/regeneration are still considered but are built into the exercises being performed, rather than treated as separate specific exercises. Full weight bearing training will provide more aggressive tissue stress, whether addressing weight bearing tissues or contractile tissues. Coordination is still a goal but more related to functional performance in more complicated diagonal patterns or tri-planar motions. Exercise intensity is progressed from coordination and local vascularity to focus more on improving endurance and strength. Emphasis on tissue repair and neurological influence are replaced with more direct influence on muscle tissue. Stimulation for hypertrophy and improved circulation will result in long-term improvements in motor performance.

Progression to Stage 3 concepts is indicated by achieving full pain free range of motion, decreased overall exercise time with increased speed of training with continued improvements in coordination and balance. Now that normal motor patterns are emerging and endurance levels are normalizing, greater loads are required to produce further neural adaptations, hypertrophy of muscles and greater tensile strength in collagen. The goals for Stage 3 exercise progression vary slightly depending on whether the hip is hypermobile or hypomobile. For hypomobility progressions, as is common with hip rehabilitation, the goal changes from increasing endurance and coordination to improving strength and stability in the gained ROM. Concentric and eccentric exercises are performed and dosed up to 75–85% 1RM for sets of 16–10 repetitions respectfully. Exercises are advanced from straight plane movements to more functional motions.

Tissue and Functional States for Stage 3

- *Full arthrokinematic and osteokinematic motion with cardinal planes*
- *Full weight bearing postures are tolerated*
- *Tissue pain only after excessive repetitions*
- *Edema has resolved*
- *Muscle guarding has resolved*
- *Palpation of primary tissues negative, provocation tests negative, trigger points may be positive to deep palpation*
- *Fair to Good coordination in planar motions*
- *Fair to Good balance/functional status*
- *May have reduced fast coordination, endurance, strength and power*

Hypermobilities progress similarly with the addition of isometric contractions at 70–85% 1RM added throughout the full range of motion, except into the pathological end range. Since the resistance is being increased for dynamic exercises during this stage, speed may need to be decreased initially to allow for development of coordination at the new load. If coordination can be maintained, then the increased speed can be maintained as well. The amount of rest allowed must also be taken into consideration when exercising at weights ranging from 70–85% of 1RM, as this will cause enough intramuscular pressure to shut off capillary flow, leading to fatigue and possible abnormal compensatory movement. This endurance/strength training level may require longer breaks for aerobic and anaerobic recovery.

Basic Outline: Stage 3 Hypomobility

- *Increase weight (60–80% of 1RM), decrease repetitions (strength)*
- *Work order: eccentric work first, followed by concentric, to stabilize new range*
- *Tri-planar motions in available range (diagonal patterns)*
- *Isometrics to fix strength in gained range*

Recommended Rest Breaks

- *Acute myositis/acute injury rest until normalization of respiration rate has occurred (aerobic)*
- *For local muscle endurance rest one minute (aerobic)*
- *For hypertrophy rest one to two minutes (glycolysis and ATP-PC)*
- *For strength and power rest two to five minutes between sets (primarily ATP-PC)*

Functional Testing

Functional testing in Stage 3 is progressed based on the acquisition of normal range of motion, resolution of pain through that range, improved balance, strength and endurance. Testing the ability to squat, lunge, hop and jump are natural progressions that are based on the end stage functional requirements of the individual patient. Testing should attempt to reflect functional or sport requirements, serving as a baseline for exercise design and prescription, as well as re-testing to demonstrate improvement. The reader is referred to the knee chapter for descriptions of these basic functional tests.

Six-Minutes Walk Test

The 6-minute walk test (6MWT) measures the distance a patient can quickly walk on a flat, hard surface in a period of six minutes. The test is not a specific test for hip or lower limb, but does evaluate the global and integrated responses of all the systems involved during exercise, including the pulmonary and cardiovascular systems, systemic circulation, peripheral circulation, blood, neuromuscular units and muscle metabolism (American Thoracic Society 2002). The test initially was designed for detection of exercise-induced asthma, cardiac stress testing and a test of cardiopulmonary fitness (Wasserman et al. 1999, Weisman and Zeballos 1994). The 6MWT was a shortened version of the 12MWT, designed to accommodate patients with respiratory disease for whom walking 12 minutes was too exhausting (Butland 1982). The 6MWT has been described as easier to administer, better tolerated, and more reflective of activities of daily living than the other walk tests (Solway et al. 2001). In some clinical situations, the 6MWT may be a better index of the patient's ability to perform daily activities than tests for peak oxygen uptake; for example, 6MWD correlates better with formal measures of quality of life (Guyatt et al. 1991). Improvement in the 6MWT after therapeutic intervention has been correlated with subjective improvement in dyspnea (Niederman et al. 1991, Noseda et al. 1989).

For assessing general exercise capacity in subjects with a greater level of disability, the time of the walking test can be reduced and still serve as an adequate baseline test. Leung et al. (2006) assessed a 2-minutes walk test (2MWT), comparing it to the 6MWT, to test exercise capacity in subjects with chronic obstructive pulmonary disease (COPD). The 2MWT proved to reliable, valid, practical and well tolerated for the assessment of exercise capacity and response to rehabilitation in patients with moderate-to-severe COPD.

Timed Stair Test

Perron et al. (2003) described Timed Stair Test (TST) for subjects with total hip arthroscopy, as a more advanced functional test then the TUG described previously. The TST consisted of a series of four subtasks: 1) standing up from a chair and walking 3 meters, 2) ascending a staircase, 3) turning around and descending stairs and 4) walking back to the chair, turning and sitting down. Their test began with subjects sitting three meters from the stair case of 13 steps, 16.5 cm high and 32 cm deep. The use of armrests, a walking aid or one of the two banisters if necessary was allowed, as well as stopping at any time to rest. Each subtask can be timed separately, as well as a total timed score. A loaded version of the test was also performed, in which the subject wore a 10kg vest. Testing for both the loaded and unloaded version correlated with previously validated 10 meter walking test.

Weight Bearing Progression

Progressing to Stage 3 is partly defined by achieving full weight bearing status. Weight bearing exercises should now include more complex tri-planar movement patterns that are consistent with retraining a specific motor deficit or directed toward a specific functional goal. Coordination is emphasized first on these more complex patterns with dosage progressing up the functional quality continuum to include endurance and strength (75–85% of 1RM). Each exercise may be dosed differently to emphasize the specific functional quality being trained.

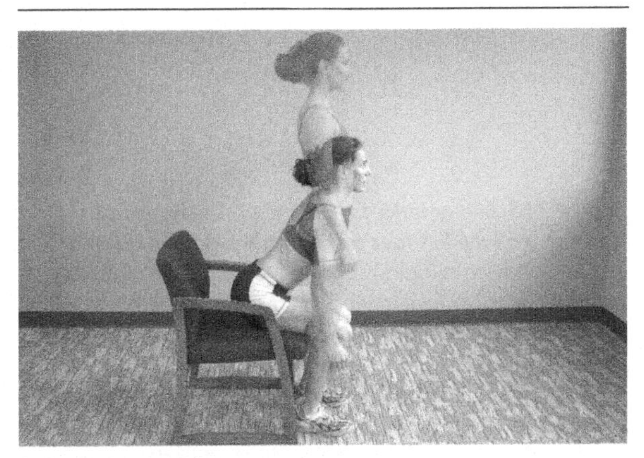

Figure 3.82: The chair squat emphasizes hip extensors, over quadriceps, with increased hip flexion and reduced anterior movement of the knee. The chair also provides an external cue of the coming in contact with the chair providing feedback for correct technique.

Choosing exercise directions or techniques, such as in prescribing specific squat or lunge activities, specific muscle strengthening and functional patterns should be built into the exercise. For example, EMG studies on older adults show that chair squats place greater demand on hip extensors compared to normal squats (Flanagan et al. 2003). The chair squat involves lowering the pelvis backward, as if sitting on a chair, rather than having the pelvis drop vertically with the knees moving forward, which would emphasize the quadriceps. The chair provides an external cue to perform this squat technique correctly.

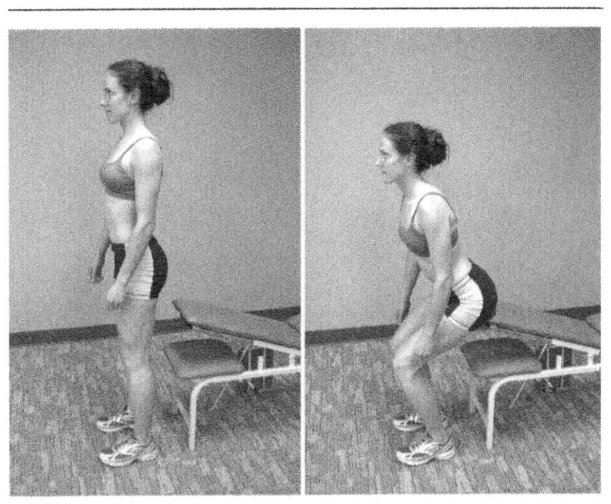

Figure 3.83a,b: Squats with posterior external cue to emphasize hip flexion with posterior motion of the pelvis rather than anterior movement of the knee. An adjustable bench can progress to a lower surface, increasing the range of motion of the squat.

Squats

Progression of exercise does not always involve adding weight. Performing at the same level of dosage but in different patterns can increase the challenge to coordination and proprioception, as well as providing a more interesting exercise program. Super set training is a weight training term referring to performing different motions while challenging the same muscle group. Super set training for bodybuilding relates more to providing different types of stress to the same muscle group for strengthening. Here the emphasis is on providing different stress not only to the muscle group, but to further challenge coordination, tissue loading and functional performance. An example would be to perform several types of squats in a three set progression. The first set may be a normal technique with the feet shoulder width apart. The second set might be performed with the feet wide to emphasize a different fiber emphasis in the gluteal or with the feet close together to work balance reduce the base of support. The third set could further challenge balance reflexes by standing on a labile surface. All three sets challenge the hip and knee extensors but place emphasis on different muscle fiber directions and with different challenges to balance reflexes.

Figure 3.84a,b,c,d: Superset with squats with 4 sets performed with different balance challenges during endurance/strength training: a) narrow base squats, b) sumo/ plie' squats with the feet wide apart, c) squats while standing on two labile surfaces and d) squats on one labile surface.

Lunges

Directions of lunging in Stage 2 may have focused on directions that were pain free or unloaded joint tissues but are now necessary to train. Multiple force exercises were described for squat and lunges with a pulley or elastic force for knee varus to assist the hip abductors, which should now be removed. A valgus force at the knee may have been used to increase the facilitation of the deep rotators and more transverse fiber orientations of the gluteus medius. Caudal resistance at the knee should now be removed, as it does not represent functional performance. Performing a lateral reach with the upper limbs, as shown in Stage 2, is recommended. An anterior or medial reach will reduce natural recruitment of the deep hip rotators making the lunge motion harder to coordinate. The basic concept is shifting from techniques that assist the pattern to ones that increase difficultly.

Initial lunge training used a step-lunge technique in which foot is set on the ground with the weight still on the back foot and then shifted onto lunging foot. Progressing to a fall out technique may now be recommended to increase joint loading forces, increasing the balance challenge, to increase eccentric requirements for deceleration and stabilization or to simulate athletic performance.

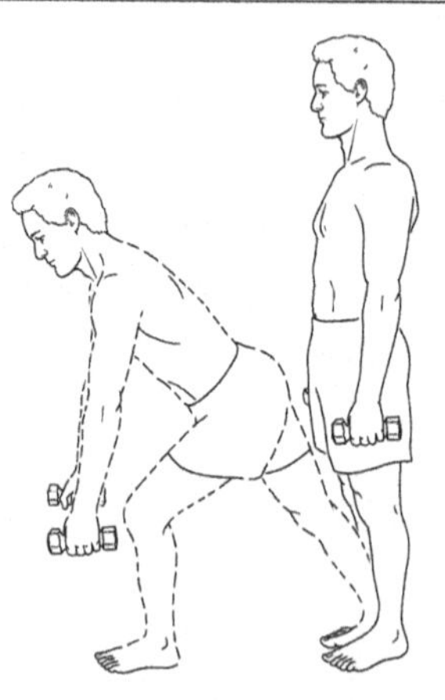

Figure 3.85: Left anterior fall out lunge. The body falls forward with the left foot contacting the floor in the end position, having to quickly decelerate and stabilize the body weight. The back heel remains on the ground, with the knee in full extension and the trunk remaining parallel to the back leg. The fall out lunge position emphasizes hip and back muscles on the back leg, while reducing the emphasis of the quadriceps on the forward leg. This same technique can be used in all directions to emphasize hip and back musculature.

Left anterior lunge with medial upper extremity reach will increase the resistance to the hip abductors by increasing the body weight and lever arm medial to the axis of the hip joint. Stage 2 involved reaching lateral to the hip to facilitate trunk rotation and to create an internal rotation moment at the pelvis. This requires increased recruitment of the deep hip internal rotators to assist with eccentric stabilization of abduction. The medial reach creates a neutral to external rotation moment at the pelvis, which does not recruit the deep hip external rotators leaving the hip abductor muscles alone to stabilize the lunge motion.

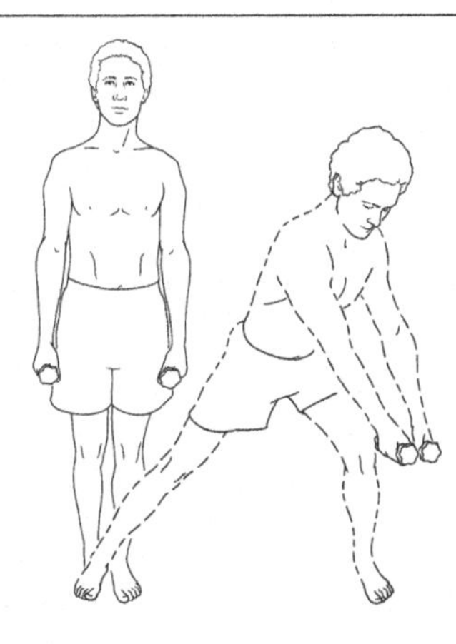

Figure 3.86: Left anterolateral lunge fall out lunge. The fall out lunge technique can be used in all directions of lunging to emphasize hip and back musculature. An upper extremity anterior reach with weights will increase the load on the hip and back extensors.

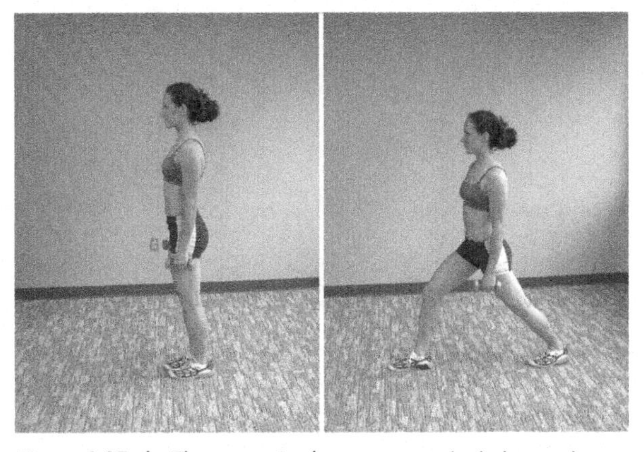

Figure 3.87a,b: The posterior lunge may assist in improving functional hip extension with facilitation of the hip extensors and reciprocal inhibition of the hip flexors. The mobilization force is minimal, versus eccentric training of the hip flexors into full hip extension, but the posterior lunge may be effective in unloading an anterior hip impingement.

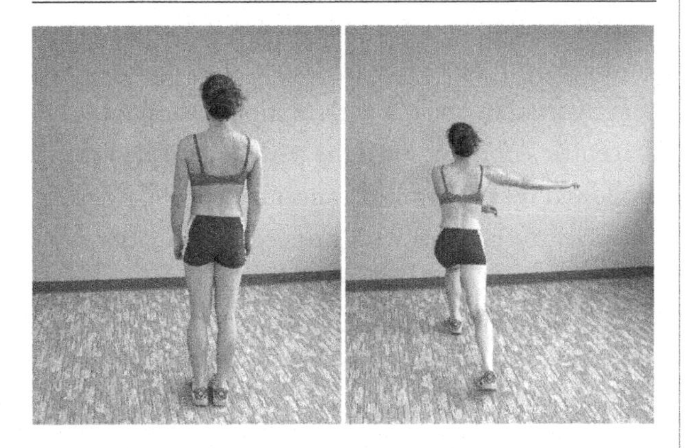

Figure 3.88a,b: Left anterior lunge with medial upper extremity reach. The medical reach will increase the load on left hip abductor muscles and increase the balance challenge.

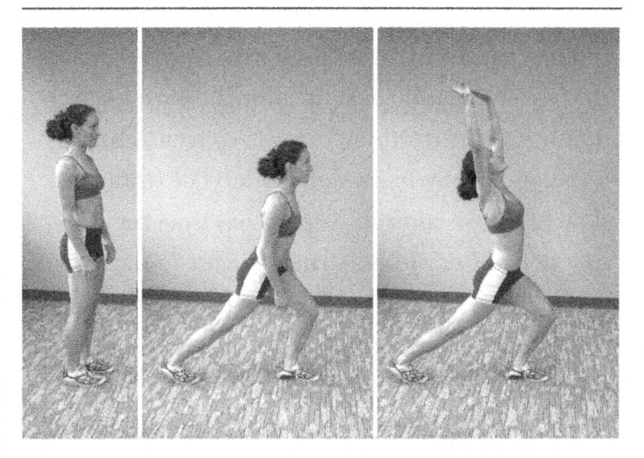

Figure 3.89a,b,c: The anterior lunge with overhead posterior reach will decrease facilitation of the hip extensors, while increasing eccentric activity of the hip flexors and lower abdominals. Emphasis is on full hip extension prior to the reach, to full lengthen the hip flexors. This is commonly used in lumbar patients that have limited hip extension with the overhead, posterior reach, limited to the pain free range.

Super set training for lunges involves several different types of lunges but working the same muscle groups. Performing two to four sets in different directions, with different upper extremity motions and with different techniques can make the training more interesting to the patient while establishing coordination in a greater variety of motions. Direct simulation of sport activity may be built into a collection of different lunge activities.

Step Training

Step training in Stage 3 should no longer require unloading devices to reduce body weight. Full body weight for step ups and step downs should be coordinated through a range of motion consistent with the normal height of steps on a staircase or curb. The basic functional requirement of ascending and descending stairs must first be achieved. If long term goals require higher level tasks or sport activities more challenging step training can be added. Increasing the step height may be necessary to simulate an activity or to work the lower extremity through a greater range of motion.

Increasing the overall work with step training is the first progression of step training. Speed was used to assist initial step training. Slow speed increased peak torque with concentric step up training and fast speeds increased peak torque with eccentric training. Reversing this concept will work against the normal neurological recruitment having fewer motor units performing the task concentrically and significantly increasing the level of work for slow eccentric control. If speed has been set at a functional level, adding hand weights to increase load to the 75–85% of 1RM range will train endurance and strength.

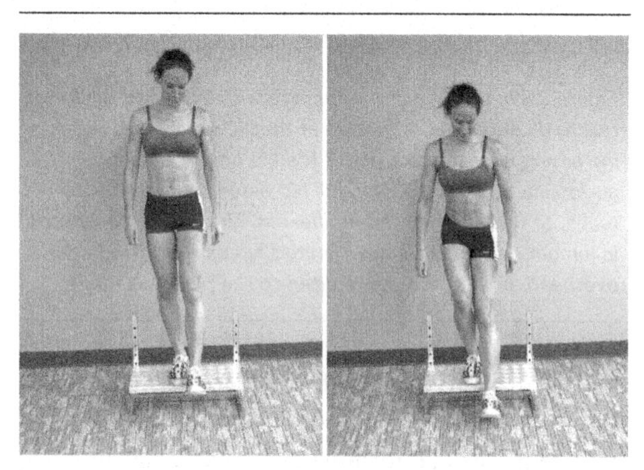

Figure 3.90a,b: Anterior step down with heel touch only requires a slow eccentric lowering of the body weight to touch the heel and immediately return concentrically.

Step training in different directions can target specific functional tasks or muscle groups. Emphasis may be on achieving full range of motion in combined tri-planar motion consistent with diagonal patterns. Combined directions, rotation of the lower extremity and arm movements can be added. More complex challenges to balance may involve cervical motions.

Figure 3.91a,b: Though all portions of the posterior hip muscles are active during step training, lateral step training will increase the force vector for the hip abductors, where anterior step ups emphasize hip extensors.

Figure 3.92a,b,c: Right lateral step up with rotation of the lower extremity will emphasize increased external rotation of the hip joint with subsequent motor control.

Figure 3.93a,b: Right anterior step up with medial arm reach will increase the lever arm of the load medial to the hip axis increasing the demand on the hip abductors.

Balance Training

Balance training should progress on a daily basis as improvements are reached. Conceptually the initial starting motions unloaded tissue or assisted muscles with reduce performance. The opposite is emphasized in Stage 3 with motions that require eccentric control toward the pathological tissues followed by a controlled concentric return. Weaker muscles, or patterns with reduced coordination, are now challenged with either or both upper limb and lower limb reaches. Emphasis should also be on matching functional requirements for activities of daily living, work or sports. Athletes may require a more aggressive progression at the end than a person wishing to only return to normal activities of daily living. The addition of labile surfaces, as described before, is utilized if functionally necessary.

Advanced Unilateral Stance Balance Assessment / Exercises

- *As above with eyes closed.*
- *Unilateral stance on a stable surface with eyes open, lower limb reaching in planes to challenge a weak motor pattern or increase load to pathological tissues.*
- *As above with eyes closed.*
- *Progress to foam and balance boards for advanced challenges, if functionally necessary or if vestibular deficits are identified.*

Gait Training

Progressing the challenge with normal walking is imperative prior to resuming more active jogging or running activities. Walking exercise for coordination and balance should no longer involve any assistive devices, upper quarter support or the eyes watching the floor for feedback. Exercises may involve many combinations of forward, backward, lateral and rotational motions. Training exercises may be specific repetitive motions in a known direction, or may involve reacting quickly to specific instruction or cues to alter direction.Lateral walking will emphasize lateral hip strengthening, and provide a balance challenge different than normal anterior walking. As with previous multidirectional training, and external force moment can increase the central core muscles for lumbopelvic and hip stabilization. A pulley force laterally attached at the ankle will emphasize muscles of the lower during the open chain performance of hip abduction. Attachment directly to the pelvis with emphasize the closed chain push off leg during lateral motion. Holding the pulley rope emphasizes the trunk muscles for more central stabilization. This may be necessary for the hip patient that has a lumbar component, or is simply weak centrally. As with the previous examples, the rope can be held bilaterally or unilaterally to emphasize different muscles groups within the trunk. Emphasis may be on the anterior or posterior muscles of the trunk, or a combination.

Figure 3.94a,b,c: Resisted left lateral walking with a pulley attached at the ankle. Emphasis is placed on the open chain abduction to the hip abductors on the left. The gluteus medius will be targeted with the abduction motion, but the tensor fascia lata will all have significant recruitment. This is not recommended with hip bursitis or "TFL syndrome."

Figure 3.95a,b,c: Resisted left lateral walking with a pulley force moment attached at the pelvis, proximal to the hip. Emphasis is placed on the closed chain abduction of the right hip, including the hip abductors and external rotators.

Figure 3.96a,b,c: Resisted left lateral crab walking with a pulley force moment attached at the ankle. Emphasis is still placed on the closed chain abduction of the right hip. The flexed position of the hip reduces the relative contribution of the tensor fascia lata, with increased emphasis on the gluteus medius as the primary abductor.

Figure 3.97a,b,c: Resisted left lateral walking with a pulley held in both hands. Emphasis is placed on training central stabilization of the trunk during lateral walking. Both hands will recruit both anterior and posterior musculature, with greater overall recruitment.

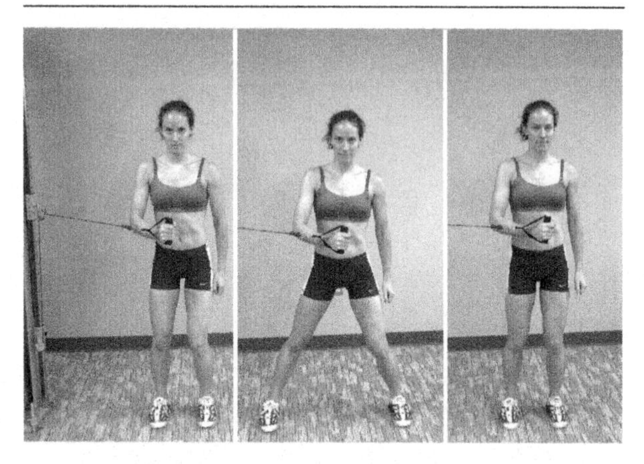

Figure 3.98a,b,c: Resisted left lateral walking with a pulley held in the right hand. Emphasis is on the anterior muscles, with a line of recruitment through the right arm, to the pectorals, the right external obliques, the left internal obliques, transverse abdominus and the left hip flexors. The right hip also has a significant increase in the hip rotators to resist the right rotation trunk moment, as well as abduction.

Figure 3.99a,b,c: Resisted left lateral walking with a pulley held in the left hand. Emphasis is placed on pushing off with the right hip and the posterior muscles of the trunk, including the right multifidi.

Section 4: Stage 4 Exercise Progression Concepts for the Hip Joint

Training Functional Goals

Stage 4 concepts apply to joints in which all acute objective findings have resolved with a moderate level of coordination through general activities of daily living. What is lacking in this stage may be more functional weight bearing tasks related to work or sport. The basic goal for Stage 4 is to restore full functional stability and coordinate tonic and phasic muscles throughout the physiological range of motion. Determining a long term, post physical therapy treatment, exercise program is necessary. Exercises in this stage should be, or should mimic, all functional requirements. It is not enough to simply perform specific therapeutic exercises with higher levels of resistance. The exercise should be specific to the patient's needs—"you get what you train" (Holten 1996)—therefore you must train the functional requirements.

Tissue Training

Training tissue repair is not forgotten throughout the progression of an exercise program as pain and edema are avoided for all exercises. A gradual increase in the loading impact of weight bearing exercises is necessary to increase tissue tolerance of weight bearing tissues. Higher levels of tension and speed will improve tolerance in tissues resisting tensile forces. Postmenopausal women lose 10–15 percent of total bone mass during the first five years after menopause (Heaney 1990). Using site specific "impact loading" exercises (walking, jogging, jumping, hopping, stair climbing) along with resistive exercises have been found effective in increasing bone mineral density in the proximal femur and femoral neck in postmenopausal women (Kelley 1998, Kohrt et al. 1997). This should be taken into consideration in the exercise prescription. The recommended dosage for impact loading exercises for bone density is two to three sets of six to eight repetitions performed for two to three weeks (Kahn 2004). Caution should be taken if the patient already has osteopenia or osteoporosis.

Strength Training

The dosage of exercises in Stage 4 should be prescribed according to the desired functional quality. Agility training and plyometrics might also be added to allow for return to a more athletic endeavor. If fast ballistic training is required it is added now after coordination has been established along with a level of strength. Proper arthrokinematic endurance, coordination and strength around a physiological axis are a prerequisite to safe aggressive weight bearing training. Higher resistance slow speed training in Stage 3 established coordination and strength. Fast ballistic movements require multi-joint phasic muscles with a reduced activation of single joint tonic muscles (Mackenzie et al. 1995, Thorstensson et al. 1985).

Directions and Motions for End Stage Functional Training

End stage exercise should have a functional focus related to activities of daily living, work and/ or sports. More complex patterns are necessary with increased resistance and speed. Simulation of normal activity is required to normalize functional coordination, tissue tolerance and improve endurance and strength.

Squat to Jump Training

Stage 4 squat training emphasizes strength, power and work/sport simulation. Added resistance to previous patterns of Stage 3 may be enough in the program design. Squatting exercises may involve lifting patterns in which the lumbar spine and lower quarter work together in straight lifts or diagonal lifting patterns. Free weight or pulleys can simulate any squat or lifting performance required for long term goals.

Progressing the squat motion to jumping may be needed for return to sport. Normal squats may continue with an emphasis on training strength at 85–95% of 1RM, but jump training would require the RM to drop down to the 30–40% of 1RM range to allow for the safe application of speed to the exercise (McBride et al. 2002, Wilson et al. 1993). Jump training may begin on a trampoline surface to reduce the loading forces on weight bearing structures and increase the time for muscular deceleration of the performance. As coordination and tissue tolerance dictates, jumping is performed on the floor. The sport may dictate if jumping should be performed with an emphasis on height or direction. Jump training is more thoroughly discussed in the knee chapter.

Example of Squat Progression from Stage 1 Through Stage 4

- *Unloaded bilateral squat*
- *Bilateral squat full body weight*
- *Squat with speed*
- *Jumping*
- *Jump with load*
- *Jump down and stop*
- *Plyometric jumping*

**Option: matrix as above in different directions*

Lunges

Stage 4 concepts involve full simulation of activities of daily living, work or sport. Resistance is increased with an emphasis on strengthening from 85–95% of 1RM. Previous lunges may be performed with added resistance to the trunk or held in the hands. Speed should be set at a functional level for the desired performance, or at a faster speed to challenge coordination and increase power. Step lunging up or down can further increase challenges in coordination, balance, range of motion, and strength. Any direction can be performed to duplicate functional tasks, increase stress to joint structures or emphasis specific muscle fibers.

Figure 3.100a,b: The Lateral Step-Up Lunge creates a greater range of motion than floor lunges further stressing the hip capsule and lengthening the hip adductors.

Prior to resuming running activities the lower extremity and spine must be trained for quick motor recruitment and stabilization of high speed impact. Anterior ballastic lunging, the hop and stop, as well as bounding, are progressions prior to resuming running. The hop and stop involves a fast lunge, or hop, with the complete transfer of weight to the lunging foot and freezing the motion for several seconds to train quick stabilization. Bounding would a more plyometric shift from eccentric deceleration to immediate concentric acceleration to hop back to the other foot. A much greater demand is placed on the hip muscles to decelerate the pelvis from adduction, internal rotation and flexion occurring excessively or too fast. The technique is not that of simply hopping from foot to foot but lunging out with a deceleration of a deeper level of flexion at the hip and knee.

Figure 3.101a,b: Left lateral lunge on labile surface. Using a labile surface can add a balance challenge to any weight bearing exercise.

Figure 3.102a,b,c: Right Lateral Hop and Stop. A lateral hop onto the right lower extremity emphasizes eccentric deceleration. The patient is instructed to establish unilateral balance stance on foot contact, followed by a one to two second hold prior to performing the next lateral hop. This can be performed in any direction, related to a specific tissue pathology, impairment or a functional requirement.

Example of Balance Progression from Stage 1 through Stage 4

- *Lower quarter balance reach*
- *Step-lunge*
- *Lunge*
- *Fall out lunge*
- *Lunge to labile surface*
- *Hop and stop*
- *Bounding*
- *Jogging*
- *Running*
- *Sprinting*

Option: matrix as above in different directions

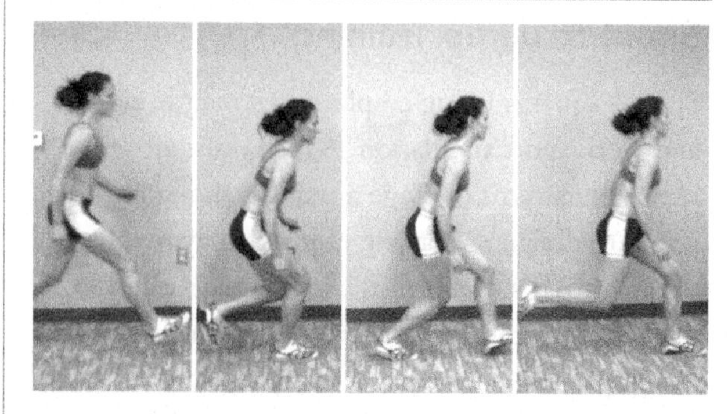

Figure 3.103a,b,c,d: Anterior bounding is more like repetitive jumping with emphasis on moving forward, rather than up or down. Stability on each heel strike is emphasized, as with the hop-and-stop exercise. Bounding begins slowly to ensure

proper coordination to avoid collapse at the hip, knee or ankle. As coordination and stability improves, both speed and distance of bounding is increased. Bounding can also be performed in multiple directions depending upon the functional needs of the patient.

Step Training

End stage step training continues with above identified patterns of dysfunction and/or functional requirements. Load is increased to 85–95 percent for pure strength training. Speed may also be increased to match speeds associated with athletic performance. Motions are performed with a ballistic eccentric deceleration and concentric acceleration. Step training can be considered a natural progression toward hopping and running activities. Higher impact with step training requires greater levels of motor recruitment in shorter periods of time. Tissue stress to weight bearing tissues reaches maximal levels with progression of ballistic step training, jumping and running.

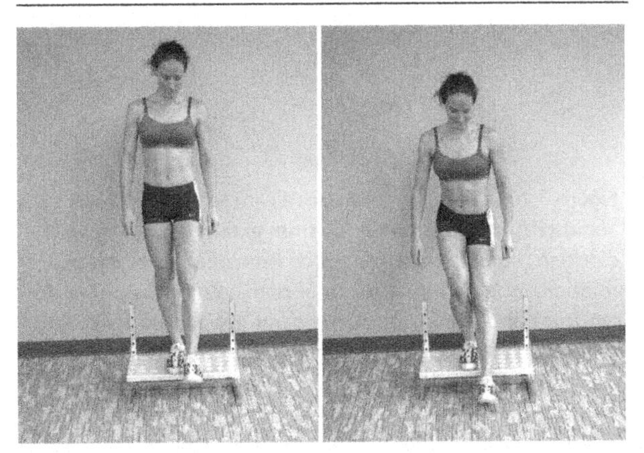

Figure 3.104a,b: The anterior step-down lunge increases the load for entire extensor pattern from the lumbar spine to the foot. Not only is functional stepping trained, but the goal may also pertain to increasing compressive tissue loading.

Upper quarter motions were added in Stage 3 to increase the level of difficulty. Performing specific athletic activities, such as catching and throwing a ball while performing step activities, can further increase the demand while focusing the mind off of the lower extremity portion of the exercise. Simply turning the neck in different directions and looking away from the primary activity will increase the level of difficulty by removing the afferent input from the eyes.

Figure 3.105a,b: Right anterior step with anterior upper extremity reach and cervical rotation left. The anterior reach will increase demand on the lumbar and hip extensor pattern. Turning the head to the left increases the balance demand by reducing visual input for the pattern. Left cervical rotation will also have a level of inhibition on the right hip external rotators by creating an external rotation stimulus to the trunk and pelvis. Where reaching lateral and looking lateral naturally increases deep hip rotator facilitation, looking away may serve to inhibit their participation, increasing demand on the hip extensor muscles.

Example Of Unilateral Step Progression From Stage 1 To 4

- *Unilateral balance*
- *Unloaded unilateral squat*
- *Unilateral balance labile*
- *Unilateral squat full body weight*
- *Step-ups*
- *Unilateral squat labile surface*
- *Unilateral hopping*
- *Hop and stops (stick)*
- *Step-up and hop*
- *Jump and stick*
- *Walking hills*
- *Hill Running*

Option: matrix as above in different directions

Figure 3.106a,b: Combinations of step and jumping can be added to a superset of step training. Stepping up with an immediate one legged hop, and controlled land on the step, requires concentric acceleration off the step with unilateral eccentric deceleration.

Balance and Functional Challenges

By Stage 4 basic functional balance has been achieved for basic activities of daily living and basic work requirements. High level sport activity may require an additional level of balance training that combines challenges to the vestibular system, ocular system and afferent receptor systems in the neck, trunk and lower limbs. Similar balance activities from previous stages can be performed with additional stress to one or more of the primary balance systems. Closing the eyes reduces the significant compensation of the visual system. Standing on foam significantly reduces the afferent input from the lower extremity increasing demand proximally in the hip, trunk, cervical spine, eyes and vestibular system. A labile surface increases the range of motion the foot and ankle must stabilize.

Upper limb and lower limb reach activities for balance can be progressed in terms of direction, speed or in combination with other movements. Upper limb reaches may have begun at shoulder height to emphasize only the transverse plane of stability. A gradual progression to hip height, knee height and eventually to the ground will increase the 3-dimensional eccentric stability and concentric return. Reaching to catch a ball or touch a target in an inconsistent pattern requires reactive activities rather than repetitive tasks in a known direction.

Returning to higher level sports may require more reaction training, as it relates to balance, coordination and strength. Balance challenges may continue to incorporate cervical motion and labile surfaces, but the added challenge of catching and throwing will provide a more reactionary activity. The patterns of catching throwing can still follow the basic stabilization pattern of concentric work away, isometric work, and eccentric motion toward.

Throwing Reaction Activities with Stabilization Emphasis for Right Hip Anterior Impingement

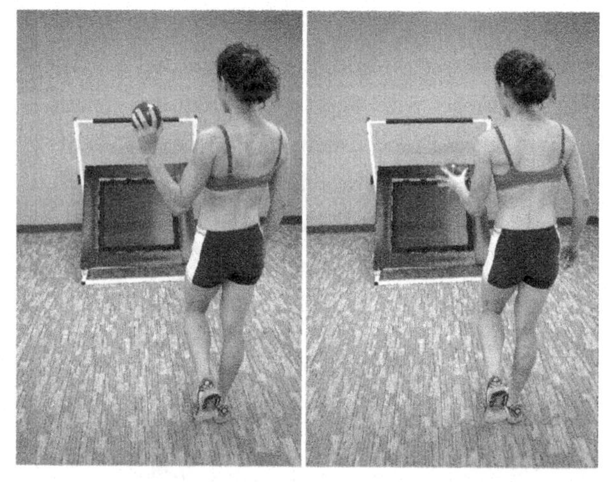

Figure 3.107a,b: Right leg balance with left lateral throw and catch. Stabilizing right hip from external rotation while catching requires eccentric action from muscles of internal rotation, including the oblique abdominals for the pelvis. As the trunk and pelvis rotate to the left, the hip joint is taken away from the tissue pathology of internal impingement. General balance can be improved with coordination for throwing and catching the ball off the rebounder in a safe environment that places little risk of tissue injury.

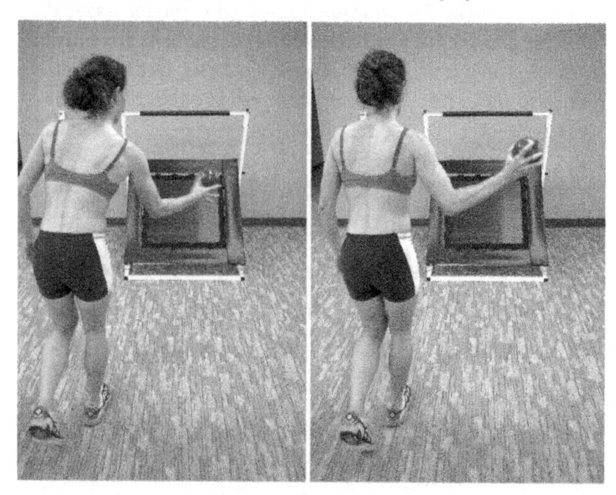

Figure 3.108a,b: Right leg balance with right arm throw and catch. After motion away from the pathology, the

next progression is isometric stabilization of the muscles that decelerate the body toward the pathology. Here the right hip external rotators, including rotators of the trunk, must prevent the pelvis moving toward the right, or relative internal rotation and flexion of the right hip. This combination of the flexion and internal rotation will irritate the anterior hip joint structures associated with anterior impingement. The exercise is initially performed slowly to ensure coordination and safe stabilization of the joint. The throw may be from elbow extension only, with emphasis on catching the ball with isometric stabilization of the hip and pelvis. The trunk and pelvis are prevented from any motion on impact of the ball in the hand. A light weight ball is used first to limit the moment that must be stabilized. Increasing the speed of throwing or the weight of the ball will increase the motor challenge for isometric stabilization.

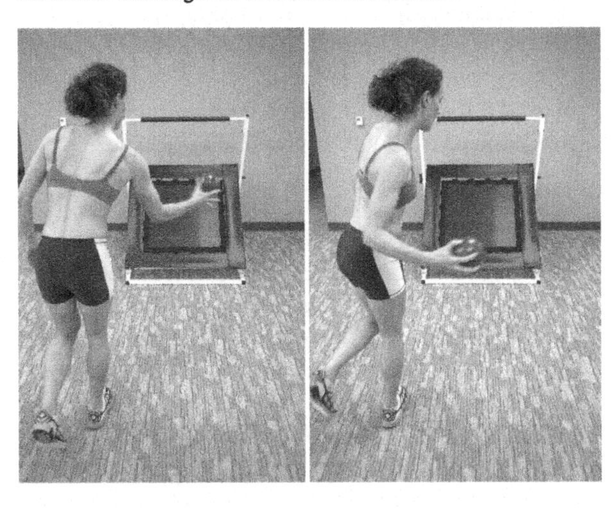

Figure 3.109a,b: Right leg balance with left lateral throw, catch and eccentric deceleration of the pelvis into flexion and internal rotation. The final stage is to work the hip extensors and external rotators, along with the trunk left rotators, to decelerate the pelvis toward the pathology. The ball is catch with a deceleration toward, but not into pain.

Gait Training to Running

Stage 4 involves the return to running and sport activity. Tissue tolerance for weight bearing activity has increased over the previous three stages to allow for aggressive weight bearing activities but running must not be initiated until adequate performance has been established with lunging, jumping, hopping and bounding activities that simulate the higher tissue impact loading of running. Significant cartilage injury may require more than six to nine months for adequate repair despite a reduction in pain and improvement in performance. Not only tissue tolerance but training coordination for running is necessary. Speed is progressed from walking, to fast walking, to jogging, to running and eventually sprinting.

How far along the progression a rehabilitation program must go depends upon the long-term functional goal, but regardless of the level of return treatment should include challenges of direction such as quick changes, cutting, and pivoting. Jogging short distances should be utilized first including backward and lateral motions. A series of changing speeds and directions is a more complete progression prior to resuming normal running activities for prolonged periods. Loading time is gradually increased with assessment of secondary joint pain after running activities. Increased pain within one to two hours of training may be associated with increased edema in the joint from excessive loading activities and should be avoided. Exercise should be dosed lower to avoid delayed onset tissue soreness (DOTS).

Summary

To achieve tissue healing and full dynamic stability of a joint there is not one protocol to perform. Depending on the condition of different tissues, a patient might be performing exercises in several of the stages at once. The functional quality being trained should serve as a guide for appropriate dosage and expected outcome in the therapeutic exercise program. With these guidelines the optimal stimulus for tissue healing and a return to functional levels can be achieved.

Bibliography

Andersson E, Oddsson L, Grundstrom H, Thorstensson A. The role of the psoas and iliacus muscles for stability and movement of the lumbar spine, pelvis and hip. Scand J Med Sci Sports 5(1):10–16, Feb, 1995.

ATS Statement. Guidelines for the six-minute walk test. Am J Respiratory Critical Care Medicine 166:111–117, 2002.

Augustsson J, Thomee R, Karlsson J. Ability of a new hop test to determine functional deficits after anterior cruciate ligament reconstruction. Knee Surg Sports Traumatol Arthrosc 12(5):350–6, Sep, 2004.

Avdic D, Skrbo A. Co-relation between risk factors of falls down and the Berg balance scale in elderly people (third age). Bosn J Basic Med Sci 3(1):49–55, Mar, 2003.

Bandy WB, Irion JM, Briggler M. The effect of static stretch and dynamic range of motion training on the flexibility of the hamstring muscles. J Orthop Sports Phys Ther 27(4):295–300, Apr, 1998.

Barbosa AR, Santarem JM, Filho WJ, Marucci Mde F. Effects of resistance training on the sit-and-reach test in elderly women. J Strength Cond Res 16(1):14–8, Feb, 2002.

Barker KL, Shamley DR, Jackson D. Changes in the cross-sectional area of the multifidus and psoas in patients with unilateral back pain: the relationship to pain and disability. Spine 15:29(22):E515–9, Nov 15, 2004.

Bernstein N. *The Co-ordination and Regulation of Movement.* London: Pergamon Press, 1967.

Bjerkeim I. Secondary dysplasia and osteoarthrosis of the hip in functional and in fixed obliquity of the pelvis. Acta Orthop Scand 45(6):873–882, 1974.

Bogle Thorbahn LD, Newton RA. Use of the Berg Balance Test to predict falls in elderly persons. Phys Ther 76(6):576–83, Jun, 1996.

Bohannon RW, Smith J, Hull D, Palmeri D, Barnhard R. Deficits in lower extremity muscle and gait performance among renal transplant candidates. Arch Phys Med Rehabil 76(6):547–551, Jun, 1995.

Bohannon RW. Alternatives for measuring knee extension strength of the elderly at home. Clin Rehabil 12(5):434–440, Oct, 1998.

Bohannon RW. Comfortable and maximum walking speed of adults aged 20–79 years: reference values and determinants. Age Ageing 26(1):15–19, Jan, 1997.

Bohannon RW. Reference values for the five-repetition sit-to-stand test: a descriptive meta-analysis of data from elders. Percept Mot Skills 103(1):215–22, Aug, 2006.

Brunt D, Greenberg B, Wankadia S, Trimble MA, Shechtman O. The effect of foot placement on sit to stand in healthy young subjects and patients with hemiplegia. Arch Phys Med Rehabil 83(7):924–929, Jul, 2002.

Buckley JP, Sim J, Eston RG, Hession R, Fox R. Reliability and validity of measures taken during the Chester step test to predict aerobic power and to prescribe aerobic exercise. Br J Sports Med 38(2):197–205, Apr, 2004.

Bullock-Saxton JE, Janda V, Bullock MI. The influence of ankle sprain injury on muscle activation during hip extension. Int J Sports Med 15(6):330–334, Aug, 1994.

Butland RJA, Pang J, Gross ER, Woodcock AA, Geddes DM. Two-, six-, and 12-minute walking tests in respiratory disease. BMJ 284(6329):1607–1608, May 29, 1982.

Campbell AJ, Borrie MJ, Spears GF. Risk factors for falls in a community-based prospective study of people 70 years and older. J Gerontol 44(4):M112–M117, Jul, 1989.

Carter ND, Kannus P, Khan KM. Exercise and the prevention of falls in older people: a systematic literature review examining the rationale and evidence. Sports Med 31(6):427–438, 2001.

Carter ND, Khan KM, Mallinson A, Janssen PA, Heinonen A, Petit MA, McKay HA. Knee extension strength is a significant determinant of static and dynamic balance as well as quality of life in older community-dwelling women with osteoporosis. Gerontology 48(6):360–8, Nov–Dec, 2002.

Cheng PT, Liaw MY, Wong MK, Tang FT, Lee MY, Lin PS. The sit-to-stand movement in stroke patients and its correlation with falling. Arch Phys Med Rehabil 79(9):1043–1046, Sep, 1998.

Chong RK, Ambrose A, Carzoli J, Hardison L, Jacobson B. Source of improvement in balance control after a training program for ankle proprioception. Percept Mot Skills 92(1):265–72, Feb, 2001.

Cibulka MT, Delitto A. A comparison of two different methods to treat hip pain in runners. J Orthop Sports Phys Ther 17(4):172–6, Apr, 1993.

Cliborne AV, Wainner RS, Rhon DI, Judd CD, Fee TT, Matekel RL, Whitman JM. Clinical hip tests and a functional squat test in patients with knee osteoarthritis: reliability, prevalence of positive test findings, and short-term response to hip mobilization. J Orthop Sports Phys Ther 34(11):676–85, Nov, 2004.

Cress ME, Schechtman KB, Mulrow CD, Fiatarone MA, Gerety MB, Buchner DM. Relationship between physical performance and self-perceived physical function. J Am Geriatr Soc 43(2):93–101, Feb, 1995.

Csuka M, McCarty DJ. Simple method for measurement of lower extremity muscle strength. Am J Med 78(1):77–81, Jan, 1985.

Cyriax JH. *Illustrated Manual of Orthopedic Medicine.* 2nd edition. London: Butterworth-Heinemann Medical, 1996.

Dangaria TR, Naesh O. Changes in cross-sectional area of psoas major in unilateral sciatica caused by disc herniations. Spine 23(8):928–31, Apr 15, 1998.

Doll C. Static stretching versus active range of motion of the triceps surae: effect on ankle dorsiflexion range of motion. DPT Dissertation: The Ola Grimsby Institute, 1999.

Drabsch T, Lovenfosse J, Fowler V, Adams R, Drabsch P. Effects of task-specific training on walking and sit-to-stand after total hip replacement. Aust J Physiother 44(3):193–198, 1998.

Duncan PW, Weiner DK, Chandler J, Studenski S. Functional reach: a new clinical measure of balance. J Gerontol 45(6):M192–7, Nov, 1990.

Elble RJ, Thomas SS, Higgins C, Colliver J. Stride-dependent changes in gait of older people. J Neurol 238:1–5, 1991.

Faugli HP. *Medical Exercise Therapy*. Laerergruppen for Medisnsk Treningsterapi AS, Norway, 1996.

Ferrandez AM, Pailhous J, Durup M. Slowness in elderly gait. Exp Aging Res 16(1–2):79–89, Spring–Summer, 1990.

Finley FR, Cody KA, Finizie RV. Locomotion patterns in elderly women. Arch Phys Med Rehabil 50(3):140–146, Mar, 1969.

Flanagan S, Salem GJ, Wang MY, Sanker SE, Greendale GA. Squatting exercises in older adults: kinematic and kinetic comparisons. Med Sci Sports Exerc 35(4):635–43, Apr, 2003.

Fredericson M, Cookingham CL, Chaudhari AM, Dowdell BC, Oestreicher N, Sahrmann SA. Hip abductor weakness in distance runners with iliotibial band syndrome. Clin J Sport Med 10(3):169–75, Jul, 2000.

Friberg O. Clinical symptoms and biomechanics of lumbar spine and hip joints in leg length inequality. Spine 8(6):643–651, Sep, 1983.

Frischknecht R. Effect of training on muscle strength and motor function in the elderly. Reprod Nutr Dev 38(2):167–174, Mar–Apr, 1998.

Fujisawa H, Takeda R. A new clinical test of dynamic standing balance in the frontal plane: the side-step test. Clin Rehabil 20(4):340–6, Apr, 2006.

Goldie PA, Bach TM, Evans OM. Force platform measures for evaluating postural control: reliability and validity. Arch Phys Med Rehabil 70(7):510–517, Jul, 1989.

Grigg P, Finerman GA, Riley LH. Joint-position sense after total hip replacement. J Bone Joint Surg Am 55(5):1016–25, Jul, 1973.

Grimsby O. Hip Chapter OMT 660. Ola Grimsby Residency Course notes, 2004.

Grimsby O. MT-4: Scientific Therapeutic Exercise Progressions, Unpublished seminar work book, 1995.

Grimsby O. Residency Course Curriculum. Manual Therapy Part I, Jan, 1991.

Gustavsson A, Neeter C, Thomee P, Silbernagel KG, Augustsson J, Thomee R, Karlsson J. A test battery for evaluating hop performance in patients with an ACL injury and patients who have undergone ACL reconstruction. Knee Surg Sports Traumatol Arthrosc 14(8):778–88, Aug, 2006.

Guyatt GH, Townsend M, Keller J, Singer J, Nogradi S. Measuring functional status in chronic lung disease: conclusions from a randomized control trial. Respir Med 85 Suppl B:17–21, Sep, 1991.

Guyatt GH, Townsend M, Keller J, Singer J, Nogradi S. Measuring functional status in chronic lung disease: conclusions from a random control trial. Respir Med 85(Suppl B):17–21, 1991.

Hageman PA, Blanke DJ. Comparison of gait of young women and elderly women. Phys Ther 66(9):1382–1387, Sep, 1986.

Headley S, Germain M, Mailloux P, Mulhern J, Ashworth B, Burris J, Brewer B, Nindl B, Coughlin M, Welles R, Jones M. Resistance training improves strength and functional measures in patients with end-stage renal disease. Am J Kidney Dis 40(2):355–364, Aug, 2002.

Heaney RP. Estrogen-calcium interactions in the postmenopause: a quantitative description. Bone Miner 11(1):67–84, Oct, 1990.

Himann JE, Cunningham DA, Rechnitzer PA, Paterson DH. Age related changes in speed of walking. Med Sci Sports Exerc 20(2):161–166, Apr, 1988.

Hoeksma HL, Dekker J, Ronday HK, Heering A, van der Lubbe N, Vel C, Breedveld FC, van den Ende CH. Comparison of manual therapy and exercise therapy in osteoarthritis of the hip: a randomized clinical trail. Rheum 51(5):722–9, Oct 15, 2004.

Hofmann DF. Arthritis and exercise. Prim Care 20:895–910, 1993.

Holbein-Jenny MA, Billek-Sawhney B, Beckman E, Smith T. Balance in personal care home residents: a comparison of the Berg Balance Scale, the Multi-Directional Reach Test, and the Activities-Specific Balance Confidence Scale. J Geriatr Phys Ther 28(2):48–53, 2005.

Holten O, Faugli HP. *Medisinsk Treningsterapi*, Ilniversitetsforlaaet. 0608 Oslo, Norway, 1996.

Hu MH, Woollacott MH. Multisensory training of standing balance in older adults: I. Postural stability and one-leg stance balance. J Gerontol 49(2):M52–M61, Mar, 1994.

Hu MH, Woollacott MH. Multisensory training of standing balance in older adults: II. Kinematic and electromyographic postural responses. J Gerontol 49(2):M62–M71, Mar, 1994.

Hughes C, Osman C, Woods AK. Relationship among performance on stair ambulation, functional reach, and Timed Up and Go tests in older adults. Issues Aging 21:18–22, 1998.

Hui SS, Yuen PY. Validity of the modified back-saver sit-and-reach test: a comparison with other protocols. Med Sci Sports Exerc 32(9):1655–9, Sep, 2000.

Ingber RS. Iliopsoas myofascial dysfunction: a treatable cause of "failed" low back syndrome. Arch Phys Med

Rehabil 70(5):382–6, May, 1989.
International Federation of Orthopaedic Manipulative Therapists. www.IFOMTorg. Definition of OMT, 2006.
Janda V. Function of Muscles in Musculoskeletal Pain Syndromes. Northeast Seminars. Tacoma, Washington, April 18–19, 1999.
Judge JO, Underwood M, Gennosa T. Exercise to improve gait velocity in older persons. Arch Phys Med Rehabil 74(4):400–6, Apr, 1993.
Juker D, McGill S, Kropf P, Steffen T. Quantitative intramuscular myoelectric activity of lumbar portions of psoas and the abdominal wall during a wide variety of tasks. Med Sci Sports Exerc 30(2):301–10, Feb, 1998.
Kaltenborn FM. *Mobilization of the Extremity Joints* (3rd edition). Oslo: Olaf Norlis Bokhandel, 1980.
Karanjia PN, Ferguson JH. Passive joint position sense after total hip replacement surgery. Ann Neurol 13(6):654–7, Jun, 1983.
Kasman GS. Surface EMG Evaluation and Feedback Training: Course Manual, 2002.
Kea J, Kramer J, Forwell L, Birmingham T. Hip abduction-adduction strength and one-leg hop tests: test-retest reliability and relationship to function in elite ice hockey players. J Orthop Sports Phys Ther 31(8):446–55, Aug, 2001.
Kelley GA. Aerobic exercise and bone density at the hip in postmenopausal women: a meta-analysis. Prev Med 27(6):798–807, Nov–Dec, 1998.
Kerr A, Rafferty D, Kerr KM, Durward B. Timing phases of the sit-to-walk movement: Validity of a clinical test. Gait Posture 26(1):11–6, Jun, 2007.
Khan KH, McKay H, Kannus P, Bailey D, Wark J, Kim Bennell K. *Physical Activity and Bone Health.* Human Kinetics, 2001.
Khan KH. Physical activity prescription for bone health across the lifespan. Lecture presented at Ola Grimsby Institute Competency Forum. San Francisco, 2004.
Kisner C, Kolby KA. *Therapeutic Exercise.* Foundation and Techniques. Philadelphia, PA: FA Davis Co, 1985.
Kohrt WM, Ehsani AA, Birge SJ JR. Effects of exercise involving predominantly either joint-reaction or ground-reaction forces on bone mineral density in older women. J Bone Miner Res 12(8):1253–61, Aug, 1997.
Kornetti DL, Fritz SL, Chiu YP, Light KE, Velozo CA. Rating scale analysis of the Berg Balance Scale. Arch Phys Med Rehabil 85(7):1128–35, Jul, 2004.
Lajoie Y, Gallagher SP. Predicting falls within the elderly community: comparison of postural sway, reaction time, the Berg balance scale and the Activities-specific Balance Confidence (ABC) scale for comparing fallers and non-fallers. Arch Gerontol Geriatr 38(1):11–26, Jan–Feb, 2004.
Leung AS, Chan KK, Sykes K, Chan KS. Reliability, validity, and responsiveness of a 2-min walk test to assess exercise capacity of COPD patients. Chest 130(1):119–25, Jul, 2006.
Lord SR, Murray SM, Chapman K, Munro B, Tiedemann A. Sit-to-stand performance depends on sensation, speed, balance, and psychological status in addition to strength in older people. J Am Geriatr Soc 57(8):M539–M543, Aug, 2002.
Mackenzie ME, Ng Gy. Investigation of progressive high speed non-weight bearing exercise to triceps surae. New England J Physio 17–19, Aug, 1995.
Mathias S, Nayak US, Isaacs B. Balance in elderly patients: the “get-up and go” test. Arch Phys Med Rehabil 67(6):387–9, Jun, 1986.
McBride JM, Triplett-McBride T, Davie A, Newton RU. The effect of heavy vs. light-load jump squats on the development of strength, power and speed. J Strength Cond Res 16(1):75–82, Feb, 2002.
McGill S. *Low Back Disorders: Evidence-Based Prevention and Rehabilitation.* Human Kinetics, Champaign, IL, 2002.
Mecagni C, Smith JP, Roberts KE, O’Sullivan SB. Balance and ankle range of motion in community-dwelling women aged 64 to 87 years: a correlational study. Phys Ther 80(10):1004–11, Oct, 2000.
Meknas K, Christensen A, Johansen O. The internal obturator muscle may cause sciatic pain. Pain 104(1–2):375–80, Jul, 2003.
Messier SP, Glasser JL, Ettinger WH Jr, Craven TE, Miller ME. Declines in strength and balance in older adults with chronic knee pain: a 30-month longitudinal, observational study. Arthritis Rheum 47(2):141–8, Apr 15, 2002.
Minor MA. Exercise in the management of osteoarthritis of the hip and knee. Arthritis Care Res 7(4):198–204, Dec, 1994.
Murray MP, Kory RC, Clarkson BD. Walking patterns in healthy old men. J Gerontol 24(2):169–178, Apr, 1969.
Nelson RT, Bandy WD. Eccentric training and static stretching improve hamstring flexibility of high school males. J Athl Train 39(3):254–258, Sep, 2004.
Netz Y, Ayalon M, Dunsky A, Alexander N. ‘The multiple-sit-to-stand’ field test for older adults: what does it measure? Gerontology 50(3):121–6, May–Jun, 2004.
Nevitt MC, Cummings SR, Kidd S, Black D. Risk factors for recurrent nonsyncopal falls: a prospective study. JAMA 261(18):2663–2668, May 12, 1989.
Newcomer KL, Krug HE, Mahowald ML. Validity and reliability of the timed-stands test for patients with rheumatoid arthritis and other chronic diseases. J Rheumatol 20(1):21–27, Jan, 1993.
Newton RA. Validity of the multi-directional reach test: a practical measure for limits of stability in older adults. J Gerontol A Biol Sci Med Sci 56(4):M248–52, Apr, 2001.
Niederman MS, Clemente PH, Fein AM, Feinsilver SH,

Robinson DA, Ilowite JS, Bernstein MG. Benefits of a multidisciplinary pulmonary rehabilitation program: improvements are independent of lung function. Chest 99(4):798–804, Apr, 1991.

Norkin C, Levangie P. *Joint Structure and Function: A Comprehensive Analysis*, 2nd ed. Philadelphia: F.A. Davis, 1992.

Noseda A, Carpiaux JP, Prigogine T, Schmerber J. Lung function, maximum and submaximum exercise testing in COPD patients: reproducibility over a long interval. Lung 167(4):247–57, 1989.

Oberg T, Karsznia A, Oberg K. Basic gait parameters: reference data for normal subjects, 10–79 years of age. J Rehabil Res Dev 30(2):210–223, 1993.

Olmsted LC, Carcia CR, Hertel J, Shultz SJ. Efficacy of the star excursion balance tests in detecting reach deficits in subjects with chronic ankle instability. J Athl Train 37(4):501–506, Dec, 2002.

Ostrosky KM, VanSwearingen JM, Burdett RG, Gee Z. A comparison of gait characteristics in young and old subjects. Phys Ther 74(4): 637–646, Jul, 1994.

Pare EB, Stern JT Jr, Schwartz JM. Functional differentiation within the tensor fasciae latae. A telemetered electromyographic analysis of its locomotor roles. J Bone Joint Surg Am 63(9):1457–1471, Dec, 1981.

Peck WA, Riggs BL, Bell NH, Wallace RB, Johnston CC Jr, Gordon SL, Shulman LE. Research directions in osteoporosis. N Eng J Med 84(2):275–282, Feb, 1988.

Perron M, Malouin F, Moffet H. Assessing advanced locomotor recovery after total hip arthroplasty with the timed stair test. Clin Rehabil 17(7):780–6, Nov, 2003.

Perry J, Weiss WB, Burnfield JM, Gronley JK. The supine hip extensor manual muscle test: a reliability and validity study. Arch Phys Med Rehabil 85(8):1345–50, Aug, 2004.

Petrella RJ, Wight D. An office-based instrument for exercise counseling and prescription in primary care. The Step Test Exercise Prescription (STEP). Arch Fam Med 9(4):339–44, Apr, 2000.

Piva SR, Fitzgerald GK, Irrgang JJ, Bouzubar F, Starz TW. Get up and go test in patients with knee osteoarthritis. Arch Phys Med Rehabil 85(2):284–9, Feb, 2004.

Podsiadlo D, Richardson S. The timed "Up & Go": a test of basic functional mobility for frail elderly persons. J Am Geriatr Soc 39(2):142–8, Feb, 1991.

Robinovitch SN, Heller B, Lui A, Cortez J. Effect of strength and speed of torque development on balance recovery with the ankle strategy. J Neurophysiol 88(2):613–20, Aug, 2002.

Ross M. Test-retest reliability of the lateral step-up test in young adult healthy subjects. J Orthop Sports Phys Ther 25(2):128–32, Feb, 1997.

Ross MD, Langford B, Whelan PJ. Test-retest reliability of 4 single-leg horizontal hop tests. J Strength Cond Res 16(4):617–22, Nov, 2002.

Rossi A, Grigg P. Characteristics of hip joint mechanoreceptors in the cat. J Neurophysiol 47(6):1029–42, Jun, 1982.

Schlicht J, Camaione DN, Owen SV. Effect of intense strength training on standing balance, walking speed, and sit-to-stand performance in older adults. J Gerontol A Biol Sci Med Sci 56(5):M281–6, May, 2001.

Skelton DA, Young A, Greig CA, Malbut KE. Effects of resistance training on strength, power, and selected functional abilities of women aged 75 and older. J Am Geriatr Soc 43(10):1081–1087, Oct, 1995.

Smidt N, de Vet HC, Bouter LM, Dekker J, Arendzen JH, de Bie RA, Bierma-Zeinstra SM, Helders PJ, Keus SH, Kwakkel G, Lenssen T, Oostendorp RA, Ostelo RW, Reijman M, Terwee CB, Theunissen C, Thomas S, van Baar ME, van 't Hul A, van Peppen RP, Verhagen A, van der Windt DA; Exercise Therapy Group. Effectiveness of exercise therapy: a best-evidence summary of systematic reviews. Aust J Physiother 51(2):71–85, 2005.

Solway S, Brooks D, Lacasse Y, Thomas S. A qualitative systematic overview of the measurement properties of functional walk tests used in the cardiorespiratory domain. Chest 119(1):256–270, Jan, 2001.

Steffen TM, Hacker TA, Mollinger L. Age- and gender-related test performance in community-dwelling elderly people: Six-Minute Walk Test, Berg Balance Scale, Timed Up and Go Test, and gait speeds. Phys Ther 82(2):128–37, Feb, 2002.

Thorstensson A, Oddsson L, Carlson H. Motor control of voluntary trunk movements in standing. Acta Physiologica Scandinavia 125(2):309–21, Oct, 1985.

Tinetti ME, Williams TF, Mayewski R. Fall risk index for elderly patients based on number of chronic disabilities. Am J Med 80(3):429–434, Mar, 1986.

Topp R, Mikesky A, Wigglesworth J, Holt W Jr, Edwards JE. The effect of a 12-week dynamic resistance strength training program on gait velocity and balance of older adults. Gerontologist 33(4):501–6, Aug, 1993.

Tropp H. Functional Instability of the Ankle Joint. Medical Dissertation. Linkoping, Sweden, 1985.

van Baar ME, Dekker J, Oostendorp RA, Bijl D, Voorn TB, Bijlsma JW. Effectiveness of exercise in patients with osteoarthritis of hip or knee: nine months' follow up. Ann Rheum Dis 60(12):1123–30, Dec, 2001.

Van Baar ME, Dekker J, Oostendorp RA, Bijl D, Voorn TB, Lemmens JA, Bijlsma JW. The effectiveness of exercise therapy in patients with osteoarthritis of the hip or knee: a randomized clinical trial. J Rheumatol 25(12):2439–9, Dec, 1998.

van Loo MA, Moseley AM, Bosman JM, de Bie RA, Hassett L. Test-re-test reliability of walking speed, step length and step width measurement after traumatic brain injury: a pilot study. Brain Inj 18(10):1041–8, Oct, 2004.

Viton JM, Atlani L, Mesure S, Franceschi JP, Massion J,

Delarque A, Bardot A. Reorganization of equilibrium and movement control strategies in patients with knee arthritis. Scand J Rehabil Med 31(1):43–8, Mar, 1999.

Viton JM, Atlani L, Mesure S, Massion J, Franceschi JP, Delarque A, Bardot A. Reorganization of equilibrium and movement control strategies after total knee arthroplasty. J Rehabil Med 34(1):12–9, Jan, 2002.

Wall JC, Bell C, Campbell S, Davis J. The Timed Get-up-and-Go test revisited: measurement of the component tasks. J Rehabil Res Dev 37(1):109–13, Jan–Feb, 2000.

Wasserman K, Hansen JE, Sue DY, Casaburi R, Whipp BJ. *Principles of Exercise Testing and Interpretation,* 3rd edition. Philadelphia: Lippincott, Williams & Wilkins; 1999.

Webright WG, Randolph BJ, Perrin DH. Comparison of nonballistic active knee extension in neural slump position and static stretch techniques on hamstring flexibility. J Orthop Sports Phys Ther 26(1):7–13, Jul, 1997.

Weiner DK, Duncan PW, Chandler J, Studenski SA. Functional reach: a marker of physical frailty. J Am Geriatr Soc 40(3):203–7, Mar, 1992.

Weisman IM, Zeballos RJ. An integrated approach to the interpretation of cardiopulmonary exercise testing. Clin Chest Med 15(2):421–445, Jun, 1994.

Whitney S, Wrisley D, Furman J. Concurrent validity of the Berg Balance Scale and the Dynamic Gait Index in people with vestibular dysfunction. Physiother Res Int 8(4):178–86, 2003.

Whitney SL, Marchetti GF, Morris LO, Sparto PJ. The reliability and validity of the Four Square Step Test for people with balance deficits secondary to a vestibular disorder. Arch Phys Med Rehabil 88(1):99–104, Jan, 2007.

Whitney SL, Wrisley DM, Marchetti GF, Gee MA, Redfern MS, Furman JM. Clinical measurement of sit-to-stand performance in people with balance disorders: validity of data for the Five-Times-Sit-to-Stand Test. Phys Ther 85(10):1034–45, Oct, 2005.

Wilson GJ, Newton RU, Murphy AJ, Humphries BJ. The optimal training load for the development of dynamic athletic performance. Med Sci Sports Exerc 25(11):1279–1286, Nov, 1993.

Winter DA, Patla AE, Frank JS, Walt SE. Biomechanical walking pattern changes in the fit and healthy elderly. Phys Ther 70(6):340–347, Jun, 1990.

Winters MV, Blake CG, Trost JS, Marcello-Brinker TB, Lowe LM, Garber MB, Wainner RS. Passive versus active stretching of hip flexor muscles in subjects with limited hip extension: a randomized clinical trial. Phys Ther 84(9):800–7, Sep, 2004.

Woerman AL. Evaluation and treatment of dysfunction in the lumbar-pelvic-hip complex. In: Donatelli R, Wooden MJ (eds), *Orthopedic Physical Therapy*, p 439. New York, NY: Churchill Livingstone, 1989.

Zwick D, Rochelle A, Choksi A, Domowicz J. Evaluation and treatment of balance in the elderly: A review of the efficacy of the Berg Balance Test and Tai Chi Quan. Neuro Rehabilitation 15(1):49–56, 2000.

NOTES:

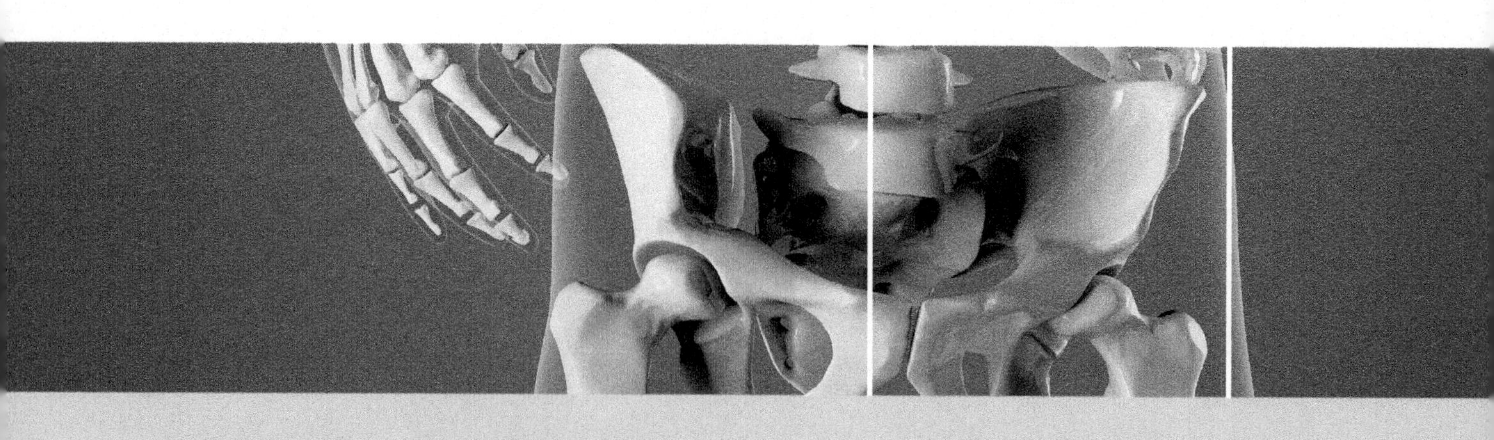

Authors: Rob Tillman, Sarah Olson, Jim Rivard, Ola Grimsby

CHAPTER 4

Exercise Rehabilitation of the Sacrioiliac Joint

Introduction

In this chapter:

Exercise for the sacroiliac joints and pelvic girdle are typically not performed in iso=lation. It is clinically difficult to separate influences of the lumbar spine and hip joints from the structural and dynamic stability of this region. This chapter will attempt to focus on exercise concepts and progressions for the pelvis. The reader is also encouraged to review the Hip and Lumbar chapters for integration into a more complete exercise program, should these areas require treatment as well.

Initial exercise will address rehabilitating the primary diagnosis of connective tissue abnormalities and/or weakness and lack of coordination of musculature related to the pelvis. Much conflict and controversy exists pertaining to the existence of sacroiliac related pathologies, but core science consistently reports motion occurring at the articulations between the iliac bones and the sacrum (Jacob et al. 1995, Smidt et al. 1995/1997, Lund et al. 1996, Wang et al. 1996). Though most of the available motions at these articulations occur in the frontal plane around a more sagittal axis, motion also occurs around vertical, anterior/posterior and oblique axes, making these joints functionally multiaxial. A detailed evaluation of the related anatomical systems must be performed to identify load sensitivity of key ligamentous and fascial structures, as well as abnormal function at the articulations between the ilia and sacrum, and the pubis. Evaluation must also include assessment of the motor system, including volume, coordination and strength. Muscles of the hip, pelvis and lumbar spine must function in proper synergy to be transducers of force between the lower limb and the upper and lower torso. It is assumed the reader has an extensive knowledge of sacroiliac joints and pelvic girdle anatomy, with a basic understanding of biomechanics. This chapter will provide limited reviews of these sciences, with emphasis on exercise application.

The sacroiliac joints (SIJ) are subjected to great shear forces. The ligamentous and capsular structures close to the instantaneous axis of motion, as well as collagen structures more distal to axis, need to be able to resist up to 4,800 N of force (Gunterberg et al. 1976).

Primary stabilizers of the SIJ are the sacroiliac ligaments. Anterior sacroiliac ligaments blend with the joint capsule and insert on the sacrum and ilium. The posterior sacroiliac or interosseous ligament is the strongest SIJ ligament and inserts dorsally on the ilium and sacrum. Avulsion of the bone from the sacrum or ilium will occur before a healthy interosseous ligament will tear. Because of their close association with the joint line, these two ligaments are taut with both nutation (forward nodding) and counter-nutation (backward nodding) of the sacrum. The sacrotuberous and sacrospinous ligaments limit posterior ilia rotation and nutation of the sacrum.

Exercise should be prescribed and dosed according to the nature of the patient presentation and pathology, as well as by the desired level of functional return. A clothing store manager and a professional football corner back may present with the same mechanism of injury, for example, a fall on the right buttocks. They may both present with bruising and discoloration of the skin, swelling and irritation in the subcutaneous tissues and musculature. Tenderness along the matrix of the dorsal SI ligaments, iliolumbar, sacrotuberous and sacrospinous ligaments as well as evidence of posterior rotation of the right innominate bone on the sacrum may also be present. The initial treatment of these two people would be close to identical for pain inhibition, tissue repair and mobilization. It may include mobilization of the ilium in an anterior direction to reduce the rotation.

Possible utilization of cryotherapy and some Stage 1 exercise activities can be performed at an intensity level below that which would cause an increase in pain. As their tissue lesions progress to the settled stage of tissue healing, a slight difference in the amount of weight these used for training will vary between the two, but their programs would still be very similar. Significant differences in their programs would occur when the clinician begins to challenge the involved tissues and musculature with activities designed to optimally promote primary tissue healing and full functional return to the pre injury state and beyond. Where the store manager may be satisfied with sitting for long periods of time without pain to perform a hobby or gardening, the professional football player would progress to aggressive plyometric exercise activities such as running with directional changes against resistance using a speed pulley, etc.

This chapter will review the various phases of rehabilitation for individuals with SIJ pathology. At times there will be overlap on some activities as the clinician progresses a patient in a program. Each patient will present with unique characteristics related to their pathologies, deficits and goals. Every exercise program should, while following the logical guidelines outlined in the chapter must be tailored to the individual for an optimal outcome.

Section 1: Stage 1 Exercise Progression Concepts for the Sacroiliac Joint

Four stages of rehabilitation for the sacroiliac dysfunction will be presented. Again, patients will need a program specifically tailored for their presentation, deficits and functional goals. Flexibility in the overall thought process by the clinician is needed for the construction and progression of a program that promotes optimal rehabilitation of patients presenting with rehabilitation needs. Some may need to enter the process of rehab with exercises taken primarily from Stage 1, while others may present with a load tolerance and functional strength base that allows them to begin with exercises in Stage 2. In cases involving elite conditioning and reactive/contact sports, patients and clients may begin with Stage 3 exercises and quickly progress to Stage 4.

The stages are defined by the patient's presentation of signs and symptoms. The stage of healing may also play a role as to what stage of exercise they can begin training. The list of impairments drives the list of exercises, as each one should address a specific impairment with a specific functional quality in mind. The specific exercise itself does not dictate which functional quality is attained; it is the dosage and design that will determine the outcome.

Potential Stage 1 Tissue States and Functional Status:

- *Reduced arthrokinematic motion*
- *Decrease in active and passive range of motion*
- *Painful joint at rest and/or with motion*
- *Abnormal respiration patterns*
- *Edema*
- *Muscle guarding*
- *Reduced coordination, timing and recruitment*
- *Poor balance/functional status*
- *Positive palpation to involved tissues*

Training Goals

The basic components of an initial exercise program are to 1) normalize joint motion/alignment, 2) provide tissue repair stimulus, 3) normalize motor patterns, 4) restore function motion and activities, 5) and finally to elevate the overall training level. Joint motion can be addressed with both passive and active treatments, though significant restrictions require passive correction prior to attempting stabilizing exercises. Normal arthrokinematic motion must be available to allow for normal range of motion, provide afferent input via mechanoreceptor firing in the surrounding joint capsule and to assist in normalizing muscle recruitment. Hypomobile joints may also benefit from mobilizing exercises to follow up more passive treatments of soft tissue work and joint mobilization. Prior to attempting to stabilize hypermobile joints, any restricted joints or facia must be normalized. This typically includes the hip joints and lumbar spine, though the thoracic spine and foot are also in the biomechanical chain.

Basic Outline for Stage 1

- *3–5 Exercises*
- *Many repetitions minimal resistance (coordination)*
  - *< 50% of 1RM for joint mobilization*
  - *< 60% of 1RM (vascularity)*
  - *< 50% of 1RM—Tissue repair, edema reduction and range of motion*
  - *< 40% of 1RM tonic fiber atrophy (hypertrophy)*
- *Low speed (coordination)*
  - *Deformity of collagen*
  - *Concentric facilitation*
- *Hypomobile: train in the outer range of available normal physiologic motion*
- *Hypermobile: train in the mid to inner range of available normal physiologic motion*
- *Selective tissue training (STT): for bone, collagen, muscle and cartilage*
- *Neurophysiologic influences: pain inhibition and coordination*

Options

- *Joint in resting position, start contraction from length tension position*
- *Concentric work only for vascularity of muscles in guarding*
  - *Use a pulley with an eccentric stop wheel to remove the eccentric phase*
  - *Weight stack rests between repetitions*
- *Artificial locking may be required to avoid motion around a non-physiologic axis*

Mobilization

Mobilization exercises attempt to improve range of motion through improving elasticity in joint capsules for increased arthrokinematic motion, or in surrounding soft tissues for osteokinematic motion. Low resistance, high repetition exercise will improve elasticity, as well as assist in normalizing muscle guarding, while prolonged holds will lead to plastic deformity of collagen. Mobilizing exercises may also address muscle guarding patterns that are responsible for restricting joint motion, limiting range or creating changes in lumbopelvic posture.

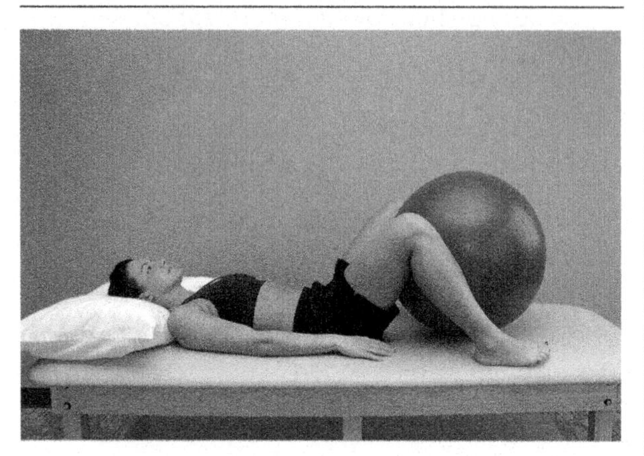

Figure 4.1: Ball squeeze for mobilization of the symphysis pubis. The patient is in supine with the hips flexed to 45° and a ball placed between the knees. The patient adducts the hips squeezing the ball to create a distraction force at the symphysis pubis with adductors. The contraction may cause a click or shift in symphasis position, but should be pain free during the performance.

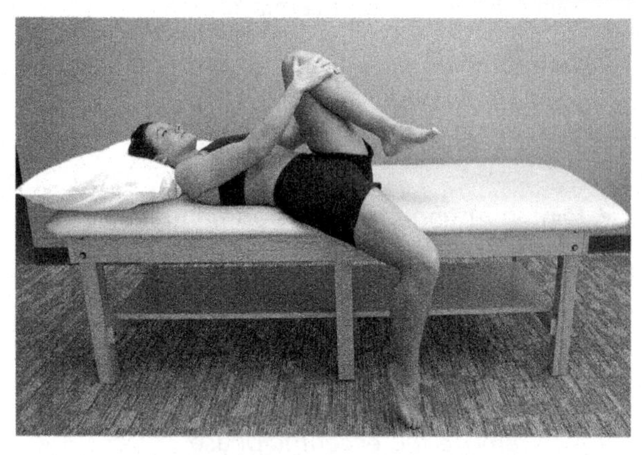

Figure 4.2: Torsion mobilization of the ilium. The same technique, sequenced differently, can be used for either an anterior or posterior mobilization of the ilium on the sacrum. To mobilize the right ilium anterior the opposite hip is first flexed, this also flexes the lumbar spine, preventing extension. The hip is then dropped off the edge of the table on the involved (right) side to create an anterior torsion mobilization. For posterior mobilization, the opposite side is first dropped off the table to lock the lumbar spine in extension. The hip is then flexed on the involved side to create a posterior torsion of the ilium. Often the mobilization is not used for osteokinematic influences of changing perceived abnormal bony alignment but is used to fire mechanoreceptors in the collagen of the joint capsule to inhibit pain and reduce muscle guarding. The neurological influence of mobilization exercises may reduce abnormal muscle guarding patterns while increasing the facilitation of normal motor patterns for lumbopelvic stability.

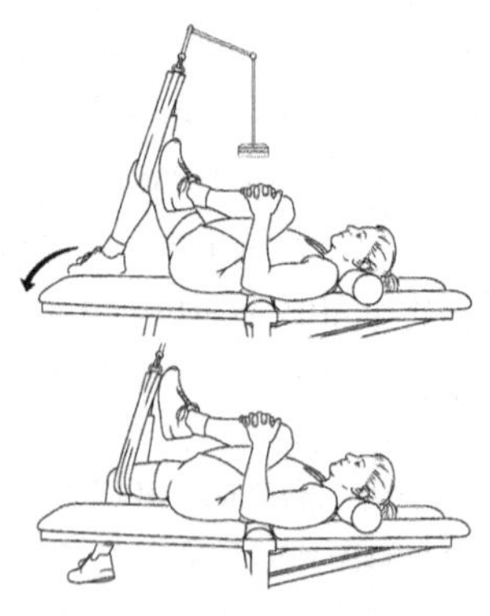

Figure 4.3: Supine assisted hip flexion. In supine, the opposite hip is flexed to lock the lumbar spine into flexion and the opposite SIJ into a relative posterior torsion. Repetitive hip flexion is performed on the involved side with the goal of reducing tone of the hip flexors. A pulley assist may be necessary to reduce the weight of the leg enough to train for local vascularity (3x24 at 60% 1RM). For patients with higher training levels, cuff weights may need to be added to achieve this level of training.

Figure 4.4: Resisted hip extension with lumbar locking – anterior mobilization of ilium Alternatively from the previous exercise, the weight may be increased on the pulley to resist hip extension, rather than assist hip flexion. Utilizing the hip extensors leads to reciprocal inhibition of the hip flexors, reducing tone allowing for an anterior torsion of the ilium.

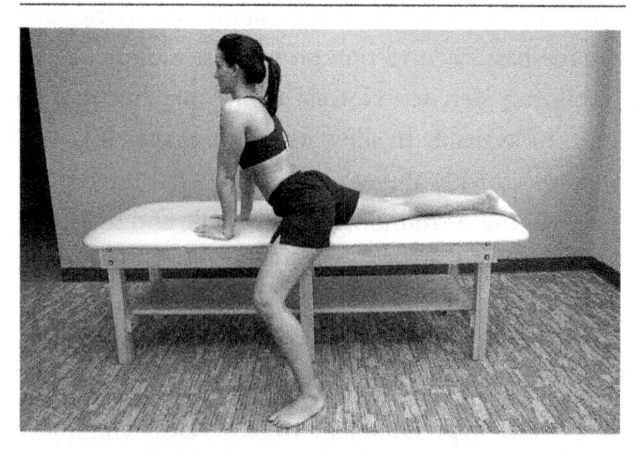

Figure 4.5: Sacral counter-nutation on right ilium. The patient is prone with the left leg off the table, flexing the hip to create a posterior torsion of the left ilium. The lumbar spine remains in neutral. The patient performs a passive press

up with the arms creating a counter-nutation of the sacrum relative to the right ilium.

Figure 4.6a,b: Chair lunge – anterior mobilization of the left ilium. The involved side is positioned with the foot back (left), while the foot of the uninvolved side is placed on a chair. The patient weight shifts forward onto the front foot creating an anterior torsion of the left ilium. Reaching overhead and/or extending the lumbar spine will bring sacrum into counter-nutation, increasing the mobilizing force.

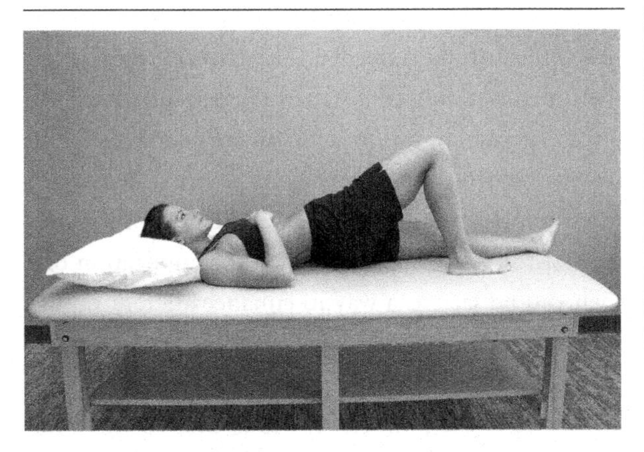

Figure 4.7: Unilateral bridge for anterior mobilization of the right ilium. In supine the involved hip only is bridged creating a relative anterior torsion of the ilium. If the knee is maintained toward the ceiling, a lengthening of the psoas also occurs. With gluteal contraction for the bridge, reciprocal inhibition of the psoas assists in the anterior ilial mobilization.

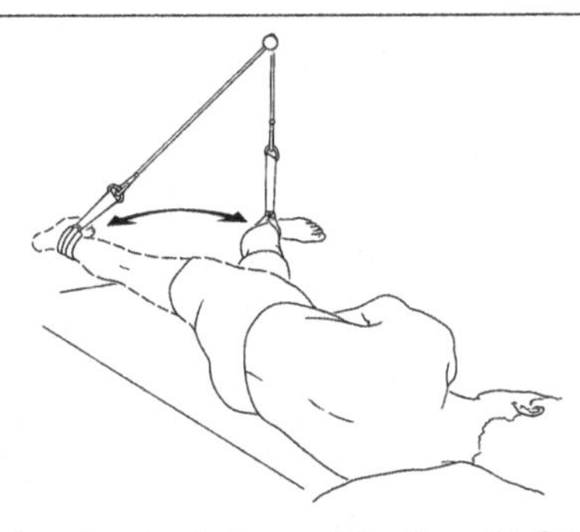

Figure 4.8: Low level anterior mobilization of left ilium on sacrum. The patient is in right side lying with the leg on the involved side in a sling. The right hip is flexed to fix the right SIJ and lumbar spine. The patient performs active left hip extension for a repetitive dynamic mobilization of the left ilium relative to the sacrum.

Figure 4.9a,b: For higher level patients an active lunge can be used for an anterior torsion mobilization. The patient lunges anterior on the opposite side leg while the involved side leg maintains knee extension and full hip extension. To increase the level of mobilization, lumbar extension and counter-nutation of the sacrum can be added by reaching overhead during the lunge.

Figure 4.10a,b: Dynamic lengthening of the iliacus and psoas muscles, with relative anterior torsion mobilization of the SIJ. The patient is in single leg kneeling on the involved side. An eccentric extension and rotation motion is performed over the ipsilateral shoulder to emphasis sacral motion. To emphasize muscle lengthening the motion is performed over the opposite shoulder.

Figure 4.11a,b: Standing dynamic lengthening of the iliacus and psoas muscles, with relative anterior torsion mobilization of the left SIJ. This is a standing progression from the previous kneeling version.

Force Locking Concepts

Counter-curve locking techniques are frequently used in SIJ training for the purposes of mobilization. Several of the previous exercises shown utilized this concept to isolate mobilizing forces to one SIJ while locking the other side, along with the lumbar spine. This concept can be used for many SIJ training exercises in different positions. It is a technique that cannot be used for all exercises and is limited in functional application. An artificial locking technique involves using a belt around the pelvis to compress and stabilize the articular surfaces. A belt to the SIJ may be more effective in stabilizing the articulations of the pelvis to allow for exercises addressing coordination and stabilization. A pelvic belt has been shown to increase SIJ stiffness both in vitro (Damen et al. 2002a) and in vivo (Vleeming et al. 1992).

The role of the pelvic belt in the treatment of subjects with pregnancy-related pelvic pain is still controversial. Clinical experience has shown that positive effects can be obtained with different positions and tensions of the belt. Pelvic belts have been found to reduce SIJ laxity (Damen et al. 2002, Mens et al. 2005), and this decrease in mobility occurred regardless of the amount of tension used to secure the belt. Placing the belt higher (at the ASIS) was found to be more effective than placing the belt at the pubic symphysis. These results are in direct conflict with Depledge et al. (2005) on the same subject. They demonstrated that the addition of a rigid or a non-rigid pelvic support belt to a pelvic floor exercise regime and advice did not add to the effects as all groups improved, regardless of whether a belt was given or not.

In one anatomical study, the mobility of the sacroiliac joint (SIJ) was significantly restricted by application of a pelvic belt with a tension of 50 N, while larger forces did not give better results. The underlying theory of the use of a pelvic belt is that the articular surfaces of the SIJ will be pressed together, which raises friction to resist shearing. However, there is no in vivo proof of this mechanical effect. The decrease of SIJ laxity values with the application of a pelvic belt is due to the position of the pelvic belt rather than the tension of the belt. Vleeming et al. (1992) found that the combination of a pelvic belt with muscle training enhanced pelvic stability. Emphasis was placed on the location of the belt, at the level of the acetabulum rather than the amount of load. Using a 100 N belt did not significantly differ from using a 50 N belt in terms of reducing sagittal rotation of the sacroiliac joints. A pelvic belt was more effective when the application was just below the anterior superior iliac spines (high position) as compared to the application the level of the symphysis (low position) (Damen et al. 2002b).

Despite limited evidence for structural support, clinical experience demonstrates that some patients are able to perform at higher levels with reduced symptoms when utilizing a belt. For these clients no clinical study is needed. The effect of the belt may not only be a structural support of the articular system, but more of an influence on motor control. If ligaments are viewed as beds of mechanoreceptors, producing afferent information to assist in motor patterns, then the application of a belt can be seen as a way to enhance afferent input. A neoprene knee brace hardly provides structural support to the knee, but the afferent input from receptors in the joint structures to those of the skin assist in providing a more stable and less painful training experience. The SI belt may serve a similar function in providing some structural stability, but also influencing afferent systems to assist in the recruitment of normal muscle patterns, which does provide dynamic stability to the pelvis.

Selective Tissue Training (STT)

Selective tissue training involves incorporating the optimal stimulus for tissue repair into the exercise chosen. Dosage is set very low to allow for a high number of repetitions, multiple times a day to encourage cellular activity for reparative processes. Patients who need to begin their exercise activities at this level often present with severe irritation and load sensitivity to the dorsal SIJ ligaments as well as the sacrotuberous, sacrospinous and iliolumbar ligaments. Related to the condition of these overstretched ligaments and connective tissues is inhibition of the motor systems, creating stability. Over time, these muscles may suffer from atrophy, with compensatory motor patterns taking over general function. Additional abnormal strain may occur to the tissues in lesion with chronic use of compensatory motor patterning. Muscle guarding from pain may be present in the tonic muscles of the pelvis (piriformis and other deep hip rotators).

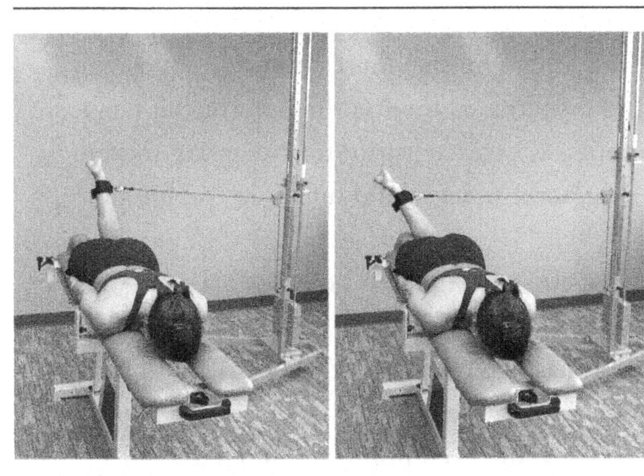

Figure 4.12a,b: Prone hip external rotation for tissue training of the sacrospinous ligament. The range of training is kept from neutral to full external rotation. Full internal rotation is avoided to prevent an end range tissue strain to the ligamentous structures. As tissue tolerance improves, range of motion is increased, rather than weight, to further improve the ligament's ability to tolerate strain.

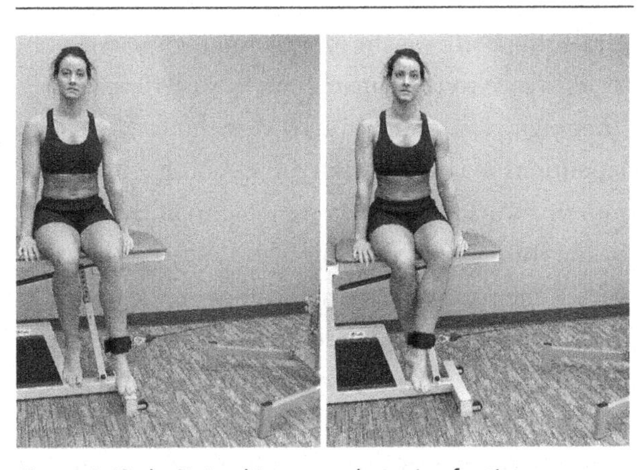

Figure 4.13a,b: Sitting hip external rotation for tissue training of the sacrotuberous ligament. Action of the deep hip external rotators will create secondary tension through the fascia and sacrotuberous ligament. The range of training is kept from neutral to full external rotation. Full internal rotation is avoided to prevent an end range tissue strain to the ligamentous structures. As tissue tolerance improves, range of motion is increased, rather than weight, to further improve the ligament's ability to tolerate strain.

Exercises can be chosen to target collagen in noncontractile structures. The sacrospinous ligament inserts on the posterior ilia and transverses medially to the lower sacrum and coccyx. The fibers of this ligament parallel the fibers of the piriformis muscle and are often damaged with attempts to aggressively stretch this muscle. Stretching of the hip external rotators is avoided, replaced by pain free repetitive contraction for tissue training.

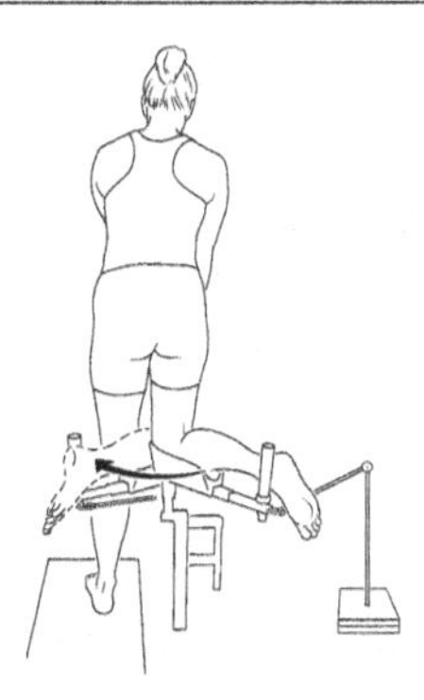

Figure 4.14: Partial weight bearing hip external rotation for tissue training of the SIJ ligamentous structures. The involved side limb is placed on a rotating stool, or the shoulder rotation trainer attachment. Resisted hip external rotation is performed in partial weight bearing to create a light tension through the SIJ ligaments for repair.

The sacrotuberous ligament arises from the ischial tuberosity and transverses superiorly to the inferior sacral border and the posterior inferior iliac spine. This ligament can be thought of as a cross, intersecting in the posterior buttock just inferior to the piriformis muscle. Many authors choose to designate an "x" or horizontal SI axis that would pass through the sacrum at the S2 level, coinciding with motion of the joint in the frontal plane. The sacrotuberous ligament is oriented in a direction that would have the greatest resistive force in checking motion of the ilium posteriorly around the x-axis. Functionally, activities loading this ligament would include sitting and hyperflexion of the hip. The "y" axis of rotation is vertical and motion around this axis could lead to tension on the sacrospinous ligament. Motions that would load the ligament would include hip internal rotation and adduction of the hip with flexion to 90 degrees. Available SI motion and loading of the ligaments and other tissues related to the SI joints are greatly related to ipsilateral hip position, as well as forces transposed through the hips proximally to the SI joints. It is of great importance to assess and clear the hips of any pathologies and issues with loading while evaluating the pelvic girdle.

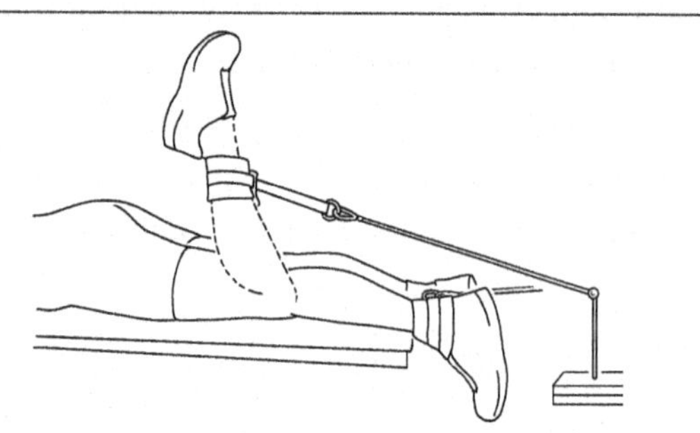

Figure 4.15: Prone hamstring curls with pulley resistance. Tissue training for the sacrotuberous ligament and sacroiliac joint ligaments via tension of the proximal hamstring tendon.

Dosage for Stage 1 Functional Qualities

Tissue Repair, Pain Inhibition, Edema Reduction and Joint Mobilization

- *Sets*: 1–5 sets or training sessions daily
- *Repetitions*: 2–10 hours during the day
- *Resistance*: Assisted exercise up to 25% 1RM
- *Frequency*: Multiple times daily

Coordination

- *Sets*: 1–5 sets or training sessions daily
- *Repetitions*: 20–50 during the day
- *Resistance*: Assisted exercise up to 50% 1RM
- *Frequency*: 1–3 times daily

Vascularity / Local Endurance

- *Sets*: 2–3 sets or training sessions daily
- *Repetitions*: 25
- *Resistance*: Assisted to 50–60% 1RM
- *Frequency*: One time daily

Muscle in Atrophy

- *Sets*: 2–3 sets or training sessions daily
- *Repetitions*: 40 or more
- *Resistance*: Assisted to 30–40% 1RM
- *Frequency*: One time daily

Coordination

Once the initial issues of restoring joint motion, reducing edema and decreasing pain have been addressed, the focus of the program shifts toward muscle performance. Coordination is the most important performance based functional quality, despite the common emphasis on strength. Often isometric work is initially involved because using muscles without moving the joint is considered safer for injured tissues, though isometric work does not significantly contribute to establishing coordination. Establishing normal muscle timing, sequencing and coordinated motion through the full range of motion is necessary prior to attempting to train endurance, strength and functional performance.

Muscle recruitment can become abnormal with altered mechanoreceptors firing in restricted joints, as well as swollen joints. Early passive and active mobilization exercises may significantly improve coordination, even prior to training, due to a normalization of mechanoreceptor firing. When muscle training for coordination is initiated, emphasis is placed on dosing the exercise in a manner in which the muscle can succeed. Using too high a level of resistance can be one factor associated with loss of coordination. For the pelvic girdle, the body weight may be too high a level of resistance to coordinate motion without pain. Training as low as 40% of 1RM with 40–50 repetitions may be necessary to establish a normal movement pattern. Often the weight, of the limb or the trunk exceeds the muscles ability to perform within the pattern, requiring compensation or abnormal movement. Unloading and pulley assist concepts allow for coordination training with a high number of repetitions. The abnormal movement pattern may correct instantly when the resistance is lowered enough. Designing the exercise to reduce load with assisted motion or reduce body weight may correct an otherwise abnormal motor pattern and allow for high repetition training. Neurological adaptation occurring during assisted training may normalize the pattern more efficiently. Once coordination is established, a progression to training local vascularity and endurance can be added.

Normal Myokinematics

The SIJ does not have a true arthrokinematic muscle system controlling the axis of motion in the same manner as most other synovial joints. Muscles crossing the joint that influence the hip and lumbar spine do play a significant role in the stability of the SIJ. The influence may be direct via placement of additional tension on collagenous structures influencing the sacrum and ilium, or more indirectly through influence on the hip or lumbar spine.

A model of SIJ function postulates that shear is prevented by friction, dynamically influenced by muscle force and ligament tension, allowing accommodation to specific loading situations (van Wingerden et al. 2004). Stiffness of the SIJ was found to significantly increase when the erector spinae, the biceps femoris and the gluteus maximus muscles were activated. In some cases, when assessing one of these muscles a significant co-contraction occurred in the other muscles. The finding that SIJ stiffness increased even with slight muscle activity supports the notion that effectiveness of load transfer from spine to legs is improved when muscle forces actively compress the SIJ, preventing shear (van Wingerden et al. 2004).

Transversely oriented fibers of the lower abdominal muscles, in combination with stiff dorsal sacroiliac ligaments, increases the compression of the sacroiliac joints and improves stability. Richardson et al. (2002) compared a drawing-in pattern of transverse abdominus facilitation with a bracing pattern on increasing stiffness of the sacroiliac joints. The draw-in pattern involves a specific contraction of the transversus abdominis through drawing in or hollowing of the abdominal wall. The brace pattern is a general contraction of all the abdominal muscles, involving isometric bracing action. Both techniques were shown to decrease laxity of the sacroiliac joints, though the authors felt the drawing-in technique had even more effect than bracing. Both strategies are likely used in the real world. The drawing-in technique can be performed with less co-activation of other abdominal muscles and multifidi (Richardson et al. 2002). The technique may be more effective for dynamic stabilization with light activity. More aggressive activity will require a higher level of muscle force and cocontraction with other muscles for lumbosacral stability, being more conducive to the bracing technique.

The thoracolumbar fascia also contributes to stability with attachments through the SIJ ligamentous structures, as well as connection into fascia of the lower limb. Contraction of the latissimus dorsi, gluteus maximus and erector muscle tenses the superficial lamina of the thoracolumbar fascia, while in some subjects the deep lamina is tensed by contraction of the biceps femoris (Vleeming et al. 1995). Tension in the posterior layer is transmitted to the contralateral side at the level of L4, or in some specimens at L2–L3 (Vleeming 1995).

Diagonal pattern training from the lower limb, through the lumbosacral region to the opposite upper limb bring together these mechanism of ligamentous, fascial and muscular control. Rotational training in the transverse plane will naturally assist in recruiting the transverse fiber direction of the transverse abdominus. Progressing to diagonal patterns, as in traditional diagonal patterns would be late stage integration of the synergy of these anatomical structures.

The pelvic floor muscles are crucial for normal body function, being termed, the "floor of the core." The pelvic floor is one part of the three functional diaphragms; the pelvic floor, the diaphragm muscle and the glottis. The pelvic floor can work antagonistically to each other with diaphragm contraction for respiration resulting in pelvic floor relaxation. The pelvic floor and diaphragm can work in synergy for bracing activities of the pelvis and trunk to stiffen the pelvis.

Abnormal Myokinematics

Mooney et al. (2001) identified in patients with SIJ pain, compared to normal patients, an abnormal hyperactivity of the gluteus muscle on the involved side with an increased activity of the latissimus on the contralateral side.

Hyper or hypotonicity of the pelvic floor musculature can be associated with any number of musculoskeletal dysfunctions. Pelvic floor muscular (PFM) coordination and function is especially crucial for urinary and fecal continence, normal sexual function, labor and delivery, and SIJ and lumbar stability. This chapter assumes the reader is familiar with normal anatomy, which can be divided into two parts: a system for bladder neck support as well as a system for sphincter closure (Ashton-Miller et al. 2001). Pelvic organs rely on their attachments to pubic bones, muscles, and connective tissue for support, and pathologies such as vaginal prolapse can occur when this support is impaired (Wei et al. 2004).

Dysfunction of the pelvic floor does not only occur in women, although most readers equate pelvic floor exercise with "Kegels" done during and post-pregnancy or later in life due to urinary incontinence. Pelvic floor dysfunction can be associated with either an increase or a decrease in tone, effecting men and women, as well as high level athletes. A "stiff" and strong pelvic floor is crucial in counteracting increased abdominal pressure, which can occur with high impact sports such as trampoline jumping, skydiving, or gymnastics. A high prevalence of incontinence (80 percent) was found in female athletes engaging in these high impact sports, as well as experiencing leaking during their sport (Eliasson et al. 2002). None of these athletes had leakage with coughing or other activities, and a pelvic floor strengthening regime was found to improve strength and decrease incontinence during trampoline jumping (Bo et al. 2004). Urinary incontinence is not caused by most types of normal exercise for women, however extreme high impact sports such as parachute jumping could actually lead to increased incidence of incontinence. A study by Davis et al. (1996) found nine previously continent women reported incontinence on completion of training. Three of the nine women had urge incontinence and six had stress urinary incontinence. The six women with stress urinary incontinence showed evidence of bladder neck hypermobility, anterior vaginal wall prolapse, and loss of urine with Valsalva maneuvers. Five of them could recall a specific traumatic event associated with a parachute jump in which pain to the pelvic region was felt and after which her urinary incontinence began. This is certainly an uncommon scenario as incontinence is typically caused by a change in tone or lack of recruitment of the PFM, which are necessary to coordinate and stabilize the spine, pelvis, hips, and lower extremity. It is interesting to note, however, that these more traumatic causes can affect continence as well. (Jiang et al. 2004).

Proper evaluation to determine the type of dysfunction is essential prior to distributing an exercise program, as a generalized protocol will not work for everyone. When planning a program for incontinence exercises it's important to assess the need first, and to have a theoretical rationale for the program to promote adherence (Alewijse et al. 2003). In a study examining differences in pelvic muscle exercise between women with stress urinary incontinence (SUI) and those without, it was found that 91 percent of those in the study with a history of gynecological surgery (hysterectomy, prolapse, etc.) were in the SUI group (Boyington et al. 2000). Taking time to question the patient regarding about past gynecological surgeries or difficulties during pregnancy may give information as to the possibility of pelvic floor dysfunction. Finding not only the correct tissue in lesion, but the area of pelvic floor weakness is also important. An exercise program based on specific anatomical defects in the anterior, middle, or posterior compartments of the vagina was found to lead to a 65 percent strength improvement in these three directional forces during coughing and straining (Petros and Skilling 2001). Knowledge of anatomy and physiology is crucial in order to give an effective pelvic floor exercise regime, and as with any other muscle group the program must be based on the principles of specificity and overload. Success depends on the ability to identify and isolate the correct muscle group (Johnson 2001). Often times it is assumed that all women suffering from incontinence or pain during or post pregnancy have decreased tone of the pelvic floor, requiring pelvic floor strengthening. However pelvic girdle relaxation has been found to be problematic for many women during pregnancy, thereby limiting ADL's due to pain secondary to hypertonus (Hansen et al. 1999). These patients would benefit more from an exercise program focused on pelvic floor relaxation and decreasing muscular tone rather than strengthening. Overall, these specific exercises given based on symptoms and diagnosis must be done as a life long process in order to maintain strength and coordination gains in the pelvic floor.

The Pelvic Floor

Physical therapists should not assume that all women are educated during pregnancy about the benefits of pelvic floor exercises and the potential for dysfunction by their obstetrician, however these issues are commonly not discussed during doctor visits with the pregnant patient. Female patient's knowledge regarding potential pelvic floor changes associated with pregnancy and delivery was polled following delivery. A total of 46.1 percent of women had received no information on kegel exercises, 51.3 percent on episiotomy, 46.6 percent on urinary incontinence, 80.6 percent on fecal incontinence, 72.8 percent on change in vaginal caliper, and 84.9 percent in neuropathy (Mclennan et al. 2005). These numbers indicate a need for education by the physical therapist, especially when dysfunction arises during or post pregnancy.

Back pain with pregnancy has been found to be as high as 49 percent, with 22–28 percent having pain from the twelfth week to delivery (Ostgaard et al. 1991). A common belief is that low back pain is normal during pregnancy and will go away after delivery. As high as 67 percent of women experience back pain directly after delivery, with 37 percent still having back pain at the follow-up examinations (Ostgaard et al. 1992). Women with previous severe low back pain during pregnancy are at a long term high risk for new episodes of severe low back pain during another pregnancy, as well as when they are no longer pregnant (Brynhildsen et al. 1998). Pregnant women who had a history of back pain prior to their pregnancy also had a higher prevalence of LBP during and post pregnancy. Younger women were found to have a higher incidence than older women, and women having their second baby were found to have more lumbar pain than women having their first child. The type of work women did prior to pregnancy, heavy labor versus a more sedentary job, did not make a difference as far as back pain during pregnancy (Turgut et al. 1998). Even six years later, many women who had LBP during pregnancy still reported having back pain. This pain improved following a physical therapy program (Ostgaard et al. 1997). The presence of pelvic pain during pregnancy can have serious detrimental effects, as 37 percent of those with pelvic pain were sick listed for an average of 12 weeks during their pregnancy (Larsen et al. 1999). Pool-Goudzwaard et al. (2005) identified 52 percent of all patients with pregnancy-related low back pain to have pelvic floor dysfunction as well. In the majority of cases (82 percent) the complaints started with low back pain, and/or pelvic pain, prior to symptoms of pelvic floor dysfunction.

Pelvic floor dysfunction can lead to an increase in muscle tone, as these muscles have been shown to stabilize the pelvic ring (Pool-Goudzwaard et al. 2004). It can also lead to a decrease in muscle tone, impacting right sided, left sided, or bilateral timing of recruitment. Timing of contraction of both sides is critical to proper stability during activity, as well as relaxation for voiding. A higher level of activity of the pelvic floor muscles can negatively influence the appropriate activity patterns during essential voluntary and reflex motor maneuvers (Pool-Goudzwaard et al. 2005). A direct impact can be expected on muscle endurance, the timing and quality of the reflexes and relaxation time of these muscles. These changes can functionally restrict the patient in many ways, including difficulty with sit to stand or rolling transfers, sports including jumping or quick changes in direction, voiding dysfunctions, constipation and sexual problems (Velde van der 1999). Pelvic floor dysfunction is also correlated with lower motor activity during coughing, but a higher activity level during pushing maneuver, as in a bowel movement (Pool-Goudzwaard et al. 2005). This means a patient with pelvic floor dysfunction may have hypotonia of the PFM during coughing which could lead to urine leaks with this activity, but may also have hypertonia with a bowel movement leading to constipation and incomplete voiding.

Pelvic Floor Training for the Male Pelvis

Though pelvic floor training is more commonly associated with females related to pregnancy and incontinence, the treatment is also used for fecal and urinary incontinence in males. Other applications include addressing premature ejaculation, erectile dysfunction and post-surgical incontinence related to prostatectomy. Several studies have specifically looked at pelvic floor function and exercise programs for men, several addressing pelvic floor muscle exercises post prostatectomy (Chang et al. 1998, Moul 1998, Porru et al. 2001). Urinary incontinence post-prostatectomy is 0.5–1.0 percent, and both pelvic floor muscle exercises, as well as electrical stimulation, were found to have decreased incontinence episodes and post-micturition dribbling. Patients who did pelvic floor muscle exercises also had better psychological and social quality of life scores, had fewer symptoms post-surgery, and had an improved grade of pelvic floor muscle contraction strength. Filocamo et al. (2005) also found a significant reduction in post-prostatectomy incontinence following a training program for pelvic floor muscles. The use of biofeedback has been shown to enhance learning of exercises for pelvic floor muscles in subjects with post-prostatectomy incontinence (Mathewson-Chapman 1997). At a minimum, avoiding rest and staying active after prostatectomy has a similar outcome to those studies showing bed rest being detrimental for low back pain. Windsor et al. (2004) found that a simple home program of moderate-intensity walking for subjects treated with radiation therapy for prostate carcinoma produced a significant improvement in physical functioning. No significant increase in fatigue was noted, as found in the control group that was instructed to rest with reduced activity. Premature ejaculation has also been treated with pelvic floor muscle training with a 61 percent success rate (La Pera G, Nicastro 1996). Claes H, Baert (1993) compared to a group of 150 male patients with erectile dysfunction and proven venous leakage randomized to either surgery or to a pelvic floor training program. Surgery was not superior to the pelvic floor training program either subjectively or objectively with 42 percent of the exercise group satisfied with the outcome and refusing surgery.

Incontinence

About 12 million adults in the United States have urinary incontinence. It's common in women over 50 years old, however it can also affect younger people, especially women who have just given birth (American Academy of Family Physicians 2005). Causes of urinary incontinence include:

- *For women, thinning and drying of the skin in the vagina or urethra, especially after menopause*
- *For men, enlarged prostate gland or prostate surgery*
- *Weakened pelvic muscles*
- *Certain medicines*
- *Build-up of stool in the bowels*
- *Inability to ambulate*
- *Urinary tract infection*
- *Problems such as diabetes or high calcium levels*

There are four types of urinary incontinence, including stress, urge, overflow, and functional. Stress incontinence is urine leakage which occurs because of sudden pressure on your lower stomach muscles, such as when you cough, laugh, lift, or

exercise. It usually occurs when the pelvic muscles are weakened, as by childbirth or surgery. This type of incontinence is more common in women.
Urge incontinence occurs when the need to urinate comes on too fast, often the body only gives a warning of seconds to minutes before urination occurs. This may not allow time to get to a toilet. This type is most common in the elderly and may be a sign of infection in the kidneys or bladder. Overflow incontinence is a constant dripping of urine caused by an overfilled bladder. There may be an inability to fully empty the bladder and may cause strain when urinating. This often occurs in men and can be caused by something blocking the urinary flow, such as an enlarged prostate gland or tumor. Diabetes or certain medications can also cause the problem. Functional incontinence occurs with normal urine control but trouble getting to the bathroom in time due to arthritis or other diseases than make ambulation difficult.

Pregnancy Related Incontinence

Incontinence during pregnancy is relatively common, with a rate of 25 percent. Prevalence of stress urinary incontinence during pregnancy has been linked to the hormone relaxin, with higher serum relaxin values actually being linked with a lower prevalence of incontinence (Kristiannsson et al. 2001)

Relaxin is more commonly known for increasing joint and ligamentous laxity during pregnancy, as it functions to allow the normally strong ligaments of the sacroiliac joint to stretch in order to allow for the passage of the baby through the pelvis. Relaxin has been shown to limit collagen production and reorganization, preventing fibrogenesis and reducing established scarring. This is important for pregnancy, but there are also new insights on potential of relaxin within the medical community to limit fibrosis with other medical conditions such as organ failure (Samuel 2005).

Along with changes in serum levels of hormones such as relaxin, bone density has also been shown to change significantly with pregnancy and lactation (Bjorklund et al. 1999). Bone density decreases significantly during and post pregnancy; however no association between bone loss and back or pelvic pain during pregnancy was found.

Several factors during labor and delivery have been studied, including choices for positioning of the mother during vaginal birth. The lateral position for delivery has been associated with having the highest rate of intact perineum (66.6 percent), while the squatting position was associated with the least favorable perineal outcome (42 percent). Deliveries overseen by obstetricians were found to have a much higher episiotomy rate with more tearing requiring sutures than did deliveries by midwives (Shorten et al. 2002).

Sacroiliac Instability

It is generally accepted that women tend to have more available motion in their SI articulations than men as well as having a greater likelihood to suffer from SI pathology. There is also an overwhelming likelihood that mechanical SI pathology will be present in the form of hypermobility rather than cases of hypomobility. Clinical observation reveals the tendency for general laxity of the dorsal sacroiliac ligaments as well as the sacrotuberous, sacrospinous and iliolumbar ligaments with female patients (deeper structures). Male patients with load sensitivity in this area present with tenderness and load sensitivity to the more superficial layers of fascia and other connective tissues.

The concept of form and force closure has been presented as a model for SIJ stability (Vleeming et al. 1990, Pool-Goudzwaard et al. 1998). Postural loading requires a combination of specific ligament and muscle forces to intrinsically stabilize the pelvis. Form closure refers to interlocking of the ridges and grooves on the joint surfaces. Roughening and increased coarseness of the auricular surface of the sacroiliac joints is viewed as a nonpathologic adaptation to the forces exerted, leading to increased stability (Vleeming et al. 1990). Force closure refers to compressive forces of structures like muscles, ligaments and fascia (Pool-Goudzwaard et al. 1998). Insufficient tension on ligaments, muscle weakness, and abnormal motor timing can reduce the compressive forces and reduce normal load transfer through the sacroiliac joints.

Increased laxity of the pelvic girdle during pregnancy has traditionally been correlated with pregnancy related pelvic pain (Abramson et al. 1934, Farbrot 1952, MacLennan 1991, Calguneri et al. 1982, Hansen et al. 1996). More recent research challenges this commonly held belief that generalized laxity is itself a cause of pelvic pain. Buyruk et al. (1999) found similar levels of sacroiliac joint stiffness between peripartum pelvic pain patients and controls, with both groups having stiff and unstiff joints. Differences were identified between the groups with regard to the relative difference between left and right sides, with asymmetry in mobility more notable in the symptomatic group.

Damen et al. (2001) also found that pregnant women with moderate or severe pelvic pain have the same sacroiliac joint laxity as pregnant women with no or mild pain. A clear relation was found between asymmetric laxity of the sacroiliac joints in patients with pregnancy-related pelvic pain. This pain includes symptoms from the posterior pelvis and/or from the pubic symphysis. Pelvic pain may be brought on by changes in the center of gravity, secondary to weight gain, creating a strain on weight bearing structures in the body (Damen et al. 2001). General laxity of collagenous tissue created by pregnancy-related hormones may also play a role. Damen et al. (2002) further defined that sacroiliac joint asymmetric laxity measured during pregnancy is predictive of the persistence of moderate to severe pregnancy-related pelvic pain well into the postpartum period.

The Manual Therapy Lesion Concept

The manual therapy lesion integrated histology, biomechanics, neurophysiology and work physiology to more completely define dysfunction and provide a road map for recovery.

Basic Stages of the Manual Therapy Lesion (Grimsby 1988)

1. Collagen/tissue Trauma:

Acute injury, surgery, degenerative joint disease, repetitive strain, postural strain, hypo/hypermobile joint. The may be an acute painful beginning to a pathological state, or a slow progression that appears to have an insidious onset related to gradual degeneration.

2. Receptor Damage:

Structural damage to the receptor, cut in neural pathway, imbedded in non-mobile capsules. Type I receptor more easily damaged than type II, as it is located more superficially on the joint capsule, with type II deeper in tissue. The pain free but restricted spinal segment will have a reduction in afferent input from lack of collagen elasticity changing motor performance within its motor pool.

3. Reduced Muscle Fiber Recruitment:

Abnormal central processing of afferent signals from cutaneous, articular and muscle/tendon receptors. Altered central feedforward facilitation to alpha motor neurons (extrafusal muscle fibers) and gamma motor neurons (intrafusal muscle fibers). End result includes altered reflex responses for proprioception and kinesthesia, motor weakness, motor delay and/or poor timing. Central spinal segmental dysfunction, both with pain and without, alters central programming for distal performance.

4. Tonic Fiber Atrophy:

Reduced motor recruitment resulting in initial lack of recruitment, leading to atrophy over time. Tonic muscle fiber initially more affected by type I mechanoreceptor loss and influences of the type IV mechanoreceptor (pain) system. Phasic muscle fiber atrophy occurring with more significant tissue damage that includes type II mechanoreceptors, and with overall reduction in higher level activity.

5. Reduced Anti-Gravity Stability:

Reduced recruitment of the tonic system results in loss of dynamic arthrokinematic control of joint motion, static postural stability/alignment, and central balance mechanisms. Central stability at the lumbar spine and pelvis is reduced, which may alter the static and dynamic alignment of the hip, knee and ankle.

6. Motion Around Non-physiologic Axis:

Loss of dynamic control increases the neutral zone of function for the joint, increasing the range of the instantaneous axis of motion. Altered mechanics for the biomechanical chain of the lower limb. Reduced tonic function creates an abnormal relationship between tonic (arthrokinematic) and phasic (osteokinematic) muscles. Compensatory motor recruitment occurs leading to abnormal mechanics.

7. Trauma/Acute Locking/Degeneration

Altered axis of motion leads to abnormal tissue stress/strain, resulting in further tissue and receptor damage. Hypermobile joints are prone to acute locking. Joint degenerative changes occur over time.

8. Pain/Guarding/Fear-anxiety of movement

The type IV mechanoreceptor responds to tissue damage with tonic reflexogenic muscle guarding and pain perception. Psychological influences of pain my lead to altered motor patterns and/or reduced effort. The overall effect is reduced tissue tolerance, abnormal joint/tissue loading, reduced afferent feedback, altered feedforward efferent drive and the possibility of reduced function.

Treatment typically focuses on pain and reduction of muscle guarding as the reflex responses from type IV mechanoreceptors register abnormal levels of deformity or chemical concentration. Resolving pain and guarding are just the beginning. Improving tissue tolerance of the injured tissues, reestablishing coordinated and stable movement around a physiological axis and returning to a higher training state are the real treatment goals. Injured tissue not only results in passive instability of joint structures, but damages mechanoreceptors that provide afferent feedback for motor recruitment. Inhibition of normal tonic arthrokinematic muscle can result from pain signals or mechanoreceptor damage. Part of stabilization training is not only to improve tissue tolerance but to improve muscle spindle sensitivity to provide additional afferent feedback to replace that from the damaged mechanoreceptors.

Motor Tests

Gross resisted testing of cardinal plane motions of the hip and trunk can be utilized to identify general directions of weakness that require training. Specific manual muscle testing may also identify isolated muscle weakness providing information for more specific exercise. These classic assessment procedures are better covered in more classic texts. Motor performance tests can provide information related to motor timing, recruitment or stabilizing effects on the pelvis. The ASLR test described below can provide the clinician with information regarding larger movement patterns and assist in exercise selection, design and prescription.

ASLR

The active straight leg raise test (ASLR) demonstrates the ability of load transfers from the lower limbs through the pelvis to the lumbar spine. The ASLR test can be helpful in identifying motor coordination dysfunction in patients with lumbar, SI, and hip dysfunctions. It can be helpful specifically for patients with an SIJ hypermobility to give guidance on which muscle group is most responsible for decreased stability at the SIJ. Studies on the ASLR focused on post pregnancy pelvic pain, and the test was found to be a valid, sensitive, and specific method to discriminate between patients who are disabled by post pregnancy pelvic pain and healthy subjects (Mens et al. 2001). It was also found to be an accurate measurement of disease severity in patients with post pelvic pain after pregnancy (Mens et al. 2002c).

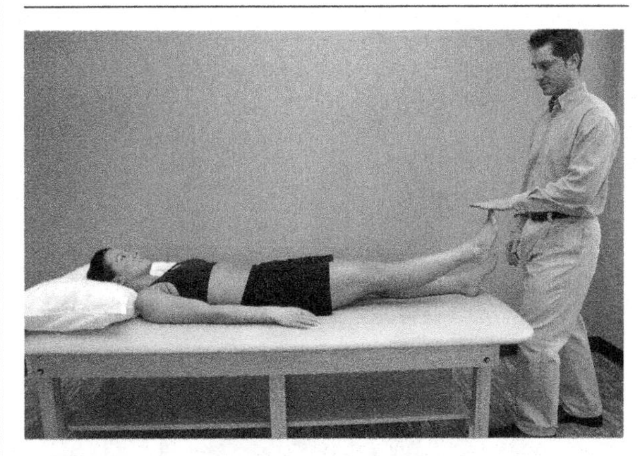

Figure 4.16: The ASLR test is performed in supine with straight legs and the feet 20 cm apart. The patient is then

instructed to "Try to raise the legs, one after the other, 20 cm off the table without bending the knee." The patient then rates the difficulty of each leg lift: Not difficult at all = 0, minimally difficult = 1, somewhat difficult = 2, fairly difficult = 3, very difficult = 4, unable to do = 5 (Mens et al. 2001). The scores for each leg are then added, giving the sum of the two scores a rating of 0–10.

The ASLR was compared in pregnant women with and without low back pain. Those with low back pain demonstrated a disturbed load transfer across the sacroiliac joints, as these women developed higher muscle activity but had significantly lower output (de Groot et al. 2006). Hip flexion force was measured at both zero cm and 20 cm of the leg raise. Sacroiliac pain has also been associated with a decrease in diaphragmatic excursion during ASLR (O'Sullivan et al. 2002).

The use of a pelvic belt to increase stability of the sacroiliac joints has been shown to improve the ASLR test (Mens et al. 1999, Mens et al. 2006). To assess the effect of improved SIJ stability, the ASLR test can be repeated while a manual compressive force is applied through the ilia, or with a belt tightened around the pelvis (O'Sullivan et al. 2002). Some of the participants tested with zero diaphragmatic motion, while several subjects also performed intermittent breath holding during the test. The addition of compression of the pelvis through the ASIS during the ASLR test also increased the diaphragmatic excursion to a level comparable with that of the comparison group (O'Sullivan et al. 2002). Identification of abnormal respiratory patterns should be an indication to incorporate respiration training with all exercises.

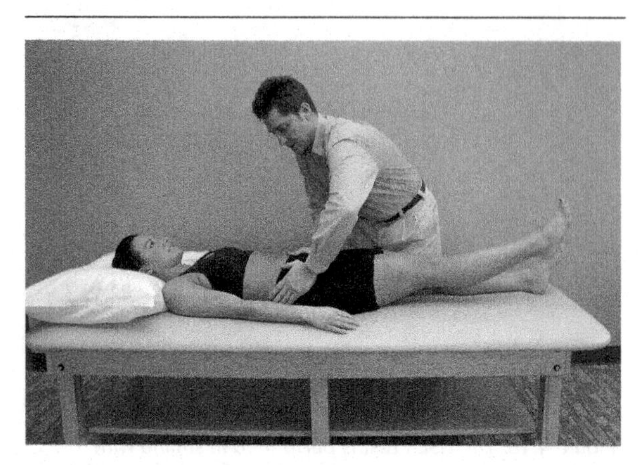

Figure 4.17: Active Straight Leg Raise test with pelvic compression anteriorly (Lee and Vleeming 2007). The therapist adds a compressive force at the ASIS to simulate dynamic stability of the transverse abdominus. Improved ease of ASLR with this compressive force is thought to indicate a need for stabilization of the anterior abdominal muscles, or more specifically the transverse abdominus.

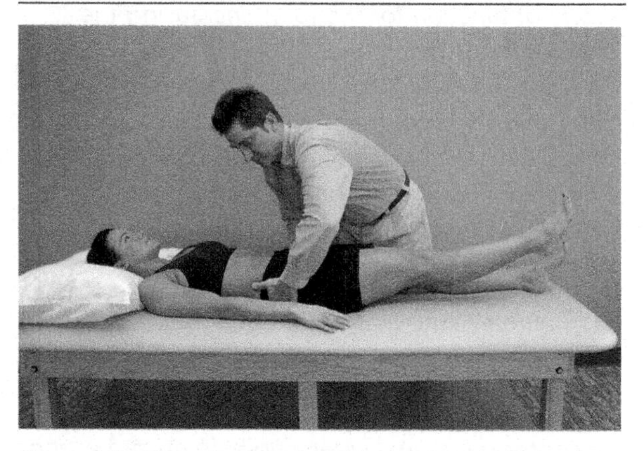

Figure 4.18: Active Straight Left Raise test with enhanced SIJ stability (Lee and Vleeming 2007). The therapist reaches around and behind the pelvis to perform a compressive force to the sacroiliac joints. If symptoms improve or there is an ease of movement with the ASLR test, then stability of the SIJ is indicated with belting and motor training.

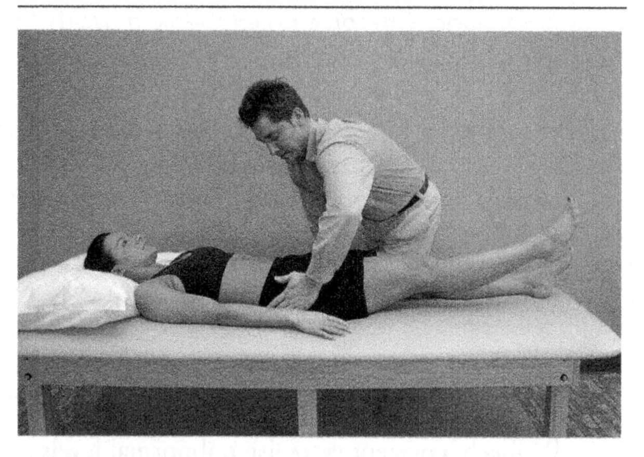

Figure 4.19: Active Straight Leg Raise test with enhanced stability of the pelvic floor. The therapist performs a compressive force at the level of the acetabulum to provide mechanical support to the pelvic floor (Lee 2003). If the test is improved, then training of the pelvic floor is indicated. If symptoms increase or test becomes more difficult this indicates abnormal elevation of pelvic floor tone requiring down training.

Prone Cross Pattern Training

The prone cross pattern test has been described by Janda (1992/1999). This test is similar to the ASLR test but performed in prone with hip extension. This test has not been directly related to improving stability of the SIJ, though the muscles involved have been described as having a positive influence. The procedure for this test is covered in the lumbar chapter, and the reader is referred to that chapter for review. The traditional prone hip extension test can also be modified with manual compression of the pelvis through the sacroiliac joints to assess potential improvements in motor performance during the test. Improvements with manual compression may add further evidence for a diagnosis of SIJ instability, but may also suggest the use of an SIJ belt to improve motor performance and to address muscles associated with SIJ and pelvic stability.

Body Position For Training

A basic challenge is to position the body during exercise to increase the demand on muscle while supporting joint structures. Initial exercises start in the position of comfort for the patient. For the hypermobile pelvis, standing is often the most stable and comfortable position. These patients may even complain of the most significant symptoms when they are lying, rolling and transferring. Standing creates stability with the sacrum forced down between the ilia. Motor facilitation is greater in standing due to postural reflexes. Muscles that may be inhibited from pain inhibition or tissue trauma may respond better in standing postures. Muscles of the hip joints, involving the gluteal muscles and piriformis, are also activated in a standing position more so than in recumbent postures. Activity for the internal oblique muscles is higher in sitting compared to lying supine, but standing activity has been shown to be facilitate even greater recruitment (Snijders et al. 1995).

Slouched sitting during exercises should be avoided to prevent abnormal tissue strain and to avoid reinforcement of bad habits. Slouching combined with a backward rotation of the pelvis in sitting results in backward rotation of the sacrum with respect to the ilium, dorsal widening of the intervertebral disc at L5–S1 and strain on the iliolumbar ligaments when protection from back muscles against lumbar flexion is absent (Snijders et al. 2004). Sitting with the hips abducted, weight anterior to the ischial tuberosities and/or use of a lumbar backrest support almost eliminates lumbosacral and sacroiliac movement.

Progression of training postures may begin in standing, progressing to sitting and lying postures, based on stability of the hypermobile pelvis. Utilizing an SIJ belt to stabilize the pelvic may allow training in any posture. Hypomobile joints, or issues of dysfunction that do not related to mobility, may begin in postures that are more functional or specific to training certain muscles.

Motions and Directions

Depending upon patient presentation and functional goals, a specific prescription of exercise and patient education can be constructed in an attempt to regain or enhance muscle coordination, ability to generate force and increase muscular volume. Activities should also supply the optimal stimulus for regeneration of injured connective tissues (ligaments etc.), while being very careful to not exercise through pain. Pain is often a key indicator that the activity is further injuring the pathological tissues, which may lead to tonic muscle guarding in the region. Predominantly tonic muscles that reflexively guard with pain include the piriformis, multifidus and quadratus lumborum. Collagen injury may also damage the matrix encapsulated and complex unencapsulated nerve endings needed to provide important afferent information for proprioception and coordination to the central nervous system.

The reciprocal relationship of the latissimus dorsi on one side and the gluteus maximus on the other side has been demonstrated anatomically. To demonstrate this relationship by muscle action, electromyographic studies were performed in 15 healthy individuals (Mooney et al. 2001). This formed the baseline for evaluation of five symptomatic patients with sacroiliac dysfunction. Abnormal hyperactivity of the gluteus muscle on the involved side and increased activity of the latissimus on the contralateral side was contrasted with normal function of the healthy individuals. All patients in the torso rotary strengthening exercise program improved in strength and return of myoelectric activity to more normal patterns (Mooney et al. 2001).

While it is of great importance to strengthen the gluteal musculature and train for the synergistic coordinated firing of the gluteus maximus and the sacral originating external rotators of the hip, integrating exercise for other trunk and hip muscles is also necessary. Exercises should include the erector spinae, the sacral multifidi, quadratus lumborum, latissimus, hamstrings, hip flexors, adductors and the entire abdominal complex. Cross pattern training of these synergistic muscles address anatomical and functional relationships of the pelvic joints with the lower limb, thoracolumbar fascia, trunk musculature and muscles of the opposite shoulder girdle. With weakness in the pelvis and lower lumbar spine, compensatory muscle recruitment may be noted in the upper thoracic spine, dorsal scapular muscles and cervical spine. Pelvic floor muscles may also need to be integrated into a more complete program for pelvic stability.

Extension: Gluteal Training

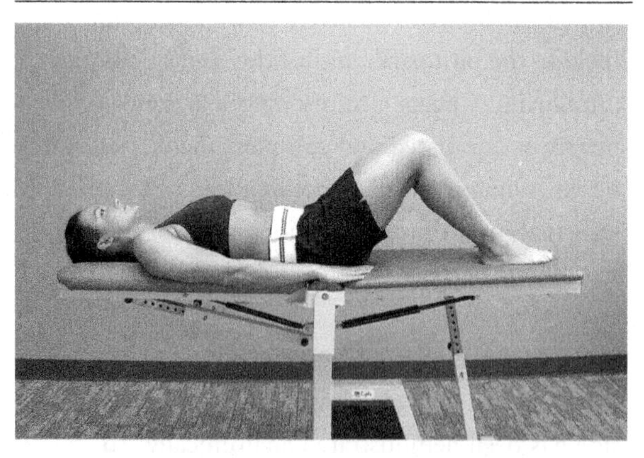

Figure 4.20: Gluteal sets. The patient is with the knees flexed and the hips abducted and externally rotated. This is the loose packed position for both the hip and sacroiliac joints. The patient will be asked to tighten or clinch the gluteus maximus muscles and hold for a one count. This firing will create a moment of muscle force causing hip extension and some external rotation. The loose packed position is also the posture that allows for the greatest recruitment of both tonic and phasic muscles within the joint system as both Type I and Type II mechanoreceptors fire to assist in recruiting motor facilitation. A SIJ belt may be utilized to increase stability of the pelvis during training. A simple progression is to continue with similar exercises without the use of the SIJ belt.

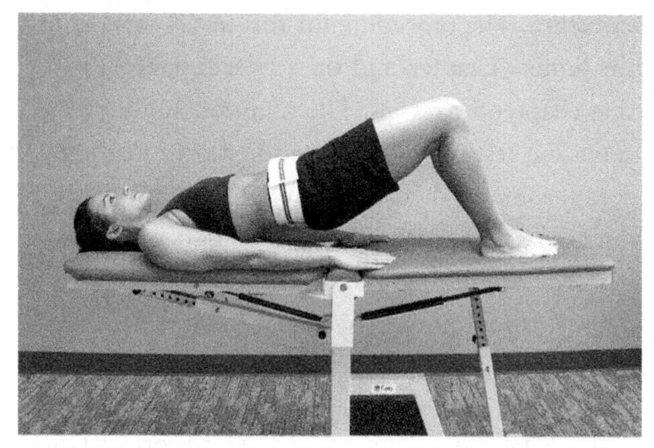

Figure 4.21: Bridging. The patient begins in the same starting position as with the gluteal sets, with the addition of lifting the buttocks from the contact surface in the pain free range, or to a point where the femur would be near in line with the torso. A one-count gluteal set is performed at the top of the bridge. Special care should be taken to keep the patient out of hyperextension of the lumbal spine. An SIJ belt may be utilized to increase stability of the pelvis during training.

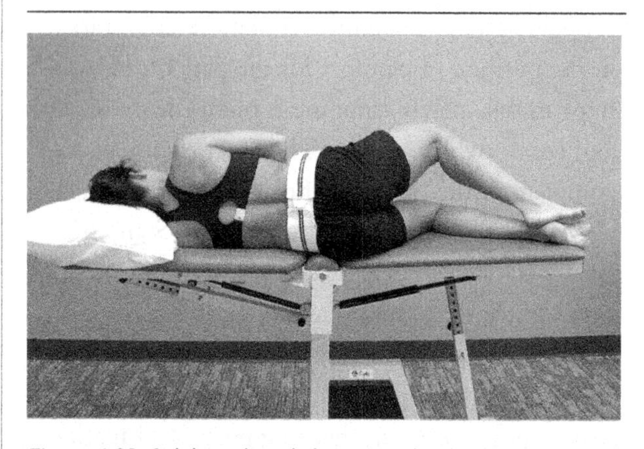

Figure 4.22: Sidelying hip abduction with a belt (Clam exercise). The Clam exercise traditionally attempts to train the gluteus medius while stabilizing the lumbar spine. Active horizontal abduction of the hip will create a relative compression moment of the SIJ. The patient is in sidelying with the knees and hips flexed to 45° with a pillow placed between the knees. The pillow is important to limit the adduction range of motion at rest. The patient abducts the hip while keeping the feet together and moderately tightening the gluteals at end range.

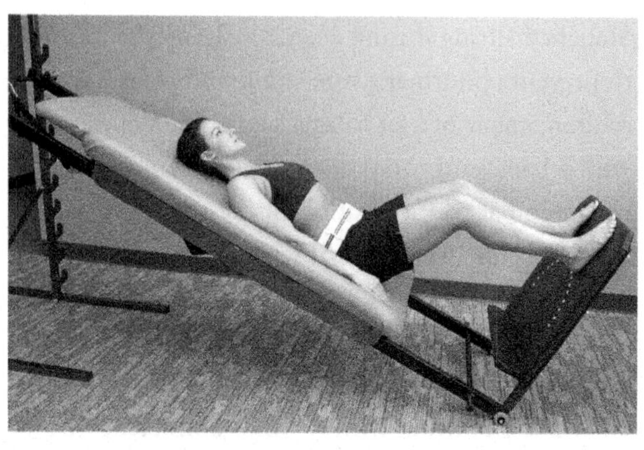

Figure 4.23: Squats – unloaded with sled. The squat motion

is a functional movement involving the major muscles of the hip and pelvis. An adjustable hip sled allows for reduced body weight with a more horizontal position, allowing high repetition coordination training. The hips are in approximately 75 degrees of flexion and slightly externally rotated with the feet slightly elevated on the footplate. Do not position the patient in a starting position with too much hip flexion, which can create a moment for force that causes posterior rotation of the ilium and overload of injured tissues. An SIJ belt may be utilized to increase stability of the pelvis during training.

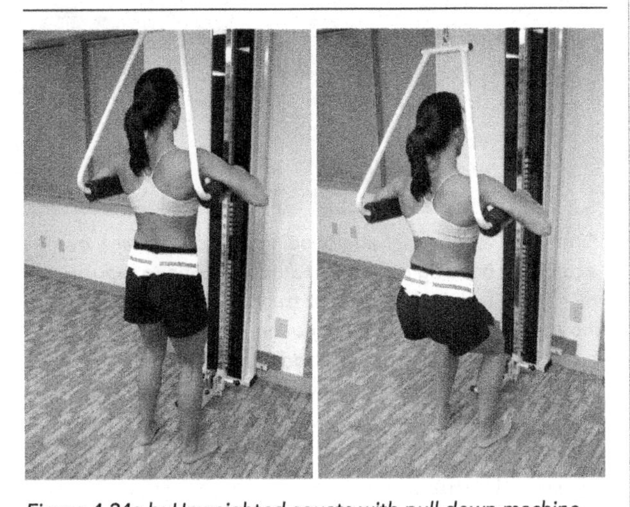

Figure 4.24a,b: Unweighted squats with pull down machine gantry allows the patient to initiate exercise in the functional standing position. The patient stands with the arms resting on the bars of the unweighting gantry. The weight is adjusted to reduce body weight to the point of no pain during the exercise and/or the desired number of repetitions can be performed. The feet will be slightly less than shoulder width apart and pointed out about 20 degrees to position the hips in more of a loose packed position and to decrease loading to the dorsal SIJ ligaments. The patient is instructed to perform a functional squat with special care in maintaining a solid base and keeping the knees in line with the feet. An SIJ belt may be utilized to increase stability of the pelvis during training. The therapist may want to use a gait training belt as the patient is first instructed in this exercise.

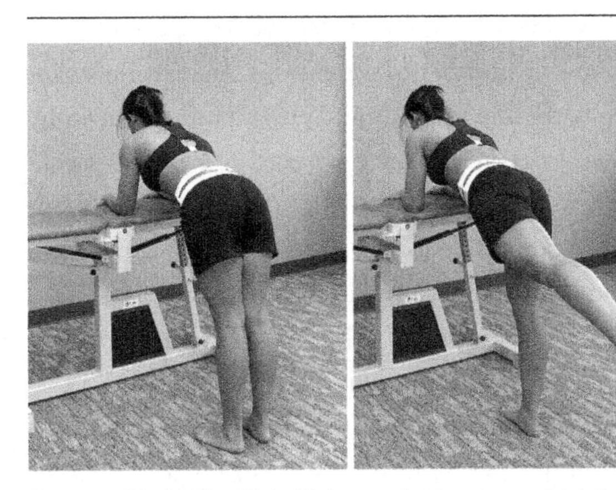

Figure 4.25a,b: The Mule Kick exercise is an open kinetic chain exercise designed to strengthen the hip extensors. The patient leans over a table and flexes the exercising hip and knee to 90 degrees. Hip and knee extension is performed to the point where the lower extremity is in line with the torso. A one count hold is performed at the out point. Performing hip extension with the leg in 15 to 20 degrees of hip external rotation will increase the rotary component of the gluteus maximus and deep hip external rotators. An SIJ belt may be utilized to increase stability of the pelvis during training.

Training the hip joint for stabilization of the plevis is not only performed in extension, but may address the abduction and rotary components of the hips as well. Below are a few examples for these motions. The reader is referred to the hip chapter for more options for early training for these motions.

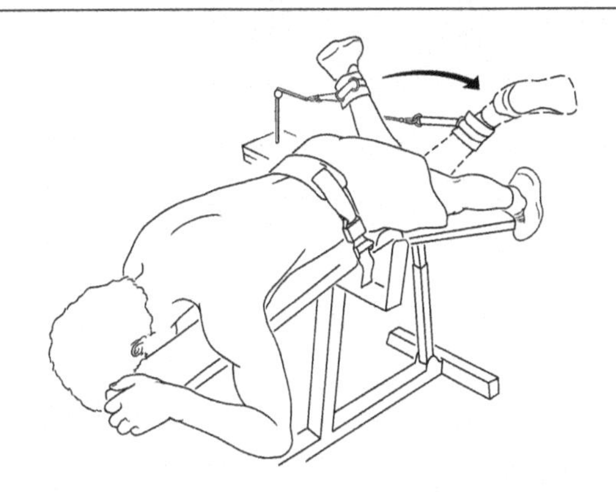

Figure 4.26: Hip external rotation in prone. Direct training to the piriformis and deep hip external rotators will assist in dynamic stabilization of the SIJ. Despite the side of symptoms, both hips should be tested and potentially trained if found weak. Initially a belt can used to stabilize the pelvis to the table, which will isolate hip motion. A progression of removing the belt requires more dynamic stabilization of the trunk muscles.

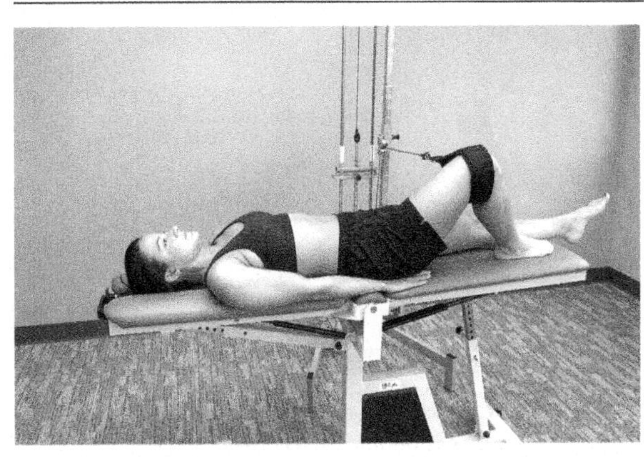

Figure 4.27: Supine hip abduction. The patient is in supine with the knee flexed to 45° and a resistance around the knee. Horizontal hip abduction is performed in this position, attempting to emphasize the gluteus medius in a supine position for training.

Side Bending

Performing exercises in side bending address

motor training of the quadratus lumborum, directly attaching to the pelvis from the rib cage, the oblique abdominals and the lumbar multifidi. Side bending motions may have cause less of a mechanical strain on a hypermobile SIJ, as opposed to motions in the sagittal plane (flexion/extension).

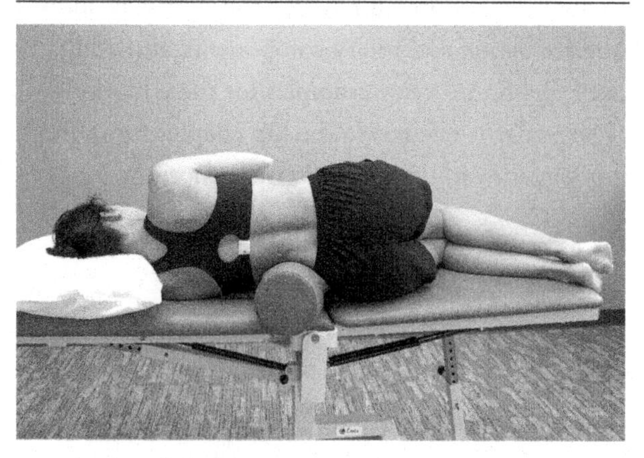

Figure 4.28: Sidelying hip hike. The hip hiking emphasizes the quadratus lumborum and lumbar multifidi. The patient is in sidelying with the hips slightly flexed. A bolster or pillow can be placed at the waist angle to decrease lumbar side bending. The patient is instructed to approximate the iliac crest to the rib cage, facilitate the quadratus lumborum. This may be very difficult for some patients with challenged proprioception and coordination to perform. The therapist may perform gentle manual resistance along the iliac crest and trochanter, as well as the rib cage and supplying the patient appropriate cuing for performance. An SIJ belt may be utilized to increase stability of the pelvis during training.

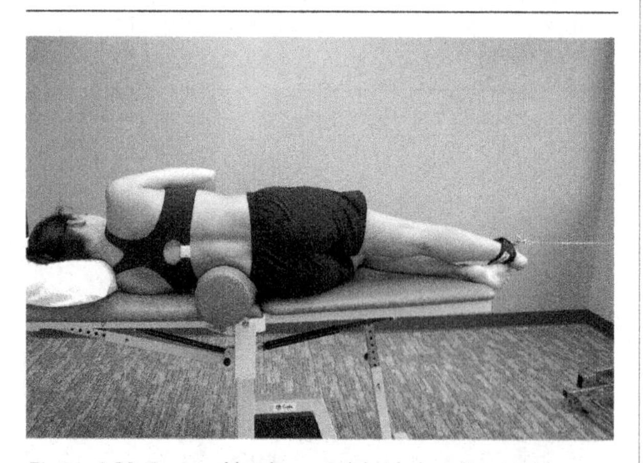

Figure 4.29: Resisted lumbar caudal side bending in side lying. The exercise is performed as previously described with the addition of a resistance from a wall pulley placed at the ankle. Providing resistance allows for training specific functional qualities in the muscles, such as local vascularity.

With pelvic instability, standing is often a position of least pain and greatest functional stability. Performing simple standing side bending training while holding a free weight at the side a basic and classic exercise for side bend training. This exercise is easily reproduced at home for early cardinal plane training, prior to combined patterns.

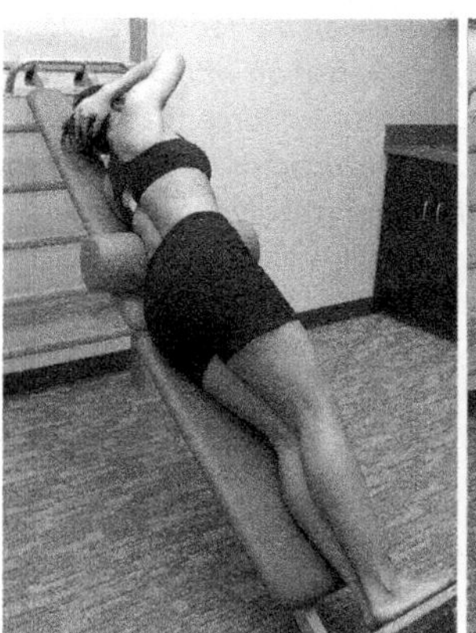

Figure 4.30a,b: Cranial side bending on a slant board, over a roll. Side bending is performed on an incline to reduce the overall weight of the upper trunk. Larger bolsters will increase the excursion of the motion. The partial weight bearing position may provide additional stability to the pelvis, as opposed to lying down postures.

Flexion Training

Training the hip flexors may also be necessary. The previous example, in the mobilization section, presented a supine hip flexion exercise with an unloading or loading option to adjust the resistance. Pain from sacroiliac pathology can lead to muscle guarding of the psoas and iliacus muscles. Muscle guarding creates a functionally "short" muscle that can effect the standing alignment and mobility of the SIJ. Passive stretching of a muscle in guarding is not as effective as repetitive exercise dosed to improve local circulation. This is most efficient around 60% of 1RM with sets of 24 repetitions. To achieve this number of repetitions, the weight of the leg may require unloading. With significant atrophy and weakness resistance may need to be even lower.

With hypermobile joints, initial range of training is from mid range to the opposite direction of the hypermobility. For the SIJ, posterior rotation of the ilium is the most common hypermobile range, as this unlocks the joint. If hypermobility is in anterior rotation of the ilium, than this end range is initially avoided. In a significantly unstable SIJ or pelvis, the range of training needs to remain in mid range, avoiding both end range excursions. In cases of posterior torsion hypermobility, the initial training range of the hip flexor muscles is from neutral to extension range at the hip joint. This emphasizes lengthening of the hip flexors. Limiting the range of hip flexion to around 60 degrees will prevent excessive strain to the SIJ ligamentous structures.

Excessive tone in the hip flexors may also be to compensate for weak or inhibited lower abdominal muscles. Sacroiliac stiffness is improved with activation of the transverse abdominus through a drawing in or bracing technique (Richardson et al. 2002). Training the lower abdominals, in isolation from the hip flexors, is performed by creating an extensor pattern in the lower limbs during a sit up performance. Simply flexing the hip joints does not prevent the hip flexors from participating during a sit up exercise. The extensor pattern of plantar flexion, knee extension and hip extension will inhibit hip flexion. With the psoas and iliacus inhibited, the lower abdominals will be the primary muscles activated with cranial trunk flexion.

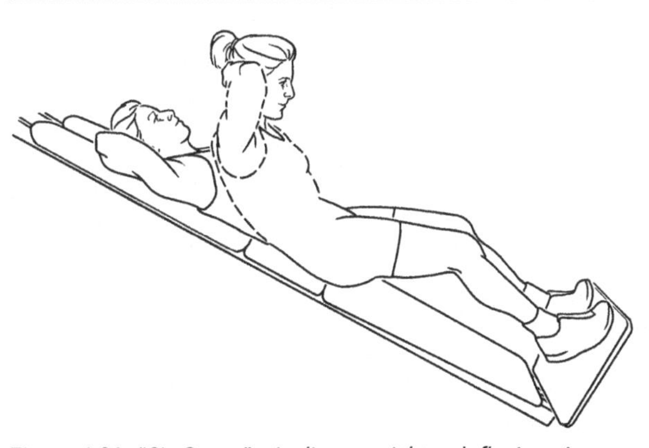

Figure 4.31: "Sit-Overs"—Incline cranial trunk flexion. An incline position with slight plantar flexion, knee flexion and hip flexion will facilitate an extensor pattern in the lower limbs. A pattern of plantar flexors, knee extensors and hip extensors will inhibit the flexor pattern including the psoas and iliacus. With cranial flexion, the lower abdominals bear most of the load. The incline board is used to create the extensor pattern but is also necessary to reduce the load to allow for 20–30 repetitions with the lower abdominals working more independently. The patient breathes out while performing a hollowing or bracing motion, and then segmentally flexes the trunk to a vertical position.

If a slant board is not available, the incline position for inhibition of the hip flexors during cranial flexion training can be duplicated with other types of equipment. For clinical training, the slant board is the most effective in providing the partial weight bearing position on a stable surface. The home or health club version can be duplicated on fitness balls or weight training benches. The fitness ball provides a limited incline position and the recoil of the ball itself may serve as too much of an assist for the concentric phase of the performance. The labile surface of the ball can be a later stage challenge.

Figure 4.32: Sit-Overs with a ball. If an incline board is not available, an attempt to create an extensor pattern for cranial flexion can be made with a fitness ball. The feet are placed against a wall in a plantar flexed position with the knees flexed for quadriceps facilitation. The trunk is angled on the ball in an incline position to facilitate the hip extensors during cranial flexion. The coordination challenge on the fitness ball may be too high for early training and cannot easily be adjusted for the level of resistance but can be a successful home option when an incline board is not available. The fitness ball approach will be easier than the slant board due to the recoil of the ball assisting the trunk motion.

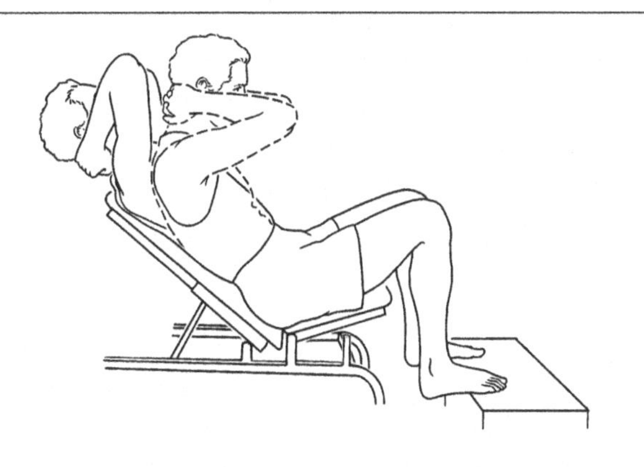

Figure 4.33: Sit-Overs with an incline weight bench. If an incline board is not available, an attempt to create an extensor pattern for cranial flexion can be made with an incline weight bench. The feet are placed on a stool or against a wall allowing for an extensor pattern through the lower quarters (the knees should be more extended than pictured). The patient is asked to push there feet into the foot stool or wall to facilitate the extensor pattern while performing cranial flexion.

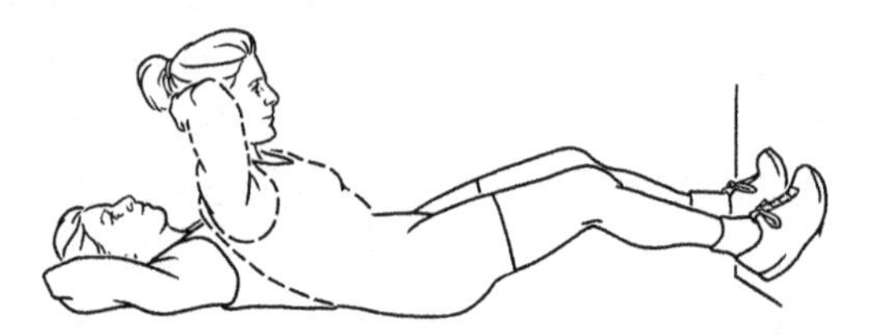

Figure 4.34: If no incline equipment is available, the basic crunch sit-up can be performed with the toes pushing into a wall. The extensor pattern will not be as strong as with an incline, but some level of inhibition of the hip flexors will take place. Cranial flexion is then performed with emphasis on initial low level setting of the transverse abdominus prior to lifting the shoulders off the floor.

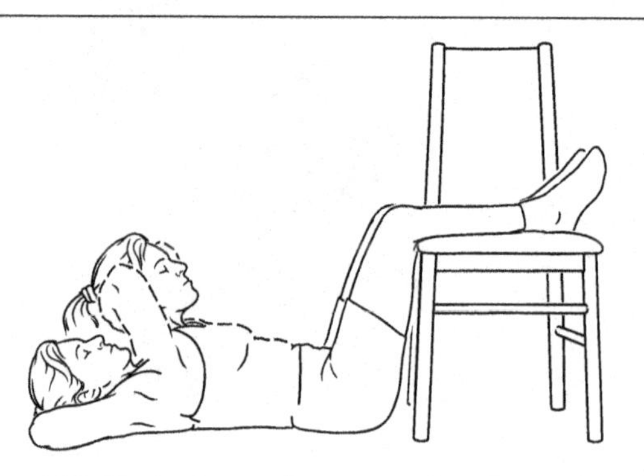

Figure 4.35: The basic crunch sit-up involves flexing the hips (here shown as the legs supported on a chair with the hips flexed to 90°). The flexed position of the hips is thought to make the hip flexors short and inhibited. The hip flexors will, however, still contract during the exercise, which may or may not be desirable. The patient is instructed to breathe out while tightening the lower abdominals and then slightly lifting the shoulders off the floor. The arms should not pull on the neck to assist the motion.

With higher levels of pain and/or impairment, early abdominal training may not allow for motion of the lumbar spine and pelvis. Indirect training through the arms may be a necessary starting point to facilitate the abdominal muscles.

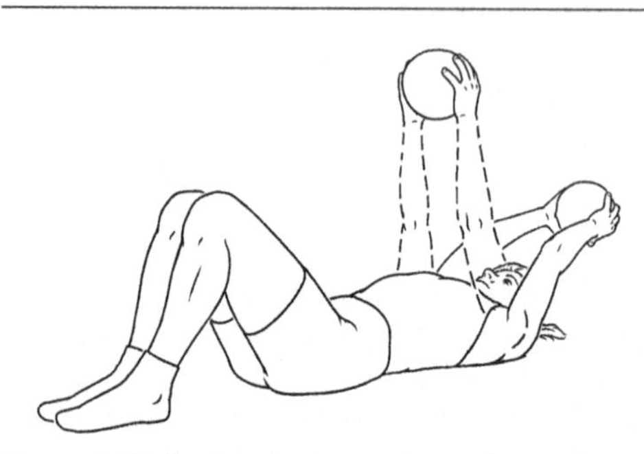

Figure 4.36: Supine arm extension with a medicine ball. The patient is in supine with the hips flexed to 45°. A medicine ball is held in both hands and lowered back above the head. The patient is instructed to breathe out, tighten the lower stomach and return the arms to a vertical position.

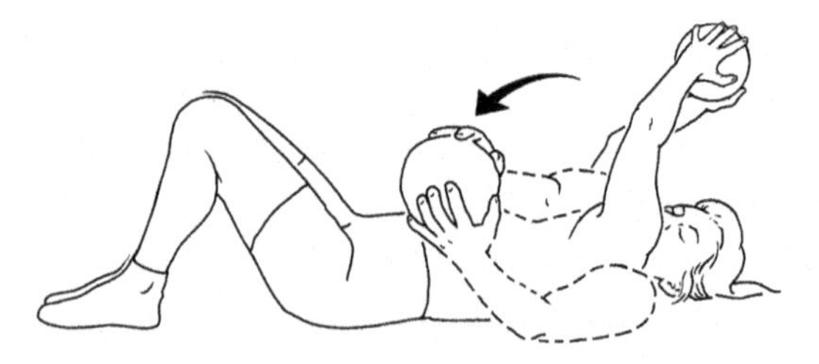

Figure 4.37: Arms motions with the medicine ball can include larger excursions or diagonal motions to facilitate the oblique abomdinals. Performing the motions with one arm can increase the challenge with asymmetrical forces through the upper trunk.

Rotation Training

As the pelvis does not have a true transverse plane motion available, rotation is trained through the lumbar spine and/or hip joints. This transition of rotary forces through the hips and pelvis to the lumbar spine is necessary for normal motion. The reader is referred to the chapters on the hip joints and lumbar spine for integration of rotational concepts and exercises. A few basic early training examples are provided below.

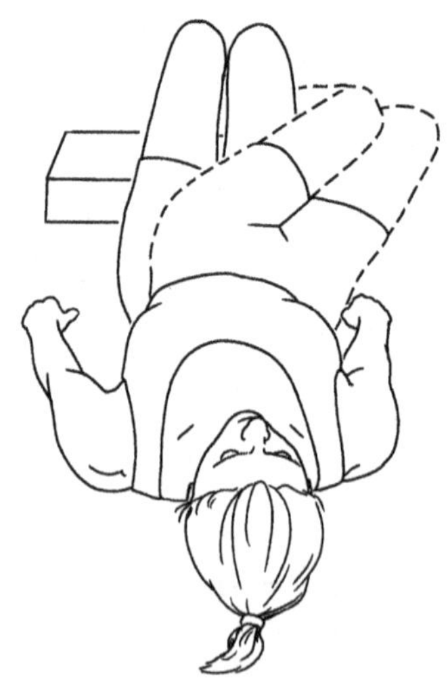

Figure 4.38: Hip Roll—Caudal supine left rotation. The patient is supine with the hips flexed to 45° and the knees at 90°. A bolster is placed under the feet, if necessary, to place the lumbar spine in neutral or slight flexion. The knees are eccentrically controlled out to the side with a concentric return to neutral. Emphasis is placed on the concentric return initiated from the lower abdominals and lumbar spine, not the hip and thigh muscles. The patient is instructed to breath out while bracing or hollowing the lumbar spine prior to initiating motion of the lower limbs. As little rotary mobility is available at each lumbar segment, the knees do not need to move far to the side to effectively train tissue and motor performance. Beginning with a "10 and 2" position on a clock is a good initial guideline, progressing in range as tolerated and/or necessary.

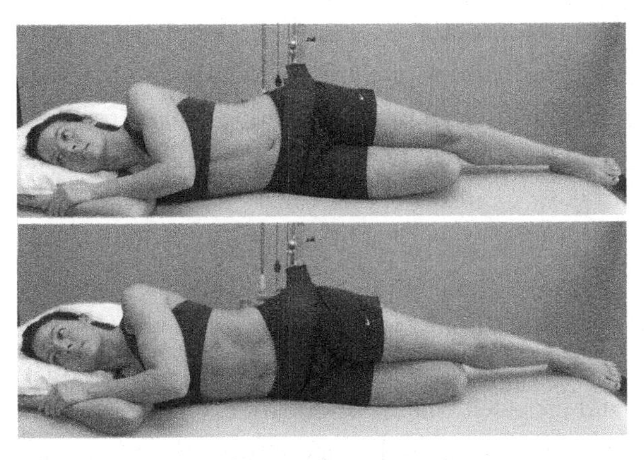

Figure 4.39a,b: Sidelying caudal rotation with posterior resistance. The patient is in side lying with the bottom leg slightly flexed for stability and the top leg straight to create an axis of motion. A bolster may be used under the foot to prevent an adduction moment at the hip joint. The posterior force moment will emphasize the anterior abdominal muscles during caudal rotation, but also creates a synergistic co-contraction of the lumbar multifidi. The force moment from the pulley is in the fiber line of the transverse abdominus, creating an external cue for recruitment of this muscles without cognitive effort.

Pelvic Floor Training

Training of pelvic floor muscles may be a primary intervention or an adjunct to stabilization training of the sacroiliac joints. Primary pelvic floor training may address urinary or fecal incontinence, as well as sexual dysfunction. Beginning with pelvic floor muscle contraction can be a helpful training technique when training the hip, abdominals or back muscles to enhance motor recruitment and dynamic stabilization of the pelvis.

It is not uncommon to find pelvic floor muscular dysfunction in patients without a primary incontinence issue or in patients who are not pregnant. As most of the pelvic floor muscles are innervated by the pudendal nerve originating from the sacral plexus, nerve roots S2–S4, lumbar and sacral dysfunction can lead to a segmental facilitation affecting the pelvic floor muscles. Pain itself can also "turn off" the pelvic floor or alter its sequence of recruitment, as increased firing of nociceptors can alter the normal firing pattern of the PFM. Changes following an MVA or surgery can also alter pelvic floor muscle tone, including lack of exercise which can decrease tone in the abdominal and pelvic floor musculature. Therefore it is important to assess the pelvic floor muscular integrity in all patients with lumbar, SI, hip, or lower extremity dysfunctions, not just those complaining of incontinence.

Studies have shown the effectiveness of pelvic floor exercise programs for both decreasing urinary incontinence and decreasing pelvic and lumbar pain during pregnancy. Specially designed postpartum pelvic floor muscle exercises have been effective in increasing pelvic floor muscle strength and reducing urinary incontinence in the immediate postpartum period (Morkved S, Bo K 1997). Compared with a control group providing usual postpartum care, the addition of specific pelvic floor exercise training as well as strategies to improve adherence significantly improved urinary continence after delivery (Chiarelli et al. 2006, Sampselle et al. 1998). In fact, compliance with pelvic muscle exercises has been found to decrease urinary stress incontinence by 85.2 percent (Siu et al. 2003). Along with pelvic floor muscle exercises, the addition of biofeedback for treatment of urinary stress incontinence has been shown to further decrease incontinence (Aksac et al. 2003, Sung et al. 2000, Dumoulin et al. 1995b). Combined treatment of pelvic floor exercise and biofeedback has been found not only to reduce urinary incontinence, but fecal incontinence as well (Pager et al. 2002). Other studies have shown pelvic floor muscle training to be effective treatment for all types of incontinence including stress, mixed, and urge incontinence, as well as for decreasing incidence of anterior vaginal prolapse (Fisher and Riolo 2004).

Incontinence is not only associated with pregnancy, but can affect women and men related to aging or back pain. Many physical therapists do not evaluate or treat the pelvic floor musculature, dismissing it as "not their area." To disregard the pelvic floor musculature in a patient a physical therapist is treating for lumbar or sacroiliac pain is as negligent as refusing to address scapular stability when treating a shoulder patient with a primary rotator cuff tear. The two are intimately related, and dysfunction in the proximal deep tonic musculature such as the pelvic floor can lead to more distal injury in much the same way that a winging or unstable scapula, eventually lead to shoulder injury.

Pelvic floor training for stress urinary incontinence, unrelated to pregnancy, has been shown to be effective in reducing urinary loss (Dougherty et al. 1993). Even a small reduction in leakage can make a big difference in quality of life to patients with incontinence, and many are not aware that treatment such as a pelvic floor exercise program can improve their quality of life. Eliasson et al. (2002) found 80 percent of trampolinists noted leaking with their sport, 51 percent of these young girls admitted to being embarrassed by their leaking, and this embarrassment can lead to changes in their quality of life. They may choose to quit their sport, a mother who leaks when jumping may chose not to attend her daughter's soccer game due to concern she will leak when she cheers, or a man who leaks with running may choose not to participate in his local fun run due to fear of leakage. Education that incontinence is not uncommon, that it can be treated, and that studies such as those mentioned have showed this treatment to be effective, can give these patients hope as well as the ability to return to a better quality of life.

When using biofeedback devices, the location of electrode placement may be important for success in treatment for urinary incontinence. External, and or internal electrodes can be useful. External placement higher on the pubic symphysis was found to be more comfortable and require less amplitude of the machine to record pressure than placement inferior to the pubic symphysis (Dumoulin et al. 1995a). Although the immediate benefits of biofeedback are clear, the long term benefits are not, therefore Peticca et al. (2002) suggested patients treated for incontinence participate in a follow up biofeedback session after one year to maintain improvement. Follow up biofeedback may not be necessary, as compliance with a pelvic floor muscle exercise program has been found to be effective in maintaining or improving continence after a successful treatment with biofeedback or electrical stimulation (Hayn et al. 2000). In a study comparing bladder training and pelvic muscle exercises, both groups were found to show improvement in urinary incontinence over the control group. Bladder training consists of a progressive program in which the interval between voluntary voiding was gradually increased. The bladder training group had decreased frequency and increased volume of urination, while the pelvic floor exercise group had increased peak and average muscle contraction (Yoon et al. 2002).

Biofeedback and subsequent pelvic floor home exercise program can be helpful not only for low level patients with incontinence, but also for higher level patients such as those who only experience leakage with jumping or other high impact activity. Biofeedback probes can be in place during jumping on the trampoline, jogging, or going up and down a stair, and can monitor pelvic floor muscular activity during these functional activities so that the patient is aware of muscular dysfunction during these activities and can monitor progress via biofeedback over time.

Providing the patient a standardized protocol checklist, so the patient can monitor their own program, optimize long-term treatment outcomes and compliance with improved long-term adherence in pelvic floor exercise programs (Alewijnse et al. 2002). Instructing patients to return for an advanced home exercise program training may also be beneficial, as progressing exercises in order to further functional gain is optimal. The addition of a general health education program has not been found to give added benefit. The addition of adding an audiotape for home pelvic exercises has been found to increase long term compliance from 65–100 percent (Gallo et al. 1997).

Long-term compliance with a pelvic floor muscle exercise program has been assessed by several studies, and short-term adherence to home exercises has been found to significantly predict long-term adherence. Women with more frequent episodes of incontinence were also found to be more likely to comply with a home program (Alewijse et al. 2003). Long term continence benefits of a pelvic floor exercise program has been found to be equal between intense in-clinic programs and a more general home exercise program. Whereas initially there was found to be a 60 percent continence improvement in an intense in-clinic exercise group compared with only a 17 percent improvement in the home exercise group, 15 years later both groups had equal long-term adherence rate and equal urinary symptoms (Bo et al. 1996). Another study showed that five years after cessation of organized training of pelvic floor muscles for incontinence, 70 percent of patients continued PFM exercises at least once weekly, 75 percent still had no leakage during a stress test, and there was an overall 70 percent satisfaction rate (Bo 2004).

Finding and isolating the pelvic floor muscles may be difficult for the patient with hypotonus or weakness. A biofeedback machine may be beneficial in order to provide cueing for proper contraction and relaxation of these muscles. If a biofeedback machine is not available, verbal cueing to localize pelvic floor muscles might be useful. The patient is cued to create tension in the muscles that stop the flow of urine, drawing tension upward to the bikini line without straining or pushing out the belly. When verbal cueing is not enough, self monitoring and tactile feedback is helpful. Females can be instructed to insert their finger or a tampon into the vaginal canal and practice contractions. Men or women can recreate the specific contraction by stopping the flow of urine. This is to be done as only a test, in order to ensure they are recreating the desired contraction. When done repeatedly, this may increase the likelihood of bladder infections.

Isolation of Pelvic Floor Musculature

Learning how to perform a "Kegel" can be difficult task, as the patient often has difficulty visualizing these muscles and/or perceiving a contraction. Motor learning techniques can involve either internal or external cues for the patient. Internal cues involve a focus within the body, including visualizing the pelvic floor musculature function. Common cues may involve an internal perception of stopping the flow of urine, saying the "p" sound to go along with contraction, saying the "t" sound with pelvic floor bulging, or visualizing picking up a small object with the muscles of the pelvic floor. External cues are more efficient for motor learning, involving a focus on the body's response to the external environment. Many patients have trouble with internal cues associated with pelvic floor training, and may benefit more from external cues or a combination of both. Biofeedback is one method of external cueing, making use of kinesthetic awareness with the probe directly on the muscles to be contracted, visual cues as the patient looks for the muscle contraction on the computer screen, and audio cues as a tone can sound when the patient reaches his or her goals. External cueing can also be achieved by sitting on a towel folded into a rectangle or triangle. The pressure of the towel on the pelvic floor provides an external cue for the muscles to push against, allowing the patient to feel their function. A large exercise ball, or on a small medicine ball can provide similar, yet softer and more diffuse feedback. The exercise may simply be to push down into the towel or ball for relaxation of the pelvis floor (bulging), or lifting up off the ball for contraction.

Pelvic floor contractions can be progressed from slow contraction to quick flicks, which require increased levels of coordination to perform. Isometric holds can also be added to fixate strength. Coordination training is emphasized with the "elevator/basement" approach to pelvic floor contraction. The patient is instructed to contract the pelvic floor and perceive an elevator rising upward. The contraction can be held at different levels of submaximal contractions to represent different "floors." More typical of the contraction performed throughout the day by this primarily tonic muscle group to avoid urination and defecation. The contraction would begin as a bulge, progressing to relaxation, minimal contraction, moderate contraction and finally a maximal contraction. Progression may include performing as many "elevator floors" as possible.

Figure 4.40: Pelvic floor training with adductor ball squeeze. Contracting the hip adductors along with the pelvic floor muscles can be done by sitting with a ball between the legs and squeezing the ball with a pelvic floor contraction. Pelvic floor training in conjunct with other bigger muscle groups may also assist in facilitation of pelvic musculature.

Figure 4.41: External cues may be beneficial in engaging pelvic floor muscles. Sitting on a towel folded into a triangle can provide pressure feedback for contracting the floor to lift up off the towel. The point of the triangle faces anterior. The feedback technique can be added to any sitting exercise to enhance awareness and feedback of pelvic floor contraction.

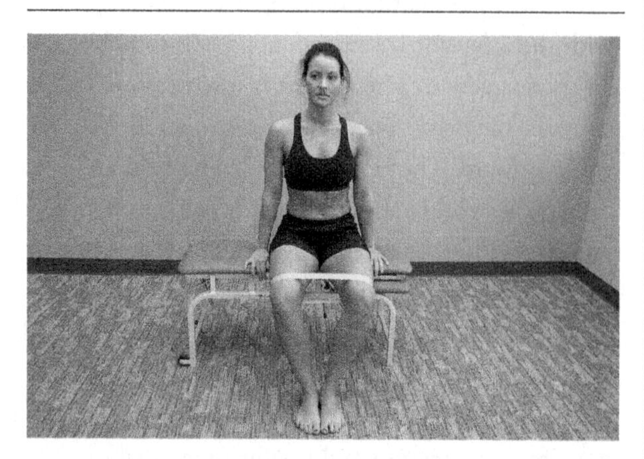

Figure 4.42: Sitting hip abduction. The patient is in sitting with an elastic band around the knees and the feet together. The knees are rolled out against the band two to three inches, emphasizing the obturator internus muscles, as the pelvic floor muscles contract. Breathing is maintained during the exercise, rather than breath holding.

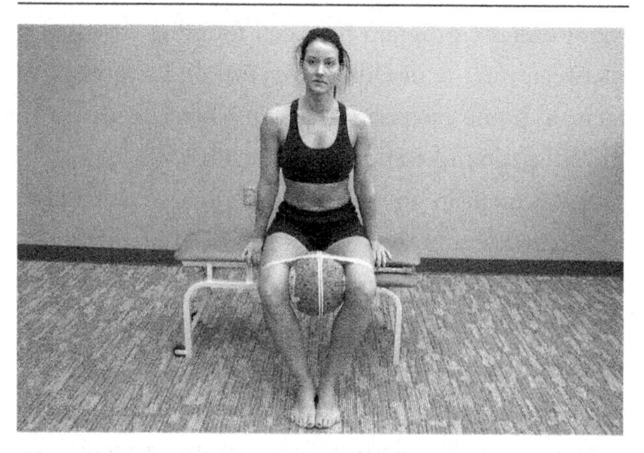

Figure 4.43: Combination hip adduction and abduction in sitting. As a progression, these two muscle groups can be combined with both a ball and elastic band, alternating between pushing in against the ball, and rolling out against the band, engaging the pelvic floor musculature simultaneously with both.

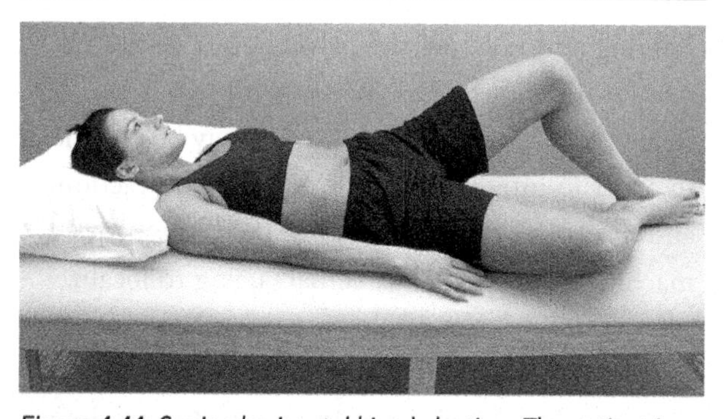

Figure 4.44: Supine horizontal hip abduction. The patient is sitting with both hips flexed to 45°, and the knees at least 90° flexed. The right hip is abducted horizontally, eccentrically controlled by the hip adductors. The motion will recruit the lower abdominal muscles to stabilize the pelvis and lumbar spine. The pelvic floor is also recruited, with an added Kegal contraction on the return to neutral.

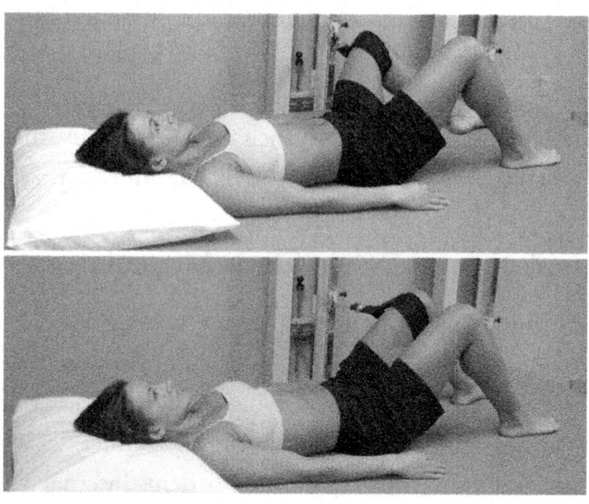

Figure 4.45a,b: Supine resisted horizontal hip adduction.

As above, horizontal hip adduction is performed against resistance to increase the recruitment of the hip adductors and pelvic floor, as well as an increased challenge for lumbosacral stabilization.

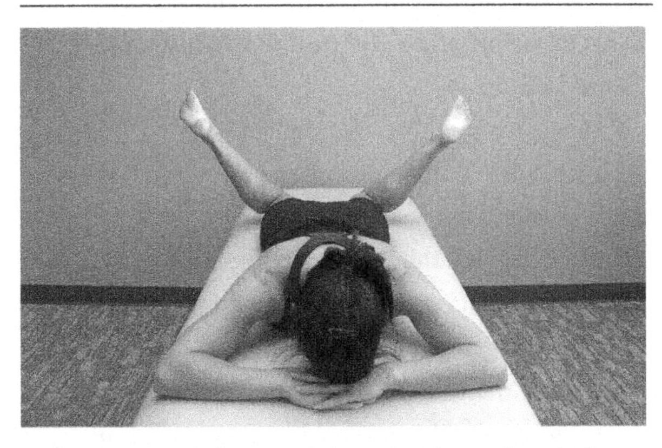

Figure 4.46: Prone hip external rotation, knees flexed. The patient is prone with the knees flexed 90°. The feet are separated, eccentrically lowering the feet to the side. The pelvic floor is relaxed, with deep inspiration, as the feet fall out to the side. On the concentric return, the hip external rotators contract in synergy with the pelvic floor muscles.

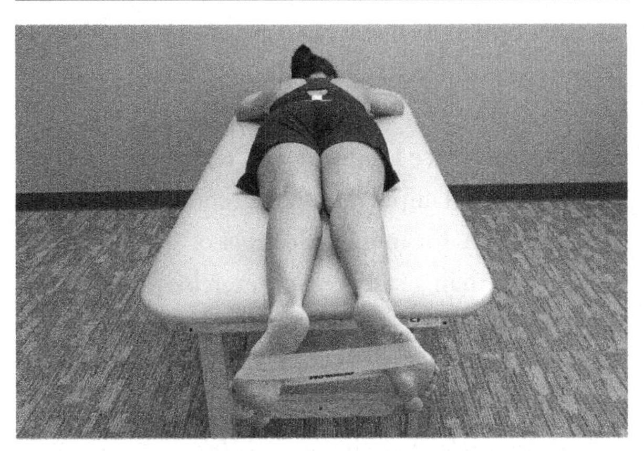

Figure 4.47: Pelvic floor contraction with prone hip external rotation prone with elastic resistance. The pelvic floor is recruited in synergy with the hip external rotators. Elastic resistance can be added to enhance the motor facilitation of the pelvic floor or to increase the challenge to the hip external rotators.

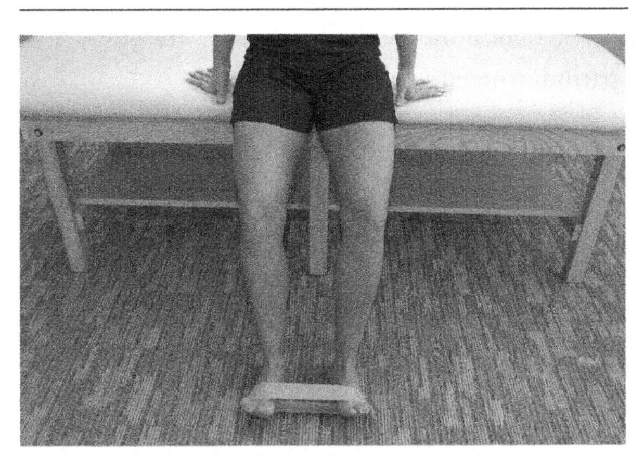

Figure 4.48: Pelvic floor contraction in sitting with hip external rotation against elastic resistance. The patient sits with a neutral spine, weight anterior to the ischial tuberosities. The pelvic floor is recruited in synergy with the hip external rotators. Elastic resistance can be added to enhance the motor facilitation of the pelvic floor or to increase the challenge to the hip external rotators. This exercise can also be performed in supine is pain is present in the lumbar spine, SIJ or coccyx with sitting postures.

Respiration with Pelvic Floor Training

The diaphragm muscle for respiration can work antagonistic to the pelvic floor diaphragm or in synergy. Both functions should be trained and restored. With normal respiration, the diaphragm contracts while the pelvic and urogenital diaphragms relax. After restoring normal diaphragmatic respiration, breathing exercise can include focusing on tightening the pelvic muscles during exhalation, then releasing with inhalation. While exhaling, pull the pelvic muscles up and in, tightening the anal and urethral openings, then relax slowly. This follows the normal respiratory synergy. Purse-lipped breathing will assist in further engaging the transverse abdominus and pelvic floor during exhalation.

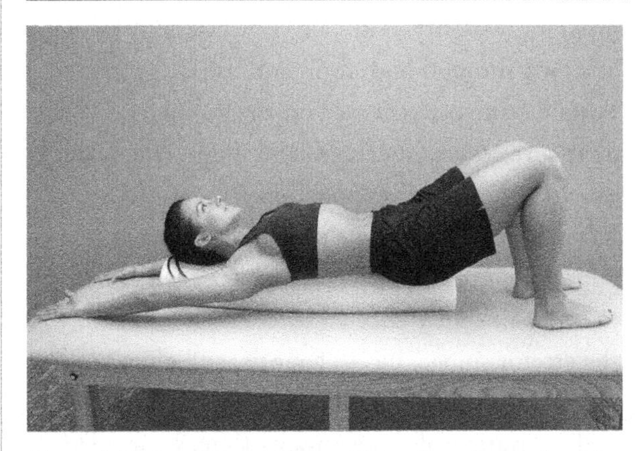

Figure 4.49: Breathing patterns with trunk motion and pelvic floor contraction. During arm elevation and trunk extension, inhalation is performed with relaxation of the pelvic floor contraction. Pulling the arms down with trunk flexion is associated with exhalation and pelvic floor contraction.

The opposite pattern can also be trained as a progression or related to functional tasks. Inhaling is performed at the same time pulling the pelvic muscles up and in, tightening the anal and urethral openings slowly. The pelvic floor muscles are released and relaxed with exhalation. More typically inhalation is associated with arm elevation and trunk extension. This can also be reversed, attempting to exhale during trunk extension patterns and inhale with flexion.

Pregnancy

Preventative exercise during pregnancy has been found to be an effective means of reducing lumbar pain during pregnancy and post-partum. Exercise during the second half of pregnancy has been found to significantly decrease the rate of low back pain and to reduce the flexibility (hypermobility) of the spine during pregnancy (Garshasbi et al. 2005).

What type of exercise is most effective for decreasing pelvic and low back pain in the pregnant woman? As with every other injury, treatment must be based on a thorough evaluation to determine the tissue in lesion, and treatment must provide optimal stimulus for repair of that tissue. Abnormal hyperactivity of the ipsilateral gluteus maximus and contralateral latissimus dorsi is commonly found with SIJ dysfunction, and this has been found to improve with a torso rotary strengthening exercise program (Mooney et al. 2001). SIJ stability is achieved through both form and force closure. Form closure occurs based on the geometry of the joint surfaces of the sacrum and ilium, which are often irregular and do not allow for significant movement. Force closure occurs due to ligamentous tension and muscular force, and has been found to be significantly effected by contraction of the erector spinae, biceps femoris, and gluteus maximus, with SIJ stiffness and stability increasing with activation of these muscles (Van Wingerden et al. 2004). Almost all studies show improvement in stability and a decrease in pain with exercise, however no difference was found in one study of pregnant women with posterior pelvic pain between those who exercised diagonal trunk muscle systems, longitudinal trunk muscle systems, and those who didn't exercise at all (Mens et al. 2000). The quality of how an exercise is performed has been found to effect the amount of muscle contraction and which muscles fire during an exercise. For example, doing a lumbar extension exercise with the pelvis stabilized by locking the lumbar spine and pelvis through increased flexion was found to increase multifidus contraction by 51 percent (San Juan et al. 2005). Simply performing the exercise without proper form or stabilization will not necessarily achieve the desired results.

As with other muscle groups, other manual therapy such as soft tissue work for the pelvic floor and joint mobilization for the SIJ may be necessary prior to beginning exercises in order to decrease muscle tone and guarding and improve joint mobility. Patients with short, painful pelvic floor musculature were found to benefit from rehabilitation including manual release of trigger points, avoiding Kegel exercises until muscle length was normalized (Fitzgerald et al. 2003). A multifaceted intervention program including soft tissue massage, biofeedback, strength training, relaxation training, a home exercise program, and patient education has also been found to improve fecal incontinence, with increased strength, endurance, and resolution of fecal incontinence following this treatment (Coffey et al. 2002). Manipulation has been found to be more effective for immediate relief of spinal and pelvic pain than a stabilization exercise program. The stabilization exercise program was still found to be effective in reducing pain and disability in chronic patients, however it was not found to be effective acutely (Ferreira et al. 2006).

Postpartum Training

The Joanna Briggs Institute (2006) has made recommendations to improve the effectiveness of pelvic floor muscle exercise programs for post-partum women, which are as follows:

- *Instruction on pelvic floor muscle contraction to ensure proper technique*
- *Recommending the exercises both pre-and post-delivery*
- *Advice and instruction for urinary incontinence*
- *At least two individual sessions incorporated during pregnancy*

- *Ensuring home programs are realistic for the time constraints of a new mother*
- *Programs should be multifaceted rather than given in printed form only*
- *Post-partum contact should be made for follow-up questions and instruction*

Stage 1 pelvic floor muscle exercises should be dosed in the same manner as any other area of the body, based on the quality the therapist is trying to achieve. Often times increased tone and guarding is present in these muscles, in which case vascularity would be the desired quality and exercise would begin at 60% of 1RM, 3–4 sets of 24 repetitions. Post-surgical patients may have atrophy of these muscles, and exercise would then begin at less than 40% 1RM, with 3–5 sets of 45 repetitions.

Determining 1RM for the pelvic floor may be difficult, as free weight and pulleys are typically not utilized in this area. The use of biofeedback may be helpful in determining proper dosage. Asking the patient to contract muscles to achieve a force target, and counting the number of repetitions the patient can achieve before fatigue which can be matched to a percentage of 1RM. If a biofeedback unit is not available, intervaginal weighted cones can be utilized to determine the length of time an isometric contraction can be held with the cone of a given weight. The use of intervaginal biofeedback or cones may also be useful for proprioceptive input, as many people cannot feel the contraction and benefit from this tactile input.

As previously noted, most important motor quality in Stage 1 is coordination. Many repetitions over time are necessary to establish a coordinated pattern. Each patient differs in the motor learning process as to how many repetitions are needed. More frequent daily training, rather than one longer duration training period resulting in fatigue, is recommended. Patients may be instructed to perform pelvic floor contractions throughout the day. Varying the position from non-weight bearing positions, sitting, or standing can increase the challenge, reinforce the movement with functional positions and prevent boredom with different training options. Body position during training has not been found to effect outcomes in women with stress urinary incontinence (SUI), as both supine only and supine combined with upright groups have been found to gain statistically significant improvements in stress urinary incontinence (Borello-France et al. 2006). It may be beneficial to begin with supine or hooklying positions in the early training stages, as the pelvis contact to the plinth helps stabilize while patient is localizing PFM contraction. The transition to include upright postures is ideal to return to functional tasks.

For the pregnant woman, education during the beginning stage of exercise is important for improving compliance. The weight of the fetus serves to increase stress on the pelvic floor muscles, which can lead to dysfunction. Many pregnant women have issues with SUI, with an added increased risk of postpartum stress SUI. Educating women that there is a decreased risk of SUI if the pelvic floor musculature is trained, is often enough to increase compliance with exercise. Initially, training the muscles in isolation with contract-relax and isometrics can give the pregnant woman several low level options for Stage 1 exercise.

Body mechanics and posture are important to discuss with pregnant women. Options for sleeping, sitting, and standing posture should be discussed. Sleeping semisupine with a pillow under the right side can be a helpful alternative to left sidelying, as decreased cardiac output by 25–30 percent in supine, related to inferior vena cava syndrome affects 3–11 percent of women. Pelvic tilts throughout the day during standing or sitting can help to counter the increased lumbar lordosis due to the increased anterior weight of the fetus. Pelvic tilt exercises have been shown to be effective in reducing ligament pain intensity and, to a lesser extent, pain duration (Andrews and O'Neill 1994). Sit to stand or side lying, log rolling, lunging, squatting, and standing to floor transfers should all be reviewed.

Performing a pelvic tilt exercise in supine may not be an option due to physician recommendations to avoid supine postures and/or patient anxiety of the position. Performing a pelvic tilt exercise in sitting can be effective in creating a flexion moment to decompress posterior lumbar structures. The sitting pelvic tilt exercise in the third trimester has been shown to be effective in reducing low back pain in woman pregnant for the first time (Suputtitada et al. 2002).

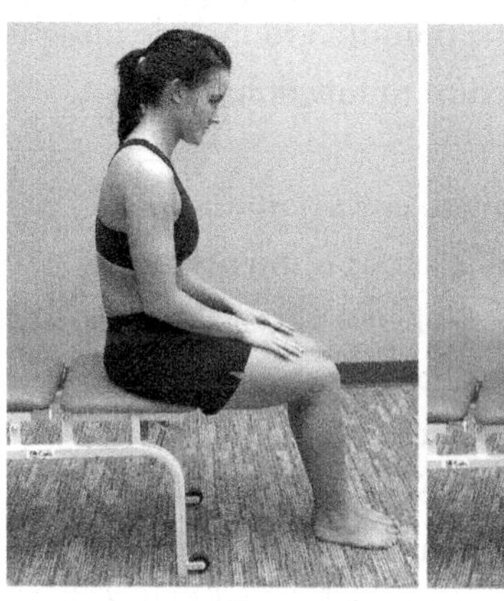
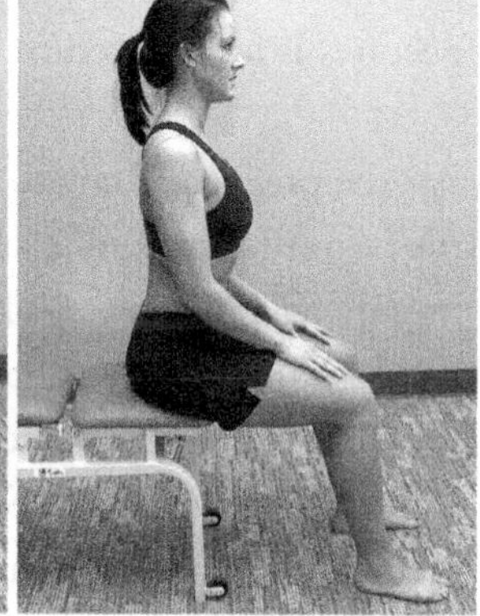

Figure 4.50a,b: The sitting pelvic tilt. The patient is sitting with the hips abducted. The pelvis is "tilted" back and forth in the pain free range, with emphasis on reversal of the lumbar lordosis into flexion.

Sometimes relaxation of the pelvic floor is the issue, not facilitation. Increased tonic guarding may cause difficulty in pelvic floor relaxation. Down-training of the pelvic floor involves having the patient gently push outward, as if trying to secretly expel a small amount of gas. The patient should avoid breath hold or pushing down with great force. The ability to relax the pelvic floor is important during intercourse, for voiding completely, as well as to relax neighboring muscles that may contribute to symptoms. It is important to balance the inner and outer pelvic musculature for maximal SI support and stability during active or resting positions.

Balance

Balance assessment and training is a basic component for rehabilitation of lower limb, pelvic and lumbar pathology. The presence of nerve fibers and mechanoreceptors in the sacroiliac ligament demonstrates central nervous system feedback for proprioception and pain. It is reasonable to speculate that the proprioceptive information is used to assist with upper body balance at this joint (Vilensky et al. 2002).

Dysfunctions not related to hypermobility allow for early incorporation of balance training. For hypermobile pelvic pathologies unilateral standing balance activities are not appropriate early in rehabilitation. The asymmetrical forces through the pelvic girdle cannot adequately be stabilized by active and passive restraints. Performing bilateral standing balance activities with eyes closed, cervical motion or using a labile surface platform are more appropriate to allow for muscular stability with symmetrical standing during balance training.

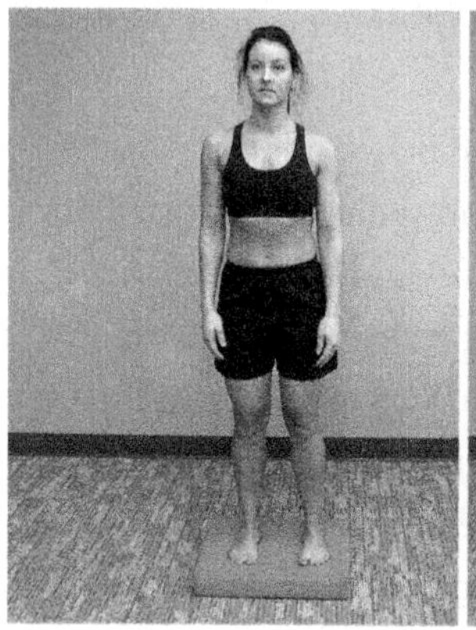
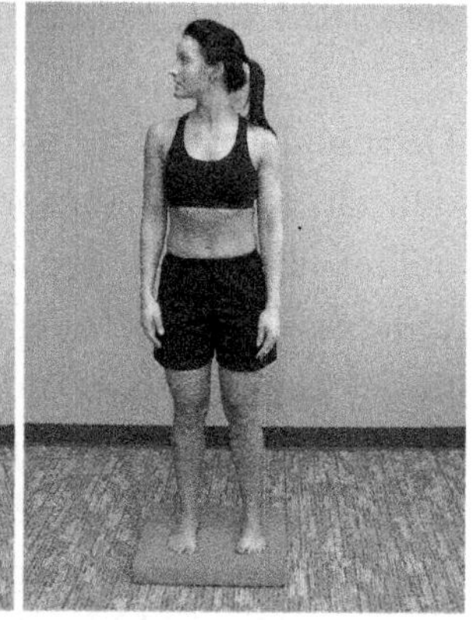

Figure 4.51a,b: Foam balance training. Basic balance training may begin with balance foam, to reduce the somatosensory input of the lower limbs, with emphasis on the trunk and vestibular system. Bilateral stance, eyes open with cervical motion is the basic starting point. Eyes closed with further challenge the vestibular system. Unilateral stance activities should only be attempted after adequate stability of the pelvis has been achieved to allow for asymmetrical force through the pelvis.

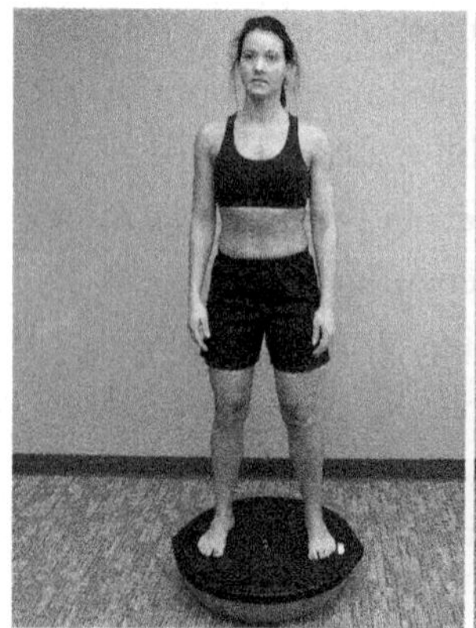
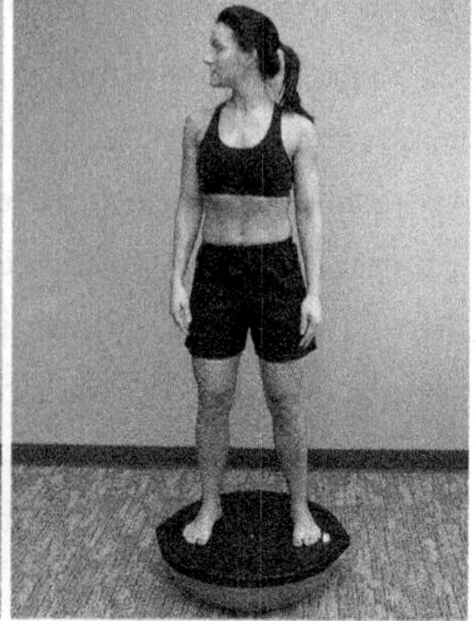

Figure 4.52a,b: Balance training on a labile surface will further challenge the somatosensory systems in the lower limbs and

pelvis. Bilateral stance with cervical motions and arm reaches are the initial starting point. Closing the eyes will increase the demand on the somatosensory and the vestibular systems. Unilateral stance activities should only be attempted after adequate stability of the pelvis has been achieved to allow for asymmetrical force through the pelvis.

The reader is referred to the lumbar chapter for more extensive options for early and later stable balance training for the pelvic girdle, as well as the hip, knee and ankle chapters.

Adjunct Treatment

Joints are not typically treated in isolation. Functional movement requires systems of joint and muscles working with complex motor patterns. Though this chapter focuses on the sacroiliac joints and pelvis, many applications for the hip and lumbar spine may need to be incorporated in to the complete rehabilitation program. Each patient is assessed for impairment with a program to address all regions of dysfunction. Issues of the lower limb and lumbar spine are addressed more specifically in their respective chapters in this text.

Respiration

Retraining normal respiration patterns may require specific attention, or be incorporated in general exercise for the pelvis. Sacroiliac pain has been associated with a decrease in diaphragmatic excursion (O'Sullivan et al. 2002). Improving respiratory function can have a positive effect on pelvic floor function. O'Sullivan et al. (2006) demonstrated that motor training for abnormal kinematics of the diaphragm and pelvic floor during the ASLR could be normalized. Improvements in pain and disability were also found to be associated with an improved ability to consciously elevate the pelvic floor.

The thoracic chapter covers more in depth issues of respiration and retraining. Basic integration involves nasal respiration for inspiration during eccentric work, with oral expiration during concentric performance. Emphasis of the breathing pattern is for anterior displacement of the lower abdomen with lateral expansion of the lower rib cage. The upper rib cage should not be involved with normal respiration during lower level exertion. When attempting to learn a new physical performance, such as a new rehabilitative exercise, a common tendency is for patients to stop breathing, using a breath holding technique. Monitoring and cueing may be required to retrain normal respiration during the exercise. Breathing patterns may also be altered based on the movement performed. Extension patterns of the trunk are associated with inspiration while flexion patterns are associated more with exhalation.

Section 2: Stage 2 Exercise Progression Concepts for the Sacroiliac Joint

Progressing to Stage 2 is not meant to imply that all exercises are progressed to a higher level. Each initial exercise was designed to improve a specific functional quality. As each functional quality is attained, the exercise is discontinued or progressed. Initial programs often involve training concepts outlined in both Stages 1 and 2. Each individual has a different list of impairments regardless of the diagnosis, and each of these should be addressed specifically. A protocol approach does not adequately address the individual variations in impairments and tolerance to training. Indications for progressing to Stage 2 include a decrease in pain, improved joint mobility, improved range of motion, resolving muscle guarding and improvement in coordination or motor learning.

Improvements in coordination may be associated with the ability to stabilize the pelvis with basic exercises and basic functional tasks. Pelvic floor training may demonstrate improvements in incontinence or specific biofeedback measurement. In a general sense, improved coordination is demonstrated with increased speed of performance.

Training Goals

The basic goals for Stage 2 are to stabilize gained range of motion, improve muscle endurance, develop fast coordination, progress the functional body position with each exercise and increase the overall volume of training. A program of eight to 12 exercises of two to three sets each is not uncommon. Speed is increased, not weight, to emphasize fast coordination and to increase the relative resistance of the exercises. The exercises from Stage 1 may be carried over, but the desired functional qualities will shift with these changes. Isometric contractions stabilization exercises may be progressed to more challenging positions and activities.

Tissue and Functional States: Stage 2

- *Decrease active range of motion more in tri-planar motions and end ranges*
- *Pain free at rest but painful with motion*
- *Partial unilateral weight bearing/loaded postures are tolerated*
- *Improved tolerance to rollingtransfer*
- *Fair coordination in basic movement patterns*
- *Fair balance/functional status*
- *Palpation of involved tissues and pressure sensitivity of guarded muscles only moderately painful*

Basic Training Goals: Stage 2

- *Increased repetitions with increased exercises (eight to 12 exercises)*
- *Increase repetitions with additional sets*
- *Increase speed, not weight*
- *IW for stabilization and strength fixation*
- *Body/limb position changes*
- *Planar motions with exercise to full range/ partial range tri-planar*
- *Remove locking or change to less aggressive joint protection*
- *Histological influence of increased lubrication with increased speed*

Endurance Training

Once coordination is established, the next goal is to elevate the training state of the tonic, arthrokinematic muscle system. The local muscles controlling joint axis and range of motion are the first priority in normalizing functional patterns. Though the SIJ does not have a direct arthrokinematic muscle system, as the majority of synovial joints, previous influences of hip, pelvic and trunk muscles have been described, which are emphasized in progressive training. Stage 1 focused on recruitment of these muscles and coordinating basic movement patterns. Emphasis now shifts toward elevating their training state. The addition of new exercises, or the addition of sets to the initial exercises, increases the overall volume of training to improve endurance. Exercises may still address coordination at lower levels of resistance, but with more complex movement patterns or postures.

Dosage for Stage 2 Functional Qualities

Coordination

- *Sets*: 1–5 sets or training sessions daily
- *Repetitions*: 20–50 during the day
- *Resistance*: Assisted exercise up to 50% 1RM
- *Frequency*: 1–3 times daily

Vascularity / Local Endurance

- *Sets*: 2–3 sets or training sessions daily
- *Repetitions*: 25
- *Resistance*: Assisted to 50–60% 1RM
- *Frequency*: One time daily

Isometric Fixation of Strength

- *Sets*: 1 set only, may be added between two dynamic sets
- *Repetitions*: 2–5
- *Resistance*: 60–80% 1RM
- *Range*: Shortened range of muscle

Isometric Stabilization

A more extensive list of isometric stabilization exercises is presented in the Lumbar chapter. Many options exist for lying, sitting and standing. Functional training should emphasize more dynamic movements while stabilizing the pelvic region, rather than only performing static exercises with isometric holding. These exercises should be used as a progressive step into more dynamic training and functional movements.

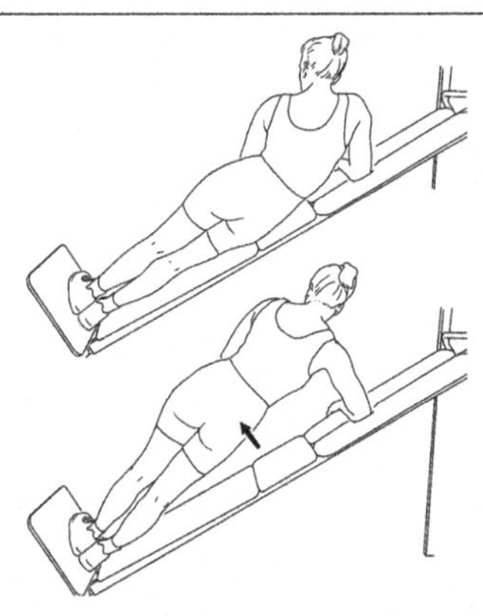

Figure 4.53: Lateral bridge with slant board. Dosing the lateral bridge may initially begin in vertical against a wall. A slant board allows specific dosing of resistance until the exercise can be performed correctly in side lying. Emphasis is placed on maintaining respiration during isometric holding.

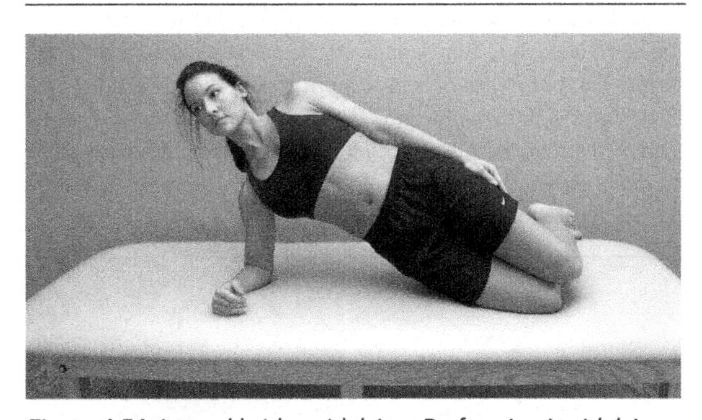

Figure 4.54: Lateral bridge sidelying. Performing in sidelying with the knees flexed also lowers the resistance and may be a progression toward performing the exercise from the feet.

Bridging exercises can be progressed to involve distal fixation with proximal motion and stabilization. The feet can initially be placed on a fixed surface, such as a step or bolster. A further challenge is to place the feet in a sling, requiring a greater level of co-activation to stabilize the proximal trunk. Initial exercises may support the lower limbs several centimeters off the floor, with a gradually progression to higher levels.

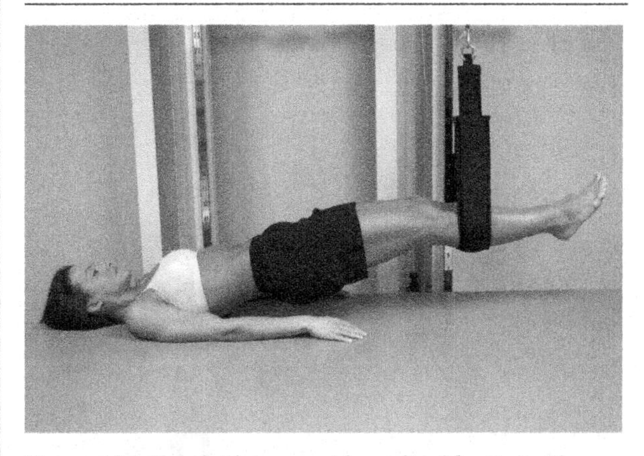

Figure 4.55: Sling bridging provides a distal fixation with proximal stabilization (Stuge and Vollestad 2007). A sling supports both legs with the trunk resting on floor. The legs can be positioned as little as one inch off the floor, with increasing heights to increase the challenge. The patient performs a lumbar bridging exercise to a neutral position, holding isometrically.

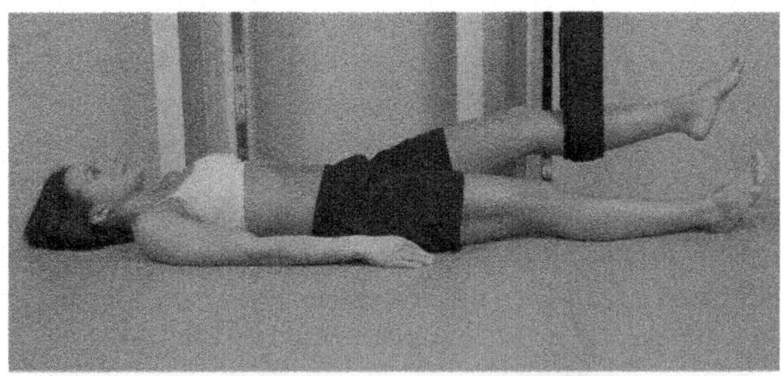

Figure 4.56a,b: Unilateral bridging with a sling. One leg is supported in a sling. The patient flexes the opposite hip until parallel with the sling leg and then performs a lumbar bridge (Stuge and Vollestad 2007).

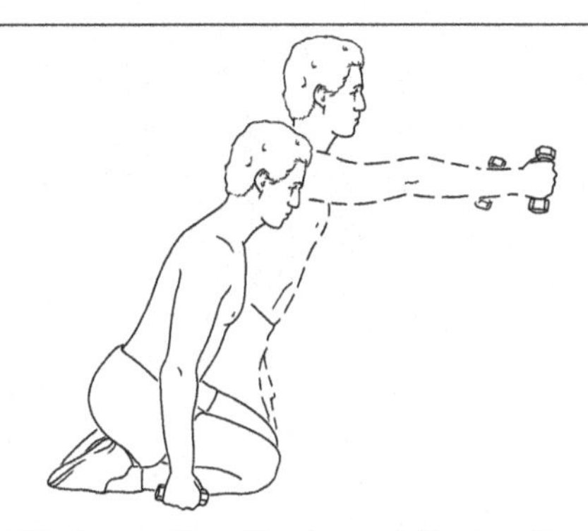

Figure 4.57: Forward kneel lumbar stabilization. From a sit-kneel position the trunk is flexed forward from the hips, while maintaining a neutral lumbar spine. The arms are then flexed forward to 90° and the knees are slowly extended. If

the psoas muscle cannot fully lengthen, an increase in lumbar lordosis occurs prior to reaching an upright kneel position. Motion is stopped when lumbar neutral can no longer be maintained, with a return to the start position.

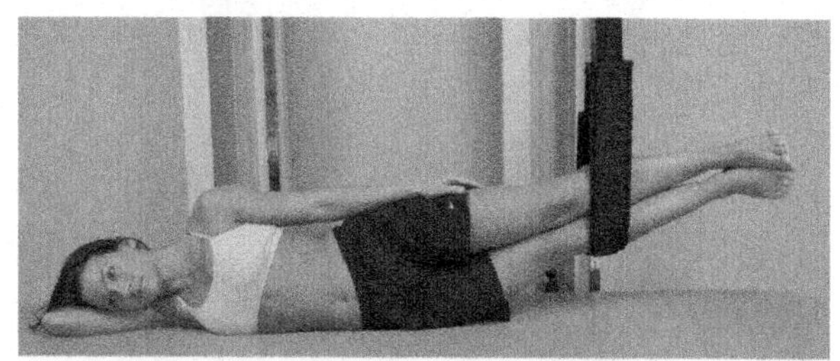

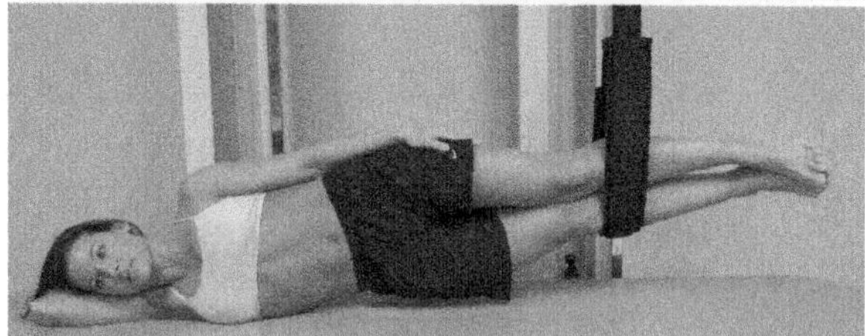

Figure 4.58a,b: Lateral bridging with proximal sling fixation. The patient begins with the pelvis resting on the table with a sling around the lower extremities. The pelvis is lifted off the table to a position of neutral side bending and flexion/extension. The exercise can be performed as an isometric hold or with small amplitude repetitive side bending. The exercise can be performed bilaterally and unilaterally, as pictured below (Stuge and Vollestad 2007). The exercise can be made more difficult by increasing the resting height of the sling, requiring an increase in side bending range. To adjust the relative resistance the sling can be placed more proximal on the thigh to reduce load or distal to increase load.

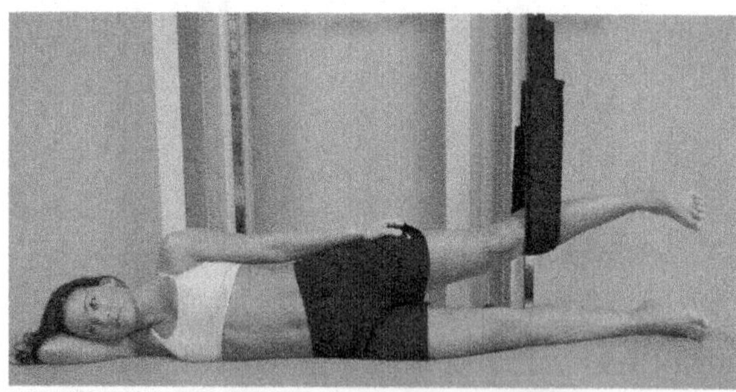

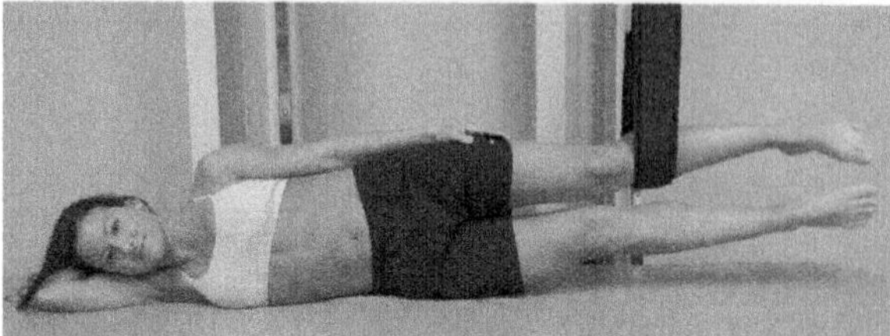

Figure 4.59a,b: Lateral bridging with unilateral lower extremity proximal sling fixation. As above, patient begins with the pelvis resting on the table but the sling is placed only around the upper leg. The pelvis is lifted off the table to a position of neutral side bending and flexion/extension. The unilateral leg fixation increases the demand of the hip adductors during the exercise, which will also increase the synergistic recruitment of the pelvic floor.

Figure 4.60: Kneeling lumbosacral stabilization with a fitness ball. From a kneeling position, the forearms are rested on a fitness ball. The ball is rolled forward placing the trunk in a more horizontal position. The patient is instructed to maintain the lumbar spine in a neutral position, not falling into an extended position.

Figure 4.61a,b: Kneeling lumbopelvic stabilization with upper limb slings. Two lines from the same sling support the arms. The patient begins in an upright kneeling position. Transferring the weight onto the sling, the patient leans forward maintaining a neutral lumbar posture. This exercise is more challenging than the fitness ball, as there is a greater freedom of movement.

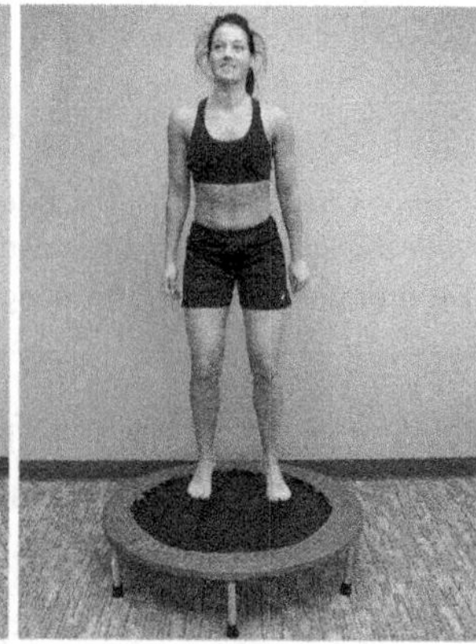

Figure 4.62a,b: Trampoline bouncing for isometric pelvic floor stabilization. Decelerating the bounce requires isometric firing of the pelvic floor to stabilize the contents of the pelvis. Light repetitive bouncing can be a simple progression for more active bladder control training.

Body Position and Motion Direction

Locking techniques are typically removed in Stage 2. The use of an SIJ belt during training may be removed with emphasis on dynamic stabilization during the same exercises and/or new more challenging exercises. Body position is progressed to more challenging positions. If dynamic stability is not possible, the use of an SIJ belt may continue until it is achieved. Lying and rolling exercises, where the form closure does not occur as in standing, may be an additional emphasis with exercises chosen.

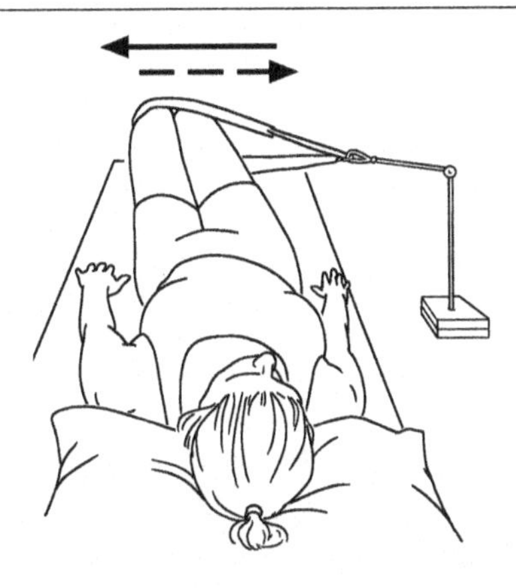

Figure 4.63: Hip Roll—Caudal supine right rotation with resistance. Supine lying with hip flexion places the SIJ close to the loose packed position requiring additional dynamic support during training. The transverse plane of motion will naturally recruit fiber directions of the transverse abdominus. Presetting the transverse abdominus can improve emphasis on central recruitment, rather than initiating the motion with the hip abductors and adductors.

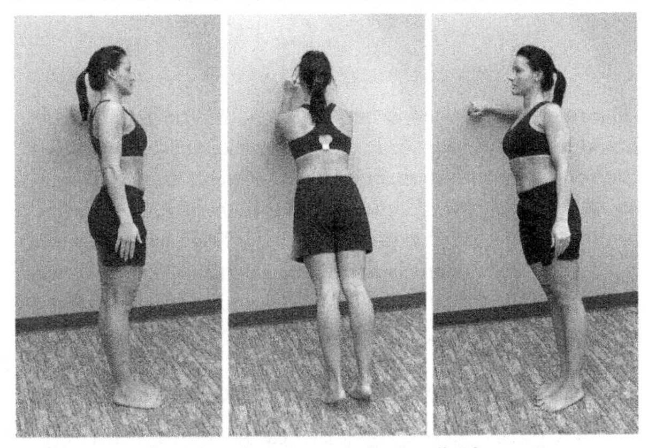

Figure 4.64a,b,c: Standing lateral plank-roll. The patient begins standing in a lateral plank position, elbow into a wall. Stabilizing the trunk, the body is rolled forward facing the wall, in a "prone" plank position on the elbows. The trunk is then again rolled up onto the opposite elbow. Breathing should continue through the transitions, rather than breath holding through the performance.

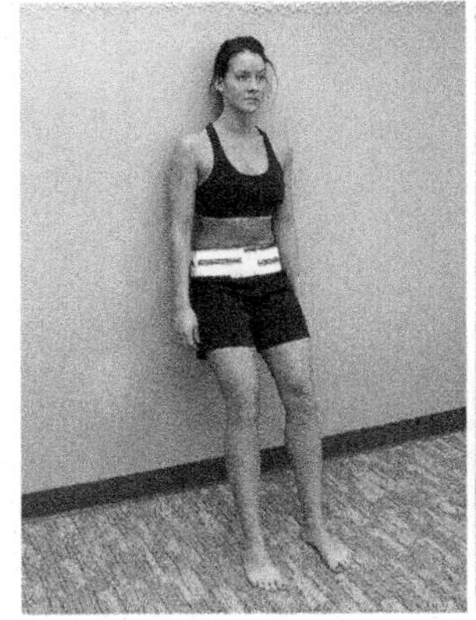
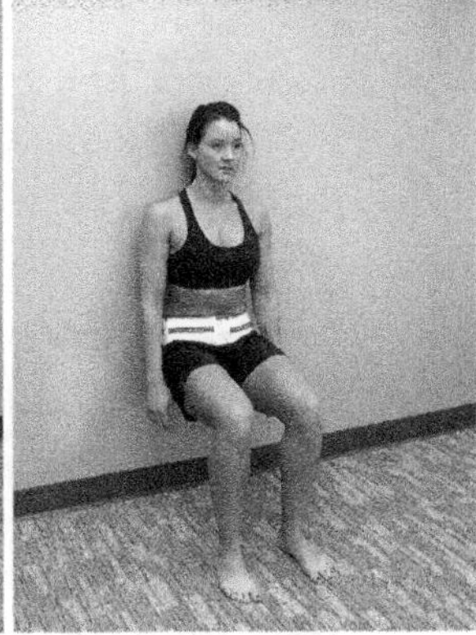

Figure 4.65a,b: Standing wall slides are a progression from the unweighted squat exercise. An SI belt is added to increase stability. The patient will start with the buttocks against a wall or slide board, with the feet slightly less than shoulder width apart and 15 to 20 degrees of external rotation. The feet will also be six to 10 inches from the wall. It is also ideal to have the patient hinge slightly forward at the hips to decrease the tendency for lumbar hyperextension. While keeping the buttocks against the wall, the patient will perform a functional squat to 90° or less and perform a one-count gluteal set at the end of the up phase of the exercise. Focus the work on the gluteals, the patient may try to compensate and use the quadriceps. This exercise works the proximal hamstrings, which are functionally important for hip extension, applying tension to the sacrotuberous ligament.

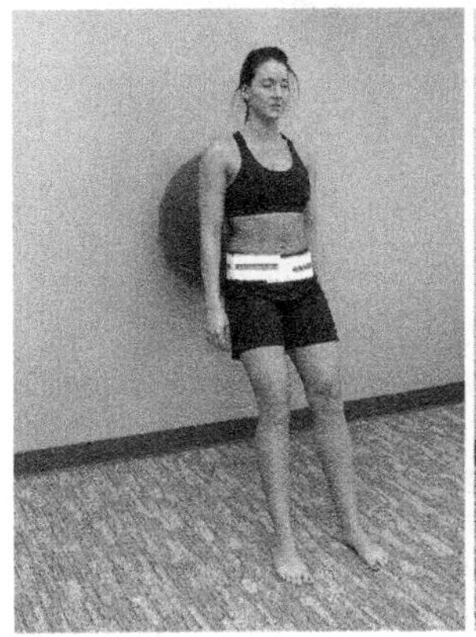
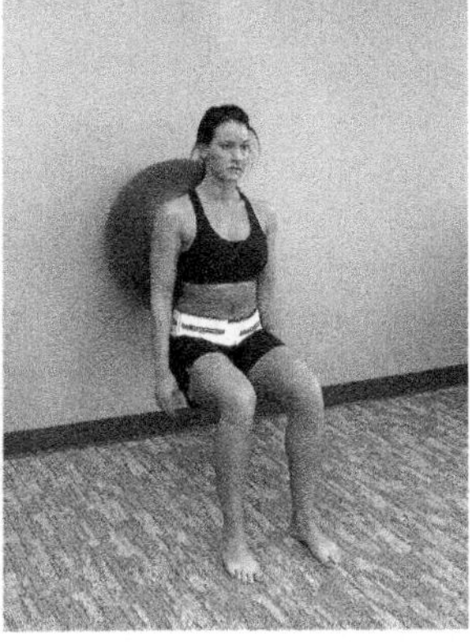

Figure 4.66a,b: Standing wall slides can also be performed with a fitness against the wall to reduce friction, as well as create a slight reduction in relative body weight.

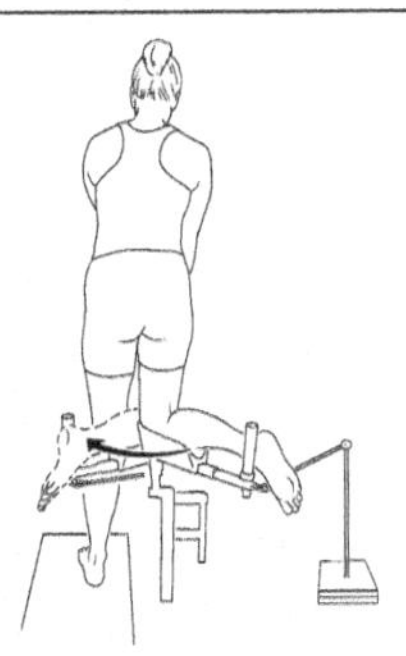

Figure 4.67: Hip resisted external rotation in standing. Performing hip rotation with the knee on a spinning stool allows for a partial weight bearing position while training hip external rotation from neutral.

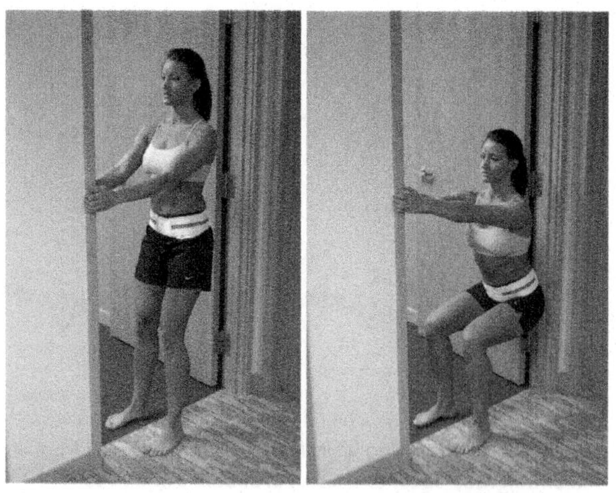

Figure 4.68a,b: Door jam squat—with SIJ belt. The patient stands in a doorway, holding onto the door jam to unload the relative body weight. This is a good home exercise version for unloaded squat training. Bilateral weight bearing through the pelvis should be more stable than non-weight bearing activities, due to form closure concepts.

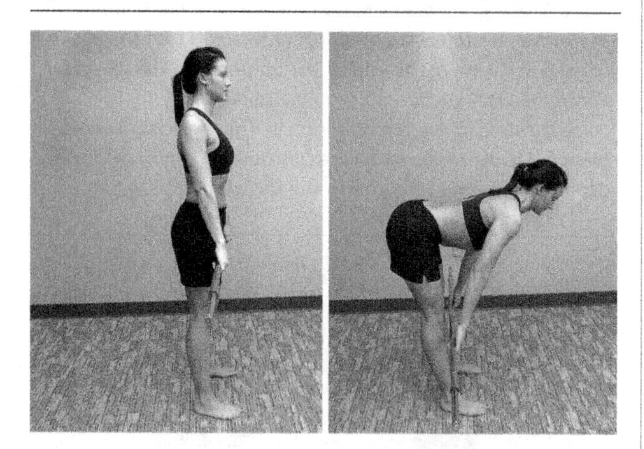

Figure 4.69a,b: Dead lift for SIJ stabilization. The patient is standing with a neutral spine, arms at the side with optional free weights or weight bar for resistance. The trunk is flexed forward at the hip joints, maintaining a neutral spine throughout the motion. Emphasis is on dynamic hamstring training placing secondary tension through the sacrotuberous ligament into the collagen of the SIJ capsule.

A significant postural challenge for the pelvis is to progress from symmetrical weight bearing through the lower limbs to training in asymmetric stance positions. Unilateral stance on the involved side will maintain the form closure, while exercises can facilitate muscles for force closure. Weight creates a natural relative posterior rotation moment of the ilium, tightening the thick posterior SIJ ligamentous structures. The non-weight bearing side of the pelvis has less stability as the lack of ground reaction forces are not present to tight the posterior ligaments, making non-weight bearing often a painful position to training in for the unstable SIJ.

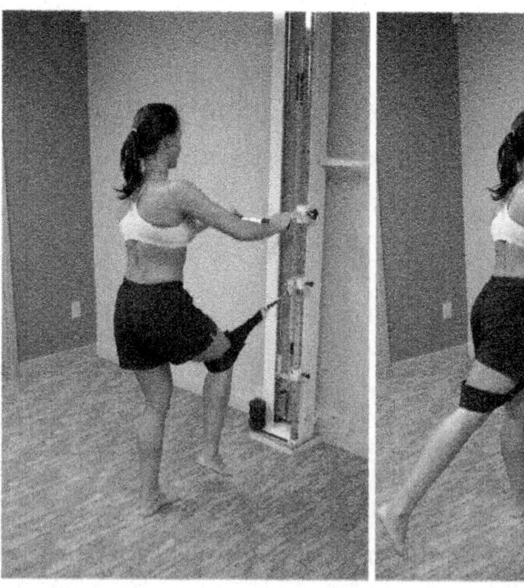

Figure 4.70a,b: Standing hip extension with pulley or elastic band around the distal thigh, with arm support to reduce the balance challenge. Initially the exercise is performed standing on the involved side, while the opposite hip performs extension. An SIJ belt may be initially used during the exercise, if needed, but is quickly removed when tolerated. A one-count hold is performed at the end point for hip extension. The working leg should be in slight abduction and external rotation at contact to full hip extension. Performing hip extension on the involved side may not be tolerated in the hypermobile SIJ, but can be performed in the hypomobile pelvis or one that requires an anterior torsion mobilization.

Figure 4.71a,b: Standing hip extension with pulley or elastic band under the foot. As in the previous exercise, the patient performs hip extension, open chain, with resistance. The pulley strap is placed under the foot to involve muscles of the entire lower quarter extension pattern. This version attempts to duplicate the step climbing in open chain. The stance leg must also work to stabilize the pelvis and lumbar spine.

These previous exercises, weight bearing leg swings, can be performed with a pulley or elastic band. The primary emphasis is the work performed by the stance leg, though the swing leg also recruits a cross pattern. The type of resistance on the swing leg, whether a pulley or elastic band, is not relevant to the motor work on the involved side.

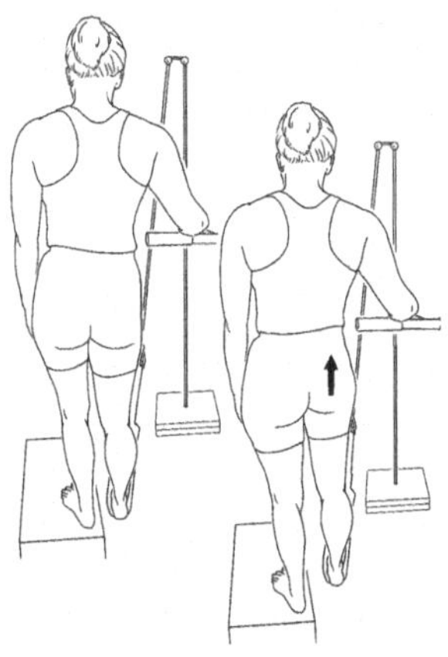

Figure 4.72a,b: Hip hiking standing – pulley assist. The hip hike emphasizes the hip abductors on the stance side, as well as the lumbar multifidi and quadratus lumborum on the opposite side. A pulley assist is used to reduce body weight, allowing for pain free coordinated motion at the desired number of repetitions. To make the exercise harder, the pulley assist weight is gradually reduced until the patient can perform the exercise with full body weight.

Figure 4.73a,b: The SIJ pivot exercise is performed with the patient standing sideways from the pulley stack and holding the resistance handle at waist height. While keeping their weight bearing on the working leg, the patient will pivot their torso and upper body around the vertical axis of the standing leg as they fire the external rotators of the standing hip. To eliminate substitution by utilization of the nonworking leg adductors, the clinician may instruct the patient to only have the toe of the off leg contacting the ground for the sake of maintained balance. At the end range of external rotation, the patient will be asked to perform a one-count hold. Do not allow the patient to overly rotate the hip of the working leg into internal rotation with the down phase of this exercise. This could cause excessive internal rotation of the hip and unwanted tension to the dorsal SIJ ligaments.

The simple horizontal pivot exercise (shown above) involves a weight bearing approach of the pelvis in pushing off the involved side with the gluteals and deep hip external rotators. The horizontal line of resistance aides in the recruitment of the transverse abdominus, as it is in the horizontal fiber direction.

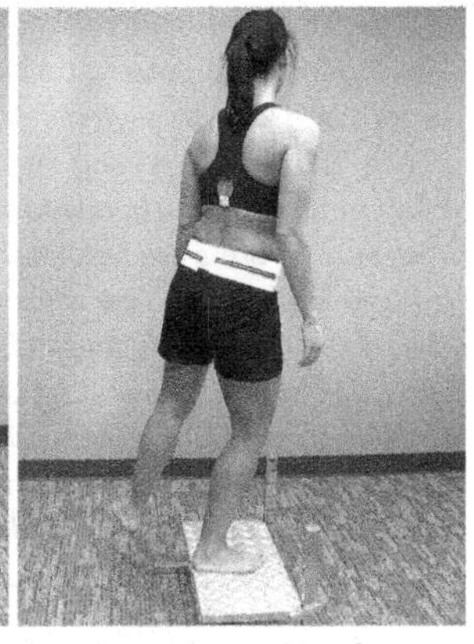

Figure 4.74a,b: Step-up with external rotation, SI belt for added stability. This exercise challenges the entire lower kinetic chain working to extend the hip and knee. The patient is instructed to step up on a variable height platform or footstool and tighten the gluteals at the end of the up phase. This will cause a natural external rotation and slight abduction of the working hip. An increase challenge to the rotational moment can be achieved by attaching an anterior force moment from a pulley line to the right anterior pelvis with a belt. As the left hip performs extension and external rotation, the pulley resistance increases the load on the external rotators. This hip abduction and external rotation pattern can also be preformed with a pulley assist, as shown in the previous pulley assist hip hike exercise.

Cross pattern training for SIJ stability recruits muscles from ipsilateral lower limb to the contralateral upper limb. Recruiting the biceps femoris creates fascial and ligamentous tension through the sacrotuberous ligament to the SIJ. The gluteal muscles and hip external rotators place an oblique and transverse orientation of tension. Ligamentous tension across the SIJ, through the thoracolumbar fascia, transmits forces to the opposite multifidi, quadratus lumborum and latissimus dorsi. Many weight bearing and non-weight bearing options are available to emphasize the recruitment of this basic motor pattern.

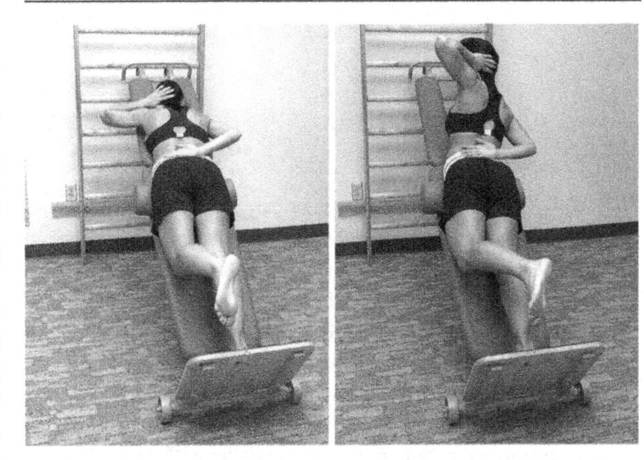

Figure 4.75a,b: Incline extension/left rotation—unilateral

stance. The slant board allows for partial unilateral weight bearing for cross pattern training. Prone extension is progressed with an asymmetrical force line through the pelvis. The patient is lying on a slant board with weight bearing through the involved side (right lower extremity shown). The arms are behind the back to minimize resistance, but the left hand can be placed behind the head to increase the relative resistance and to facilitate the left rotation of the trunk. Right SIJ stabilization training is achieved with facilitation of the right hamstrings, hip extensors/external rotators, bilateral lumbar multfidi, through the thoracolumbar fascia into the left latissimus dorsi.

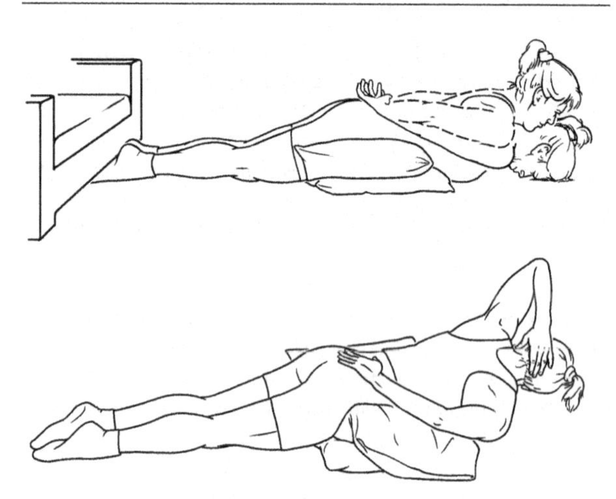

Figure 4.76a,b: Home versions for prone trunk extension can be achieved on the floor with pillows under the lumbosacral areas to create flexion. The feet can be anchored, for more aggressive training, or left free. Both feet can be anchored for initial symmetrical training of the pelvis and lumbar spine. Progressions may include unilateral anchoring on the involved side with or without trunk rotation away.

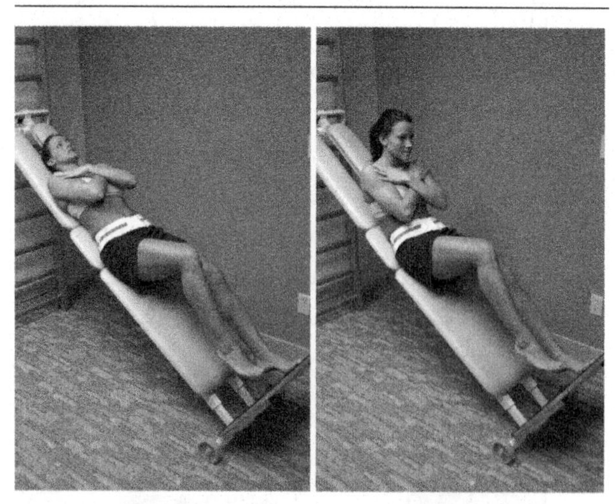

Figure 4.77a,b: The incline sit-over can also be progressed with the unilateral weight bearing concept. The patient is lying on an incline board with weight bearing through the involved side (left), the opposite foot held off the footplate. The left lower extremity extensor pattern is facilitated by maintaining the weight on the forefoot (plantarflexors), with the knee slightly flexed (quadriceps), as well as slight hip flexion (gluteals and hip external rotators). The extensor pattern reduces the participation of the hip flexors (psoas and iliacus), with increased emphasis on the lower abdominals. The asymmetrical force line increases the challenge for lumbopelvic stabilization.

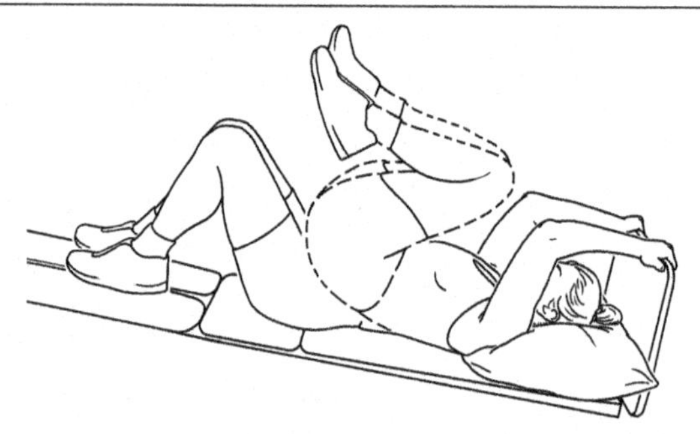

Figure 4.78: Decline hip flexion. Hip flexion in the decline position reduces the overall load, with gravity providing a slight assist. Arm support superiorly will facilitate the latissimus muscles, assisting to stabilize the lumbodorsal fascia, improving abdominal recruitment. Emphasis is on a smooth transition from hip flexor muscles to the lower abdominals without a pause or catch in the middle.

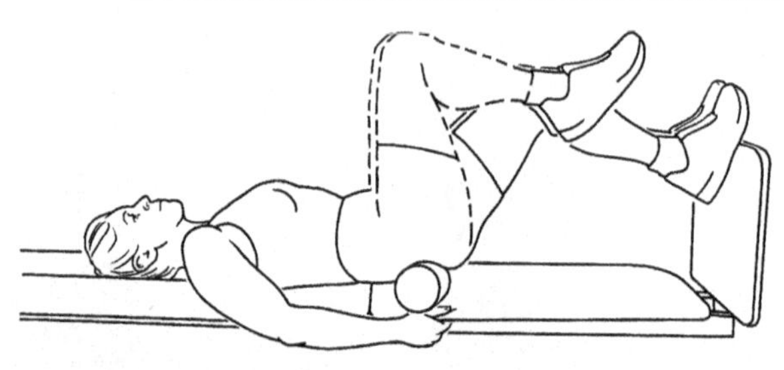

Figure 4.79: As a progression from above, hip flexion is performed horizontally. Here a small bolster is placed under the sacrum to create flexion at the lumbosacral juncture. If motor control cannot stabilize the lumbosacral juncture during hip extension, allowing an extension moment in the lumbar spine, the roll provides an external block (artificial locking) to prevent extension. As motor control improves, the bolster is removed (coordinative locking). The foot rest on the slant board also serves to limit the range of hip extension, reducing the torque in the lumbar spine.

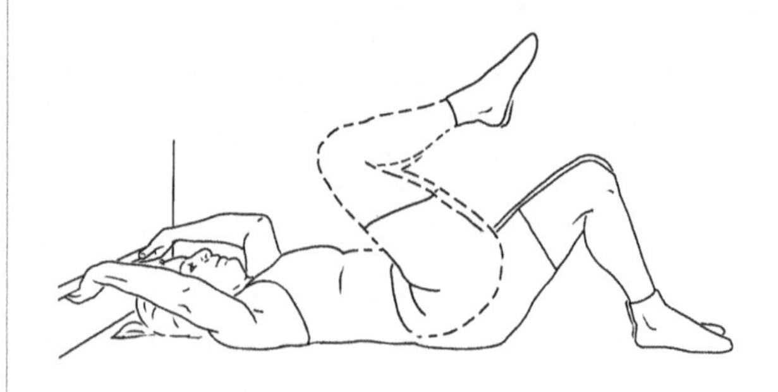

Figure 4.80: Hip flexion (caudal flexion) from the floor. The arms are again stabilized to improve motor recruitment through the lumbodorsal fascia. Maintaining knee flexion through the motion will reduce the torque in the lumbar spine, as opposed to extending the knees. The patient is instructed to breathe in while extending the legs, and breathe out during the flexion phase.

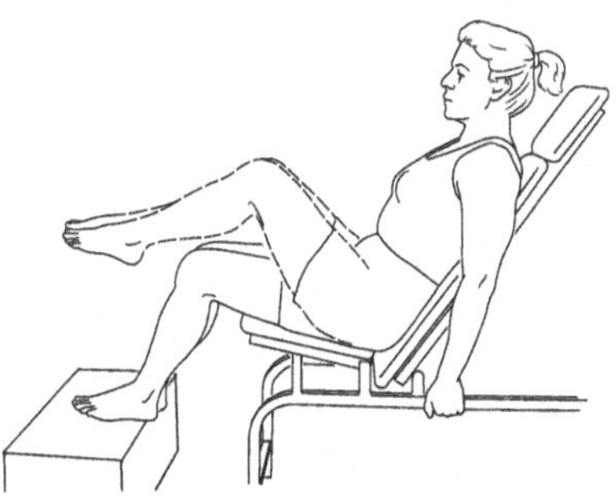

Figure 4.81: Incline hip flexion. The patient is sitting on an incline bench with the arms fixed. The knees remain flexed as the hips are flexed, lifting the feet off the box. This version of caudal flexion will emphasize the hip flexors over the lower abdominals. Range of hip extension is reduced by the sitting position, but this approach may be necessary in patients with greater impairment with difficulty transferring from recumbent postures.

Lunge training offers a weight bearing option for cross pattern training. The lunge is typically performed emphasizing the knee; the trunk remains vertical with little muscle back muscle recruitment. The fall out lunge emphasizes muscles of the front hip and lumbar spine. The back knee remains extended with the heel on the ground. The weight is transferred to the front foot with the trunk inclined forward incline with the back leg. A greater amount of work should be felt in the front hip, rather than the knee, and the lumbar spine. The lunge creates an asymmetrical force across the pelvis, requiring greater demand of the motor system for stabilization then the symmetrical squat.

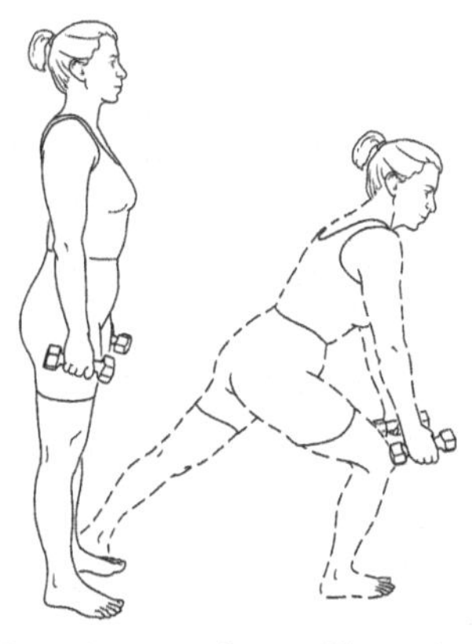

Figure 4.82: Fall out lunge. A forward lunge is performed with the back in neutral, the back knee remains in extension with the heel on the ground. The trunk remains in line with the lower limb, with the nose over the front foot. The lumbar extensors work isometrically to hold the trunk against gravity. When performed correctly the front hip feels most of the work, with little effort at the knee. When performed incorrectly the trunk remains vertical with an extended lumbar spine and the work felt at the knee, not the hip.

Figure 4.83a,b: Right anterior fallout lunge (involved SIJ) with left arm anterior reach—dumbbell resistance. Working the posterior cross pattern from the right hamstrings, right gluteals, across to the lumbar multifidi and into the left shoulder girdle.

Figure 4.84a,b: Straight arm row with pulley, free weight or elastic band. The patient is standing with a neutral spine and the hips and knees lightly flexed. Bilateral shoulder flexion is performed against resistance while the trunk is stabilized from the extension moment maintaining neutral. A free weight approach for home does not provide the same lumbar extension torque moment as a pulley or elastic band.

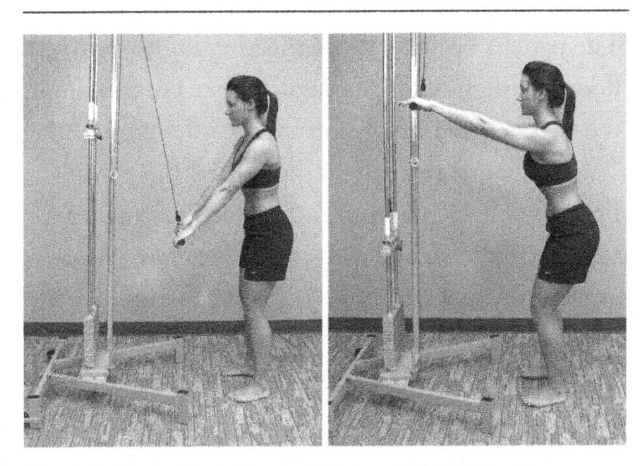

Figure 4.85a,b: Straight arm pull down. The patient is standing with a neutral spine and the hips and knees lightly flexed. Shoulder extension is performed, eccentrically first from the thighs to 90° of elevation, with a quick concentric return. Emphasis is on maintaining a neutral lumbar spine and pelvis during the elevation moment.

Pelvic Floor Training

Typically dysfunction during pregnancy is associated with hypermobility secondary to increases in relaxin levels to provide ligamentous laxity to allow the passage of the baby during delivery. Relaxin concentrations increase during the first 14 weeks of pregnancy, a gradual decrease was found from week 14 to 24, remaining more constant during the last part of pregnancy (Petersen et al. 1995). Thus Stage 2 exercises may be progressed consistently with the hypermobility model, adding isometric holds at this stage. The patient can tighten the pelvic floor muscles to 50–75 percent of their maximum, and focus on not allowing the tension to lessen for the entire count of 10–15 seconds. Percentage of tension as well as time can increase as tolerated.

Speed of contraction may be increased during this phase, with a "quick flick" contraction done with the same dosage as Stage 1. This improves the function of the fast acting muscle fibers primarily of the urogenital diaphragm and external sphincter muscles. These fibers are important for the prevention of leaking during coughing and sneezing or any sudden increase in intra-abdominal pressure. As the pelvic floor muscles are tonic and should be contracted only ten percent of their maximum throughout the day with normal activity to avoid leakage, the "elevator and basement" exercise may be helpful to allow the patient to distinguish between full and slight contraction. The patient begins with contracting to the "first floor" by slightly tightening the outer portion of the vagina and anus. Tightening a little more to the "second floor," the patient should feel the pelvic floor muscle lifting up slightly. This slight increase in contraction can be taken up to as many "floors" as possible, then gradually released back through the floors to neutral, maintaining a normal breathing pattern throughout. Going up the elevator focuses on concentric muscle contraction, going back down is eccentric, and both should be incorporated in a pelvic floor exercise training program. This exercise can be done in isolation initially, then progressed to doing this exercise with more functional activity such as with squats or straight arm pull-down.

The incidence of diastasis rectus abdominus (DRA) is 37 percent for first deliveries, 62 percent multiparas (Bursch 1987), which may occur during the second and third trimesters, during the second stage of labor, or during the postpartum period. Data suggests that many DRAs do not resolve spontaneously postpartum, and this may contribute to pubic symphysis and pelvic joint laxity. It may hinder posture, trunk stability, respiration, strength, visceral support, pelvic floor facilitation, and delivery of the fetus. Exercise for DRA consists of abdominal crunches or sit-ups with the patient manually pushing the sides of the rectus abdominus together prior to doing the abdominal curl.

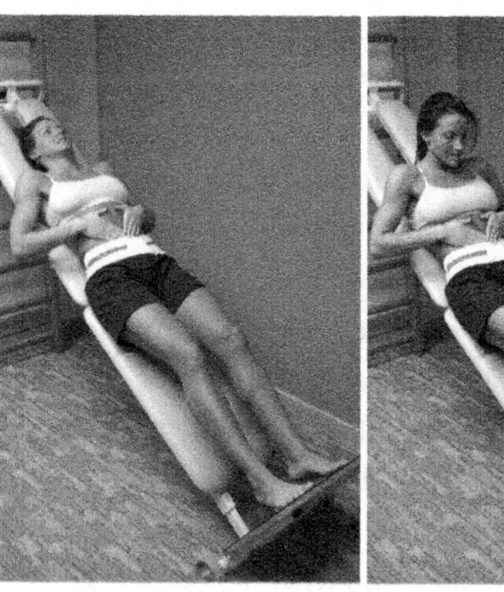

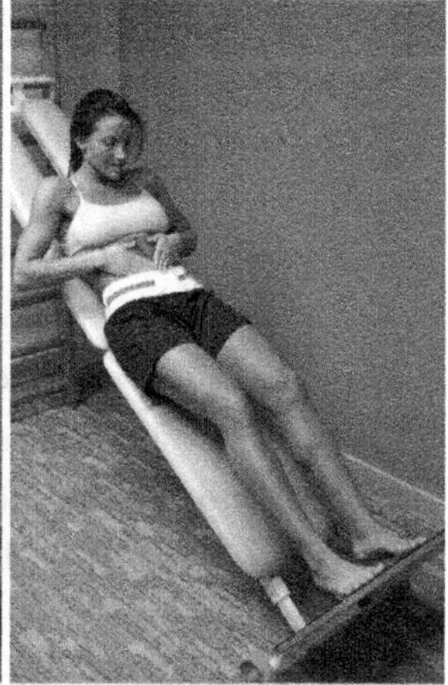

Figure 4.86a,b: Abdominal crunch training for diastasis rectus abdominus. The patient places a midline force on the diastasis with the hands while performing a crunch sit-up. Emphasis is on performing a slow, low tension contraction that avoids separation of the diastasis. Using a slant board to reduce the load of the exercise may be more effective in controlling the exercise without separation of the DRA. Manual assist at the DRA is gradually removed as stabilization is achieved.

Pelvic Floor Training with Functional Exercise

Pelvic floor contraction creates a counternutation moment of the sacrum. Incorporating a counternutation concept with exercise design can be used to enhance pelvic floor contraction with functional activities. The sit-to-stand transfers require a forward lean of the trunk to initiate standing. The pelvic floor relaxes with nutation of the sacrum, and contracts with counternutation upon standing. Relaxation of the pelvic floor occurs with nutation as the patient returns to a sitting position, and contraction occurs again with full sitting to maintain postural stability.

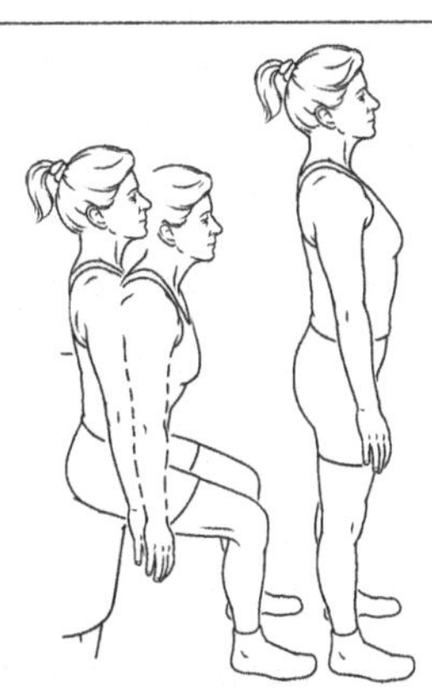

Figure 4.87a,b: Pelvic floor contraction with sit-to-stand transfer. The patient begins in sitting and is instructed to contract the pelvic floor, with the optional addition of the transverse abdominus muscle, and then move to a standing position. The patient should also be reminded to breathe during the exercise, rather than perform breath holding with the pelvic floor contraction.

Figure 4.88: Pelvic floor training in quadruped. Relaxation of the pelvic floor occurs with nutation as the patient leans back into a leaning position, and contracts as the sacrum counternutates as the patient returns to full quadruped.

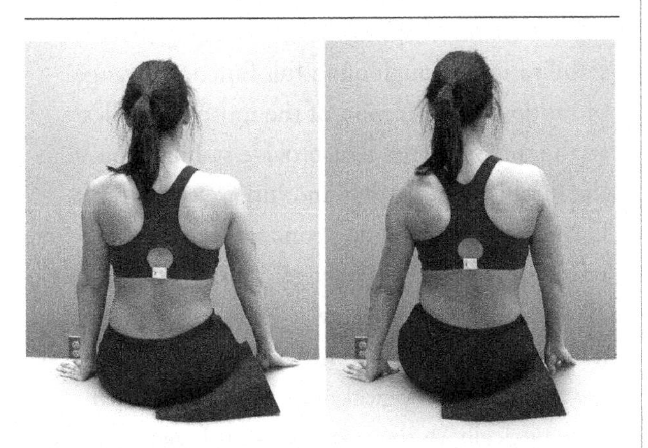

Figure 4.89a,b: Caudal lumbar sidebending with pelvic floor contraction. The patient is sitting on a towel under one ischial tuberosity. Contraction of the pelvic floor occurs as the patient sidebends the lumbar spine lifting the pelvis off the table.

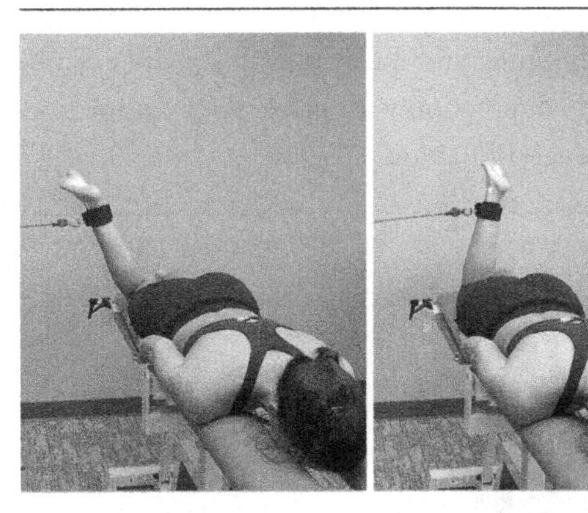

Figure 4.90a,b: Pelvic floor training with synergistic hip external rotation. The obturator internus and piriformis are part of the pelvic wall, which has direct fascial connections to the pelvic diaphragm. Initial setting of the pelvic floor and transverse abdominis is followed by hip external rotation. Resistance from a pulley can further enhance the synergy, but the weight is set on coordination of the pelvic floor and abdominals. Heavier weight set for the strength of the hip external rotators will often exceed the ability of the trunk and pelvis to stabilize the motion.

Figure 4.91: Plie squats, performed with abducted and externally rotated hips, will enhance contraction of the pelvic floor in synergy with the hip adductors. The pelvic floor is relaxed during the eccentric lowering phase, with pelvic floor contraction just prior to the concentric return.

Figure 4.92: Lateral lunges can also be used to facilitate a more global pattern of the hip adductors with the pelvic floor. The pelvic floor must contract to stabilize on foot impact, relaxing after full push-off.

High level activities can be incorporated into pelvic floor training, as many patients have incontinence with running, jumping, or quick cutting turns with sports. Jogging or jumping on a trampoline, ballistic lunging, hopping and quick cutting activities can be trained in conjunction with pelvic floor contraction for higher level function.

Figure 4.93a,b: Light jumping on a trampoline requires a contraction on foot impact to stabilize the pelvic floor. The downward force of the pelvic contents on the pelvic floor muscles on impact provides a mechanical tension on the muscles for a stretch-reflex facilitation. Deceleration of jumping is a functional task for eccentric training of the pelvic floor muscles, decelerating the abdominal contents.

Section 3: Stage 3 Exercise Progression Concepts for the Sacroiliac Joint

Progression to Stage 3 is defined by resolution of primarily acute tissue pain, symptoms are more generalized discomfort or exacerbations are related to excessive activity. Edema and muscle guarding have resolved. Full cardinal plane range of motion is a signal to move to Stage 3 concepts with more aggressive full range diagonal patterns. The patient may lack full active range of motion in more complex movement patterns but basic levels of recruitment, timing and coordination have been normalized. Patient complaints relate more to functional limitations with activities of daily living, work or sports. If joint restrictions were present at the beginning stages, they have normalized with an emphasis shifting toward stabilization. If the initial finding was of joint hypermobility than the progression focuses on dynamic stabilization through the full range of physiological motion. Locking techniques and SI belting are removed, placing further demand on dynamic stabilization.

Tissue and Functional States—Stage 3

- *Full arthrokinematic and osteokinematic motion in cardinal planes*
- *Full weight bearing/loaded postures are tolerated*
- *Joint may be painful with excessive repetitions*
- *Edema and muscle guarding resolved*
- *Palpation of primary tissues and provocation tests are negative*
- *Fair to Good coordination in planar motions*
- *Fair to Good balance/functional status*
- *May have reduced fast coordination, endurance, strength, power*

Training Goals

As basic clinical sign and symptoms are resolving, Stage 3 focuses primarily on improving motor performance, elevating the training state and returning to normal functional activities. Resistance is increased to improve strength. Coordination may still be part of the training program, but this relates to more complex and functional patterns of movement. Emphasis is on achieving dynamic stabilization through out a full functional range of motion. Sensitization of the muscle spindles in the tonic muscles that provide stabilization to the lumbar spine, pelvis and hip are emphasized. Eccentric biased training concepts for dynamic stabilization involve a change in the work order. The body is positioned away from the pathological range, first working eccentrically to decelerate toward the pathological range and then return concentrically to the start position. Length tension relationships are also adjusted for eccentric work.

Basic Outline: Stage 3

- *Increase weight (60–80%)/decrease repetitions (strength)*
- *Change work order to Eccentric-Concentric to stabilize new range*
- *Tri-planar motions in available range (diagonal patterns)*
- *Isometrics to fixate strength in gained range*

Eccentric Stabilization Concepts

Stabilization training begins as isometric performance of the pelvic muscles bracing the pelvis. But true stabilization training involves dynamic performance while controlling the instantaneous axis of motion of the joint through full range of motion. Movement toward a direction of instability is controlled with agonists and antagonists working in synergy. The agonist performs concentrically, moving the body part toward the pathological range, while the antagonist decelerates the motion eccentrically in preparation for a concentric return. Emphasis is placed on training this eccentric function to decelerate motion by changing the work order. Concentric to eccentric performance is switched to eccentric to concentric performance. The joint is positioned away from the direction of pathology, moving eccentrically first toward the pathology and then concentrically back to the start position. There should be no pause between the transitions from eccentric to concentric work, with emphasis on a quick transition. As a plyometric performance, as in jump training, the stretch shortening cycle is utilized with a quick transition between eccentric and concentric work.

For the unstable SIJ in posterior torsion, forward bending would unlock the joint, requiring increased motor performance to control motion. Eccentric emphasis would involve beginning in neutral, moving the trunk anterior into flexion with eccentric control of the lumbar extensors, with a concentric return to neutral. This concept of range of training and work order could be performed in any posture or position. For the opposite example of instability with anterior torsion, then extension of the spine would be the hypermobile range for the SIJ. The pattern of training begins in flexion or neutral with eccentric work of the flexor muscles decelerating the trunk and pelvis into extension with a concentric return.

Strength Training

Dynamic stabilization first requires coordination. This includes appropriate facilitation and timing of the arthrokinematic muscles controlling the joint axis. Once normal patterns have been established, resistance increases to build strength. Mechanoreceptor damage associated with pathology reduces the normal afferent feedback to improve proprioception and coordination. Sensitizing the muscle spindle with strength training boosts the afferent feedback during the motor performance to achieve dynamic stability. Increasing the resistance in Stage 3 up to 80% of 1RM improves the muscle spindle sensitivity to improve afferent feedback with training and function.

Body Position for Training

Stage 3 should represent normal functional positions, movement and patterns. Working toward functional patterns, and away from specific "rehabilitation" type exercises should be emphasized. Each patient will have a different functional demand, which should direct the position, motions and directions that are being trained through Stages 3 and 4.

Motions and Directions

A progression to asymmetrical weight bearing places greater demand on the passive and contractile elements associated with pelvic stability. Unilateral stance requires balance and stability at the hip joint, along with the contralateral lumbar spine. Asymmetrical forces may be applied through the upper limb, lower limb or both. Many options are available and can address functional activities or athletic simulation.

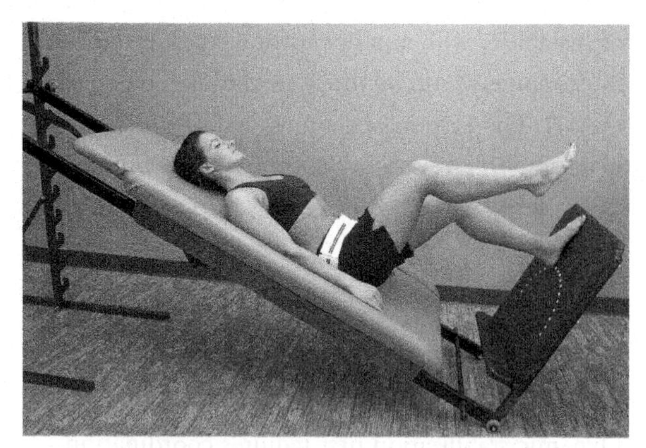

Figure 4.94: Unilateral squat—unloaded. A progression for pelvic training is from symmetrical weight bearing to asymmetrical activities. The asymmetrical force is transmitted from the right hip, through the pelvis, crossing in the lower lumbar region to the opposite side of the trunk. A belt can be initially worn to assist in passively stabilizing the pelvis during the exercise, but the first progression should be removal of the belt, requiring dynamic stabilization.

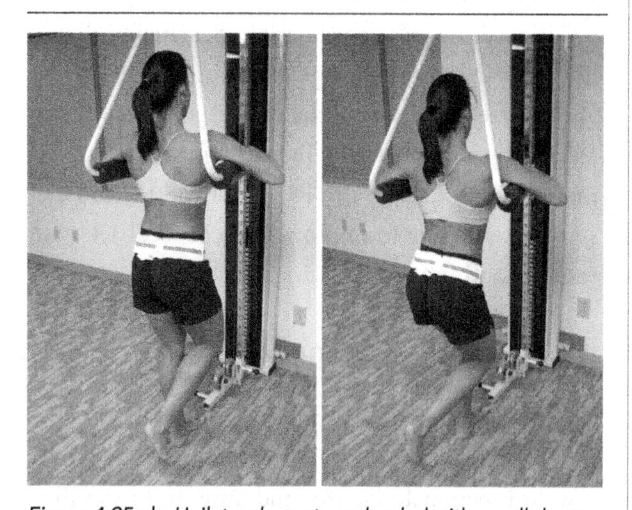

Figure 4.95a,b: Unilateral squat—unloaded with a pull down machine. The patient is instructed to fix the shoulders on the gantry while performing a unilateral squat. The motion should begin with the pelvis moving posterior, to engage the muscles of the hip, rather than the knees moving forward.

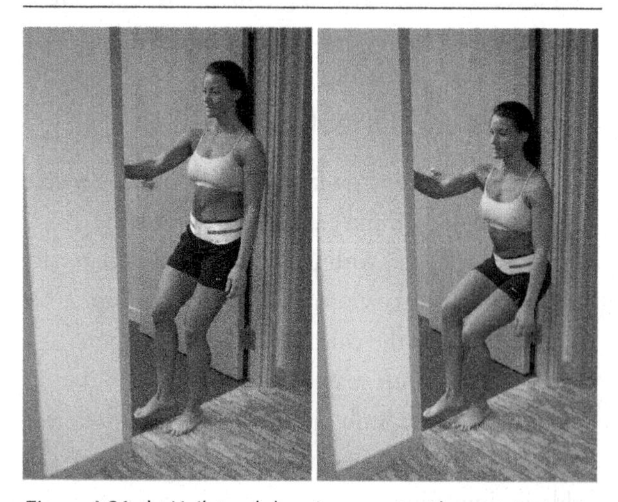

Figure 4.96a,b: Unilateral door jam squat with opposite upper extremity support. An SIJ belt is optional. The patient stands in a doorway on the involved limb (left pictured), holding the door jam with the opposite upper extremity. The unilateral squat will increase the demand on the left gluteal muscles, the pelvic floor, the right lumbar multifidi, the right quadratus lumborum and latissimus dorsi.

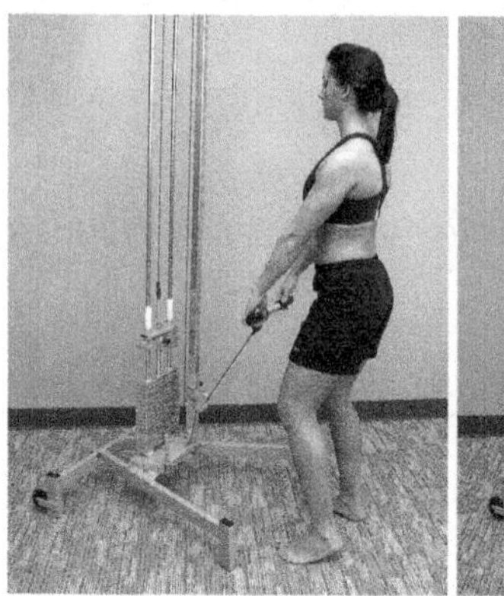

Figure 4.97a,b: Straight arm rowing with toe touch standing (left foot toe touch shown). The rowing motion is performed with unilateral stance on the involved side, with toe touch on the opposite foot for balance control. Increased demand is placed on the right hip and lower quarter, as well as the cross pattern across the SIJ to the left lumbar spine.

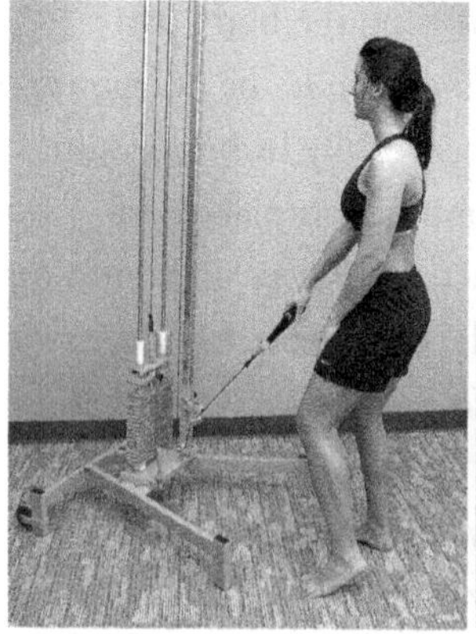

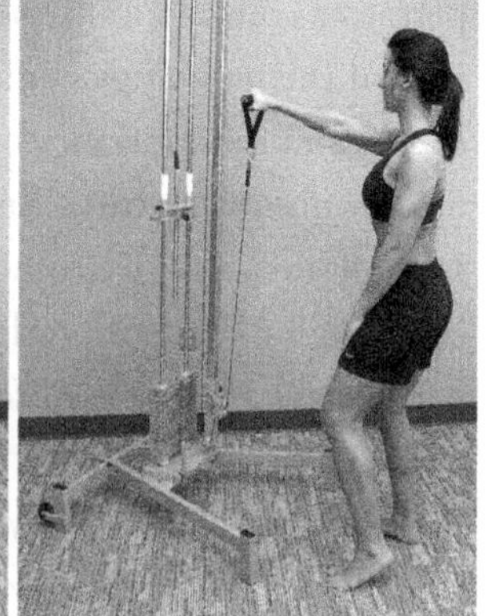

Figure 4.98a,b: Cross pattern single arm row, toe touch standing. Again, the patient is standing on the involved SIJ with toe touch standing to stabilize balance. Shoulder flexion of the opposite arm emphasizes the cross pattern from the stance leg to the opposite side of the trunk and shoulder.

The vertical pull shown above, or using a free weight, emphasizes the extension pattern from lower limb, through the trunk to the opposite shoulder in a primary extension pattern. Later progressions place the pulley line more medial to the shoulder to produce a diagonal pattern, involving a more three dimensional recruitment pattern. Performing the same motion holding a horizontal pulley line will emphasize a rotary, or transverse plane, of motor recruitment (transverse abdominus, lumbar multifidus and hip rotators).

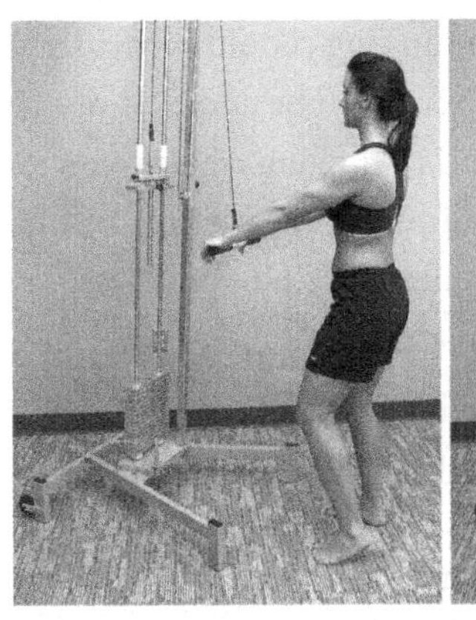

Figure 4.99a,b: Straight arm pull down, toe touch stance. The patient is standing on the involved lower limb, with the toe touch support on the opposite limb while maintaining a neutral spine. Shoulder extension is performed, eccentrically first from the thighs to 90° of elevation, with a quick concentric return. Emphasis is on maintaining a neutral lumbar spine and pelvis during the elevation moment, rather than initiating the movement with excessive lumbar extension or swinging the pelvis anterior.

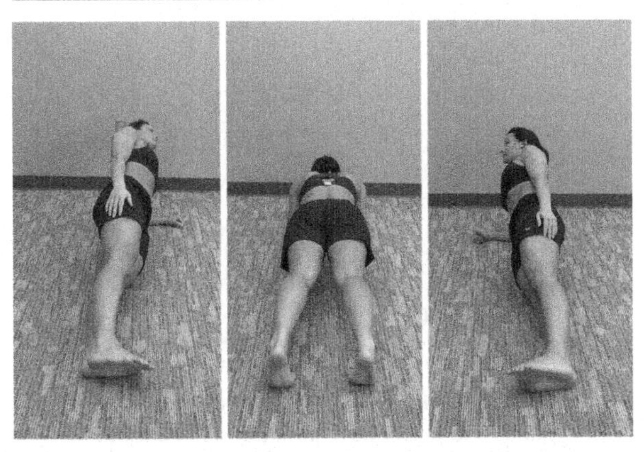

Figure 4.100a,b: Lying lateral plank-roll. The patient begins in a sidelying bridge position, either from the knees or feet (pictured). Stabilizing the trunk, the body is rolled forward to a prone plank position on elbows. The trunk is then again rolled up onto the opposite elbow. Breathing should continue through the transitions, rather than breath holding.

Figure 4.101: Prone plank on a labile surface. A progression of plank training may involve the use of different labile surfaces to increase the dynamic balance challenge through the arms.

Figure 4.102: Prone plank on labile surface with alternate leg lifts. A further challenge to central stabilization is to alternately life one leg while maintaining alignment of the pelvis to the shoulders.

Despite the common emphasis on the lower abdominals (transverse abdominus) with sit-up training, endurance and strength training of the hip flexors is often necessary in lumbopelvic dysfunction. Performing sit-up exercises with the feet first, facilitating a lower quarter flexion pattern, will emphasize the hip flexors over the lower abdominal muscles. Weakness in the hip flexors with lumbar and pelvic dysfunction is often overlooked, with emphasis on the transverse abdominus. Inhibition of the hip flexors is created with pelvic pain as well as upper lumbar segmental dysfunction. Motor function must be tested in each patient with each deficit addressed, rather than assuming a motor pattern of dysfunction based on a given diagnosis or symptom pattern.

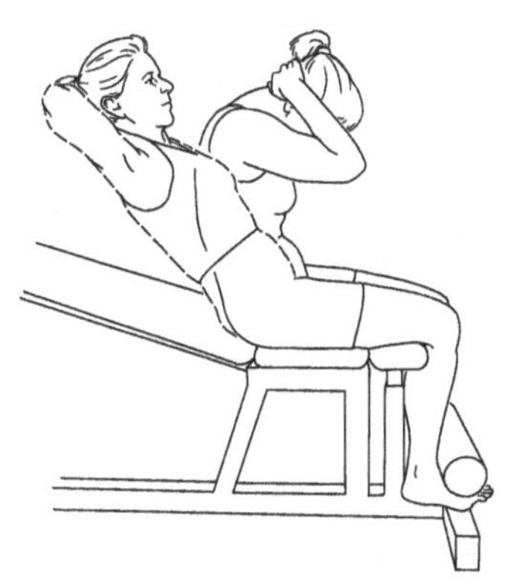

Figure 4.103: Cranial hip flexion through the trunk in sitting. The patient is in sitting on a mobilization bench with the feet fixed. The patient maintains a neutral lumbar spine while the trunk is lowered back, creating hip extension, then flexed forward. The patient is instructed to breathe in during the eccentric phase back and breathe out during the flexion phase forward.

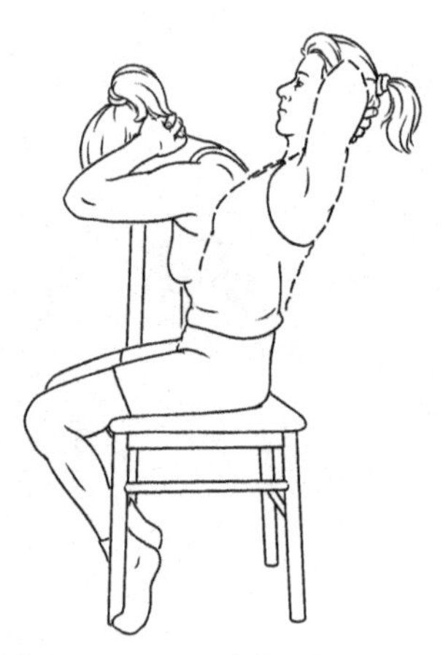

Figure 4.104: A home version of the above exercise involves sitting in a chair, anchoring the feet around the chair legs. The patient is instructed to maintain a neutral lumbar spine while the trunk is moved back and forth around an axis through the hip joints.

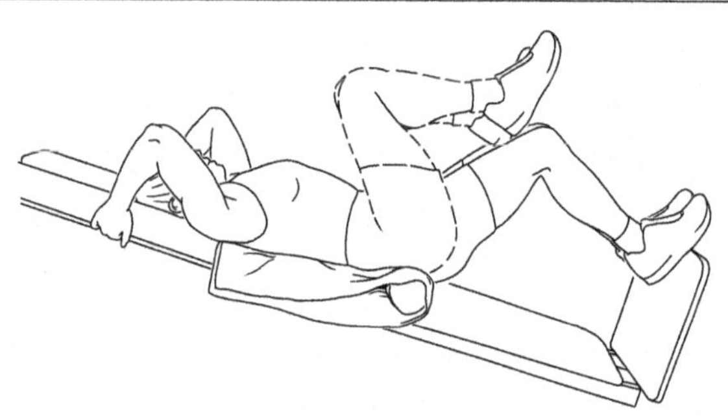

Figure 4.105: Incline hip flexion (caudal trunk flexion). The patient is lying on an incline board with the arms fixed above the head. The ankles and knees are held together as the hips are flexed. Pelvic floor contraction with lower abdominal contraction sets the lumbar stability prior to lifting the hips. To assist stabilization of lumbodorsal from hyperextension, a bolster is placed under the sacrum (artificial locking) to maintain a relative flexed position. A towel is wrapped around the bolster and placed under the trunk to hold the bolster in place as the sacrum flexes off the bolster.

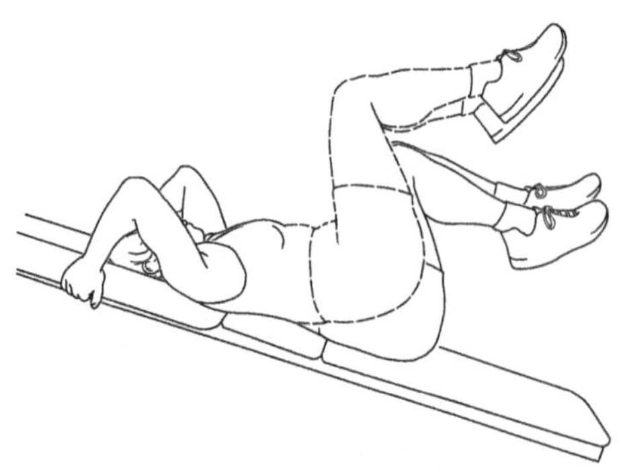

Figure 4.106: As a progression from above, the bolster is removed requiring the patient to dynamically stabilize the lumbosacral juncture.

Returning to more athletic activities requires training for more ballistic impact with weight bearing training. A transition from squatting and lunging to jumping and running should be made prior to resuming sport activity. The addition of speed, rather than weight, to lunging activity creates a demand for faster recruitment and strength of hip and pelvic muscles. Plyometric training involves eccentric deceleration followed by a concentric explosive propulsion. Emphasis is placed on a fast transition between eccentric work and concentric work. The patient should start the activity with a slow transition to learn the exercise and develop coordination. Speed is gradually increased towards a more ballistic activity and tolerance improves if coordination is maintained.

Figure 4.107: Ballistic lunging. The patient lunges forward quickly, lands and performs a quick transition back to the start position. The ballistic lunge can be performed at different angles, related to sport requirements. Even runners performing primarily a forward movement benefit from lateral and angular training for stability.

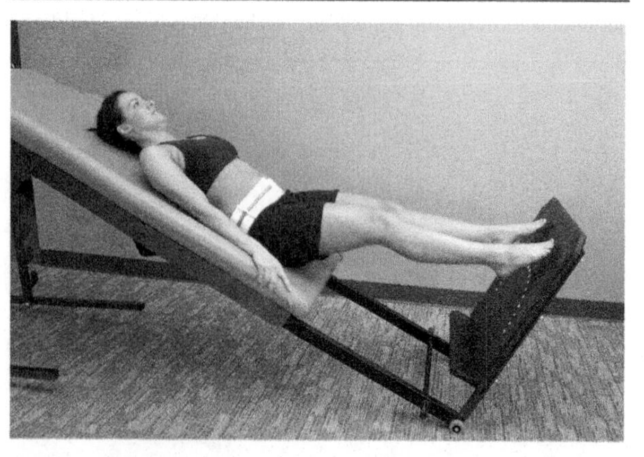

Figure 4.108: Light jumps or "hops" can be performed using the a Vigor Gym® or Total Gym®. They would begin with both feet at the same level focusing on a firm push off or hop from approximately 30 degrees of knee flexion to push off. Jump training is performed at a lower RM than previous squat training, between 30–40% of 1RM.

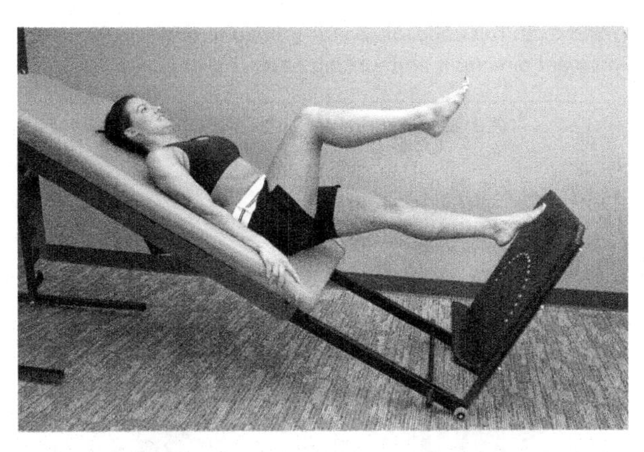

Figure 4.109: Progressing from the unloaded jumping, unilateral unloaded jumping increases the challenge to the lumbopelvic region. Initially an SIJ belt can assist with stabilization of a hypermobile pelvis, but is lateral removed as dynamic stabilization improves. Eventual progression to full weight bearing may be initiated on a small trampoline to reduce impact forces, prior to jumping on the ground.

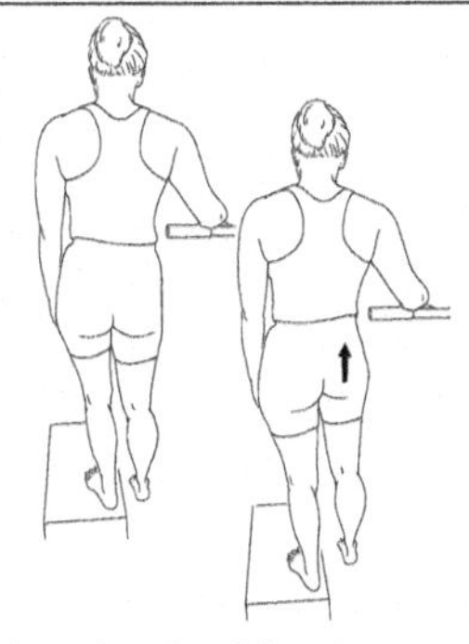

Figure 4.110a,b: Standing hip hike. A progression from Stage 2 assisted hip hiking, the motion is now performed with full body weight. The hip hike emphasizes the hip abductors on the stance side, as well as the lumbar multifidi and quadratus lumborum on the opposite side. To increase the weight challenge, arm support can be with the left, holding a dumbbell in the right hand to increase the relative weight medial to the axis of abduction/adduction of the left hip.

Figure 4.111a,b: Weight bearing horizontal hip abduction. The patient stands on the involved side (left), the arm supported, trunk horizontal with the opposite hip extended. The trunk remains in alignment with the pelvis, as the hip performs external rotation on the fixed femur. Instructing the patient to internally rotate the right leg (point toes toward the floor) on lowering and externally rotate (point the toes toward the ceiling) will assist in emphasizing rotation of the pelvis on the femur, rather then extending the left hip without rotational motion of the pelvis.

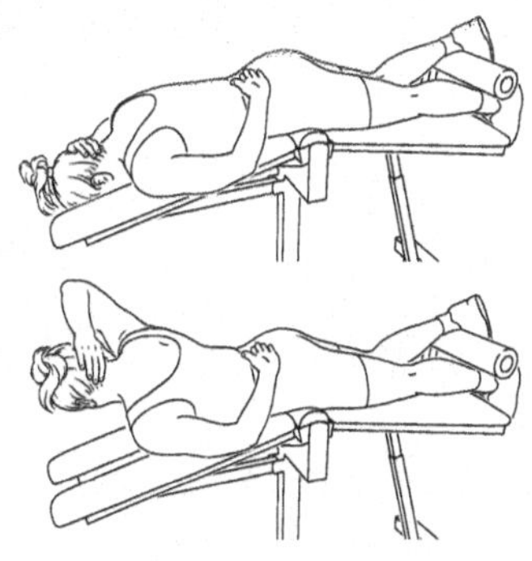

Figure 4.112a,b: Prone lumbar extension/rotation with cross pattern stabilization of the lower limb. The left leg is anchored with lumbar extension and right rotation. Right SIJ stabilization training is achieved with facilitation of the right hamstrings, hip extensors/external rotators, bilateral lumbar multifidi, thoracolumbar fascia into the left latissimus dorsi.

Figure 4.113: Lunging—lateral lunge. The patient is instructed to step laterally. Alignment of the hip, knee and ankle is maintained, with the body weight over the lower limb. Emphasis is placed on posterior motion the pelvis to engage the hip and lower back, as opposed to maintaining a vertical trunk which emphasizes the quadriceps.

Figure 4.114a,b: Multidirectional lateral lunge training with right arm resistance from a lateral force moment. The pulley recruits a diagonal line of muscle from the right arm, crossing anteriorly to the left hip. Anterior musculature is emphasized, including the transverse abdominus and hip flexors. Greater recruitment is achieved through the left side internal obliques, right external obliques, transverse abdominus and hip flexors to control the rotary vector from the pulley.

Figure 4.115a,b: Multidirectional lateral lunge training with left arm resistance from a right lateral force moment. The pulley recruits a line diagonal line of muscle from the left arm, crossing the lumbodorsal fascia, into the right hip. Posterior musculature is emphasized, including the lumbar multifidi and the deep hip rotators. The primary resistance for the right hip is to push off for the lunge, but further stabilization is required to control the rotary vector caused by the pulley.

Figure 4.116: Ballistic training may involve running anterior (pictured) or lateral against resistance. A speed pulley or sport cord attached to the waist during running will increase the motor demand of the lower limbs, but also the core muscles of stabilization of the pelvis and lumbar spine.

Figure 4.117: Gait training with asymmetrical posterior pelvic resistance, pulley fixed to the right ASIS. A pulley is attached to the right ASIS via a waist belt. The patient walks forward. The posterior force moment from the pulley increases the recruitment of the lower abdominals to counteract the resistance. This concept can be applied to differing force moment directions and walking backward or sideways.

Figure 4.118a,b: Plyoball toss overhead. The patient is in a one leg kneeling position, matching any pelvic dysfunction present. The ball is tossed and caught overhead. The trunk and hip flexors are recruited to stabilize an extension moment at the lumbosacral juncture. A change in body position or direction of toss/catch can be utilized to challenge different variables or pathologies.

Figure 4.119a,b,c: Left lateral walking with ankle resistance, adductor emphasis. Pulley resist at the ankle increases the hip adductor, pelvic floor and anterior lower abdominal recruitment during lateral walking. Options: Lateral walking with ankle resistance, abductor emphasis. Medial resistance will increase recruitment of the muscles of hip abduction and external rotation. Lateral walking with a partial squat position ("crab walk") will reduce the tensor fascia lata recruitment, increasing emphasis on the gluteus medius.

Providing external resistance through the arms or trunk emphasize trunk musculature. Providing a caudal resistance through the lower extremities increases the recruitment in pelvic, hip and thigh musculature. Resistance applied at the ankle can be added to lunging and gait exercises to increase the need for proximal stabilization.

Pelvic Floor Training

At this stage, exercises can become more functional, incorporating pelvic floor contractions with lunges, squats, with exercises done at 80 percent of 1RM. Isometric contractions can be performed for the pelvic floor and at the same time, isometrics throughout the range are performed for gluts, multifidi, erector spinae, quadratus lumborum, and abdominals. Side and prone planks can be initiated in an elevated position and progress to horizontal on the floor. Bridge isometrics can be done with pelvic floor contraction. Bridging can be progressed to include one leg extension in the bridge position.

Functional, diagonal planes can now be incorporated with diagonal lifting and multi-plane lunges. Balance activities such as single leg standing can be added and progressed to reaching. Balance exercises include gluteus medius training, an important outer pelvis muscle that contributes to support of the hip and pelvis. For pre or post partum patients, abdominal and hip work can lead to functional training for child care. Preparation for lifting and carrying the baby can include the aforementioned muscle groups as well as progressing biceps and triceps exercises, trunk flexion and extension, and squats to 80% 1RM.

Section 4: Stage 4 Exercise Progression Concepts for the Sacroiliac Joint

Stage 4 is defined by resolution of primary clinical signs and symptoms. The patient has achieved full mobility and coordination with dynamic stability. Exercises include coordination of tonic and phasic muscle systems working in synergy through the entire range of motion, around a physiological axis.

Tissue and Functional States—Stage 4

- *Full active and passive range of motion*
- *Pain free joint motion at a significant level of exercise*
- *Good coordination*
- *Limitations in endurance and strength with functional performance*

Training Goals

Training goals focus on education for long term independent exercise and return to function. Work and sport simulation should be built into the program. The patient should have returned to these activities by now, at least at a reduced level. Emphasis is on elevation of the overall training state and improving endurance and strength.

Basic Outline Stage 4

- *Tri-planar motion through full range of motion around the physiological axis (coordination, endurance, strength and hypertrophy)*
- *Strength, speed, power, endurance, hypertrophy (80–85% of 1RM)*
- *Functional exercises and retraining for activities of daily living, sport and job activities*

Resistance

Training for endurance and strength should be no less than 75% of 1RM, up to 85% of 1RM. As weight is initially increased, speed may decrease slightly to allow for coordination at the new level of resistance. As performance improves, speed will naturally increase, as seen in earlier stages marking improved coordination. Higher resistance will increase the level of tissue stress and strain, further improving the tissue tolerance with more functional activities. Higher loads will also increase hypertrophy of muscle, making more permanent changes in strength and performance.

Pelvic Floor Training

Pelvic floor exercises can also be progressed to a high end functional level. Performing pelvic floor contractions during activities that provide the most aggressive challenge are necessary to train for sports, exercise or functional activities. Pelvic floor contractions can be performed just prior to coughing, sneezing, or jumping. All of these activities are associated with bladder incontinence in the beginning stage of training. Simulation of activities such as picking up toys, pushing the stroller, vacuuming, sweeping, carrying, and lifting, using squat, lunge, or golfers lift method can be performed. These can be performed as specific exercises, but are more important as a functional integration of pelvic floor contraction during normal daily performance of these activities. Sports athletes who have pain with kicking a soccer ball for example, may be instructed to contract the pelvic floor muscles prior to their kick and may find the activity pain free with the added stability. Older patients who have trouble getting out of bed may benefit from PFM contraction prior to getting out of bed or transferring to the toilet for decreased pain or leakage. At this stage any painful task or functional activity which was previously causing leakage should be able to be directly trained with consideration for body mechanics and pelvic floor contraction during the activity.

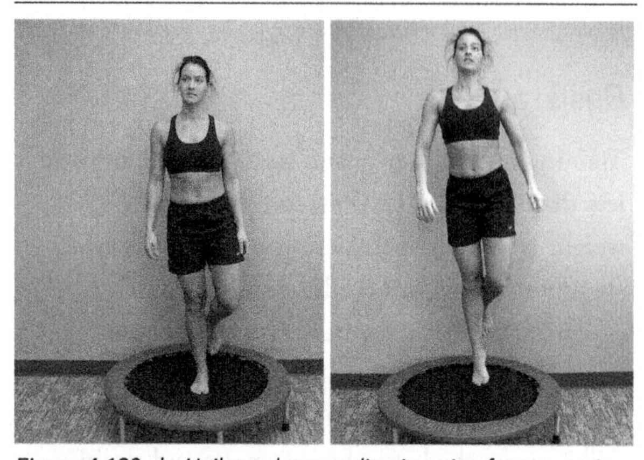

Figure 4.120a,b: Unilateral trampoline jumping for eccentric pelvic floor stabilization as well as asymmetrical lumbar spine and hip training. A progression from earlier stages of light bilateral jump training may include an increase in the amplitude of the jump, unilateral jumping, as well as tandem and wide foot positions.

Strength Training

Increasing strength may be a function of added weight or added speed. Whether to progress with weight or speed is more determined by the activity or the functional requirement of the patient. Weight for strength training is around 80–85% of 1RM for a slow or normal speed performance. Ballistic training with speed, such as jump training, requires lower resistance from 30–40% of 1RM.

End Stage Exercises

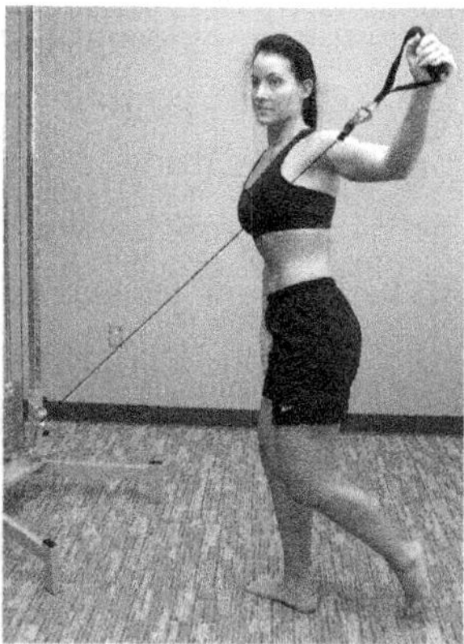

Figure 4.121a,b: Single leg stance (involved SIJ), opposite arm diagonal row. As a progression from Stage 3 exercises, further challenge is created by removing the toe touch support, requiring both balance and strength to maintain pelvic and trunk position.

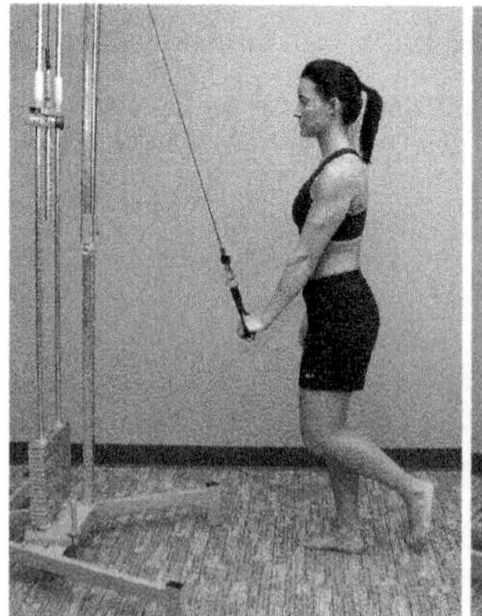

Figure 4.122a,b: Straight arm pull down, unilateral stance. The patient is standing on the involved lower limb, while maintaining a neutral spine. Unilateral shoulder extension is performed with the opposite arm, eccentrically first from the thighs to 90° of elevation, with a quick concentric return. Emphasis is on maintaining a neutral lumbar spine and pelvis during the elevation moment.

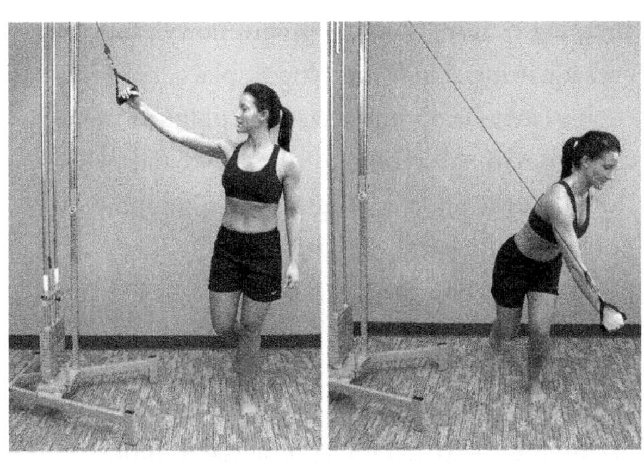

Figure 4.123a,b: Range of elevation can be progressed with unilateral stance reaching overhead and lateral. Balance on the lower limb is required while the lower abdominals stabilize the lumbar spine in neutral during the reach performance.

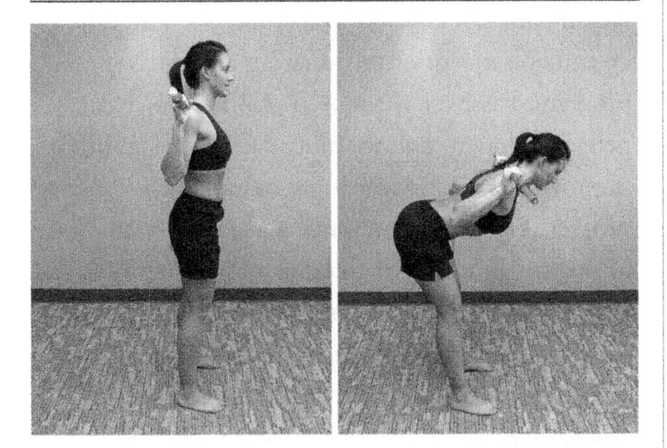

Figure 4.124a,b: Dead lift with shoulder bar for resistance. As a progression from previous dead lift training with the arms hanging the weight forward, placing the bar on the shoulders creates a longer lever arm of resistance to increase the torque on the lumbar spine, gluteal muscles and hamstrings. This progression is not necessary for all patients, but those returning to athletic performance or heavier lifting work may benefit from this strength training approach. Stabilization is achieved through the hamstrings, sacrotuberous ligament, across the SIJ to the lumbar spine.

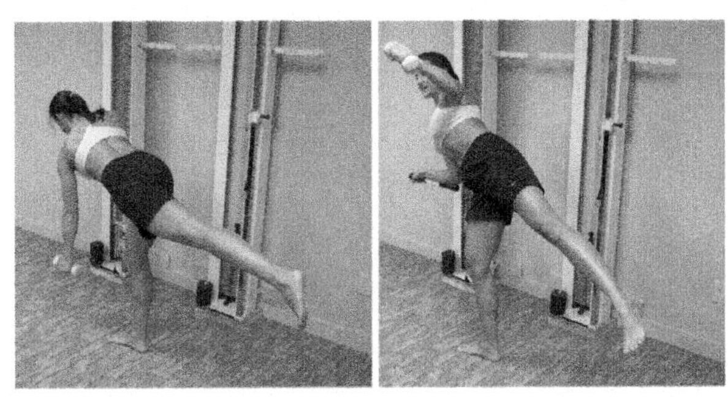

Figure 4.125a,b: Cross pattern dumbbell row. The ipsilateral hip is held in extension with the trunk horizontal and the arm resting on a weight bench. The contralateral knee is slightly flexed with the upper limb performing free weight rowing. A progression from isometric stabilization involves the addition of trunk rotation or hip extension to create more dynamic motion of the lumbopelvic region.

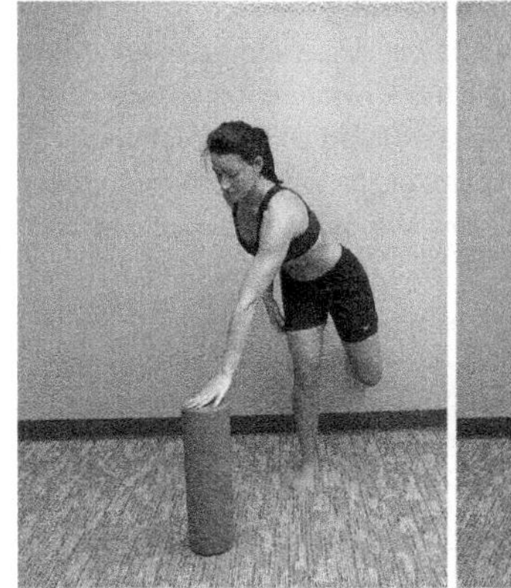
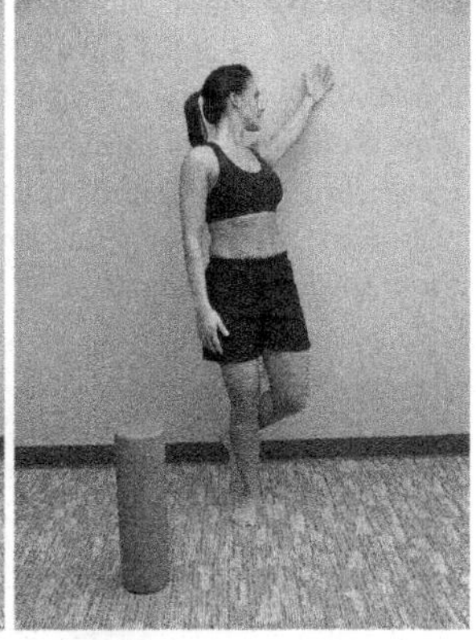

Figure 4.126a,b: Unilateral stance on the involved limb, anterolateral reach into flexion/rotation right, followed by posteriomedial reach into extension rotation left. Hip and pelvic stabilization through unilateral stance diagonal patterns combines central balance training with strength training of the lower limb and pelvis. This is a functional progression of basic lunging exercises.

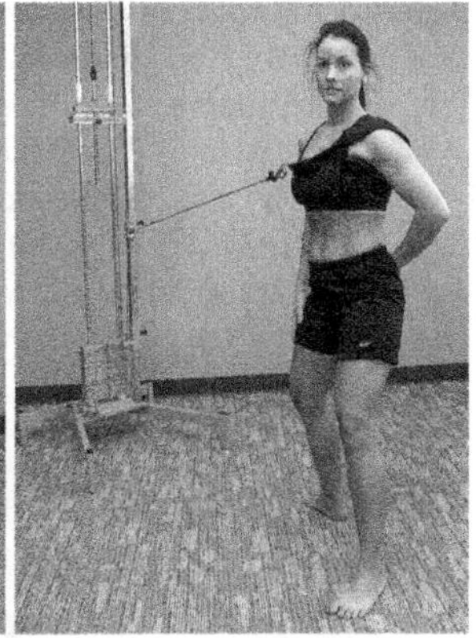

Figure 4.127a,b: The pivot twist exercise is a combination of the SIJ pivot exercise and standing lumbar rotation against resistance. The client starts with the resistive strap over the uninvolved side shoulder (left side pictured). The knees and hips are slightly flexed with the feet shoulder width, the involved side hip (right) extends and external rotates while rising up and rotating the torso away. The lumbar spine must be stabilized on the pelvis while the hip right hip performs an extension, abduction and external rotation moment.

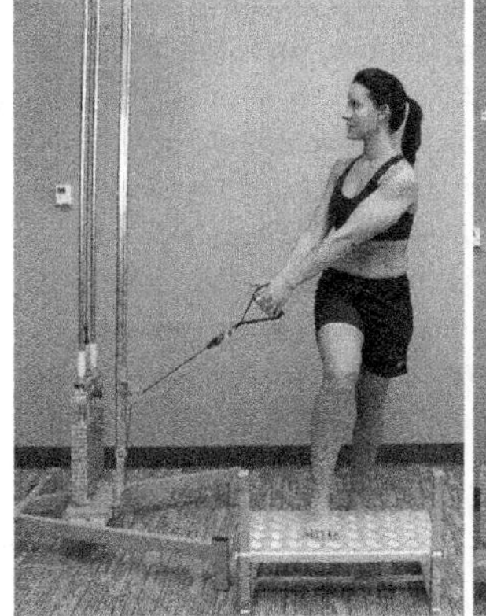
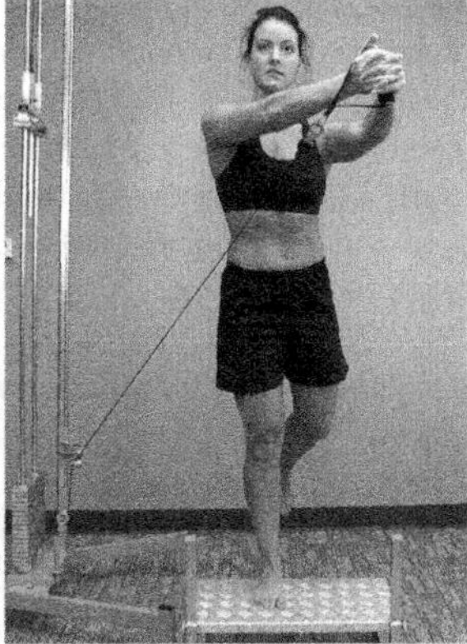

Figure 4.128a,b: Multidirectional right step-up with a right inferiolateral force moment. An extension moment for the hip

and trunk is created with gravity during the step-up motion. The right lateral force increases the extension moment from the inferior vector and a right trunk rotational moment. The arms swing through to the left, creating an left rotational trunk pattern, increasing activity of the right lumbar multifidi, right hip deep external rotators, the oblique abdominals and the transverse abdominus.

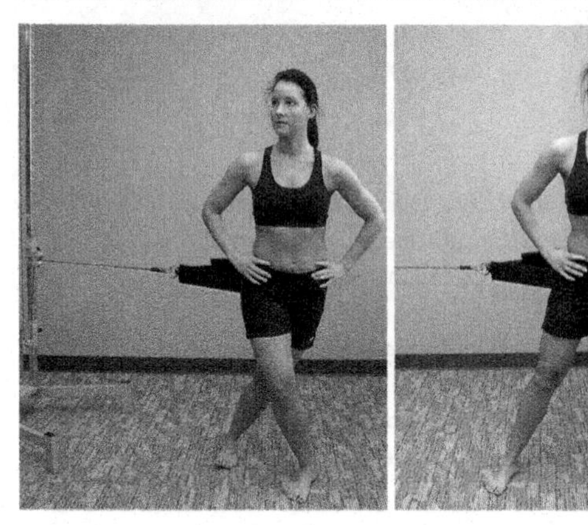

Figure 4.129a,b: Cross over lateral walking with resistance. The cross over step aggressively challenges the integrity of the SI articulation with extreme moments of force and range. Motion around the "x" axis occurs with hip flexion while aggressive "y" axis motion occurs with adduction of the hip. This is an important activity for athletes participating in reactive sports and their ability to perform this should be a key indicator as to if they are ready to participate.

Figure 4.130a,b,c: Right Lateral Hop and Stop. A lateral hop onto the right lower extremity emphasizes eccentric deceleration. The patient is instructed to establish unilateral balance stance on foot contact, followed by a one to two second hold prior to performing the next lateral hop. This is a transitional exercise for lunging to running is hopping. The hop-and-stop requires the body to be stabilized between each successive hop. An anterior hop is performed into the involved side. The patient must stabilize and maintain unilateral stance balance upon landing. Until the patient can perform a stable landing, without valgus collapse of the lower limb or excessive shake in the lower limb, a return to running is not suggested. The hop and stop and be performed in any direction in the 360° plane of the floor, matching directions of dysfunction and/or sport requirement.

Jumping training is also a progression of lunging and squatting. Light jumping in place, even jump roping, may be an initial starting point to train more functional stability of the pelvis and pelvic floor. For the athlete, progressing to jumping in horizontal direction, jumping for height or box jumping can be added as functionally necessary.

Summary

Training for pelvic stability and pelvic floor function requires integration of the primary diagnosis and impairments with function of neighboring joint systems. The pelvis cannot be treated in isolation. Passive manual therapy techniques, not covered in this exercise text, are often necessary prior to initiating an exercise program. Passive joint mobilization to normalize pelvic alignment and mobility can make significant improvements in motor function, allowing a faster progression through the training stages. A review of the hip and lumbar chapters of this text will also provide additional insights and ideas on how to develop an individualized exercise program that addresses overall function of the hips, pelvis and lumbar spine.

Bibliography

Abramson D, Roberts SM, Wilson PD. Relaxation of the pelvic joints in pregnancy. Surg Gynecol Obstet 58:595–613, 1934.

Aksac B, Aki S, Karan A, Yalcin O, Isikoglu M, Eskiyurt N. Biofeedback and pelvic floor exercises for the rehabilitation of urinary stress incontinence. Gynecol Obstet Invest 56(1):23–7, 2003.

Albert H, Godskesen M, Westergaard J. Evaluation of clinical tests used in classification procedures in pregnancy-related pelvic joint pain. Eur Spine J 9(2):161–6, Apr, 2000.

Alewijnse D, Mesters I, Metsemakers J, van den Borne B. Predictors of long-term adherence to pelvic floor muscle exercise therapy among women with urinary incontinence. Health Educ Res 18(5):511–24, Oct, 2003.

Alewijnse D, Mesters IE, Metsemakers JF, van den Borne BH. Program development for promoting adherence during and after exercise therapy for urinary incontinence. Patient Educ Couns 48(2):147–60, Oct–Nov, 2002.

Alewijnse D, Metsemakers JF, Mesters IE, van den Borne B. Effectiveness of pelvic floor muscle exercise therapy supplemented with a health education program to promote long-term adherence among women with urinary incontinence. Neurourol Urodyn 22(4):284–95, 2003.

American Academy of Family Physicians. www.familydoctor.org 8/2005

Andrews CM, O'Neill LM. Use of pelvic tilt exercise for ligament pain relief. J Nurse Midwifery 39(6):370–4, Nov–Dec, 1994.

Ashton-Miller JA, Howard D, DeLancey JO. The functional anatomy of the female pelvic floor and stress continence control system. Scand J Urol Nephrol Suppl (207):1–7, 2001.

Berghmans LC, Frederiks CM, de Bie RA, Weil EH, Smeets LW, van Waalwijk van Doorn ES, Janknegt RA. Efficacy of biofeedback, when included with pelvic floor muscle exercise treatment, for genuine stress incontinence. Neurourol Urodyn 15(1):37–52, 1996.

Bjorklund K, Naessen T, Nordstrom ML, Bergstrom S. Pregnancy-related back and pelvic pain and changes in bone density. Acta Obstet Gynecol Scand 78(8):681–5, Sep, 1999.

Bo K, Kvarstein B, Nygaard I. Lower urinary tract symptoms and pelvic floor muscle exercise adherence after15 years. Obstet Gynecol 105(5 Pt 1):999–1005, May, 2005.

Bo K, Talseth T. Long-term effect of pelvic floor muscle exercise 5 years after cessation of organized training. Obstet Gynecol 87(2):261–5, Feb, 1996.

Bo K. A tailored pelvic floor exercise program commenced immediately post-partum promotes continence. Aust J Physiother 48(4):317, 2002.

Bo K. Adherence to pelvic floor muscle exercise and long-term effect on stress urinary incontinence. A five-year follow-up study. Scand J Med Sci Sports 5(1):36–9, Feb, 1995.

Bo K. Pelvic floor muscle exercise and urinary incontinence—train yourself continent! Tidsskr Nor Laegeforen 120(29):3583–9, Nov 30, 2000.

Bo K. Urinary incontinence, pelvic floor dysfunction, exercise and sport. Sports Med 34(7):451–64, 2004.

Borello-France DF, Zyczynski HM, Downey PA, Rause CR, Wister JA. Effect of pelvic-floor muscle exercise position on continence and quality-of-life outcomes in women with stress urinary incontinence. Phys Ther 86(7):974–86, Jul, 2006.

Boyington AR, Dougherty MC. Pelvic muscle exercise effect on pelvic muscle performance in women. Int Urogynecol J Pelvic Floor Dysfunct 11(4):212–8, 2000.

Brynhildsen J, Hansson A, Persson A, Hammar M. Follow-up of patients with low back pain during pregnancy. Obstet Gynecol 91(2):182–6, Feb, 1998.

Bursch SG. Interrater reliability of diastasis recti abdominis measurement. Phys Ther 67(7):1077–9, Jul, 1987.

Buyruk HM, Stam HJ, Snijders CJ, Lameris JS, Holland WP, Stijnen TH. Measurement of sacroiliac joint stiffness in peripartum pelvic pain patients with Doppler imaging of vibrations (DIV). Eur J Obstet Gynecol Reprod Biol 83(2):159–63, Apr, 1999.

Calguneri M, Bird HA, Wright V. Changes in joint laxity occurring during pregnancy. Ann Rheum Dis 41(2):126–8, Apr, 1982.

Chang PL, Tsai LH, Huang ST, Wang TM, Hsieh ML, Tsui KH. The early effect of pelvic floor muscle exercise after transurethral prostatectomy. J Urol 160(2):402–5, Aug, 1998.

Chiarelli P, Cockburn J. Promoting urinary continence in women after delivery: randomised controlled trial. BMJ 324(7348):1241, May 25, 2002.

Claes H, Baert L. Pelvic floor exercise versus surgery in the treatment of impotence. Br J Urol 71(1):52–7, Jan, 1993.

Coffey SW, Wilder E, Majsak MJ, Stolove R, Quinn L. The effects of a progressive exercise program with surface electromyographic biofeedback on an adult with fecal incontinence. Phys Ther 82(8):798–811, Aug, 2002.

Damen L, Buyruk HM, Guler-Uysal F, Lotgering FK, Snijders CJ, Stam HJ. The prognostic value of asymmetric laxity of the sacroiliac joints in pregnancy-related pelvic pain. Spine 27(24):2820–4, Dec 15, 2002a.

Damen L, Buyruk HM, Guler-Uysal F, Lotgering FK, Snijders CJ, Stam HJ. Pelvic pain during pregnancy is associated with asymmetric laxity of the sacroiliac

joints. Acta Obstet Gynecol Scand 80(11):1019–24, Nov, 2001.
Damen L, Spoor CW, Snijders CJ, Stam HJ. Does a pelvic belt influence sacroiliac joint laxity? Clin Biomech (Bristol, Avon) 17(7):495–8, Aug, 2002b.
de Groot M, Pool-Goudzwaard AL, Spoor CW, Snijders CJ. The active straight leg raising test (ASLR) in pregnant women: Differences in muscle activity and force between patients and healthy subjects. Man Ther Dec 22, 2006.
Depledge J, McNair PJ, Keal-Smith C, Williams M. Management of symphysis pubis dysfunction during pregnancy using exercise and pelvic support belts. Phys Ther 85(12):1290–300, Dec, 2005.
Dougherty M, Bishop K, Mooney R, Gimotty P, Williams B. Graded pelvic muscle exercise. Effect on stress urinary incontinence. J Reprod Med 38(9):684–91, Sep, 1993.
Dumoulin C, Seaborne DE, Quirion-DeGirardi C, Sullivan SJ. Pelvic-floor rehabilitation, Part 2: Pelvic-floor reeducation with interferential currents and exercise in the treatment of genuine stress incontinence in postpartum women—a cohort study. Phys Ther 75(12):1075–81, Dec, 1995a.
Dumoulin C, Seaborne DE, Quirion-DeGirardi C, Sullivan SJ. Pelvic-floor rehabilitation, Part 1: Comparison of two surface electrode placements during stimulation of the pelvic-floor musculature in women who are continent using bipolar interferential currents. Phys Ther 75(12):1067–74, Dec, 1995b.
Eliasson K, Larsson T, Mattsson E. Prevalence of stress incontinence in nulliparous elite trampolinists.Scand J Med Sci Sports 12(2):106–10, Apr, 2002.
Faugli HP. *Medical Exercise Therapy*. Laerergruppen for Medisnsk Treningsterapi AS, Norway, 1996.
Farbrot E. The relationship of the effect and pain of pregnancy to the anatomy of the pelvis. Acta Radiol 38(5):403–19, Nov, 1952.
Ferreira P. Specific stabilising exercise improves pain and function in women with pelvic girdle pain following pregnancy. Aust J Physiother 50(4):259, 2004.
Ferreira PH, Ferreira ML, Maher CG, Herbert RD, Refshauge K. Specific stabilisation exercise for spinal and pelvic pain: a systematic review. Aust J Physiother 52(2):79–88, 2006.
Filocamo MT, Li Marzi V, Del Popolo G, Cecconi F, Marzocco M, Tosto A, Nicita G. Effectiveness of early pelvic floor rehabilitation treatment for post-prostatectomy incontinence. Eur Urol 48(5):734–8, Nov, 2005.
Finckenhagen HB, Bo K. The effect of pelvic floor exercise on stress urinary incontinence. Tidsskr Nor Laegeforen 118(13):2015–7, May 20, 1998.
Fisher K, Riolo L. Evidence in physical therapy. Phys Ther 84(8):744–753, 2004.
FitzGerald MP, Kotarinos R. Rehabilitation of the short pelvic floor. II: Treatment of the patient with the short pelvic floor. Int Urogynecol J Pelvic Floor Dysfunct 14(4):269–75, Oct, 2003.
FitzGerald MP, Kotarinos R. Rehabilitation of the short pelvic floor. I: Background and patient evaluation. Int Urogynecol J Pelvic Floor Dysfunct 14(4):261–8, Oct, 2003.
Gallo ML, Staskin DR. Cues to action: pelvic floor muscle exercise compliance in women with stress urinary incontinence. Neurourol Urodyn 16(3):167–77, 1997.
Garshasbi A, Faghih Zadeh S. The effect of exercise on the intensity of low back pain in pregnant women. Int J Gynaecol Obstet 88(3):271–5, Mar, 2005.
Gordon D, Luxman D, Sarig Y, Groutz A. Pelvic floor exercise and biofeedback in women with urinary stress incontinence. Harefuah 136(8):593–6, 660, Apr 15, 1999.
Grimsby O. Neurophysiological view points on hypermobilities. The Nordic Group of Specialists. Manual Therapy 2:2–9, 1988.
Gunterberg B, Romanus B, Steiner B. Pelvic strength after major amputation of the sacrum—an experimental study. Acta Orthopedica Scandinavica 47:635–642, 1976.
Hansen A, Jensen DV, Larsen E, Wilken-Jensen C, Petersen LK. Relaxin is not related to symptom-giving pelvic girdle relaxation in pregnant women. Acta Obstet Gynecol Scand 75(3):245–9, Mar, 1996.
Hansen A, Jensen DV, Wormslev M, Minck H, Johansen S, Larsen EC, Wilken-Jensen C, Davidsen M, Hansen TM. Symptom-giving pelvic girdle relaxation in pregnancy. II: Symptoms and clinical signs. Acta Obstet Gynecol Scand 78(2):111–5, Feb, 1999.
Hayn MA, Greco SJ, Capuano K, Byrnes A. Compliance with pelvic floor exercise program: maintaining bladder symptom relief. Urol Nurs 20(2):129–31, Apr, 2000.
Hoek van Dijke GA, Snijders CJ, Stoeckart R, Stam HJ. A biomechanical model on muscle forces in the transfer of spinal load to the pelvis and legs. J Biomech 32(9):927–33, Sep, 1999.
Ishiko O, Hirai K, Sumi T, Tatsuta I, Ogita S. Hormone replacement therapy plus pelvic floor muscle exercise for postmenopausal stress incontinence. A randomized, controlled trial. J Reprod Med 46(3):213–20, Mar, 2001.
Jacob HA, Kissling RO. The mobility of the sacroiliac joints in healthy volunteers between 20 and 50 years of age Clin Biomech (Bristol, Avon) 10(7):352–361, Oct, 1995.
Janda V. Function of Muscles in Musculoskeletal Pain Syndromes. Course workbook. Northeast Seminars. Tacoma, Washington, April 18–19, 1999.
Janda V. Treatment of chronic low back pain. Journal of Manual Medicine 6:166–168, 1992.
Jiang K, Novi JM, Darnell S, Arya LA. Exercise and urinary incontinence in women. Obstet Gynecol Surv 59(10):717–21; quiz 745–6, Oct, 2004.

Joanna Briggs Institute. A pelvic floor muscle exercise programme for urinary incontinence following childbirth. Nurs Stand 20(33):46–50, Apr 26–May 2, 2006.

Johnson VY. How the principles of exercise physiology influence pelvic floor muscle training. J Wound Ostomy Continence Nurs 28(3):150–5, May, 2001.

Klein MC. In the literature: pushing in the wrong direction. Birth 33(3):251–3, Sep, 2006.

Kristiansson P, Svardsudd K, von Schoultz B. Back pain during pregnancy: a prospective study. Spine 21(6):702–9, Mar 15, 1996.

La Pera G, Nicastro A. A new treatment for premature ejaculation: the rehabilitation of the pelvic floor. J Sex Marital Ther 22(1):22–6, Spring, 1996.

Larsen EC, Wilken-Jensen C, Hansen A, Jensen DV, Johansen S, Minck H, Wormslev M, Davidsen M, Hansen TM. Symptom-giving pelvic girdle relaxation in pregnancy. I: Prevalence and risk factors. Acta Obstet Gynecol Scand 78(2):105–10, Feb, 1999.

Lee D. Course notes: The Pelvis–Restoring Function, Relieving Pain. Northeast Seminars, Seattle, Washington, May 18–19, 2003.

Lee D, Vleeming A. An integrated therapeutic approach to the treatment of pelvic girdle pain. In: Vleeming A, Mooney V, Stoeckart R. Movement, stability and lumbopelvic pain. Integration of research and therapy. 2nd Edition. Churchill Livingstone, 2007.

Leung MW, Wong BP, Leung AK, Cho JS, Leung ET, Chao NS, Chung KW, Kwok WK, Liu KK. Electrical stimulation and biofeedback exercise of pelvic floor muscle for children with faecal incontinence after surgery for anorectal malformation. Pediatr Surg Int 22(12):975–8, Sep, 2006.

Lund PJ, Krupinski EA, Brooks WJ. Ultrasound evaluation of sacroiliac motion in normal volunteers. Acad Radiol 3(3):192–6, Mar, 1996.

MacLennan AH. The role of the hormone relaxin in human reproduction and pelvic girdle relaxation. Scand J Rheumatol 88(Suppl): 7–15, 1991.

Mathewson-Chapman M. Pelvic muscle exercise/ biofeedback for urinary incontinence after prostatectomy: an education program. J Cancer Educ 12(4):218–23, Winter, 1997.

McLennan MT, Melick CF, Alten B, Young J, Hoehn MR. Patients' knowledge of potential pelvic floor changes associated with pregnancy and delivery. Int Urogynecol J Pelvic Floor Dysfunct 17(1):22–6, Jan, 2006.

Mens JM, Damen L, Snijders CJ, Stam HJ. The mechanical effect of a pelvic belt in patients with pregnancy-related pelvic pain. Clin Biomech (Bristol, Avon) 21(2):122–7, Feb, 2006.

Mens JM, Snijders CJ, Stam HJ. Diagonal trunk muscle exercises in peripartum pelvic pain: a randomized clinical trial. Phys Ther 80(12):1164–73, Dec, 2000.

Mens JM, Vleeming A, Snijders CJ, Koes BW, Stam HJ. Reliability and validity of the active straight leg raise test in posterior pelvic pain since pregnancy. Spine 26(10):1167–71, May 15, 2001.

Mens JM, Vleeming A, Snijders CJ, Koes BW, Stam HJ. Validity of the active straight leg raise test for measuring disease severity in patients with posterior pelvic pain after pregnancy. Spine 27(2):196–200, Jan 15, 2002a.

Mens JM, Vleeming A, Snijders CJ, Ronchetti I, Ginai AZ, Stam HJ. Responsiveness of outcome measurements in rehabilitation of patients with posterior pelvic pain since pregnancy. Spine 27(10):1110–5, May 15, 2002b.

Mens JM, Vleeming A, Snijders CJ, Ronchetti I, Stam HJ. Reliability and validity of hip adduction strength to measure disease severity in posterior pelvic pain since pregnancy. Spine 27(15):1674–9, Aug 1, 2002c.

Mens JM, Vleeming A, Snijders CJ, Stam HJ, Ginai AZ. The active straight leg raising test and mobility of the pelvic joints. Eur Spine J 8(6):468–73, 1999.

Mooney V, Pozos R, Vleeming A, Gulick J, Swenski D. Exercise treatment for sacroiliac pain. Orthopedics 24(1):29–32, Jan, 2001.

Morkved S, Bo K. The effect of postpartum pelvic floor muscle exercise in the prevention and treatment of urinary incontinence. Int Urogynecol J Pelvic Floor Dysfunct 8(4):217–22, 1997.

Moul JW. Pelvic muscle rehabilitation in males following prostatectomy. Urol Nurs 18(4):296–301, Dec, 1998.

Ostgaard HC, Andersson GB Postpartum low-back pain. Spine 17(1):53–5, Jan, 1992.

Ostgaard HC, Andersson GB, Karlsson K. Prevalence of back pain in pregnancy. Spine 16(5):549–52, May, 1991.

Ostgaard HC, Zetherstrom G, Roos-Hansson E. Back pain in relation to pregnancy: a 6-year follow-up. Spine 22(24):2945–50, Dec 15, 1997.

O'Sullivan PB, Beales DJ, Beetham JA, Cripps J, Graf F, Lin IB, Tucker B, Avery A. Altered motor control strategies in subjects with sacroiliac joint pain during the active straight-leg-raise test. Spine 27(1):E1–8, Jan 1, 2002.

O'sullivan PB, Beales DJ. Changes in pelvic floor and diaphragm kinematics and respiratory patterns in subjects with sacroiliac joint pain following a motor learning intervention: A case series. Man Ther 12(3):209–18, Aug, 2007.

Pager CK, Solomon MJ, Rex J, Roberts RA. Long-term outcomes of pelvic floor exercise and biofeedback treatment for patients with fecal incontinence. Dis Colon Rectum 45(8):997–1003, Aug, 2002.

Petersen LK, Vogel I, Agger AO, Westergard J, Nils M, Uldbjerg N. Variations in serum relaxin (hRLX-2) concentrations during human pregnancy. Acta Obstet Gynecol Scand 74(4):251–6, Apr, 1995.

Peticca L, Pietroletti R. Combined pelvic floor rehabilitation. Tech Coloproctol 6(3):203, Dec, 2002.

Petros PP, Skilling PM. Pelvic floor rehabilitation in the female according to the integral theory of female urinary incontinence. First report. Eur J Obstet Gynecol Reprod Biol 94(2):264–9, Feb, 2001.

Pool-Goudzwaard AL, Hoek van Dijke G, Gurp van M, Mulder P, Snijders CJ, Stoeckart R. Contribution of pelvic floor muscles to stiffness of the pelvic ring. Clin Biomech 19(6):564–571, Jul, 2004.

Pool-Goudzwaard AL, Slieker ten Hove MC, Vierhout ME, Mulder PH, Pool JJ, Snijders CJ, Stoeckart R. Relations between pregnancy-related low back pain, pelvic floor activity and pelvic floor dysfunction. Int Urogynecol J Pelvic Floor Dysfunct 16(6):468–74, Nov–Dec, 2005.

Pool-Goudzwaard AL, Vleeming A, Stoeckart R, Snijders CJ, Mens JM. Insufficient lumbopelvic stability: a clinical, anatomical and biomechanical approach to 'a-specific' low back pain. Man Ther 3(1):12–20, Feb, 1998.

Porru D, Campus G, Caria A, Madeddu G, Cucchi A, Rovereto B, Scarpa RM, Pili P, Usai E. Impact of early pelvic floor rehabilitation after transurethral resection of the prostate. Neurourol Urodyn 20(1):53–9, 2001.

Richardson CA, Snijders CJ, Hides JA, Damen L, Pas MS, Storm J. The relation between the transversus abdominis muscles, sacroiliac joint mechanics, and low back pain. Spine 27(4):399–405, Feb 15, 2002.

Sampselle CM, Miller JM, Mims BL, Delancey JO, Ashton-Miller JA, Antonakos CL. Effect of pelvic muscle exercise on transient incontinence during pregnancy and after birth. Obstet Gynecol 91(3):406–12, Mar, 1998.

Samuel CS. Relaxin: antifibrotic properties and effects in models of disease. Clin Med Res 3(4):241–9, Nov, 2005.

San Juan JG, Yaggie JA, Levy SS, Mooney V, Udermann BE, Mayer JM. Effects of pelvic stabilization on lumbar muscle activity during dynamic exercise. J Strength Cond Res 19(4):903–7, Nov, 2005.

Shorten A, Donsante J, Shorten B. Birth position, accoucheur, and perineal outcomes: informing women about choices for vaginal birth. Birth 29(1):18–27, Mar, 2002.

Siu LS, Chang AM, Yip SK, Chang AM. Compliance with a pelvic muscle exercise program as a causal predictor of urinary stress incontinence amongst Chinese women. Neurourol. Urodyn 22(7):659–63, 2003.

Smidt GL, McQuade K, Wei SH, Barakatt E. Sacroiliac kinematics for reciprocal straddle positions. Spine 20(9):1047–54, May 1, 1995.

Smidt GL, Wei SH, McQuade K, Barakatt E, Sun T, Stanford W. Sacroiliac motion for extreme hip positions. A fresh cadaver study. Spine 22(18):2073–82, Sep 15, 1997.

Snijders CJ, Hermans PF, Kleinrensink GJ. Functional aspects of cross-legged sitting with special attention to piriformis muscles and sacroiliac joints. Clin Biomech (Bristol, Avon) 21(2):116–21, Feb, 2006.

Snijders CJ, Hermans PF, Niesing R, Spoor CW, Stoeckart R. The influence of slouching and lumbar support on iliolumbar ligaments, intervertebral discs and sacroiliac joints. Clin Biomech (Bristol, Avon) 19(4):323–9, May, 2004.

Snijders CJ, Slagter AH, van Strik R, Vleeming A, Stoeckart R, Stam HJ. Why leg crossing? The influence of common postures on abdominal muscle activity. Spine 20(18):1989–93, Sep 15, 1995.

Stapleton DB, MacLennan AH, Kristiansson P. The prevalence of recalled low back pain during and after pregnancy: a South Australian population survey. Aust N Z J Obstet Gynaecol 42(5):482–5, Nov, 2002.

Stevens VK, Vleeming A, Bouche KG, Mahieu NN, Vanderstraeten GG, Danneels LA. Electromyographic activity of trunk and hip muscles during stabilization exercises in four-point kneeling in healthy volunteers. Eur Spine J Eur Spine J 16(5):711–8, May, 2007.

Stremler R, Hodnett E, Petryshen P, Stevens B, Weston J, Willan AR. Randomized controlled trial of hands-and-knees positioning for occipitoposterior position in labor. Birth 32(4):243–51, Dec, 2005.

Stuge B, Vollestad NK. Important aspects for efficacy of treatment with specific stabilizing exercises for postpartum pelvic girdle pain. In: Vleeming A, Mooney V, Stoeckart R. *Movement, Stability and Lumbopelvic Pain. Integration of Research and Therapy.* 2nd Edition. Churchill Livingstone, 2007.

Sung MS, Hong JY, Choi YH, Baik SH, Yoon H. FES-biofeedback versus intensive pelvic floor muscle exercise for the prevention and treatment of genuine stress incontinence. J Korean Med Sci 15(3):303–8, Jun, 2000.

Suputtitada A, Wacharapreechanont T, Chaisayan P. Effect of the "sitting pelvic tilt exercise" during the third trimester in primigravidas on back pain. J Med Assoc Thai 85 Suppl 1:S170–9, Jun, 2002.

Turgut F, Turgut M, Cetinsahin M. A prospective study of persistent back pain after pregnancy. Eur J Obstet Gynecol Reprod Biol 80(1):45–8, Sep, 1998.

van Wingerden JP, Vleeming A, Buyruk HM, Raissadat K. Stabilization of the sacroiliac joint in vivo: verification of muscular contribution to force closure of the pelvis. Eur Spine J 13(3):199–205, May, 2004.

Velde van der F. Psychophysiological investigation of the pelvic floor. The mechanism of vaginismus. Thesis, Free University of Amsterdam, 1999.

Vilensky JA, O'Connor BL, Fortin JD, Merkel GJ, Jimenez AM, Scofield BA, Kleiner JB. Histologic analysis of neural elements in the human sacroiliac joint. Spine 27(11):1202–7, Jun 1, 2002.

Vleeming A, Buyruk HM, Stoeckart R, Karamursel S, Snijders CJ. An integrated therapy for peripartum pelvic instability: a study of the biomechanical effects of pelvic belts. Am J Obstet Gynecol 166(4):1243–7,

Apr, 1992.

Vleeming A, de Vries HJ, Mens JM, van Wingerden JP. Possible role of the long dorsal sacroiliac ligament in women with peripartum pelvic pain. Acta Obstet Gynecol Scand 81(5):430–6, May, 2002.

Vleeming A, Mooney V, Stoeckart R. *Movement, Stability and Lumbopelvic Pain. Integration of Research and Therapy*. 2nd Edition. Churchill Livingstone, 2007.

Vleeming A, Pool-Goudzwaard AL, Stoeckart R, van Wingerden JP, Snijders CJ. The posterior layer of the thoracolumbar fascia. Its function in load transfer from spine to legs. Spine 20(7):753–8, Apr 1, 1995.

Vleeming A, Stoeckart R, Volkers AC, Snijders CJ. Relation between form and function in the sacroiliac joint. Part I: Clinical anatomical aspects. Spine 15(2):130–2, Feb, 1990.

Vleeming A, Volkers AC, Snijders CJ, Stoeckart R. Relation between form and function in the sacroiliac joint. Part II: Biomechanical aspects. Spine 15(2):133–6, Feb, 1990.

Wang M, Bryant JT, Dumas GA. A new in vitro measurement technique for small three-dimensional joint motion and its application to the sacroiliac joint. Med Eng Phys 18(6):495–501, Sep, 1996.

Wei JT, De Lancey JO. Functional anatomy of the pelvic floor and lower urinary tract. Clin Obstet Gynecol 47(1):3–17, Mar, 2004.

Windsor PM, Nicol KF, Potter J. A randomized, controlled trial of aerobic exercise for treatment-related fatigue in men receiving radical external beam radiotherapy for localized prostate carcinoma. Cancer 101(3):550–7, Aug 1, 2004.

Yoon HS, Song HH, Ro YJ. A comparison of effectiveness of bladder training and pelvic muscle exercise on female urinary incontinence. Int J Nurs Stud 40(1):45–50, Jan, 2003.

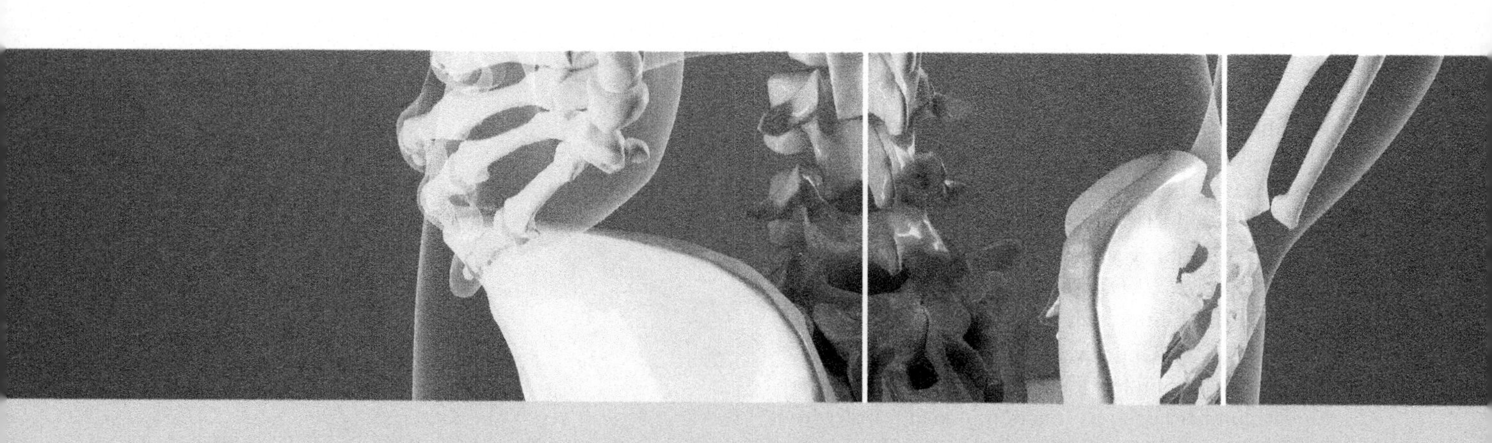

Authors: Alec Kay, Ola Grimsby, Jim Rivard, Dan Washeck, Art Schwarcz

CHAPTER 5

Exercise Rehabilitation of the Lumbar Spine

Introduction

In this chapter:

Treatment of the lumbar spine has been described as "the medical catastrophe of the 20th century" (Waddell 2000). Exercise has been applied for therapeutic purposes for thousands of years, and the treatment of the lumbar spine is no exception. Unlike many of the other topics evaluated by the Cochrane Study (Hayden et al. 2006), the report showed some evidence of benefit for exercise when used with patients experiencing chronic low back pain. Unfortunately, in reviewing the randomized trials, it is clear that the methodology used did not place subjects into subgroups. This resulted in very heterogeneous populations of subjects with low back pain being treated with specific exercise approaches. It is not reasonable to expect good outcomes if the same exercise protocol is applied to a group of patients with low back pain that have a variety of diagnoses and degree of acuity. Looking at the studies, it was generally found that exercise for low back pain has minimal to no effect for acute pain, with good evidence for benefiting patients with chronic pain (Hart et al. 2006, Hayden et al. 2006, Hayden et al. 2005a/b/c)

It is easy to predict only marginal outcomes with exercises for back pain if they are not specific to a patient's diagnosis, and this needs to be the focus of future scientific inquiry. After detailed tissue differentiation, exercise can be applied to not only improve symptoms and function, but to assist in facilitating tissue repair. Tissue differential diagnosis is challenging, yet it should remain a goal with each patient to better establish a treatment program. Even if an evaluation fails to identify a specific tissue, many tissues can be ruled out and any contraindications can be identified. Also, a pattern of symptoms with compression or tension can be established to assist in designing a safe program to improve tissue tolerance. Much of the literature addressing the application of exercise in the treatment of lumbar pain has identified two main properties: muscle strength or motor control. The application of scientific therapeutic exercise progressions (STEP) addresses these two issues, as well as tissue regeneration, resolving muscle guarding, reducing fear and avoidance behaviors, acute pain control, initiating functional training, increasing endurance and edema reduction.

Exercise is not applied alone to the spine, but is a part of a more complete treatment program involving modalities, soft tissue work, joint mobilization, exercise and prophylaxis. Exercise and manual therapy have been shown to be more effective at reducing pain levels than exercise alone (Geisser et al. 2005). The individual patient may require more emphasis on one of these interventions over another to control symptoms, but exercise is the primary intervention in improving overall performance. Benefit is noted from combining manual therapy with exercise for low back pain patients (Bronfort et al. 1996, Niemisto et al. 2003). Rasmussen-Barr et al. (2003) found no difference in treating low back pain patients with manual therapy versus stabilization exercises. Stabilization training, however, seemed to be more effective in terms of reducing the need for recurrent treatment periods. Conversely, Aure et al. (2003) found manual therapy to be more advantageous than exercise therapy. Improving joint mechanics with passive mobilization techniques prior to exercise can assist in normalizing afferent feedback to influence motor patterns. Segmental mobilization techniques in the lumbar spine have been shown to improve recruitment of the superficial abdominal muscles in subjects with chronic low back pain, though no improvements were found in the transverse abdominis muscle (Ferreira et al. 2007). Improved recruitment of the transverse abdominis following spinal manipulation has been reported in a case study (Gill et al. 2007). In a follow up controlled trial, Ferreira et al. (2007b) assessed outcomes for general exercise, specific motor control exercise or spinal manipulation for chronic non-specific back pain. A six and 12 month follow up found motor control exercise and spinal manipulative therapy slightly better than general exercise for short-term function and perception of effect, but not better medium or long-term effects. As all three approaches have evidence for positive outcomes, the emphasis on passive manual techniques, specific motor training and general exercise should be determined on an individual basis, rather than assuming one is superior for every patient. An overall treatment incorporates passive manual treatments, exercise and education. Manual therapy, as defined in this text, includes exercise therapy. For every degree of mobility improvement through passive treatment, comes the need for active treatment to stabilize this new range.

Wand et al. (2004) demonstrated the benefit of an assess/advise/treat model of care, offering better outcomes, than an assess/advise/wait model of care. Early intervention consisted of biopsychosocial education, manual therapy, and exercise. In the short-term, intervention is more effective than just advising the patient to stay active. Early intervention leads to a more rapid improvement in function, mood, quality of life, and general health. The timing of intervention also has an affect on the development of psychosocial features. When treatment was provided later, the same psychosocial benefits were not achieved. Though physician advice to stay active and return to work does have psychosocial benefits, a complete treatment program involving education, manual care and exercise can be more effective and more long lasting. Early intervention is also more cost effective due to an earlier return to work (Wright et al. 2005).

The lumbar spine is inherently unstable (Lucas and Bresler 1961), with the upright spine devoid of musculature only being able to support two kilograms of vertical load. Normal weight of the head and torso would cause buckling unless adequate muscular support is available. Looking at exercise to address low back pain, it is important to be able to see the body both globally and at a cellular, or tissue, level. From a mechanical standpoint, the human body is built so that the lumbar spine is a transition point between weight bearing through the feet and the active head, neck and trunk. As the lumbar spine endows the pelvic girdle with movement at L5/S1, it is also a natural point of stress. The requirements of coordination, strength, and flexibility are apparent in observation of the union of the stable pelvis with the relatively large moment forces from the head, shoulder, upper limbs, and torso. The lumbar spine has a higher number of Type II mechanoreceptors, compared to Type I, which are associated with recruitment of phasic muscle systems. Strength is crucial, but without the appropriate timing of contraction and coordination, recurrence has a significantly higher chance of occurring, as much as a six times greater chance of recurrence was measured two and three years after implementation of an exercise program (Hides et al. 2001).

This chapter will attempt to outline a thought process for exercise dosage and progression to address a continuum of clinical requirements from pain and mobility up through stability and function. It is assumed the reader is familiar with spinal anatomy and biomechanics, as well as earlier theoretical chapters in this book series. A more clinical approach is taken on how to apply work physiology concepts in a rehabilitative setting with conceptual outlines as well as specific examples.

Bed Rest or Stay Active

Bed rest and immobilization as the treatment for acute non-specific low back pain has come under question from a diverse accumulation of evidence based scientific research. Immobilization and bed rest for acute low back pain was based on the early understanding of the medical disease model. The thought was that early physical activity might increase pain, aggravate inflammation, prevent healing and eventually lead to chronic pain. John Hunter (1794) is credited as being one of the first to propose therapeutic rest as a treatment for wounds and inflammation based on the understanding of healing at that time (Waddell 1998). Later, Hugh Owens Thomas (1874), considered the father of English Orthopedics, applied the concept of therapeutic rest to the newly emerging field of orthopedics. This became the orthopedic rationale for acute injuries and is still described today as a basic principle of orthopedic treatment (Waddell 1998). Therapeutic rest has also been widely applied to acute cases of low back pain. According to the National Institutes of Health (NIH), low back pain ranks as the second most common symptom-related reason for seeing a physician, and is the most common and most expensive cause of work-related disability in the United States.

With the discovery of the herniated (or ruptured) disc in 1932, many physicians began to assume that most pain in the back was the result of this new problem. In addition, Nachemson (1966) discovered that the lumbar intradiscal pressures significantly change from supine (25–35 percent body weight) to over 300 percent in weight bearing positions. This information further emphasized the need to rest lying down. The thought was that laying and rest would allow the disc relief so that it may "go back in," resolving the back pain (Waddell 1998). An increase in intradiscal pressure was mistakenly taken as a negative thing, rather than a natural stimulus for tissue maintenance. Cartilage pressure also increases with weight bearing, but progressive weight bearing training is a mainstay of rehabilitation of cartilage pathologies, not bed rest.

By the mid 1980's, more and more research was published on bed rest and immobilization for acute back pain. Two key randomized control studies came out that challenged the idea of therapeutic rest for acute back pain. Deyo et al. (1996) showed that patients who were advised to take two days bed rest returned to work sooner than those advised to take seven days. Gilbert et al. (1985) showed those who had no bed rest returned to normal activities faster than those who had four days' bed rest. Waddell et al. (1988) did a systematic literature search for all randomized clinical trials of acute back pain comparing the effects of bed rest to advice by the medical practitioner to maintain activity levels. Ten trials of bed rest and eight trials of advice to stay active were identified. Consistent findings showed that bed rest is not an effective treatment for acute low back pain but may delay recovery. Advice to stay active and to continue ordinary activities results in a faster return to work, less chronic disability, and fewer recurrent problems (Waddell 1997). This review also found that early mobilization leads to fewer recurrent problems and less sick leave over the next one to two years (Waddell 1998). Studies have also shown that proper explanation and reassurance by the medical practitioner can have positive effects on patients' beliefs about back pain and satisfaction with healthcare (Deyo and Diehl 1996).

Patients with acute low back pain have a more rapid recovery when continuing ordinary activities within the limits permitted by the pain than either bed rest or back-mobilizing exercises (Malmivaara et al. 1995). van Tulder et al. (1988) published a systematic review of randomized controlled trials of the most common interventions used for acute and chronic non-specific low back pain. A rating system was used to assess the strength of the evidence based on the quality of the randomized controlled trials, the relevance of the outcome measures, and the consistency of the results. The most common treatments were categorized as to whether there was strong evidence, some evidence or contradictory evidence of their effectiveness. Strong evidence was found for advice to continue ordinary activity as normally as possible for nonspecific acute low back pain (van Tulder et al. 1998).

All the available evidence suggests that staying as active as possible and staying at work or returning to work as soon as possible can give faster symptomatic recovery and lead to less chronic disability and less time off work. Patient education and reassurance of the benefit of early pain free mobilization and return to activity by ones' health practitioner might be one of the most beneficial interventions for quicker recovery of acute low back pain. Staal et al. (2003) performed an international comparison of low back pain guidelines as it pertains to advice, return to work strategy, and treatment. A general agreement was found on advice that low back pain is a self limiting condition, and that remaining at work or an early (gradual) return to work, if necessary with modified duties, should be encouraged. Evidence suggests that prescribed exercise does not increase the recurrence rate of back pain in those with a history of acute, sub-acute, recurrent, or chronic low back pain. Several randomized controlled trials have demonstrated a significant decrease in the number of recurrences of back pain in subjects randomized to exercises, compared to controls (Donchin et al. 1990, Soukup et al. 1999, Hides et al. 2001).

The only time bed rest is prescribed is in the presence of nerve root compression. For the lumbar spine, this involves a loss of sensation, reduced deep tendon reflexes and reduced myotome findings. Bed rest is prescribed with compression to allow GAG to be broken down by enzymes, taking pressure off of the nerve. The half life of GAG is 1.7 to 7 days, in which time the clinical tests should move from a compressed state, to a mixed state, signaling the time to begin exercise. The mixed state involves at least one of the three tests moving from a state of compression to normal or hyperreactive state. As pain increases from a state of nerve compression to an irritated state, it cannot be used as a gauge for improvement. The patient should actually be warned that their pain might increase as nerve function improves. Continued clinical testing of the nerve function will allow the clinician to determine whether the treatment is having an overall positive effect. Too aggressive treatment can result in further tissue injury, leaking of more GAG from the disc and the nerve moving from an irritated state to a compressed state. Pain levels would decrease, but the state of tissue injury has actually regressed.

Nerve Root Pathology

	Sensory	Motor	Reflex	Pain
Nerve Root Irritation	+	+	+	+
Mixed Stage	+/-	+/-	+/-	+/-
Nerve Root Compression	-	-	-	-

Table 5.1: Nerve conduction assessment. True nerve compression results in reduced clinical tests for sensation, reflexes and myotomes. This is accompanied with a relative reduction in pain signals, compared to the mixed and irritated states. The irritated state is defined by hyperesthesia,

hyperreflexia and muscle spasm. The mixed state involves any combination of normal, irritated or compressed signs.

Exercise Outcomes

Much is still to be discovered as to the effectiveness of different kinds of exercises for lumbar patients. The effectiveness of specific exercises to achieve set training goals is possible, but the overall effect on pain and disability cannot be adequately addressed with one exercise. Specific dosage of exercise can determine the functional qualities obtained, such as coordination, endurance and strength. Rehabilitation is not achieved through obtaining one functional quality, but rather through a continuum of resolving acute symptoms through restoration of overall function.

Generalized Fitness Programs

Literature reviews for exercise and low back pain have generally shown that exercise is effective for chronic low back pain, however limited evidence exists for acute back pain. The effectiveness of general fitness programs, rather than individualized physical therapy exercise programs, has been assessed. General fitness programs have demonstrated effective outcomes (Klaber Moffett et al. 2000, Frost et al. 1995/1998), as well as being cost effective (Carr et al. 2005). These programs appear most effective on low grade, nonspecific low back pain without significant impairment. These studies typically exclude patients that have had recent physical therapy, predominant sciatica with leg pain, recent significant surgery, spinal surgery, presence of a neurological condition or systemic condition (such as rheumatoid arthritis), inability to get off the floor unaided, or pregnancy. More specific programs are likely necessary, addressing individualized needs that are not met through general exercise programs. Klaber Moffett et al. (2004) were able to demonstrate a positive outcome in patients with fear-avoidance behavior with utilization of a generalized back fitness program, however the control group consisted of patients treated only by a general practitioner. More specific programs, with guided exercise, would also be effective in addressing fear avoidance issues.

Koumantakis et al. (2005a/b) compared general endurance exercise with supplemental stabilization training, concluding that physical exercise alone, and not the exercise type, was the key determinant for improvement in this patient group. Contrary to this, O'Sullivan et al. (1997) found specific exercises statistically significant in reducing pain intensity and improving functional disability levels for patients with chronically symptomatic spondylolysis or spondylolisthesis.

Aggressive Generalized Back Exercise

More aggressive, yet still generalized protocols and exercise programs have also been assessed. These programs have been deemed appropriate for more significant back pain, stemming from conditions such as lumbar disc degeneration, herniated nucleus pulposus, spinal stenosis, facet syndrome, Grade I–II spondylolisthesis, spondylolysis, myofascial pain and postoperative fusion patients or laminectomy patients (Cohen and Rainville 2002). Contraindication for aggressive back exercise is noted for more significant pathologies including: medical instability; lesions best treated with surgery; severe osteoporosis; fracture; tumor, Cauda Equina and Conus Medullaris Syndromes, progressive neurological deficit; spinal instability; Grade III–IV spondylolisthesis; visceral/systemic pathology and spondyloarthropathy (Cohen and Rainville 2002). Exercises in this program included Cybex back extension, Roman chair hyperextension, lumbar crate lifting, pull-down machine, Cybex rotary torso machine and the Multihip weight machine. Training sessions lasted from one to two and half hours, with two to three sessions per week.

Prospective and retrospective analysis of studies utilizing aggressive exercise as treatment for patients with chronic low back pain reveal significant improvements within a six to eight week period. Trunk flexibility has been shown to improve by 20 percent, trunk strength and lifting capacities by 50 percent and endurance by 20 to 60 percent (Hazard et al. 1989, Mayer et al. 1985, Risch et al. 1993, Brady et al. 1994, Curtis et al. 1994, Mayer et al. 1994, Pollock et al. 1989). Pain-related disability was reduced by 50 percent (Fairbank et al. 1980), on average, and pain severity by 30 percent (Hartigan et al. 1994). Successful completion of exercise in the presence of chronic pain, from a cognitive standpoint lessens fear and concern, improves self-efficacy and confidence for performing daily activities, resulting in reduced disability (Rainville et al. 1993, Dolce et al. 1986).

Exercise is a primary intervention for physical therapists in the treatment of back pain. Research utilizing generalized fitness or generalized stabilization approaches for back pain should have limited results. This chapter will attempt to incorporate more general and traditional exercise approaches for back pain, but more importantly will attempt to illustrate the potential for specifically dosed exercise programs addressing identified impairments and tissue pathology.

Section 1: Stage 1 Exercise Progression for the Lumbar Spine

Despite limited evidence for exercise for acute low back pain, much can be accomplished early in rehabilitation to control symptoms and improve impairments, while protecting injured tissue. A lack of evidence in the literature may reflect a lack of research on more specific training for acute issues, rather than a lack of effectiveness. Faster resolution of acute symptoms more quickly transitions the patient to a subacute stage, in which evidence does exist supporting the positive effects of exercise. The concepts and techniques outlined in the STEP curriculum provide safe options, for early training. The exercises, dosage and progression concepts outlined in this chapter are an extension of Medical Exercise Therapy (MET) principles originating from the 1960's (Faugli and Holten 1996). Emphasis may be more on passive manual therapy techniques in the early stage, but this section will also provide many active options as well.

Stage 1 Progression Concepts

The basic components of an initial exercise program are to 1) normalize joint motion, 2) provide tissue repair stimulus, 3) resolve muscle guarding, 4) normalize motor patterns (coordination), 5) improve function and finally 6) to elevate the overall training level. Pain is indirectly improved by addressing these basic building blocks to function. Joint motion is facilitated first with both passive and active treatment. Arthrokinematic motion must be available to allow for normal range of motion, provide afferent input via mechanoreceptor firing in the surrounding joint capsule and to assist in normalizing muscle recruitment. Hypomobile joints require mobilizing exercises, while hypermobile joints limited by muscle guarding require exercise to normalize tone. For tissue training and improving motor function, additional exercises are dosed around the general parameters of high repetition with minimal resistance and slow speed.

Potential Tissue States and Functional Status: Stage 1

- *Reduced arthrokinematic motion*
- *Decrease in active and passive range of motion*
- *Painful joint at rest and/or with motion*
- *Abnormal respiration patterns*
- *Pain with weight bearing*
- *Edema with palpable temperature (if in the superficial joint)*
- *Muscle guarding at rest locally and with distal referral (Active trigger points with satellites and referred patterns)*
- *Poor coordination*
- *Poor balance/functional status*
- *Positive palpation to involved tissues*
- *All higher level functions of endurance, strength and power are reduced*
- *Sympathetic hyperactivity*

Exercises in Stage 1 are commonly associated with the functional qualities of pain reduction, edema reduction, resolution of muscle guarding, stimulation of metabolic activity of the injured tissue(s) for repair/regeneration, neurological adaptation to improve coordination and range of motion. Several or all of these functional qualities may be achieved in combination with simple pain free movements. Deciding which of these qualities to emphasize first will be influenced by a patient's pathology, stage of healing, surgical intervention, training level and general health, as well as any other specific limitation of the diagnosis. The state of the tissues and the functional level of each patient are assessed to determine necessary starting points, required training goals and to establish the appropriate dosage of training. All impairments are addressed specifically through a properly dosed exercise program for the functional quality that will reverse the abnormal condition. A shot gun approach for every low level functional quality is avoided and a more focused program is designed.

For rehabilitative exercise there is no difference between the athlete and the non-athlete, the young or old, the male or female, other than the dosage of the exercise. Even overweight elderly subjects have demonstrated positive training results for the lumbar spine and extremities (Vincent et al. 2006). Exercises are dosed in a patient specific manner, based on the stage of healing in the tissue and the training state. Each exercise is tested for resistance, repetitions, coordination and speed, then adjusted to the specific functional quality desired.

Many Repetitions / Dosage

Initial training emphasizes high numbers of repetitions with minimal resistance. Safe, pain free repetitive motion will improve all Stage 1 functional qualities. Specific parameters for dosage are outlined below, with the patient's performance and tolerance being the true gauge of how much should be initially achieved. Each of these functional qualities will be more specifically addressed in this section. Pain assessment and delayed onset tissue soreness (DOTS) will guide the clinician in modifying the program. Any pain during the exercise is avoided to prevent further tissue damage, increased muscle guarding or altered motor patterns due to motor reflexes from the pain. The introduction of additional pain during training would suggest abnormal tissue deformity or significant tissue ischemia, neither of which is a desired goal. Pain within several hours of a training session suggests excessive levels of stress to tissue, resulting in an inflammatory response. Development of pain or excessive stiffness the next morning would suggest unnecessary tissue stress or muscle soreness. This should also be avoided in the early stages. In later stages, when tissue tolerance has improved, post exercise soreness may relate more to muscle strain, that may be a tolerable level associated with higher levels of training.

Tissue Repair / Edema Resolution / Pain Inhibition

An acute ankle sprain is initially actively dosed with high repetitions of low resistance exercise to address pain, edema and muscle guarding. If dosed safely and performed within the available tissue tolerance level, improved healing should result. Depending on any contraindication present, this concept is no different for the spine. A patient with an acute strain of superficial fibers of the annulus fibrosus has a similar list of health issues, which can be tackled with both passive and active measures. High repetition, pain free movement recruits joint mechanoreceptors that will inhibit pain and muscle guarding. Movement will facilitate edema reduction and provide modified tension in the line of stress, promoting fibroblast activity related to tissue repair. Repetitive motion will also begin to address issues of motor recruitment, timing and coordination. Psychological benefits are also achieved by providing the patient with a safe environment to move, a positive exercise experience, and some personal sense of control over their situation. Rather than prescribing bed rest and limited activity, feeding into fear and anxiety, an active approach is prescribed that is safe and effective. Signs of over training need to be strictly monitored. Indicators of exercise exceeding tissue tolerance include increased pain, increased muscle guarding and any worsening of neurological symptoms.

Dosage for Stage 1 Functional Qualities

Tissue Repair, Pain inhibition, Edema Reduction and Joint Mobilization

- *Sets*: 1–5 sets or training sessions daily
- *Repetitions*: 2–6 hours during the day
- *Resistance*: Assisted to 25% of 1RM
- *Frequency*: Multiple times daily

Coordination

- *Sets*: 1–5 sets or training sessions daily
- *Repetitions*: 20–50 during the day
- *Resistance*: Assisted to 50% of 1RM
- *Frequency*: One to three times daily

Vascularity / Local Endurance

- *Sets*: 2–3 sets or training sessions daily
- *Repetitions*: 25
- *Resistance*: Assisted to 60% of 1RM
- *Frequency*: One time daily

Muscle in Atrophy

- *Sets*: 2–3 sets or training sessions daily
- *Repetitions*: 40 or more
- *Resistance*: Assisted to 40% of 1RM
- *Frequency*: One time daily

The dosage for tissue repair, edema resolution and pain inhibition is similar. Repetitive pain free motions are performed with minimal to zero resistance, stimulating tissue and cell repair. Just as an acute ankle sprain is moved repetitively for hours, so is the acute lumbar injury. The key is to begin with a comfortable posture, first moving in the least painful plane, and progressing to the most painful plane but remaining in the pain free range. Tissue training targets fibroblasts for type I collagen in the annular fibers of the disc, facet capsules and ligaments, while production of type II collagen is stimulated in the chondrocytes in the nucleus pulposus and facet cartilage. Annular fibers are more specifically targeted with modified tension in the line of stress with rotation, but this would also include flexion, extension and side bending. Compression and decompression for chondrocytes stimulation in the nucleus is also achieved with intradiscal pressure changes through rotation, but would also progress to all planes of motion and combined planes. Moving from lying postures to more erect postures will also increase the level of compression, improving the repair process (Nachemson 1966).

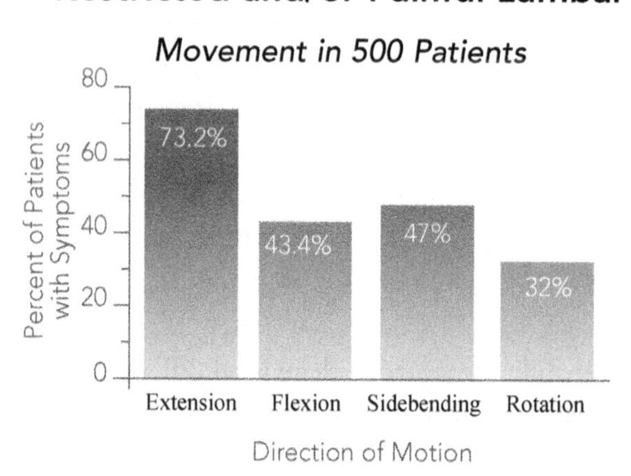

Table 5.2: Holten tested 500 patients with no exclusion criteria to document which directions were most and least tolerated. Rotation was found to be the most tolerated with extension producing pain in the greatest number of patients (Faugli and Holten 1996). The initial direction chosen for tissue training is based on the direction of tolerance by the patient, not by statistical descriptions from research.

In a more basic sense, injured tissues need modified motion to stimulate repair without causing further tissue injury. Motion can be performed in one plane, or all three planes, depending upon the type of tissue injury and the severity. Matching the individual's directional preference for pain free motion can significantly and rapidly decrease pain, the use of medication and overall improvement in all other outcomes (Long et al. 2004). The body should be placed in the most demanding position a patient can tolerate, though often initial training is in lying postures to reduce the tissue load. As tissue tolerance improves, similar exercises should be performed in more challenging postures, rather than adding increased resistance to recumbent exercises. Repetitive movements are recommended on almost an hourly basis with shorter time periods to avoid tissue irritation.

Dosage for these basic functional qualities is typically less than 25% of 1RM, so fatigue is typically not a limiting factor. The patient performs these exercises throughout the day at a level of tissue strain that does not create pain, swelling or stiffness. The program is organized so the patient can perform several hours of pain free exercise daily. Inflammation secondary to overstraining tissue may take several hours to become symptomatic. Should delayed onset tissue soreness (DOTS) occur, then the level of training is decreased.

Exercise Examples for Lumbar Tissue Repair, Pain and Edema:

- *Side lying caudal rotation*
- *Side lying cranial rotation*
- *Hip rolls supine*
- *Cat and camel*
- *Pelvic tilts*
- *Decline single and double knee to chest*

Tissue repair models of exercise are most important for acute injury and post surgical patients. Chronic pain patients may no longer have acute, specific tissue damage requiring tissue training, but

	Body Position	*Plane of Motion*	*Time and Frequency*	*Speed*	*Exercise Description*
Tissue Repair	1) Non weight bearing (WB), 2) partial WB, 3) full WB, 4) loaded WB, may vary on body region or tissue	1) Alternate plane from pain, 2) in plane of pain but in the opposite direction, 3) in plane of pain toward but not into pain	1) < 15 minutes with 5–10 sessions daily, 2) <30–60 minutes 1–3 sessions daily	Slow speed progressing to moderate speed based on pain and coordination.	Specific and controlled motion progressing to functional motions.
Edema Reduction	1) Recumbent postures 2) Dependant postures	1) Motions at joints proximal and distal, 2) Controlled motions at joint in least painful planes	1) At least 15 minutes with 5–10 sessions daily, 2) At least 30–60 minutes 1–3 sessions daily	Slow speed progressing to moderate speed based on coordination, pain and preventing an increase in temperature or redness	Specific and controlled cardinal plane motion progressing to functional tri-planar motions
Pain Inhibition	1) Non Weight Bearing (WB), 2) to partial WB, 3) to full WB, 4) to loaded WB	1) Distraction at joint, 2) motions at joints distal and proximal, 3) at joint but away from pain, 4) toward pain but not into pain	Perform daily when pain is present of <10 minutes – as frequently as is effective	Slow speed progressing to moderate speed based on pain and coordination	Specific and controlled motion

Table 5.3: General guidelines for acute training.

general exercise will still address improvement in tissue stress and strain tolerance. Improved joint mechanics and motor performance from exercise will also reduce excessive or abnormal loads to tissue. Patients should be encouraged to move as much as possible with daily activity and with specific exercises. Bed rest may address the fear and anxiety of the patient or clinician, but will only delay tissue healing and functional restoration.

Literature Review on Recommendations to Surgeons for Lumbar Post-Operative Care (McGregor et al. 2007):

- *There is a lack of evidence for restricting post-operative activity*
- *Most restrictions are probably from anxiety or uncertainty of the clinician or patient*
- *Most restrictions are unnecessary*
- *Restrictions may delay recovery and return to work*

Patients are uncertain of what they can and cannot do:

- *Knowing what to expect can facilitate recovery*
- *"Let pain be your guide" is an unhelpful approach*

Better outcome is achieved with early activation:

- *Most recovery occurs in the first 2-3 months*
- *Function and activities of daily living may improve with early rehabilitation programs*

Early return to work produces better outcomes and faster recovery

- *Early return to work is not harmful and may be helpful*
- *Most pain and disability is noted in those that do not go back to work*

Mobilization

Not every patient has joint restriction, but when limited joint movement is present it must be corrected prior to attempting to gain any improvement in motor performance. Muscle performance cannot significantly improve if the hinge is not moving properly around a normal physiological axis. Assessment of joint mobility must also go beyond the lumbar spine, assessing the thoracic spine, pelvis and lower limb joints. Any restriction in the biomechanical chain will affect force transmission and motor performance. Treating only the area of pain often results in missing significant pain free impairments that are part of the primary symptom complaint. In some cases, simply restoring joint motion normalizes motor recruitment and patterns due to improved afferent input from joint mechanoreceptors.

Specific mobilization exercises involve the use of a segmentally specific wedge or block, but cannot be as specific as passive manual techniques. Dosage for mobilizing exercise is more classically initiated at one set of greater than 40 repetitions at <50% of 1RM to improve elasticity of collagen and lubrication of cartilage. Progressions with additional sets and number of repetitions are based on the patient's response. Multiple sets may be performed during the day, or during training sessions; rest breaks between sets to restore oxygen are not necessary when training collagen, as with training muscle. Plastic deformity of restricted collagen in joint capsules is trained with similar resistance but with only two to five slow repetitions holding 10–20 seconds. Time, not force, is more effective in creating plastic changes. The clinical setting can provide segmentally specific mobilization exercises. Home exercise may not be as specific but is still important to maintain and augment improvements made in the clinic.

The dosage of the exercise, not the exercise itself, determines the outcome. For example, the simple supine hip roll exercise can be dosed differently to achieve pain inhibition, tissue repair, joint mobilization or motor performance. An exercise should not be considered to have a specific purpose, it is the dosage associated with it that determines the outcome.

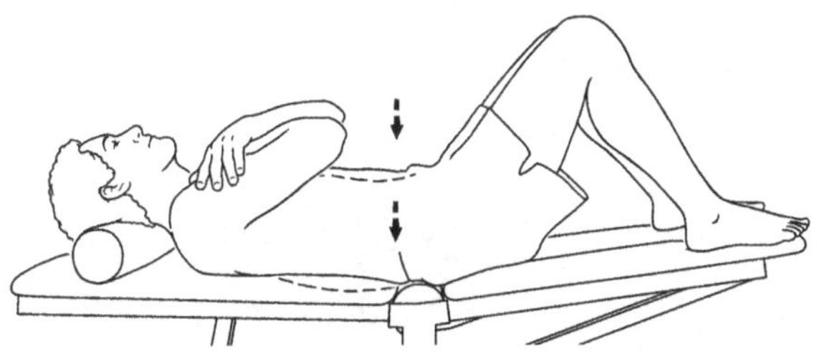

Figure 5.1: Pelvic tilt—caudal lumbar flexion mobilization with the lower abdominals. A towel or bolster can be placed under the sacrum to increase the excursion of flexion. The patient is instructed to relax the lower limbs to avoid pushing the feet into the floor to create the pelvic motion. Emphasis is on joint motion for tissue stress and may not be on any particular motor recruitment pattern of the lower abdominals. Extension can be performed with emphasis also on gaining extension range or improving tissue tolerance to extension.

Figure 5.2:a,b: Cat-Camel self-mobilization. Tissue stress in the sagittal plane can also be improved in quadruped by caudally flexing and extending the back. The amount of excursion is typically greater than with the supine pelvic tilts.

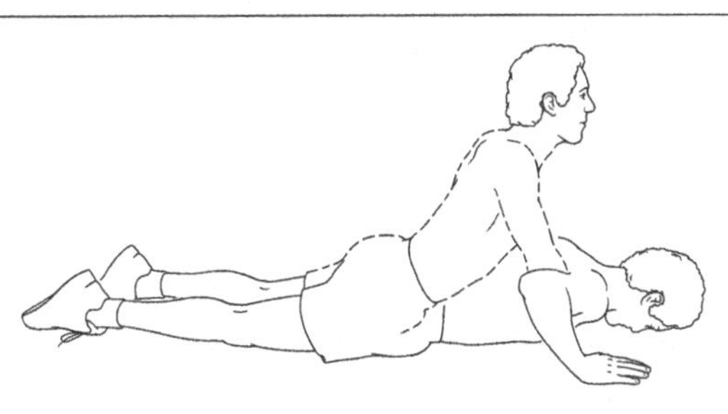

Figure 5.3: Press Up—passive lumbar extension mobilization. The patient is instructed to relax the back muscles and perform the press up emphasizing muscles of the shoulder girdle. A pillow can be placed under the pelvis to reduce the extension range if pain or neurological compromise occurs. The mobilization can be performed with repetitive motion or as a sustained stretch. The press up is intended to be passive maneuver, not active extension, to load the tissues at end range for improved mobility and symptoms (McKenzie 2001). The press-up mobilization is contraindicated with nerve root pathologies due to foraminal closure with the maneuver.

Figure 5.4: Quadruped lumbar side bending mobilization. From a quadruped position, the patient is instructed to side bend the lumbar spine by attempting to look at their hip joint while pulling the pelvis around.

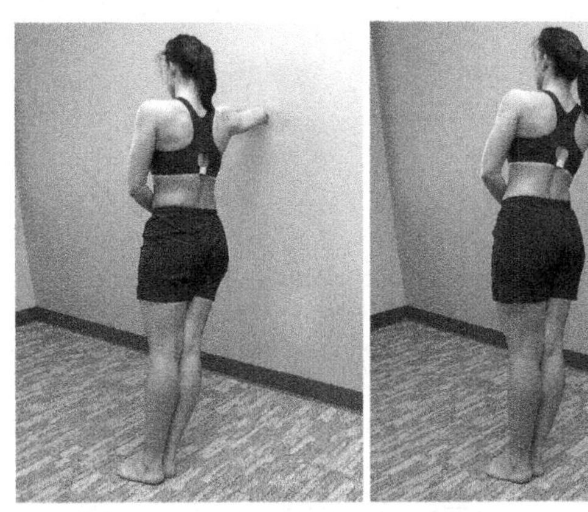

Figure 5.5a,b: Caudal lumbar side bending mobilization. With the feet together and the elbow supporting the body against the wall, the pelvis is lowered slowly toward the wall into the pain free range. The feet can be moved further from the wall, or the arm can support the body further away from the wall, to increase the excursion of the mobilization.

Dosage for Mobilization Exercises

- *Elasticity: one set 30–50 repetitions, one to three times daily.*
- *Plasticity: 10–15 second holds into mobilization, three to five repetitions, one time daily.*

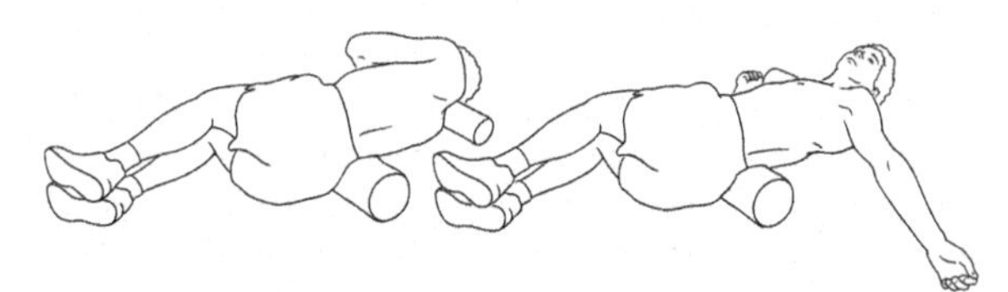

Figure 5.6a,b: Left lumbar rotational mobilization in right side bending for maximal left foraminal opening. The hips and knees are flexed to 90° with the lumbar spine in right side bending. The trunk is rotated back with the arm reaching backward. The exercise can be performed repetitively by swinging the arm for elasticity, or as a prolonged hold of 20–30 seconds for plasticity changes in collagen and muscle stretching. A free weight can also placed in the left hand to increase the mobilizing force into rotation. Emphasis is placed on the eccentric phase into relative left rotation to lengthen the right multifidi (and other muscles of producing right rotation) and adjacent connective tissues.

Figure 5.7a,b: Cranial left rotation mobilization with the lumbar spine in neutral. The hips and knees are flexed to 90° with the lumbar spine in neutral. The trunk is rotated back with the arm reaching backward.

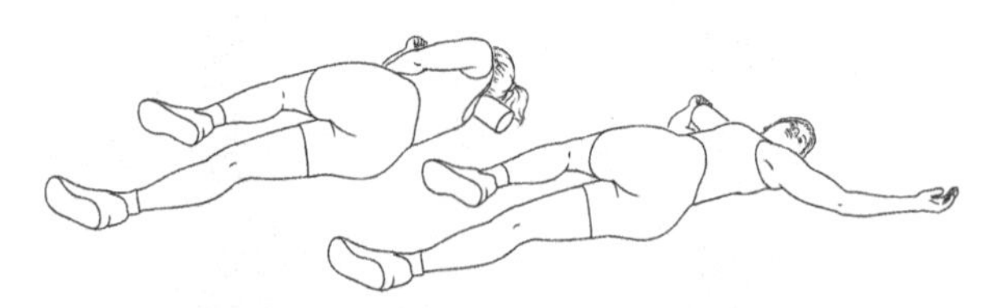

Figure 5.8a,b: Left rotation mobilization—caudal and cranial force moment. A progression is to increase the flexion of the top leg creating caudal rotation prior to moving the upper trunk and arm.

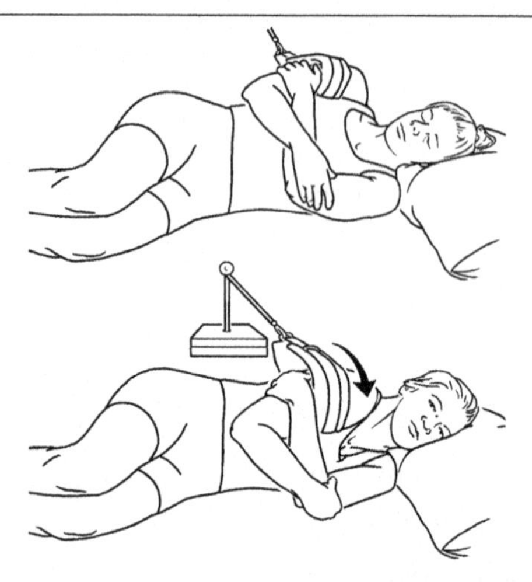

Figure 5.9a,b: Right rotation mobilization—resisted in side lying. The patient is in side lying with the resistance coming posteriorly. The patient rotates concentrically to the left, with emphasis on a slow eccentric return into right rotation.

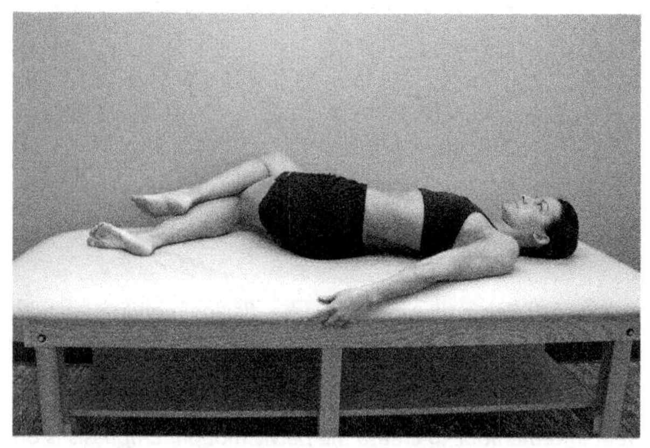

Figure 5.10: Caudal left rotation mobilization with legs crossed. The patient is instructed to cross the right hip over the left, then rotate the knees to the right (relative left rotation). The crossed legs will increase the stretch emphasis to the hip joint and tensor fascia lata.

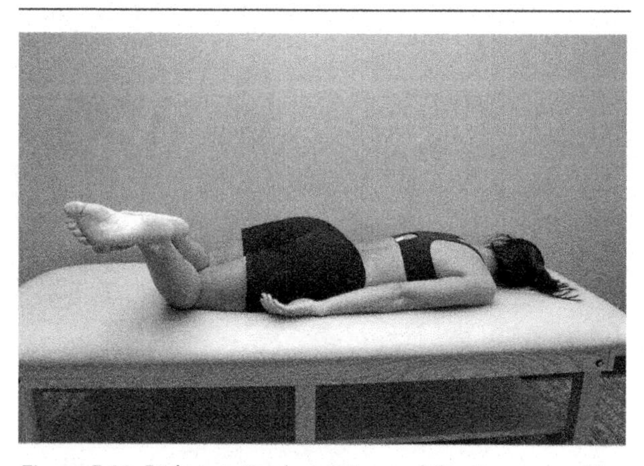

Figure 5.11: Right rotation/extension mobilization prone. The patient lies in prone with the knees flexed to 90°. The lumbar spine is in neutral, without a pillow under the stomach. The feet drop off to the right, creating a relative right rotation of the lower lumbar segments.

Figure 5.12a,b: Slump flossing of lumbar dural and neurovascular structures. This approach is designed to floss the dura and neurovascular structures, not stretch the dura. The patient is in sitting with the lumbar spine flexed

(slumped). The neck is in flexion to place tension on the cranial portion of the dura, while the knee is flexed to place slack on the caudal portion. The neck is extending, reducing cranial tension, while the knee is extended to increase caudal tension. The dura and neurovascular structures are, in this way, flossed through their surrounding tissues.

Figure 5.13: Dural flossing in supine. The lower extremity is placed up against a wall, with the knee extended. The neck is alternately flexed and returned to neutral to creating a flossing force.

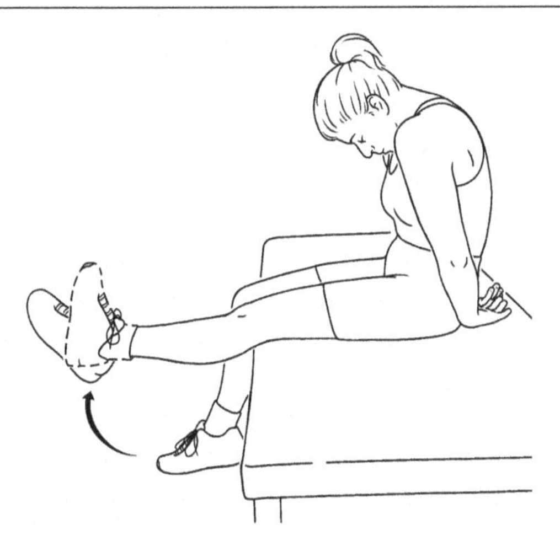

Figure 5.14: Sciatic neural flossing with tibial nerve (L4,5; S1 to S3) emphasis. The tibial nerve descends inferiorly through the popliteal space, passing between the heads of the gastrocnemius muscle to the dorsum of the leg, as the posterior tibial nerve, and into the ankle and foot. As the posterior tibial nerve traverses under the flexor retinaculum at the tarsal tunnel, it is subject to possible compression (i.e., tarsal tunnel syndrome). The patient is instructed to flex the neck and lumbar spine. As the medial and lateral plantar nerves course along the plantar surface of the foot, the tibial nerve is stretched by dorsiflexing and everting the ankle. The flossing may be performed by first dorsiflexing the ankle and extending the knee. Alternatively, the leg can be extended with the flossing from alternately dorsiflexing and plantar flexing the ankle.

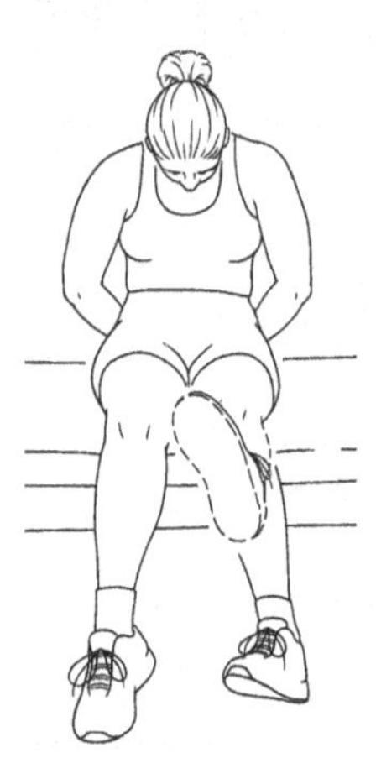

Figure 5.15: Sciatic neural flossing with common peroneal emphasis. The common peroneal nerve (L4,5; S1,2) lies posterior to the proximal fibular head. The patient is instructed to flex the neck and lumbar spine. To place tension to the common peroneal nerve, the hip is flexed and medially rotated, the knee is extended, and the ankle is plantar flexed and inverted. The goal is to achieve functional gliding of the common peroneal nerve, not a stretch.

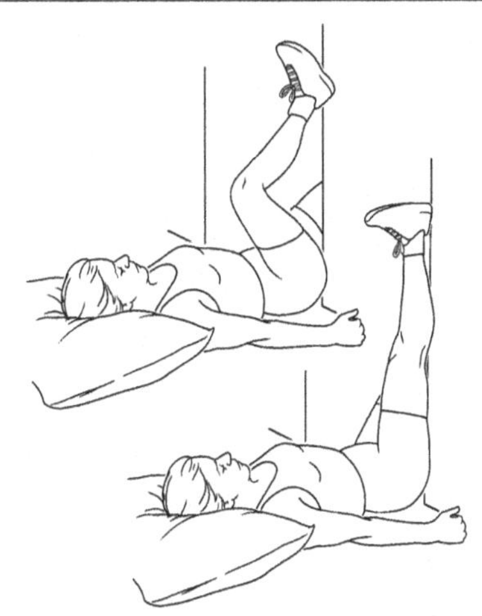

Figure 5.16a,b: Sciatic nerve flossing with heel slides on the wall. The patient is supine with pelvis against the wall and the involved heel on the wall, knee flexed. The heel is slide up and down the wall to provide a flossing force to the sciatic nerve. The ankle can first be positioned in dorsiflexion and eversion to emphasize the tibial portion, or plantar flexion and inversion to emphasize the common peroneal portion.

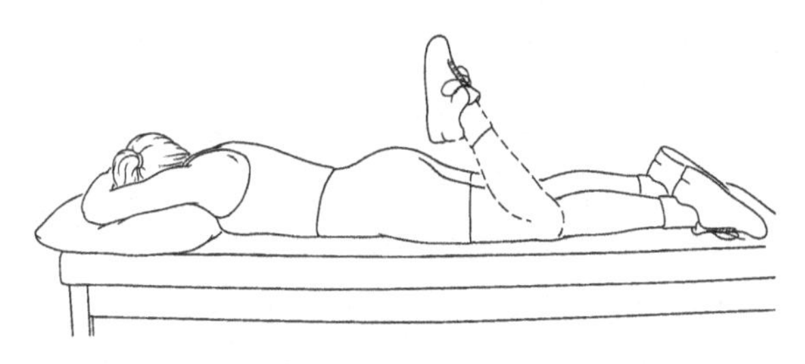

Figure 5.17: Femoral nerve flossing. The femoral nerve is a branch of the lumbar plexus, formed by the ventral primary rami of L1, L2, L3, part of L4, and possibly T12, continuing medial to the knee as the saphenous nerve. The knee is

progressively flexed to increase femoral nerve tension, along with stretching of the quadriceps femoris muscle. Hip extension can be added by first placing a pillow under the thigh. The lateral femoral cutaneous nerve can be stretched by adding hip adduction. The saphenous nerve is emphasized with hip in extension, abduction, and lateral rotation while extending the knee and dorsiflexing/everting the ankle.

Locking Techniques

Locking techniques with exercise provide better positioning, preventing motion from occurring in areas of contraindication and assisting in mobilizing specific joints or tissues. A hypomobile joint often causes a compensatory hypermobility elsewhere. The hypermobile joint is typically the area of pain complaints, while the hypomobile joint can be asymptomatic. Protecting the hypermobility of the joint while improving motion at the hypomobility, requires more specific exercise design than the general mobilizing exercises previously listed.

The utilization of locking techniques often allows early, more aggressive training models, than would otherwise be tolerated or safe. Locking techniques protect injured tissues, allowing muscles over the area to be trained safely and for higher level movement patterns or body position. For the purpose of mobilization, locking techniques may be utilized to focus all forces into restricted tissues, preventing more normal areas from participating in the exercise. Options for locking techniques include artificial, ligamentous (counter curves), joint locking (coupled force locking) and coordinative locking. For Stage 1, the most common forms of locking used are artificial and joint locking.

Artificial Locking

Artificial locking utilizes external belts, bolsters, wedges, benches, etc., to block movement from occurring into a range that is contraindicated. One of the most basic techniques is to have the weight stack on a pulley system come to rest to block any further range of resistance. Often with spinal treatment a specific block is necessary to allow for a more segmentally specific approach to training, either preventing motion from occurring at a specific segment or focusing for into an area.

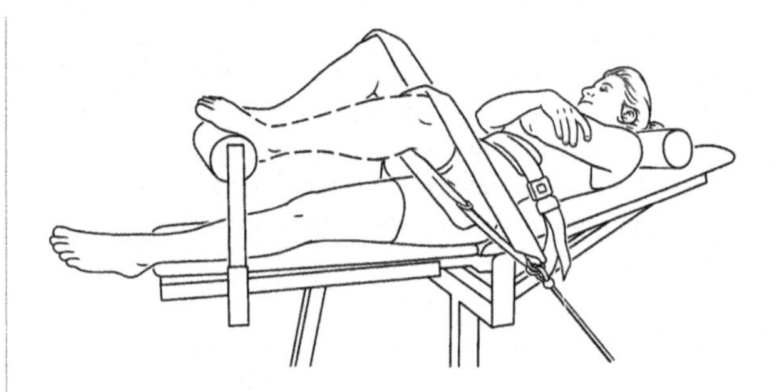

Figure 5.18: Artificial Locking—caudal right rotation mobilization with locking of cranial segments. A mobilization wedge is placed at the level of the belt under the cranial segment that is to be blocked. Relative right rotation is performed from below with the cranial segments locked from the motion. The right lower limb horizontally abducts eccentrically into right rotation as a repetitive motion followed by a sustained hold of 20–30 seconds. This exercise is indicated for restriction from L3 to L5/S1 without any adjacent hypermobile segments. This exercise is best utilized to mobilize a stiff lower lumbar spine by focusing forces to the restricted area.

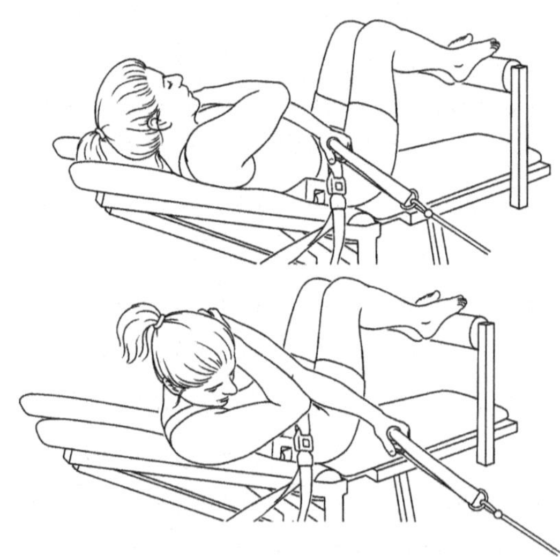

Figure 5.19a,b: Cranial left rotation mobilization. A mobilization wedge is placed under the right side at the level of the caudal segment. A belt around the trunk and angle bench will fixate the caudal segment and artificially lock the rest of the trunk. The left arm reaches down and across the body eccentrically into left rotation as a repetitive motion followed by a sustained hold of 20–30 seconds. This exercise is indicated more for thoracolumbar restriction without any adjacent hypermobile segments.

Ligamentous Locking (Counter Curve Locking)

Ligamentous locking techniques focus motion by taking up all the slack in the connective tissue around the area to be locked, while the specific joint or region is allowed to move. Because all of the slack in the collagen is taken up, this type of locking is contraindicated in the presence of a hypermobile joint. Taking up all of the collagen slack in a hypermobile joint while exercising muscle or another joint region, will further deform collagen. Ligamentous locking is indicated to decompress painful joint structures during an exercise, such as cartilage or a nerve root and is most effective when performing exercises to mobilize joint restrictions in the absence of any compensatory hypermobilities.

Figure 5.20: Thoracic extension mobilization with lumbar ligamentous locking in flexion. The hips are flexed creating lower lumbar flexion, taking up collagen slack in facet capsules and interspinous ligaments, preventing extension into a painful or contraindicated range. An artificial lock in the thoracic spine allows an active extension motion to create a general mobilization to the mid thoracic region. A more vertical position reduces the mobilizing force, while a more decline position will increase the relative weight of the body to increase the mobilizing force.

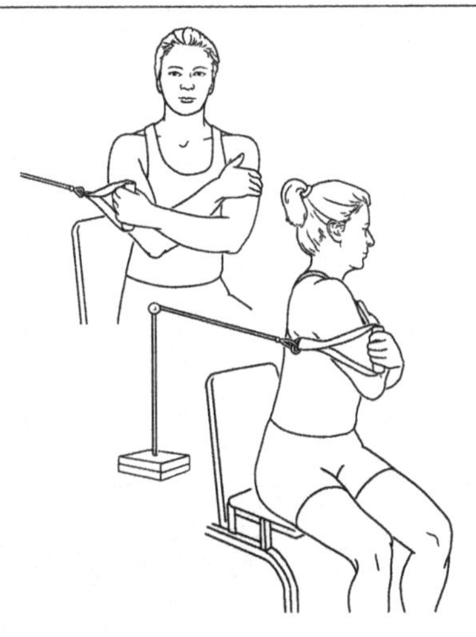

Figure 5.21a,b: Seated active rotation mobilization of L2/3 to the right, ligamentous locking from below. The patient is sitting with lumbar flexion up to L3. A posterior resistance from a pulley is held in the hand or wrapped around the trunk. Left rotation is performed with emphasis on a slow eccentric phase to the end or right rotation range. For higher segments flexion counter-curve can be increased by placing the feet up on a stool, further flexing the hips. For lower segments abduct the hips and forward tilting the pelvis.

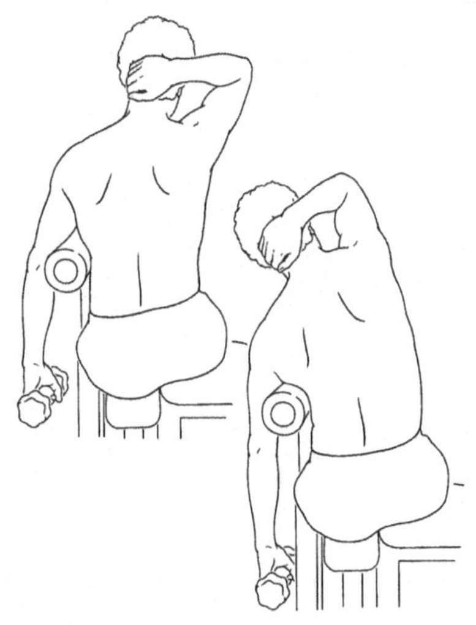

Figure 5.22a,b: Left cranial side bending with counter-curve locking (ligamentous locking) from below. The segment to be mobilized is at the level of the bolster. The lumbar spine is positioned in side bending to the right from below. The caudal locking will prevent the lumbar spine from participating in the right side bending, protecting a potential right lumbar hypermobility.

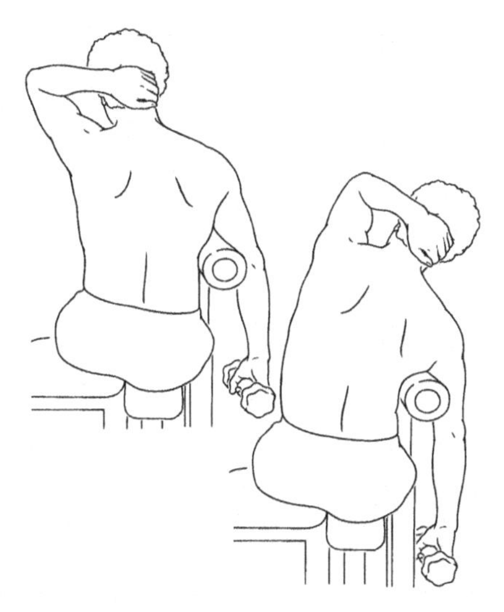

Figure 5.23a,b: Right cranial side bending with counter-curve locking—home version. If a drop seat is not available, placing folded towels under the contralateral ischial tuberosity can create caudal side bending for locking exercises.

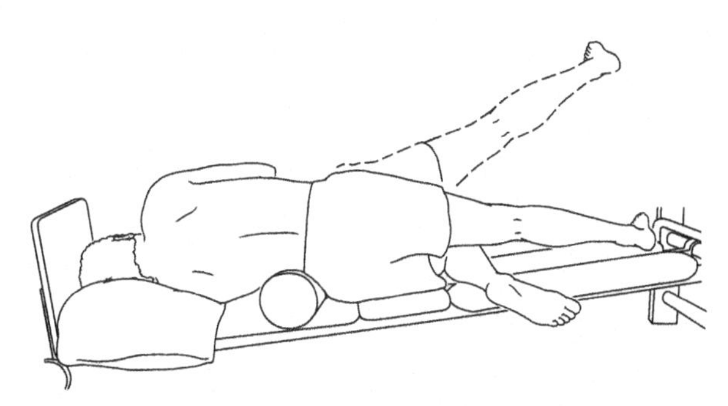

Figure 5.24: Caudal right side bending from performing hip abduction with a left side bending counter curve from above with a bolster.

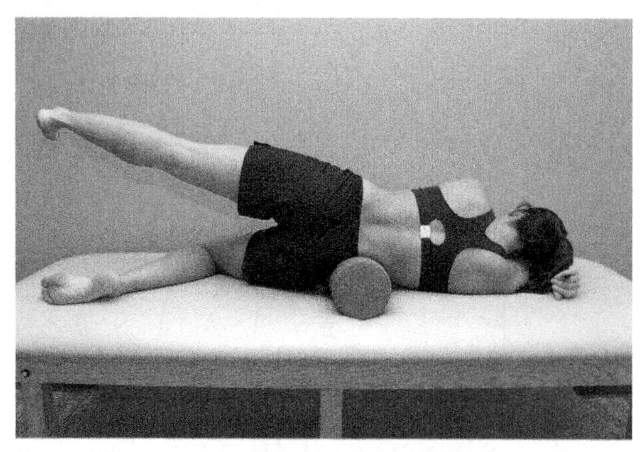

Figure 5.25: As previous, caudal right side bending with hip abduction, left side bending counter curve. Without a slant board, resistance can be decreased by flexing the knee.

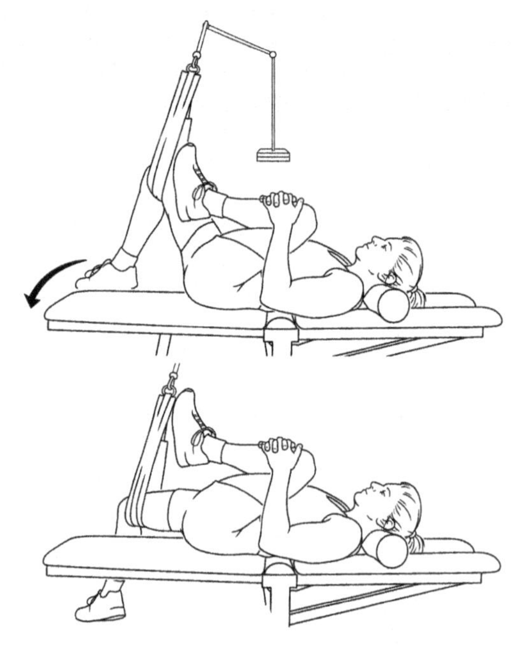

Figure 5.26a,b: Hip extension mobilization with lumbar counter curve locking. A hypomobile hip joint in extension may result in a compensatory hypermobility at L5/S1 to extension. The lumbar spine is ligamentously locked in flexion, with hip extension performed. Emphasis can be on training the hip flexors with eccentric return into hip extension. The exercise shown is an assist to the hip flexors, lightening the weight of the lower limb to allow for 30–40 mobilizing repetitions. Depending on the training level of the patient, the weight of the leg may be tolerable for high repetitions, or even the use of cuff weights to the ankle may be appropriate if the patient is still be able to complete 30–40 repetitions.

Joint Locking (Coupled Force Locking)

Ligamentous locking requires normal collagen structures to tolerate end range tension during locking procedures. Joint locking, or coupled force locking, locks joints in the opposite direction and utilizes the natural spine coupled forces to attain facet opposition. With a hypermobility, compression of the facets prevents motion of the vertebra with collagen on slack. Coupled forces dictate which direction rotation and side bending occur in the spine while in flexed or extended positions. Coupled force locking with exercise was introduced to the Medical Exercise Therapy (MET) curriculum by Ola Grimsby. Traditional lumbar coupled force descriptions involved rotation and side bending occurring in the same direction in flexion and the opposite in neutral or extension. Variations do occur in different individuals, which can be tested prior to designing a specific exercise utilizing coupled force locking techniques. Volume 1, chapter 16, of this text series provides a more thorough discussion of coupled forces and how to test them for the lumbar and lower cervical regions.

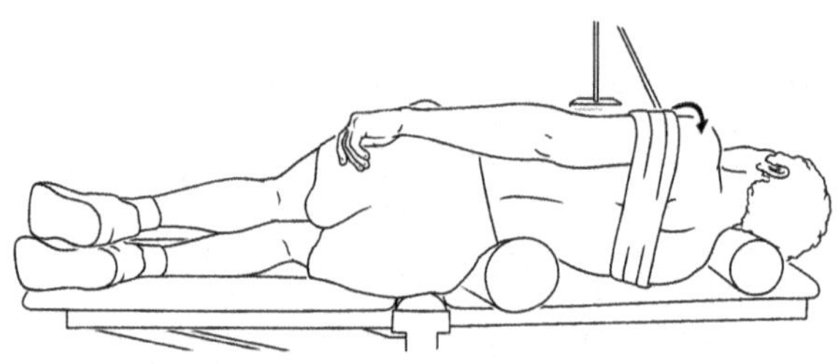

Figure 5.27: Cranial left rotation with locking in the mid lumbar spine—stabilizing exercise for muscles of right rotation. The bolster placed at L2/3 creates a right side bending with a flexed lumbar spine. Normal coupled forces would induce a right rotation at L2/3, locked and not participating in left rotation occurring in the upper lumbar spine. This technique would be used to protect the L2/3 segment from moving into a left rotation hypermobility.

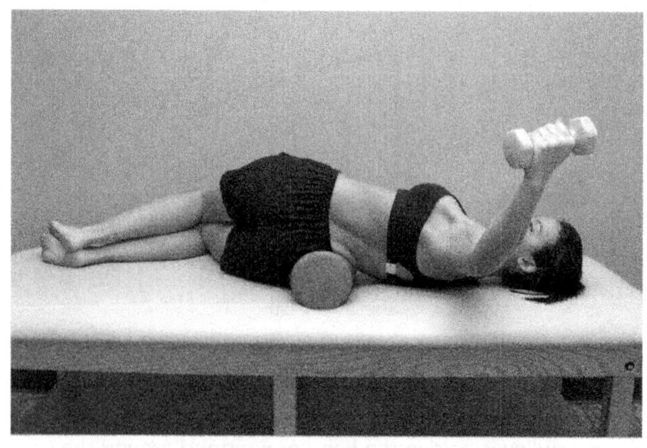

Figure 5.28: Cranial left rotation with locking in the mid lumbar spine—mobilizing exercise for left rotation. The bolster placed at L2/3 creates a right side bending with a flexed lumbar spine, creating right rotation at L2/3. Left rotation mobilization is performed eccentrically in the upper lumbar spine, with locking of the mid and lower lumbar spine.

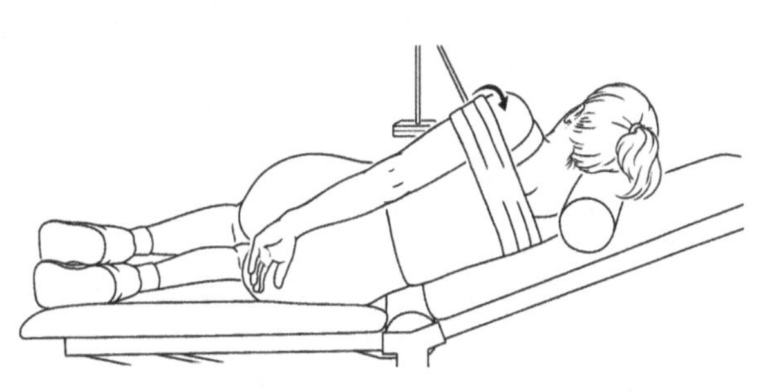

Figure 5.29: The hypermobile segment that needs to be protected in this case is located at L4/5, the level of the bend in the table. It needs to be protected from left rotation so it is kept in neutral and left side bent. Coupled forces will lock the spine from rotating to the left even though there is left rotation cranially. The same exercise can emphasize joint mobilization into left rotation by placing the resistance on the posterior side, allowing eccentric training into left rotation.

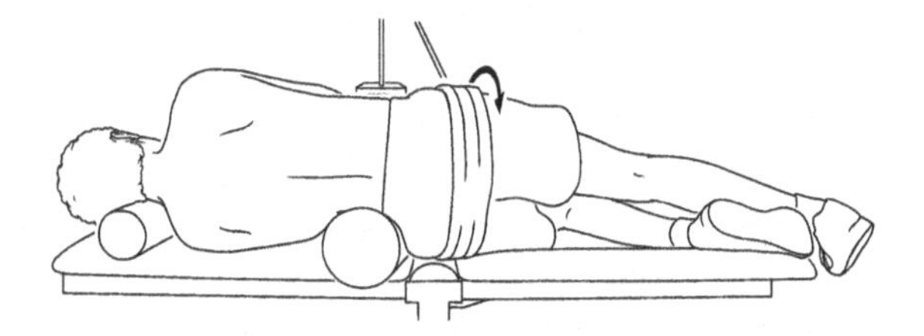

Figure 5.30: Left side lying caudal left rotation. The spinal locking occurs at the apex of the bolster, in this case L3/4. The exercise involves left rotation (always described using the cranial vertebral segment as a reference) from below but since the L3/4 segment is left side bent in neutral, coupled forces will rotate it to the right. The left rotation will not occur at the locked level due to facet opposition.

Coordinative Locking

Coordinative locking involves controlled, coordinated motion, while actively preventing movement into a range of motion or joint region that is contraindicated. As a higher level of coordination is required, this type of locking is typically used as a progression in later stages from the above, more aggressive locking approaches. In a more general sense, all exercises in Stage 1 involve a certain amount of coordinative locking, as the patient is required to perform all exercises in a range that can be coordinated and pain free. In Stage 1, coordinative locking is more typically used when one point in the range negatively affects a specific structure. The final goal is always to be able to perform motion without the assistance of locking. The level of locking is gradually reduced until the patient can preform normal functional movements.

Figure 5.31: Caudal lumbar extension over a bolster. Extension can be trained from a position of full flexion. In the presence of a contraindication for end range extension, the patient is cued to use coordinative locking to prevent extension from occurring past neutral, working from full flexion to neutral only. This would be indicated in cases of lower lumbar hypermobility in extension, spondylolisthesis, nerve root pathology or for any reason requiring avoiding end range lumbar extension.

Figure 5.32: As a progression from the previous exercise, caudal lumbar extension performed with lifting the lower extremities off the bolster for added resistance.

Coordination / Motor Learning

As joint mobility is improved by joint mobilization and resolving edema, exercise can begin to focus on functional qualities that more directly train the motor system. Basic qualities include coordination, motor learning and endurance. Resolving tonic muscle guarding should occur first as this is a cause of pain, reduced range of motion and altered movement patterns. The tonic muscle system is the arthrokinematic system responsible for controlling the axis of motion, which is fundamental to training coordination. Understanding which muscles serve this purpose and their functions, as well as a complete knowledge of passive and dynamic constraints to motion makes exercise design specific and logical. Coordination can be disrupted for many reasons. The Manual Therapy Lesion outlined below describes a global and local influence of pathology on motor performance. An integration of deficits in biomechanics, neurophysiology, biochemical and psychology are integrated into a model of defining impairment. The broad view of dysfunction provides a logical approach to exercise design and progression.

Treatment addressing the manual therapy lesion first attempts to reduce pain and muscle guarding. Numerous passive treatments and modalities can be used to this end, but removing pain and guarding is not really treatment, only a necessary step to allow for more intervention to resolve specific impairment. Treatment must focus on the reduced tissue tolerance to stress and strain, abnormal joint mechanics, reduced mechanoreceptor function altering motor patterns and the resulting functional loss. Exercise provides the repetitive motion for tissue repair. Motor impairment is first addressed by attempting to improve afferent feedback in the presence of damaged mechanoreceptors. The use of cutaneous receptors, through taping techniques, will also increase afferent input to improve motor patterns for both central programming and distal motor performance. Emphasis, however, is on progressive training of the arthrokinematic tonic muscle system in attempting to sensitize local muscle spindles to provide adequate afferent feedback to normalize motor patterns. Nitz and Peck (1986) found the lumbar multifidus to have a significantly greater spindle density compared with the semispinalis, while no difference between the spindle densities of the cervical or thoracic multifidus and the semispinalis were noted. Muscle spindle density in arthrokinematic muscles such as the multifidi function as 'kinesiological monitors' generating important proprioceptive feedback to the central nervous system (Peck et al. 1984). Proprioceptive function is located throughout the lumbar spine. Proprioceptive training must be an important part of a complete exercise program.

Basic Stages of the Manual Therapy Lesion (Grimsby 1988)

1. ***Collagen/Tissue Trauma:*** *Acute injury, surgery, degenerative joint disease, repetitive strain, postural strain or hypo/hypermobile joint. This may be an acute injury or a slow progression that appears to have an insidious onset related to gradual degeneration.*

2. ***Receptor Damage:*** *Afferent signals are lost or altered from direct structural damage to mechanoreceptors, a loss or restriction in the neural pathway, or reduced feedback from receptors imbedded in non-mobile capsules. Type I mechanoreceptors are more easily damaged than type II, as they are located more superficially on the joint capsule, while type II mechanoreceptors are deeper in tissue. The pain free but restricted spinal segment will have a reduction in afferent input from lack of collagen elasticity changing motor performance within its motor field.*

3. ***Reduced Muscle Fiber Recruitment:*** *Abnormal central processing of afferent signals from cutaneous, articular and muscle/tendon receptors. Altered central feedforward facilitation to alpha motor neurons (extrafusal muscle fibers) and gamma motor neurons (Intrafusal muscle fibers). End result of includes altered reflex responses for proprioception and kinesthesia, motor weakness, motor delay and/or poor timing. Central spinal segmental dysfunction, both with and without pain, alters central motor programming.*

4. ***Tonic Fiber Atrophy/Loss of Phasic Power:*** *Reduced motor recruitment resulting in initial lack of recruitment, leading to atrophy over time. Tonic muscle fiber, the multifidi in the cervical spine, is initially more affected by type I mechanoreceptor loss and influences of the type IV mechanoreceptor (pain) system. Phasic muscle fiber atrophy, or inhibition, may occur with more significant tissue damage that includes type II mechanoreceptors, and with overall reduction in higher level activity.*

5. ***Reduced Anti-Gravity Stability:*** *Reduced recruitment of the tonic system results in loss of dynamic arthrokinematic control of joint motion, static postural stability/alignment and central balance mechanisms. Central stability at the lumbar spine and pelvis is reduced, which may alter the static and dynamic alignment of the hip, knee and ankle.*

6. ***Motion Around Non-Physiologic Axis:*** *Loss of dynamic control increases the neutral zone of function for the joint, increasing the range of the instantaneous axis of motion. Altered mechanics for the biomechanical chain of the lower limb. Reduced tonic function creates an abnormal relationship between tonic (arthrokinematic) and phasic (osteokinematic) muscles. Compensatory motor recruitment occurs, leading to abnormal mechanics.*

7. ***Trauma/Acute Locking/Degeneration:*** *Altered axis of motion leads to abnormal tissue stress/strain, resulting in further tissue and receptor damage. Hypermobile joints are prone to acute locking. Joint degenerative changes occur over time.*

8. ***Pain/Guarding/Fear-Anxiety of Movement:*** *The type IV mechanoreceptor responds to tissue damage with pain signals and tonic reflexogenic muscle guarding. Psychological influences of pain my lead to altered motor patterns and reduced effort. Overall effect is reduced tissue tolerance, abnormal joint/ tissue loading, reduced afferent feedback, altered feedforward efferent drive and reduced function (Grimsby 1988).*

Early emphasis for coordinating the abdominals involves recruitment of the more tonic portions: the transverse abdominis (TrA) and obliquus internus (OI) (Wohlfart et al. 1993). The two basic techniques used to recruit these muscles are abdominal bracing and hallowing (Richardson et al. 1992). Bracing involves a contraction in which the abdominals flare laterally, while hallowing draws the umbilicus toward the spine (Richardson et al. 1992). It has been argued that hallowing can destabilize the spine by narrowing the base of support of the guy-wires and some authors have gone as far as claiming TrA isolation creates dysfunctional spines and "these unfortunate patients become paralyzed by their own hyper-analysis of what their transverse is doing" (McGill 2004). As a part of the normal tonic arthrokinematic system, these muscles are involuntarily recruited prior to active loading of the spine or the movement of an extremity (Hodges and Richardson 1997). Consistent with the Manual Therapy Lesion, the more tonic muscles of the spine are affected with low back pain (Hides et al. 1994, Hodges and Richardson 1996).

From a clinical standpoint, muscle isolation training in early motor learning is not emphasized during rehabilitation. Having appropriate tonic muscle recruitment is imperative but no single muscle or muscles are the most important. Focus is placed on timing and recruitment within a normal pattern of movement. The key muscles for dynamic stabilization during functional tasks may change throughout the task, depending upon the pattern of movement being performed.

Barnett and Gilleard (2005) compared different types of abdominal sit-ups to determine which type more preferentially recruited the TrA and OI. Performing an abdominal hallowing and curl was the most effective technique. For a majority of subjects, performing a brace and hold, followed by a curl or a hallowing with a rotational curl also recruited the TrA and OI first. A straight curl, without an initial hallowing or bracing, selected the rectus abdominis first.

Eversull et al. (2001) investigated and described the neuromuscular neutral zones (NNZ) in the lumbar spine in terms of activity patterns at different lumbar levels, and rate of tension/elongation of viscoelastic structures. The authors defined NNZ as "the lumbar displacement or tension threshold below which muscles remain reflexively inactive." The NNZ concept describes the complex and adaptive sensorimotor feedback loop originating from afferent receptors located in viscoelastic tissue

(Eversull et al. 2001). The sensorimotor feedback system needs to be tapped into for optimal recovery. With intact collagen and receptors, the NNZ functions in a small range. As tissue and receptor damage occur, greater displacement occurs prior to afferent feedback for motor support. The NNZ has been defined in the feline lumbar spine as 5–15 percent of the maximal physiological range of motion. Displacements greater than 5–15 percent trigger activity in the lumbar multifidi, which contribute to segmental stability. The NNZ is a different concept compared to what has previously been described as the Neutral Zone Concept (Panjabi 1992). The Neutral Zone Concept refers only to the passive constraints of collagen and the influence of muscles in maintaining the joint motion within physiological limits but does not take into account the neurophysiological implications of afferent feedback systems.

The other side of the coin, in terms of afferent input from mechanoreceptors improving dynamic stability in the spine, is the fact that desensitization of the mechanoreceptors induced by cyclical loading of passive tissue might lead to a significant loss of stiffening of the lumbar spine from muscle contraction, leaving the spine unstable and unprotected (Solomonow et al. 1999). These results demonstrated a destabilizing effect of cyclical deformation in the passive tissues of the lumbar spine. The authors speculated that this change in recruitment was evidence of a decrease in protective muscular reflex caused by laxity in the passive structures of the spine. This can be appreciated clinically in the systemic hypermobile or local hypermobile patient where they have difficulty recruiting muscles due to decreased firing of mechanoreceptors. When this occurs muscle spindles can be sensitized with progressive training leading to improvements in motor control. Twenty minutes of sustained static flexion exposes the spine to decreased muscular stabilizing forces, muscle spasm and impaired tension-relaxation status of viscoelastic structures (Jackson et al. 2001). Demonstrated with EMG muscle activity was five percent of its initial value within three to four minutes of sustained flexion. The recovery pattern of reflexive muscular activity after continuous passive, cyclic loading of tissue requires at least 15 to 20-minute rest after a 50-minute episode of cyclic loading to produce most of the possible recovery of reflexive muscular activity. Full recovery was not expected until after two hours of rest (Gedalia et al. 1999).

Stimulation of the mechanoreceptors in the supraspinous ligament will create stabilizing contraction in the lumbar spine multifidi (Solomonow et al. 1998). Receptors in discs, interspinous ligaments, anterior and posterior longitudinal ligaments and facet joints may also contribute to this recruitment. The authors identified a reflex arc of EMG activity in the lumbar multifidi (LM) from mechanoreceptors. They concluded that therapy for muscle strengthening had the potential to improve spinal stability and to decrease pathological situations associated with episodes of spinal instability. The decreased lumbosacral position sense in the patient group might have been due to an altered paraspinal muscle spindle afferent and central processing of this sensory input (Brumagne et al. 2000). The lack of return of multifidus after injury, and the wasting that has been found, might be partly due to the muscle spindles' role in reflex inhibition. The muscle spindle input from LM was crucial for accurate positioning of the pelvis and lumbosacral spine in a sitting position (Brumagne et al. 1999).

As with other regions in the body, in the spine, coordination and motor learning represent two separate but equally important functional qualities that are addressed though out exercise progressions. Motor learning involves the cognitive process of acquiring a new skill. With lumbar patients and clinical practice this is most often a muscle isolation approach. Coordination training may involve this but the focus is to dose exercise at a level that allows for many repetitions to obtain neurological adaptation or coordination. This is accomplished by having the patient focus on the motion itself. To achieve proper coordination all aspects of dosage are relevant including: appropriate resistance, angle of resistance and appropriate repetition dosage. Both approaches can be effective and should not be looked at as one being superior over another. The pathology will determine which approach is most effective, along with the patient's learning style and their body awareness. Matching the exercise style to the patient should be more effective than attempting to force the patient into only one way of approaching a clinical problem.

Coordination requires cooperation from two different stabilizing systems: the global (more phasic) and the local systems (more tonic). These two systems differ histologically, biomechanically, metabolically and neurologically. The muscles in the lumbo-pelvic region that would be considered local would include the following: lumbar multifidus (Bogduk 1992), posterior fibers of the psoas (Bogduk 1992), the medial fibers of the quadratus lumborum (McGill 1992), the transverses abdominis and some fibers of the internal oblique abdominals (Snijders 1995). These local, or arthrokinematic muscles of the spine, are more involved with stability at the segmental level and are generally located closer to the axial skeleton and axis of motion. More tonic, arthrokinematic muscles, also possess a greater proprioceptive function, obviated by the significantly greater density of muscle spindles (Nitz 1986, Peck et al. 1987, Peck et al. 1984).

The global, or osteokinematic muscles are active in relationship to the direction of movement that they create. For example, the rectus abdominis is active with trunk flexion while the TrA is active with flexion, extension and extremity movement. Coordinated segmental motion is the result of tonic muscles co-contracting in an early feed forward, non-directional manner, while the phasic muscles are direction specific more responsible for moving the bone or body segment (osteokinematic motion).

The local arthrokinematic muscles possess a greater predominance of Type I, or slow twitch muscle fiber, making them more oxidative from a metabolism standpoint. Characteristically this group of muscles, particularly the LM and TrA, is active regardless of the direction of motion (Pauly 1966), and show electromyographic activity prior to the primary mover creating osteokinematic movement (Hodges et al. 1997a/b/c, Moseley 2002).

Concepts involving altered recruitment have been presented prior to the research on the TrA. Janda (1977) described the dysfunctional motor system of a patient with weak abdominals and shortened back extensors. He provides an example of the tightness in the extensors inhibiting the abdominals and treatment of abdominal strengthening being futile. In this case, addressing the abnormally high tone in the erector spinae is necessary before it is possible to get normal muscle facilitation in the abdominals (Janda 1978).

In addition to the examples of recruitment changes and loss of coordination in pathology, there is impressive evidence of segmental atrophy following collagen and receptor damage in the psoas and the LM. Dangaria et al. (1998) found a significant correlation between cross-sectional areas of the psoas and the level of disc herniation. There is very predictable atrophy ipsilateral to unilateral sciatica due to disc herniation. Hides et al. (1994) examined the cross-sectional area of the lumbar multifidus in 26 patients with acute unilateral low back pain and 51 normal subjects to assess the differences between the groups. They used real-time ultrasound throughout the lumbar spine and detected statistically significant evidence of lumbar multifidus wasting ipsilateral to the side of their symptoms in patients with acute unilateral low back pain. The scans revealed that marked asymmetry occurred in each patient and was isolated to one vertebral level. The speed of onset and localization indicated that this wasting was not the result of disuse atrophy, and they theorized that it was the result of pain or other local inhibition. In follow up studies, based on these subjects, the author determined that without exercise, the recovery of the LM was not spontaneous (Hides et al. 1996). In addition, with one and three year follow ups, it was determined that the subjects in the control group suffered an increased number of recurrent episodes of low back pain with greater levels of severity compared to the exercise group (Hides et al. 2001). During the first year, the patients in the control group were 12.4 times more likely to experience recurrence of LBP compared to the patients in the exercise group. In years two and three the control group was still 5.9 times more likely to suffer recurrences of LBP than the exercise group. It is possible that the muscle wasting measured in this study was present prior to the onset of acute low back pain, potentially predisposing the group toward low back injury. A follow up study with a porcine model alleviated this possibility, finding segmental multifidus wasting within 72 hours following disc injury (Hodges et al. 2006).

Subjects with spinal pathology can present with many different examples of muscle inhibition and dysfunction. To focus excessively on individual muscles is to miss the big picture of stability combined with coordination. It has been determined that there are many different muscles that provide stability to the spine depending on the loads imposed (Kavcic et al. 2004). All of the muscles are vital players in the symphony that is ultimately responsible for producing spinal stability and coordination.

Abnormal Motor Patterns

Identifying movement pattern dysfunction is a necessary step in identifying which types of exercises to prescribe. General active, passive and resisted testing may elicit basic directional information and indicate the muscle groups in dysfunction. Directions of motions that are limited and/or painful actively, but are notably improved with passive lumbar range of motion testing, are suggestive of motor deficits. Often limitations in motor performance are an issue of dosage; the level of resistance of the trunk is too high for the functional level of the muscles involved leading to compensation. Applying basic unloading principles to spinal training can significantly improve the range that can be trained.

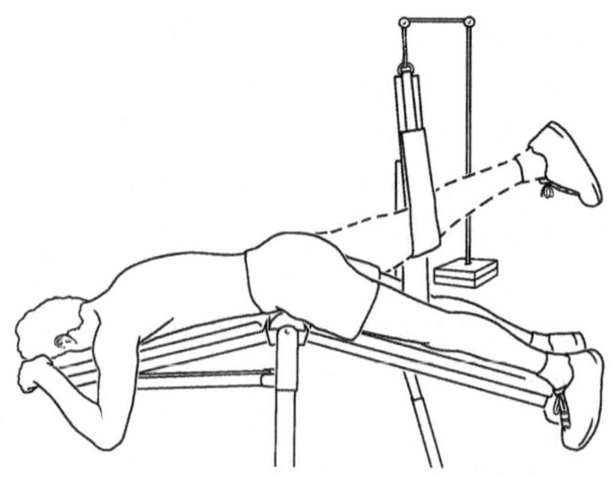

Figure 5.33: Abnormal cross pattern firing with prone hip extension may be a function of too high a resistance. Reducing the weight of the lower limb with the pulley assist may elicit a more normal firing sequence of the involved muscles, allowing for high repetition coordination training.

Figure 5.34: Unloaded squat training can also be used as a form of unloading trunk flexion. Tissue pain into flexion may be avoided with an unloading gantry and the reduced load may be within the performance level of the trunk muscles to begin high repetition training for coordination.

Figure 5.35: Extension/rotation assist with a wall pulley. When pain is present returning to neutral from a flexed position a pulley can reduce the weight of the trunk assisting the trunk extensors for a return to neutral.

Changes in Motor Function with Low Back Pain

Structural Changes:

- *Histological changes in muscle fiber atrophy (Lehto et al. 1989, Mattila et al. 1986).*
- *Reduced cross sectional area of LM at level of symptoms, and to a much lesser degree above and below the level of symptoms, due to pain inhibition (Stokes et al. 1992b, Hides 1994).*
- *Denervation of the LM with instability (Sihvonen and Partanen 1990).*
- *Slow twitch hypertrophy with fast twitch atrophy in pathological LM on side of symptoms in acute and chronic LBP (Stokes et al. 1992a).*
- *Positive correlation between paraspinal muscles atrophy and self-reported disability in CLBP (Alaranta et al. 1993).*
- *Atrophy of the LM with lumbar radiculopathy (Campbell et al. 1998).*
- *Higher percentage of type I fibers and lower percentage of Type IIA and IIB (Bajek et al. 2000).*
- *The relative percentage of slow twitch fibers increase both with age and pathology (Jowett et al. 1975).*

Performance Changes with LBP:

- *Greater fatigability of lumbar extensors (Cassisi et al. 1993, Kumar et al. 1995, Mayer et al. 1985, Shirado et al. 1995).*
- *Increased fatigue speculated with higher risk of injury (Sparto et al. 1997).*
- *Reduced strength from smaller unilateral cross sectional area (CSA) and the lower percentage of type II fiber in the LM (Zhao et al. 2000).*
- *Extensor weakness with sciatica (McNeil et al. 1980).*
- *Impaired postural stability (Luoto et al. 1998, Nies and Sinnott 1991).*

Continued...

- *Proprioceptive dysfunction and LBP, not only peripheral problems but also impaired functioning of the central nervous system (Luoto et al. 1996).*
- *Proprioception deficits in the lumbar spine, not compensated by any other mechanisms outside of the lumbar spine (Gill and Callaghan 1998).*
- *Abnormal neuromuscular control of paraspinal recruitment with chronic LBP (Grabiner et al. 1992)*
- *Motor control dysfunctions of trunk muscles with LBP might result in a decreased stiffening of the spine leading to excessive mobility, or instability (Hodges and Richardson 1996).*
- *Abnormal dynamic motor control responses (King et al. 1988).*
- *Reduced utilization of deeper abdominal muscles (O'Sullivan et al. 1997).*

Global Changes with LBP:

- *Central balance deficits chronic unilateral LBP (Byl and Sinnott 1991, Kuukkanen and Malkia 2000, Alexander and LaPier 1988, Nies and Sinnott 1991, Mok et al. 2004).*
- *Loss of proprioception and postural control (Leinonen et al. 2003).*
- *Mechanoreceptor in the joint capsule or ligaments of the SIJ contribute to proprioceptive function and motor recruitment for lumbar stability (Indahl et al.1999).*
- *Reduced seated postural control (Radebold et al. 2001).*
- *Slower reaction time, decision time and choice reaction time (Taimela et al. 1993).*

Bracing

In cases of spinal instability, external bracing or ortheses, have been used to control symptoms. Celestini et al. (2005) compared the use of bracing with a second group that received bracing and stabilizing exercises. Both groups had some overall improvement, but the group with bracing and exercise had a higher level of improvement in symptom reduction, neuromuscular control and lifestyle. Over time this group used less medication, relied less on the lumbar brace and tended to resort to using the home exercise program to assist in controlling symptoms.

> *Post Surgical Changes in Motor Performance:*
>
> - *Potential denervation atrophy resulting from damage to the dorsal rami in patients undergoing posterior lumbar surgical procedures (Kawaguchi et al. 1996, Sihvonen et al. 1993, Zoidl et al. 2003).*
> - *Decrease in trunk strength following discectomy (Kahanovitz et al. 1989, Hakkinen et al. 2003a/b).*
> - *Strong correlation between multifidus muscle dysfunction and poor functional outcome with recurrence of low back pain after disc surgery (Rantanen et al. 1993, Sihvonen et al. 1993).*
> - *Loss of proprioception and postural control did not improve following discectomy, requiring exercise therapy post operatively (Leinonen et al. 2003).*
> - *Poorer surgical outcomes correlation with size of Type II muscle fibers of LM, more fatty deposits in LM and more denervation of LM (Rantanen et al. 1993).*

The vast majority of patients with low back pain do not require surgery, but there are appropriate indicators of the need for surgery: safe guarding nerve roots, fixation of fractures, correction of deformity and to fuse pathologically unstable segments. Changes in muscle structure and function cannot, however, be corrected with surgery (Filiz et al. 2005). Surgery can even have a detrimental effect on motor function. Muscle atrophy may result from long-term retraction of weak muscles during surgery and the severity of pain in the postoperative period (Gejo et al. 1999, Kawaguchi et al. 1996). Significant support is noted for improvements in outcomes regarding pain, disability and functional recovery with exercise regimens after discectomy (Alaranta et al. 1986, Brennan et al. 1994, Danielsen et al. 2000, Dolan et al. 2000, Donceel et al. 1999, Filiz et al. 2005, Hakkinen et al. 2003b, Johannsen et al. 1994, Manniche et al. 1993a/b, Ostelo et al. 2003, Postacchini et al. 1978, Skall et al. 1994, Yilmaz et al. 2003, Choi et al. 2003).

Motor Testing

Testing motor performance is a helpful step in prescribing exercise. Identifying reduced motor performance supports the use of more specific selection criteria for the initial exercise program. Testing may not only involve the lumbar spine, extremity and balance performance can also be included. Motor testing is not diagnostic, but may assist in the development of clinical predictive rules for the use of certain types of exercises. For example, in the case of lumbar instability, Hicks et al. (2005) identified testing criteria that best predicted a successful outcome with the prescription of stabilization exercises. The key physical tests were identified as straight leg raise, lumbar mobility testing, aberrant motions during lumbar range of motion and the prone instability test. Individual tests may provide specific information as to which muscles should be trained, or what functional quality is lacking (i.e., timing, endurance or strength). Many basic tests are available to assess performance. Each may involve a different or unique emphasis as to what exactly is being tested.

Side Support Test

Figure 5.36: The patient is positioned in side lying, legs extended, resting on the lower elbow for support with the top foot in front of the lower foot (McGill 1998). The patient

is instructed to lift the hips off the table with only the elbow and feet remaining in contact, holding the position as long as possible. The test is performed bilaterally, scored as time held per side. This test is simply a timed motor performance test to establish a level of function, or to track improvement with training. Poor performance of the lower abdominal is seen by the patient flexing the hips and trunk slightly and/ or rotation of the pelvis or upper thoracic spine. The ability to breathe diaphragmatically during the test is also assessed, as the patient should be able to breathe normally with the diaphragm during isometric work of the lower abdominals.

Extensor Endurance Test

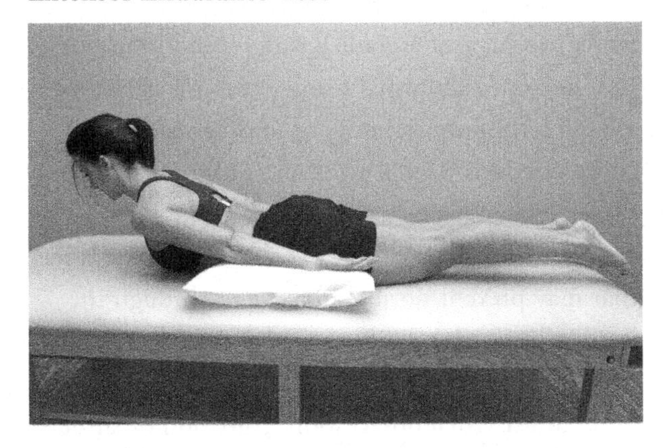

Figure 5.37: In prone, the patient is asked to hold the sternum off the floor for as long as possible (McGill 1998). A small pillow is placed under the lower abdomen to decrease the lumbar lordosis, the cervical spine is maintained in maximum flexion and the pelvis is stabilized through gluteal contraction. The performance time is recorded in seconds, with the patient is asked to hold this position as long as possible, not to exceed 5 minutes. This test is simply a timed motor performance test to establish a level of function, or to track improvement with training. Muscle bulk of the gluteal muscles and lumbar multifidi can also be assessed during the test, with atrophy identified visually or through palpation.

Active Sit-Up Test

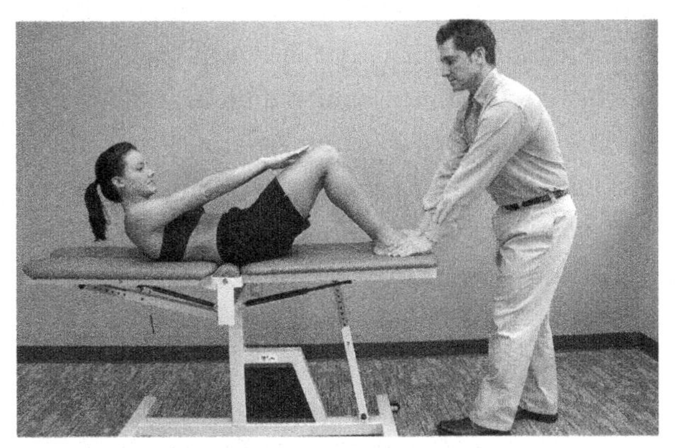

Figure 5.38: The patient is positioned in supine with the knees flexed 90°. The soles of the feet are flat on the floor with the examiner holding them down with one hand. The patient reaches up with the fingertips of both hands to touch (not hold) both knees (Waddell et al. 1992). If the patient cannot maintain this position for five seconds, the test is positive. This test is simply a timed performance to establish a level of function, or to track improvement with training.

Active Bilateral SLR Test

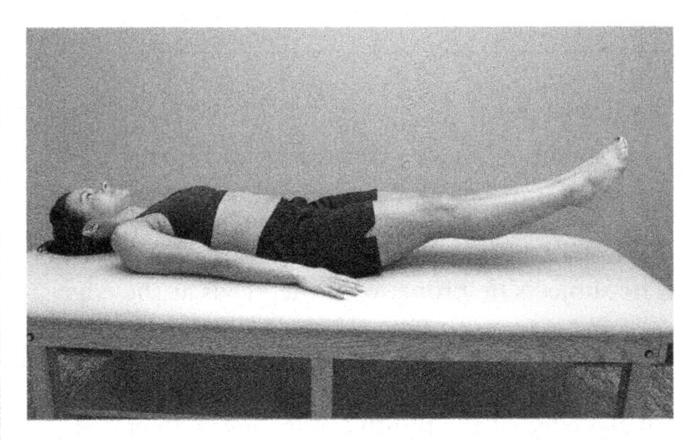

Figure 5.39: In supine, the patient is asked to lift both legs together six inches (15.24 cm) off the examining surface and hold for five seconds (Waddell et al. 1992). Both heels and calves should not touch the examining surface during the test. The test is positive if the patient cannot maintain this position for five seconds. This test is simply a timed motor performance test to establish a level of function, or to track improvement with training.

Prone Instability Test

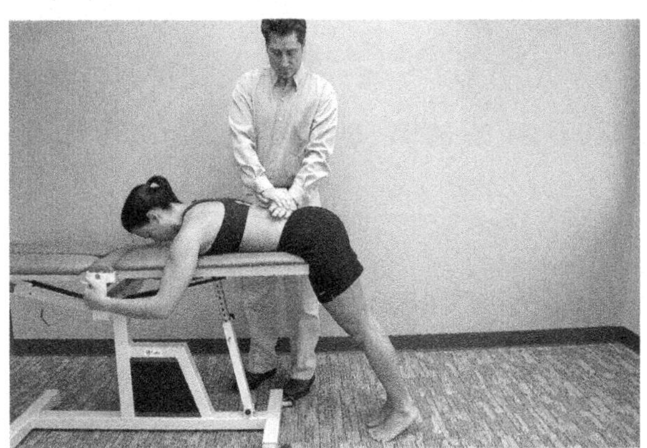

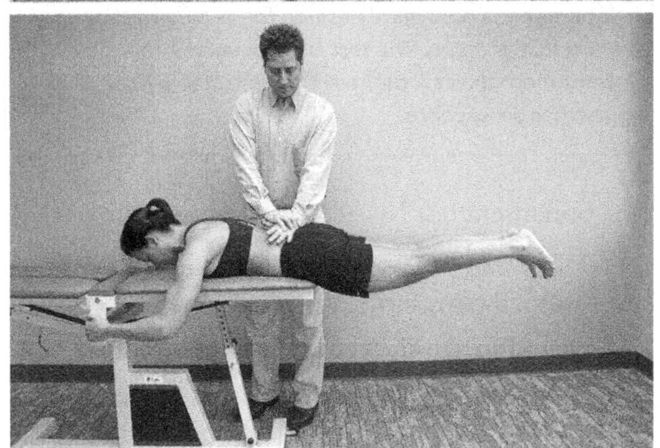

Figure 5.40a,b: The patient lies with the trunk on the examining table, flexed at the hips, with the feet resting on the floor. The patient relaxes while the examiner applies posterior to anterior pressure to the lumbar spine, recording any pain provocation (McGill et al. 1999). The patient then

lifts both legs off the floor, holding the table to maintain position while posterior compression is applied again to the lumbar segments. The test is positive if pain present during the resting test position subsides in the second position while muscles are contracted. This test is an indicator that enough muscle function is present to control an unstable segment.

Sorensen Test

The Sorensen Test is a timed test for isometric capacity of the trunk extensors ability to sustain an antigravity position (Biering-Sorensen et al. 1984). The subjects lie prone, with the pelvis at the edge of a plinth and the lower limbs fixed. The body is maintained in a horizontal position with the trunk off the edge of the plinth for as long as tolerated. The test is terminated if the torso deviates more than six degrees from the stable position for longer than six seconds.

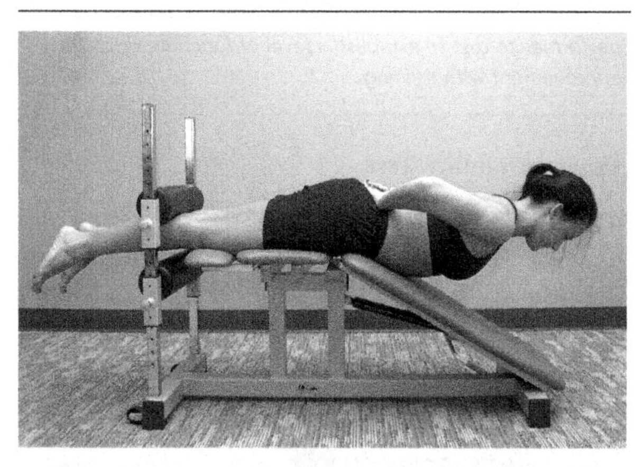

Figure 5.41: The Sorensen Test: timed test for isometric capacity of the trunk extensors ability to sustain an antigravity position (Biering-Sorensen et al. 1984). The subject lies prone, with the pelvis at the edge of a plinth and the lower limbs fixed. The subjects are instructed to maintain their bodies in a horizontal position for as long as they can tolerate that position. The test is terminated if the torso deviates more than six degrees from the stable position for longer than six seconds.

The Spremsem test has been used to measure endurance for lumbar extensors (Chok et al. 1999), but is actually measuring isometric endurance. To improve isometric endurance, strength training is required, not endurance training or isometric holding. Isometric endurance is limited by blood flow. As intramuscular pressure increases, from the tension of muscle contraction, circulation decreases. With contraction below 20–25 percent of maximum, circulatory compromise does not occur, and the activity can be sustained. If the body weight exceeds this percentage, circulatory compromise occurs, causing fatigue with isometric holding. Repetitive strength training, rather than isometric holding, will increase the lifting capacity. A fixed body weight would then fall closer to the relative 20–25 percent of maximum contraction, allowing for longer isometric holding times.

Lumbar Extension Cross Patterns

More subtle alterations in motor patterns of the trunk can prevent a gradual progression in motor training as an abnormal firing pattern is occurring related to improper timing of muscles and/or compensations. Visual assessment of movement may identify more global issues of limited mobility or strength through the hips, pelvis and thoracic spine that may prevent normal movement through the lumbar spine. Subtle changes in timing or limited recruitment of the lumbar multifidi may require more direct visual observation, palpation or surface EMG to identify. Janda (1999) described a series of abnormal firing patterns of the hip, pelvis and lumbar spine with a simple prone hip extension test. Janda's theory stated that with a prone hip extension of the right hip, the recruitment order for posterior muscle groups should occur in the following order: right hamstring, right gluteus maximus, left lumbar erector spinae, right lumbar erector spinae, left thoracolumbar erector spinae and lastly right thoracolumbar erector spinae (Janda 1992). Vogt and Banzer (1997) also felt a consistent firing pattern was noted with prone hip extension but in a different order: ipsilateral lumbar erector spinae, semitendinosus, contralateral lumbar erector spinae, tensor facia latae and gluteus maximus. Bullock-Saxton et al. (1993) compared hip extension in pain free subjects concluding that muscle onset times were almost simultaneous. Lehman et al. (2004) performed a similar EMG study on muscle firing patterns for prone hip extension in 14 asymptomatic subjections, finding no consistent order of activation for the biceps femoris, contralateral erector spinae and ipsilateral erector spinae. The authors also concluded that the prone hip extension test is not sufficient for a diagnostic test due to physiological variation, feeling that an overlap between normal and potentially abnormal activation patterns may exist.

The clinical application of the prone hip extension test is somewhat lost in these previous investigations. The intention of the test is not to be diagnostic. Identifying weakness in muscle, altered firing patterns or gross muscle impairment is not diagnostic, but simply a clinical impairment finding from specific motor tests. True tissue diagnosis is made from a thoughtful process of evaluating all the tissues with the available tests, eliminating those tests that are negative and coming to a hypothesis for a pathological description.

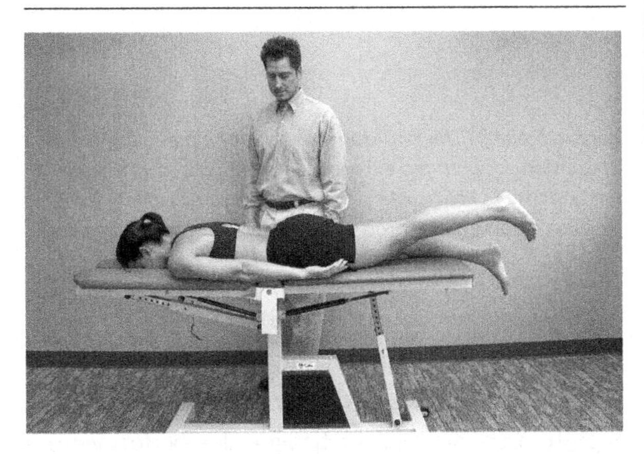

Figure 5.42: Prone hip extension test. The patient lies prone with the arms at the side and the simple instruction of lifting the leg while keeping the knee straight. The normal firing pattern for right hip extension would be 1) right hamstrings, 2) right gluteus maximus 3) left lower lumbar multifidi, 4) right lower lumbar multifidi, 5) left upper lumbar multifidi, 6) right upper lumbar multifidi, with no participation of thoracic paraspinal muscles or dorsal scapular muscles (Janda 1999).

The value of the prone hip extension test, in a more simplistic sense, is to observe gross motor performance. The order of muscle recruitment may not be as significant as notable asymmetries in performance and compensatory firing of upper thoracic and dorsal scapular musculature. Poor coordination of anterior abdominal muscles, including the transverse abdominis, may lead to an extension or a shear moment in the lumbar spine or loss of rotational stability during the test. Late activation gluteus maximus may be significant when coupled with weakness in muscle testing or weight bearing functional testing. The actual timing between muscle activation is measured in milliseconds and may not relate to the actual gross motor performance. Hodges and Richardson (1997) attempted to make a similar claim in timing issues related to the transverse abdominis firing prior to lifting the arm and when in dysfunction the muscle fired late. Again the timing is in milliseconds and the clinical relevance must incorporate all other evaluative findings. These types of observations and tests may assist in guiding which muscles are emphasized during coordination training and may simply be used as retests for performance improvement. Observation of abnormal motor performance is not diagnostic but should be used in assisting exercise design. Abnormal firing patterns may resolve after normalization of joint restriction, resolution of muscle guarding or resolution of pain (Grimsby 1991, Janda 1997).

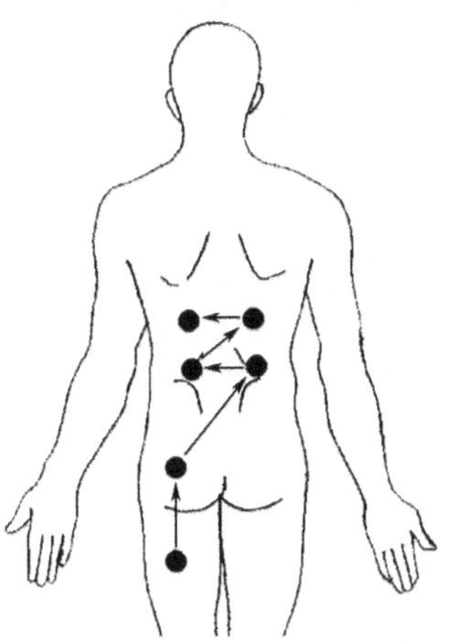

Figure 5.?: Normal pelvic cross pattern recruitment for the prone hip extension (Janda 1999).

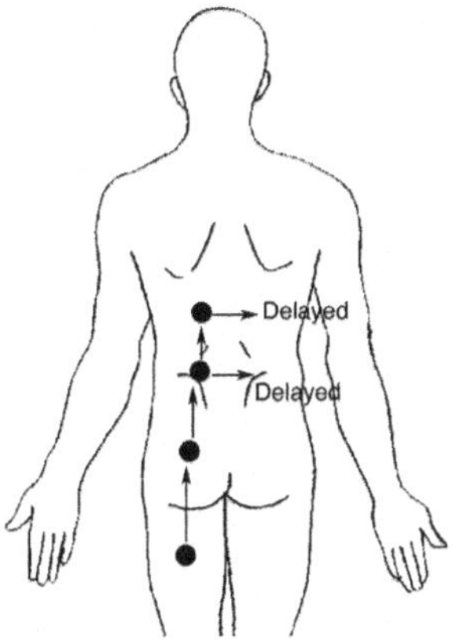

Figure 5.43: An example of an abnormal pelvic cross pattern as described by Janda (1997). Firing order is altered to: 1) hamstrings, 2) gluteals, 3) ipsilateral lower multifid, 4) ipsilateral upper lumbar multifidi, 5) contralateral lower lumbar multifidi (delayed) and 6) contralateral upper lumbar multifidi (delayed).

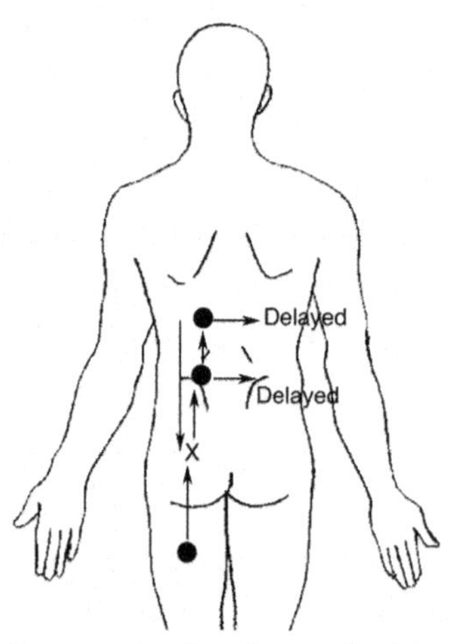

Figure 5.44: An example of an abnormal pelvic cross pattern as described by Janda (1997). Firing order is altered to: 1) hamstrings, 2) ipsilateral lower multifidi, 3) ipsilateral upper multifidi, 4) contralateral multifidi (delayed), 5) contralateral multifidi (delayed), 6) absent or delayed gluteals. Absent or reduced gluteus maximus strength and activation is postulated to decrease the efficiency of gait and is associated with chronic low back pain (Janda 1992, Janda 1996).

Abnormal Recruitment Pattern 3
Hamstring
Ipsilateral upper multifidi
Contralateral multifidi (delayed)
Contralateral multifidi (delayed)
Absent gluteal muscles and Multifidi
Late firing of ipsilateral multifidi

Abnormal Recruitment Pattern 4
Hamstrings
Absent gluteals
Ipsilateral lower multifidi
Ipsilateral upper multifidi
Absent/diminished contralateral muscle recruitment
Ipsilateral of upper trapezius

Abnormal Recruitment Pattern 5
Hamstrings
Gluteals
Ipsilateral lower multifidi
Ipsilateral upper multifidi
Contralateral multifidi (delayed)
Contralateral multifidi (delayed)
Contralateral upper trapezius

Abnormal Recruitment Pattern 6
Hamstrings
Gluteals
Ipsilateral lower multifidi
Ipsilateral upper multifidi
Contralateral multifidi (delayed)
Contralateral multifidi (delayed)
Pelvis rotates (right) as contralateral multifidi and transverse abdominis can not stabilize with left rotation moment

Using the prone hip extension test does require the clinician to at least consider the motor timing and coordination during a motor performance. Certainly many alterations in the pattern of firing may exist in normal subjects, but an even greater variation can be seen in symptomatic patients. The classic bird-dog exercise is so commonly used for early back training. The true functional applications of this exercise can be argued but if an abnormal firing pattern is occurring, this exercise may continue to reinforce abnormal movement.

Figure 5.45a,b: The bird-dog exercise for hip and lumbar extension. Limited hip extension mobility or the presence of an abnormal pelvic cross pattern (as previously described) can reinforce abnormal motor recruitment of lumbar musculature.

Comparing performance on both sides may assist in determining if abnormal motion is occurring. Providing assistance to the exercise may create a significant change in the firing order, performance and pain levels suggestive of the firing pattern being a contributing factor to dysfunction. Principles of STEP involve dosing exercises below 40% of 1RM when emphasizing coordination training to allow for a higher number of repetitions. The resistance of the leg during prone hip extension may exceed 1RM for some of the participating muscles, requiring compensation of other muscles or altered order of recruitment. Shoulder elevation with motor dysfunction is associated with abnormal firing of scapular muscles and the rotator cuff leading to abnormal elevation of the scapula and humeral head. Providing a pulley assist to lighten the weight of the arm can instantly normalize the coordination in the performance. The same unloading concept can be applied to spinal training. Dosing the resistance to a lower level may allow for a normal pain free pattern for high repetition training for coordination. Once the pattern is coordinated a faster progression to endurance and strength training can be made.

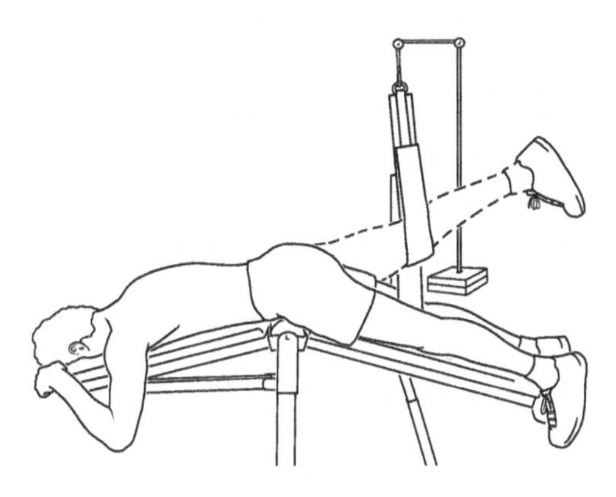

Figure 5.46: A pulley assist for prone hip extension reduces the weight of the limb allowing for a more normal firing pattern without compensation. With reduced load (<60% of 1RM) a high number of repetitions can be performed for coordination training and motor learning.

Deep Abdominals / Hip Flexors

Anterior muscles of the trunk contribute to lumbar stability and function. Weakness, or inhibition, of these muscles is not uncommon in both chronic and acute low back pain. The lumbar multifidus, quadratus lumborum and erector spinae are tonic muscles, frequently in reflexive spasm or guarding in the presence of low back pain. Muscle spasm of the posterior extensors can lead to reciprocal inhibition of the anterior deep abdominals. The deep abdominal muscles include the external oblique, internal oblique and the transverse abdominals. The greater the level of pain and subsequent guarding of the posterior tonic extensors, the more difficult it is to recruit these anterior flexor muscles. For this reason, treatment should initially focus on resolving the cause of pain and reducing the posterior muscle before trying to recruit and retrain the abdominal muscles (Basmajian 1979). Frequently, resolution of pain and guarding will lead to normal, or significantly improved, recruitment. Difficulties with motor learning for up-training previously inhibited muscles may resolve when the source of inhibition is reduced.

The ability for the transverse abdominis muscle (TrA) to provide adequate stability to the vertebral column is hard to quantify, as the muscle is paper thin and not directly attached to lumbar vertebra. Indirect influence on vertebral motion is transmitted through the thoracolumbar fascia or related to increasing intraabdominal pressure.

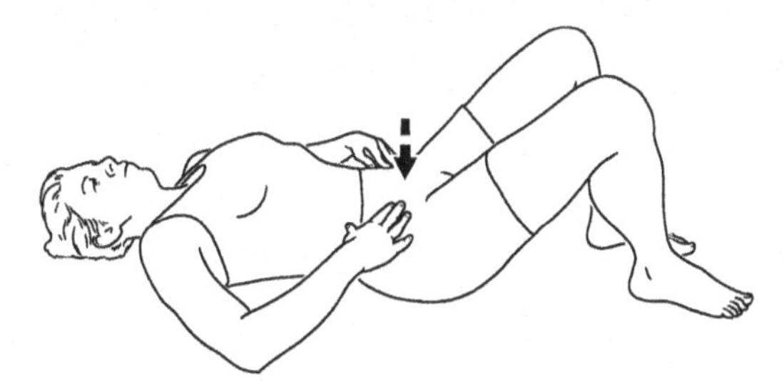

Figure 5.47: Abdominal hollowing. The patient is instructed to pull the umbilicus in toward the spine.

Some rehabilitation programs attempt to teach the patient to isolate this muscle from the other deep abdominal. The process of motor learning for isolating individual muscles is difficult and counter-intuitive. Isolating an individual muscle is not consistent with normal movement patterns and is not necessary in the overall program of stabilization training for the spine. Providing a mild deep abdominal contraction is a component of normal function and has obvious clinical benefits in retraining spinal stability. This approach is crucial in Stage II when isometrics are introduced with a hypermobility or to protect a joint from moving around a non-physiological axis with a hypomobility. Complimentary research has supported a gentle isometric bracing, rather than a hollowing, of the deep abdominals without placing more responsibility on the TrA (McGill 2001). Bracing involves only a tightening of the abdominal wall without a change in the wall profile. Hollowing involves pulling the anterior abdominal wall posterior, sinking the stomach toward the spine. According to McGill (2001), this reduces stability in the spine. Bracing is easier for patients to learn, as it is a more natural motor response. Bracing should be a precursor to spinal motion but not as an aggressive, hard, isometric performance. Normal stability requires as little as 15 percent of a maximum voluntary contraction for stability. Excessive bracing creates too much stiffness, reducing necessary motion for the delicate modulation for movement. It is imperative that adequate coordination ensures that the muscles are modulated so that the spine can change its stiffness to match the demands of internal and external forces on the body.

Dysfunction of the TrA should not be considered in isolation, but as a breakdown in an overall movement pattern. Weakness, inhibition and/or poor timing of the TrA can be a component of dynamic instability in the spine, but should be considered in the larger motor pattern involving many muscles throughout the trunk and pelvis. Rather than muscle isolation, early emphasis is placed on coordination, or the execution of a particular skill, of the lumbopelvic region. Rather than learning the motor skill of isolating a muscle unnaturally, the focus is on recruiting the muscle in a normal pattern of movement. Motor learning refers to the process of acquisition, or acquiring the skill and can be achieved quickly over a period of hours or may require days, months, or years. Motor learning has three phases: acquisition, consolidation and retention (Blazquez et al. 2004). The cognitive processes of motor learning involve verbal instruction, imitation, imagery and mental practice (Annett 1994).

The goal of achieving coordination for new movement patterns leads to consideration of the best way a patient can learn. The approach for recruiting the TrA focuses on internal attention, which can be detrimental to the performance of a well-learned skill as well as learning new skills (McNevin et al. 2000). Attention focused internally toward the lower abdominals will disrupt a normal bracing maneuver by the entire lower trunk and degrade the learning process of the transfer or lift. An exercise set up with movement goals that accomplish the desire recruitment pattern will improve coordination more efficiently. Some level of internal awareness of body position and abdominal tightening may be necessary, but directing full attention internally toward specific muscles can disrupt the execution of automated skills and can have a degrading effect on the learning of new skills (NcNevin et al. 2000).

Hollowing is an appropriate technique for motor learning for isolation of the TrA, but not necessarily for normal stability or function. Bracing is the more natural and functional approach for the abdominal contribution toward dynamic stability. In some cases an initial approach directed internally toward isolation of the TrA with hollowing type exercises may be a necessary first step in a progression. An approach of bracing should, however, be the emphasis and eventual progression for all spinal stabilization training. Exercises can be designed to naturally recruit the deep anterior muscles without an internal focus but through an external resistance that naturally recruits these muscles. With the focus more on performance of an activity that naturally recruits certain fiber directions, motor learning can address functional movements and occur more quickly. Care should be taken to avoid directing exercise too much toward learning specific rehabilitation exercises versus learning more functional movements.

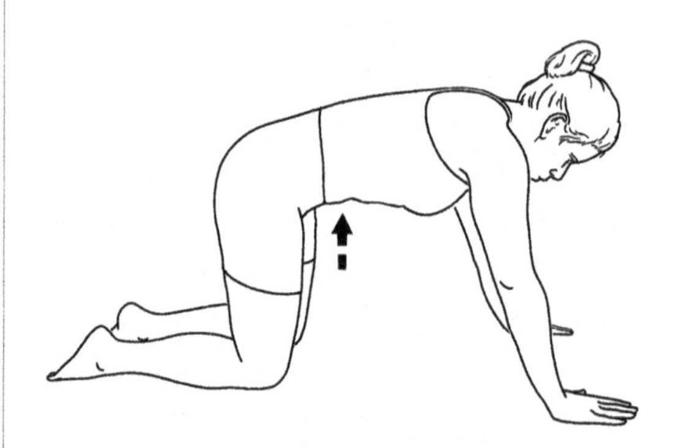

Figure 5.48: TrA hollowing performed in quadruped. The weight of the organs on the abdomen wall provides feedback for contraction of the TrA muscle. The patient is instructed to exhale and pull the umbilicus toward the spine. The can be practiced in alternative postures, such as sitting and standing.

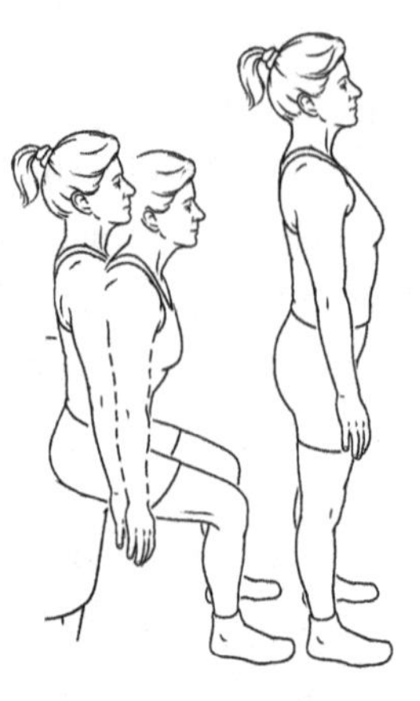

Figure 5.49: Stabilization training during the functional motion of transferring from sit to stand. Whether TrA hallowing or abdominal bracing is preferred, practice can involve many other functional motions such as rolling transfers and lifting.

The deep hip flexors also play a key roll in spinal stability. Weakness in the psoas and iliacus often relate to spinal dysfunction. Iliacus and psoas activation is not consistent with spinal movement but primarily hip flexion (Juker et al. 1998). The iliacus has been shown to stabilize the pelvis in contralateral hip extension during standing, while the psoas is selectively involved in contralateral trunk loading situations, requiring stabilization of the spine in the frontal plane (Andersson et al. 1995). The iliacus also produces hip flexion but working alone without the psoas creates an anterior pelvic tilt, forcing the spine into lumbar extension. The psoas counteracts this anterior tilting moment. A weak or inhibited psoas can lead to excessive anterior tipping or shearing. The psoas provides minimal spinal segment stability and shear stiffness but only in the presence of hip flexion torque (McGill 1998).

In some cases, excessive tone is present in the hip flexors, preventing normal movement and creating secondary pain from muscle guarding. An elevation in tone of the hip flexors may be directly related to pain and spinal facilitation or related to compensation for weak lower abdominals. These muscles often test short with muscle length tests but this is largely due to elevated resting tone. Directly training the hip flexors (as above) may be helpful with acute guarding to increase circulation and reduce tone but if they are over active, compensating for lower abdominal weakness, than a combination of down training the hip flexors while up-training the lower abdominals is most efficient.

Reducing hip flexor contribution to cranial trunk flexion (sit ups) has largely focused on the position of the hip. The more common approach is to flex the hip to shorten the hip flexors in attempt to make them actively insufficient. The position of the hips and lower quarter is not as important as whether an extensor pattern or flexor pattern is facilitated in the lower limbs during the movement (Janda 1996). Hooking the feet to stabilize the lower limbs recruits a flexor pattern of ankle dorsiflexors, knee flexors and hip flexors (psoas and iliacus). Sit-ups performed in this position emphasize the hip flexors rather than the lower abdominals. Creating an extensor pattern would more effectively inhibit the hip flexors to allow for trunk flexion to be performed to a greater extent by the lower abdominals.

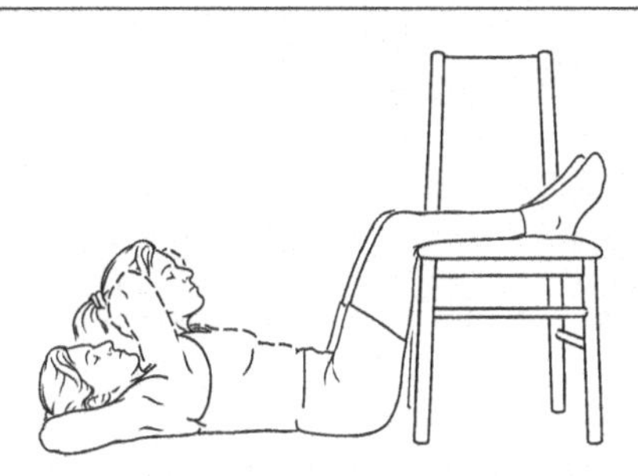

Figure 5.50: Crunch sit-ups with the hips flexed. Performing cranial flexion with the hips in a flexed position attempts to reduce recruitment of the hip flexors during the motion. This approach is largely ineffective and will continue to facilitate the hip flexors. The movement also emphasizes the cranial portion of the rectus abdominus, which has little contribution to spinal stabilization.

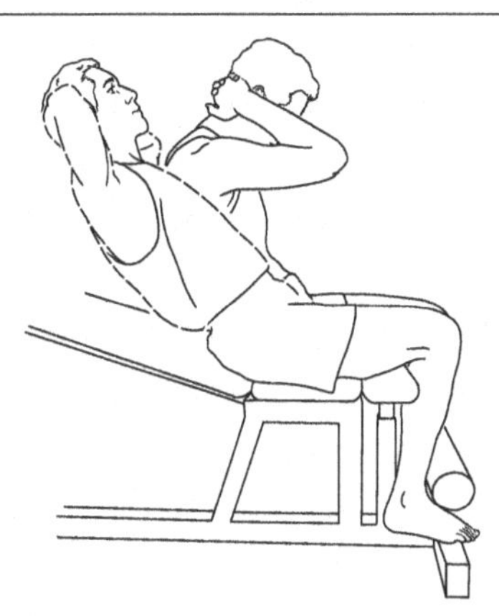

Figure 5.51: Cranial flexion with lower limb fixation. Any block of the lower limbs in supine, incline or sitting will facilitate a flexor pattern increasing emphasis of the hip flexors and reducing emphasis of the abdominals.

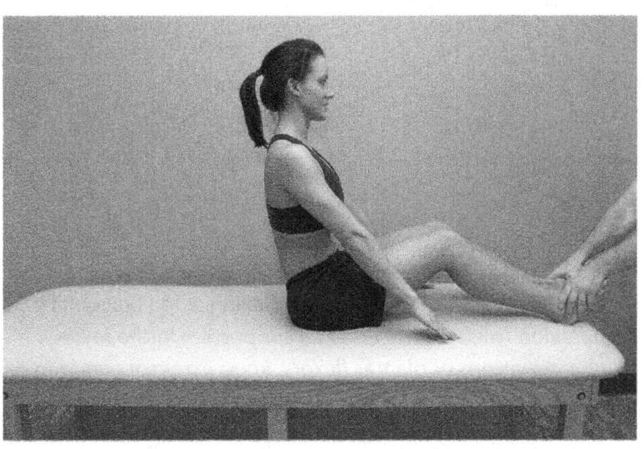

Figure 5.52: Testing can determine the contribution of the hip flexors to the sit-up performance (Janda 1996). In supine, the operator fixes the ankles in dorsiflexion. This will facilitate a

flexion pattern of dorsiflexion, knee flexion and hip flexion to assist the abdominals.

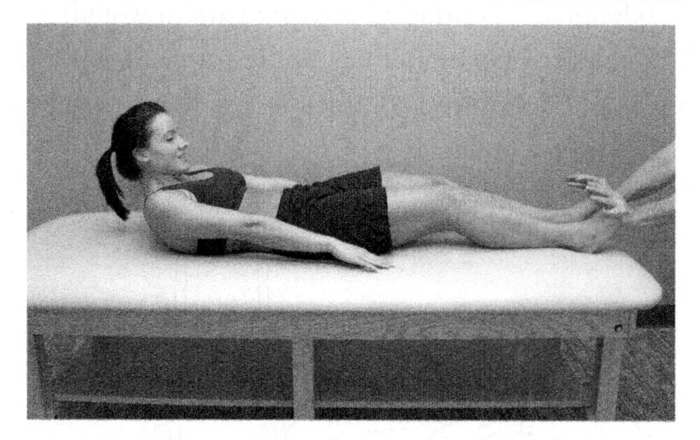

Figure 5.53: Sit-up test continued. A comparison is then made to the height of the sit-up, or ease of movement, with resisted plantar flexion. An extension pattern of plantar flexion, knee extension and hip extension inhibits the facilitation of the hip flexors, as the abdominals are forced to contribute the majority of torque for performance. A patient with weak abdominals will often be able to perform a full sit-up with a flexion block, but be unable to even rise off the table with a plantarflexion fixation.

Training the lower abdominals with reduced hip flexor contribution involves a position creating a lower limb extension pattern. This position is best achieved with a slant board, allowing for a relative reduction in the weight of trunk, while allowing for the lower limbs to support the body with a primary extension pattern. The angle of the slant board is increased until the patient can coordinate a sit-up motion, and raised higher until the desired number of repetitions can be achieved.

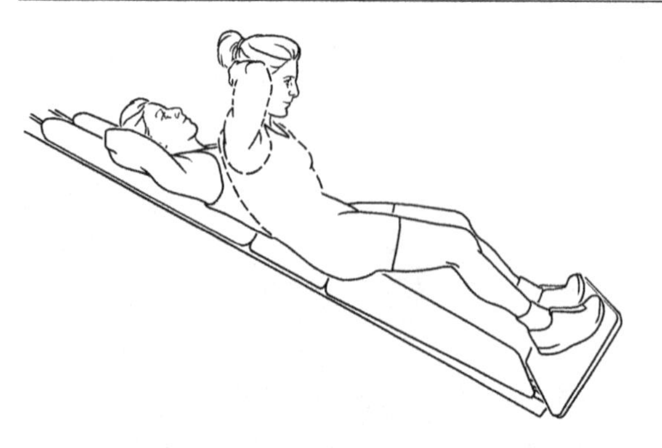

Figure 5.54: "Sit-Overs"—Incline cranial trunk flexion. An incline position with slight plantar flexion, knee flexion and hip flexion will facilitate an extensor pattern in the lower limbs. A pattern of plantar flexors, knee extensors and hip extensors will inhibit the flexor pattern including the psoas and iliacus. With cranial flexion the lower abdominals bear most of the load. The incline board is used to create the extensor pattern but is also necessary to reduce the load to allow for 20–30 repetitions with the lower abdominals working more independently. This down training approach of the hip flexors is particularly effective in the presence of a hypermobile spondylolisthesis or disc pathology with shearing that are prone to excessive extension.

Figure 5.55: Sit-Overs with a ball. If an incline board is not available, an attempt to create an extensor pattern for cranial flexion can be made with a fitness ball. The feet are placed against a wall in a plantar flexed position with the knees flexed for quadriceps facilitation. The trunk is angled on the ball in an incline position to facilitate the hip extensors during cranial flexion. The coordination challenge on the fitness ball may be too high for early training and cannot easily be adjusted for the level of resistance but can be a successful home option when an incline board is not available. The fitness ball approach will be easier than the slant board due to the recoil of the ball assisting the trunk motion.

Motions and Directions

The initial movement directions of the Stage 1 exercises are chosen based on contraindications, tissue presentation, presence of pain, direction of joint restriction or hypermobility, muscles in guarding and/or muscle performance issues. Selecting specific exercises and directions should follow a logical thought process. Hypomobile joints are mobilized toward the end range, where hypermobile joints may be protected while mobilizing adjacent restrictions and/or muscles.

Some authors suggest that either flexion or extension biased exercises are of equal benefit (Elnaggar et al. 1991). Motion should be considered in all planes related not only to the pathology but functional requirements. Combined motions in diagonal patterns are necessary with daily activities, work and sport. Rehabilitation should have the end goal in mind for progressing exercises through all planes of motion. A thorough evaluation should make the initial choices for motion self-evident. Long et al. (2004) demonstrated that matching subjects' directional preference and pain free planes of motion significantly and rapidly decreased pain, medication use and improvements in all other outcomes. In this study, subjects were compared to two other groups, one of which performed exercises in the opposite direction of the preferred motions. The third group performed evidence-based exercises commonly prescribed, including multidirectional, midrange lumbar exercises and stretches for the hip and thigh muscles.

Descarreaux et al. (2002) compared the effectiveness of two home exercise programs at decreasing disability and pain related to subacute and chronic nonspecific low back pain. A specific (individualized) exercise program was compared with a program of commonly prescribed exercises for low back pain. Both groups demonstrated some improvement, but only the members of the group who received specific exercises significantly reduced their level of pain and disability.

A tissue based approach to exercise is to first move away from pain, or in the direction of preference of the patient. Motion is safely performed away from injured tissues but repetitive motion provides cellular stimulus for repair. When muscle guarding is present, exercises are chosen to train into the guarded pattern, emphasizing local vascularity and resolving abnormal tone. For the lumbar multifidi in guarding, this would be a pattern of extension, ipsilateral side bending and contralateral rotation. Low level training to bring in more oxygen allows for full elongation of the muscle to improve segmental motion in the range of flexion, contralateral side bending and ipsilateral rotation. When specific motor deficits in timing, coordination, endurance or strength are discovered, these muscles are selectively trained to improve their function. Emphasis on training muscles for dynamic stabilization is then considered. For every increase in range of motion gained, exercise is performed to stabilize this new range.

Hypomobile Joint Concepts

Active movements for mobilizing hypomobile ranges involve slow movement with a low demand on coordination. Emphasis is not on concentric work toward the restriction, but rather concentric work away, with subsequent eccentric work toward the restriction. A slow eccentric motion toward the restriction, addressing both contractile and noncontractile tissue that may be involved in the restriction is incorporated. In the example of mobilizing shoulder abduction, it is more effective to perform adduction with an eccentric return into elevation, than to perform active abduction toward the restriction. For the spine, the same pattern is performed concentrically away from the restriction, with an eccentric return toward the restriction. The motion is toward the end range but kept in the pain free range to prevent an increase in muscle guarding, adding to the limitation in motion.

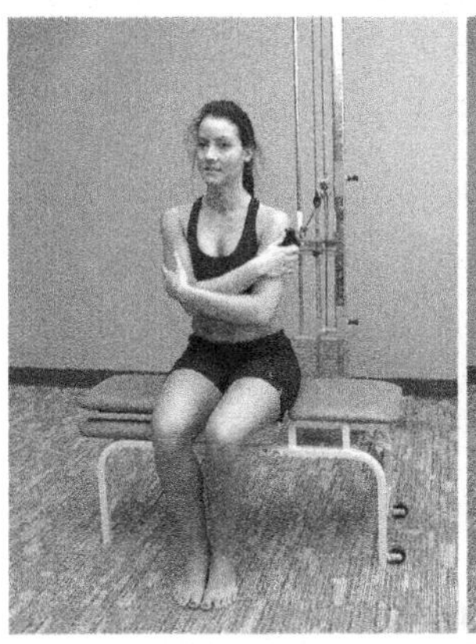
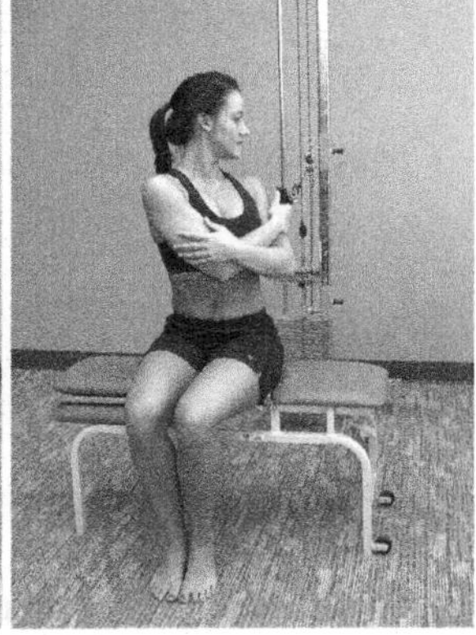

Figure 5.56a,b: Left rotation mobilization of the thoraco-lumbar region. Sitting reduces the coordination needed to stabilize the pelvis in standing. Left rotation is performed with an emphasis on a slow eccentric return into right rotation and may even involve a sustained hold.

Hypermobile Joint Concepts

Hypermobile joints lack the passive collagen structures to limit motion, as well as having reduced neural recruitment of muscles associated with stabilization. It is initially crucial to avoid exercising into the end range collagen tension in the direction of the hypermobility. The range of training is from a mid length of collagen tension toward a shortened range, avoiding end range strain. Whether the pathology involves a hypermobility facet, unstable disc, nerve impingement or just pain, the range of motion avoids end range strain to these tissues. Motion begins from neutral, moving away from the contraindication. Exercises may be initiated in a different plane than the primary plane of contraindication, but when in the same plane, motion is away from pain. Repetitive motion in the same plane as the pathology will provide the optimal stimulus for tissue repair but the end range of tension is avoided to prevent further pain or excessive tissue strain. As tissue tolerance and motor control improves, the range of training will gradually increase.

Basic Sequencing of Direction of Tissue Training:

- *Movement is initiated in the pain free planes of motion.*
- *Movement away from pain in painful plane of motion.*
- *Hypomobile: movement away with eccentric return toward end range.*
- *Hypermobile: movement away with eccentric return to neutral only.*
- *Movement toward pain, but not into painful range.*

Dynamic Stabilization

Dynamic stabilization implies that the motor system is functioning properly to control the axis of motion of a joint or joint system. The primary local muscle for segmental stabilization of the lumbar segments is the multifidus muscle. This predominantly tonic, arthrokinematic muscle, has the appropriate lever arms to control normal motion and abnormal shear. Eccentric function of the right multifidi will decelerate flexion, right rotation and left side bending. With rotation, bilateral multifidi are working in synergy, one side concentric and one side eccentric (Donisch and Basmajian 1972). This is why these muscles are considered to function as dynamic ligaments. Despite being on opposite sides of the axis of motion, the opposing multifidi are seen as synergists, not antagonist. Synergy with the tonic transverse abdominis and obliquus internus are also part of the normal movement synergy for coordinated and stabilized motion. The remaining portions of the abdominals, the long extensors and muscles of the hip coordinate the osteokinematic motion of the trunk.

The axial attachments and vertical orientation of the LM leave it poorly suited to contribute to the motion of lateral side bending or rotation in terms of providing torque. The muscles best suited for trunk rotation are the oblique abdominals. Because of the flexion moment associated with these abdominals, the erector spinae might be involved as an antagonist to flexion. In a similar function, the authors speculated the LM were acting as "anti-flexors" to balance the anterior sagittal rotation produced by the flexors.

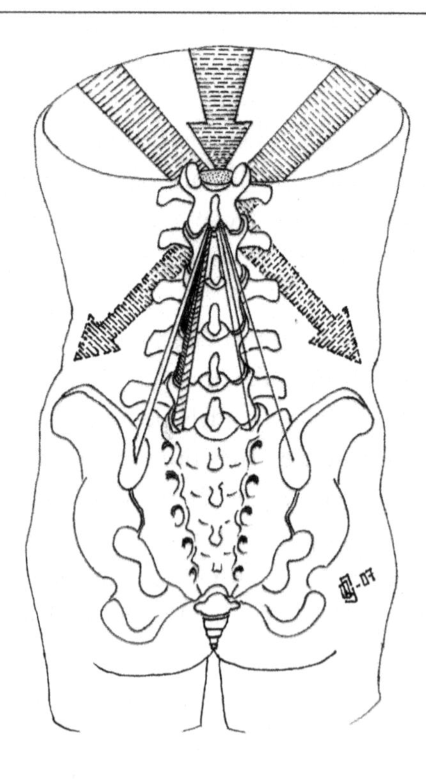

Figure 5.57: This drawing helps illustrate how the right and left LM will be active during rotation. In the example of left trunk rotation, the primary mover would be the right external oblique abdominal and the left internal oblique abdominal. Because of their fiber orientation, their contraction would provide a flexion moment in addition to left rotation. As the multifidi are primarily extensors, they will fire bilaterally to counter the flexion moment of the abdominals. The right LM would fire concentrically to assist with rotation while the left LM fire eccentrically. The deep multifidi serve the arthrokinematic function of controlling the segmental motion while the osteokinematic abdominals rotate the trunk.

In the clinical example of a hypermobile lumbar segment in flexion and right rotation, initial training focuses on the right side multifidus. This muscle will be targeted as it is not only inhibited due to receptor damage and pain from the pathology, but is also the primary stabilizer of the segment into the pattern of hypermobility. Initial training focuses on concentric work (CW) to recruit the muscle and resolve guarding. The range of training avoids the hypermobile range of the joint, while training any or all of the three motions of the multifidi. Training is from neutral to end range left rotation, from neutral to extension and/ or from neutral to right side bending. The patient's tolerance or pathology determines the plane(s) to be initially trained. Performing repetitive motion in these limited ranges provides a safe level of tissue stress and strain to maintain and repair collagen and cartilage of the right facet, ligaments and portions of the disc under load.

If these same motions are performed from both a cranial and a caudal direction than at least six exercises can be safely performed acutely without concern of overstretching damaged collagen, while providing a high number of repetitions for pain inhibition, tissue repair, vascularity and coordination. This three dimensional example is for the purpose of outlining the training direction concept. For different clinical conditions, focusing on only one plane, or a combination of only two planes, may be effective. The patient's presentation and evaluation findings should guide the selection of directions and ranges to train. This example focuses on the posterior muscles, but incorporating the tonic contributions of the transverse abdominis and hip musculature would also be built into the program as well.

The Manual Therapy Lesion refers to altered or lost feedback due to mechanoreceptor damage from the primary pathology. This in turn, reduces the normal recruitment of the tonic muscles that control the axis of motion for a spinal segment. A major component of achieving dynamic stabilization is sensitizing the remaining afferent receptors in the systems to aide with motor control. Sensitizing the muscle spindles in the local tonic muscles is part of the progressive training process. Initial emphasis on concentric work away from the pathology focuses on recruitment, timing and resolution of guarding. Isometric was not initially trained as it reduces circulation, does not provide tissue stimulus and is not efficient for developing coordination. Stage 2 adds isometric work (IW) to allow an increase in resistance, and improved efficiency at developing sensitization of muscle spindles. Isometric work is initially performed in the shortened position of the multifidi, later progressed to the mid and lengthened range.

Figure 5.58a,b: Concentric work from neutral to left rotation, emphasizing the right side multifidus for Stage 1 training of a right rotation hypermobility.

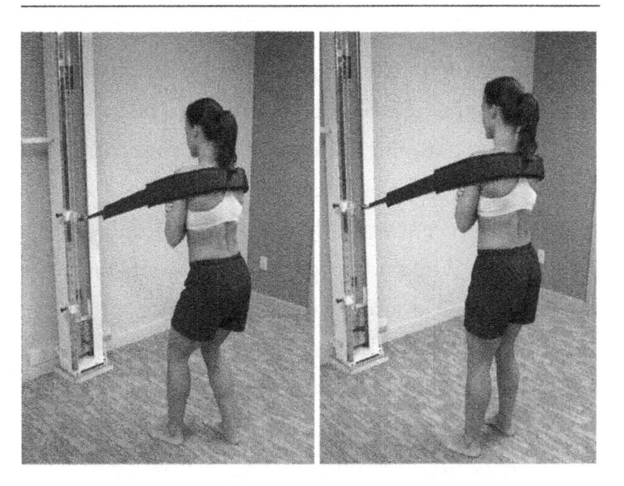

Figure 5.59a,b: Isometric work of the right multifidus for left rotation: Stage 2 training of a right rotation hypermobility. The spine is positioned in a range from neutral rotation to full rotation way (left) from the pathology (right rotation hypermobility). In the beginning position(a) the weight stack is at rest with one foot forward. The patient steps back lifting the weight stack while maintaining the trunk position. After a hold time the patient steps forward resting the weight stack.

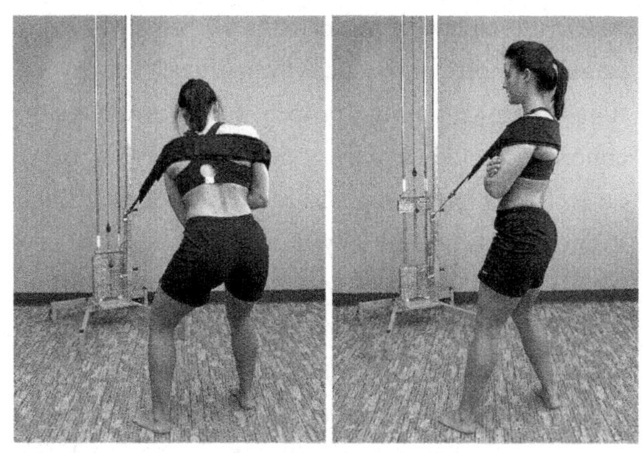

Figure 5.60a,b: Eccentric work from left rotation to right rotation, but not into the pathological range, Stage 3. Emphasis is on eccentric deceleration of right rotation with the right side multifidi working eccentrically.

Figure 5.61a,b: Concentric work from left rotation to right rotation with a one to two second pause just prior to the pathological range. This can be added in Stages 2–4. Emphasis is on the left side multifidus, with an end range hold for co-contraction and stabilization.

The next step in the progression is to emphasize coordinated eccentric muscle function, decelerating toward the pathological range with a concentric return. This is the emphasis in Stage 3. To emphasize eccentric work (EW) the training order is changed. The body is positioned away from the pathological range with the resistance set for eccentric work toward the pathological range and a concentric return. The work order changes from CW-EW of Stage 1 and 2, to EW-CW. Motion is eccentrically controlled toward, but not into, the pathological range with concentric return to the start position. Speed is initially slow to develop coordination and then can be increased to match the functional demand. Additional IW is performed closer to the pathological range to fix strength where the joint is less stable. In this example of right rotation hypermobility, the right side multifidus is emphasized through this training of CW away, IW toward and EW toward. The opposing side multifidus also needs to be trained, working concentrically toward right rotation. The trunk is moved toward the pathological range, but not into it, with a one to two second pause in this position before returning to the start position. This isometric pause at the end assists in the training of recruiting additional motor units at the pathological range where they are needed. Remember that because of their arthrokinematic nature, both sides multifidi are always working, so this is performing a short duration co-contraction for stabilization.

The goal of Stage 4 is to achieve a coordinated synergy between all tonic and phasic muscles, around a physiological axis through a full range of motion. Exercises should involve more diagonal patterns, or functional patterns. Simulation of work, sport or activities of daily living should be built into the program. Basic dynamic stabilization has been achieved by Stage 4, with emphasis shifting from coordination for stabilization to elevation of endurance, strength and power in this newly stabilized range.

Summary of Basic Stabilization Progression

- *Concentric work away (Stage 1)*
- *Isometric work (Stage 2)*
- *Eccentric work toward (Stage 3)/ Concentric work toward (Stage 2-4)*
- *Functional motion patterns through full range (Stage 3 and 4)*

Hundreds of options exist for back exercises. The specific exercises chosen will dictate which muscles are working, the functional quality trained, specific dosage for the functional quality desired and the range of motion trained. Understanding dosage and progression concepts allows the clinician to customize, modify or invent exercises specific to the individual patient presentation. Below are just a few basic examples of common exercises but the reader should focus on the dosage and progression concepts and apply them to the hundreds of exercises available. The exercise chosen does not determine the training outcome, it is the dosage and exercise design that determine the outcome.

Extension Training

Stage 1 emphasizes active motion for recruitment and coordination. Whenever tolerated, emphasis is on motion of the lumbar spine, rather than direct isometric work in the spine or indirect through extremity motion. Resistance is reduced to allow for high repetitive motion with pain or excessive fatigue. Fatigue with repetitive back extension has been shown to result in up to 75 percent reduction in erector spinae muscle force, with compensatory elevation in the latissimus dorsi and external oblique muscles (Sparto and Parnianpour 1998).

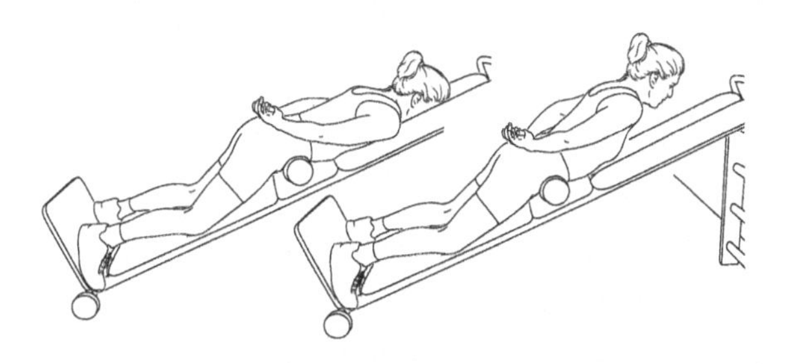

Figure 5.62: Cranial extension on slant board. An incline board reduces the weight of the trunk to allow for training in the 40–60% of 1RM range for vascularity and coordination. In the example of flexion hypermobility or pain, the board artificially limits the flexion. A larger roll will increase the amount of available flexion.

Caudal extension allows for direction motion at the lower lumbar spine for mobilization or local muscle training with reduced load. When extension is painful, as potentially with spinal stenosis or with nerve root pathologies, the range of training is limited from caudal flexion to neutral, avoiding painful end range extension. A roll under the pelvic can position the involved segment or region in flexion, with the active motion returning to neutral only. As mobility improves, the extension range can be increased until pain free end range extension is achieved. Exercise for spinal stenosis has been recommended as a form of conservative treatment (Bodack and Monteiro 2001). Flexion exercises are more commonly advocated for spinal stenosis because of the spinal canal and neuroforaminal narrowing produced by lumbar extension (Fast 1988, Fritz et al. 1997, Sikorski 1985), although some authors have also advocated using lumbar flexion and extension exercises (Nagler and Bodack 1993). The direction of motion should not be set by the pathology, but dictates more the range that is trained. The therapist must have the ability to design an exercise in any direction that is pain free and can be dosed for what ever training quality that is desired while maintaining the safety of the injured tissue(s).

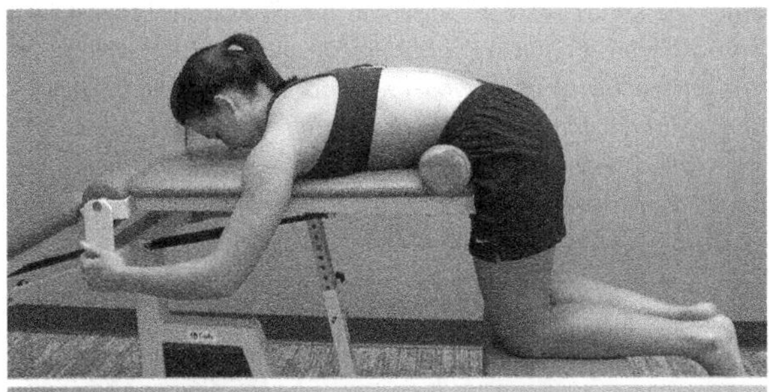

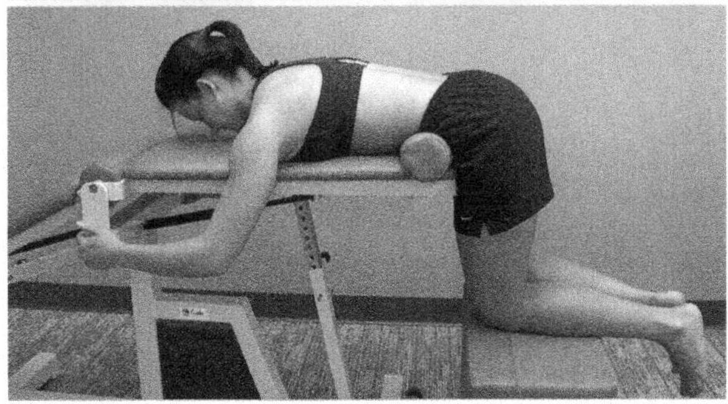

Figure 5.63a,b: Prone caudal extension, short lever with pelvic motion only. As above but with caudal motion, concentric extension away from the hypermobility will target the lumbar multifidi for early vascularity and coordination training. The range is maintained with coordinative locking by the patient from neutral to full extension, while avoiding the hypermobile range of full flexion. Supporting the knees on a bolster and/or putting more of the trunk on the table, to shift the axis of the motion caudal reduces the overall load.

Flexion Training

Active flexion initially addresses mobility, pain, tissue tolerance and/or coordination. Previously, concepts regarding lower abdominal recruitment were addressed with motor control issues specific to their recruitment. Here the emphasis is on gross motion of flexion in an attempt to improve tolerance and performance. Isometric stabilization exercises are performed at a later stage, when the emphasis is only on improving motor performance.

Whether initiating motion caudally or cranially, the goal is to perform segmental flexion through all lumbar segments.

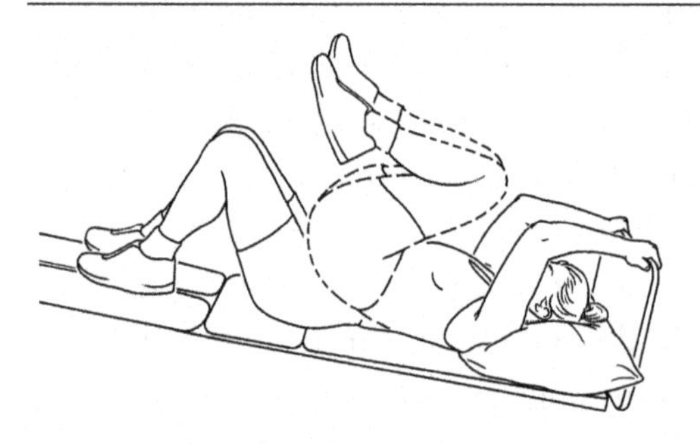

Figure 5.64: Double knees to chest on a decline board. With the knees and feet together, the hips are flexed with a smooth transition into lumbar flexion. Arm fixation will recruit the latissimus muscles, which will improve the facilitation of the abdominal muscles. The eccentric return also focuses on controlled segmental motion of the lumbar spine. The angle of the board is increased until the motion can be performed correctly for the desired number of repetitions.

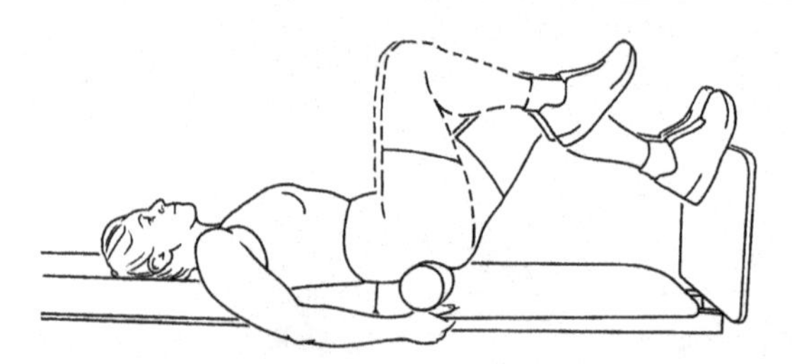

Figure 5.65: Double knees to chest on a decline board, with sacral counter-nutation bias. A roll is positioned under the distal sacrum to bias lumbar flexion. This can be effective for a deep lordotic curve and/or significant limited flexion mobility. The exercise emphasis joint motion, rather than muscle performance. The exercise is performed as above, flexing the hips with a smooth transition into lumbar flexion. A block of the feet also limits extension range and torque during the eccentric return.

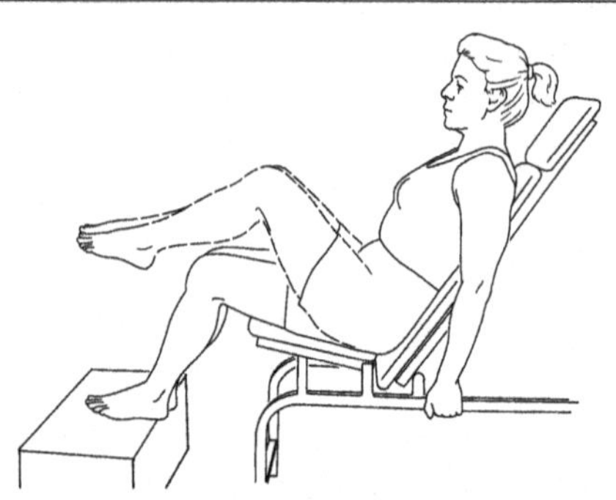

Figure 5.66: Basic progression of caudal flexion with emphasis on hip flexor training. The patient sits with a slight incline, feet supported on a bolster. The hips are flexed in unison, held for one second and then the feet are returned to the bolster. Fixing the hands on the bench with improve contraction of the abdoinals by increasing stiffness of the thoracolumar fascia through the latissimus muscles.

Spinal compression increases with activation of the psoas during hip flexion torque. Pathologies in which compression with flexion is a precaution, such as advanced osteoporosis, can still be trained with reduced forces in a non-weight bearing posture. Caudal flexion supine with bilateral hip flexion focuses on a smooth transition between the hip flexors and lower abdominals. Motion should not stop at the end of hip flexion range nor should there be a pause before transitioning to flexion through the lumbar spine. A supine position may still be too heavy a resistance to perform a coordinated pattern. Performing hip flexion and caudal trunk flexion in a declined position may be necessary to reduce load to the muscles and compression to the spine. Progression through later stages would involve moving toward horizontal and eventual incline positions. Standing hip flexion training may also be performed unilaterally with cuff weights or pulleys strapped to the ankle.

Side Bending Training

Initial motions of training are determined by patient tolerance. Side bending may be an available motion to train. The lumbar multifidi and quadratus lumborum work more arthrokinematically, while the erector spinae, abdominals and latissimus perform the osteokinematic trunk motion. Side bending caudally may be more effective for mobilizing restricted motion, as motion occurs directly at the limited segments. For intraforaminal or compression pathologies, motion can be controlled from neutral to the contralateral side to avoid tissue provocation.

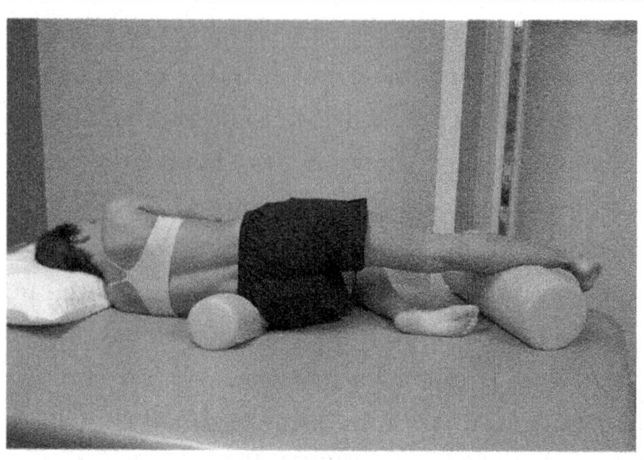

Figure 5.67: Caudal side bending over a roll. A bolster is placed under the lumbar spine to create a side bending.

A hip hike is performed from below. The top leg can be supported on a rolling bolster to support the leg in line with the trunk and remove friction from dragging the leg. To increase the resistance for side bending, a pulley can be added at the ankle with a horizontal force moment.

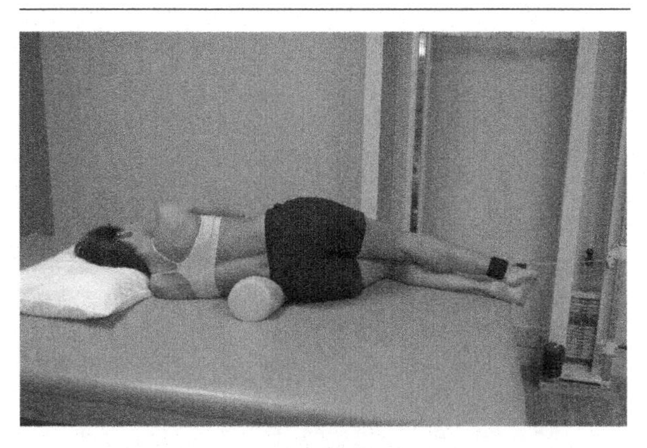

Figure 5.68: Resisted lumbar caudal side bending in side lying. The exercise is performed as previously described with the addition of a resistance from a wall pulley at the ankle.

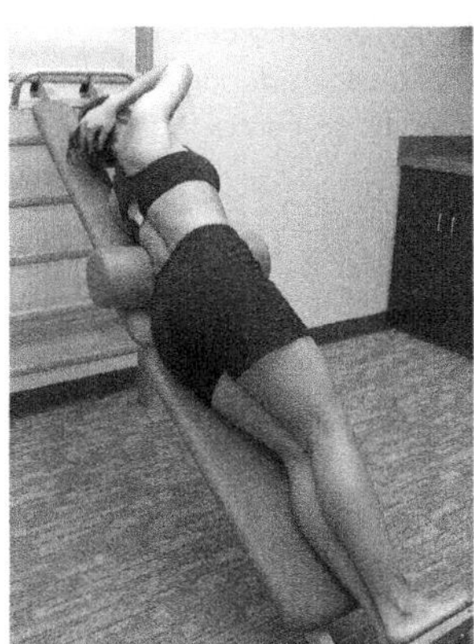

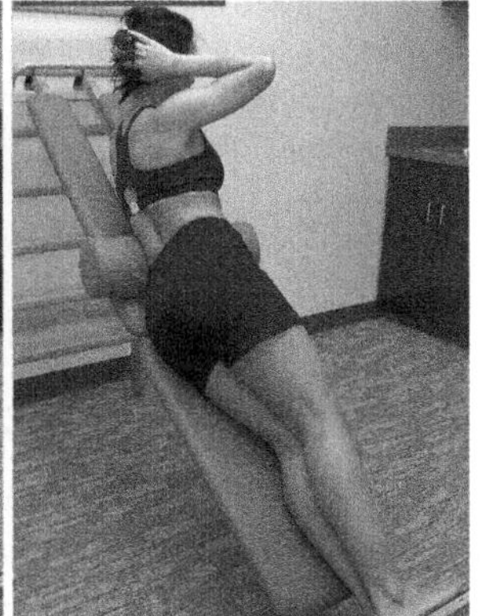

Figure 5.69a,b: Cranial side bending on a slant board, over a roll. Cranial side bending can also be performed over a bolster. A slant board can be used to reduce the overall load to allow for high repetition motion.

Rotation Training

Rotation is a key motion to train for motor recruitment and tissue tolerance. Rotation is more often the least painful motion in lumbar pathology (Faugli/Holten 1996), often making it one of the easiest directions to start training. With horizontal positioning, rotary motions can safely provide the optimal stimulus for tissue repair to all segmental tissues of the spine. Though the multifidus muscle is a primary extensor, it is the key arthrokinematic stabilizer during rotary motion. The horizontal fibers of the transverse abdominis are naturally facilitated in a stabilizing function for rotary motion and would be recruited in the normal spine prior to movement as a pre-movement stabilizing contraction (Hodges and Richardson 1997).

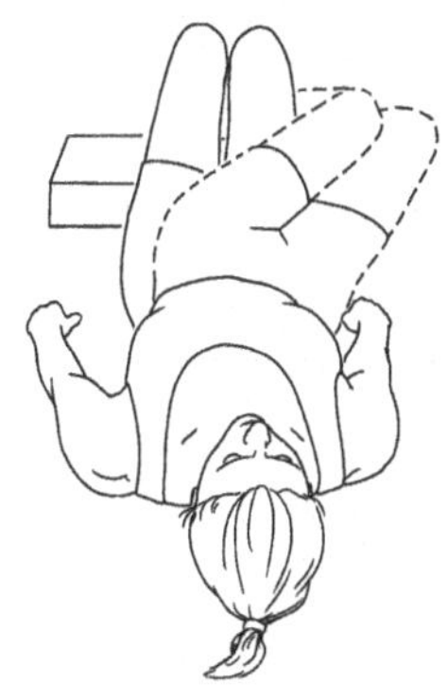

Figure 5.70: Hip Roll—Caudal supine left rotation. The patient is supine with the hips flexed to 45° and the knees at 90°. A bolster is placed under the feet, if necessary, to place the lumbar spine in neutral or slight flexion. The knees are eccentrically controlled out to the side with a concentric return to neutral. Emphasis is placed on the concentric return initiated from the lower abdominals and lumbar spine, not the hip and thigh muscles. The patient is instructed to breath out while bracing or hollowing the lumbar spine prior to initiating motion of the lower limbs. As little rotary mobility is available at each lumbar segment, the knees do not need to move far to the side to effectively train tissue and motor performance. Beginning with a "10 and 2" position on a clock is a good initial guideline, progressing in range as tolerated and/or necessary.

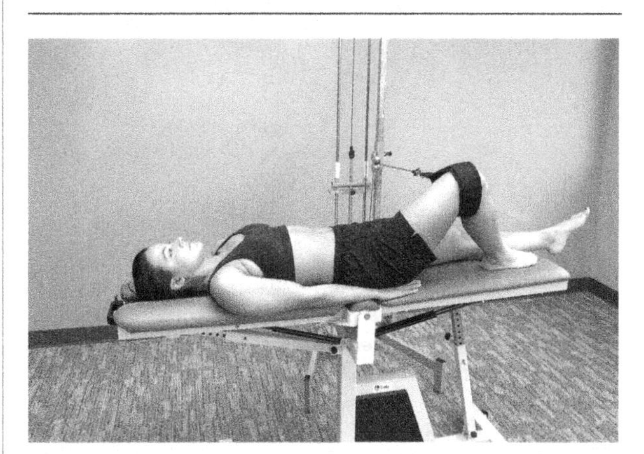

Figure 5.71: Supine hip abduction (horizontal abduction). The patient is in supine with resistance placed around the knee. The patient performs hip abduction, with emphasis on the gluteus medius. Increasing the hip flexion will reduce the contribution of the tensor fascia lata. The lumbopelvis juncture can be allowed to rotation, for a mobilizing force, or be held still for a stabilizing for at the lumbosacral juncture.

Sidelying cranial rotation is a commonly used early form of rotation training, as the position can easily be adjusted for safety and comfort. The locking section of this chapter offered several examples for protecting tissues by using bolsters and coupled forces to allow early and safe training. Cranial rotation in side lying will emphasize the multifidus on the bottom side, though both sides are always working in synergy. Rotation is not pure, as the axis for motion is no longer the central spine, but the contact of the body on the table. Progressing to sitting and standing postures will allow for a more pure rotational motion and motor recruitment.

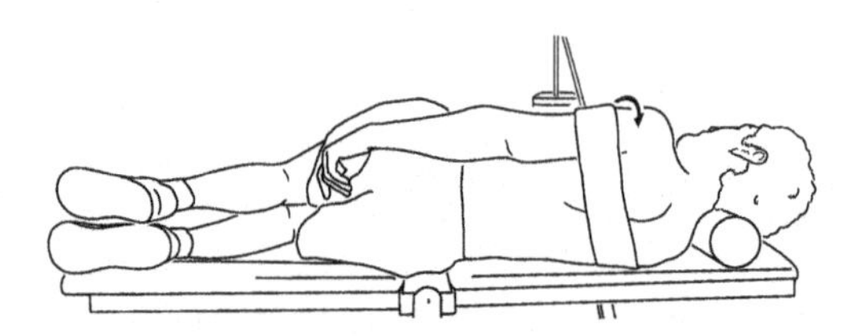

Figure 5.72: Sidelying cranial rotation with anterior resistance. Left rotation (pictured) will emphasize the right side multifidus. The hips and knees are flexed to stabilize the position and should not move during the exercise. The patient lies on the strap to fix one end, while the other end is attached to the pulley. To avoid shoulder and/or cervical irritation, the strap can be placed directly over the trunk, under the arm. A free weight can be used, rather than a pulley, but the muscle recruitment changes to the opposite side after passing vertical.

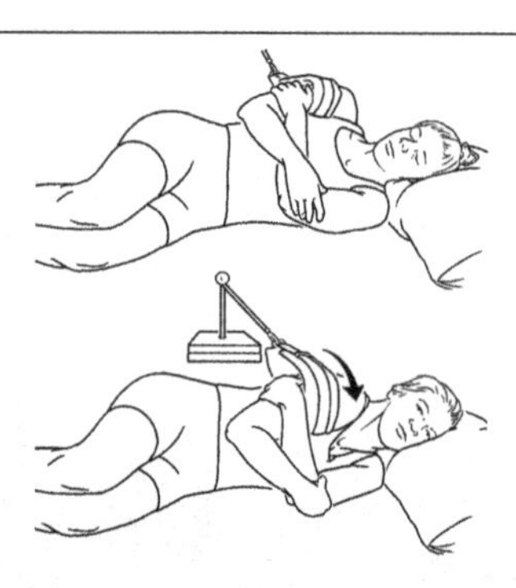

Figure 5.73: Sidelying cranial rotation with anterior trunk muscle emphasis. The exercise is similar to the previous one, with the resistance changed to the posterior side. This will emphasize the anterior muscles, as well as joint mobilization on the eccentric phase.

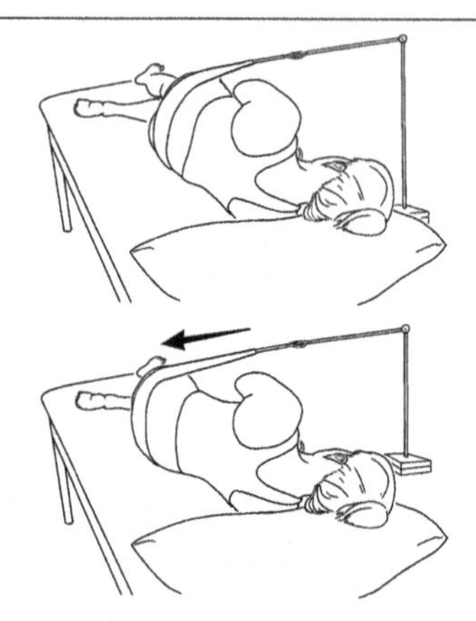

Figure 5.74: Sidelying caudal rotation with anterior resistance. Caudal rotation emphasizes direct motion in the lumbar region, with greater recruitment of the left side multifidus, with relative right rotation (pictured – motion listed for cranial segment). The top leg remains extended to form an axis of motion, while the bottom leg is flexed for stability. Facilitation of the lower abdominals (transverse abdominis) is enhanced with a horizontal resistance in the fiber direction, and with the initial bracing prior to initiating the motion.

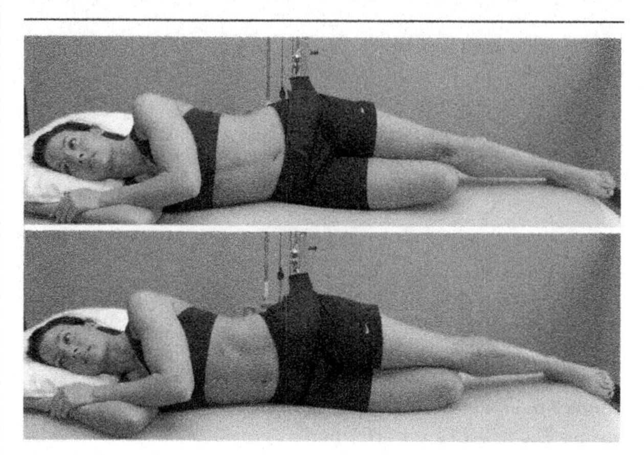

Figure 5.75a,b: Sidelying caudal rotation with posterior resistance. The exercise is performed as previously shown with resistance anterior. A posterior force moment will further emphasize the anterior abdominal muscles during caudal rotation, but also creates a synergistic co-contraction of the lumbar multifidi.

Local Muscle Endurance

Hyperactivity of paraspinal muscles in chronic low back pain has been attributed to reflex-spasm response to pain (Collins et al. 1982); however, support for this theory is considered somewhat ambiguous (van Dieen et al. 2003). Marras et al. (2001) has shown a greater compressive load on the spine associated with higher levels of paraspinal muscle activity. Excessive spinal loading has been suggested to impede tissue self-repair (Kumar 1990) and decreasing load tolerance over time (Brinckmann et al. 1988, Callaghan and McGill 2001). Elevated resting paraspinal muscle activity shown in individuals with chronic low back pain is one such mechanism of increasing load, prolonging tissue recovery (Chiou et al. 1999, Ambroz et al. 2000, Lariviere et al. 2000, Marras et al. 2001). Healey et al. (2005) demonstrated that subjects with elevated paraspinal muscle activity from chronic low back pain had increased compression on the intervertebral disks and diminished ability to recover the height lost through loaded exercise. Rodacki et al. (2003) found similar results in pregnant women with chronic low back pain; the control group recovered 93.8 percent of stature that was lost during exercise, whereas the chronic low back pain group recovered only 54.4 percent. Resolving muscle guarding is considered an important step to normalize tissue loading, allowing for repair processes, but also for reducing symptoms, improving mobility and normalizing motor patterns. This can be achieved by dosing exercises specifically to improve local circulation and endurance. Endurance training of lumbar extensor muscles has been shown to reduce symptoms and improve function in the initial three weeks of training (Chok et al. 1999).

Pain signals for any lumbar pathology result in an increase in sympathetic efferent response, causing vasoconstriction, reducing microcirculation to tonic muscles in guarding. These tonic muscles rely on oxygen as a primary fuel source. Ischemia associated with chronic muscle guarding is best treated with active exercise dosed to increase local circulation, while not producing additional energy depletion. Too low of a resistance does not require an increase in local circulation, while too high a resistance increases intramuscular pressure reducing circulation. Between the two extremes is the range of training to maximize local circulation. A reduction in the blood flow starts as early as 30 percent of MVC and by 70 percent of MVC the blood flow is likely entirely cut off, although quadriceps circulation has been cutoff at only 20 percent of MVC (Edwards 1972). Maintaining blood flow is necessary for muscle protein synthesis (Biolo et al. 1995). In the presence of chronic muscle guarding active training is performed to the tonic muscles in guarding dosed in a range from 55–65% of 1RM to improve local circulation.

Pure concentric training, with removal of the eccentric phase, is effective in improving circulation. Concentric exercise has been shown to increase capillary density not associated with eccentric work (Hather et al. 1991). Performing concentric and eccentric work together with the muscle constantly loaded maintains an intramuscular pressure that results in a reduction in blood flow. The use of an eccentric stop wheel on a wall pulley with early training for vascularity can improve the efficiency of training. If an eccentric stop wheel is not available, allowing the weight stack to come to rest with each repetition, and the muscle to fully relax, will improve local blood flow through the muscle during training. In severe guarding, resting one to two seconds between each repetition will be more effective in oxygen restoration.

Figure 5.76: The eccentric-stop wheel attachment for the wall pulley spins for the concentric phase only. During the eccentric phase the wheel is blocked from turning, creating a friction to lower the weight and remove the need for muscle contraction during the weight lower phase.

The lumbar multifidi are typically the primary tonic muscle in guarding with acute lumbar pathology. The muscle can be selectively trained for local vascularity to increase local circulation and provide oxygen for fuel for recovery. For example, the right multifidi performs extension, right side bending and left rotation. All three of these motions can be performed caudally and/or cranially. All of these motions can be trained with any number of exercise options with the dosage set for vascularity (60% of 1RM for three sets of 24 repetitions). Other commonly guarded muscles such as the piriformis and the psoas can be addressed with a similar approach for vascularity to provide oxygen and normalize resting tone. Specific exercises chosen may relate to the least painful planes of motion, safest for participating tissues, easiest to coordinate, easiest to reproduce at home or related to any number of contraindications for motion in certain planes. With proper exercise design and

Endurance Training General Guidelines	Direction of Training	Resistance	Sets and Repetitions	Rest Intervals	Speed	Frequency
Local Muscle Endurance	Break down pattern into simple uni-planar motions progression to tri-planar motion.	60% of 1RM	1 set of 30 repetitions or 3 sets of 24 repetitions	Recovery of respiration rate If no change in respiration occurs, rest up to 1 minute	Slow to moderate matching respiration rate	1x daily 5x/weekly

Table 5.4: Dosage to emphasize endurance/local muscle vascularity.

dosage muscle guarding associated with ischemia can be resolved within the treatment session resulting in a significant improvement in mobility and reduction in symptoms. As muscle guarding is resolved the neurological inhibition of the opposing antagonistic muscles reduces which may result in a significant improvement in their performance for early coordination training. For examples, resolving muscle guarding in the multifidi, a primary extensor muscle, may improve weakness or poor coordination in the antagonistic lower abdominals without direct training of the anterior muscles.

Respiration

Normalizing respiratory patterns is a fundamental component of lumbar spine rehabilitation. Breathing patterns have an impact on oxygenation for reducing pain levels, energy during performance and recovery. With attachment to the spine, the diaphragm also contributes to the dynamic stability of the lumbar spine. The trunk has three functional diaphragms: the glottis, the diaphragm muscle and the pelvic floor. All three work in synergy as a co-contraction during isometric stability of the spine. With normal respiration, the diaphragm contracts downward, creating a negative pressure to fill the lungs. The pelvic floor is reflexively inhibited during contraction of the diaphragm. During high level sport activity, the diaphragm may be working with increased intensity and frequency, but the pelvic floor may also need to contract at a high level for spinal stability. These opposing concepts are easily coordinated in the normal spine, but can breakdown in the pathological state.

Initial training may simply be to emphasize diaphragmatic respiration, without accessory muscles of respiration contributing. Though thought of as a simple task, it is often impaired in the chronic back pain patient. Proper breathing patterns should be established prior to more complicated coordinative, endurance or strengthening exercises to provide proper oxygen and to avoid reinforcing abnormal patterns of respiration. When a patient is attempting to learn a new exercise with a coordination challenge, and mental challenge, the first compensation is to stop breathing during this learning process. Cueing the patient to breath is frequently necessary during the acquisition phase of a new skill.

Figure 5.77a,b: Diaphragmatic respiratory training. In supine, the patient places one hand on the lower abdomen and one hand on the sternum. The patient is instructed to feel the abdominal hand rise during inspiration, without motion of the sternum. Breathing through the nose during inspiration and out the mouth during exhalation may enhance the performance and learning process.

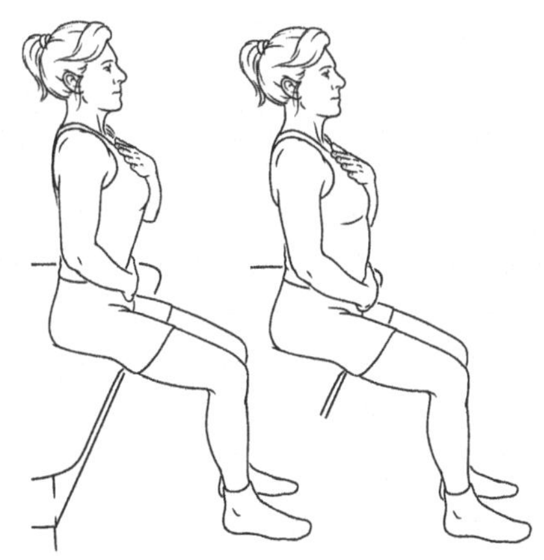

Figure 5.78a,b: Diaphragmatic respiratory training in sitting. Reinforcing the diaphragmatic breathing pattern should take place in different postures, to transition the normal breathing pattern to other functional tasks and exercises.

Progressing breathing concepts into training exercises is typically taught as performing exhalation during concentric work (lifting) and inspiration during eccentric work (lowering). This approach is appropriate for most early exercises. When training the trunk, however, the breathing pattern may be altered to follow the biomechanical motions of the trunk. Inspiration is performed during extension patterns of the thoracic and lumbar spines with exhalation during flexion patterns.

Balance

Balance training for lumbar and cervical patients is not common practice despite balance deficits being a common finding (Nies and Sinnott 1991). Several studies have reported reduced balance control with chronic low back pain (Alexander and LaPier 1988, Nies and Sinnott 1991). When a task involves increased complexity and removal of visual information, low back pain patients demonstrate increased postural sway compared to pain free controls (Mientjes and Frank 1999). Failure rate in balance testing has been shown to be four times that of a normal person (Mok et al. 2004). A visual dependency with reduced hip strategies results in reduced proprioception and postural control.

Balance deficits have also been reported in patients with peripheral neuropathy (Ducic et al. 2004). This may include diabetic neuropathy as a complicating factor in back patients with diabetes. Nerve root pathologies would have obvious influence on afferent signals affecting balance. Somatosensory deficits due to tissue and receptor damage may play a less obvious role in balance deficit in the presence of normal neurological conduction of lumbar nerve roots.

Assessment of lumbar and cervical patients should begin without shoes on a flat stable surface. Tissue injury in the spine or periphery can affect posture due to abnormal signals from the somatosensory system. Mechanoreceptor damage will alter afferent input affecting normal motor recruitment. Conflicting afferent input may be created in the presence of pain and muscle guarding. Mechanoreceptors in the joint capsule at rest report an absence of joint motion while muscle spindles, in the state of acute or chronic muscle spasm, report tension or motion. This confusion can manifest in many ways: altered movement patterns, altered postural tone, altered kinesthesia, vertigo and loss of balance or dizziness.

Initial Dosage for Balance Training

- *Balance—30 seconds to two minutes per exercise.*
- *Greater than five minutes of training on to three times daily.*

Balance improvements in chronic low back pain can be enhanced with specific and customized exercise programs (Kuukkanen and Malkia 2000). A progression of balance testing for the spine is performed with assessment of trunk sway and perceived stability of the patient. Cervical movements during the testing, even for the lumbar patient, assess the ability to integrate postural reflexes associated with somatosensory system of the cervical spine. Identifying specific directions of cervical movements that create balance deficits can assist in determining retraining exercises.

Patients with chronic low back pain have been shown to have impaired psychomotor speed and, among females, impaired postural control (Luoto et al. 1996). Postural control reaction time improved without specifically training it. This implies that LBP caused not only peripheral problems but also impaired functioning of the central nervous system (feedforward mechanisms). A correlation between proprioceptive dysfunction and LBP was also established. A slight difference was found between the groups in the slow corrections of posture, suggesting that the central nervous system's processing of the responses was somewhat different in the patients with LBP (Luoto et al. 1998). The differences found in this study suggested that the differences in postural corrections might be non-cortical. Impaired postural stability is considered one factor in the multi-factorial problem of chronic back pain.

Suggested Progression of Balance Assessment for the Spine:

- *Bilateral stance with the eyes closed*
- *Bilateral stance with the eyes closed with cervical rotation, flexion and extension*
- *Unilateral stance with eyes open*
- *Unilateral stance with eyes closed*
- *Unilateral stance with the eyes closed with cervical rotation, flexion and extension*
- *Progress to foam and balance boards for advanced challenges, if functionally necessary or if vestibular deficits are identified.*

Balance training can begin with simple bilateral and unilateral stance. Closing the eyes will remove visual feedback, with emphasis on the vestibular and somatosensory systems. Balance foam reduces somatosensory input, emphasizing vestibular function, as well as visual if the eyes are open. The use of wobble boards or wobble shoes more effectively challenges the somatosensory system, than the balance foam. Later stage progressions in balance will also involve upper and lower limb reach activities. This emphasizes a change in center of gravity with a return to neutral, rather than labile a static posture on a labile surface.

Adjunct Training

Treatment should focus on the person, not the diagnosis. Intervention addressing only the region of pain may miss important impairments in other regions that are a part of the overall pathology. The biomechanical model addresses joint mobility and range of motion in all joints in the biomechanical chain. This involves the sacroiliac joints, entire lower limb, thoracic and cervical function. Restricted hip mobility in extension places in increased torque at the lumbosacral juncture during many functional motions and patterns. Loss of mobility of the thoracic spine in extension and/or a forward head position of the spine can alter overall posture of the lumbar spine. Many causes for reduced mobility exist, with many potential areas distal to the lumbar spine. Evaluation and passive treatments may address these, but exercise may also need to directly address some of these impairments. The other chapters in this text offer mobilizing exercises and concepts for these regions.

The neurophysiological model addresses movement patterns, with muscles not only crossing the lumbar spine, but affecting the trunk or lower limbs. Inhibition of normal movement patterns leads to compensation in posture, movement and performance. Central balance training is also a component of this model. Normalizing balance and stability of the lower limb with heel contact during gait and/or running can significantly reduce the impact load to the lumbar spine. For the manual therapist, the hip is part of the functional unit of the pelvis and lumbar spine. Addressing motor deficits in the lower limb, with emphasis on the hip, may be a primary need for the back patient.

Myofascial Stretching Techniques

Addressing limited mobility of muscle or fascia in Stage 1 focuses on passive manual therapy techniques. Stretching exercises are not commonly prescribed in the presence of pain and muscle guarding. The facilitated segment model teaches us that a painful segment will elevate the resting tone of muscles innervated by that level. Pathologies affecting the L4, L5 and S1 nerve roots result in an elevation of the resting tone of the hamstrings, or a reduced ability to lengthen. This is considered a symptom, not a primary cause of low back pain. Fisk (1979) demonstrated a single lumbar spinal manipulation resulted in an immediate significant increase in hamstring length, without any direct stretching to the muscle. Stage 1 addresses the reasons for pain, which in turn will reduce the neural drive for tone, improving functional hamstring length. More specific stretching concepts are applied later in the exercise progression.

Home Exercise

Establishing a home exercise program is necessary to achieve early training goals. Selective tissue training, pain inhibition and coordination all require hundreds of repetitions, preferably multiple times daily. In later stages, frequency and duration of training may decrease, but adherence to a set program is still necessary to achieve meaningful long term outcomes.

Providing specific instructions is necessary in teaching home exercise, but more detailed information does not necessarily improve adherence (Raynor 1998). Lack of time on the part of the patient is one of the most common reasons for lack of adherence (Sluijs et al. 1993, Dean et al. 2005). It is important for patients to identify time to carry out the prescribed exercise, but physical therapists must recognize that they have a role to play in helping patients manage their time (Dean et al. 2005). Working with the patient to identify their life priorities can assist the therapist in designing individually tailored programs in a way that does not interfere with these priorities. Exercises programs can be designed around work schedules, childcare issues, a current fitness program or built into daily activities. Creating rapport is also critical for establishing a therapeutic relationship that promotes adherence to home exercise programs (Dean et al. 2005).

Friedrich et al. (2005) determined that motivational strategies to improve adherence to home exercise does improve disability outcomes for chronic low back pain, compared to home exercise without motivational approaches. Techniques used included providing clear instructions, emphasizing the importance of regular and consistent exercise for reducing the pain and likelihood of recurrent episodes, enhancing each patient's internal locus of control, providing positive feedback, reward and punishment strategies, written contract, posting written contract at home and maintaining a daily exercise log.

Patients that are expected to independently adjust equipment and perform exercises during supervised sessions demonstrate improved adherence to long-term home exercise programs (Hartigan et al. 2000). Requiring independence in the clinical setting provides patients with an opportunity to practice and master exercise techniques.

Section 2: Stage 2 Exercise Progression for the Lumbar Spine

Stage 2 progressions are determined by improvement in functional qualities being trained in Stage 1. Often there is an overlap between exercises addressing both Stage 1 and 2 concepts within one training session. Indications for progression include a decrease in pain, improved joint mobility, improved range of motion, resolving muscle guarding and improvement in coordination or motor learning. Improvements in coordination may be demonstrated with an increased speed with a given exercise. The basic goals for Stage 2 are to stabilize the gained range of motion, improve muscle endurance, develop faster coordination, progress the functional body position with each exercise and increase the overall volume of training. A program of 8–12 exercises, of two to three sets each, is not uncommon. Speed is increased, not weight, to emphasize fast coordination and to increase the relative resistance of the exercises. The exercises from Stage 1 may be carried over, but the desired functional qualities will shift with these changes. Isometric contractions within the mid to inner range of motion are also introduced to help fix strength. The following are examples of Stage 2 exercises dosed for specific functional qualities.

Tissue and Functional States: Stage 2

- *Decreased active range of motion in tri-planar motions and end ranges*
- *Pain free at rest but painful with motion*
- *Partial weight bearing/loaded postures are tolerated*
- *Fair coordination in basic movement patterns*
- *Fair balance/functional status*
- *Palpation of involved tissues and pressure sensitivity of guarded muscles only moderately painful*

Basic Training Goals: Stage 2

- *Increased repetitions with increased exercises (5 to 10 exercises).*
- *Increase repetitions with additional sets (endurance).*
- *Increase speed (strength/endurance).*
- *Combined concentric and eccentric work for further tissue tension accommodation.*
- *Body/limb position changes from recumbent to more dependent.*
- *Planar motions with exercise to full range/ partial range tri-planar.*
- *Remove locking or change to less aggressive types of locking.*
- *Histological influence of increased lubrication with increased speed.*

Increase Repetitions

The most basic progression to an exercise program is the addition of more repetitions to the program to further increase tissue training qualities and to further enhance performance. Tissue repair and coordination are a function of high repetitions in training. Adding sets and exercises to the program increases the overall number of repetitions. Initial training directions are matched to the patient's tolerance. The next progression is to then begin training in the primary direction of pain, but not into pain. As the overall time of training increases, appropriate dosage and rest breaks are necessary to avoid excessive fatigue and to allow enough reserve energy for recovery and tissue repair.

Increase Speed / Not Weight

Frequently in physical therapy practice, increasing the weight with an exercise is often the first approach toward progressing an exercise. As coordination improves speed is naturally increased. This increase in speed creates greater resistance through inertia, making it unnecessary to increase the weight. Obtaining full, pain free range of motion should be the first emphasis, prior to increasing the level of difficulty. This may also include achieving full range in more challenging positions. Flexion with supine training is not the same performance as in standing. Once full range has been achieved, increasing speed of training, not increasing the resistance, creates an increased level of challenge. The addition of speed is the addition of inertia. Inertia, with acceleration and deceleration, is a force, and is therefore an added resistance. As an example, running requires a greater level of work than walking, though the weight of the body is the same for both performances. Speed will also further challenge coordination, which is speed specific. Initial training may have been below a functional level, with the increase in speed reflecting more of a functional demand.

Overall changes in body position, number of sets and repetitions, along with changes in speed provide a more aggressive training program. Each training session should involve a progression of at least one variable in at least one exercise. Progression is not from one list of exercise for Stage 1 to a second list of exercise for Stage 2. Each exercise is progressed as improvements are attained. Even acute patients with a low training state will benefit from a high volume, low intensity program. Comparing patients with moderate or hard physical occupations to those with sedentary/light job functions, Hansen et al. (1993) found that those with sedentary/light job functions did better with intensive back exercises.

Isometric Work

For every degree of mobility created, dynamic stabilization is necessary. Cortical drive (feedforward) and proprioceptive afferent input (feedback) establish normal movement patterns for stability. With afferent receptor damage associated with pathology, afferent signals are altered along with normal recruitment patterns. The muscle spindle provides afferent feedback that can help substitute for damaged joint mechanoreceptors. Biasing the muscle spindle through progressive training is a part of establishing dynamic stabilization with enhanced motor control.

Isometric work (IW) is associated with a higher level of resistance to improve sensitivity to stretch in the muscle spindles. Oxygen compromise with IW is no longer a concern. A higher peak torque is noted with IW than CW, allowing greater resistance for training. Joint motion does not occur with IW, making heavier resistance safe, which might otherwise not be coordinated with CW. The tonic muscle system associated with arthrokinematic control is still emphasized, using heavier weights with hold times for at least six to ten seconds with heavy resistance and up to 30 seconds with lighter.

The joint is initially positioned in mid range, where some level of structural stability exists. In later stages the joint will be positioned closer to the pathological range. More aggressive approaches to this utilization of IW are necessary in the hypermobile joints, while the hypomobile joint may achieve dynamic stability more rapidly due to less tissue and receptor damage. Isometric training is also associated with fixation of strength. Performing an isometric hold assists in creating a neurological fixation of a motor pattern, or specific muscle recruitment. As range of motion is gained, isometric hold times in the new range can fix muscle strength in this new range, improving stabilization more quickly than performing only high repetitions as lighter loads.

The application of IW may involve replacing one of the three dynamic sets with an isometric set. The second set is often replaced with one isometric set at a higher resistance, followed by the third dynamic set at the original lower resistance. Only one to three holds are performed with emphasis on neurological adaptation, not structural changes in muscle tissue.

Dosage for Isometric Work

- *Isometric hold in mid range as second set between first and third dynamic sets, 80% of 1RM 6–10 second hold.*
- *Two to five isometric holds can be performed toward pathological range*

More classic examples of back exercises involving co-contraction of trunk muscles for isometric stabilization are added in Stage 2. The trunk is stabilized against gravity, external resistance or during extremity movements. These exercises do not address the primary functional need of the patient as defined by Stage 1 criteria related to tissue repair, mobility, resolving muscle guarding and coordination. The focus is more on heavier load co-activation with isometric holding. Resistance is often high, with hold times of only 10 seconds. This type of dosage is more consistent with strength training, as opposed to coordination or endurance training, which are necessary qualities to attain prior to strength training.

Early dosage for strength training may account for some of the studies assessing exercise for acute low back pain having minimal positive effects. Strength qualities in larger phasic muscles does not address recruitment and coordination of tonic, arthrokinematic, muscle. Pure isometric exercises with higher loads are of more significant benefit in Stage 2, providing many options for tonic and phasic muscles to work together with higher loads.

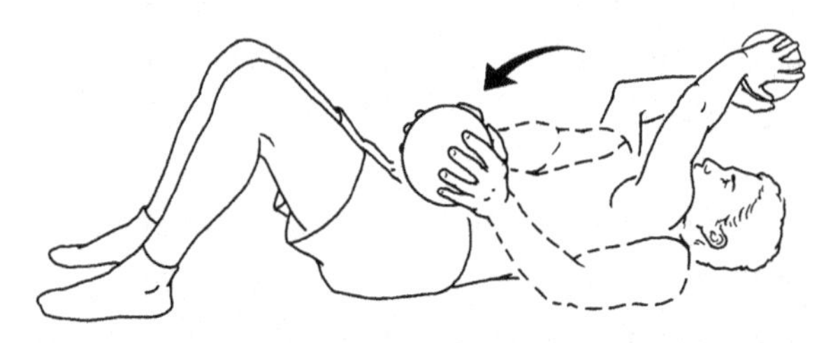

Figure 5.79: Isometric stabilization with supine overhead arm movements. The lumbar spine is stabilized in neutral while the upper extremities reach straight overhead or in diagonals. Resistance can be with a free weight, medicine ball, pulley or elastic resistance.

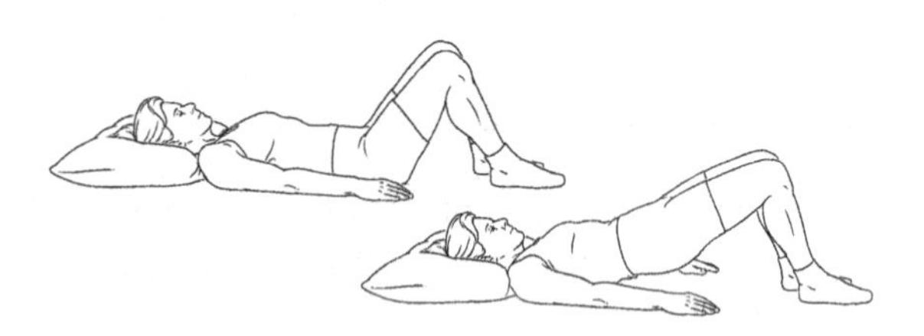

Figure 5.80a,b: Lumbar stabilization with bridging. The supine bridge exercise is more effective at lumbar muscle recruitment than the gluteal muscles (Kasman 1998). The bride can be modified with holding isometrically for time, bracing while lifting the heels, sliding the heels or performing leg lifts.

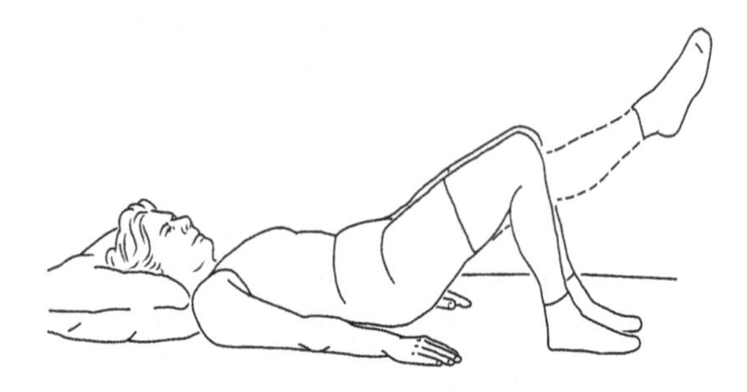

Figure 5.81a,b: Lumbar stabilization while performing leg lifts. The bridge is performed with the lumbar spine held in a neutral position. The patient lifts one heel off the floor several inches without allowing the pelvis to drop or rotate. If this is can be coordinated the heel is lifted and the is extended to increase the challenge for stabilization of the lumbar spine and pelvic girdle.

Bridging exercises can be progressed to involve distal fixation with proximal motion and stabilization. The feet can initially be placed on a fixed surface, such as a step or bolster. A further challenge is to place the feet in a sling, requiring a greater level of co-activation to stabilize the proximal trunk. Initial exercises may support the lower limbs several centimeters off the floor, with a gradually progression to higher levels.

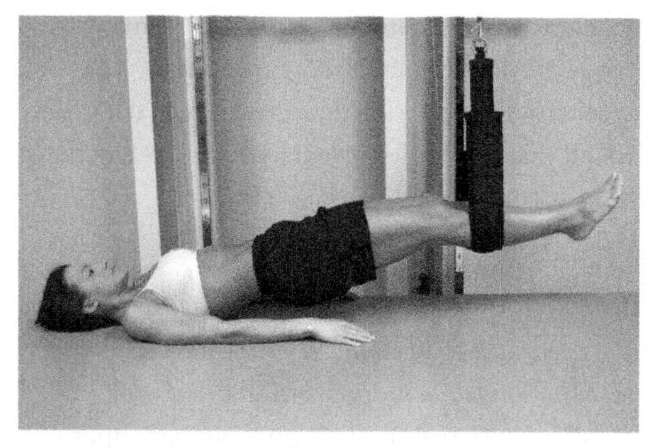

Figure 5.82a,b: Sling bridging provides a distal fixation with proximal stabilization (Stuge and Vollestad 2007). A sling supports both legs with the trunk resting on floor. The legs can be positioned as little as one inch off the floor, with increasing heights to increase the challenge. The patient performs a lumbar bridging exercise to a neutral position, holding isometrically.

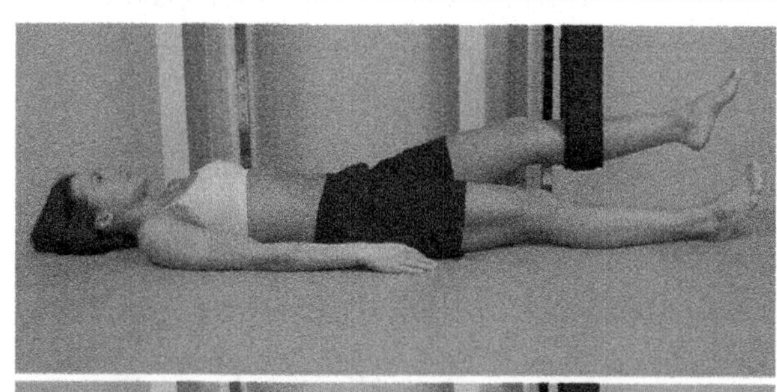

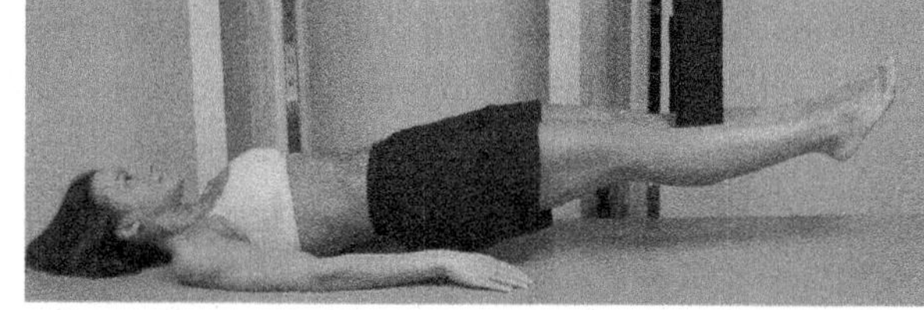

Figure 5.83a,b: Unilateral bridging with a sling. One leg is supported in a sling. The patient flexes the opposite hip until parallel with the sling leg and then performs a lumbar bridge (Stuge and Vollestad 2007).

Figure 5.84: Lateral plank with increased challenge of the weight bearing arm being fully extended. If the patient is able to hold this position, the opposite arm is raised toward the ceiling.

Figure 5.85: Bridging on a fitness ball, knees flexed. The upper thoracic spine and shoulders are supported on a fitness ball with the knees flexed. The pelvis is elevated to take the lumbar spine from flexion to neutral. The exercise emphasizes the lumbar extensor muscles with secondary influence on the hip extensors. In this position the patient can also be instructed to lift one heel off the ground while maintaining a neutral lumbar spine and avoiding rotation of the pelvis.

Figure 5.86: Bridging on a fitness ball, knees extended. The lower legs and ankles are supported on a fitness ball with the neck, shoulder and upper thoracic spine on the floor. The pelvis is elevated to take the lumbar spine from flexion to neutral. The exercise emphasizes the lumbar extensor muscles with secondary influence on the hamstrings and gluteal muscles.

Figure 5.87: Kneeling stabilization with upper limb support on a fitness ball. The forearms are positioned on a fitness ball, while in an upright kneeling position. Weight is transferred onto the ball rolling it forward. The patient leans forward maintaining a neutral lumbar posture while maintaining normal respiration.

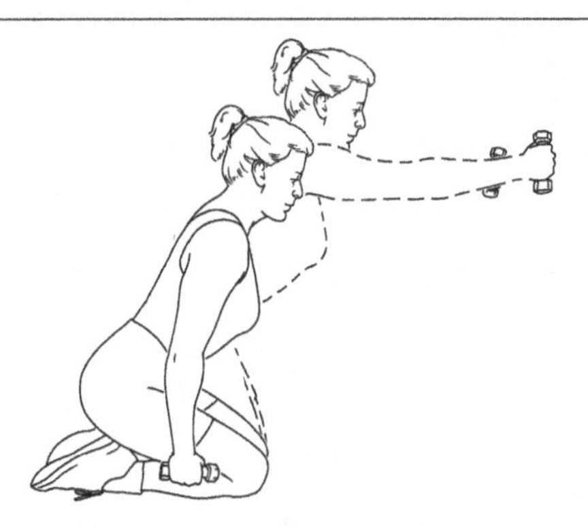

Figure 5.88: Forward kneel lumbar stabilization. From a sit-kneel position the trunk is flexed forward from the hips while maintaining a neutral lumbar spine. The arms are then flexed forward to 90°and the knees are slowly extended. If the psoas muscle cannot fully lengthen, an increase in lumbar lordosis occurs prior to reaching an upright kneeling position. Motion is stopped when neutral cannot be maintained.

Figure 5.89a,b: Kneeling stabilization with upper limb slings. Two lines from the same sling support the arms. The patient begins in an upright kneeling position. Transferring the weight onto the sling, the patient leans forward, maintaining a neutral lumbar posture.

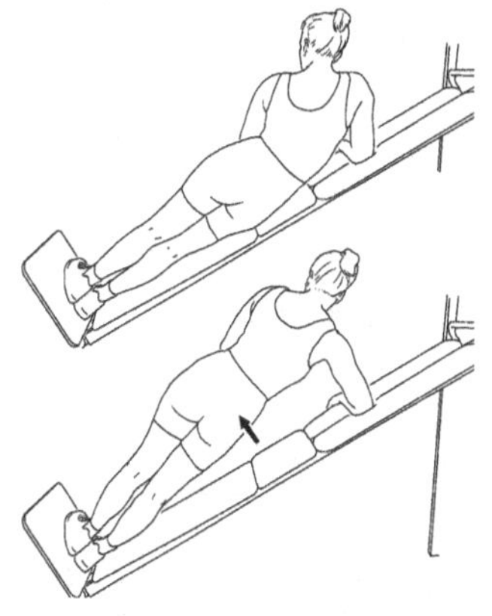

Figure 5.90a,b: Lateral bridge. Isometric stabilization of the lumbar spine is performed at an incline with the elbows supported on a wall. A slant board allows specific progressing dosing of increased resistance working toward a horizontal position in side lying. The shoulders, ribcage and pelvis are maintained in alignment.

Primary Muscle Group	Exercise	Criteria for Progression
Transversus Abdominus	Abdominal bracing	30 repetitions with 8-s hold
	Bracing with heel slides	20 repetitions per leg with 4-s hold
	Bracing with leg lifts	20 repetitions per leg with 4-s hold
	Bracing with bridging	30 repetitions with 8-s hold, then progress to 1 leg
	Bracing in standing	30 repetitions with 8-s hold
	Bracing with standing row exercise	20 repetitions per side with 6-s hold
	Bracing with walking	
Erector spinae/ Multifidus	Quadruped arm lifts with bracing	30 repetitions with 8-s hold on each side
	Quadruped leg lifts with bracing	30 repetitions with 8-s hold on each side
	Quadruped alternate arm and leg lifts with bracing	30 repetitions with 8-s hold on each side
Quadratus Lumborum	Side support with knees flexed	30 repetitions with 8-s hold on each side
	Side support with knees extended	30 repetitions with 8-s hold on each side
Oblique Abdominals	Side support with knees flexed	30 repetitions with 8-s hold on each side
	Side support with knees extended	30 repetitions with 8-s hold on each side

Table 5.5: Stabilization exercises with suggested criteria for progression, as described by Hicks (2005).

Performing abdominal training with trunk flexion exercises increases compressive forces, being too aggressive for early training in pathologies related to disc or vertebral compression. Use of the lateral bridge exercises is a safe substitute to reduce loading while providing significant recruitment of the abdominal muscles. The side bridging exercise has been shown to produce up to 100 Nm of abdominal torque with much lower spinal loading (Axler and McGill 1997).

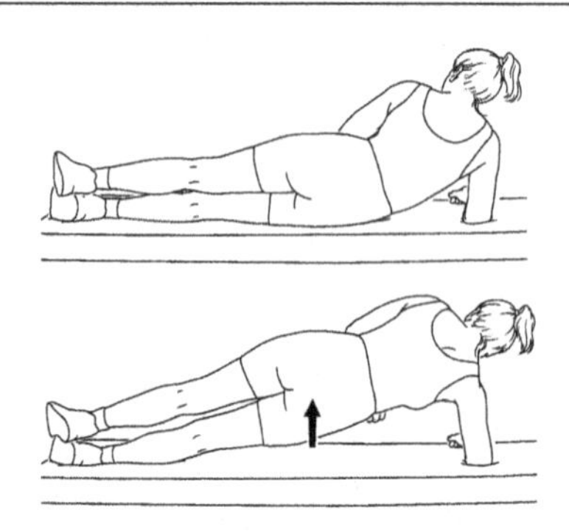

Figure 5.91a,b: Lateral bridge is progressed from an incline position to a horizontal lateral plank position. To reduce difficulty both knees can be flexed, or just the bottom leg, but the knees are kept inline with the pelvis and shoulders. Separating the feet also increases the base of support to reduce the need for rotational stabilization. To increase the difficulty the patient can horizontally rotate the top arm forward and backward. Turning the cervical spine may also increase the rotary stabilization challenge of the trunk.

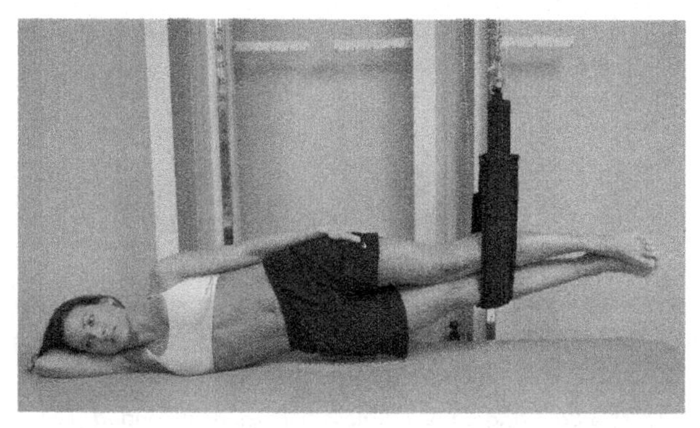

Figure 5.92: Lateral bridging with proximal sling fixation. The exercise can be performed bilaterally and unilaterally (Stuge and Vollestad 2007).

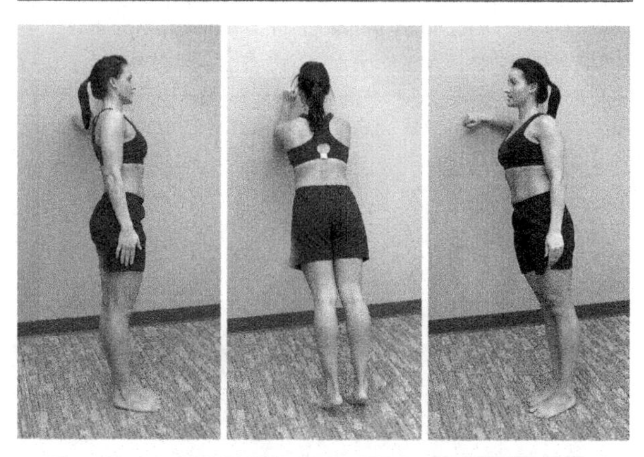

Figure 5.93a,b,c: Wall-Rolls. A progression of the plank exercises is to perform both lateral and relative prone plank positions can be performed in a rolling sequence in standing against a wall (McGill 2002). The lumbar spine is held in neutral during the change in positions.

Figure 5.94: The incline prone bridge (plank) for lumbar bracing training. The shoulders, ribcage and pelvis are maintained in alignment. The patient should be instructed to maintain normal breathing, rather than breath holding. Flexing at the hips, lifting the pelvis up, is a common compensation for weak lower abdominals. The patient is instructed to lower the pelvis in line with the shoulders and feet, unless pain is reproduced. Raising the pelvis may initially be used to assist the patient, with a gradual progression to horizontal. Incline positions using a wall, table or slant board allow for a correct position during lower training states.

Figure 5.95: Incline prone bridge on a fitness ball. The ball can provide a slight incline, to reduce the relative load compared to a prone position. The labile surface provides an additional challenge for lumbopelvic stabilization.

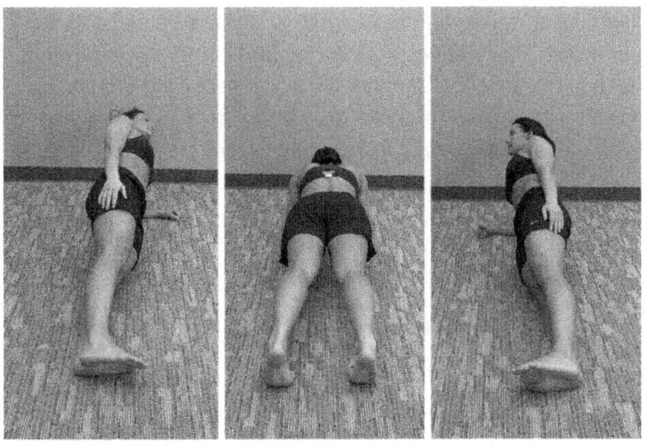

Figure 5.96a,b,c: Side bridge-rolls prone. A progression of the plank exercises is to perform both lateral and prone plank positions in a rolling sequence. The lumbar spine is held in neutral during the change in positions. The patient is instructed to lift the elbow slowly off the wall, rather than push off. The motion should be initiated from the lower trunk, rather than pushing off with the shoulder girdle.

Figure 5.97: Isometric stabilization training of the trunk with overhead quick press. The patient stands with the lumbar spine in neutral and slightly flexed knees. A bar is pressed overhead with quick motions. Emphasis is placed on the trunk remaining stable during the motion.

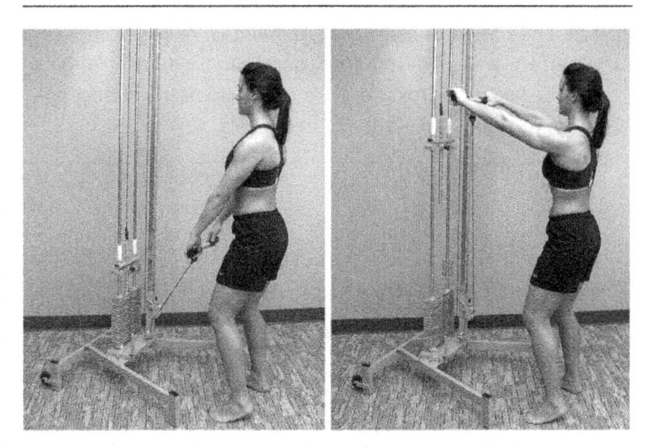

Figure 5.98a,b: Standing arm row. The patient stands with the lumbar spine in neutral and slightly flexed knees. A pulley handle is lifted bilaterally while maintaining a neutral lumbar spine. Lack of stability results in an extension moment of the lumbar spine as the arms are raised. The exercise can also be performed with free weights, though the torque on the back is not as significant. Elastic resistance will produce a greater torque at the end range, but shoulder limitations may require adjustment in tension that provides little resistance at the beginning of the motion. Resistance should be kept light, with a limited arm swing, to avoid excessive torque in the lumbar spine and reduce the risk of disc injury.

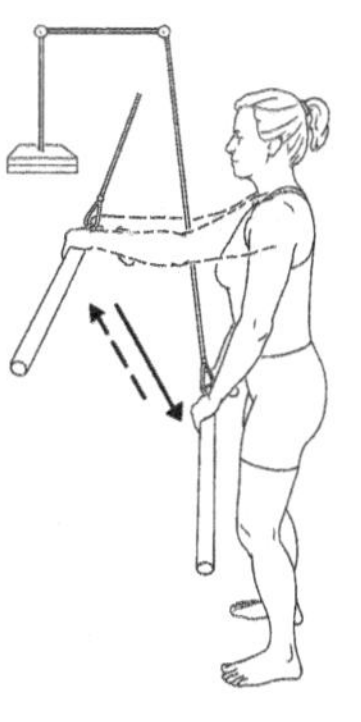

Figure 5.99: Standing pull down, lumbar stabilization. The

patient is standing with the lumbar spine in neutral with the knees flexed. With the lat bar starting on the thighs, the bar is raised to shoulder height and returned. The lumbar spine is held in neutral by the abdominals, avoiding the extension moment at the top position of the bar.

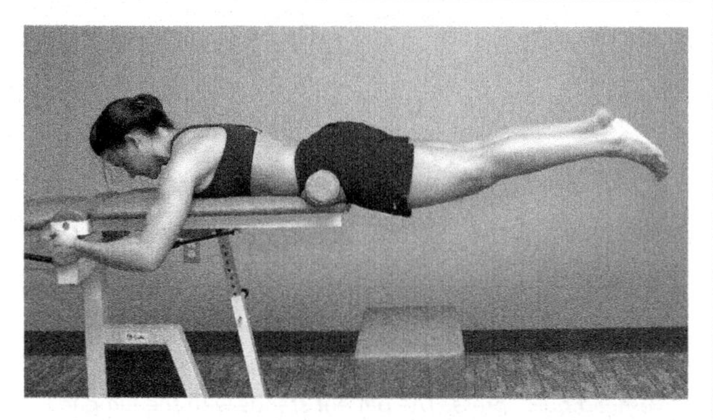

Figure 5.100: Caudal extension-isometric hold. The patient begins prone with the feet on the floor, or the knees resting on a bolster, while the arms are secured on the bench. The legs are lifted to horizontal, in line with the trunk, and held in place. The patient is instructed to maintain respiration during the performance. An isometric contraction is held as long as possible, up to one minute, to improve isometric strength fixation for the lumbar and hip extensors.

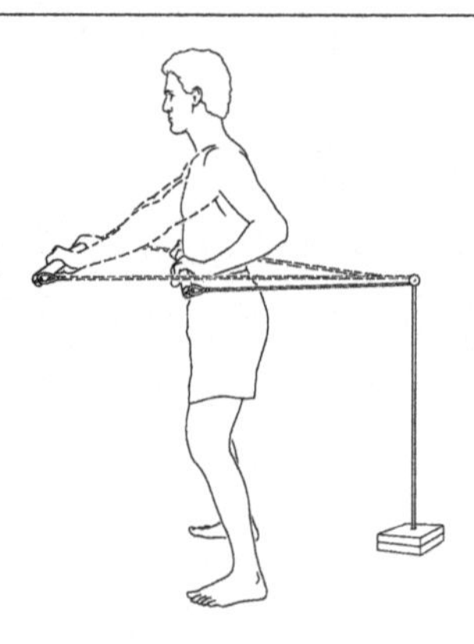

Figure 5.101: Isometric lumbar stabilization with pushing and pulling motions. The patient stands with the lumbar spine in neutral, hips and knees slightly flexed. Pushing and pulling motions are performed without allowing motion in the spine. Resistance can be with elastic bands or a bar attached to a double pulley.

Figure 5.102a,b: Fall out lunge. A forward lunge is performed with the back in neutral, the back knee remains in extension. The trunk remains in line with the lower limb, with the nose over the front foot. The lumbar extensors work isometrically to hold the trunk against gravity. When performed correctly the front hip feels most of the work, with little effort at the knee. When performed incorrectly the trunk remains vertical with an extended lumbar spine and the work felt at the knee, not the hip.

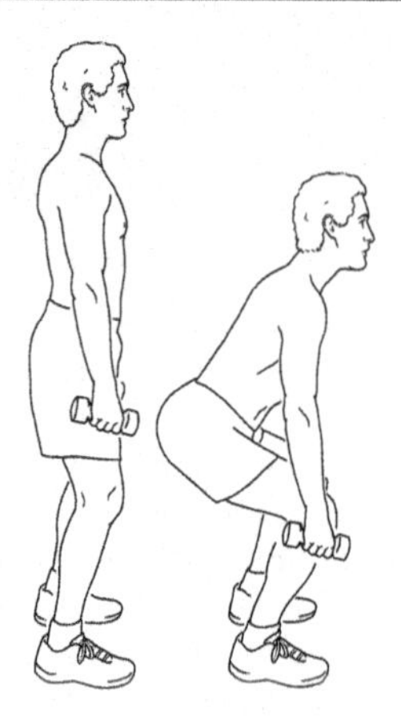

Figure 5.103a,b: The hack squat initially is performed with isometric stabilization of the lumbar spine, without flexion of the lumbar segments. Abdominal bracing stabilizes the lower lumbar segments during a functional squat motion.

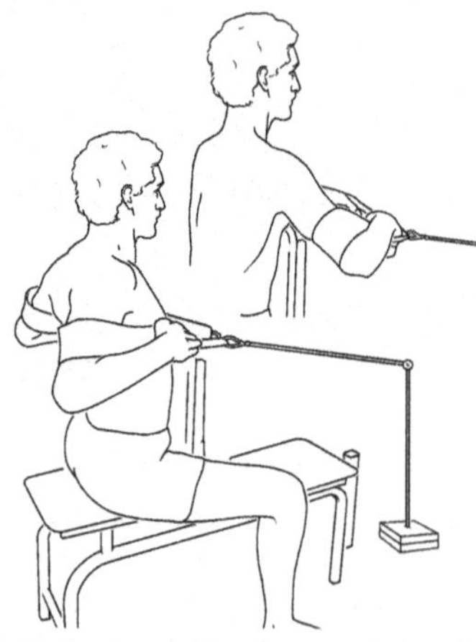

Figure 5.104a,b: Sitting rowing for isometric back extensor training. Sitting removes the need to coordination the pelvis and lower limbs with back stabilization required in standing positions. The patient sits with the pelvis rotated anterior to achieve a neutral lumbar spine. Cable rowing or elastic resistance is used for a rowing motion for the upper limbs while the lumbar spine maintains a neutral position.

Motion and Directions

From the standpoint of tissue tolerance and pain, the basic motions of Stage 1 are progressed in terms of range of motion and by adding the directions that were avoided. Full range of motion and coordination are the goal of Stage 2, allowing progression to more functional strength training in later stages. Initial exercises are progressed in terms of range of motion and/or body position. A more significant progression may be to train in the same range, but in a more challenging body position. Lying down postures should be progressed to partial weight bearing, then sitting or standing, with an emphasis on continued improvement in coordination and tissue tolerance. Examples of more aggressive exercise are listed below, but many options are possible. Progression of motion and direction is a continuum of low-level tissue training toward the end stage functional demand, and are therefore dictated by the individual patient.

Extension

Back extension exercises, whether with active motion or isometric holding, have long been shown to be beneficial in the treatment of low back pain (Davies et al. 1979). Cranial lumbar extension is a classic exercise for training the lumbar extensor muscles. The Roman chair exercise has been utilized in both the clinic and health club settings for back training. This exercise does not isolate the lumbar extensors, but is coupled with the hip extensors. Even the biceps femoris muscles are connected via the sacrotuberous ligament and thoracolumbar fascia (Vleeming et al. 1995). The importance of the gluteus maximus and the biceps femoris to the force production during trunk movement has been demonstrated (Clark et al. 2002, Kankaanpaa et al. 1998, Vleeming et al. 1995). Clark et al. (2003) studied the recruitment patterns of the lumbar and hip extensors as it relates to fatigue during Roman chair extension, finding that the maximal degree of lumbar muscle activation to be around 85 percent. They also observed a reduced recruitment of the lumbar paraspinal musculature occurring at approximately 55 percent of maximal fatigue with a concomitant increase in hip extensor muscle activity. Robinson et al. (1992) has also reported similar decrements in lumbar paraspinal muscle activity with fatiguing trunk extension exercise. Clark et al. (2002) suggested that dynamic Roman chair exercise, due to muscular failure and fatigue, does not allow for maximal muscle activation of the lumbar paraspinal muscles. This statement may be true for strength training in isolation for the lumbar extensors, but is effective for training the functional synergy between the lumbar spine and hip extensors. The exercise can also be adjusted to reduce the load, allowing for coordination and endurance training, reducing the early fatigue of the lumbar extensors.

To adjust the amount of resistance with back extension exercises the axis of motion can be set more cranial. The peak angle of the table is set more cranial on the trunk to reduce the amount of body weight on the cranial side of the axis of movement. The arms can be placed behind the back shifts more body weight caudally, reducing the load, or behind the head (as pictured) to increase the load. A further reduction was achieved with the Stage 1 option shown on a slant board to position the body more vertically. An alternative approach to reducing weight is to adjust the table more vertically.

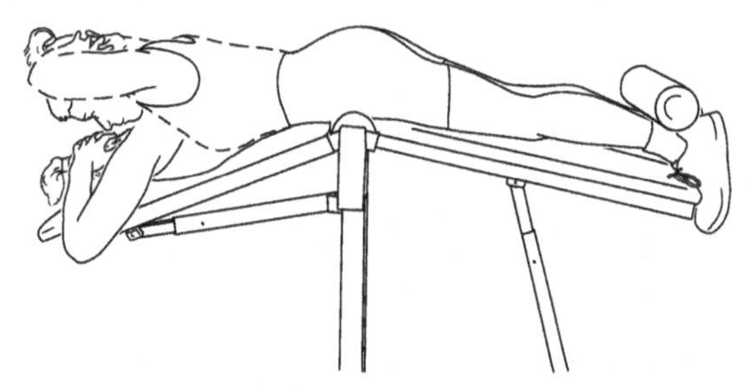

Figure 5.105: Roman chair lumbar extension. The functional synergy between the lumbar and hip extensors can be trained in prone, with cranial extension. Internal rotation of the lower limbs will emphasize lumbar extensor muscles, while external rotation will increase the contribution of the gluteal muscles. Modified Roman chair extension: to reduce contribution from the hip extensors, more specifically the biceps femoris, the fixating roll can be moved to the upper thigh. This will increase the contribution of the lumbar extensor muscles during the exercise.

When using a Roman chair for lumbar extension, placing the hips in internal rotation places the gluteal muscles at a mechanical disadvantage, increasing the multifidus recruitment by 18 percent (Mayer et al. 2002). External rotation of the hips (toes pointed 45 degrees outward) created a 39 percent increase in gluteal recruitment, compared to the internally rotated position. Hamstring recruitment did not change with different hip positioning. Hip rotation can easily be adjusted for cranial extension exercises, with internal rotation emphasizing back muscles, while external rotation will emphasize the hip musculature.

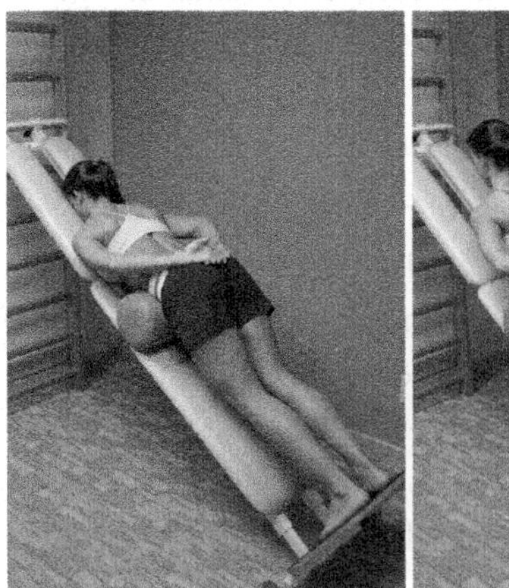

Figure 5.106a,b: Back extension in a more vertical position to reduce the load for endurance training.

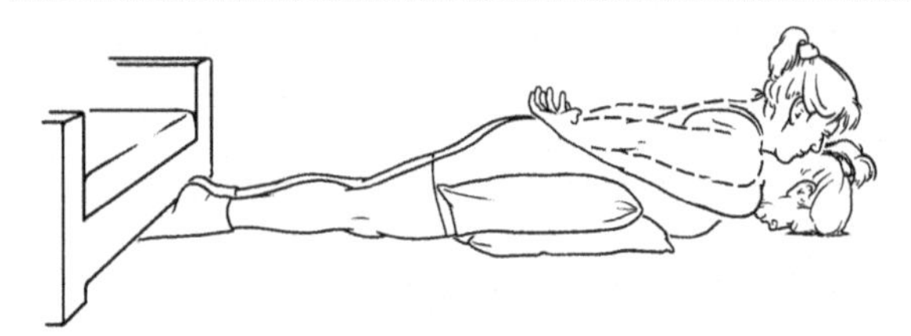

Figure 5.107: Home exercise version of cranial lumbar extension. A pillow is placed under the lumbar spine to reduce the lordosis. The feet can be anchored under a piece of heavy furniture. The patient exhales during the concentric lift, inhaling during the eccentric return.

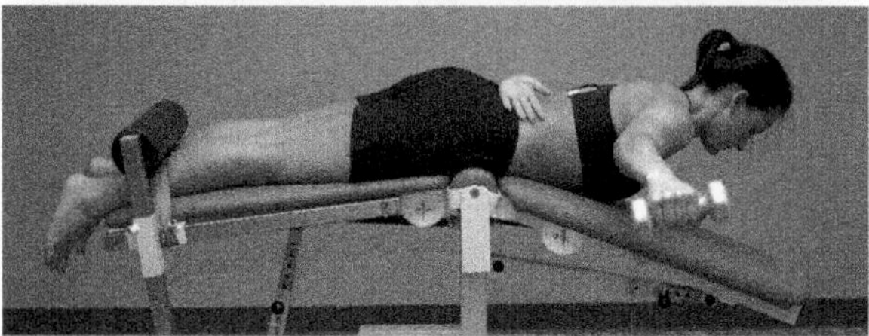

Figure 5.108a,b: Multidirectional cranial extension. Prone extension is performed as above, but one arm is held out to the side to place a rotational moment on the spine. The right arm to the side produces a left rotation moment that must be stabilized during the primary motion of extension.

Figure 5.109: Multidirectional cranial extension. Prone extension is performed as above, but one arm is held out to the side to place a rotational moment on the spine. The right arm to the side produces a left rotation moment that must be stabilized during the primary motion of extension.

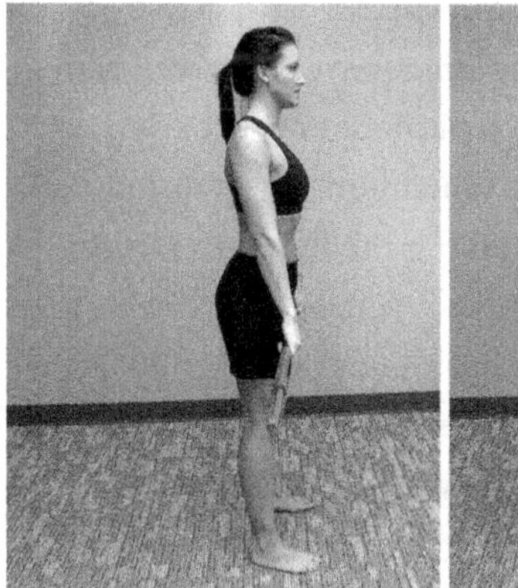

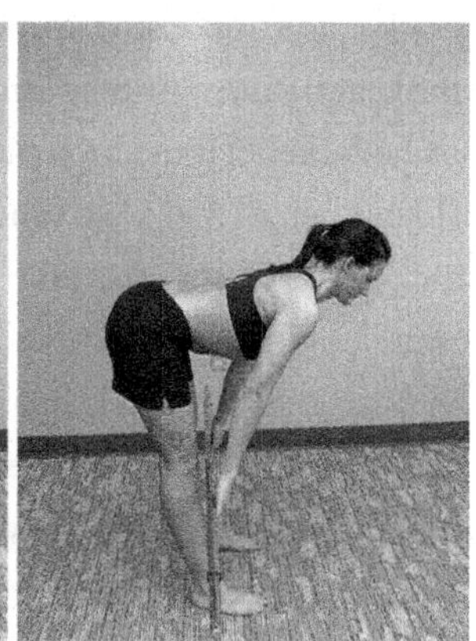

Figure 5.110a,b: Dead lift with shoulder bar for resistance. As a progression from previous dead lift training with the arms hanging the weight forward, placing the bar on the shoulders creates a longer lever arm of resistance to increase the torque on the lumbar spine, gluteal muscles and hamstrings. This progression is not necessary for all patients, but those returning to athletic performance or heavier lifting work may benefit from this strength training approach. Resistance through the hamstrings, sacrotuberous ligament, across the SIJ to the lumbar spine assists in the dynamic stabilization of the pelvis.

If concern exists over minimizing the contribution of the hip extensors during lumbar extension, changing to a sitting position is an effective option. Sitting significantly increases the compression to the lumbar discs, and is therefore not recommended early in rehabilitation of acute disc pathology. For disc pathologies, it may be more appropriate to wait until Stage 3 or 4 training to utilize sitting postures for cranial extension training.

Figure 5.11: Seatted lumbar cranial extension. A roll is placed in the lower abdomen to increase the reversal of the lumbar lordosis in the flexed position. Extension begins cranially from the upper thoracic spine with a segmental progression to the lower lumbar spine. Pressure is maintained anteriorly on the roll during extension to emphasize lumbar extension, rather than hip extension with a return to vertical. The cervical spine remains in neutral throughout the exercise. Placing the feet further forward will increase the amount of lumbar flexion, locking the lower lumbar segments from extension during the exercise. This may be appropriate if extension is painful in the lower lumbar spine, or the emphasis is on gaining mobility in the thoracic spine. Moving the feet back will allow motion to occur down to the sacrum.

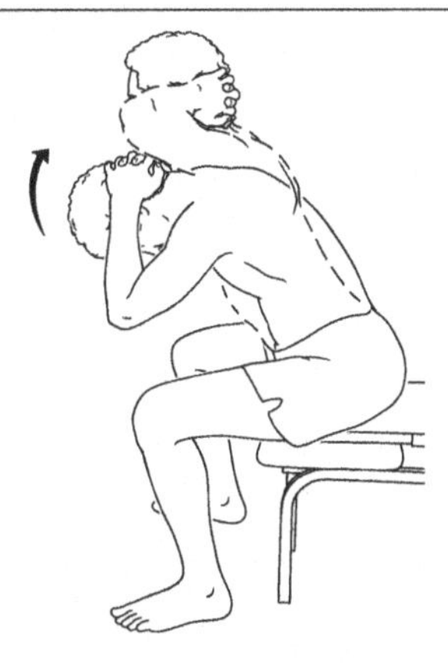

Figure 5.112: Seated cranial extension without anterior block. As coordination improves, the anterior block is removed with the patient performing cranial extension segmentally. Hip extension is avoided by not returning to a vertical position, but stopping the motion after full spinal extension is achieved.

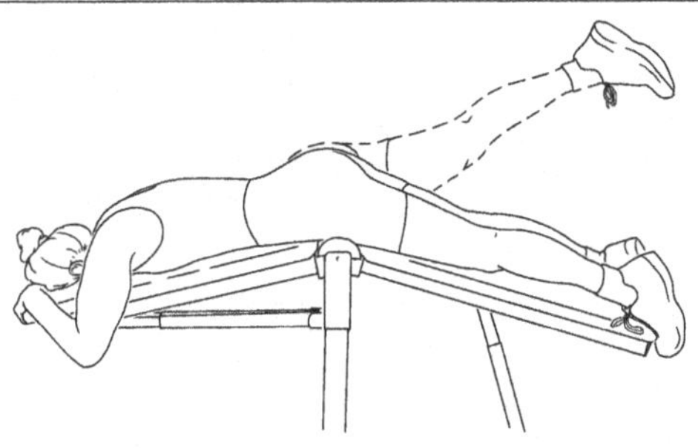

Figure 5.113: Caudal extension with unilateral limb extension. A fixed surface is more beneficial for earlier coordination training and strength training. Performing in quadruped, without any support, can allow for excessive lumbar extension during the motion. The patient must be able to dynamically control the lumbar extension moment with the abdominal muscles. Weakness in the lower abdominals and lumbar multifidi may lead to pelvic rotation, seen as the ASIS on the ipsilateral side of the leg lift moving down toward the table. A towel roll can be placed under the ASIS to passively stabilize the pelvis and assist the motion. The towel can also be placed on the opposite side with the patient instructed to push down into the towel to fix the lower abdominals prior to lifting the leg.

Figure 5.114: Caudal extension with unilateral limb extension on labile surface. Once coordination is achieved on a stabile surface, performing the same exercise on a fitness ball provides an increased stabilization challenge.

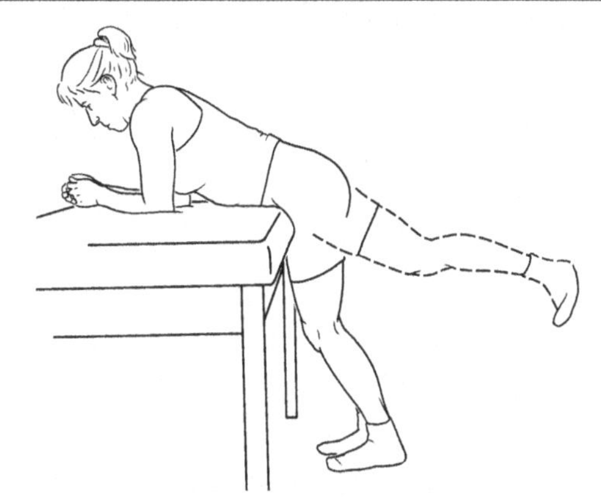

Figure 5.115: Mule Kick. Hip extension for back training can also be performed in standing with the upper body supported on a table. This position will limit extension in the lumbar spine, as the back is flexed with the hip extending from a position of lumbar flexion to lumbar neutral.

Flexion

Flexion training in Stage 2 may progress in range of motion and slight increases in speed to increase the relative resistance. Brief isometric holds during the mid phases of sit-up exercises can also increase the amount of muscle work, as well as challenging active respiration during an isometric hold.

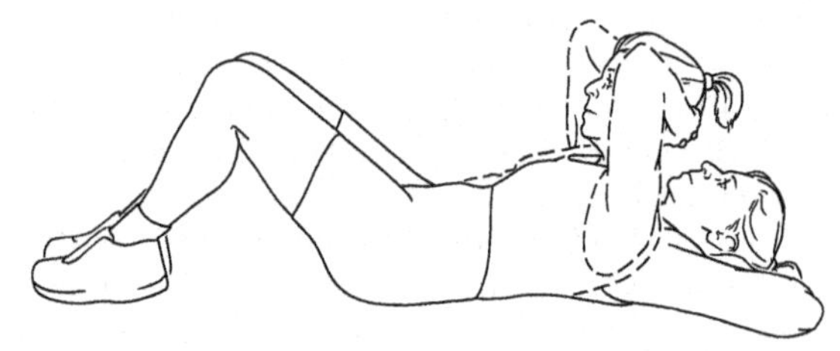

Figure 5.116: Modified Crunch Sit-up. As with the previous slant board examples for cranial flexion, the hip flexor contribution can be reduced by creating an extensor pattern through the lower limbs. The feet are plantar flexed into the wall, along with hip and knee extension. This extensor pattern will reduce recruitment of the hip flexors, allowing emphasis on the lower abdominals. The patient continues to perform a bracing or hollowing prior to motion to normal motor patterns.

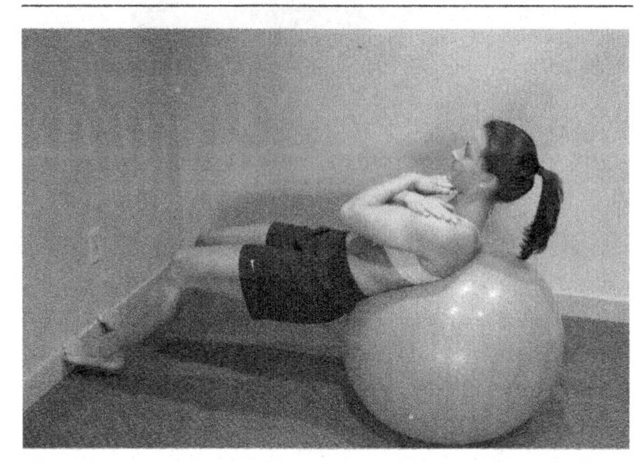

Figure 5.117: Sit-overs on ball. As with the previous slant board examples for cranial flexion, the hip flexor contribution can be reduced by creating an extensor pattern through the lower limbs. When a slant board is not available, lying on a ball with the body on an incline, instead of horizontal, can reduce the load. The feet are plantar flexed into the wall, along with hip and knee extension. This extensor pattern will reduce recruitment of the hip flexors, allowing emphasis on the lower abdominals. The patient continues to perform a bracing or hollowing prior to motion to normal motor patterns. The ball will not be as challenging for the abdominals in terms of flexing the trunk, but will increase the core challenge to stabilize the exercise.

Figure 5.118: Spine shoulder extension (eccentric elevation) on a fitness ball. The patient is supine on a fitness ball with the lumbar spine maintained in neutral. With a bar or free weights the arms are elevated over the head.

Figure 5.119: Free weights allow for asymmetrical arm motion, introducing a rotary components to lumbar stabilization challenges.

Flexion training may focus on coordination of a larger pattern. Performing short-range diagonals into a flexion pattern with a wall pulley can begin the process of establishing coordination for a full functional pattern in later stages. A lighter pulley weight may also serve to unload the weight of the trunk during the eccentric phase of flexion, assisting with the muscles of trunk extension. Heavier loads will emphasize abdominal and hip flexor activation.

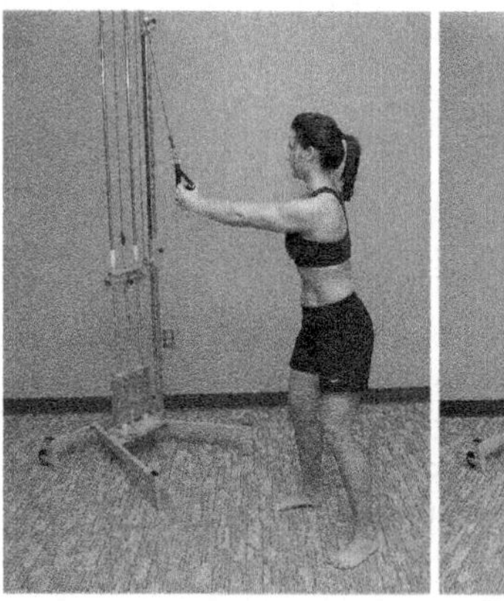

Figure 5.120a,b: Lumbar flexion diagonal with pulley assistance or resistance. Range of motion is kept short, less than from shoulder to knees. A lighter pulley resistance will assist the lumbar extensor muscles during the motion, while heavier resistance will emphasize flexor muscles of trunk and hip for resisted flexion trainings.

Side Bending

Progressing side bending exercises to partial and full weight bearing represents a more functional training position to involve not only back muscles, but also contributions from the hip musculature. Weight bearing is also a progression for improving tissue stress and strain tolerance.

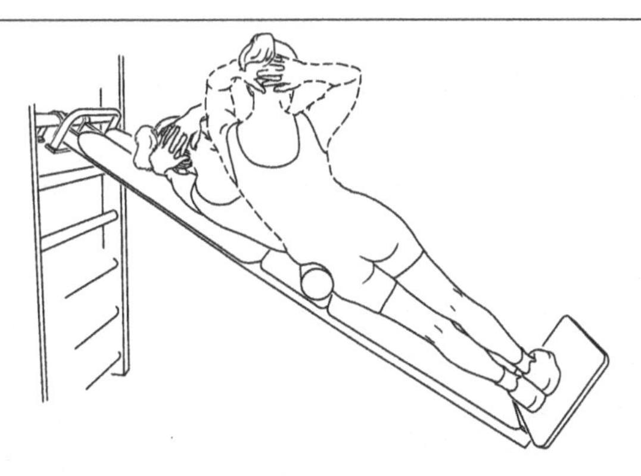

Figure 5.121: Cranial side bending on slant board to reduce resistance. A bolster for the axis of motion can be set at the rim of the pelvis to involve the entire lumbar spine, or set at a specific spinal level to prevent side bending caudal to the roll. Increasing the angle and/or moving with bolster more cranially will reduce the load.

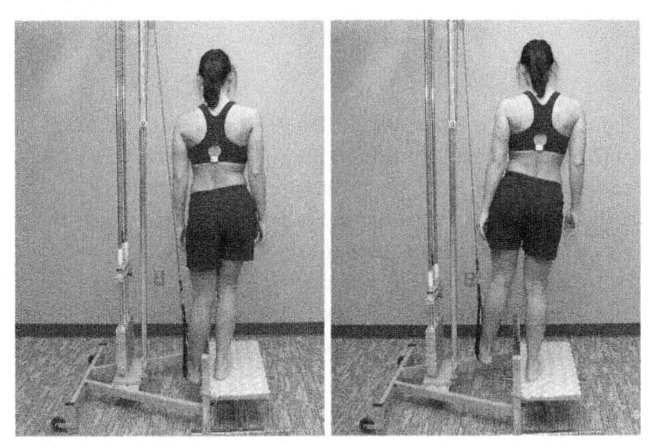

Figure 5.122a,b: Hip hiking standing – pulley assist. The hip hike works primarily the hip abductors on the stance side and the quadratus lumborum on the opposite side. A pulley assist will reduce the body weight to allow for pain free coordinated motion.

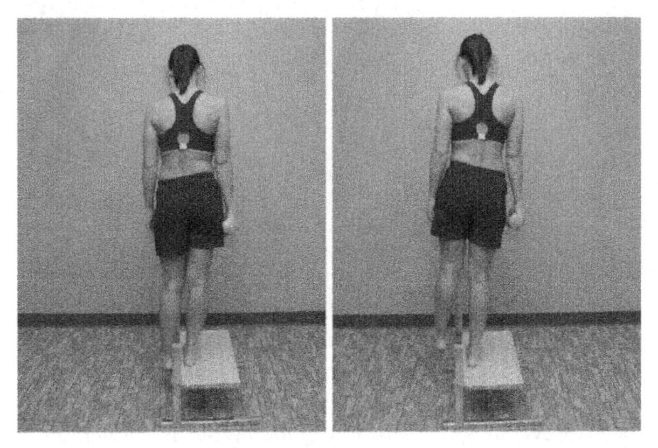

Figure 5.123a,b: Hip hiking standing – free weight assist. To assist the motion a pulley can produce a force cranially on the stance side, or a free weight can be held in the hand on the opposite side of the joint axis.

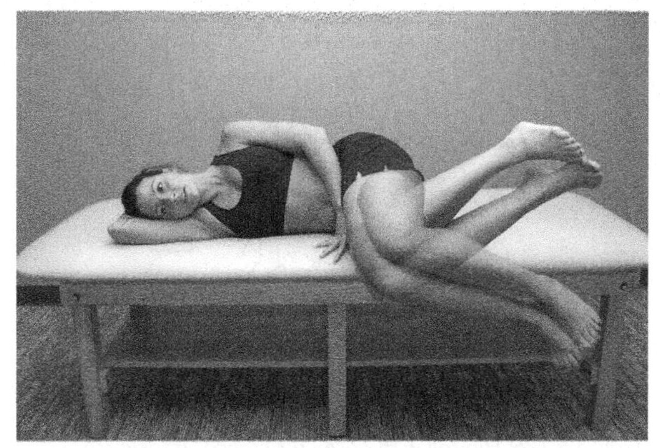

Figure 5.124: Lumbar side bending in sidelying. The hips and knees are flexed to 90°, with the mid thigh on the edge of the table. The feet are rotated from the floor to the ceiling to create a side bending moment in the lumbar spine.

Figure 5.125: Lateral bridge on fitness ball. From a sidelying position, with the feet and lower legs on a fitness ball, the pelvis is lifted off the floor. The pelvis should remain in alignment with the shoulders, with the lumbar spine in neutral. The labile surface of the ball increases the challenge for stabilization.

Rotation

With improvement in acute symptoms, tissue tolerance and mobility, rotation training focuses more on progressing motor performance. If recruitment, timing and coordination are established, using a higher resistance with the same exercises will develop qualities within the muscles, such as increased capillaries for endurance or improve strength qualities. Non-weight bearing training with increased resistance emphasize muscle qualities without increasing tissue strain. Progressing to weight bearing exercises places more demand on tissue tolerance, as well as advancing toward more functional activities. The process of beginning work or sport simulation begins at a low level.

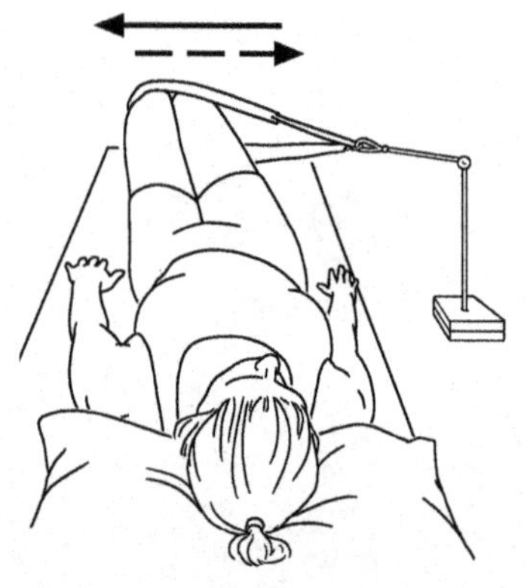

Figure 5.126: Hip roll with pulley resistance. Transverse resistance from the pulley will recruit the oblique abdominal muscles, as well as the multifidus for arthrokinematic control of the lumbar segments. In a normal spine, the transverse abdominis will be recruited as well. The more phasic oblique abdominals provide the primary torque for rotation. Eccentric control occurs as the knees move to the right, with emphasis on the concentric return beginning at the more cranial segments in the lower thoracic spine, moving toward the lower lumbar segments, with the pelvis moving last. If the motion is initiated at the knees, rather than the lower thoracic spine, then emphasis is placed on hip muscles of adduction and abduction.

Figure 5.127a,b: Resistance for the supine hip roll can be created by crossing the knees, moving body weight lateral.

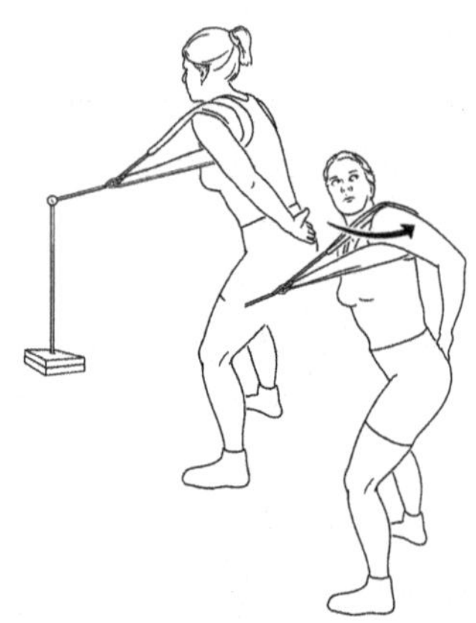

Figure 5.128a,b: Standing left cranial rotation, pulley resistance. The patient stands facing the pulley with the knees and hips slightly flexed. Left rotation emphasizes the right multifidus, though both sides work in synergy and the oblique abdominals are the prime mover. Motion should be initiated with the eyes and head turning, with the trunk following through.

Adjusting the resistance is only one way to modify and exercise. The flexibility of a wall pulley allows for any force vector during an exercise. Performing the same standing trunk rotation can have different emphasis depending upon what vector the pulley is set at and how the body reacts to that force.

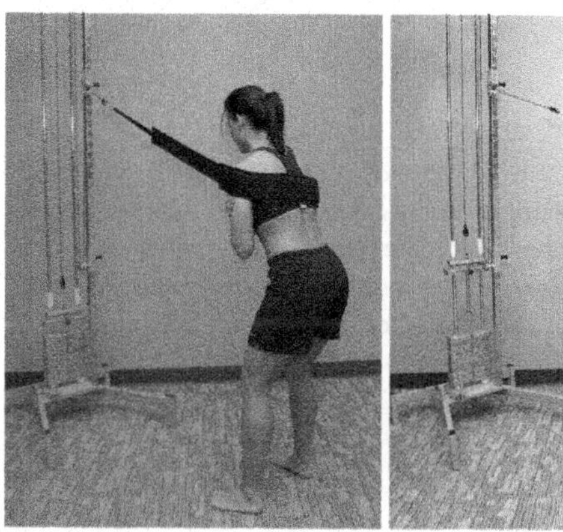
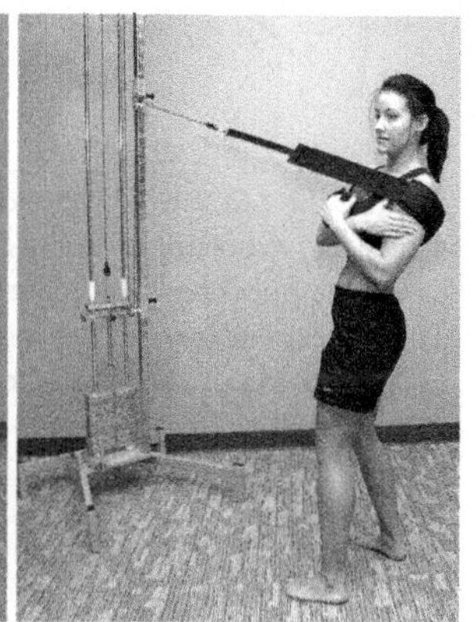

Figure 5.129a,b: Standing left trunk rotation with a cranial resistance inclined up and forward. The force moment creates a motor response for extension and rotation but the incline vector increases the relative range of extension that is performed, as opposed to pure rotation. This option may be desired in a stiff back attempting to gain range into extension. Another reason for this selection may be a functional requirement of combined extension and rotation, such as a tennis serve.

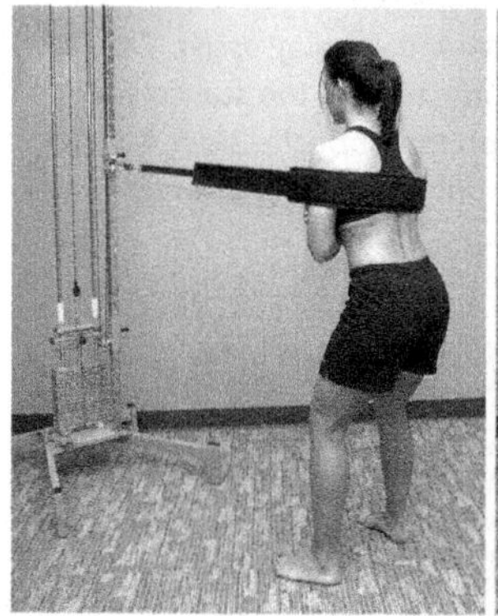

Figure 5.130a,b: Standing left trunk rotation with a horizontal resistance. A horizontal force moment will emphasize pure rotation of the trunk with little to sagittal plane motion in flexion or extension. From a motor standpoint, this setup may be desired to recruit muscles of rotation while reducing the demand for muscles of extension (i.e., recruiting the multifidi and transverse abdominus for rotation while minimizing the contribution of the long trunk extensors). Additionally, the emphasis may simply be on improving gross rotation range. Lastly, progressing to strength training for trunk rotation to address functional requirements.

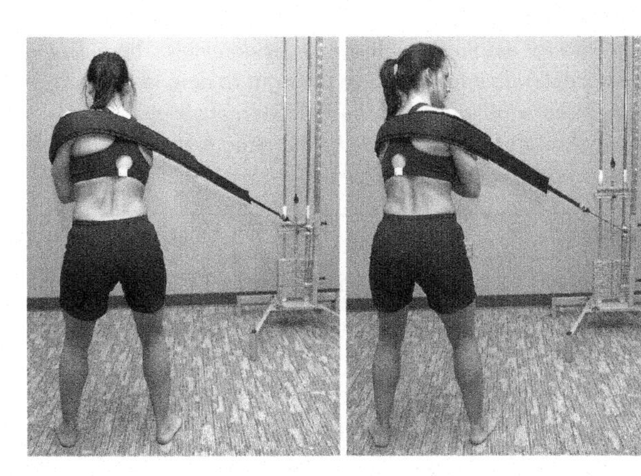

Figure 5.131a,b: Standing right trunk rotation with a force vector to emphasize recruitment of the lumbar left lumbar multifidi (LM). The pulley line is set at a vector to naturally recruit the LM, without cognitive effort. The "external cue" for training can improve the facilitation of LM that may be in a state of atrophy or inhibited by pathology. The force line is set at a decline to address the extension moment of the LM. The line is lateral to the right to emphasize the left side bending component of the LM. A pure rotational motion is performed to emphasize the rotational vector of the multifidi without recruiting the long extensors of the trunk.

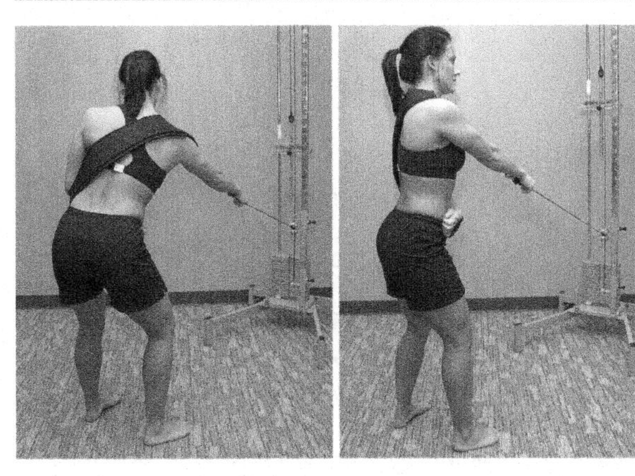

Figure 5.132a,b: Standing right trunk rotation using a wrap around strap attachment. The end of the belt is secured by the left hand in front of the pelvis. The belt wraps around the trunk in a vector similar to the left side lumbar multifidi. A combined extension and right rotation moment is performed.

The exercise itself does not determine the training outcome. It is the dosage and the exercise design that will determine the training outcome. The same exercise can be dosed anywhere on the continuum from mobilization, to strength and to power. Adjusting the weight and repetitions will determine which of these functional qualities is attained with training. Beyond simple dosage concepts, the design of the exercise can be adjusted to emphasize gross range of motion, specific muscle fiber recruitment or pertain to specific functional patterns. Functionally, a co-contraction of anterior and posterior muscles occurs, with both tonic arthrokinematic and phasic osteokinematic muscles working in synergy. Anterior resistance will place greater emphasis on the posterior muscles, while a posterior resistance will emphasize anterior muscles.

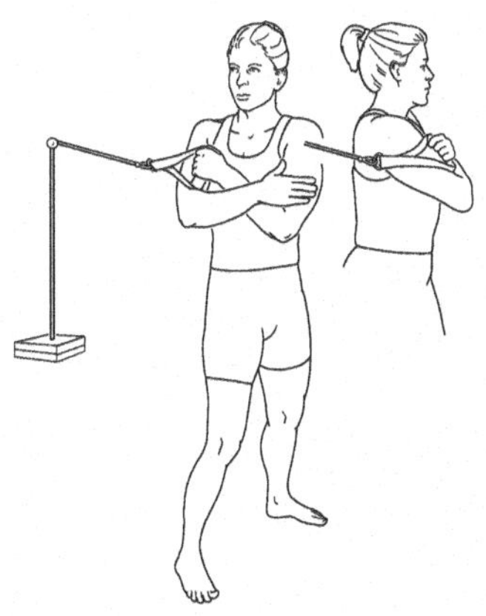

Figure 5.133a,b: Standing trunk rotation, posterior pulley resistance. The patient stands with the back to the pulley, knees and hips slightly flexed, arms crossed and holding on to a pulley handle. The lower limbs remain stable while a cranial rotation movement is performed.

With the pulley line set perpendicular to the trunk, emphasis is on range of motion, gross mobilization or training general functional patterns of rotation. The same motion can be performed, but the movement of the arm created by the line of the pulley can be adjusted to facilitate specific muscle fibers. Internal attention focusing on a person's own body movement can be detrimental to the performance of a well-learned skill as well as in learning new skills (McNevin et al. 2000). Having the patient try to feel specific muscle recruitment during the training process can slow the learning process. External attention, focusing on the body's reaction to an external force or object is a more efficient approach. Having the patient attempting to focus on firing specific lumbar multifidi muscles during the rotational motion is not as efficient as setting the pulley line to act as an external resistance that the body has to respond to with specific muscle recruitment. This is done all the time with free weights placed in the line of fiber direction to recruit muscle with simple cardinal plane motions. Attempting to recruit the multifidi requires the pulley line to be set in a vector that will recruit the muscle as an extensor, side bender and rotator. As the pulley line comes from the opposite side of the axis of motion, resistance is set perpendicular to the fiber direction, rather than in line with the fiber direction, creating a natural recruitment emphasis of the lumbar multifidi during rotation.

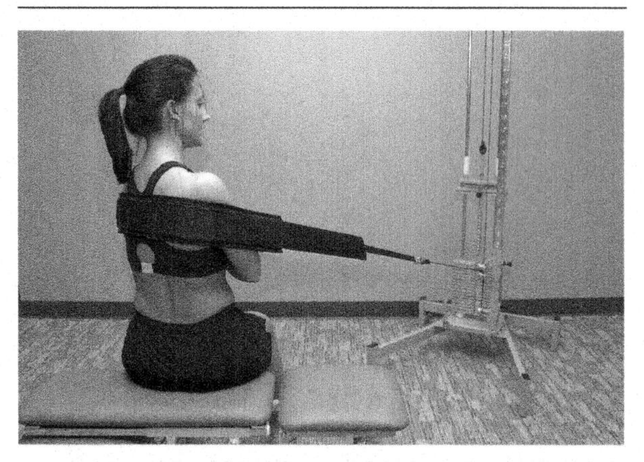

Figure 5.134: Standing left cranial rotation with pulley resistance for right multifidus emphasis. The pulley line is set inferior and to the left to create a moment arm that naturally requires increase recruitment of the right multifidus muscles.

Figure 5.135a,b: Standing trunk rotation with pulley line adjusted for emphasis of the transverse abdominis. A pelvis high horizontal resistance is in line with the fibers of the transverse abdominis. The patient performs a cranial rotation moment fixating the pelvis and hips.

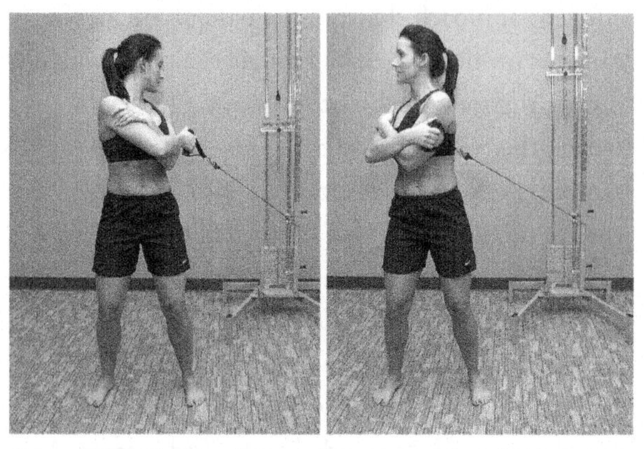

Figure 5.136a,b: Standing trunk rotation with pulley line adjusted for emphasis on the lower abdominals. The pulley is set posterior, inferior and to the right to emphasize the right internal obliquus and the left external obliquus. The transverse abdominis will still be recruited first as the initial tonic stabilizer.

Figure 5.137a,b: Sitting cranial rotation with posterior resistance through a bar. The patient is in sitting with a neutral lumbar spine. A bar is held anteriorly, attached to a pulley. Cranial rotation is performed while stabilizing the lumbar spine. The bar increases the lever arm, creating a greater torque moment through the shoulders and trunk.

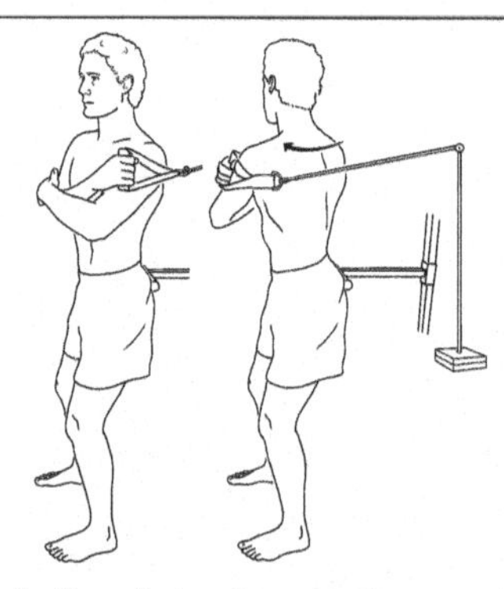

Figure 5.138a,b: For all standing rotation exercises, pelvic stabilization through the lower limbs is required. If the patient cannot initially control the pelvis and lower limbs an external block with a pulley T-bar across the sacrum removes the stabilization requirement. Perch-sitting on a tall bench will also serve as an external stabilization, which also allows extension/rotation exercises facing the pulley. As coordination improves, the support for the pelvis is removed. A horizontal line of resistance will emphasize more transverse fiber directions of spinal muscles.

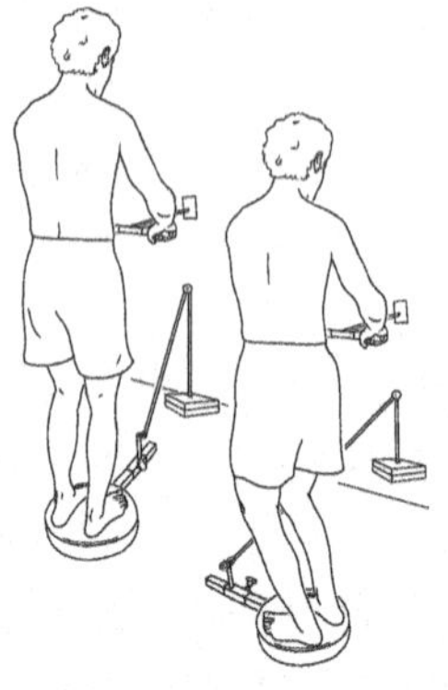

Figure 5.139a,b: Caudal rotation on a rotation plate. The patient is standing on a rotation plate, knees slightly flexed and the lumbar spine in neutral. The arms stabilize the

shoulders and upper trunk through a T-bar attached to the wall pulley. Motion is initiated by tightening the lower abdominals and beginning the rotary motion from the pelvis, not the feet.

Locking

A basic progression concept for Stage 2 is to move the body to a more challenging position while training. Stage 1 typically involves more recumbent positions. Utilizing locking techniques may allow for a faster progression into weight bearing, or more functional position. Some level of improvement in tissue tolerance and coordination, achieved in Stage 1, is usually necessary prior to progressing to using locking techniques in weight bearing postures.

Rather than changing body position, an alternative progression might be changing the locking procedure to a less restrictive version requiring more demand on the patient to coordinate the movement safely. If the patient has developed adequate coordination, artificial and spinal locking may be removed with a progression to coordinative locking. For example, a hypermobility into right rotation of L5/S1 may have been locked into left rotation using coupled force locking during exercises. As a progression, the coupled force locking is removed, allowing the joint to participate in the exercise. Coordinative locking involves eccentric coordination of the trunk left rotators (oblique abdominals and right side multifidi) into right rotation. The patient controls the motion toward the pathological range. This concept may apply to exercises in lying, sitting or standing postures. This change in locking is an important progression and it is never necessary to retreat to artificial locking once coordinative locking has been achieved.

If a patient is ready to progress to Stage 2 but still lacks the necessary coordination, spinal or artificial locking may still be necessary, particularly if the patient does not have full pain free range of motion. If locking is needed in this stage, it is generally tolerated well in a weight bearing position. Below are examples of specific exercises using different locking techniques.

Joint Locking (Coupled Force Locking)

As discussed in Stage 1, joint locking utilizes coupled forces to lock spinal segments from motion with facet compression. With hypermobile segments, ligamentous locking (countercurve) techniques cannot be used, as taking up slack of collagen in hypermobile segments may cause additional tissue damage. The opposite direction of facet compression with coupled forces is the safe alternative. Joint locking techniques can be utilized in sitting, allowing a progression of tissue loading from side lying postures.

Figure 5.140a,b: Seated cranial right rotation with locking from below in flexion, left side bending/left rotation. In this case the L4/5 and L5/S1 segments are in flexion from below, by adducting the hips and increasing the range of hip flexion. Left lumbar side bending from below, from an adjustable bench on the left or sitting on folded towels on the left, leads to left rotation of the lower lumbar spine. The right rotation from above will not include the locked and protected segments from below. *Variations in coupled forces may require the use of extension, rather than flexion, to create left rotation from below. Lumbar extension can be achieved by abducting the hips and anteriorly tipping the pelvis.*

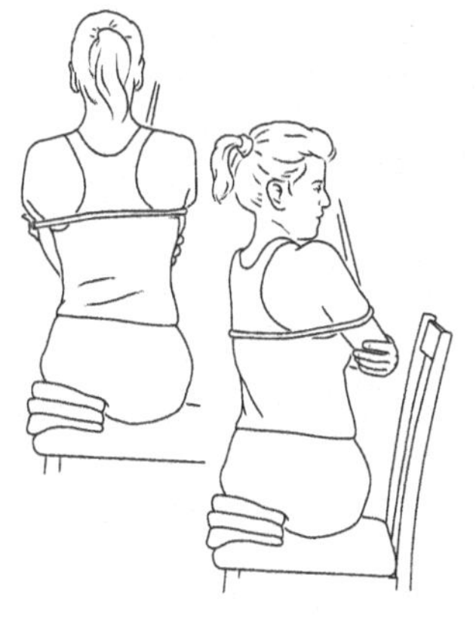

Figure 5.141a,b: Home version or clinic version of locking using towels to create a caudal side bending left. With classic coupled forces, flexing from below with side bending left will create a left rotation. With the lower lumbar segments held in left rotation by the coupled force action, they cannot move with cranial right rotation.

Ligamentous Locking (Counter Curve Locking)
With lower lumbar contraindications to exercise, using counter curve locking concepts may allow for a faster progression into more aggressive body positions or exercises. Counter curves may involve one plane of motion where opposite curves are performed caudal and cranial. Different planes of training can also be effective in locking motion; i.e., flexion in the lower lumbar spine while training rotation from above. Taking up collagen slack in flexion from below will reduce the available range for rotation coming from above.

Ligamentous locking is also practical for using one plane of motion to tighten and protect segments while training in another plane of motion. One example in the lumbar spine would be flexing from below, taking up collagen tension in the posterior elements (supraspinous and interspinous ligaments, ligamentum flavum, facet joint capsules). This tension would lock the spine from participating in other directions of movement, rotation for example. It may be necessary to palpate the spinous processes to assess segmental motion to ensure that the desired locking is taking place.

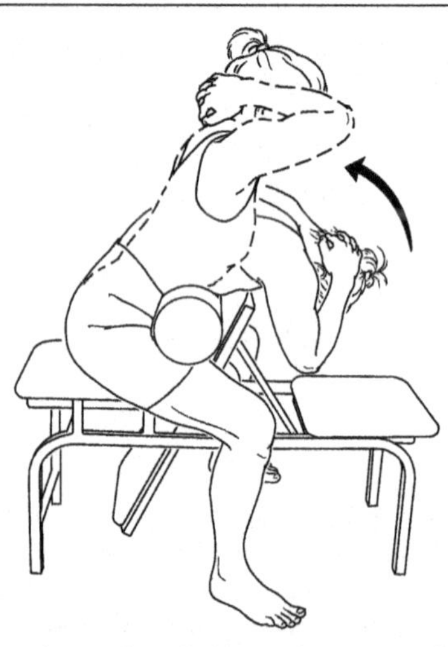

Figure 5.142: Flexion from below and extension from above. This example would suit the patient with a contraindication concerning extension in the lower lumbar and needing extension training in the upper lumbar segments. The patient is instructed to reach the elbows forward and up to initiate a cranial extension moment.

Figure 5.143: Performed as above without the bolster in the abdomen. This is a common home exercise version performed on a chair. The patient continues to perform a caudal-to-cranial motion with a caudal flexion counter-curve.

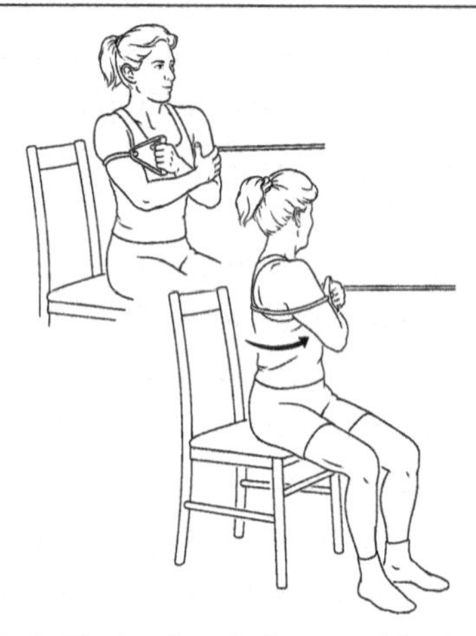

Figure 5.144a,b: Flexion from below by adducting the hips with knees together will provide ligamentous locking. To flex higher segments in the lumbar spine the feet can be placed on a stool to increase the degree of hip flexion and secondary lumbar flexion. Training from above into right or left rotation either from anterior or posterior resistance can be localized above the level of ligamentous locking.

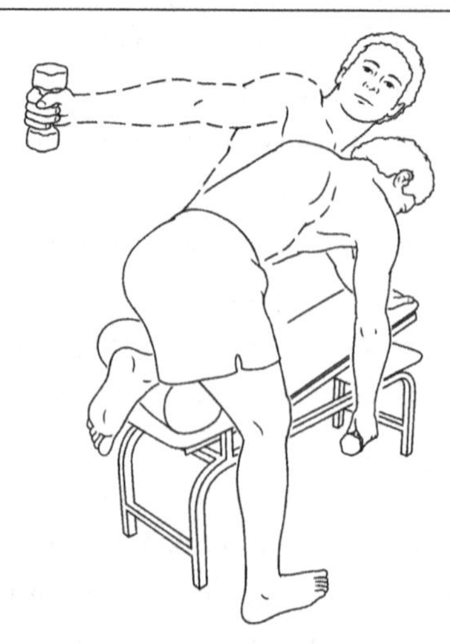

Figure 5.145: Cranial rotation training in extension with lumbar locking with flexion from below. The degree of flexion from below can be controlled by the degree of hip flexion on the left in this example. The locking occurs with the counter curve of flexion from below, preventing the spine from participating in rotation from above.

Caudal coupled force locking techniques can be performed in standing, though they are much more difficult to design. The simplest version is to apply basic countercurves from below to facilitate motion or block motion. By standing with the feet staggered, one forward and one backward, a caudal rotation is created rotating the pelvis. This approach is only effective for the lower lumbar region. If this rotation matches the rotation from above, then motion is facilitated. This is used most commonly when trying to mobilize the full end range from above and below. If the caudal rotation is set opposite to that occurring from above, then there is a block of rotation at the lower lumbar segments.

Figure 5.146a,b: Cranial left rotation mobilization. The left foot forward rotates the pelvis to the right, which is a relative left rotation (named for top segment). As the motion is performed eccentrically from above into left rotation, all segments move to the full end range. This position is most effective for mobilizing restricted segments to full end range, or decompressing pathologies that are painful with facet compression in rotation (facet DJD, spondylolisthesis, entrapments).

Figure 5.147a,b: Cranial left rotation with caudal countercurve locking of the lower lumbar segments. The left foot is backward, rotating the pelvis to the left, which is a relative right rotation (named for top segment). As the motion is performed eccentrically from above into left rotation motion cannot occur at the lower lumbar segment that is held in a position of left rotation. For this locking to be effective, the patient must actively stabilize the pelvis, not allowing rotation. This position is effective for any contraindication to rotation during training.

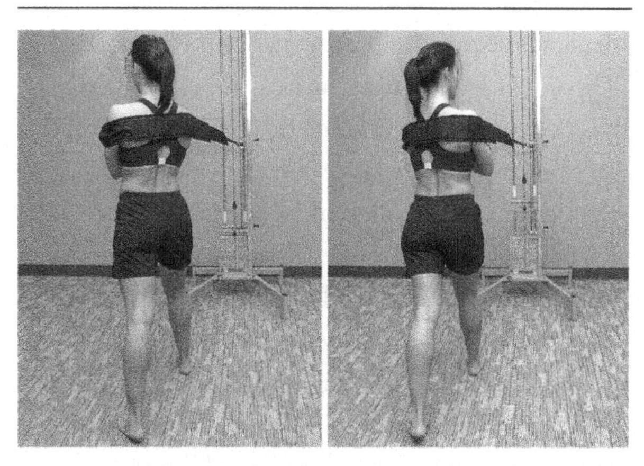

Figure 5.148a,b: Cranial right rotation with staggered feet to emphasize range of motion of the lower lumbar spine in right rotation. The right foot forward rotates the pelvis to the left, which is a relative right rotation. Slack is taken up from below with the cranial right rotation motion reaching the full end range for trunk right rotation.

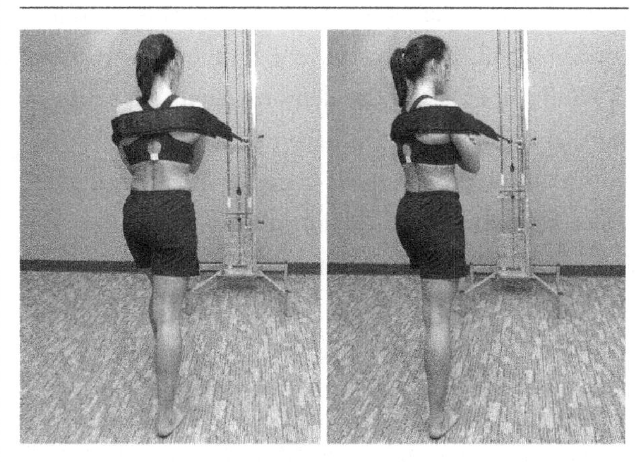

Figure 5.149a,b: Cranial right rotation with caudal locking in left rotation countercurve. The patient is standing facing the pulley with the right foot backward, the pelvis is rotated to the right, creating a relative left rotation of the caudal lumbar segments. Right rotation is performed cranially with motion stopping at the caudal counter curve of left rotation.

With standing countercurve locking the easiest progression is to move the foot position gradually toward an even alignment of the feet, reducing the relative counter rotation from below. As the position moves toward a normal stance position, a greater demand is placed on the motor system for dynamic control of the motion. Stage 3 will attempt to remove all locking, other than coordinative locking, which by definition is dynamic motor control for stabilization.

Balance

Balance exercises can be progressed as necessary to return to a level of safety, or work to even more challenging levels, should the patient's functional demands require high end balance performance. The use of labile sitting surfaces for clinical exercise has been suggested (Manniche et al. 1991, Norris 1995). The use of a labile surface for balance training may be enough of an added challenge for previous Stage 1 exercises for bilateral and unilateral stance. Cervical motions or upper limb reach activities can also be performed with a labile surface. A labile sitting surface, such as a gym ball or air cushion, theoretically creates an increased challenge for muscles of lumbar stabilization.

The sitting position itself, however, removes the contributions of the hip joints and pelvis to overall lumbar mobility and stability. Standing requires a much greater demand than sitting on this complex system of muscles crossing the lumbopelvic region. O'Sullivan et al. (2006) demonstrated that labile sitting did not result in increased EMG activity of the lumbar multifidi, transverse fibers of the internal oblique and the iliocostalis lumborum. As sitting on a labile surface does not represent a functional requirement for most patients, and does not create a significant muscular challenge, it may not be the most efficient progression for stabilization training.

Standing may represent a greater functional challenge. Snijders et al. (1995) found that the activity for the internal oblique muscles is higher in sitting compared to lying supine, but standing activity was even greater. Standing may represent a more functional position, as well as a more natural posture for recruiting the tonic components of the abdominals for stabilization training. Standing on a labile surface does create a postural sway challenge, requiring synergistic action of the muscles of the lower limb, pelvis and lumbar spine to maintain a center of gravity. Not all labile surfaces emphasize the same things. Balance examples below outline variation is standing challenges for different emphasis of training.

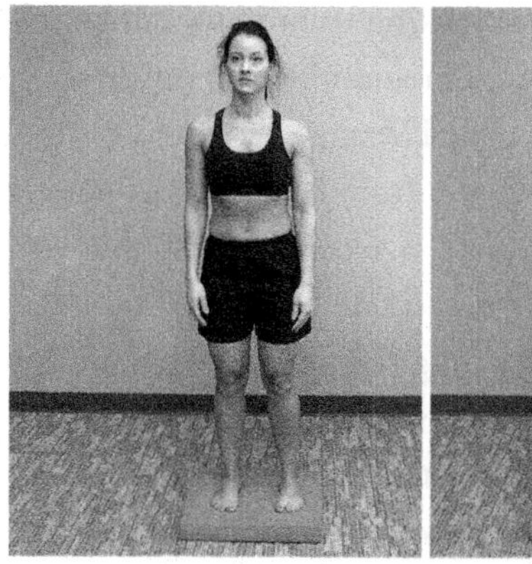
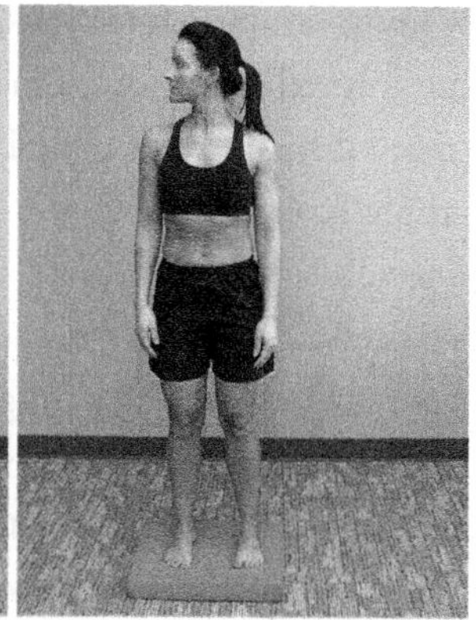

Figure 5.150a,b: Standing balance on foam with cervical rotation. Balance foam is often misinterpreted as a challenge to lower limb and pelvic balance. Though more difficult than standing on a flat surface, foam emphasizes the vestibular system. Foam reduces the somatosensory information from the feet and lower limb, requiring greater integration from afferent input from the vestibular system.

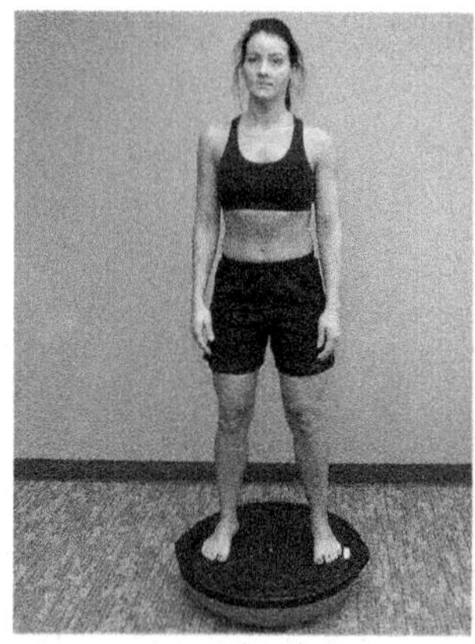
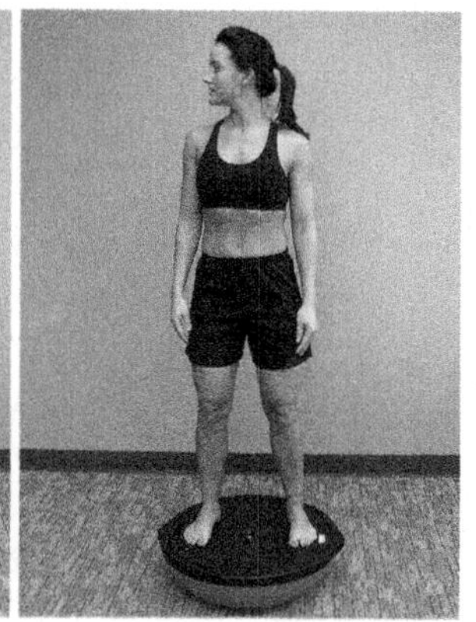

Figure 5.151a,b: Balance training on a BOSU with cervical rotation. Any air cushion platform provides a solid flat surface to allow for contribution of the lower limb somatosensoary system, while providing a labile challenge to postural reflexes. This can be performed bilaterally as shown or with two smaller platforms for each individual foot. Cervical motions add a greater challenge to postural reflexes from the head and neck. Upper quarter activities, such as reaching, can also increase the challenge by moving the center of gravity away from the body and back again. Throwing and catching activities requires a reaction, rather than a planned motor performance, further challenging postural reflexes.

Static balance training may be less functional than balance activities that include movement of the extremities. Upper extremity motions during balance training focuses the patient on extremity movements while postural reflexes function at a more subconscious level. Excessive bracing during standing balance activity is an abnormal motor response to instability. This can be monitored by palpating the lumbar multifidi during balance training activities. As arm motions move through cardinal planes, the lumbar multifidi should contract as relax in response to different movement vectors. An abnormal response would be bilateral multifidi splinting throughout the activity.

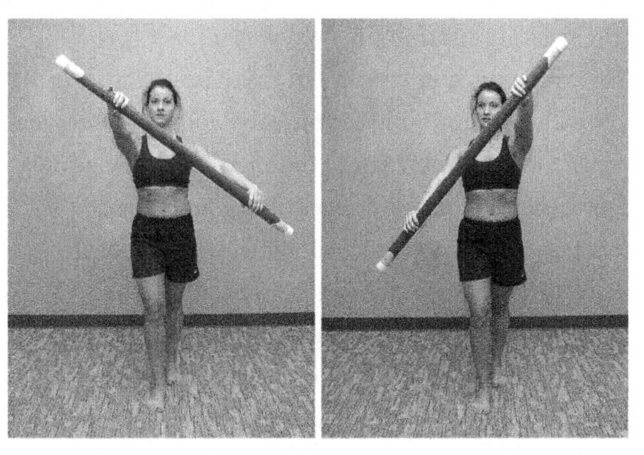

Figure 5.152a,b: Frontal plane motion with pulse bar during heel-toe stance. The bar is moved back and forth in the frontal plane while balance is maintained. The feet may begin in a normal shoulder width position, progressing to the heel-toe position. A light wooden dowel, weight bar or pulse bar can be used. The pulse bar has rolling weights inside that roll to the end creating an end position thrust requiring more of a reactive response for lumbar stabilization.

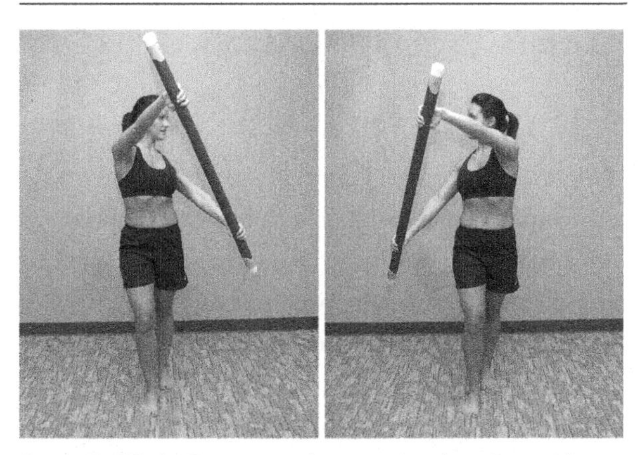

Figure 5.153a,b: Transverse plane rotational motions with a pulse bar, feet in heel-toe position.

Figure 5.154a,b: Sagittal plane motions with a pulse bar, feet in shoulder width position.

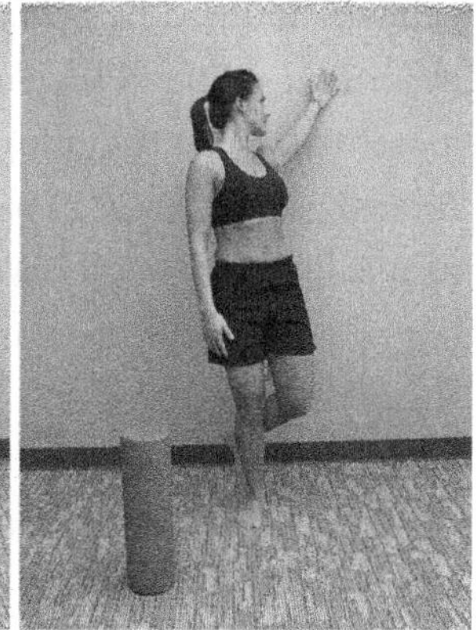

Figure 5.155a,b: Balance reach with unilateral stance in diagonal pattern: anterior-inferior-lateral to superior-medial-posterior. The patient is given an anterior target (bolster) and posterior target (wall) to reach and touch. The opposite foot is to remain off the ground during the reach and the return. The patient is instructed to flex at the hip joint, maintaining a relatively neutral lumbar spine.

Aerobic Exercise

Stage 2 is marked by a reduction in acute tissue pain, with improved tolerance to stress and strain. As soon as the back tolerates normal standing and walking, an aerobic component to rehabilitation should be added. Mannion et al. (2001) found low impact aerobics to be an effective intervention for chronic low back pain patients. Subjects had a better one-year follow up than patients treated with three months of standard physical therapy. Even aerobic leg cycling for people with chronic low back pain has been shown to reduce pressure pain perception for up to 30 minutes (Hoffman et al. 2005). Simple walking will provide a mechanical stress to the lumbar spine to improve general circulation, as well as build low-level endurance. Normalizing segmental mobility or hip mobility is a precursor to beginning an effective walking program. Limited hip extension will cause an excessive shear or torque at the L5/S1 juncture. In this case, walking may be an irritant and slow the rehabilitation process. Lamoth et al. (2004) found that induced pain and fear of pain have subtle effects on erector spinae EMG activity during walking. The global patterns of EMG activity and trunk kinematics were found to be unaffected. Arendt-Nielsen et al. (1996) demonstrated unilateral alterations in EMG activity in chronic low back pain patients, as well as controls in which pain was induced by bolus injection of 5 percent hypertonic saline. Musculoskeletal pain modulates motor performance during gait via reflex pathways. Altered gait observed in patients with low back pain is likely a complex evolved consequence of a lasting pain, rather than a simple immediate effect (Lamoth et al. 2004).

Walking should begin at a normal to slow speed, with normal stride length and arm swing. Swinging the arms is a natural counter balance that reduces forces across the lumbar spine. Holding on to hand rails on a treadmill can change this normal pattern and be less effective. Initially, holding on may be for safety purposes, which can be a necessary compromise. Pathologies that do not tolerate extension, or have pain with compression, should initially avoid down hill walking. Walking down hill biases the lumbar spine in extension and may aggravate these conditions. Conversely, using a slight incline on a treadmill may bias flexion, making a walking program more tolerable.

Walking may begin daily for as little as 10 minutes. Time can be increased as long as no adverse symptoms result. Increasing up to 20–30 minutes should be the goal, but may be progressed gradually over weeks. With poor coordination of the muscles for lumbar stability, or a poor training state, fatigue in the lumbar spine may occur during the walking program, but not be perceived by the patient. As muscles fatigue, greater forces are transmitted into the non-contractile tissues, which may irritate the primary tissue pathology. Any level of increased symptoms during walking, right after walking or the next morning are signs of too long a duration of training. When tissue tolerance allows, walking provides a multitude of positive stimuli to the lumbar spine. The act of walking provides a gentle pumping of lumbar and abdominal muscles to aid in the removal of waste products from guarded tissue. The facets joint cartilage can benefit from the compression/ decompression and gliding associated with rotation and slight flexion and extension. The non-contractile connective tissues, including the annular fibers of the disc can receive an optimal stimulus of modified tension to stimulate fibroblast activity and increased tissue strength.

The elliptical trainer is a common aerobic machine found in health clubs. This machine allows for the walking motion but removes impact forces by removing heel strike. This may be effective in allowing for training degenerative joint pathologies in the lower limbs. Some back patients may also perceive less irritation when using an elliptical trainer, versus normal walking. But by removing heel strike on the elliptical trainer, the normal forces that assist in recruiting muscles in the hip and lumbar spine during walking are reduced. This change in motor firing may aggravate the back patient, and it should not be assumed that the lower impact load from the elliptical is a better training environment than normal walking.

Section 3: Stage 3 Exercise Progression for the Lumbar Spine

Progression to Stage 3 concepts is again done one exercise at a time, as specific functional qualities are attained. By this time in the treatment cycle complaints of pain have been replaced with a more generalized discomfort, except possibly with excessive range or activity. Edema and muscle guarding have resolved with Stage 2 training. Full cardinal plane range of motion is a signal to move to Stage 3 concepts, with more aggressive full range diagonal patterns. The patient may still lack full active range of motion in more complex movement patterns but basic levels of recruitment, timing and coordination have normalized. Complaints usually relate more to an inability or limitation to activities of daily living, work duties or participating in sports. If joint restrictions were present at the beginning stages, they have normalized with an emphasis shifting toward stabilization. If the initial finding was of joint hypermobility than the progression focuses on dynamic stabilization through the full range of physiological motion.

Tissue and Functional States- Stage 3

- *Full arthrokinematic and osteokinematic motion with cardinal planes*
- *Full weight bearing/loaded postures are tolerated*
- *Joint may be painful with excessive repetitions*
- *Edema has resolved*
- *Muscle guarding has resolved*
- *Palpation of primary tissues negative, provocation tests negative, trigger points may be positive to deep palpation*
- *Fair to good coordination in planar motions*
- *Fair to good balance/functional status*
- *May have reduced fast coordination, endurance, strength, power*

As basic clinical signs and symptoms are resolving, Stage 3 focuses primarily on improving motor performance, elevating the training state and returning to normal functional activities. Resistance is increased to improve strength. Coordination may still be part of the training program, but this relates to more complex and functional patterns of movement. Eccentric concepts for dynamic stabilization can affect the design and work order of each exercise.

Basic Outline: Stage 3 Hypomobility

- *Increase weight (60%–80% of 1RM), decrease repetitions (strength)*
- *Eccentric/Concentric exercises to stabilize new range*
- *Tri-planar motions in available range (PNF patterns)*
- *Isometrics to fix strength in gained range*

Dosage

As strength training becomes more of an emphasis, resistance is increased with a subsequent reduction in repetitions. Though the initial focus for motor performance was on timing, coordination and endurance, studies have shown that patients with chronic low back pain have deficits in trunk strength (Addison and Schultz 1980, Brady et al. 1994, Kahanovitz et al. 1989, Mayer et al. 1985, Mayer et al. 1989, McNeill et al. 1980, Nachemson and Lindh 1969, Novy 1999, Smith 1985). Emphasis of training remains on the posterior muscles, as the loss of extensor strength is greater than that of flexor strength (Addison and Schultz 1980, Kahanovitz et al. 1989, Mayer et al. 1989, McNeill 1980). Strengthening in Stage 3 should reflect a functional need of not only anterior and posterior trunk muscles, but also arm and leg strength as well.

Isometric Stabilization

Isometrics are used in Stage 3 into the hypermobile range. This allows higher dosages of strength to be utilized with lower coordinative demands. Isometric training may include the entire trunk with bracing activities, such as performing plank exercises or upper extremity weight training. Additionally, isometric work may specifically focus on the lumbar multifidi in vectors of instability for more segmentally specific stabilization training. This will help to fix strength and sensitize muscle spindles of the LM for greater dynamic stability.

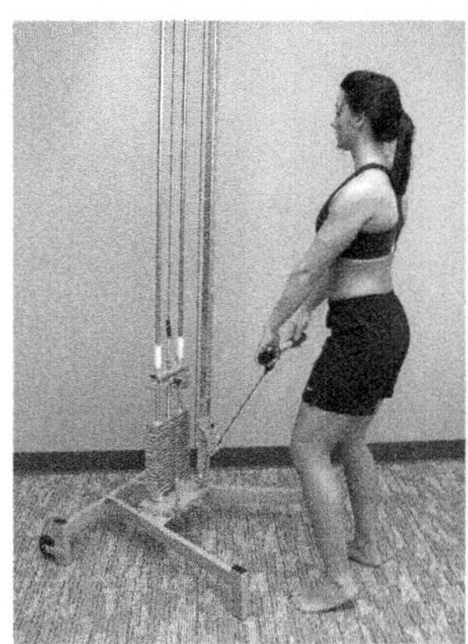

Figure 5.156a,b: Straight arm rows with unilateral stance, or toe-touch stance. The lumbar spine must perform a co-contraction, with emphasis on the posterior musculature, to stabilize from the extension moment created by resisted elevation. Unilateral stance challenges balance and creates a cross pattern from the stance limb, across the pelvis to the opposite side of the lumbar spine.

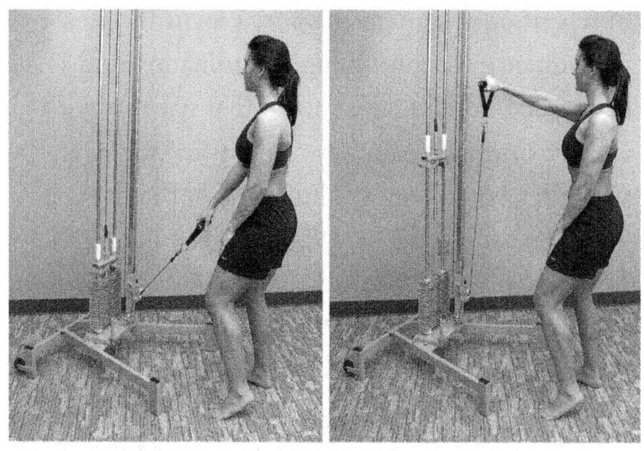

Figure 5.157a,b: Unilateral straight arm row with opposite unilateral stance, or toe-touch stance. Performing the resisted elevation with one arm completes the cross pattern from the left foot to the right shoulder.

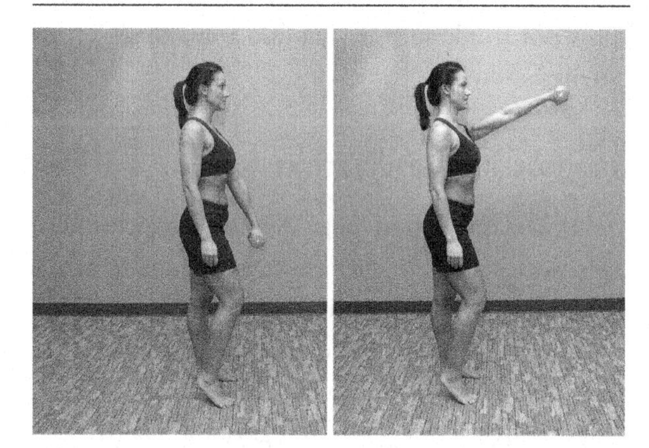

Figure 5.158a,b: Shoulder flexion for lumbar stabilization training, unilateral stance position. Lifting the opposite arm will facilitate a diagonal cross pattern for training the stance leg gluteals, cross over to the opposite side multifidi to the opposite shoulder. Lifting the ipsilateral arm (left arm pictured) will be more challenging to the non-stance leg, as synergistic action of the gluteals does not assist the lumbar multifidi in stabilizing the pelvis.

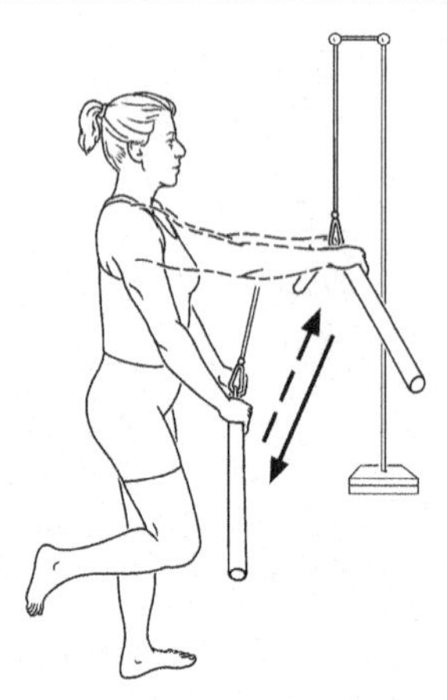

Figure 5.159: Unilateral standing lat pull for lumbar stabilization, as a progression from bilateral stance. The patient is standing with the lumbar spine in neutral with the stance knee slightly flexed. Starting with the pull down bar on the thighs, the bar is raised to shoulder height and returned. The lumbar spine is held in neutral by the abdominals, avoiding the extension moment at the top position of the bar. The unilateral stance creates a cross pattern, or torsion, from the rib cage to the pelvis. To increase the torsional moment, the exercise can also be performed using only the opposite upper extremity.

Figure 5.160: Ballistic military press for lumbar stabilization. The patient stands with a neutral lumbar spine, slightly flexed hips and knee. The bar is repeatedly pressed overhead at a fast rate, requiring quick responses from the lumbar spine to avoid an extension moment on the lift phase.

Figure 5.161: Lateral plank, straight arm with the addition of hip abduction.

Figure 5.162: Prone plank stabilization can be progressed by lifting one hand, lifting one foot, reaching forward or lateral,

extending the hip or placing the hands on separate labile surfaces (shown).

Figure 5.163: Multiple resistance prone plank stabilization training with lateral trunk motion. When lifting an extremity, common compensation is to shift the center of gravity to the contralateral side prior to extremity movement. A manual force, or pulley strap, can be used to prevent the contralateral trunk shift by requiring muscle stabilization toward the side of extremity motion.

Eccentric Stabilization

Stabilization should not be considered an isometric performance of holding a joint position, but a dynamic performance of controlling the joint's instantaneous axis of motion through full range of motion. Movement toward a direction of instability is controlled with agonists and antagonists working in synergy. The agonists perform concentrically, moving the body part toward the pathological range, while the antagonist decelerates the motion eccentrically in preparation for a concentric return. Emphasis is placed on training this eccentric function to decelerate motion by changing the work order. Concentric to eccentric performance is switched to eccentric to concentric performance. The joint is positioned away from the direction of pathology, moving eccentrically first toward the pathology and then concentrically back to the start position. There should be no pause between the transition from eccentric to concentric work. As a plyometric performance, as in jump training, the stretch shortening cycle is utilized with a quick transition between eccentric and concentric work.

The most basic application of this concept is the simple hack squat motion. The exercise begins from a neutral standing position. A squat motion is performed, involving eccentric deceleration into flexion controlled by the posterior lumbar multifidus and erector spinae. At the end of the squat motion, a quick transition of concentric work is made with a return to the neutral starting position. This is a simple sagittal plane version of this concept. This concept of changing the work order moving toward the pathology can be applied to any plane of dysfunction and progress to combined planes with PNF diagonal patterns or more functional patterns.

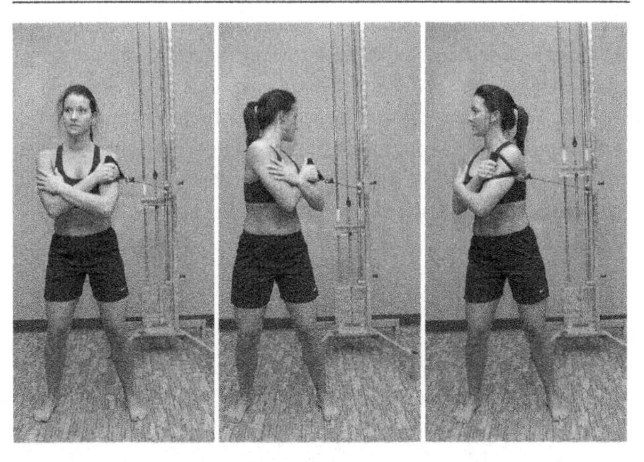

Figure 5.164a,b,c: Eccentric stabilization training for emphasizing the left side lumbar multifidi decelerating left rotation in the presence of a hypermobility to the left. The start position is right rotation (shortened position of the left LM), away from the direction of the hypermobility. Eccentric deceleration is performed toward left rotation, but not into the pathological or painful range. The transition from eccentric to concentric work should occur without a pause.

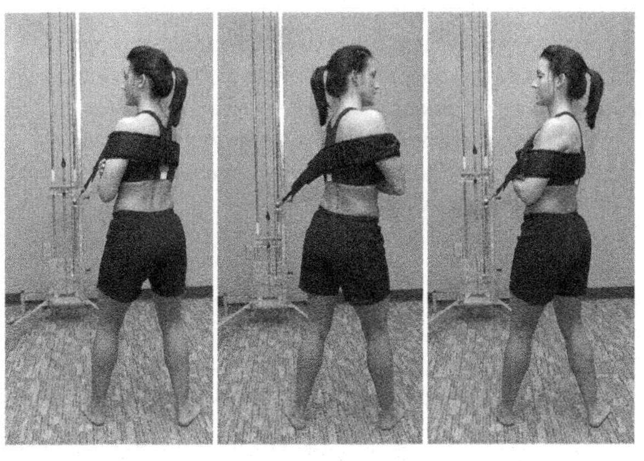

Figure 5.165a,b,c: Eccentric stabilization training emphasizing the right LM for rotation stabilization for a right rotation hypermobility. The start position is left rotation (shortened position of the right LM), opposite of the pathological range of the right rotation hypermobility. Eccentric deceleration is performed toward right rotation, but not into the pathological or painful range, followed by a concentric return. The transition from eccentric work at the end of right rotation to concentric work to the left is quick without a pause.

Figure 5.166a,b,c: Eccentric stabilization training for flexion right rotation stabilization, emphasizing the right LM. The start position is extension and left rotation (relatively shortened position of the right LM), opposite of the pathological range. Eccentric deceleration is performed toward flexion and right rotation, but not into the pathological or painful range, followed by a concentric return to extension and left rotation. The transition from eccentric to concentric work should occur without a pause.

Figure 5.167a,b,c: Eccentric stabilization training for extension right rotation stabilization, emphasizing anterior abdominal muscles and the right LM. The start position is in flexion left rotation, opposite of the pathological range. Eccentric deceleration is performed toward extension and right rotation followed by a concentric return to extension and left rotation. The transition from eccentric to concentric work should occur without a pause.

Figure 5.168: Cranial trunk rotation with a weight bar for eccentric stabilization training of a left rotation hypermobility. The trunk starts in a right rotated position accelerating toward left rotation. Eccentric deceleration to the left is followed by a quick acceleration back toward the right rotated start position.

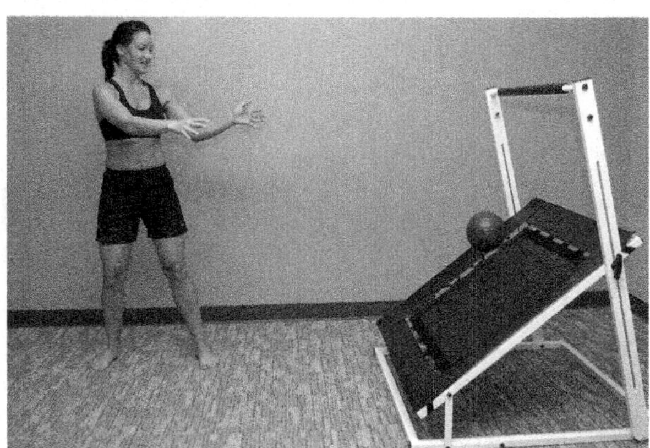

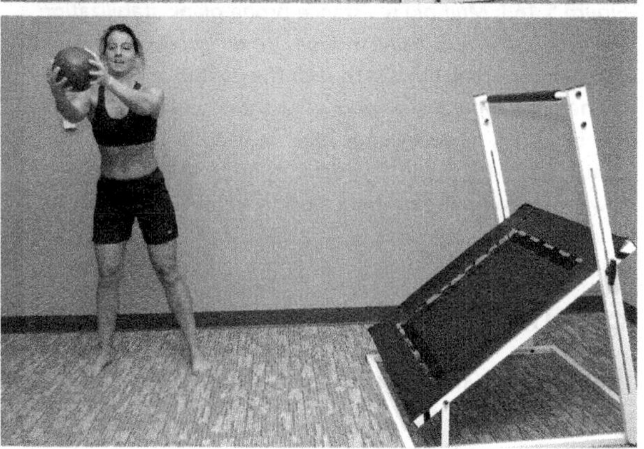

Figure 5.169a,b: Plyoball toss for rotational stabilization training, eccentric emphasis. The patient stands lateral to a rebounder tossing the ball into a left rotation pattern. The ball is caught away from the body with outstreteched arms requiring an eccentric deceleration toward right rotation (hypermobility in right rotation). The patient is instructed to catch the ball stabilizing the lumbar spine and pelvis from excessive right rotation. The concept can be performed in multiple patterns, following the pattern of instability.

Figure 5.170a,b: Sitting eccentric stabilization training for left rotation with a weight bar. The trunk starts in a right rotated position accelerating toward left rotation. Eccentric deceleration to the right is followed by a quick acceleration back toward the right rotated start position.

Figure 5.171a,b: Cranial right rotation sitting on fitness ball, anterior muscle emphasis. Sitting on a labile surface will increase the challenge to the trunk to maintain postural stability. As the sitting posture removes the hips and lower extremities from the activity, and the ball provides support and limits motion of the lower lumbar segments, this approach emphasizes upper lumbar and thoracic stabilization.

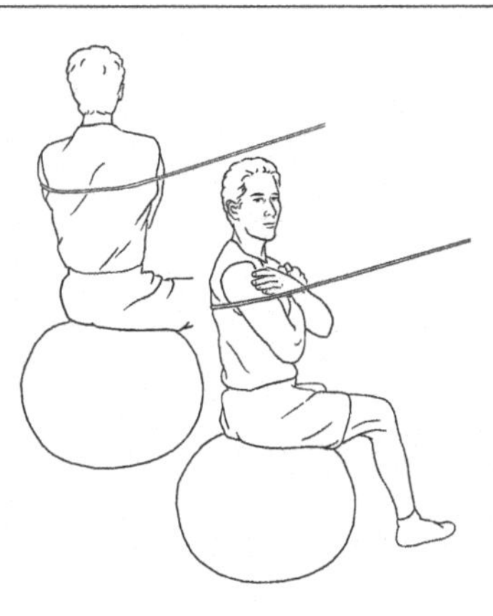

Figure 5.172a,b: Cranial right rotation sitting on fitness ball, posterior muscle emphasis. Sitting on a labile surface will increase the challenge to the trunk to maintain postural stability. As the sitting posture removes the hips and lower extremity from the activity, and the ball provides support and limits motion of the lower lumbar segments, this approach will likely emphasize upper lumbar and thoracic stabilization.

Motions and Directions

General back strengthening and conditioning offers any number of exercise options. The main emphasis should be to get off the floor and focus more on functional exercises and motions that address the individual's needs. Direction of training may be specific, as in the case of stabilization training for hypermobility/instability, or may involve general conditioning. Increased resistance, with increased body position challenges and range of motion increases, make Stage 3 training more difficult but more functionally beneficial.

Extension

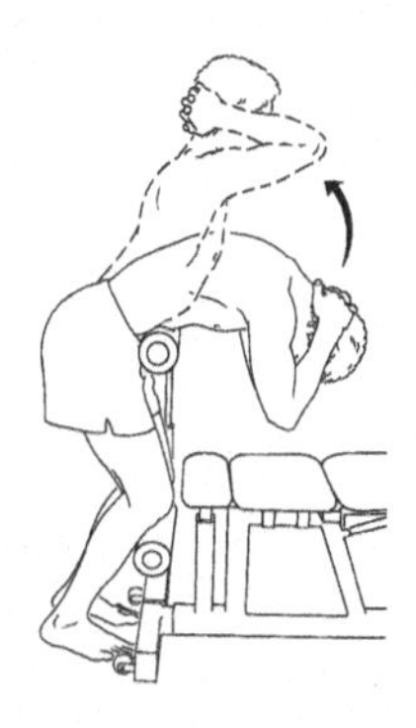

Figure 5.173: Standing cranial extension with pelvic block – modified Roman Chair. This is eccentric to concentric work order first moving eccentrically into the hypermobility of flexion, but not the pathological range of motion and then performing a concentric return. This is done in a more upright position to challenge the functional coordination. The speed would begin slow and progress as coordination improved. The exercise could be progressed by adding tri-planar movement of rotation and side bending along with flexion.

Figure 5.174: Cranial trunk extension with posterior pelvic fixation. If a bench is not available for an anterior fixation, the wall can be used to support the pelvis for training in the clinic or at home. Eccentric flexion should emphasizes cranial motion at each segment with slight anterior tipping of the pelvis with the return emphasizing movement of each segment, rather than the spine moving as a straight unit.

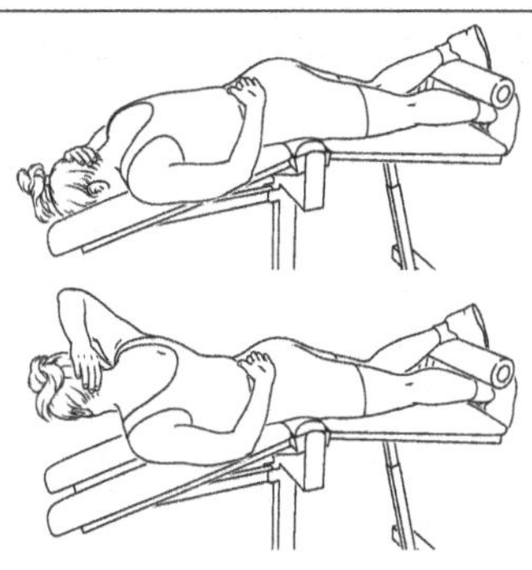

Figure 5.175a,b: Cranial extension combined with right rotation. This example brings in the concept of adding tri-planar movement in Stage 3. A cross pattern from the left hamstrings, across the lumbar spine to the opposite shoulder is created. This example is for a hypermobility into flexion and left rotation and the work order emphasis would start in extension and right rotation and eccentrically move into flexion and left rotation.

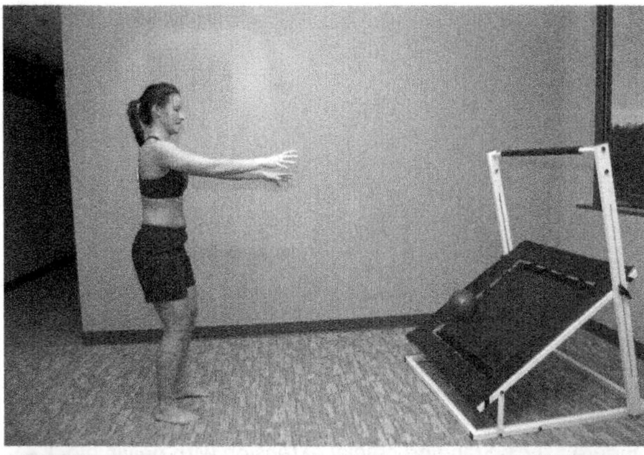

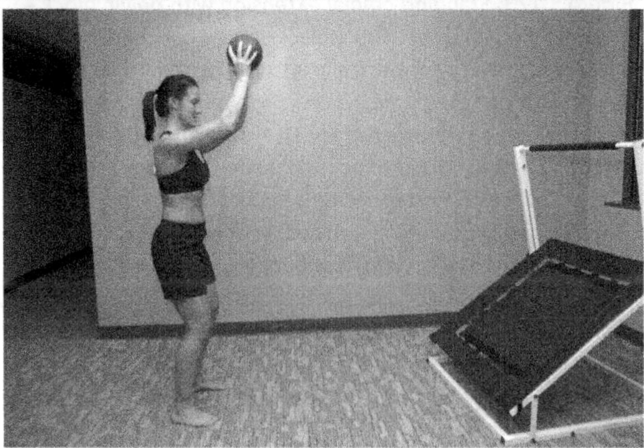

Figure 5.176a,b: Plyoball toss anterior for extension stabilization training of the lumbar spine. The patient stands facing the rebounder, holding the ball overhead. The ball is tossed forward maintaining a stable spine. The patient is first instructed to catch the ball while maintaining a straight and neutral lumbar spine, preventing an extension moment from occurring. This is the isometric phase of the stabilization progression. Once the patient has demonstrated mastery of the isometric phase a progression to eccentric deceleration is added. The patient is instructed to catch the ball and allow a slight extension motion in the lumbar spine. The hands catch the ball allowing it to move over with eccentric stabilization of a slight extension motion in the back.

Flexion

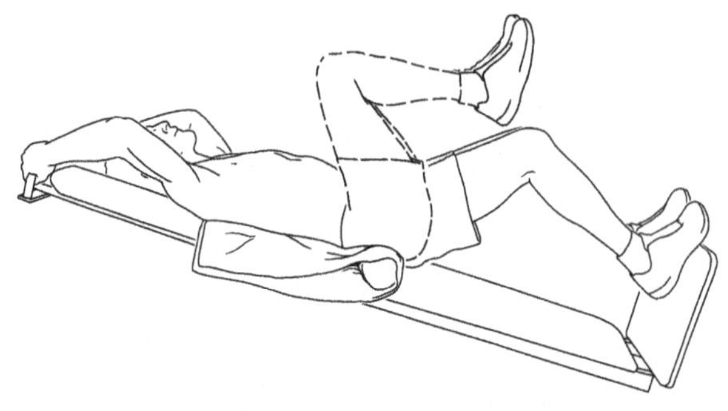

Figure 5.177: Caudal flexion, incline board. Stabilization through the arms will improve abdominal facilitation. A roll under the sacrum to bias flexion is not always necessary, but serves to assist the eccentric control of lordosis when lowering the legs. As coordination and strength improve, the roll can be removed. The work order emphasizes eccentric work followed by concentric work.

Figure 5.178: Step lunge with overhead stabilization. The patient stands with the lumbar spine in neutral and a posterior resistance in both hands from a pulley or elastic resistance. An anterior fall out lunge is performed while maintaining a neutral lumbar spine. This exercise is an aggressive challenge to the lower abdominals to prevent the extension moment created by the resistance. The fall out lunge position also creates a challenge to the spinal extensors, and posterior hip muscles on the forward leg, to hold the position against gravity.

Figure 5.179a,b: Step lunge with unilateral overhead resistance provides a flexion and rotary torque moment. Emphasis is placed on the fall-out lunge position, with the trunk flexed forward at the hip joint to be in line with the extended back leg, to increase the gluteal and lumbar recruitment. The pulley line increases the anterior abdominal recruitment during a motion that is more typically recruiting the posterior hip and trunk muscles.

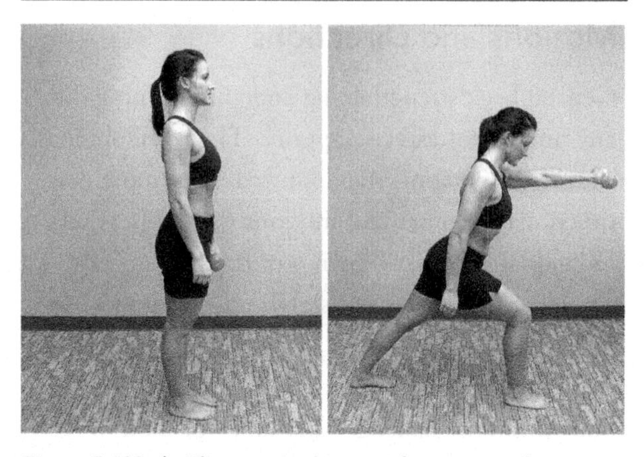

Figure 5.180a,b: The anterior lunge with upper reach using free weights. Again, emphasis of the movement is the right

hip muscles and posterior trunk. The anterior reach creates a greater extension moment, increasing the stabilizing demand on the low back and hip extensors. Performing the motion with only the opposite arm reaching forward (pictured) will also introduce a rotary moment requiring increased recruitment of the deep external rotators of the right hip, as well as the rotary components of the abdominals.

Side Bending

Basic lumbar side bending training can be performed standing with active side bending. The patient is standing with a neutral lumbar spine and the knee slightly flexed. A dumbbell is held at the side for resistance to the contralateral muscles of side bending. To emphasize mobility to the same side, the opposite arm rests at the side. To emphasize lengthening soft tissues on the opposite side, including fascial tightness in the latissimus dorsi, the hand is placed behind the head.

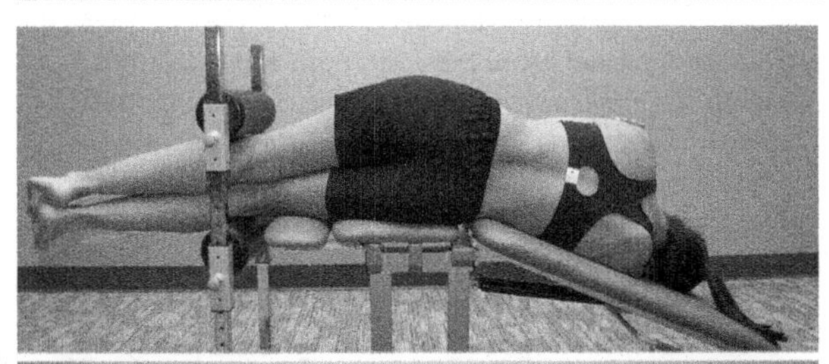

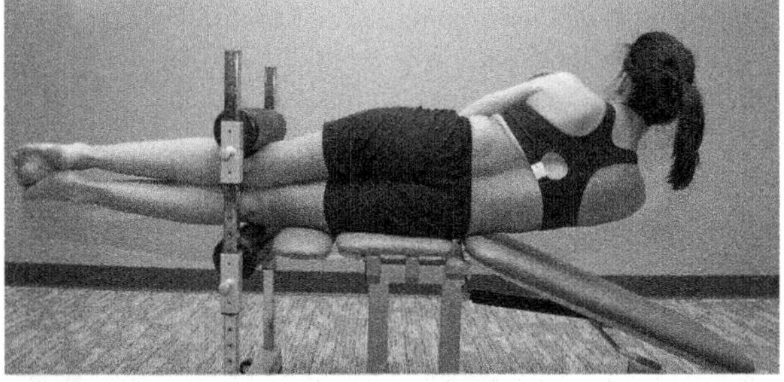

Figure 5.181a,b: Cranial side bending. Cranial side bending is progressed from vertical and incline to the most challenging position in horizontal.

Figure 5.182: Lateral lunge. This exercise dose not directly challenge lumbar side bending, but does train lateral hip strength. Improved lateral hip stability will assist stabilization of the pelvis in the frontal plane. Free weight and arm motions can be added for additional challenge.

Rotation

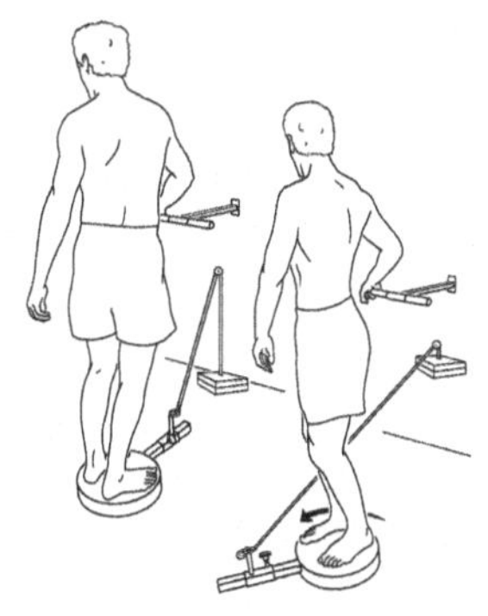

Figure 5.183a,b: Standing caudal right rotation (relative to cranial segment) with the rotation plate. The patient stands with the lumbar spine in neutral, slight flexion at the hip and knees. The shoulders are stabilized through the arm preventing upper trunk motion. The movement is initiated at the lumbar spine, not at the ankles.

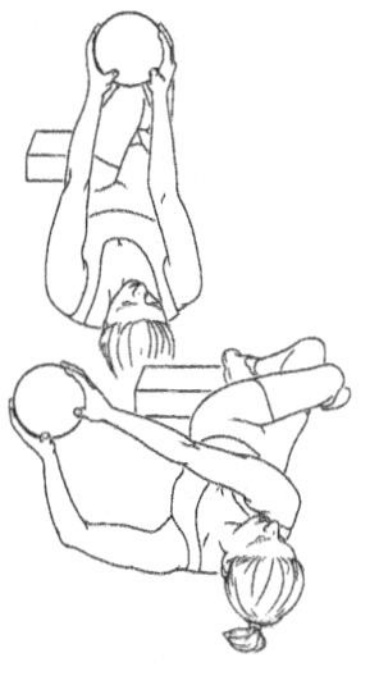

Figure 5.184a,b: Caudal hip rotation can be progressed with an upper quarter weight moving in opposite direction of the knees. The leg is crossed over to increase the resistance by moving weight laterally. An additional challenge can be added by holding both feet off the ground while performing the exercise.

Figure 5.185: Standing right rotation with bar attached to pulley. Using a bar attachment increases the lever arm to increase the challenge for lumbar stabilization, as well as increase erector spinae recruitment in the thoracic spine. Eccentric emphasis continues by starting in full right rotation, decelerating into left rotation, followed by concentric return to the right.

Removal of Locking

By Stage 3 all coupled force locking, counter curve locking and artificial locking are removed. Emphasis is on coordinative locking; the patient dynamically controls the exercise without going into the painful or pathological range. Standing locking exercises performed in Stage 2, with a staggered foot stance to create caudal locking, are progressed by removing locking with a parallel foot stance requiring dynamic control.

Functional Balance Training

By Stage 3 the patient should have already achieved significant improvement in balance activities for basic activities of daily living. Should a higher level of balance be required for sport performance, continued balance challenges can be added. Performing all previous exercises with labile surfaces, eyes closed or both can provide a significant challenge. A balance component may be added to a previous exercise or movement pattern. For example, a hack squat could be performed for lifting training, but rather than standing on the floor, the feet may be on a labile surface. Lunges may be changed to have the foot land on a labile surface, rather than the floor. These challenges can be changed with each set of a three set exercise, having a variation in demand to reduce the monotony of the exercise. A balance challenge will reduce the ability to use a heavier load. If the real goal is strengthening for activities such as squatting or lunging, then a firm surface with heavier weights is a better functional choice.

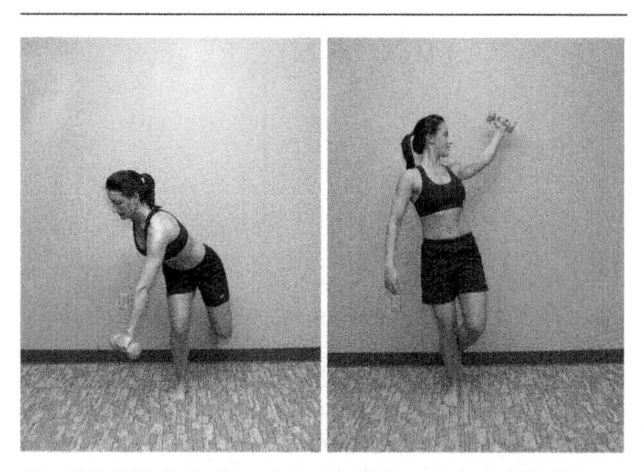

Figure 5.186a,b: Unilateral stance, balance reach diagonal patterns with free weight resistance. Emphasis is placed on mobility, strength and stabilization through the hip joint with the pelvis moving on the femur. The lumbar spine remains neutral while the hip flexes, adducts and internally rotations with the hand reaching forward. The hip is then extended abducted and externally rotated with the hand reaching back toward the wall. The addition of the free weight increases the demand on extensor muscles of the hip and trunk.

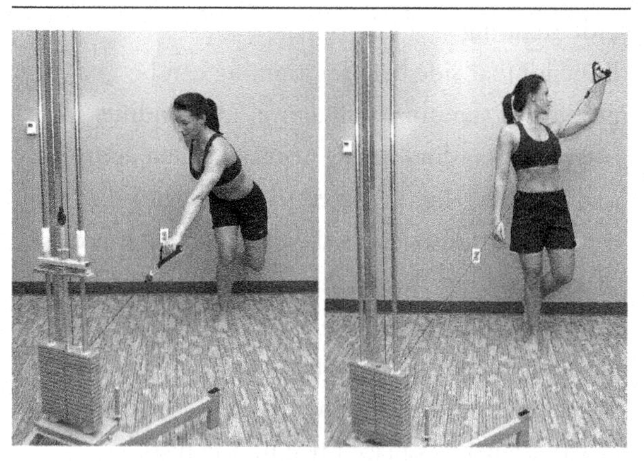

Figure 5.187a,b: Unilateral stance, balance reach diagonal patterns with pulley resistance. As with the previous exercise, emphasis is placed on mobility, strength and stabilization through the hip joint with the pelvis moving on the femur. A diagonal pattern is performed in the range that can be coordinated and stabilized. The lumbar spine remains neutral while the hip flexes, adducts and internally rotations with the hand reaching forward. The hip is then extended abducted and externally rotated with the hand reaching back toward the wall. Compared to using a free weight, the pulley will provide an additional rotational vector of resistance, with slightly less of an extension moment.

Figure 5.188a,b: Squat reach with a pulley while standing on a labile surface (BOSU pictured). Maintaining a neutral lumbar spine, the patient performs a squat while reaching forward toward the pulley. Standing back up straight, the pulley handle is pulled in toward the lower abdomen. The exercise should first be coordinated on a stable flat surface prior to progressing to a labile surface. The addition of the labile surface will further challenge the postural reflexes for combined lower extremity, pelvis and trunk stabilization training. The labile surface will emphasize postural control over strengthening. When moving to a labile surface the resistance must be decreased.

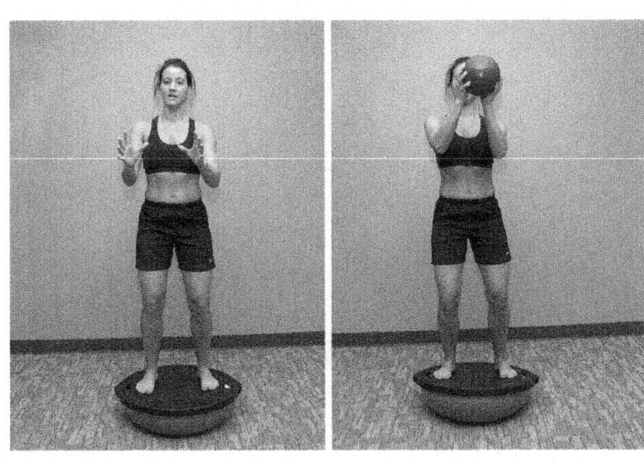

Figure 5.189a,b: Catch and throw activities on a labile surface (BOSU). The patient stands on a labile surface while performing reactive exercises, i.e., catching and throwing a ball. Concentration on the activity focuses attention off the balance requirement of the exercise, attempting to train postural reflexes for lumbar and lower extremity stability.

Functional Exercises

Continuing the progression of more complex movement patterns, exercises should begin to duplicate more required functional tasks, work or sport performance. Issues of work order, speed and range of training may not be as relevant as training normal motion. Exercises can be customized to mimic specific work related tasks, such as lifting, pushing or pulling. More sport specific exercises may begin with functional patterns related to the sport performance, with eventual progression to sport simulation.

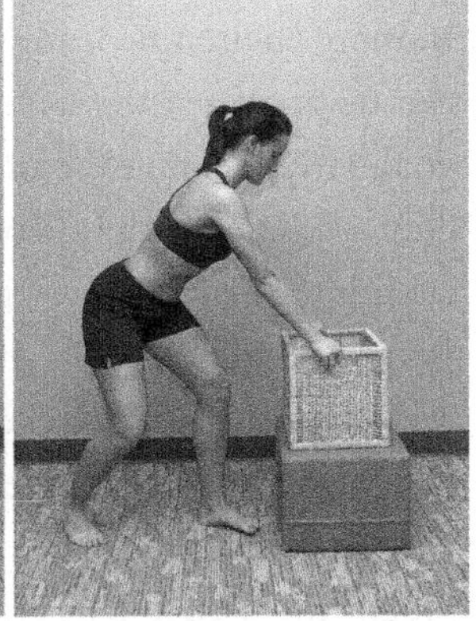

Figure 5.190a,b: Crate lifting. Functional lifting patterns for job requirements can be trained with free weights or crate lifting. The patient is instrucred in proper lower extremity use for lifting, including flexing at the knees and hips. Depending upon the lifting pattern being utilized, emphasis may also be on pivoting through the feet and lower extremities, as opposed to twisting through the trunk. Previous training may emphasize on the squat and pivot prior to adding lifting.

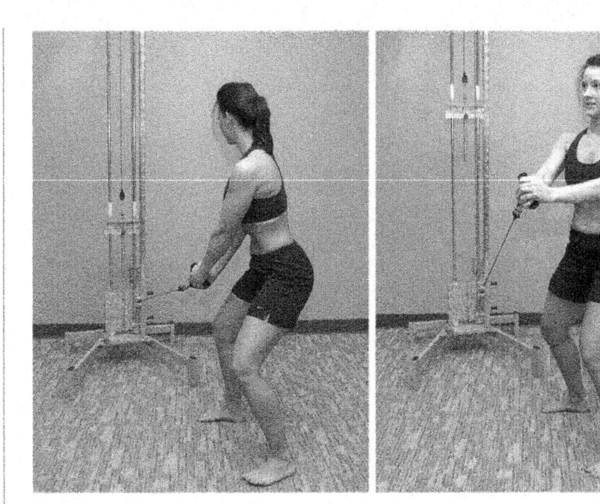

Figure 5.191a,b: Diagonal patterns—50% range. As a progression from horizontal rotation to full diagonal lifting patterns, the pulley bracket can be lowered to knee height. The diagonal pattern is performed through half the range of motion. Emphasis may be placed on proper weight shifting, maintaining foot (left) pronation during the initial push-off, lifting with the hips and following through with the arms. As coordination is achieved, the bracket can be gradually lowered to floor level for full range of motion training.

Figure 5.192: Functional pushing motions with isometric lumbar stabilization using a double pulley. Work simulation for pushing motions can be performed with stabilization through the lower limbs or while walking forward, as in pushing a cart.

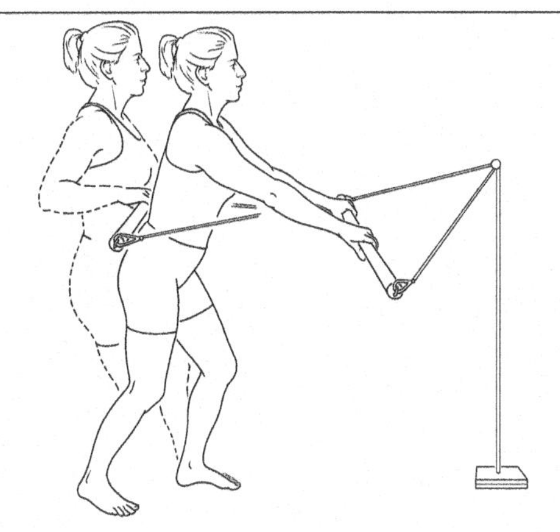

Figure 5.193: Functional pulling motions with isometric lumbar stabilization using a double pulley. Work simulation for pulling motions can be performed with stabilization through the lower limbs or while walking backward, as in pulling a cart.

Figure 5.194: Standing elastic rowing. Elastic resistance may be used for home training of functional pulling or pushing motions with isometric lumbar stabilization.

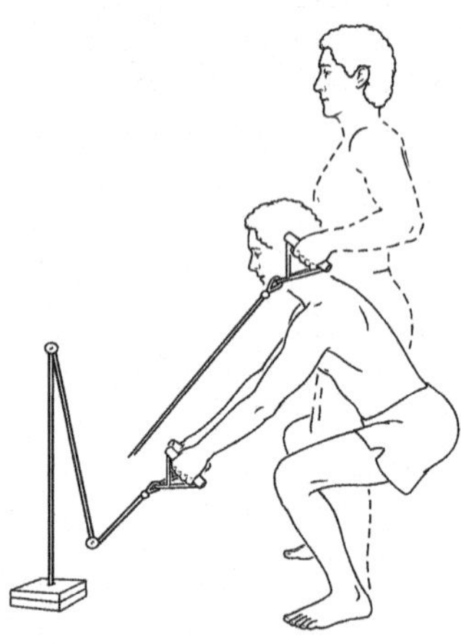

Figure 5.195: Deep squat and lift with pulley or cable system. The patient is instructed to maintain a neutral lumbar spine while performing a deep squat and anterior reach toward the pulley. Returning to a standing position is initiated with hip and knee extension, not trunk extension.

Figure 5.196a,b: Anterior lunge with anterolateral reach using free weights. The patient performs a fallout lunge, with the back knee extended, heel on the ground and trunk in line with the back leg. Free weights are lifted anterior to challenge the extensors and then rotated out laterally to provide a rotary challenge to the trunk and right hip.

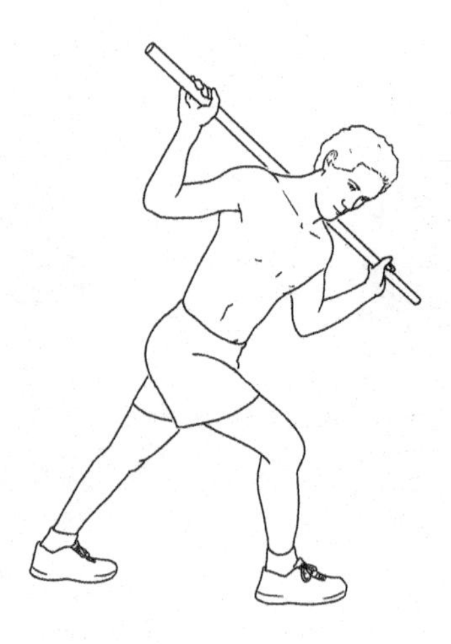

Figure 5.197: Anterior fallout lunge with trunk left rotation – with a weight bar.

Exercise Classes

Stage 3 is marked by resolution of acute pain, improved tissue tolerance, normalization of range of motion and joint mobility. Coordination of basic movement patterns have been established, so training goals focus on more advanced stabilization and improving general conditioning. Exercises are transitioned into more functional movements. Patients may have worked through Stages 1 and 2 to be ready for Stage 3 concepts. Chronic back pain patients may begin therapy already at this level of function, and can bypass some of the early training concepts and move into more of a general conditioning program for the low back. Utilizing an exercise class or circuit program provides a group training environment in which several patients can be trained at the same time. Klaber Moffett et al. (1999) found the exercise class setting to be clinically effective and have better outcomes than traditional general physician management. The program was also more cost effective. Soukup et al. (1999) used the Mensendieck approach from Scandinavia that combines exercises and education in the prevention of low back pain. The program taught the participants ergonomic principles for movements of daily activities and improved their knowledge related to prevention of low back pain. The group setting is ideal for education issues in which a group can be trained at the same time, and benefit from each other in the learning process. Yoga has also been shown to improve function and reduce chronic low back pain, with benefits persisting for at least several months (Sherman et al. 2005). This study, however, compared yoga to a self-care book rather than a more specific lumbar rehabilitative program. Galantino et al. (2004) performed a controlled trial of Yoga reported improvements in flexibility, balance and disability, though findings were not statistically significant.

Physical therapy treatment involves more specific education and exercise training than general exercise classes. Biomechanical deficits and abnormal movement patterns are not best solved in generalized training programs. But once these clinical findings are resolved with specific manual and active training, the patient can be progressed into a more generalized exercise program that is consistent with their long term goals, time commitments, convenience and personal interest.

Section 4: Stage 4 Exercise Progression for the Lumbar Spine

Stage 4 is defined by resolution of primary clinical signs and symptoms. The patient has achieved full mobility and coordination with dynamic stability. Exercises include coordination of tonic and phasic muscle systems working in synergy through the entire range of motion around a physiological axis. Emphasis is on elevation of the overall training state, improving endurance and strength. Higher end power training may be necessary for more athletic goals.

Tissue and Functional States - Stage 4

- *Full active and passive range of motion*
- *Pain free joint motion with exercise*
- *Good coordination*
- *Limitations in endurance and strength with functional performance*

Resistance

Training for endurance and strength should be no less than 75% of 1RM, and up to 90% of 1RM for athletes. Increasing resistance is not only a factor of muscle performance, but must also consider tissue tolerance. Collagen and cartilage can take up to a year or longer to fully repair. Excessive loading early during rehabilitation may be tolerated by the motor system, but the underlying tissues may not yet be ready for high compression and/or tension associated with heavy resistance training. As weight is initially increased, speed may decrease slightly to allow for coordination at the new level of resistance. As performance improves, speed will naturally increase, as seen in earlier stages marking improved coordination. Higher resistance will increase the level of tissue stress and strain, further improving the tissue tolerance with more functional activities. Higher loads will also increase hypertrophy of muscle, making more permanent changes in strength and performance.

Basic Outline Stage 4

- *Tri-planar motion through full range of motion around the physiological axis (coordination, endurance, strength, hypertrophy).*
- *Endurance, speed, hypertrophy, strength and power—80% of 1RM.*
- *Functional exercises and retraining for activities of daily living, sport and job activities.*

Motions and Directions

In Stage 4 the exercises are functional and task specific. The long-term goals of the patient will determine the types of exercises in this stage. Long-term goals for return to activities of daily living, work or sport should predetermine the type of movements that should be trained by Stage 3. Similar patterns should already have been added in Stage 3 with a simplified form through reducing normal range of motion, speed, resistance or planes of motion involved. If the end goal is be able to pick up the newspaper in the morning without needing to call the neighbor for assistance, you can imagine the Stage 4 exercises would look different than for someone returning to work lifting in a warehouse or performing pole vaulting.

All directions should be worked with coordination of concentric and eccentric contractions of tonic (arthrokinematic) and phasic (osteokinematic) muscle groups through the entire functional range of motion. Higher-level training should emphasize functional tasks but also address the key elements for the long term stabilization needs of the individual. One "best" exercise does not exist, but should be matched to the initial tissue injury and motor control issues.

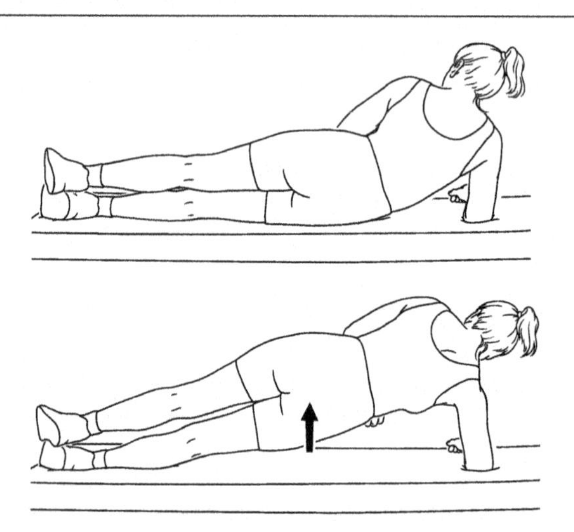

Figure 5.198a,b: Respiration emphasis: Lateral bridge after aerobic activity. Isometric bracing is coordinated with antagonistic breathing challenge. Emphasis is not only on stabilizing the position but maintaining normal respiration, rather than breath holding, during the exercise. Performing the exercise immediately following aerobic training, in which the respiration rate is elevated, will further challenge the ability to stabilize with lower abdominal contraction while allowing for a diaphragmatic respiration pattern. A breathing challenge during stabilization training can cause the supporting musculature of the spine to drop to inappropriately low levels in some people (McGill et al. 1995).

Strength Training

As with all functional qualities, strength training is a function of dosage and not related to a particular set of exercises. Resistance is increased to 80% 1RM or greater with a reduction in repetitions. Exercises in this stage should involve multiple large muscle groups through greater ranges of motion. Increased time of rest breaks may be necessary for proper muscle recovery between sets, from up to one minute or greater.

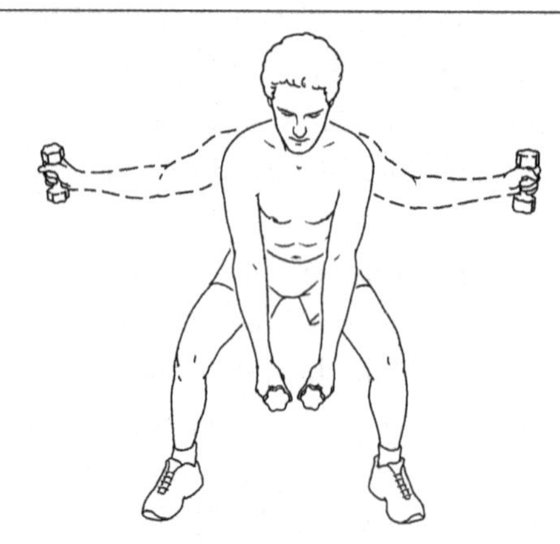

Figure 5.199: Forward bent shoulder fly. The patient is standing with the knees slightly flexed, the lumbar spine is in neutral while the trunk is flexed forward at the hips 30–60°. The lumbar spine remains in neutral while the arms perform horizontal abduction.

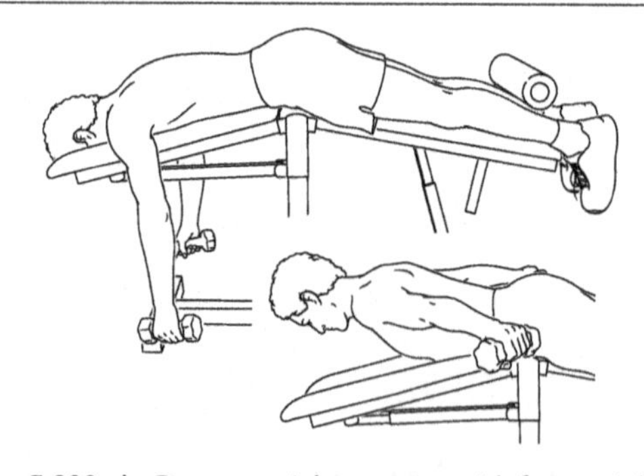

Figure 5.200a,b: Prone cranial extension with free weight resistance. The patient is prone with the feet anchored, the hips in slight flexion. The shoulders perform a resisted extension motion during trunk extension. The angle of the arms can be changed to horizontal abduction during the exercise. As a progression, unilateral arm motions in elevation will create extension and trunk rotation.

Figure 5.201a,b: Squat-overhead row. Progressing upright rowing includes starting from a squat position then extending the hips and knees while lifting the bar overhead.

Figure 5.202: Anterior lunge with resisted trunk rotation. Attaching a bar to a pulley system will emphasize rotation with a horizontal resistance. Using a heavier free weight bar, without a pulley or cable system, emphasizes the extension moment from gravity.

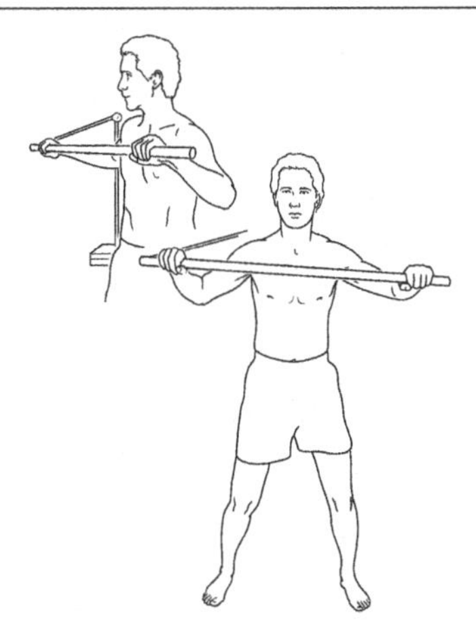

Figure 5.203a,b: Cranial rotation with long lever arm. The patient is standing with their back to the pulley, the lumbar spine in neutral, with the knees and hips slightly flexed. Trunk rotation is performed while maintaining fixed shoulder girdles, pelvis and lower limbs.

Figure 5.204: Standing trunk rotation with a double pulley, bar attachment. A bar is placed behind the shoulders with alternate rotational motions requiring a co-contraction to stabilize the extension moment while performing rotation.

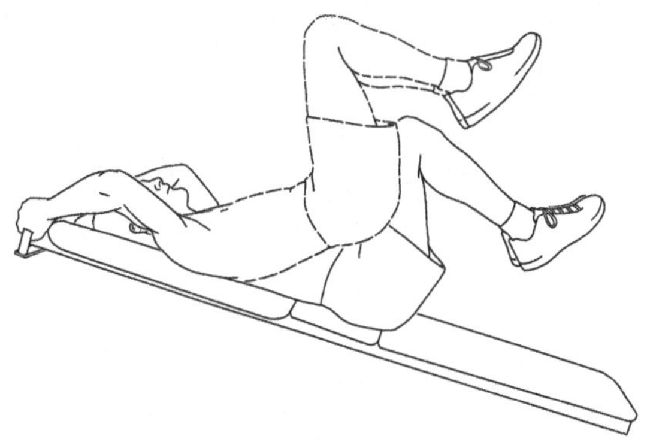

Figure 5.205: End stage training caudal flexion on an incline board. The arms are fixated as the hips are flexed to a vertical position, performed with a smooth transition from hip flexors to lower abdominals. With the thighs in a vertical position, the back is extended lifting the knees toward the ceiling and then lowering the pelvic back down on the board.

Functional Movement Patterns

More functional and global patterns are necessary when returning to higher-level jobs and athletic activity. If the spine is not challenged to full range functional patterns, the patient is at risk of re-injury when returning to these activities. The tissues involved require progressive stress and strain through the functional ranges in which they will be challenged. The motor system requires a training progression to establish coordination through the full range of motion around a physiological axis. As this is achieved, progressive training involves strength training to elevate the training level to above the initial level that the patient was injured.

Figure 5.206a,b: Functional pushing while walking forward. A speed pulley with two lines attached to a bar allows for functional simulation of pushing, as in training flight attendants to push a cart while walking forward. The trunk is maintained in neutral throughout the performance.

Figure 5.207: Pushing exercise as above with a double pulley, sport cord or elastic resistance. Elastic resistance will increase the challenge with a increasing resistance toward the end of the motion.

Figure 5.208: Changing the angle of the arms and the vector of pulley to above the head will increase the length of the lever arm from the resistance to the lumbar spine. This arm position creates greater involvement of the pectoral and upper abdominal muscles.

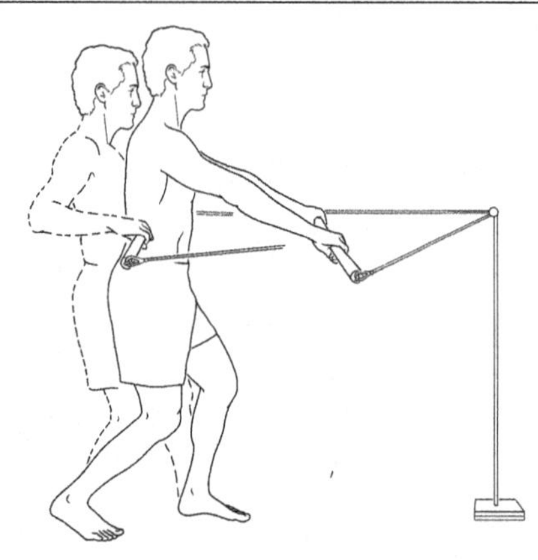

Figure 5.209: Functional pulling exercise for lumbar stabilization training. A speed pulley with two lines attached to a bar, allowing for functional simulation of pulling while walking backward. The lumbar spine is maintained in neutral throughout the performance, avoiding a flexion moment in the trunk or at the hips.

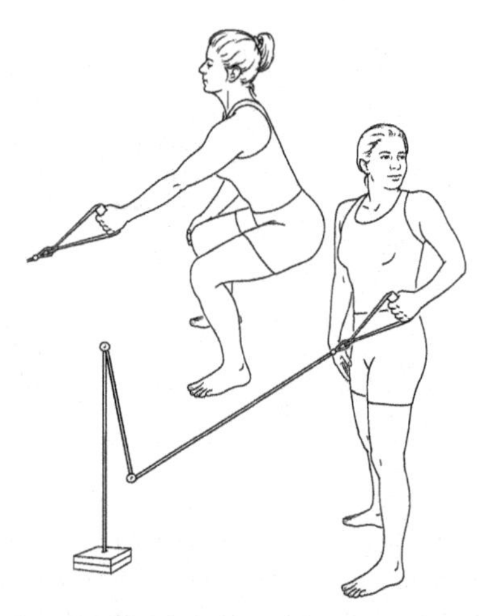

Figure 5.210a,b: Unilateral squat lift with trunk rotation. A further challenge for the squat lift motion is to perform the exercise unilaterally with rotation of the trunk.

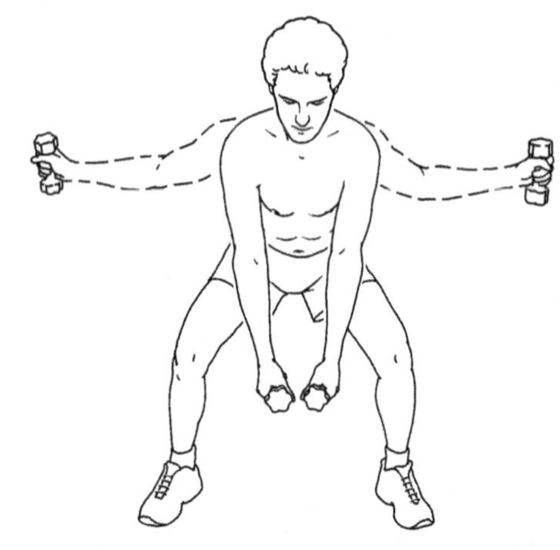

Figure 5.211: Decline shoulder flys. The patient is standing with the trunk flexed forward at the hips, with the lumbar spine in neutral. Free weights are lifted out to the side.

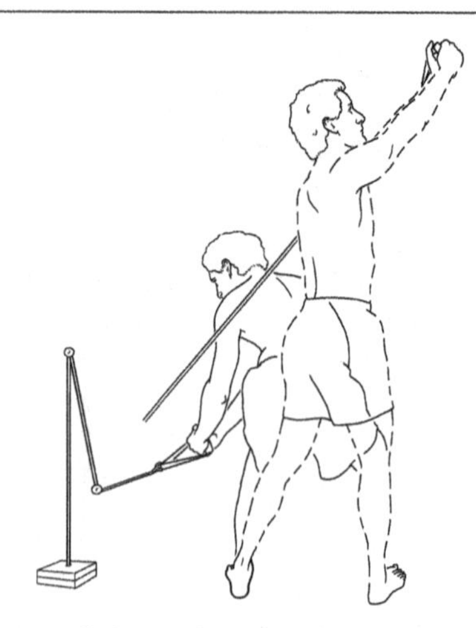

Figure 5.212: Full diagonal patterns from extension and right rotation pattern of the trunk. Emphasis is placed on maintaining the left foot in neutral or pronation to allow for push off with the left lower limb during the lifting moment. Allowing the left heel to roll into supination while bending and reaching to the left does not allow the lower leg to push off, increasing the torque moment on the low back.

Figure 5.213a,b: Full diagonal trunk pattern from extension and left rotation to flexion and right rotation. Emphasis is on pivoting through the hip joints to minimize the amount of motion in the lower lumbar spine.

Figure 5.214a,b: Trunk diagonal pattern of extension and right rotation with bilateral stance and unilateral left arm motion resisted by a pulley. The full range diagonal pattern can be performed through one upper extremity to increase the rotary torque through the trunk and to assist with load transfer through the shoulder to the pelvis. This pattern will emphasize anterior abdominal strength for trunk rotation.

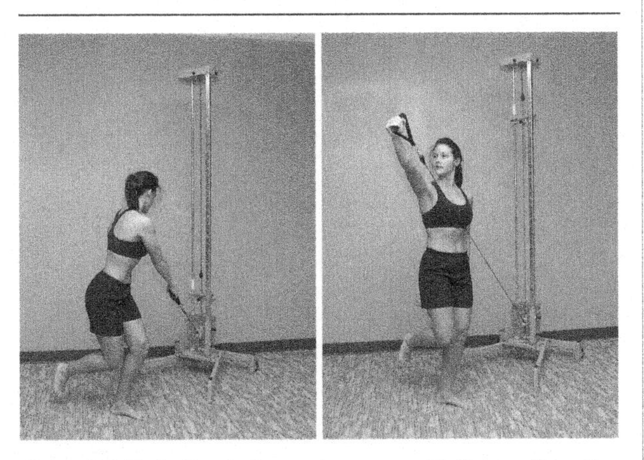

Figure 5.215a,b: Trunk diagonal pattern with right unilateral stance and unilateral right arm motion resisted by a pulley. Unilateral stance will further challenge balance and stabilization from the lower extremity and pelvis through the lumbar spine to the shoulder. The right arm diagonal pattern will emphasize posterior muscles of the trunk and deep hip external rotators of the left lower extremity.

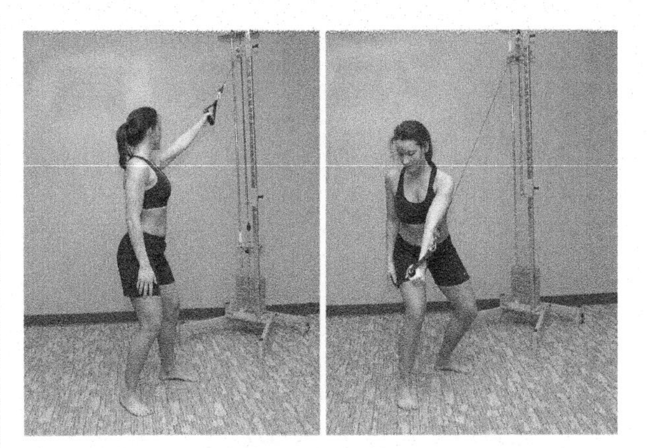

Figure 5.216a,b: Trunk diagonal pattern into flexion and right rotation, bilateral stance and unilateral left arm motion resisted by a pulley. Anterior abdominal and hip flexor muscles are emphasized, with force transfer from the shoulder into the trunk.

Figure 5.217a,b: Trunk diagonal pattern into flexion and right rotation, bilateral stance and unilateral right arm motion resisted by a pulley. The arm pattern emphasizes posterior lumbar muscles for rotation, while the flexion pattern emphasizes anterior abdominals and hip flexors.

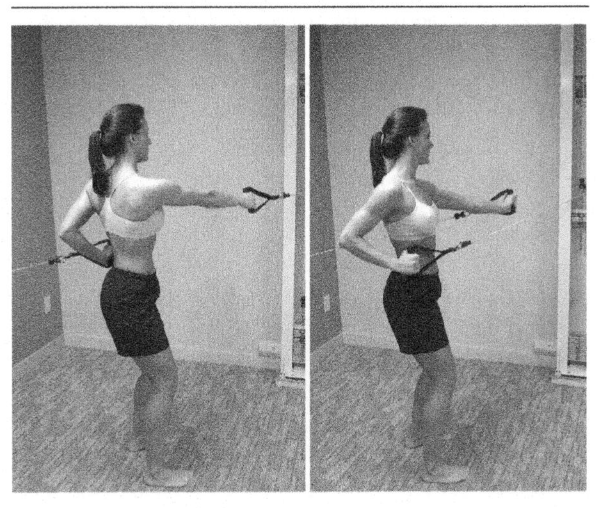

Figure 5.218a,b: "Push-me, Pull-me's" synergistic right trunk rotation multiple resistance. The patient is standing with the lumbar spine in neutral and the hips and knees slightly flexed. The left arm pushes forward as the right arm pulls backward. The trunk is allowed to rotate during the motion.

Figure 5.219a,b: "Pull and Punch". The patient pulls back with the right hand while punching forward with the left. Weight is transferred to the front foot during the movement.

Figure 5.220: Fallout lunge with barbell resistance. The fallout lunge is performed as previously described, with the back knee straight and the trunk flexed forward at the hips. A weight bar is placed on the shoulders to add resistance for strength training.

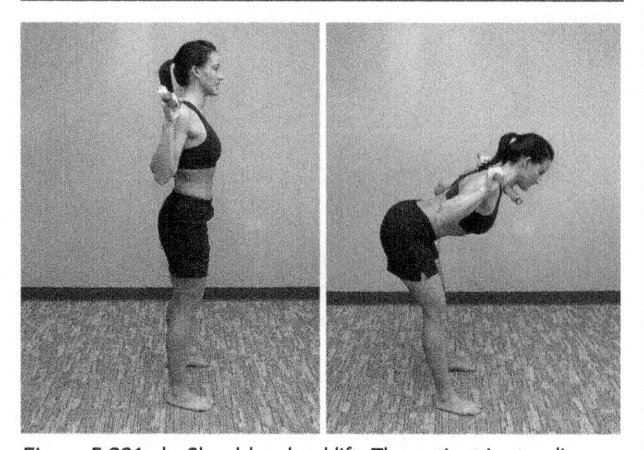

Figure 5.221a,b: Shoulder dead lift. The patient is standing with a neutral spine and a bar on the shoulders. The lumbar spine remains in neutral as the pelvis is flexed forward at the hip joints. The knees remain slightly flexed. This is an aggressive strengthening exercise for the hamstrings, hip extensors and lumbar spine.

Exercise Post Treatment

Regular exercise after outpatient active treatment is related to better outcome in the long-term regarding both the recurrence of chronic low back pain and work absenteeism (Taimela et al. 2000). For chronic low back pain, benefit from a home training program have been shown to be as effective as the supervised dynamic strength muscle training program, yielded lasting improvement after at least one year of adherence. The adherence rates, however, were much better when the training was supervised at the start (Bentsen et al. 1997). Ljunggren et al. (1997) found no difference in sick leave comparing one year of supervised physical therapy exercise with a home exercise program. The home program however, did involve an initial training phase with physical therapy, a follow-up visit every six weeks to adjust the program and two telephone follow-ups. Motivation and program adherence can be enhanced by clinical visits, but the frequency of treatment can certainly be reduced, with emphasis on independent long-term training.

Bibliography

Addison R, Schultz A. Trunk strengths in patients seeking hospitalization for chronic low-back disorders. Spine 5(6):539–44, Nov–Dec, 1980.

Alaranta H, Hurme M, Einola S, Kallio V, Knuts LR, Torma T. Rehabilitation after surgery for lumbar disc herniation: Results of a randomized clinical trial. Int J Rehabil Res 9:247–257, 1986.

Alexander KM, LaPier TL. Differences in static balance and weight distribution between normal subjects and subjects with chronic unilateral low back pain. J Orthop Sports Phys Ther 28(6):378–83, Dec, 1998.

Ambroz C, Scott A, Ambroz A, Talbott EO. Chronic low back pain assessment using surface electromyography. J Occup Environ Med 42:660–9, 2000.

Anderson GB. Epidemiological features of chronic low-back pain. Lancet 354(9178): 581–5, Aug 14, 1999.

Andersson E, Oddsson L, Grundström H, Thorstensson A. The role of the psoas and iliacus muscles for stability and movement of the lumbar spine, pelvis and hip. Scand J Med Sci Sports 5(1):10–6, Feb, 1995.

Annett J. The learning of motor skills: sports science and ergonomics perspectives. Ergonomics 37(1):5–16, Jan, 1994.

Arendt-Nielsen L, Graven-Nielsen T, Svarrer H, Svensson P. The influence of low back pain on muscle activity and coordination during gait: a clinical and experimental study. Pain 64(2):231–40, Feb, 1996.

Aure OF, Nilsen JH, Vasseljen O. Manual therapy and exercise therapy in patients with chronic low back pain: a randomized, controlled trial with 1-year follow-up. Spine 28(6):525–31, Mar 15, 2003.

Axler CT, McGill SM. Low back loads over a variety of abdominal exercises: searching for the safest abdominal challenge. Med Sci Sports Exerc 29(6):804–11, Jun, 1997.

Barnett F, Gilleard W. The use of lumbar spinal stabilization techniques during the performance of abdominal strengthening exercise variations. J Sports Med Phys Fitness 45(1):38–43, Mar, 2005.

Basmajian JV. *Muscles Alive, Their Functions Revealed by Electromyography*. Williams & Wilkins, 1979.

Bendix AF, Bendix T, Labriola M, Boekgaard P. Functional restoration for chronic low back pain. Two-year follow-up of two randomized clinical trials. Spine 23(6):717–25, Mar 15, 1998.

Bendix T, Bendix A, Labriola M, Haestrup C, Ebbehoj N. Functional restoration versus outpatient physical training in chronic low back pain: a randomized comparative study. Spine 25(19):2494–500, Oct 1, 2000.

Bentsen H, Lindgarde F, Manthorpe R. The effect of dynamic strength back exercise and/or a home training program in 57-year-old women with chronic low back pain. Results of a prospective randomized study with a 3-year follow-up period. Spine 22(13):1494–500, Jul 1, 1997.

Bergmark A. Stability of the lumbar spine. A study in mechanical engineering. Acta Orthop Scand Suppl 230:1–54, 1989.

Biering-Sorensen F. Physical measurements as risk indicators for low-back trouble over a one-year period. Spine 9(2):106–19, Mar, 1984.

Biolo G, Maggi SP, Williams BD, Tipton KD, Wolfe RR. Increased rates of muscle protein turnover and amino acid transport after resistance exercise in humans. Am J Physiol 268(3 Pt 1):E514–20, Mar, 1995.

Bodack MP, Monteiro M. Therapeutic exercise in the treatment of patients with lumbar spinal stenosis. Clin Orthop Relat Res (384):144–52, Mar, 2001.

Bogduk N, Pearcy M, Hadfield G. Anatomy and biomechanics of psoas major. Clin Biomech 7:109–19,1992.

Brady S, Mayer TG, Gatchel RJ. Physical progress and residual impairment quantification after functional restoration. Part II: isokinetic trunk strength. Spine 19(4):395–400, Feb 15, 1994.

Brennan GP, Shultz BB, Hood RS, Zahniser JC, Johnson SC, Gerber AH. The effects of aerobic exercise after lumbar microdiscectomy. Spine 19(7):735–739, Apr 1, 1994.

Brinckmann P, Biggemann M, Hilweg D. Prediction of the compressive strength of human lumbar vertebrae. Spine 14(6):606–10, Jun, 1989.

Brinckmann P, Biggemann M, Hilwig D. Fatigue fracture of human lumbar vertebrae. Clin Biomech (Bristol, Avon) 3(Suppl 1):1–23, 1988.

Bronfort G, Goldsmith CH, Nelson CF, Boline PD, Anderson AV. Trunk exercise combined with spinal manipulative or NSAID therapy for chronic low back pain: a randomized, observer-blinded clinical trial. J Manipulative Physiol Ther 19(9):570–82, Nov–Dec, 1996.

Brumagne S, Cordo P, Lysens R, Verschueren S, Swinnen S. The role of paraspinal muscle spindles in lumbosacral position sense in individuals with and without low back pain. Spine 25(8):989–994, Apr 15, 2000.

Brumagne S, Lysens R, Swinnen S, Verschueren S. Effect of paraspinal muscle vibration on position sense of the lumbosacral spine. Spine 24(13):1328–1331, Jul 1, 1999.

Byl NN, Sinnott P. Variations in balance and body sway in middle-aged adults: subjects with healthy backs compared with subjects with low-back dysfunction. Spine 16(3):325–330, Mar, 1991.

Callaghan JP, McGill SM. Intervertebral disc herniation: studies on a porcine model exposed to highly repetitive flexion/extension motion with compressive force. Clin Biomech (Bristol, Avon) 16(1):28–37, Jan, 2001.

Calmels P, Jacob JF, Fayolle-Minon I, Charles C, Bouchet JP, Rimaud D, Thomas T. Use of isokinetic techniques

vs standard physiotherapy in patients with chronic low back pain. Preliminary results. Ann Readapt Med Phys 47(1):20–7, Feb, 2004.
Carey TS. Review: exercise therapy reduces pain and improves function in chronic but not acute low-back pain. ACP J Club 144(1):12–3, Jan–Feb, 2006.
Carr JL, Klaber Moffett JA, Howarth E, Richmond SJ, Torgerson DJ, Jackson DA, Metcalfe CJ. A randomized trial comparing a group exercise programme for back pain patients with individual physiotherapy in a severely deprived area. Disabil Rehabil 27(16):929–37, Aug, 2005.
Cassisi JE, Robinson ME, O'Conner P, MacMillan M. Trunk strength and lumbar paraspinal muscle activity during isometric exercise in chronic low-back pain patients and controls. Spine 18(2):245–51, Feb, 1993.
Celestini M, Marchese A, Serenelli A, Graziani G. A randomized controlled trial on the efficacy of physical exercise in patients braced for instability of the lumbar spine. Eura Medicophys 41(3):223–31, Sep, 2005.
Chiou WK, Lee YH, Chen WJ. Use of the surface EMG coactiviational pattern for functional evaluation of trunk muscles in subjects with and without low-back pain. Int J Ind Ergon 23:51–60, 1999.
Choi G, Raiturker PP, Kim MJ, Chung DJ, Chae YS, Lee SH. The effect of early isolated lumbar extension exercise program for patients with herniated disc undergoing lumbar discectomy. Neurosurgery 57(4):764–72, Oct, 2005.
Chok B, Lee R, Latimer J, Tan SB. Endurance training of the trunk extensor muscles in people with subacute low back pain. Phys Ther 79(11):1032–42, Nov, 1999.
Cholewicki J, McGill SM. Mechanical stability of the in vivo lumbar spine: implications for injury and chronic low back pain. Clin Biomech (Bristol, Avon) 11(1):1–15, Jan, 1996.
Cholewickin J, Simons APD, Radebold A. Effects of external trunk loads on lumbar spine stability. J Biomechanics 33(11):1377–1385, Nov, 2000.
Clark BC, Manini TM, Mayer JM, Ploutz-Snyder LL, Graves JE. Electromyographic activity of the lumbar and hip extensors during dynamic trunk extension exercise. Arch Phys Med Rehabil 83(11):1547–52, Nov, 2002.
Clark BC, Manini TM, Ploutz-Snyder LL. Derecruitment of the lumbar musculature with fatiguing trunk extension exercise. Spine 28(3):282–7, Feb 1, 2003.
Cohen AM. The role of exercise in the treatment of postural low back pain. Wis Med J 58(2):121–6, Feb, 1959.
Cohen I, Rainville J. Aggressive exercise as treatment for chronic low back pain. Sports Med 32(1):75–82, 2002.
Collins GA, Cohen MJ, Maliboff BD, Schandler SL. Comparative analysis of paraspinal and frontalis EMG, heart rate and skin conductance in chronic low back pain patients and normals to various postures and stress. Scand J Rehabil Med 14(1):39–46, 1982.
Cooper RG, St Clair Forbes W, Jayson MI. Radiographic demonstration of paraspinal muscle wasting in patients with chronic low back pain. Br J Rheumatol 31(6):389–394, Jun, 1992.
Curtis L,Mayer TG, Gatchel RJ. Physical progress and residual impairment quantification after functional restoration. Part III: isokinetic and isoinertional lifting capacity. Spine 19(4):401–5, Feb 15, 1994.
Dangaria T, Naesh O. Changes in cross-sectional area of psoas major muscle in unilateral sciatica caused by disc herniation. Spine 23(8):928–931, Apr 15, 1998.
Danielsen JM, Johnsen R, Kibsgaard SK, Hellevik E. Early aggressive exercise for postoperative rehabilitation after discectomy. Spine 25(8):1015–1020, Apr 15, 2000.
Dankaerts W, O'Sullivan P, Burnett A, Straker L. Altered patterns of superficial trunk muscle activation during sitting in nonspecific chronic low back pain patients: importance of subclassification. Spine 31(17):2017–23, Aug 1, 2006.
Danneels LA, Vanderstraeten GG, Cambier DC, Witvrouw EE, De Cuyper HJ. CT imaging of trunk muscles in chronic low back pain patients and healthy control subjects. Eur Spine J 9(4):266–272, Aug, 2000.
Davies JE, Gibson T, Tester L. The value of exercises in the treatment of low back pain. Rheumatol Rehabil 18(4):243–7, Nov, 1979.
De Benedittis G, Petrone D, De Candia N. Effect of the cervical reflex on the posture of normal subjects. Balance measurement study. Boll Soc Ital Biol Sper 67(3):303–9, Mar, 1991.
Dean SG, Smith JA, Payne S, Weinman J. Managing time: an interpretative phenomenological analysis of patients' and physiotherapists' perceptions of adherence to therapeutic exercise for low back pain. Disabil Rehabil 27(11):625–36, Jun 3, 2005.
Descarreaux M, Normand MC, Laurencelle L, Dugas C. Evaluation of a specific home exercise program for low back pain. J Manipulative Physiol Ther 25(8):497–503, Oct, 2002.
Deyo RA, Diehl AK, Rosenthal M. How many days of bed rest for acute back pain? A randomized clinical trial. New Eng J Med 315(17):1064–1070, Oct 23, 1996.
Dolan P, Greenfield K, Nelson RJ, Nelson IW. Can exercise therapy improve the outcome of microdiscectomy? Spine 25(12):1523–1532, Jun 15, 2000.
Dolce JJ, Crocker MF, Moletteire C, Doleys DM. Exercise quotas, anticipatory concern and self-efficacy expectancies in chronic pain: a preliminary report. Pain 24(3):365–72, Mar, 1986.
Donceel P, Du Bois M, Lahaye D. Return to work after surgery for lumbar disc herniation. A rehabilitation-oriented approach in insurance medicine. Spine 24(9):872–876, May 1, 1999.

Donchin M, Woolf O, Kaplan L, Floman Y. Secondary prevention of low-back pain. A clinical trial. Spine 15(12):1317–1320, Dec, 1990.

Donisch EW, Basmajian JV. Electromyography of deep back muscles in man. Am J Anat 133(1):25–36, Jan, 1972.

Ducic I, Short KW, Dellon AL. Relationship between loss of pedal sensibility, balance, and falls in patients with peripheral neuropathy. Ann Plast Surg 52(6):535–40, Jun, 2004.

Edwards RH, Hill DK, McDonnell M. Myothermal and intramuscular pressure measurements during isometric contractions of the human quadriceps muscle. J Physiol 224(2):58P–59P, Jul, 1972.

Elnaggar IM, Nordin M, Sheikhzadeh A, Parnianpour M, Kahanovitz N. Effects of spinal flexion and extension exercises on low-back pain and spinal mobility in chronic mechanical low-back pain patients. Spine 16(8):967–72, Aug, 1991.

Eversull E, Solomonow M, Zhou E, Baratta R, Zhu M. Neuromuscular neutral zones sensitivity to lumbar displacement rate. Clin Biomech 16(2):102–113, Feb, 2001.

Fairbank JC, Couper J, Davies JB, O'Brien JP. The Oswestry low back pain disability questionnaire. Physiotherapy 66(8):271–3, Aug, 1980.

Fast A. Low back disorders: Conservative management. Arch Phys Med Rehabil 69(10):880–891, Oct, 1988.

Faugli HP, (Holten). *Medical Exercise Therapy.* Laerergruppen for Medisnsk Treningsterapi AS, Norway, 1996.

Ferreira ML, Ferreira PH, Hodges PW. Changes in postural activity of the trunk muscles following spinal manipulative therapy. Man Ther 12(3):240–8, Aug, 2007.

Ferreira ML, Ferreira PH, Latimer J, Herbert RD, Hodges PW, Jennings MD, Maher CG, Refshauge KM. Comparison of general exercise, motor control exercise and spinal manipulative therapy for chronic low back pain: A randomized trial. Pain 131(1–2):31–7, Sep, 2007.

Filiz M, Cakmak A, Ozcan E. The effectiveness of exercise programmes after lumbar disc surgery: a randomized controlled study. Clin Rehabil 19(1):4–11, Jan, 2005.

Fisk JW. A controlled trial of manipulation in a selected group of patients with low back pain favoring one side. New Zealand Med J 90(645):288–291, Oct, 1979.

Friedrich M, Gittler G, Arendasy M, Friedrich KM. Long-term effect of a combined exercise and motivational program on the level of disability of patients with chronic low back pain. Spine 30(9):995–1000, May 1, 2005.

Fritz JM, Erhard RE, Vignovic M. A nonsurgical treatment approach to patients with lumbar spinal stenosis. Phys Ther 77(9):962–973, Sep, 1997.

Frost H, Klaber Moffett J, Moser J, Fairbank J. Evaluation of a fitness programme for patients with chronic low back pain. BMJ 310(6973):151–154, Jan 21, 1995.

Frost H, Lamb S, Klaber Moffett J, Fairbank J, Moser J. A fitness programme for patients with chronic low back pain: 2 year follow-up of a randomised controlled trial. Pain 75(2–3):273–279, Apr, 1998.

Galantino ML, Bzdewka TM, Eissler-Russo JL, Holbrook ML, Mogck EP, Geigle P, Farrar JT. The impact of modified Hatha yoga on chronic low back pain: a pilot study. Altern Ther Health Med 10(2):56–9, Mar–Apr, 2004.

Garshasbi A, Faghih Zadeh S. The effect of exercise on the intensity of low back pain in pregnant women. Int J Gynaecol Obstet 88(3):271–5, Mar, 2005.

Gedalia U, Solomonow M, Harris M. Biomechanics of increased exposure to lumbar injury caused by cyclic loading: Part 2. Recovery of reflexive muscular stability with rest. Spine 24(23):2461, Dec 1, 1999.

Geisser ME, Wiggert EA, Haig AJ, Colwell MO. A randomized, controlled trial of manual therapy and specific adjuvant exercise for chronic low back pain. Clin J Pain 21(6):463–70, Nov–Dec, 2005.

Gejo R, Matsui H, Kawaguchi Y, Ishihara H, Tsuji H. Serial changes in trunk muscle performance after posterior lumbar surgery. Spine 24(10):1023–28, May 15, 1999.

Gilbert JR, Taylor DW, Hildebrand A. Clinical trial of common treatments for low back pain in family practice. British Med J 291(6498):791–794, Sep 21, 1985.

Gill KP, Callaghan MJ. The measurement of lumbar proprioception in individuals with and without low back pain. Spine 23(3):371–7, Feb 1, 1998.

Gill NW, Teyhen DS, Lee TI. Improved contraction of the transversus abdominis immediately following spinal manipulation: A case study using real-time ultrasound imaging. Man Ther 12(3):280–285, Aug, 2007.

Grimsby O. Neurophysiological view points on hypermobilities. The Nordic Group of Specialists. J Manual Therapy 2:2–9, 1988.

Gustavsen R. *Training Therapy: Prophylaxis and Rehabilitation*. Renate Streeck, 1993.

Hakkinen A, Kuukkanen T, Tarvainen U, Ylinen J. Trunk muscle strength in flexion, extension, and axial rotation in patients managed with lumbar disc herniation surgery and in healthy control subjects. Spine 28(10):1068–1073, May 15, 2003a.

Hakkinen A, Ylinen J, Kautiainen H, Airaksinen O, Herno A, Tarvainen U, Kiviranta I. Pain, trunk muscle strength, spine mobility and disability following lumbar disc surgery. J Rehabil Med 35(5):236–240, Sep, 2003b.

Hansen FR, Bendix T, Skov P, Jensen CV, Kristensen JH, Krohn L, Schioeler H. Intensive, dynamic back-muscle exercises, conventional physiotherapy, or placebo-control treatment of low-back pain. A randomized, observer-blind trial. Spine 18(1):98–108, Jan, 1993.

Hart L. Exercise therapy for nonspecific low-back pain: a meta-analysis. Clin J Sport Med 16(2):189–90, Mar, 2006.

Hartigan C, Rainville J, Sobel JB, Hipona M. Long-term exercise adherence after intensive rehabilitation for chronic low back pain. Med Sci Sports Exerc 32(3):551–7, Mar, 2000.

Hartigan C, Sobel JB, Rainville J. Functionally oriented rehabilitation in low back pain: changes in pain scores. 9th Annual Meeting of the North American Spine Society; Minneapolis, Oct 19–20, 1994.

Hather BM, Tesch PA, Buchanan P, Dudley GA. Influence of eccentric actions on skeletal muscle adaptations to resistance training. Acta Physiol Scand 143(2):177–85, Oct, 1991.

Hayden JA, van Tulder MW, Malmivaara A, Koes BW. Exercise therapy for treatment of non-specific low back pain. Cochrane Database Syst Rev (3):CD000335, Jul 20, 2005a.

Hayden JA, van Tulder MW, Malmivaara AV, Koes BW. Meta-analysis: exercise therapy for nonspecific low back pain. Ann Intern Med 142(9):765–75, May 3, 2005b.

Hayden JA, van Tulder MW, Tomlinson G. Systematic review: strategies for using exercise therapy to improve outcomes in chronic low back pain. Ann Intern Med 142(9):776–85, May 3, 2005c.

Hayden JA, van Tulder MW, Malmivaara A, Koes BW. Exercise therapy for treatment of non-specific low back pain. Cochrane Database of Systematic Reviews, Issue 4, 2006.

Hazard RG, Fenwick JW, Kalisch SM, Redmond J, Reeves V, Reid S, Frymoyer JW. Functional restoration with behavioral support. A one-year prospective study of patients with chronic low-back pain. Spine 14(2):157–61, Feb, 1989.

Healey EL, Fowler NE, Burden AM, McEwan IM. Raised paraspinal muscle activity reduces rate of stature recovery after loaded exercise in individuals with chronic low back pain. Arch Phys Med Rehabil 86(4):710–5, Apr, 2005.

Hermann KM, Barnes WS. Effects of eccentric exercise on trunk extensor torque and lumbar paraspinal EMG. Med Sci Sports Exerc 33(6):971–7, Jun, 2001.

Hicks GE, Fritz JM, Delitto A, McGill SM. Preliminary development of a clinical prediction rule for determining which patients with low back pain will respond to a stabilization exercise program. Arch Phys Med Rehabil 86(9):1753–62, Sep, 2005.

Hides JA, Jull GA, Richardson CA. Long-term effects of specific stabilizing exercises for first-episode low back pain. Spine 26(11):E243–E248, Jun 1, 2001.

Hides J, Richardson CA, Jull GA. Magnetic resonance imaging and ultrasonography of the lumbar multifidus muscle: Comparison of two different modalities. Spine 20(1):54–58, Jan 1, 1995.

Hides JA, Richardson CA, Jull GA. Multifidus muscle recovery is not automatic after resolution of acute, first-episode low back pain. Spine 21(23):2763–9, Dec 1, 1996.

Hides JA, Richardson CA, Jull GA. Multifidus muscle rehabilitation decreases recurrence of symptoms following first episode low back pain. In: Proceedings of the National Congress of the Australian Physiotherapy Association, Brisbane, 1996.

Hides JA, Stokes M, Saide M, Jull GA, Coopers D. Evidence of lumbar multifidus muscle wasting ipsilateral to symptoms in patients with acute/subacute low back pain. Spine 19(2):165–172, Jan 15, 1994.

Hides JA, Jull GA, Richardson CA. Long-term effects of specific stabilizing exercises for first-episode low back pain. Spine 26(11):E243–8, Jun 1, 2001.

Higbie EJ, Cureton KJ, Warren GL 3rd, Prior BM. Effects of concentric and eccentric training on muscle strength, cross-sectional area, and neural activation. J Appl Physiol 81(5):2173–2181, Nov, 1996.

Hodges P, Holm AK, Hansson T, Holm S. Rapid atrophy of the lumbar multifidus follows experimental disc or nerve root injury. Spine 31(25):2926–33, Dec 1, 2006.

Hodges PW, Kaigle Holm A, Holm S, Ekstrom L, Cresswell A, Hansson T, Thorstensson A. Intervertebral stiffness of the spine is increased by evoked contraction of transversus abdominis and the diaphragm: in vivo porcine studies. Spine 28(23):2594–2601, Dec 1, 2003.

Hodges PW, Richardson CA. Contraction of the abdominal muscles associated with movement of the lower limb. Phys Ther 77(2):132–144, 1997a.

Hodges PW, Richardson CA. Feedforward contraction of transversus abdominis is not influenced by the direction of arm movement. Exp Brain Res 114(2):362–370, Nov, 1997b.

Hodges PW, Richardson CA. Inefficient muscular stabilization of the lumbar spine associated with low back pain: A motor control evaluation of transverse abdominis. Spine 21(22):2640–2650, Nov 15, 1996.

Hodges PW, Richardson CA. Relationship between limb movement speed and associated contraction of the trunk muscles. Ergonomics 40(11):1220–30, Nov, 1997c.

Hoffman MD, Shepanski MA, Mackenzie SP, Clifford PS. Experimentally induced pain perception is acutely reduced by aerobic exercise in people with chronic low back pain. J Rehabil Res Dev 42(2):183–90, Mar–Apr, 2005.

Hunter J. A *Treatise on the Blood, Inflammation and Gun-Shot Wounds*. Nicol, London, 1794.

Jackson M, Solomonow M, Zhou B, Baratta R, Harris M. Multifidus EMG and tension-relaxation recovery after prolonged static lumbar flexion. Spine 26(7):715–723, Apr 1, 2001.

Janda V. Evaluation of Muscular Imbalance (Ch.6). In: Liebenson C, Editor. *Rehabilitation of the Spine: A Practitioner's Manual*. Baltimore, Lippincott, Williams

& Wilkins, pp. 97–112, 1996.

Janda V. Function of Muscles in Musculoskeletal Pain Syndromes. Course workbook. Northeast Seminars. Tacoma, Washington, April 18–19, 1999.

Janda V. Muscles, central nervous motor regulation and back problems. Korr I. editor. *The Neurobiologic Mechanisms in Manipulative Therapy*. Plenum Press, New York. Pp. 27–41, 1978.

Janda V. Treatment of chronic low back pain. J Manual Medicine 6:166–168, 1992.

Johannsen F, Remvig L, Kryger P, Beck P, Lybeck K, Larsen LH, Warming S, Dreyer V. Supervised endurance exercise training compared to home training after first lumbar diskectomy: A clinical trial. Clin Exp Rheumatol 12(6):609–614, 1994.

Jowett RL, Fidler MW, Troup JD. Histochemical changes in the multifidus in mechanical derangements of the spine. Orthop Clin North Am 6(1):145–61, Jan, 1975.

Juker D, McGill S, Kropf P, Steffen T. Quantitative intramuscular myoelectric activity of lumbar portions of psoas and the abdominal wall during a wide variety of tasks. Med Sci Sports Exerc 30(2):301–10, Feb, 1998.

Kahanovitz N, Viola K, Gallagher M. Long term strength assessment of postoperative discectomy patients. Spine 14(4):402–403, Apr, 1989.

Kankaanpaa M, Taimela S, Laaksonen D, Hanninen O, Airaksinen O. Back and hip extensor fatigability in chronic low back pain patients and controls. Arch Phys Med Rehabil 79(4):412–7, Apr, 1998.

Kasman GS, Cram JR, Wolf SL. *Clinical Applications in Surface Electromyography: Chronic Musculoskeletal Pain*. Aspen Publishers, Inc., 1998.

Kavcic N, Grenier S, McGill SM. Determining the stabilizing role of individual torso muscles during rehabilitation exercises. Spine 29(11):1254–65, Jun 1, 2004.

Kavcic N, Grenier S, McGill SM. Quantifying tissue loads and spine stability while performing commonly prescribed low back stabilization exercises. Spine 29(20):2319–29, Oct 15, 2004.

Kawaguchi Y, Matsui H, Tsuji H. Back muscle injury after posterior lumbar spine surgery. A histologic and enzymatic analysis. Spine 21(8):941–4, Apr 15, 1996.

Klaber Moffett J, Frost H. Back to fitness programme. The manual for physiotherapists to set up the classes. Physiotherapy 86(6):295–305, 2000.

Klaber Moffett JA, Carr J, Howarth E. High fear-avoiders of physical activity benefit from an exercise program for patients with back pain. Spine 29(11):1167–72, Jun 1, 2004.

Kofotolis N, Sambanis M. The influence of exercise on musculoskeletal disorders of the lumbar spine. J Sports Med Phys Fitness 45(1):84–92, Mar, 2005.

Koumantakis GA, Watson PJ, Oldham JA. Supplementation of general endurance exercise with stabilisation training versus general exercise only. Physiological and functional outcomes of a randomised controlled trial of patients with recurrent low back pain. Clin Biomech (Bristol, Avon) 20(5):474–82, Jun, 2005a.

Koumantakis GA, Watson PJ, Oldham JA. Trunk muscle stabilization training plus general exercise versus general exercise only: randomized controlled trial of patients with recurrent low back pain. Phys Ther 85(3):209–25, Mar, 2005b.

Kumar S, Dufresne RM, Van Schoor T. Human trunk strength profile in flexion and extension. Spine 20(2):160–8, Jun 15, 1995.

Kumar S. Cumulative load as a risk factor for back pain. Spine 15(12):1311–6, Dec, 1990.

Kuukkanen TM, Malkia EA. An experimental controlled study on postural sway and therapeutic exercise in subjects with low back pain. Clin Rehabil 14(2):192–202, Apr, 2000.

Lamoth CJ, Daffertshofer A, Meijer OG, Lorimer Moseley G, Wuisman PI, Beek PJ. Effects of experimentally induced pain and fear of pain on trunk coordination and back muscle activity during walking. Clin Biomech (Bristol, Avon) 19(6):551–63, Jul, 2004.

Lariviere C, Gagnon D, Loisel P. The comparison of trunk muscles EMG activation between subjects with and without chronic low back pain during flexion-extension and lateral bending tasks. J Electromyogr Kinesiol 10(2):79–91, Apr, 2000.

Leggett S, Mooney V, Matheson LN, Nelson B, Dreisinger T, Van Zytveld J, Vie L. Restorative exercise for clinical low back pain. A prospective two-center study with 1-year follow-up. Spine 24(9):889–98, May 1, 1999.

Lehto M, Hurme M, Alaranta H, Einola S, Falck B, Jarvinen M, Kalimo H, Mattila M, Paljarvi L. Connective tissue changes of the multifidus muscle in patients with lumbar disc herniation. An immunohistologic study of collagen types I and III and fibronectin. Spine 14(3):302–309, May, 1989.

Leinonen V, Kankaanpaa M, Luukkonen M, Kansanen M, Hanninen O, Airaksinen O, Taimela S. Lumbar paraspinal muscle function, perception of lumbar position, and postural control in disc herniation-related back pain. Spine 28(8):842–848, Apr, 2003.

Ljunggren AE, Weber H, Kogstad O, Thom E, Kirkesola G. Effect of exercise on sick leave due to low back pain. A randomized, comparative, long-term study. Spine 22(14):1610–6, Jul 15, 1997.

Long A, Donelson R, Fung T. Does it matter which exercise? A randomized control trial of exercise for low back pain. Spine 29(23):2593–602, Dec 1, 2004.

Lucas D, Bresler B. Stability of the ligamentous spine. Biomechanics Laboratory, Berkeley: University of California, Tech Rep #40, 1961.

Luoto S, Aalto H, Taimela S, Hurri H, Pyykko I, Alaranta H. One-footed and externally disturbed two-footed postural control in patients with chronic low back pain and healthy control subjects. A controlled study with follow-up. Spine 23(19):2081–9, Oct 1, 1998.

Madeleine P, Prietzel H, Svarrer H, Arendt-Nielsen L. Quantitative posturography in altered sensory conditions: a way to assess balance instability in patients with chronic whiplash injury. Arch Phys Med Rehabil 85(3):432–8, Mar, 2004.

Maher CG, Latimer J, Hodges PW, Refshauge KM, Moseley GL, Herbert RD, Costa LO, McAuley J. The effect of motor control exercise versus placebo in patients with chronic low back pain. BMC Musculoskelet Disord 6:54, Nov 4, 2005.

Malmivaara A, Hakkinen U, Aro T, Heinrichs ML, Koskenniemi L, Kuosma E, Lappi S, Paloheimo R, Servo C, Vaaranen V, et al. The treatment of acute low back pain—bed rest, exercises, or ordinary activity? N Engl J Med 332(6):351–5, Feb 9, 1995.

Manniche C, Asmussen K, Lauritsen B, Vinterberg H, Karbo H, Abildstrup S, Fischer-Nielsen K, Krebs R, Ibsen K. Intensive dynamic back exercises with or without hyperextension in chronic back pain after surgery for lumbar disc protrusion. A clinical trial. Spine 18(5):560–567, Apr, 1993a.

Manniche C, Lundberg E, Christensen I, Bentzen L, Hesselsoe G. Intensive dynamic back exercises for chronic low back pain: a clinical trial. Pain 47(1):53–63, Oct, 1991.

Manniche C, Skall HF, Braendholt L, Christensen BH, Christophersen L, Ellegaard B, Heilbuth A, Ingerslev M, Jorgensen OE, Larsen E. Clinical trial of postoperative dynamic back exercises after first lumbar discectomy. Spine 18(1):92–97, Jan, 1993b.

Mannion AF, Muntener M, Taimela S, Dvorak J. Comparison of three active therapies for chronic low back pain: results of a randomized clinical trial with one-year follow-up. Rheumatology (Oxford) 40(7):772–8, Jul, 2001.

Marras WS, Davis KG, Ferguson SA, Lucas BR, Gupta P. Spine loading characteristics of patients with low back pain compared with asymptomatic individuals. Spine 26(23):2566–74, Dec 1, 2001.

Mattila M, Hurme M, Alaranta H, Paljarvi L, Kalimo H, Falck B, Lehto M, Einola S, Jarvinen M. The multifidus muscle in patients with lumbar disc herniation. A histochemical and morphometric analysis of intraoperative biopsies. Spine 11(7):732–738, Sep, 1986.

Mayer JM, Verna JL, Manini TM, Mooney V, Graves JE. Electromyographic activity of the trunk extensor muscles: effect of varying hip position and lumbar posture during Roman chair exercise. Arch Phys Med Rehabil 83(11):1543–6, Nov, 2002.

Mayer T, Tabor J, Bovasso E, Gatchel RJ. Physical progress and residual impairment quantification after functional restoration. Part I: lumbar mobility. Spine 19(4):389–94, Feb 15, 1994.

Mayer TG, Gatchel RJ, Kishino N, Keeley J, Capra P, Mayer H, Barnett J, Mooney V. Objective assessment of spine function following industrial injury. A prospective study with comparison group and one-year follow-up. Spine 10(6):482–93, Jul–Aug, 1985.

Mayer TG, Smith SS, Keeley J, Mooney V. Quantification of lumbar function. Part 2: Sagittal plane trunk strength in chronic low-back pain patients. Spine 10(8):765–72, Oct, 1985.

Mayer TG, Vanharanta H, Gatchel RJ, Mooney V, Barnes D, Judge L, Smith S, Terry A. Comparison of CT scan muscle measurements and isokinetic trunk strength in postoperative patients. Spine 14(1):33–6, Jan, 1989.

McGill SM, Childs A, Liebenson C. Endurance times for low back stabilization exercises: clinical targets for testing and training from a normal database. Arch Phys Med Rehabil 80(8):941–4, Aug, 1999.

McGill S, Juker D, Kropf P. Quantitative intramuscular myoelectric activity of quadratus lumborum during a wide variety of tasks. Clin Biomech (Bristol, Avon). 1996 Apr;11(3):170–172, Apr, 1996.

McGill S. *Low Back Disorders: Evidence-Based Prevention and Rehabilitation.* Human Kinetics, 2002.

McGill S. *Ultimate Back Fitness and Performance.* Wabuno Publishers, Ontario, p. 167, 2004.

McGill SM, Sharratt MT, Seguin JP. Loads on spinal tissues during simultaneous lifting and ventilatory challenge. Ergonomics 38(9):1772–92, Sep, 1995.

McGill SM. Low back exercises: evidence for improving exercise regimens. Phys Ther 78(7):754–64, Jul, 1998.

McGregor AH, Burton AK, Sell P, Waddell G. The development of an evidence-based patient booklet for patients undergoing lumbar discectomy and un-instrumented decompression. Eur Spine J 16(3):339–46, Mar, 2007.

McKenzie R. Spine 26(16):1829–31, Aug 15, 2001. Letter Re: van Tulder et al, Exercise therapy for low back pain. Spine 25:2784–96, 2000.

McNeill T, Warwick D, Andersson G, Schultz A. Trunk strengths in attempted flexion, extension, and lateral bending in healthy subjects and patients with low-back disorders. Spine 5(6):529–38, Nov–Dec, 1980.

McNevin NH, Wulf G, Carlson C. Effects of attentional focus, self-control, and dyad training on motor learning: implications for physical rehabilitation. Phys Ther 80(4):373–85, Apr, 2000.

Mientjes MI, Frank JS. Balance in chronic low back pain patients compared to healthy people under various conditions in upright standing. Clin Biomech (Bristol, Avon) 14(10):710–6, Dec, 1999.

Miyazaki A, Sako T, Taniguchi Y, Nagamine T, Morinaga H. Exercise therapy for back pain. Seikei Geka 19(11):873–9, Oct, 1968.

Moffett JK, Torgerson D, Bell-Syer S, Jackson D, Llewlyn-Phillips H, Farrin A, Barber J. Randomised controlled trial of exercise for low back pain: clinical outcomes, costs, and preferences. BMJ 319(7205):279–83, Jul 31, 1999.

Mok NW, Brauer SG, Hodges PW. Hip strategy for balance control in quiet standing is reduced in people with low back pain. Spine 29(6):E107–12, Mar 15, 2004.

Moseley GL, Hodges, PW, Gandevia SC. Deep and superficial fibers of the lumbar multifidus are differentially active during voluntary arm movements. Spine 27(2):E29–36, Jun 15, 2002.

Nachemson A. The load on lumbar discs in different positions of the body. Clin Orthop 45:107, 1966.

NachemsonAL, Lindh M. Measurement of abdominal and back muscle strength with and without low back pain. Scand J Rehabil Med 1(2):60–5, 1969.

Nagler W, Bodack MP: Management Options in Lumbar Spinal Stenosis. In Ernst E, Jayson MIV, Pope MH, Porter RW (eds). Advances in Idiopathic Low Back Pain. Vienna, Blackwell-MZV, 292–297, 1993.

Niemisto L, Lahtinen-Suopanki T, Rissanen P, Lindgren KA, Sarna S, Hurri H. A randomized trial of combined manipulation, stabilizing exercises, and physician consultation compared to physician consultation alone for chronic low back pain. Spine 28(19):2185–91, Oct 1, 2003.

Nies N, Sinnott PL. Variations in balance and body sway in middle-aged adults. Subjects with healthy backs compared with subjects with low-back dysfunction. Spine 16(3):325–30, Mar, 1991.

Nitz A, Peck D. Comparison of muscle spindle concentrations in large and small human epaxial muscles acting in parallel combinations. Am Surg 52(5):273–277, May, 1986.

Norris CM. Spinal stabilization: an exercise program to enhance lumbar stabilization. Physiother 81:138–145, 1995.

Nourbakhsh MR, Arab AM. Relationship between mechanical factors and incidence of low back pain. J Orthop Sports Phys Ther 32(9):447–60, Sep, 2002.

Novy DM, Simmonds MJ, Olson SL, Lee CE, Jones SC. Physical performance: differences in men and women with and without low back pain. Arch Phys Med Rehabil 80(2):195–8, Feb, 1999.

Ostelo RW, de Vet HC, Waddell G, Kerckhoffs MR, Leffers P, van Tulder M. Rehabilitation following first-time lumbar disc surgery: A systematic review within the framework of the Cochrane collaboration. Spine 28(3):2090–218, Feb1, 2003.

O'Sullivan P, Dankaerts W, Burnett A, Chen D, Booth R, Carlsen C, Schultz A. Evaluation of the flexion relaxation phenomenon of the trunk muscles in sitting. Spine 31(17):2009–16, Aug 1, 2006.

O'Sullivan P, Dankaerts W, Burnett A, Straker L, Bargon G, Moloney N, Perry M, Tsang S. Lumbopelvic kinematics and trunk muscle activity during sitting on stable and unstable surfaces. J Orthop Sports Phys Ther 36(1):19–25, Jan, 2006.

O'Sullivan PB, Phyty GD, Twomey LT, Allison GT. Evaluation of specific stabilizing exercise in the treatment of chronic low back pain with radiologic diagnosis of spondylolysis or spondylolisthesis. Spine 22(24):2959–67, Dec 15, 1997.

O'Sullivan PB. Lumbar segmental 'instability': clinical presentation and specific stabilizing exercise management. Man Ther 5(1):2–12, Feb, 2000.

Panjabi MM. The stabilizing system of the spine. Part II. Neutral zone and instability hypothesis. J Spinal Disord 5(4):390–397, Dec, 1992.

Panjabi MM. Clinical spinal instability and low back pain. J Electromyogr Kinesiol 13(4):371–379, Aug, 2003.

Pauly JE. An electromyographic analysis of certain movements and exercises: some deep muscles of the back. The Anat Record 155(2):223–234, Jun, 1966.

Peck D, Buxton D, Nitz A. A comparison of spindle concentrations in large and small muscles acting in parallel combinations. J Morphology 180(3):243–252, Jun, 1984.

Peck D, Buxton D, Nitz A. In: Hnik P, Soukup T, Vejsada R, Selena J, eds. A proposed mechanoreptor role for the small redundant muscles which act in parallel with large prime movers. Mechanoreceptors: Development, structure and function. International Symposium on Mechanoreceptors. Prague, Czechoslovakia, New York, NY, Plenum Press, pp. 377–382, 1987.

Petersen T, Kryger P, Ekdahl C, Olsen S, Jacobsen S. The effect of McKenzie therapy as compared with that of intensive strengthening training for the treatment of patients with subacute or chronic low back pain: A randomized controlled trial. Spine 27(16):1702–9, Aug 15, 2002.

Pollock ML, Leggett SH, Graves JE, Jones A, Fulton M, Cirulli J. Effect of resistance training on lumbar extension strength. Am J Sports Med 17(5):624–9, Sep–Oct, 1989.

Postacchini F, Montanaro A. Early mobilisation and functional re-education in the post operative treatment of prolapsed lumbar disc. Ital J Orthop Traumatol 4(2):231–236, Aug, 1978.

Rainville J, AhernDK, Phalen L. Altering beliefs about pain and impairment in a functionally oriented treatment program for chronic low back pain. Clin J Pain 9(3):196–201, Sep, 1993.

Rantanen J, Hurme M, Falck B, Alaranta H, Nykvist F, Lehto M, Einola S, Kalimo H. The lumbar multifidus muscle five years after surgery for a lumbar intervertebral disc herniation. Spine 18(5):568–574, Apr, 1993.

Raty HP, Kujala U, Videman T, Koskinen SK, Karppi SL, Sarna S. Associations of isometric and isoinertial trunk muscle strength measurements and lumbar paraspinal muscle cross-sectional areas. J Spinal Disord 12(3):266–270, Jun, 1999.

Raynor DK. The influence of written information on patient knowledge and adherence to treatment. In: Myers LB, Midence K, editors. *Adherence to Treatment in Medical Conditions* (chap. 4). Amsterdam: Harwood Academic Publishers; 1998.

Richardson C, Jull G, Hodges P, Hides J. *Therapeutic Exercise for Spinal Segmental Stabilization in Low Back Pain*. London, England: Churchill Livingstone, 1999.

Richardson CA, Jull GA, Toppenberg R, Comerford M. Techniques for active lumbar stabilization for spinal protection: a pilot study. Aust J Physiother 38:105–112, 1992.

Risch SV, Norvell NK, Pollock ML, Risch ED, Langer H, Fulton M, Graves JE, Leggett SH. Lumbar strengthening in chronic low back pain patients. Physiologic and psychological benefits. Spine 18(2):232–8, Feb, 1993.

Robinson ME, Cassisi JE, O'Connor PD, MacMillan M. Lumbar iEMG during isotonic exercise: chronic low back pain patients versus controls. J Spinal Disord 5(1):8–15, Mar, 1992.

Rodacki C, Fowler NE, Rodacki AL, Birch K. Stature loss and recovery in pregnant women with and without low back pain. Arch Phys Med Rehabil 84(4):507–12, Apr, 2003.

Selkowitz DM, Kulig K, Poppert EM, Flanagan SP, Matthews ND, Beneck GJ, Popovich JM Jr, Lona JR, Yamada KA, Burke WS, Ervin C, Powers CM. The immediate and long-term effects of exercise and patient education on physical, functional, and quality-of-life outcome measures after single-level lumbar microdiscectomy: a randomized controlled trial protocol. BMC Musculoskelet Disord 7:70, Aug 25, 2006.

Shaughnessy M, Caulfield B. A pilot study to investigate the effect of lumbar stabilisation exercise training on functional ability and quality of life in patients with chronic low back pain. Int J Rehabil Res 27(4):297–301, Dec, 2004.

Sherman KJ, Cherkin DC, Erro J, Miglioretti DL, Deyo RA. Comparing yoga, exercise, and a self-care book for chronic low back pain: a randomized, controlled trial. Ann Intern Med 143(12):849–56, Dec 20, 2005.

Shirado O, Ito T, Kaneda K, Strax TE. Concentric and eccentric strength of trunk muscles: influence of test postures on strength and characteristics of patients with chronic low-back pain. Arch Phys Med Rehabil 76(7):604–11, Jul, 1995.

Sihvonen T, Herno A, Paljarvi L, Airaksinen O, Partanen J, Tapaninaho A. Local denervation atrophy of paraspinal muscles in postoperative failed back syndrome. Spine 18(5):575–581, Apr, 1993.

Sikorski JM. A rationalized approach to physiotherapy for low back pain. Spine 10(6):71–79, Jul–Aug, 1985.

Sjostrom H, Allum JH, Carpenter MG, Adkin AL, Honegger F, Ettlin T. Trunk sway measures of postural stability during clinical balance tests in patients with chronic whiplash injury symptoms. Spine 28(15):1725–34, Aug 1, 2003.

Skall FH, Manniche C, Nielsen CJ. Intensive back exercises 5 weeks after surgery of lumbar disk prolapse. A prospective, randomized multicenter trial with a historical control group. Ugeskr Laeger 156(5):643–6, Jan 31, 1994.

Sluijs EM, Kok GJ, van der Zee J. Correlates of exercise compliance in physical therapy. Phys Ther 73(11):771–782, Nov, 1993.

Smith SS, Mayer TG, Gatchel RJ, Becker TJ. Quantification of lumbar function. Part 1: Isometric and multispeed isokinetic trunk strength measures in sagittal and axial planes in normal subjects. Spine 10(8):757–64, Oct, 1985.

Snijders C, Vleeming A, Stoeckart R, Mens JMA, Kleinrensink GJ. Biomechanical modeling of sacroiliac joint stability in different postures. Spine: State Art Rev 9:419–32, 1995.

Snijders CJ, Slagter AH, van Strik R, Vleeming A, Stoeckart R, Stam HJ. Why leg crossing? The influence of common postures on abdominal muscle activity. Spine 20(18):1989–93, Sep 15, 1995.

Solomonow M, Zhou B-E, Baratta R, Lu Y, Harris M. Biomechanics of increased exposure to lumbar injury caused by cyclic loading: Part I. Loss of reflexive muscular stabilization. Spine 24(23):2426–2434, Dec 1, 1999.

Solomonow M, Zhou B-E, Harris M, Lu Y, Baratta R. The ligamento-muscular stabilizing system of the spine. Spine 23(23):2552–2562, Dec 1, 1998.

Soukup MG, Glomsrod B, Lonn JH, Bo K, Larsen S. The effect of a Mensendieck exercise program as secondary prophylaxis for recurrent low back pain. A randomized, controlled trial with 12-month follow-up. Spine 24(15):1585–91, Aug 1, 1999.

Sparto PJ, Parnianpour M, Marras WS, Granata KP, Reinsel TE, Simon S. Neuromuscular trunk performance and spinal loading during a fatiguing isometric trunk extension with varying torque requirements. J Spinal Disord 10(2):145–56, Apr, 1997.

Sparto PJ, Parnianpour M, Reinsel TE, Simon S. The effect of fatigue on multijoint kinematics and load sharing during a repetitive lifting test. Spine 22(22):2647–54, Nov 15, 1997.

Sparto PJ, Parnianpour M, Reinsel TE, Simon S. The effect of fatigue on multijoint kinematics, coordination, and postural stability during a repetitive lifting test. J Orthop Sports Phys Ther 25(1):3–12, Jan, 1997.

Sparto PJ, Parnianpour M. Estimation of trunk muscle forces and spinal loads during fatiguing repetitive trunk exertions. Spine 23(23):2563–73, Dec 1, 1998.

Staal JB, Hlobil H, van Tulder MW, Waddell G, Burton AK, Koes BW, van Mechelen W. Occupational health guidelines for the management of low back pain: an international comparison. Occup Environ Med

60(9):618–26, Sep, 2003.

Staal JB, Rainville J, Fritz J, van Mechelen W, Pransky G. Physical exercise interventions to improve disability and return to work in low back pain: current insights and opportunities for improvement. J Occup Rehabil 15(4):491–505, Dec, 2005.

Stuge B, Vollestad NK. Important aspects for efficacy of treatment with specific stabilizing exercises for postpartum pelvic girdle pain. In: Vleeming A, Mooney V, Stoeckart R. *Movement, Stability and Lumbopelvic Pain. Integration of Research and Therapy.* 2nd Edition. Churchill Livingstone, 2007.

Taimela S, Diederich C, Hubsch M, Heinricy M. The role of physical exercise and inactivity in pain recurrence and absenteeism from work after active outpatient rehabilitation for recurrent or chronic low back pain: a follow-up study. Spine 25(14):1809–16, Jul 15, 2000.

Thomas HO. *Contributions to Medicine and Surgery.* Lewis, London, 1874.

Torstensen TA, Ljunggren AE, Meen HD, Odland E, Mowinckel P, Geijerstam S. Efficiency and costs of medical exercise therapy, conventional physiotherapy, and self-exercise in patients with chronic low back pain. A pragmatic, randomized, single-blinded, controlled trial with 1-year follow-up. Spine 23(23):2616–24, Dec 1, 1998.

van Dieen JH, Selen LP, Cholewicki J. Trunk muscle activation in low-back pain patients, an analysis of the literature. J Electromyogr Kinesiol 13(4):333–51, Aug, 2003.

van Tulder M, Malmivaara A, Esmail R, Koes B. Exercise therapy for low back pain: a systematic review within the framework of the Cochrane collaboration back review group. Spine 25(21):2784–96, Nov 1, 2000.

van Tulder, Koes BW, Bouter LM. Conservative treatment of acute and chronic nonspecific low back pain. A systematic review of randomized controlled trials of the most common interventions. Spine 22(18):2128–56, Sep 15, 1997.

Vincent KR, Braith RW, Vincent HK. Influence of resistance exercise on lumbar strength in older, overweight adults. Arch Phys Med Rehabil 87(3):383–9, Mar, 2006.

Vleeming A, Pool-Goudzwaard AL, Stoeckart R, van Wingerden JP, Snijders CJ. The posterior layer of the thoracolumbar fascia. Its function in load transfer from spine to legs. Spine 20(7):753–8, Apr 1, 1995.

Waddell G, Burton AK. Concepts of rehabilitation for the management of low back pain. Best Pract Res Clin Rheumatol 19(4):655–70, Aug, 2005.

Waddell G, Feder G, Lewis M. Systematic review of bedrest and advice to stay active for acute low back pain. Bri J Gen Practice 47(423):647–652, Oct, 1997.

Waddell G, Somerville D, Henderson I, Newton M. Objective clinical evaluation of physical impairment in chronic low back pain. Spine 17(6):617–28, Jun, 1992.

Waddell G. AAOMPT: American Academy of Orthopaedic Manual Physical Therapy. 6th Annual. Charlottesville, NC. Oct 19–22, 2000.

Waddell G. *The Back Pain Revolution.* Churchill Livingston, 1998.

Wand BM, Bird C, McAuley JH, Dore CJ, MacDowell M, De Souza LH. Early intervention for the management of acute low back pain: a single-blind randomized controlled trial of biopsychosocial education, manual therapy, and exercise. Spine 29(21):2350–6, Nov 1, 2004.

Wohlfart D, Jull GA, Richardson CA. The relationship between the dynamic and static function of abdominal muscles. Aust J Physiother 39:9–13, 1993.

Wright A, Lloyd-Davies A, Williams S, Ellis R, Strike P. Individual active treatment combined with group exercise for acute and subacute low back pain. Spine 30(11):1235–41, Jun 1, 2005.

Yilmaz F, Yilmaz A, Merdol F, Parlar D, Sahin F, Kuran B. Efficacy of dynamic lumbar stabilization exercise in lumbar microdiscectomy. J Rehabil Med 35(5):163–167, Jul, 2003.

Zoidl G, Grifka J, Boluki D, Willburger RE, Zoidl C, Kramer J, Dermietzel R, Faustmann PM. Molecular evidence for local denervation of paraspinal muscles in failed-back surgery/postdiscotomy syndrome. Clin Neuropathol 22(2):71–77, Mar–Apr, 2003.

Lightning Source UK Ltd.
Milton Keynes UK
UKOW07f0455230917
309727UK00002B/55/P

9 780578 015583